t CRITICAL VALUES

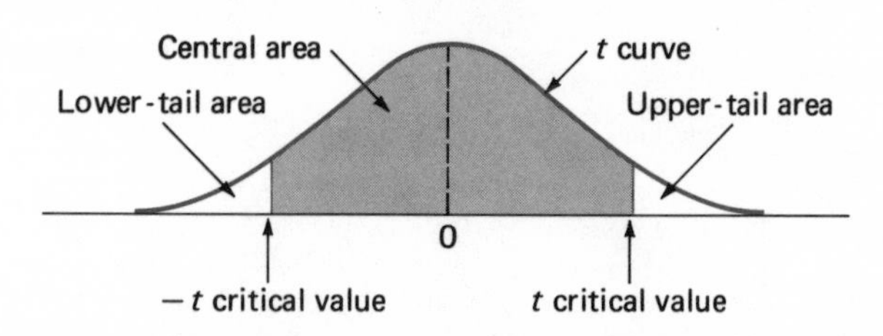

Central Area Captured	.80	.90	.95	.98	.99	.998	.999
Confidence Level	80%	90%	95%	98%	99%	99.8%	99.9%
Degrees of Freedom							
1	3.08	6.31	12.71	31.82	63.66	318.31	636.62
2	1.89	2.92	4.30	6.97	9.93	23.33	31.60
3	1.64	2.35	3.18	4.54	5.84	10.21	12.92
4	1.53	2.13	2.78	3.75	4.60	7.17	8.61
5	1.48	2.02	2.57	3.37	4.03	5.89	6.86
6	1.44	1.94	2.45	3.14	3.71	5.21	5.96
7	1.42	1.90	2.37	3.00	3.50	4.79	5.41
8	1.40	1.86	2.31	2.90	3.36	4.50	5.04
9	1.38	1.83	2.26	2.82	3.25	4.30	4.78
10	1.37	1.81	2.23	2.76	3.17	4.14	4.59
11	1.36	1.80	2.20	2.72	3.11	4.03	4.44
12	1.36	1.78	2.18	2.68	3.06	3.93	4.32
13	1.35	1.77	2.16	2.65	3.01	3.85	4.22
14	1.35	1.76	2.15	2.62	2.98	3.79	4.14
15	1.34	1.75	2.13	2.60	2.95	3.73	4.07
16	1.34	1.75	2.12	2.58	2.92	3.69	4.02
17	1.33	1.74	2.11	2.57	2.90	3.65	3.97
18	1.33	1.73	2.10	2.55	2.88	3.61	3.92
19	1.33	1.73	2.09	2.54	2.86	3.58	3.88
20	1.33	1.73	2.09	2.53	2.85	3.55	3.85
21	1.32	1.72	2.08	2.52	2.83	3.53	3.82
22	1.32	1.72	2.07	2.51	2.82	3.51	3.79
23	1.32	1.71	2.07	2.50	2.81	3.49	3.77
24	1.32	1.71	2.06	2.49	2.80	3.47	3.75
25	1.32	1.71	2.06	2.49	2.79	3.45	3.73
26	1.32	1.71	2.06	2.48	2.78	3.44	3.71
27	1.31	1.70	2.05	2.47	2.77	3.42	3.69
28	1.31	1.70	2.05	2.47	2.76	3.41	3.67
29	1.31	1.70	2.05	2.46	2.76	3.40	3.66
30	1.31	1.70	2.04	2.46	2.75	3.39	3.65
40	1.30	1.68	2.02	2.42	2.70	3.31	3.55
60	1.30	1.67	2.00	2.39	2.66	3.23	3.46
120	1.29	1.66	1.98	2.36	2.62	3.16	3.37
z critical values	1.28	1.645	1.96	2.33	2.58	3.09	3.29
Level of significance for a *two*-tailed test	.20	.10	.05	.02	.01	.002	.001
Level of significance for a *one*-tailed test	.10	.05	.025	.01	.005	.001	.0005

Statistics

THE EXPLORATION AND ANALYSIS OF DATA

Statistics

THE EXPLORATION AND ANALYSIS OF DATA

JAY DEVORE
California Polytechnic State University,
San Luis Obispo

ROXY PECK
California Polytechnic State University,
San Luis Obispo

WEST PUBLISHING COMPANY
St. Paul New York Los Angeles San Francisco

Copyediting: Linda Thompson
Cover and Interior Design: Bruce Kortebein, The Design Office
Cover Art: Frank Stella, *Star of Persia 1,* copyright © Gemini G.E.L. 1967.
Technical Illustrations: J&R Services, Inc.
Composition: Interactive Composition Corporation

50 West Kellogg Boulevard
P.O. Box 64526
St. Paul, MN 55164-1003

Printed in the United States of America

Library of Congress Cataloging-in-Publication Data
Devore, Jay L.
Statistics, the exploration and analysis of data.

Bibliography: p.
Includes index.
1. Mathematical statistics. I. Peck, Roxy L.
II. Title.
QA276.D48 1986 001.4'22 85-20340
ISBN 0-314-93172-4
1st Reprint—1986

To Carol, Allie, and Teri

To my parents, Lucelle and Doyle

CONTENTS

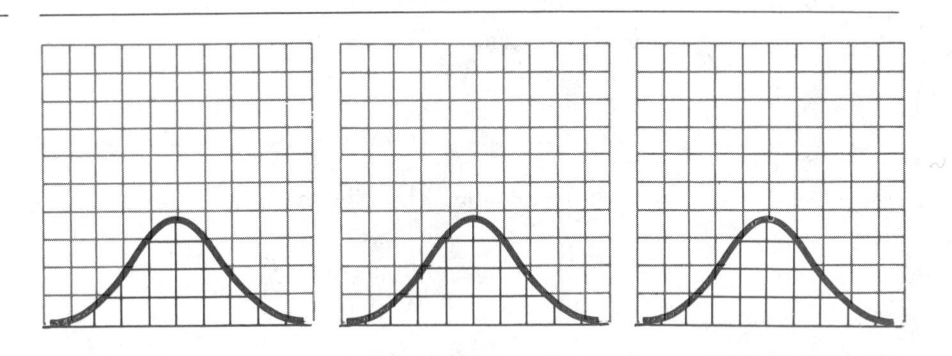

*Optional Section. After Chapter 8, coverage and ordering will depend on individual instructors' tastes, so no sections have been marked as optional.

PREFACE

Statistics: The Exploration and Analysis of Data is intended for use as a textbook in introductory statistics courses at two and four year colleges and universities. We believe that the following special features of our book distinguish it from other texts.

Features

A Traditional Structure with a Modern Flavor The topics included in almost all introductory texts are here also. However, we have interwoven some new strands which reflect current and important developments in statistical analysis. These include stem-and-leaf displays, boxplots, transformations, residual analysis, normal probability plots, and distribution-free confidence intervals. The organization gives instructors considerable flexibility in deciding which of these topics to include in a course.

The Use of Real Data Many students are skeptical of the relevance and importance of statistics both to their own interests and goals and to the greater concerns of our society. Contrived problem situations often reinforce this skepticism. A strategy that we have employed successfully to motivate students is to present examples and exercises which involve data extracted from journal articles, newspapers, and other published sources. Most examples and exercises in the book are of this nature. They cover a very wide range of disciplines and subject areas, but addressing the statistical questions posed does not require familiarity with the various problem settings.

Mathematical Level and Notational Simplicity One year of high school algebra constitutes sufficient mathematical background for reading and understanding the material presented herein. However, students at this level often have much difficulty mastering the mathematical notation and arguments used in many introductory texts. We want students to focus on concepts without having to grapple with formula and symbol manipulation. To achieve this, we have sometimes used words and phrases in place of symbols. We hope that this makes the exposition more accessible to those who are apprehensive about their mathematical skills.

The Use of Computer Output The ability of the computer to perform many computations and operations (such as ordering, grouping, and drawing pictures) very quickly has stimulated the development of much new statistical methodology. The wide availability of statistical computer packages such as

MINITAB, BMDP, SAS, and SPSS has made it much easier for investigators to analyze their own data using old or new methods. Statistical software for microcomputers has also proliferated. To highlight the role of the computer in contemporary statistics, we have included sample output from the aforementioned packages throughout the book. In addition, numerous exercises contain data which could be analyzed by a statistical package, though use of the book does not presuppose access to any such package.

The Role of Probability All too often students find probability difficult, remain unconvinced of its relevance to inference, and let their difficulties color their attitude to the remainder of the course. Our treatment of probability is informal, intuitive, and restricted to those aspects which bear most directly on the logic of inference.

Pedagogical Aids There are a great many worked examples. An exercise set appears at the end of each section, and a supplementary set concludes each chapter. Boxes are used to call out important definitions, concepts, and procedures. A key concepts list summarizes the contents of each chapter. In addition, a student's solutions manual, instructor's manual, transparency masters, and a test bank are available from the publisher.

Topic Coverage

Our book can be used in courses as short as one quarter or as long as one year in duration. Particularly in shorter courses, an instructor will need to be selective in deciding which topics to include and which to set aside. The book divides naturally into four major sections: descriptive methods (Chapters 2-4), probability material (Chapters 5-6), the basic one- and two-sample inferential techniques (Chapters 7-10), and more advanced inferential methodology (Chapters 11-14). We have joined a growing number of books in including an early chapter (Chapter 4) on descriptive methods for bivariate numerical data. This early exposure raises questions and issues which should stimulate student interest in the subject, and is also advantageous for those teaching courses in which time constraints preclude covering advanced inferential material. However, this chapter can easily be skipped until the basics of inference have been covered, and then combined with Chapter 11 for a unified treatment of regression and correlation.

With the possible exception of Chapter 4, Chapters 1-8 should be covered in order. We anticipate that most will then continue with the two-sample and paired data material of Chapters 9 and 10, though regression could be covered before either of these. Analysis of variance (Chapter 13) and/or categorical data analysis (Chapter 14) can be discussed prior to the regression material of Chapters 11-12. In addition to flexibility in the order in which chapters are covered, material in some sections can be skipped entirely, just in part, or postponed. The following commentary identifies these sections.

Chapter 1: The basic terminology is essential, but no class time need be devoted to discussing the examples at the end of this chapter.

Chapter 2: The first four sections should be covered, but the relation between sample and population distributions (Section 5) and transformations (Section 6) are optional topics.

Chapter 3: Sections 1 and 2 contain core material, but Section 3, and in particular the discussion of boxplots, can easily be skipped.

Chapter 4: As discussed earlier, this material can be covered to any desired extent at this point. Whenever regression and correlation are presented, residual plots (Section 4) and transformations (Section 5) are optional.

Chapter 5: The first four sections cover standard topics that should be part of any introductory course. Section 5 discusses the use of normal probability plots to check for normality of a population distribution. This section can be skipped entirely, or alternatively a very brief qualitative summary can be given. The binomial probability distribution is obviously of great interest and utility in its own right. However, because our focus is on inference, we view it as an optional topic. A self contained discussion of the properties on which large-sample inferences concerning population proportions are based appears in Chapter 6 (in the context of sampling distributions). Those who wish to cover the binomial probability calculations of Section 6 can do so either before or after the normal distribution material in Section 4.

Chapter 7: Once the first section on large-sample confidence intervals for a population mean has been discussed, the next two sections can be covered in either order or skipped (postponed) in favor of immediately proceeding to Chapter 8 on hypothesis testing.

Chapter 8: Sections 1 and 2 present the basic concepts of hypothesis testing, and Sections 3, 5, and 6 describe the standard one-sample test procedures for a population mean and proportion. Section 4 discusses P-values, which have appeared with increasing frequency in recent years in summaries of statistical analyses. Although it is possible to omit this material, we do not recommend doing so. For those who believe that the determination of type II error probabilities is slighted in introductory texts, the optional Section 6 includes a discussion of how β can be obtained for several important tests.

Chapter 9: After discussing the sampling distribution of a difference between two independent variables in Section 1, the usual two-sample z and t tests and confidence intervals appear in Sections 2-4. The distribution-free (non-parametric) Wilcoxon rank-sum test and associated confidence interval are introduced in Section 5. This material can be covered at this point, deleted entirely, or discussed in conjunction with other distribution-free methods from the last sections of Chapters 10 and 13.

Chapter 10: Section 1 discusses the advantages of a paired experiment, after which the paired t test and confidence interval are presented in Section 2. Wilcoxon's signed-rank test and the associated confidence interval are given in Section 3; as with the rank-sum procedures, this material can be omitted or postponed.

Chapter 11: The first three sections present the simple linear regression model and related inferential procedures. If Chapter 4 has not already been covered, appropriate material from that chapter should be integrated into the present chapter. Section 4 presents inferences concerning the population correlation coefficient; prerequisite material appears in Section 4.2. A major focus of exploratory data analysis has been the study of residuals from a fitted model. Section 5 discusses the use of residuals to check the adequacy of the simple linear regression model; this material is mentioned only briefly in the next chapter.

Chapter 12: Multiple regression analysis is frequently not part of an introductory course. For those who have the time and inclination, our focus on concepts and the use of computer output should make this material accessible and useful. Section 1 discusses various models, including the use of interaction and dummy predictors. It is not necessary to cover all these models before moving on to the methods of analysis in Sections 2 and 3. Various aspects of model building, including variable selection and multicollinearity, are considered in the last section.

Chapter 13: Section 1 presents the fundamental ideas of single-factor ANOVA. This is supplemented by a discussion of sums of squares, computations, and ANOVA tables in Section 2. Once past these sections, the remaining sections can be covered in almost any order. For example, one might choose to skip Section 3 on multiple comparisons, discuss randomized block experiments (Section 4), and omit the last two sections on two-factor and distribution-free ANOVA.

Chapter 14: The first section focuses on summarizing bivariate categorical data in a two-way frequency table. It is entirely descriptive in nature, and could be discussed in combination with the Chapter 4 material prior to probability and inference. The remaining sections present the usual test procedures based on the chi-squared distribution.

Acknowledgments

Many people have made valuable contributions to the preparation of this book. We have derived great benefit from many discussions with our colleague John Groves, who also prepared several of the supplements. Jim Daly and Joyce Curry-Daly were invaluable in reviewing examples and checking calculations. Numerous constructive criticisms and suggestions came from the following reviewers:

Norma Agras-Innes (Miami-Dade Community College)
James K. Baker (Jefferson Community College)
Calvin P. Barton (Stephen F. Austin University)
William H. Beyer (University of Akron)
Ben P. Bockstege (Broward Community College)
Fred Booth (Northern Virginia Community College)
John Brevit (Western Kentucky University)
Carl Cuneo (Essex Community College)
Shirley Dowdy (West Virginia University)
Art P. Dull (Diablo Valley College)
Josephine L. Gervase (Manchester Community College)
Lloyd A. Iverson (University of Oklahoma)
Eric Lubot (Bergen Community College)
Stanley M. Lukawecki (Clemson University)
David R. Lund (University of Wisconsin-Eau Claire)
Jeff Mock (Diablo Valley College)
Ron Morgan (West Chester University)

Julia A. Norton (California State University, Hayward)
Mary Parker (Austin Community College)
D.S. Purohit (Claflin College)
Larry J. Ringer (Texas A & M University)
Richard D. Semmler (Northern Virginia Community College)
Bill Stines (North Carolina State University)
Ara B. Sullenberger (Tarrant County Junior College)
David C. Trunnell (Xavier University)
Ann Watkins (Los Angeles Pierce College)

Typing chores were admirably handled by Charty Beam, Pat Fleischauer, Debbie Paxson, Lygia Peck, and Ellen Stier. We can't say enough about how pleasurable it was to work with our editors Peter Marshall, Theresa O'Dell, Mark Jacobsen, and Linda Thompson, as well as others at West Publishing Co. Finally, the support of our families, friends, and colleagues made our task much easier.

JAY DEVORE
ROXY PECK

A Note to the Student

IN ALL LIKELIHOOD, you have started reading this book because it is the text for an introductory statistics course required of all students in your major. You may well be thinking to yourself that if it weren't for this requirement, you wouldn't be enrolled in a statistics course and could then spend your time in more interesting and productive ways. Perhaps you are even somewhat apprehensive about your ability to do well in the course, since you've probably heard through the grapevine that mastering statistics requires some facility for mathematical reasoning and manipulation. If you are indeed ambivalent about studying statistics and a bit fearful of what lies ahead, please realize that these feelings are shared by many other students. We hope to lay these fears to rest in short order by convincing you that statistics is important for gaining a better understanding of the world around you, relevant to your particular interests and field of study, and accessible even if you have a very modest mathematical background. To this end, the book emphasizes concepts and an intuitive presentation of the core methodology used in a wide variety of applications. Statistics does rest on a mathematical foundation, but we have tried to keep the notation and mathematical development simple. We hope the result is a friendly and informal survey that will help you in various ways long after the course is finished.

The key to success in your statistics course, as in so many endeavors, is to start with a positive attitude and resolve to invest a reasonable amount of time and effort. It won't always be easy and may occasionally be frustrating. (We ourselves sometimes get quite frustrated when attempting to learn new material.) But with the right attitude and commitment of your resources, we think that understanding, enjoyment, and a sense of accomplishment will quickly follow.

Statistics

THE EXPLORATION AND ANALYSIS OF DATA

1 What Is Statistics?

INTRODUCTION

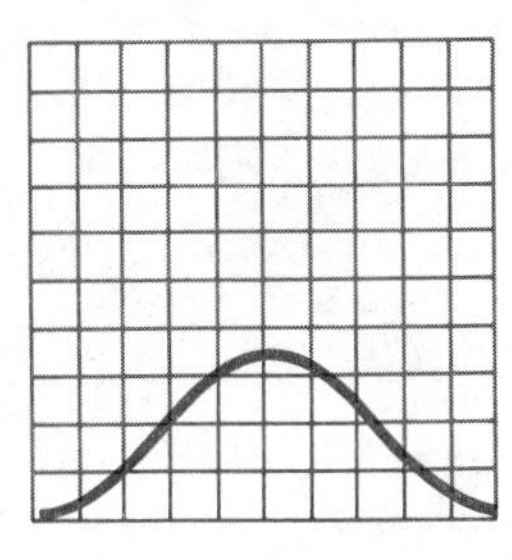

COLLECTIONS of facts and figures are constantly encountered both in the pursuit of professional goals and in everyday life. **Statistics** is the scientific discipline that provides methods to help us make sense of such data. Statistical methodology is being employed with increasing frequency in the social sciences, natural sciences, and applied sciences, such as engineering, business, medicine, and agriculture, to aid in organizing, summarizing, and drawing conclusions from data. The pervasiveness of statistical analysis in reports and studies done by investigators in many different subject areas has led educators in these areas to recognize the importance of statistical literacy. An exposure to statistical reasoning and the most widely used methods for analyzing data is now viewed by many as an integral part of a college education.

Some Basic Elements

For hundreds of years individuals have been using rudimentary statistical tools to organize and summarize data. You undoubtedly have a passing acquaintance with some of these—bar charts, tabular displays, various plots of economic data, averages and percentages, and results of polls, to name a few. Methods for organizing and summarizing data to aid in effective presentation and increased understanding constitute a branch of the discipline called **descriptive statistics.** Chapters 2–4 contain an introduction to the most commonly encountered and useful descriptive methods, many of which are then used repeatedly through the remainder of the book. Some of the more recently developed descriptive techniques fall under the heading of what statisticians now refer to as *exploratory data analysis* (EDA). These include ingenious plots and visual displays designed to uncover some of the subtler relationships in the data and reveal more clearly any hidden messages and patterns. As examples, we discuss stem-and-leaf displays in Chapter 2, box plots in Chapter 3, and straightening transformations and residual plotting in Chapter 4.

One might ask what else there is to statistics aside from the organization and summarization of data. In a word, the answer is *generalization.* Frequently, the individuals, objects, or measurements gathered by an investigator come from a much larger collection. It is this larger collection about which the investigator wishes to draw conclusions. A pollster who interviews some registered voters in a particular state wants to use the resulting data to make statements about the opinions of all registered voters in that state. When a drug is administered by a medical researcher to some patients suffering from a certain disease, data resulting from this clinical trial will be used to infer the effects of the drug if applied to all (present and future) sufferers. The sales figures that result from test marketing a new product in selected locations provide information about the product's likelihood of success in all possible locales. A manufacturing team that tests some prototype tires incorporating a new tread design will use the test data to draw conclusions about characteristics of all possible tires of this type that might be produced.

DEFINITION	The entire collection of individuals, objects, or measurements about which information is desired is called the **population** of interest. A **sample** consists of a part or subset of the population selected in some prescribed manner.

The second major branch of statistics, **inferential statistics,** involves generalizing from a sample to the population from which it was selected. That is, the objective of inferential statistics is to use sample information in order to draw conclusions about a population or several populations. For example, a television cable company might want information about the level of customer satisfaction with respect to programming, reception, and service. It would be very expensive and time-consuming to contact every subscriber (resulting in a *census* of the population). An obvious alternative is to select a representative sample of subscribers and use their responses to make inferences about the entire subscriber population. Similarly, a representative sample of homes in a large metropolitan area that have been sold recently would yield information about various characteristics of the population consisting of all recently sold homes—the average sale price, percentage financed by an FHA mortgage, and so on.

In both the cable television and home sales examples, there was a concrete, existing population. Sometimes the sample must be viewed as having been selected from a conceptual or hypothetical population, one that does not actually exist in a concrete sense. When a sample of prototype tires is tested for tread wear, the population consists of all tires that could be produced by using the same manufacturing process that yielded the prototypes. A sample of observations on crop yield when a new fertilizer is used on selected fields comes from the hypothetical population of all possible measurements that might result from repeated use of the fertilizer under similar conditions.

One obvious reason for sampling is the savings in time and expense associated with a census of the entire population. Another reason is that data collection may be destructive. One could subject every light bulb in a large batch to a life test to learn about average lifetime, but then there would be nothing left to sell or use. Even when resources allow for a complete examination of all population members, working with a sample may permit the use of fewer, more highly trained investigators, who can exercise great care in gathering and processing the data. The U.S. census contains some amusing examples of errors that can occur when the volume of data to be processed is very large, such as 13-year-old children who have been married five times.

With very few exceptions, inferential methods have been developed only in the present century, primarily in the last 40 years. Nevertheless, inferential statistics is now regarded as the more important branch of statistics. Our development of inferential methods begins in Chapter 6 and continues through the remainder of the book.

The material that bridges the gap between descriptive statistics in the early chapters and the later inferential methodology is called *probability*. Chapter 5 and most of Chapter 6 are devoted to an introduction to this material. The role of probability and its relationship to inference can be glimpsed by considering a simple population consisting of balls in an urn. Some of these balls are red and the remaining balls are green, but they are identical in every other way. The balls are mixed up, and someone selects 25 balls without looking. Once the percentage of red balls in the urn has been specified (e.g., 40% red), probability methods allow us to assess the likelihood that the sample contains between 10 and 15 red balls, or that at least 15 red balls appear in the sample. That is, in probability the nature of the population is specified. Questions concerning the composition of the sample can then be asked and answered. Inference, on the other hand, begins with a description of what is in a particular sample. The sample of 25 balls might have 12 red balls (48% of the sample). Does this sample evidence strongly suggest that the population percentage of red balls is between 40% and 50%? If there were prior grounds for believing that at most 40% of the population consisted of red balls, does the sample data strongly contradict this prior belief? These and other types of questions about the population can be answered by using appropriate inferential methods.

In probability one uses known facts about the population to reason out what might occur when a sample is selected from the population. Statistical inference involves reasoning inductively from the results of a particular sample back to the population from which the sample was drawn.

Once we understand what a sample drawn from any specified population might contain (probability), it will be much easier to reason from a given sample back to the population (statistical inference).

Some Examples of How Statistics Is Used

The best way to appreciate the scope and power of statistics is to look at examples. We'll present many examples in a variety of subject areas throughout the book, but here are a few preliminary ones that we hope will spark your interest.

EXAMPLE 1

An airline passenger must frequently travel on planes from several different airlines in order to get from one city to another. The revenue from the passenger's ticket must then be split among the airlines according to various characteristics of the individual route segments. Similarly, when freight is shipped by rail from an origin to a destination, several different railroads are often used. Again, the money paid by shippers must be split among the different companies. One way to determine how to divide the collected revenue is to examine each and every ticket or shipping bill and calculate the split for each one according to prescribed formulas. This census of the population is obviously time-consuming and expensive. In recent years statistical methods have been used to estimate the true revenue split based on a sample of tickets or shipping bills.

An interesting study of the effectiveness of such sampling techniques is discussed in the article "How Accountants Save Money by Sampling," which appears in *Statistics: A Guide to the Unknown* (see the references at the end of this chapter). The Chesapeake and Ohio Railroad considered the population of roughly 23,000 shipping bills for shipments made during a particular 6-month period. Each bill was for a shipment made partly with C&O and partly with one other railroad. One sampling method, called (simple) *random sampling,* amounts to regarding the 23,000 bills as balls in an urn and selecting a specified number. Because the amount of revenue due the C&O was related to the total charge on the bill, an alternative sampling method called *stratified sampling* was used to ensure that bills with large total charges were adequately represented in the sample. The resulting estimate of revenue due C&O was $64,568. A population census revealed that the true amount owed to C&O was $64,651, so the difference between the actual and estimated amounts was only $83! While sampling won't always produce results this accurate, studies of this sort argue convincingly for the use of statistical methodology in such situations. ■

EXAMPLE 2

Results of opinion polls are constantly reported by the news media and included in advertising. The design and execution of such polls constitute a very fertile area of application for statistical methodology. Many polls are political in nature. Some give information concerning how individuals feel about specific issues such as taxes, abortion, or an officeholder's performance. Other polls forecast election results.

Some election polls done in the 1930s and 1940s were so far off the mark that they cast shadows over the value and veracity of polling for many years. Fortunately, these debacles also helped stimulate the development of more refined methods to ensure accuracy. One of the most publicized polling disasters was the *Literary Digest* presidential poll of 1936, in which this now-defunct magazine forecast that Landon would overwhelmingly defeat Roosevelt. This projection was based on roughly 2.4 million responses. Why did the results from a sample this large yield such an erroneous conclusion? On the basis of available evidence, there were two primary culprits: (1) a sampling design guaranteed to produce a sample that was unrepresentative of the voting population, and (2) failure of most individuals in the sample to respond to the survey.* Voters in the sample were selected primarily from lists of automobile and telephone owners. These individuals tended to be relatively better off than those in the entire population, and such people were and still are more likely to vote Republican. So Republicans were substantially overrepresented in the sample. The other major problem was that, whereas the sample consisted of roughly 10 million individuals, only about 2.4 million of these sent in responses. Subsequent studies by political scientists, market researchers, and others have established that respondents to a survey differ from nonrespondents not only in that they responded but typically in many other ways. That is, even if the sample itself is representative of the entire population, those in the sample who voluntarily respond are often not representative.

How do modern-day polls avoid these pitfalls? First of all, the best polls use methods based on probability to select a sample. Provided that the sample size is large enough, such methods make it very unlikely that characteristics of the sample will differ much from those of the population. The Gallup poll, which uses fewer than 4000 respondents, has a 2% margin of error. Roughly speaking, this means that only rarely will the sample percentage favoring a particular candidate differ by more than 2% from the corresponding population percentage. In Chapter 7 we'll develop the notion of a margin of error more fully for a (simple) random sample; although the Gallup sample is more complicated (primarily to keep down the cost of polling), the idea is the same.

The problem of nonresponse is more difficult to deal with. Many polling organizations eschew mail surveys in favor of direct contact of those selected for the sample. Well-trained interviewers employ various techniques to elicit responses, including repeated visits or calls to track down hard-to-locate individuals. In addition, certain methods can be employed during the data-analysis stage to correct for possible nonresponse bias (these involve giving more weight to responses from individuals who were hard to contact). Two other more subtle difficulties also deserve mention. The first is that the population

*The article "The Literary Digest Poll: Making of a Statistical Myth," *The American Statistician* (1976): 184–85 discounts the importance of (1), but other sources disagree with this assessment. An excellent expository discussion appears in Chapter 19 of *Statistics* by Freedman, Pisani, and Purves (listed in the chapter references). See also "Opinion Polling in a Democracy" in *Statistics: A Guide to the Unknown.*

of interest consists of all those who actually vote rather than all who are registered to vote. The Gallup poll includes some seemingly peripheral questions, whose role is to enable the pollsters to assess the likelihood that any particular respondent will vote on election day. Responses from those deemed unlikely to vote can then be downweighted. Secondly, responses can be heavily influenced by the way in which questions are worded. For example, the question "Do you favor higher taxes?" would typically result in fewer favorable responses than "Do you favor higher taxes to fund better schools?" Carrying out a good survey requires not only expert statistical advice but also input from psychologists, sociologists, and others to ensure that questions accurately reflect issues of interest. ■

EXAMPLE 3

Polio is a relatively unknown disease today, but up until several decades ago it was a substantial public health threat. The use of the Salk polio vaccine in the late 1950s was the driving force behind the decreased incidence rate. When the vaccine was first developed, its effectiveness had to be tested, and this was done in an extremely large experiment carried out in 1954. In order to have a firm basis for judgment, both a treatment group (children receiving the vaccine) and a control group (children who received no treatment) were needed. To ensure that experimental subjects were as alike as possible except with respect to treatment status, all were first, second, or third graders whose parents had given permission for their children to receive the vaccine.

The children in the control group actually received a shot, but it was a *placebo*—a substance whose appearance was identical to that of the vaccine but which was known to have no effect as far as preventing polio was concerned. Vials were coded so that only personnel at the disease control center and not those actually giving or receiving shots knew whether any particular child received the vaccine or the placebo. This is an example of a **double-blind experiment,** one in which neither the experimental subject nor the person directly administering to the subject knows whether any particular subject is in the treatment or control group. Such an experimental design is strongly advocated by statisticians because it prevents both obvious and subtle biases from influencing the results. The outcome of the Salk experiment was an impressive victory for the vaccine. Each group consisted of roughly 200,000 children. Paralytic polio, the most virulent form, was contracted by 110 children in the control group and only 33 vaccinated children. Statistical analysis showed that a discrepancy of this magnitude would be extremely unlikely in the absence of any vaccine effect.

Why were the sample sizes so large? Think of a related situation in which there are two populations, each consisting of some red and some green balls in an urn. To see if each population contains the same percentage of red balls, a sample of balls is selected from each urn. If red balls constitute between 10% and 90% of each urn, relatively moderate sample sizes will usually reveal any difference in population percentages that might have practical significance. However, suppose the percentage of red balls in each urn is known to be at most .01% (1 in 10,000). Then, even if the percentages differ, samples of several hundred or even several thousand will look almost identical. Very large samples are now required to reveal any difference. The incidence rate of paralytic polio without any treatment was believed to be roughly 1 in 10,000; halving this rate would have important practical consequences.

You might wonder about the ethics of not giving the vaccine to every child, but remember that its effectiveness had not been demonstrated. Several papers in *Statistics: A Guide to the Unknown* (see the chapter references) describe studies of innovative medical treatments and social programs. It is far from true that the innovation is always successful, so it is not necessarily desirable to be in the treatment group. One other important point here concerns the efficacy of using a control group and randomizing (by coin tossing, for example) to decide in which group to place each individual subject. A few years ago a gastric freezing technique for treatment of ulcers was introduced. The first study involved no controls and reported a dramatic improvement after treatment. However, a more carefully designed double-blind experiment done several years later suggested that the freezing technique was without merit. Evidently patients in the first study were responding to the appearance of treatment rather than to the treatment itself, what psychologists call the *placebo effect*. Similarly, a number of experiments have been carried out to study the portacaval shunt, a technique for treating patients with acute cirrhosis of the liver. Experiments that involved no controls or controls selected in a nonrandomized fashion produced results very favorable to the technique, while six randomized, controlled experiments were substantially less encouraging (see page 217 of "Assessing Social Innovations: An Empirical Base for Policy" in *Statistics and Public Policy,* listed in the chapter references, for more detail). These and other examples have led statisticians and many researchers to be quite skeptical of results from experiments which lacked controls or randomization in allocating subjects to the treatment or control groups. ■

EXAMPLE 4

Probability and statistical methods are being employed with increasing frequency in both legal settings and criminology studies. One experiment that had a substantial impact on our legal system was the Manhattan bail project. It had been thought that the imposition of bail and threat of forfeiture was a critical factor in getting those accused of crimes to appear for trial. Some investigators hypothesized that individuals with close ties to their communities would be unlikely to flee even without having posted bail. They developed a scoring system to identify those judged suitable for release without bail. A large sample of such defendants was then selected and randomly split into a treatment group and a control group. For each member of the treatment group, a recommendation to release without bail was made to the court; no such recommendation was made for members of the control group. This experimental design permitted the investigators to see whether the act of recommendation itself affected case dispositions.

One rather unexpected result was that roughly 60% of those in the treatment group were acquitted or had their cases dismissed, whereas this happened to only 26% of the control group. This is analogous to drawing a sample of balls from each of two different urns and finding that 60% of the balls in the first sample and only 26% of those in the second sample are red. Statistical analysis can be used to show that with the sample sizes used, it is highly unlikely that such an extreme difference in sample percentages could result when the two urns have identical percentages of red balls. In the context of the bail project, the observed differential treatment associated with a recommendation to release cannot be explained by chance alone. Furthermore, the way in which the two groups were formed—random allocation of individuals who were all judged deserving of release without bail—makes it plausible that the observed difference can be attributed to the treatment rather than anything else.

A very gratifying result of the experiment was that fewer than 1% of those released as described failed to show for trial. As a result of this success, release without bail has become much more widespread in recent years. In addition, as noted in a paper describing the experiment ("The Manhattan Bail Project: Its Impact on Criminology and the Criminal Law Process," *Texas Law Review* (1964–1965): 319–31), the project was significant in its impact on the legal profession because of "the spur and stimulus it has given to experimentation and study in criminal law throughout the country."

Many criminology studies are observational in nature rather than being based on randomized controlled experiments. In an observational study the investigator does not assign individuals to the two (or more) groups to be compared but instead simply observes the results after the groups have been formed. Often each subject chooses his or her own group. An interesting example was described in the paper "Does it Pay to Plead Guilty? Differential Sentencing and the Functioning of Criminal Courts" (*Law and Society Review* (1981–82): 45–69). Roughly 53% of 191 defendants in San Francisco who pleaded guilty to robbery were sent to prison, whereas about 88% of the 64 defendants who were found guilty after pleading innocent were sent to prison. This difference in sample percentages is too large to be explained by chance alone (as we show in Chapter 9).

It is tempting to attribute the difference in type of sentence to the nature of the plea—i.e., to conclude that there is a causal relationship. This conclusion would be justified if we knew that the two groups were similar in all ways except for the treatments (guilty or not-guilty pleas). A random allocation of subjects to groups helps ensure this. Here the accused themselves chose their pleas, so the allocation to treatments was not random. The two groups might well differ with respect to characteristics other than the plea, and these other characteristics might have caused the differential sentences. More evidence is needed before an observed difference in the nature of sentencing can safely be attributed to type of plea. In general, conclusions drawn from randomized controlled studies have more solid foundations than those drawn from observational studies. ■

EXAMPLE 5

The size and spread of many plant and animal populations precludes examining all population members to gain desired information. One characteristic of particular interest to animal ecologists is the number of animals in a given population. Several different capture-recapture methods have been proposed for estimating the population size from sample data. The simplest such method involves capturing an initial sample of animals and tagging or otherwise marking each one. These animals are then released and given time to mix with the remainder of the population. A second sample of animals is then obtained and the number of tagged animals in this recapture sample is determined. The reasoning that leads to an estimate of population size is that the fraction of tagged animals in the recapture sample ought to be roughly the fraction of tagged animals in the population (i.e., that the sample should be representative of the population). Suppose that 100 animals were initially tagged and that 40 of the 200 animals in the recapture sample were found to be tagged. Then 20% of the recapture sample was tagged, so we estimate that 20% of the entire population was tagged. Since there are 100 tagged animals in the population, the estimated population size is 500 (because 100 is 20% of 500). The article "The Plight of the Whales" (in *Statistics: A Guide to the Unknown* in the chapter references) discusses estimating the number of blue whales. As you can imagine, tagging a whale is not such an easy task!

Some populations are composed of individuals from a number of different species. After obtaining a sample from the population and counting the number of species represented in the sample, an ecologist might wish to estimate the total number of species in the population (or, equivalently, the number of unobserved species, since total equals observed plus unobserved). The paper "Estimating the Number of Unseen Species: How Many Words Did Shakespeare Know" (*Biometrika* (1976): 435–47) gives a new twist to this problem. Shakespeare's known works comprise 884,647 total words. There are 846 words that he used more than 100 times—commonly used words such as *and, to,* and *the.* At the other extreme, 14,376 words appeared exactly once. The cited paper contains much more data of this sort. The total number of different words that Shakespeare is known to have used is 31,534 (this seems very large, but *state* and *states* are different words, for example, as are *ate, eat, eats, eaten*). With each distinct word representing a species, 31,534 species were observed. Surely Shakespeare knew some words that don't appear in his known writings. What was the size of his vocabulary? The authors of this paper posed a related question: Suppose that works consisting of an additional 884,647 words were discovered; how many words not previously used would appear in these works? The statistical methods developed in the paper yielded an estimate of 11,430 new words. An estimate of 35,000 was also given as a conservative lower bound for the total number of words that Shakespeare knew but did not use in his known works—an estimated lower bound on the total number of unseen species. ■

EXAMPLE 6

A *system* can be defined as any collection of components that interact with one another in the course of accomplishing a specified task or tasks. Our world is full of systems, some of them rather simple in nature and others quite complex. Systems analysts, those who are able to design and maintain systems that operate efficiently, are in great demand. Simulation techniques, which make heavy use of probability and statistics, allow an analyst to study various characteristics and configurations of a system before it is actually constructed. The resulting information can then be used to select a design that optimizes system behavior.

Consider as an example an inventory-control system for a company that distributes a single product to customers. Units of this product can be ordered from the manufacturer only at the beginning of a month. When an order is placed, there is a fixed charge of $400 plus a $40 charge per unit ordered (e.g., a 20-unit order costs $400 + $800 = $1200). In addition, each unit in inventory incurs a holding charge of $1 per day (for rental of storage space, maintenance, insurance, etc.). Customers pay $100 for each unit of the product. Assume that once the inventory level falls to zero in a given month, subsequent requests in that month constitute lost sales (no backlogging of demand is allowed). Thus frequent small orders are expensive because of the fixed charge in ordering, but so are infrequent large orders that result in high holding costs. A good inventory policy must strike a balance between these two extremes. The company is considering the following policy: If the inventory level at the beginning of the month is less than 20, order enough units to bring the level up to 100; otherwise order nothing (this type of policy is widely used by businesses).

The company faces uncertainty (randomness) both in the amount of time that elapses between successive customer requests and in the number of units requested by any customer. Suppose for simplicity that either 1, 2, or 3 days can elapse between successive requests, and that a customer will request either one, two, three, or four units. A simulation can be carried out with the aid of two urns filled with balls, as pictured. Each stage of the simulation consists of two steps:

1. Draw a ball from the first urn, note its number, and replace it.
2. Repeat step 1 for the second urn.

The number on the ball selected from the first urn determines the elapsed time until the next customer makes a request, and the number on the second ball determines how many units that next customer requests. The simulation can be carried out for any desired time period, such as 120 months, by repeating steps 1 and 2 until a sufficient amount of time has elapsed.

Urn 1:
Time between requests

Urn 2:
Number of units requested

The composition of the first urn implies that elapsed time between requests is equally likely to be 1, 2, or 3 days. Because 4 of the 10 balls in the second urn are marked 1, there is a 40% chance that a customer will request just one unit. Similarly, there is a 30% chance that two units will be requested, and so on (in Chapter 5, we replace these informal percentages with probabilities). The systems analyst can alter the composition of balls in either urn and study the consequences. Very importantly, the analyst can consider the effect of changing the inventory policy through replacing the numbers 20 and 100, which define the policy, with other numbers. This allows for identification of a policy that maximizes profit.

Simulations are not usually done by using balls in urns as described above. Instead, simulation is done on a computer. A computer can be programmed to generate a sequence of random numbers consistent with any desired distribution of numbered balls in urns. There are even special-purpose programming languages for simulation, such as GPSS or SIMSCRIPT, that will automatically repeat the simulation as many times as desired and then compute various summary quantities that describe system behavior. In this way inferences about how a system might behave in the real world can be based on data from a simulation. Simulation studies of complex systems give an important basis for decision making in industry, business, and government. ■

Some individuals regard conclusions based on statistical analysis with a great deal of suspicion. Extreme skeptics, usually speaking out of ignorance, characterize the discipline as a subcategory of lying, something used for deception rather than for positive ends. That is obviously not our view. We believe that statistical methods, used intelligently, constitute a set of powerful tools for gaining insight into the world around us. As with other powerful tools, there is the potential for misuse and abuse. We hope that our book will help you to understand the logic of statistical reasoning, prepare you to apply statistical methods correctly, and enable you to recognize when others are not doing so. By the end of your course, we trust that our perspective will also become yours.

KEY CONCEPTS*

Descriptive statistics Numerical, graphical, and tabular methods for organizing and summarizing data. (p. 2)

Population The entire collection of individuals or measurements about which information is desired. (p. 2)

Sample A part of the population selected for study. (p. 2)

Inferential statistics Methods for generalizing from a sample to the population from which the sample was selected. (p. 2)

REFERENCES

Fairley, William, and Frederick Mosteller, ed. *Statistics and Public Policy*. Reading, MA: Addison-Wesley, 1977. (Interesting articles focusing on how statistical methods are used in studying public policy questions and issues.)

Freedman, David, Robert Pisani, and Roger Purves. *Statistics*. New York: W. W. Norton, 1978. (The first two chapters contain some interesting examples of both well-designed and poorly designed experimental studies.)

Moore, David. *Statistics: Concepts and Controversies*. San Francisco: W. H. Freeman, 1978. (Contains an excellent chapter on the advantages and pitfalls of experimentation and another one in a similar vein on sample surveys and polls.)

Tanur, Judith, ed. *Statistics: A Guide to the Unknown*. San Francisco: Holden-Day, 1978. (Short articles by a number of well-known statisticians and users of statistics, all very nontechnical, on the application of statistics in various disciplines and subject areas.)

*Key concepts in this and later chapters are usually listed in the order in which they appear in the text.

2 Tabular and Pictorial Methods for Describing Data

INTRODUCTION

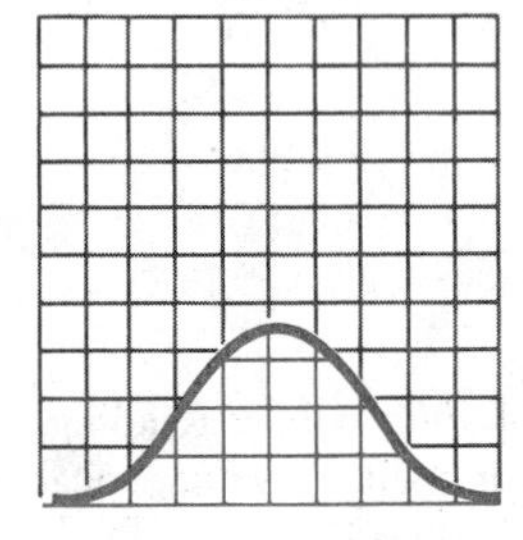

WHEN WE MAKE observations on or record characteristics of individuals or objects in some collection, the result is a body of data. Sometimes data is gathered in order to address specific issues and questions; on other occasions, the nature of the information contained in a set of data may not be obvious to those who requested or gathered the data. In either case, an important first step in extracting information and gaining insight is to organize the data and perhaps summarize it so that its salient features are more clearly revealed. Methods for organization and summarization depend on the nature of the data in hand, so we will first distinguish between several different types of data. The remainder of the chapter introduces some useful tabular and pictorial methods for describing and exploring data.

2.1 Types of Variables and Data

A statistical investigation usually stems from a desire to study various attributes of individuals or objects in some collection. For example, much effort has been expended in studying how various characteristics of an automobile affect its fuel efficiency (measured in miles per gallon under specified conditions). These characteristics include type of transmission, type of engine, number of cylinders, number of transmission speeds, horsepower, and quarter-mile time. There are only two types of transmission (two categories), manual and automatic. The attribute engine type is also categorical in nature, with the three possible categories being rotary, V-shaped, and straight. The number of cylinders is numerical rather than categorical, with possible response values of 4, 6, or 8. Quarter-mile time is also a numerical characteristic, but there are a great many possible values rather than just two or three; the same is true of fuel efficiency.

We will use the word **variable** for any characteristic that varies or changes when moving from individual to individual or object to object in a collection under study. For the attributes of automobiles just mentioned, transmission type and engine type are examples of **categorical variables,** since each automobile will fall into one of the two transmission-type categories and into one of the three engine-type categories. The variables horsepower, number of cylinders, quarter-mile time, and fuel efficiency are examples of **numerical variables;** the value of each of these variables for a particular automobile is a number. In a different setting, when examining prospective jurors for a murder trial, attorneys may seek information on the categorical variables sex, job classification, and attitude toward capital punishment (five possible responses, from strongly favor to strongly opposed), as well as on the numerical variables age, number of family members under 21, and number of years of schooling completed.

Whenever we make observations on a variable for some collection of individuals or objects, we call the result a **data set.** If we determine the transmission type for each of 10 automobiles and the results are (with *A* denoting automatic and *M* manual) *A, M, M, A, M, M, M, M, A,* and *A,* the collection of 10 observations is a **categorical data set.** The 10 quarter-mile times (in seconds) 12.1, 9.2, 11.0, 10.4, 12.3, 9.8, 12.0, 11.0, 10.5, and 12.1 constitute a **numerical data set.** Thus the nature of the variable on which we make observations determines whether the data set is categorical or numerical.

Two Types of Numerical Variables

With a numerical variable and numerical data, it is convenient to make a further distinction. Visualize a number line (Figure 1) for locating values of the numerical variable being studied. To every possible number (2, 3.125, −8.12976, etc.), there corresponds exactly one point on the number line. Now suppose that the variable of interest is the number of cylinders of an automobile engine. The possible values 4, 6, and 8 are identified in Figure 2(a)

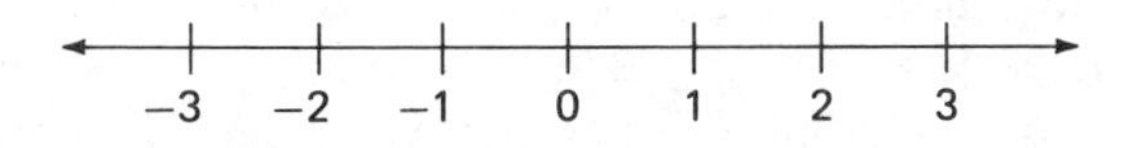

FIGURE 1 A NUMBER LINE

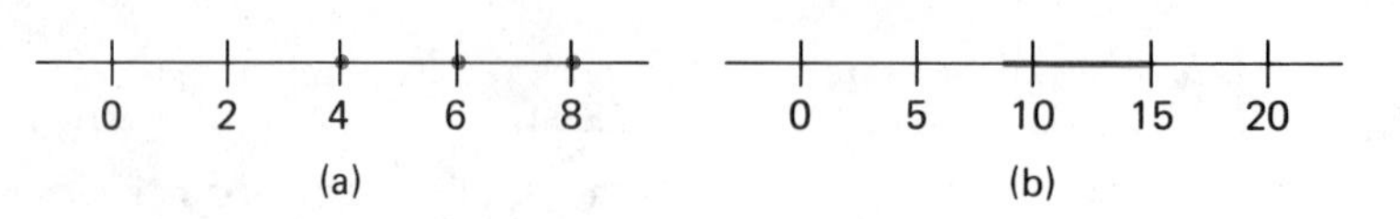

FIGURE 2 POSSIBLE VALUES OF A VARIABLE
(a) Number of Cylinders
(b) Quarter-mile Time

by the dots at the points marked 4, 6, and 8. These possible values are isolated from one another on the line; about any possible value, we can place an interval that is short enough so that no other possible value is included in the interval. On the other hand, the line segment in Figure 2(b) identifies a plausible set of possible values for the variable quarter-mile time. Here the possible values comprise an entire interval on the number line, and no possible value is isolated from other possible values.

DEFINITION

A numerical variable is **discrete** if its set of possible values, when pictured on the number line, consists only of isolated points. A numerical variable is **continuous** if its set of possible values comprises an entire interval on the number line.

The most frequently encountered type of discrete variable is a **counting variable,** one which arises from counting the number of occurrences of some event or attribute—the number of courses for which a student is registered, the number of admissions to a hospital during a given month, the number of dollars won or lost by a player in a Las Vegas blackjack game, etc. Possible values of a counting variable are integers, often including zero, and occasionally even negative integers (a profit of \$5 would be a value of 5, while a loss of \$5 would be a value of -5). Clearly the integer points are isolated on the number line.

Examples of continuous variables include time necessary to complete a task, density of a chemical compound, commuting distance, elevation (with a location below sea level, such as Death Valley, yielding a negative value), and body temperature. In theory, the value of a continuous variable can be determined to any desired degree of accuracy, such as a distance of 1.275 miles or a temperature of 98.63°, while a counting variable cannot assume any value between 1 and 2 or between 10 and 11, etc. In practice, limitations of measuring instruments usually impose a limit on the accuracy with which we can measure values, so that observations will not be spread continuously on the number line. Nevertheless, the distinction between discrete and continuous variables is useful when studying models for how values of a variable are distributed within a population.

Univariate, Bivariate, and Multivariate Data

Frequently, an investigator will wish to focus attention on just one variable. If the value of a single variable is determined for each individual or object in a collection, the result is a **univariate** data set. The univariate data set could be categorical, as with the set of responses resulting from determining the brand of calculator owned by each student in a statistics class, or numerical, as with the set of values of daily maximum ozone level for a period of 30 consecutive days in downtown Los Angeles.

Many investigations are carried out to study the relationship between two variables. To see how vocabulary size changes with age, researchers might select a number of children of different ages and determine the vocabulary size (number of words) of each child. Each observation in the data set would be a pair of values, such as (4, 1225) or (6.5, 1530), where the first number in each pair is the value of the variable age and the second number is the value of the variable vocabulary size. A data set consisting of pairs of values of two variables is referred to as **bivariate** data. Both variables could be numerical, as in the above example, or both could be categorical (type of siding (brick, wood, stucco) and type of roof (shingle, shake, rock, tile) for a collection of houses), or one variable could be categorical and the other numerical, as with brand of car and fuel efficiency.

With the increasing availability of sophisticated computer programs for analyzing data, it is now possible to analyze complicated data sets consisting of observations on more than two variables (sometimes many more); the objective is usually to describe relationships between the variables. A **multivariate** data set is one in which each observation consists of a value for three or more variables. If, for each automobile in a sample, we determined the type of transmission, number of cylinders, horsepower, quarter-mile time, and fuel efficiency, each observation would specify the value of each of these five variables; one observation might be (A, 6, 225, 12.8 18.6), while another might be (M, 4, 130, 14.0, 27.5).

We will spend much of our time looking at how to summarize and analyze univariate data sets, where the concepts and methods can most easily be appreciated without being overwhelmed by notation and manipulation. Chapter 4 presents selected descriptive methods for bivariate data, and toward the end of the book we look at inferential methods for bivariate and multivariate data.

EXERCISES

2.1 Classify each of the following variables as either categorical or numerical. For those that are numerical, determine whether they are discrete or continuous.

a. Brand of personal computer purchased by a customer

b. State of birth for someone born in the United States

c. Price of a textbook

d. Concentration of a contaminant (micrograms/cm^3) in a water sample

e. Zip code (Think carefully about this one.)

f. Actual weight of coffee in a 1-pound can

2.2 For the following numerical variables, state whether each is discrete or continuous.

a. The number of checks that bounce received by a grocery store during a given month

b. The amount by which a 1-pound package of ground beef decreases in weight (because of moisture loss) before purchase

c. The number of New York Yankees during a given year who will not play for the Yankees during the following year

d. The number of students in a class of 35 who have purchased a used copy of the text book
e. The length of a 1-year-old rattlesnake
f. The altitude of a location in California selected randomly by throwing a dart at a map of the state
g. The distance from the left edge at which a 12-inch plastic ruler snaps when bent sufficiently to cause a break
h. The price per gallon paid by the next customer to buy gas at a particular station

2.3 For each of the following situations, give some possible data values that might arise from making observations as described.

a. The country of manufacture for each of the next 10 automobiles to pass through a given intersection is noted.
b. The grade-point average for each of the 15 seniors in a statistics class is determined.
c. The number of gas pumps in use at each of 20 gas stations at a particular time is determined.
d. The actual net weight of each of 12 bags of fertilizer having a labeled weight of 50 lbs. is determined.
e. Fifteen different radio stations are monitored during a 1-hour period and the amount of time devoted to commercials is determined for each one.
f. The brand of breakfast cereal purchased by each of 16 customers is noted.
g. The number of defective tires is determined for each of the next 20 automobiles stopped for speeding on a certain highway.

2.2 Stem-and-Leaf Displays

It is difficult to extract information from a data set without first doing some preliminary organization. The organization and display of data may be an end in itself, or it may be the first step in a search for interesting questions, relationships, and pathways for further analyses. A very nice way to organize numerical data without expending much effort is to construct a **stem-and-leaf display.** Each number in the data set is broken into two pieces; the leading digit or digits form the **stem** and the trailing digit or digits form the **leaf.** For a data set consisting of readings on carbon monoxide level (parts per million, or ppm), we could choose stems and leaves so that the value 13.3 had stem 13 and leaf .3; an alternative choice would result in stem 1 and leaf 3.3. Once a choice of stem values is made, the display is obtained by listing these values vertically and then listing the leaves out to one side.

EXAMPLE 1

The May 1983, *San Luis Obispo Telegram-Tribune* reported the accompanying data on 1982 average per capita income for each of the 50 states and the District of Columbia (in hundreds of dollars, so a value of 85 corresponds to an income between $8500 and $8599).

State	Income	State	Income	State	Income	State	Income
Alabama	$ 85	Illinois	$121	Montana	$ 97	Rhode Island	$107
Alaska	152	Indiana	101	Nebraska	104	South Carolina	84
Arizona	102	Iowa	105	Nevada	117	South Dakota	95
Arkansas	83	Kansas	114	New Hampshire	107	Tennessee	88
California	125	Kentucky	88	New Jersey	130	Texas	113
Colorado	117	Louisiana	100	New Mexico	89	Utah	87
Connecticut	136	Maine	90	New York	123	Vermont	94
Delaware	117	Maryland	121	North Carolina	90	Virginia	110
District of Columbia	143	Massachusetts	119	North Dakota	104	Washington	116
Florida	108	Michigan	110	Ohio	107	West Virginia	88
Georgia	95	Minnesota	110	Oklahoma	107	Wisconsin	104
Hawaii	116	Mississippi	77	Oregon	103	Wyoming	119
Idaho	92	Missouri	101	Pennsylvania	109		

Good choices for stem values are 7 (for $7000), 8, 9, . . . , 15, so that Arizona's income of 102 has stem 10 and leaf 2. Figure 3(a) gives the first 10 entries in the display (through Florida), and the entire display appears in Figure 3(b). Notice that the leaves for any particular stem are not ordered; to do so would take time and accomplish little.

Stem	Leaves
7	
8	5, 3
9	
10	2, 8
11	7, 7
12	5
13	6
14	3
15	2

(a) Stems: Thousands
Leaves: Hundreds

Stem	Leaves
7	7
8	5, 3, 8, 9, 4, 8, 7, 8
9	5, 2, 0, 7, 0, 5, 4
10	2, 8, 1, 5, 0, 1, 4, 7, 4, 7, 7, 3, 9, 7, 4
11	7, 7, 6, 4, 9, 0, 0, 7, 3, 0, 6, 9
12	5, 1, 1, 3
13	6, 0
14	3
15	2

(b) Stems: Thousands
Leaves: Hundreds

FIGURE 3 CONSTRUCTING A STEM-AND-LEAF DISPLAY FOR THE PER CAPITA INCOME DATA
(a) The First 10 Leaves
(b) The Entire Display

The display is quite helpful in getting a feel for how the values are distributed on the income scale. Per capita income ranges from 77 to 152, but the bulk of the values have stems of 8, 9, 10, and 11. Mississippi's value, 77, is the only one that stands out as a relatively low value. There are two values that are quite a bit above the rest, 152 and 143 (Alaska, where the fraction of wage-earning adults is high, and the District of Columbia, where many well-paid government employees live). Two more values, 136 and 130, are somewhat above the main body of data. A typical, or central, value appears to be one with stem 10. There are a few more relatively high than relatively low incomes, but there is also a reasonable amount of symmetry about the typical value. ■

Using Repeated Stems to Stretch a Display

If the number of stem values is too small relative to the size of the data set, the display may be too compressed to give an informative picture of how data values are distributed. This would certainly be true of a data set consisting of 100 SAT scores, all between 400 and 799, if the stems are chosen as 4, 5, 6, and 7. Alternatively, with too many possible stems relative to the number of data values, the display would be too stretched out to give useful information. With the stems 40, 41, . . . , 79 (40 of them) and 100 SAT scores, typically no stem would have more than three or four leaves. Sometimes an effective way of obtaining an informative display is to repeat stem values twice, once for low-valued leaves (e.g., 0–4 or 00–49) and the second time for high-valued leaves (e.g., 5–9 or 50–99).

EXAMPLE 2

Figure 4 gives a stem-and-leaf display of advertised prices for a sample of 50 used Japanese subcompact cars (*Los Angeles Times,* May 8, 1983). The last digit of price has been deleted, so a stem of 2 and a leaf of 98 corresponds to a price between $2980 and $2989.

```
0 || 95
1 || 90,30,80,70,50,39,50,50,99,50,75,79,80,10,99,39,80
2 || 98,98,90,49,55,60,85,00,39,55,00,90,99,39,68,48
3 || 99,99,20,75,28,92,80,19
4 || 88,39,48                          Stems: Thousands
5 || 20,49,30,30,87                    Leaves: Tens
```

FIGURE 4 STEM-AND-LEAF DISPLAY FOR USED-CAR PRICES

All the prices between 100 and 299 are compressed into just two stems, making it difficult to see how these values are spread out in this $2000 range. In Figure 5 we display the same data with each stem repeated twice. The stem 2*l* is for prices between 200 and 249, and the stem 2*h* is for prices between 250 and 299. This display is substantially less symmetric in appearance than the display of Figure 3(b). A typical value here would be one with stem 2*h*, and there are quite a few values that are a fair distance above any typical value. The display has two peaks, one at the stem 1*h* and one at the stem 2*h*; many cars of this type have advertised prices just below $2000 or $3000.

```
0h || 95
1l || 30,39,10,39
1h || 90,80,70,50,50,50,99,50,75,79,80,99,80
2l || 49,00,39,00,39,48
2h || 98,98,90,55,60,85,55,90,99,68
3l || 20,28,19
3h || 99,99,75,92,80
4l || 39,48
4h || 88
5l || 20,49,30,30             Stems: Thousands
5h || 87                      Leaves: Tens
```

FIGURE 5 STEM-AND-LEAF DISPLAY FOR USED-CAR PRICES USING REPEATED STEMS ■

Computer-Generated Stem-and-Leaf Displays

The computer is a very powerful tool for doing statistical analysis, because it can perform routine data organization and arithmetic calculations many times as fast as these tasks can be done by hand. Most such analysis is done using prepared packages of statistical computer programs. A package of this sort is one in which the user takes advantage of a program that has already been written. The user need only enter the data properly and then give the computer a command that causes the desired operation to be performed. The most frequently used packages are MINITAB, BMDP (Biomedical Computer Programs), SAS (Statistical Analysis System), and SPSS (Statistical Package for the Social Sciences). Almost all the methods of analysis discussed in this book can be carried out using any one of these four packages.

Figure 6 pictures a MINITAB stem-and-leaf display for the used-car price data of Example 2. Whereas the leaves in Figure 5 are two-digit numbers, MINITAB drops the second digit and uses only one-digit leaves. Unfortunately, the user has no say in this—MINITAB itself chooses both the stems and leaves. It did decide to use repeated stems, with . replacing h and * replacing l. Additionally, within each stem the leaves are automatically ordered from smallest to largest.

```
STEM-AND-LEAF DISPLAY OF PRICE

     +0. 9
      1* 1333
      1. 5555777888999
      2* 003344
      2. 5566899999
      3* 122
      3. 78999
      4* 34
      4. 8
      5* 2334
      5. 8
```

FIGURE 6 A COMPUTER-GENERATED STEM-AND-LEAF DISPLAY USING MINITAB

Using a Stem-and-Leaf Display for Comparison

Frequently an analyst has two groups of data and wishes to see if they differ in some fundamental way. A comparative stem-and-leaf display, in which the leaves from one group are listed to the right of the stem values, whereas those from the other group extend out to the left, can give preliminary visual impressions and insights.

EXAMPLE 3

The Institute of Nutrition of Central America and Panama (INCAP) has carried out extensive dietary studies in various parts of Central America. One such study, reported on in the paper "The Blood Viscosity of Various Socioeconomic Groups in Guatemala" (*Amer. J. of Clinical Nutrition* (Nov. 1964):303–07) determined values of various physiological characteristics for several groups of Guatemalans. The stem-and-leaf display pictured in Figure 7 gives serum total cholesterol values (mg/l) both for a sample of high-income urban individuals and for a sample of low-income rural Indians. The first sample contains one value, 330, that is far above the rest of the data. Rather than extend the stems to capture this value at the bottom of the display, it is marked at the bottom with the symbol HI. This is routinely done in computer-generated displays for unusually high or low values.

A first impression is one of great variability of the cholesterol values in each sample considered separately. For low-income rural Indians, values run from 95 to 231, a range of $231 - 95 = 136$. Disregarding for the moment the one outlying value on the high end of display, the range for high-income urban individuals is $284 - 133 = 151$, not greatly different from the range

I	Stem	II
	9	5
	10	8,8
	11	5,4
	12	9,9,4
3,4	13	5,1,6,6,1,9
	14	0,6,4,5,2,3,8,3,4,2
5	15	2,8,7,2,5,8
	16	6,5,2
9,5,0	17	5,4,3,2,1
1,4,8,9	18	0,9,1
9,7,0,6	19	2,4,7
1,5,4,5,5,0,1,0,6	20	4
4,7	21	
2,7,8,7,2	22	3,6,0
4,6,4,9	23	1
2,9,4,1	24	
2	25	
	26	
9,3	27	
HI: 330 4,4,4	28	

Stems: Tens
Leaves: Ones

FIGURE 7 COMPARATIVE STEM-AND-LEAF DISPLAY OF SERUM TOTAL CHOLESTEROL VALUES
(I) Urban Guatemalans
(II) Low-Income Rural Indian Guatemalans

for group II. A reasonable central value for group I is one with a stem of 20, and for group II a central value is one in the high 140s or low 150s. The shapes of the two sides of the display are rather similar, each rising to a peak near a central value and then declining. Roughly speaking, the main difference between the two groups is in location. If we push the display for group II down five or six stem values (50 or 60 cholesterol units), the two halves would be quite similar. The suggested explanation for this difference in location is the presence of more fats and more saturated fats in group I diets. A formal statistical analysis would yield more precise information on the size of the shift in location. ■

EXERCISES

2.4 The Bureau of Justice's Statistics Bulletin on jail inmates for 1982 reported the following inmate population sizes for 40 of the smaller federal prisons:

644	512	448	730	401	450	419	647	792	885
501	458	755	569	417	405	509	440	402	624
603	599	791	407	433	559	777	856	492	400
484	554	634	553	723	565	424	417	524	468

Using stems 4, 5, 6, 7, and 8, construct a stem-and-leaf display for this data.

2.5 The accompanying values are rental rates per foot for boat storage at the 19 marinas in Marina del Rey and the 17 marinas at the Los Angeles–Long Beach Harbor (*Source: Los Angeles Times,* June 5, 1983).

Marina del Rey				Los Angeles–Long Beach			
$6.37	$6.60	$6.27	$6.49	$4.60	$4.75	$4.70	$8.75
$6.64	$6.82	$7.16	$6.45	$4.50	$5.40	$6.00	$6.00
$5.60	$5.95	$4.50	$6.60	$6.50	$6.00	$5.00	$5.00
$6.00	$6.82	$7.04	$5.30	$5.50	$4.35	$4.50	$5.20
$7.05	$7.05	$6.96		$4.95			

Construct a comparative stem-and-leaf display for rent per foot for the two areas. What conclusions concerning differences between the two locations can you draw from the stem-and-leaf display?

2.6 The January 1983 issue of *Consumer Reports* published the 7-day yields (in percent, for the week ending November 10, 1982) for the 36 largest money market funds.

8.5	8.7	6.7	9.9	6.2	7.5	8.9	9.0	9.1	9.1	9.2	9.5
9.0	9.5	9.3	8.7	9.0	8.2	9.6	9.6	10.0	9.0	7.4	7.5
9.5	8.9	8.8	9.7	8.9	9.4	8.7	9.5	9.2	9.2	9.3	9.8

Construct a stem-and-leaf display for this data. You may want to use repeated stems to obtain a more informative display.

2.7 The Los Angeles Board of Education has enacted a policy that prohibits students who do not have a C average from participating in extracurricular activities. The *Los Angeles Times* (May 17, 1983) reported the percentage of students who were declared ineligible for 47 Los Angeles high schools. Figures were reported separately for athletes and nonathletes.

	Percent Ineligible									
Athletes	27	12	15	15	21	15	17	14	21	27
Nonathletes	24	25	48	17	3	3	14	18	52	22
Athletes	15	10	36	19	29	35	16	13	17	18
Nonathletes	4	14	29	18	13	17	14	23	25	15
Athletes	13	28	24	40	35	10	16	37	26	18
Nonathletes	15	6	34	48	45	12	38	28	29	34
Athletes	8	21	17	21	20	15	39	37	10	23
Nonathletes	14	18	17	7	24	34	44	13	16	14
Athletes	14	12	16	8	29	17	9			
Nonathletes	15	20	18	9	30	25	8			

Construct a comparative stem-and-leaf display of the percent of ineligible students for athletes and nonathletes. Based on your stem-and-leaf display, do you think there is evidence that the percentage of students disqualified tends to be smaller for nonathletes than for athletes? Justify your answer.

2.8 The 1982 American Statistical Association proceedings of the Social Statistics Section included a paper titled "The Cost of Tenure/Promotion." In that paper the authors reported the age for each tenured faculty member at the University of Richmond in 1980.

34	26	48	28	32	34	45	49	41	55	39	30	32
34	42	43	53	59	66	59	55	52	46	66	50	47
52	48	35	44	43	33	34	41	35	34	43	49	56
52	58	48	50	60	54	53	41	36	40	33	30	39
30	34	45	34	59	42	44	43	46	34	65	37	49
34	48	36	53	42	43	32	59	43	33	28	67	31
37	39	43	54	58	45	44	52	44	43	44	48	42
38	35	43	34	34	40	34	37	36	52	43	46	58
48	45	43	47	56	49	34	35	42	44	33	30	37
41	57	57	59	50	58	53	59	62	43	53	43	57
34	37	43	44	39	49	39	37	54	38	59	41	44
32	37	58	70	62	63							

Construct a stem-and-leaf display for this data set. Use your stem-and-leaf display to determine the proportion of the faculty who will be eligible to retire during the next 5 years (assume that a person becomes eligible to retire at age 65).

2.9 The paper "Registration and Voting" in *Statistics: A Guide to the Unknown* (see the Chapter 1 references) gave voter registration information for 104 cities. Given below is the percent of the voting-age population that was registered to vote in 1960 for each of the cities.

96.4	92.6	92.5	92.0	92.0	91.9	91.2	90.5	90.4	89.4
89.3	88.4	87.7	87.4	87.3	87.3	87.0	85.8	85.1	84.9
84.7	84.0	83.2	83.2	83.0	82.4	82.4	82.0	81.9	81.7
81.4	81.3	81.2	81.1	81.1	81.0	81.0	80.9	80.4	79.8
79.6	79.4	79.4	79.3	79.2	78.9	78.8	77.6	77.3	77.1
77.1	77.0	77.0	76.9	75.8	75.6	75.1	74.9	74.0	73.9
73.8	73.6	72.4	71.9	71.5	71.2	70.7	70.7	70.6	69.9
69.7	68.8	68.5	68.4	68.1	68.0	67.7	67.7	66.6	65.8
65.7	65.0	64.7	62.2	61.8	61.4	61.2	60.0	60.0	59.2
59.0	55.9	55.6	54.9	48.4	48.3	46.5	43.6	42.6	39.1
38.0	35.0	33.8	32.1						

Construct a stem-and-leaf display for this data set. How would you describe this distribution of values?

2.3 Frequency Distributions

A stem-and-leaf display can be constructed only for numerical data. We also want to have a way of displaying data from a categorical sample. In addition, because the stem-and-leaf display lists all values in the data set, it can require too much space when the data set is large. A frequency distribution is useful for summarizing even a very large data set in a compact way. The easiest case is when the data is categorical, so we'll start there, then look at count data, and finally consider data on a continuous variable.

Frequency Distributions for Categorical Data

The **frequency distribution** for a categorical data set is obtained by tallying the responses category by category and then recording the count, or **frequency,** for each category. For example, we might wish to examine data from a survey of motorcycle owners in which each owner in a sample was asked which brand he or she had first owned. Possible responses include Honda, Yamaha, Kawasaki, Suzuki, and Harley-Davidson. There are other brands, but since not many people own one of these others, we can include all such responses in a single catchall category, Other. Sample responses might then be *Y, H, Y, K, H, H,* . . . , resulting in a Honda frequency of 24, a Yamaha frequency of 21, and so on.

Even more important than the frequency itself is the relative frequency, or proportion, of all responses in the data set that fall into a category. For example, if there are 24 Honda owners in a sample of size 100, the relative frequency for the Honda category is 24/100 = .24 (or 24%).

DEFINITION

The **relative frequency** for any particular category is the proportion of all responses in the data set that fall in the category:

$$\text{relative frequency} = \frac{\text{frequency}}{\text{total number of responses}}$$

EXAMPLE 4

The sample of individuals to be included in a survey is often obtained from a list of voters, a telephone directory, or other master list. An attempt is then made to contact each individual in the sample. Unfortunately, not all individuals can be reached on the first attempt or even on future attempts. The paper "I Hear You Knocking But You Can't Come In: The Effects of Reluctant Respondents and Refusers on Sample Survey Estimates" (*Sociological Methods and Research* (Aug. 1982):3–32) reported on a study of how the composition of the group actually contacted changed as the number of attempts to contact increased. Table 1 presents a frequency distribution for labor force categories both after the first attempt to contact and after the tenth (and final) attempt to contact those who had not already been interviewed on earlier attempts. Of the 1365 individuals in the target sample, 260 refused to be interviewed and 56 could never be reached. The relative frequencies should have a sum of 1 but don't here because of the effects of rounding. Similarly, the percentages should add to 100. The relative frequencies and corresponding percentages differ substantially in the two distributions. For example, only 28.6% of those contacted on the first attempt work full time, but this increases to 52.4% in the final sample. People in different categories of the labor force might have divergent opinions on issues such as welfare reform and government unions. A sample that was not representative, such as the one consisting of all individuals contacted on the first attempt, might then provide misleading information. This potential for misleading information is why the issue of nonresponse is of such concern to pollsters.

TABLE 1

FREQUENCY DISTRIBUTION FOR CATEGORY OF LABOR FORCE
(a) After One Attempt to Contact Individuals in the Selected Sample
(b) After 10 Attempts to Contact Individuals in the Selected Sample

	(a)			(b)		
Labor Force Status	Frequency	Relative Frequency	Percent	Frequency	Relative Frequency	Percent
Full time	67	.286	28.6	550	.524	52.4
Part time	24	.103	10.3	113	.108	10.8
Looking or laid off	12	.051	5.1	50	.048	4.8
Retired	53	.226	22.6	138	.132	13.2
Not working	78	.333	33.3	198	.189	18.9
Total	234	.999	99.9	1049	1.001	100.1

■

Frequency Distributions for Count Data

An automobile manufacturer might want information on the number of visits to dealers made by new-car owners during the warranty period for the purpose of correcting defects. A sample of owners would yield some who had made no visits, some who had made just one visit, others who had made two visits, and so on. Instead of categories, the data now consist of possible values of the count variable *number of visits*. A frequency distribution would then display the possible values and the number of times each occurred in the data set (frequency of each value). As with categorical data, dividing each frequency by the total number of values in the data set (total frequency) yields the relative frequencies.

EXAMPLE 5

A sample of 708 bus drivers employed by public corporations was selected and the number of traffic accidents in which each was involved during a 4-year period was recorded ("Application of Discrete Distribution Theory to the Study of Noncommunicable Events in Medical Epidemiology," *Random Counts in Biomedical and Social Sciences,* ed. G. P. Patil, Pennsylvania State University Press, University Park, PA., 1970). A listing of all 708 values would look something like this: 3, 0, 6, 0, 0, 2, 1, 2, 4, 1, ... The largest sample value was 11. (You might think that a driver with 11 accidents would be an ex-driver, but there was no information as to who was at fault in each accident).

The frequency distribution (Table 2) shows that 117 of the 708 drivers sampled, a proportion of .165, or 16.5% of the sample, had no accidents. We might want to know what proportion of the drivers had at most 1 accident. Since at most one means either 0 or 1, we add the corresponding relative frequencies: .165 + .222 = .387, or 38.7%. Similarly, the proportion of drivers involved in at most 5 accidents is .165 + .222 + · · · + .062 = .944. The proportion of drivers involved in at least 6 accidents is the sum of relative frequencies for the count values 6, 7, . . . , 11, which is .030 + · · · + .001 = .054. This is also the proportion with more than 5 accidents, since for a count variable *at least* 6 and *more than* 5 are equivalent statements.

TABLE 2

FREQUENCY DISTRIBUTION FOR NUMBER OF ACCIDENTS BY BUS DRIVERS

Number of Accidents	Frequency	Relative Frequency
0	117	.165
1	157	.222
2	158	.223
3	115	.162
4	78	.110
5	44	.062
6	21	.030
7	7	.010
8	6	.008
9	1	.001
10	3	.004
11	1	.001
	708	.998

Every one of the 708 drivers sampled had either at most 5 accidents or at least 6 accidents, so adding the proportions for these two conditions should give 1. Once one of these two proportions is known, the other can be calculated by subtracting the known proportion from 1, so

$$\begin{pmatrix}\text{proportion of drivers with}\\ \text{at least 6 accidents}\end{pmatrix} = 1 - \begin{pmatrix}\text{proportion of drivers with}\\ \text{at most 5 accidents}\end{pmatrix}$$
$$= 1 - .944 = .056$$

This answer differs slightly from the earlier one of .054 because in rounding we obtained relative frequencies that had a sum of .998 rather than 1.000. ■

Frequently a data set contains a few large values separated by quite a bit from the bulk of the observations. For example, consider adding two more drivers, one with 16 accidents and one with 21 accidents, to the 708 drivers of Example 5. Rather than list individual count values all the way to 21, we might stop listing at 10 and add one further category, *at least 11* (often written ≥ 11). Then 3 of the 710 observations (the 11, the 16, and the 21) would belong in this category. Table 3 presents the resulting computer-generated frequency distribution.

TABLE 3

COMPUTER-GENERATED FREQUENCY DISTRIBUTION FOR NUMBER OF ACCIDENTS USING SPSS

CATEGORY LABEL	CODE	ABSOLUTE FREQ	RELATIVE FREQ (PCT)
0	0	117	16.5
1	1	157	22.1
2	2	158	22.3
3	3	115	16.2
4	4	78	11.0
5	5	44	6.2
6	6	21	3.0
7	7	7	1.0
8	8	6	.8
9	9	1	.1
10	10	3	.4
AT LEAST 11	11	3	.4
	TOTAL	710	100.0

Frequency Distributions for Continuous Data

A frequency distribution for categorical data necessitated a listing of the possible categories. For count data, we used count values in place of categories. The difficulty with continuous data, such as observations on miles per gallon for a sample of 100 automobiles, is that there are no natural categories. Before we can compute and list frequencies, we need something analogous to categories. The way out of this dilemma is to define our own categories. For miles-per-gallon data, suppose that we mark off some intervals on a horizontal miles-per-gallon measurement axis, as pictured in Figure 8. Each data value should fall in exactly one of these intervals. If the smallest observation were 25.3 and the largest were 29.8, we might use intervals of width .5, with the first one starting at 25.0 and the last one ending at 30.0. The resulting intervals are called **class intervals,** or just **classes.** The classes play the same role that the categories played earlier, with frequencies and relative frequencies tabulated as before.

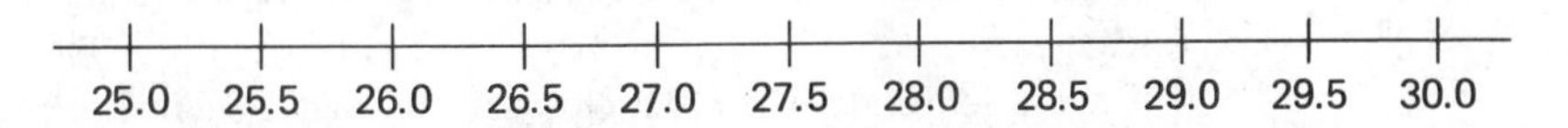

FIGURE 8 SUITABLE CLASS INTERVALS FOR MILES-PER-GALLON DATA

There is one further difficulty: Where should we place an observation such as 27.0, which falls on a boundary between classes? Our convention will be to define intervals so that such an observation is placed in the upper rather than the lower class interval. Thus in our frequency distribution, a typical class will be 26.5 – < 27.0, where the symbol < is a substitute for the phrase *less than*. The observation 27.0 would then fall in the class 27.0 – < 27.5.

EXAMPLE 6

An interesting problem in microbiology is to determine the lengths of macromolecules in order to obtain information about molecular weights. The paper "Estimation of the True Length of Broken Molecules" (*Biometrics* (1982): 201–13) reported data on segment lengths of a certain type of denatured RNA molecule observed with the aid of an electron micrograph. A frequency distribution appears in Table 4. The class intervals all have width .20. The first one begins at .05 rather than 0, since very small fragments cannot be counted using the electron micrograph. An interesting aspect of the frequency distribution is that the frequencies decline in a reasonably smooth fashion through the interval 1.65 – < 1.85, but then there is a substantial jump in the next two intervals. The authors provided an explanation for this jump.

TABLE 4

FREQUENCY DISTRIBUTION FOR SEGMENT LENGTHS OF RNA MOLECULES (MICRONS)

Class	Class Interval	Frequency	Relative Frequency
1	.05–< .25	298	.2326
2	.25–< .45	211	.1647
3	.45–< .65	149	.1163
4	.65–< .85	116	.0906
5	.85–<1.05	84	.0656
6	1.05–<1.25	78	.0609
7	1.25–<1.45	61	.0476
8	1.45–<1.65	60	.0468
9	1.65–<1.85	56	.0437
10	1.85–<2.05	63	.0492
11	2.05–<2.25	79	.0617
12	2.25–<2.45	20	.0156
13	2.45–<2.65	5	.0039
14	2.65–<2.85	1	.0008
		1281	1.0000

We can now add relative frequencies for various individual class intervals to obtain proportions associated with wider intervals on the measurement axis. A segment whose length is less than .85 belongs in one of the first four class intervals, so the proportion of segments in the sample whose lengths are less than .85 is the sum of relative frequencies for these four intervals: .2326 + .1647 + .1163 + .0906 = .6042, or roughly 60%. The proportion of segments whose lengths are *not* less than .85 (that is, whose lengths are at least .85) is then 1 − .6042 = .3958, or roughly 40%. This is also the result if we add relative frequencies for the last 10 class intervals (beginning with .85–<1.05). Similarly, the proportion of segment lengths between 1.25 and 2.25 is the sum of relative frequencies for the five intervals beginning with

1.25–<1.45 and ending with 2.05–<2.25: .0476 + · · · + .0617 = .2490, or roughly 25%. ■

There are no strict guidelines for selecting either the number of class intervals or the interval lengths. Using a few relatively wide intervals will bunch the data, whereas using a great many relatively narrow intervals may spread the data out too much over the intervals, so that no interval contains more than a few observations. Neither type of distribution will give an informative picture of how values are distributed over the range of measurement. Generally speaking, with a small amount of data one should use relatively few intervals, perhaps between 5 and 10, whereas with a large amount of data a distribution based on from 15 to 20 (or even more) intervals is often recommended. Two people making reasonable and similar choices for number of intervals, width, and starting point should obtain very similar pictures of the data.

Class Intervals of Unequal Width

Figure 9 pictures a data set in which there are a great many observations concentrated near one another at the center of the set and just a few outlying, or stray, values both below and above the main body of data. If a frequency distribution is based on short intervals of equal width, a great many intervals will be required to capture all observations, and many of them will contain no observations (zero frequency). On the other hand, only a few wide intervals will capture all values, but then most of the observations will be grouped into a very few intervals. Neither choice will yield an informative distribution. In such a situation it is best to use a few relatively wide class intervals at the ends of the distribution and some shorter intervals in the middle.

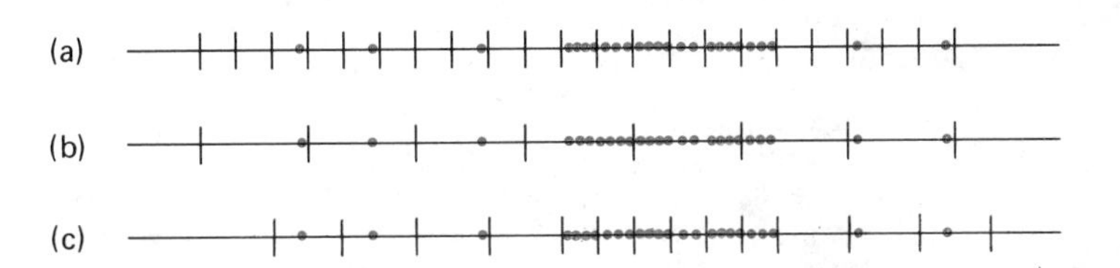

FIGURE 9 THREE DIFFERENT CHOICES OF CLASS INTERVALS FOR A DATA SET WITH OUTLIERS
(a) Many Short Intervals of Equal Width
(b) A Few Wide Intervals of Equal Width
(c) Intervals of Unequal Width

EXAMPLE 7

Often individuals are asked to give correct values of characteristics such as age, weight, and years of education. How accurate are these self-reported values? The paper "Self-Reports of Academic Performance" (*Sociological Methods and Research* (Nov. 1981):165–85) discussed a study of how accurately SAT scores and grade-point averages were reported (many exploratory techniques, including stem-and-leaf displays, were used in the paper). For each student in a sample, the difference between reported grade-point average and actual grade-point average was calculated. A positive difference corresponds to a reported value that was larger than the actual value, whereas a negative value corresponds to underreporting the actual value. Most sample observations were quite near zero, but there were some rather gross errors in both directions. The largest difference was +1.75 (delusions of grandeur?) and the smallest was −1.07 (an extreme case of modesty?).

Table 5 displays a frequency distribution for one choice of class intervals. The two extreme intervals are quite wide, while the middle intervals are narrower. Because one tends to focus on frequencies without considering the effects of unequal widths, the distribution may give some misleading impressions about how values are distributed on the measurement axis. For example, the jump in frequency in going from the twelfth to the thirteenth interval (31 to 47) occurred only because the thirteenth interval is much wider than the twelfth and not because of a "spike" in the distribution. The next section introduces a picture, called a *histogram,* as a way of conveying stronger and more accurate impressions concerning a distribution. Of particular interest is the fact that more than 10% (9.0 + 1.5 + .8 = 11.3) of the students overreported their grade-point average by at least .5, while only 1% (.2 + .8) underreported by at least that much. The data certainly casts doubt on the wisdom of any policy that uses such self-reported values to make substantive decisions (such as using self-reported high school grade-point averages in making college-admission decisions).

TABLE 5

FREQUENCY DISTRIBUTION FROM SPSS FOR ERRORS IN REPORTED GRADE-POINT AVERAGE

CATEGORY LABEL	CODE	ABSOLUTE FREQ	RELATIVE FREQ (PCT)
-1.5-<-1.0	1	1	.2
-1.0-<- .5	4	4	.8
- .5-<- .4	3	7	1.3
- .4-<- .3	4	11	2.1
- .3-<- .2	5	18	3.4
- .2-<- .1	6	51	9.7
- .1-< 0.0	7	110	21.0
0.0-< .1	8	99	18.9
.1-< .2	9	73	13.9
.2-< .3	10	34	6.5
.3-< .4	11	27	5.1
.4-< .5	12	31	5.9
.5-< 1.0	13	47	9.0
1.0-< 1.5	14	8	1.5
1.5-< 2.0	15	4	.8
		-------	-------
	TOTAL	525	100.0

■

Cumulative Relative Frequencies

Interesting information can often be obtained by adding together, or accumulating, the relative frequencies from the lowest count value or class interval to some intermediate value or interval. For example, the proportion of students at a college who had made at most 3 visits to a health center would be the sum of the proportions with 0, 1, 2, or 3 visits. Similarly, starting from a frequency distribution for playing times of phonograph records with 19–<20 as one class interval, the proportion of records whose playing times were less than 20 minutes would be the sum of the relative frequencies for this interval and all intervals below it on the measurement axis.

For count data (observations on a count variable), the **cumulative relative frequency** for any specified count value is the proportion of observations in the data set that are at most the specified value. It is computed by adding together the relative frequencies for the specified value and all smaller values. The cumulative relative frequencies are usually displayed next to the relative frequencies, each being the sum of relative frequencies in that row and all rows above it. Frequency distributions produced by the common statistical computer packages give cumulative relative frequencies as well as the relative frequencies themselves.

EXAMPLE 8

Is it really the case, as it might seem to an unsuccessful and frustrated angler, that 10% of those fishing reel in 90% of the fish caught? More generally, how is the number of fish caught distributed among those who are trying to catch them? Table 6 presents data from a survey of 911 anglers done during a particular time period on the lower Current River in Canada ("Fisherman's Luck," *Biometrics* (1976):265–71). Over 50% of the fishermen—56.53%, to be exact—caught nothing at all.

TABLE 6

FREQUENCY DISTRIBUTION FOR SIZE OF CATCH AMONG 911 ANGLERS

Number of Fish Caught	Number of Anglers (Frequency)	Relative Frequency	Cumulative Relative Frequency
0	515	.5653	.5653
1	65	.0714	.6367
2	60	.0659	.7026
3	66	.0724	.7750
4	53	.0582	.8332
5	55	.0604	.8936
6	27	.0296	.9232
7	25	.0274	.9506
8	25	.0274	.9780
9	20	.0220	1.0000
	911	1.0000	

The proportion of anglers who caught at most 1 fish is .5653 + .0714 = .6367, the cumulative relative frequency for the count value 1. Similarly, the cumulative relative frequency for the count value 5 is .8936 (the sum of relative frequencies for the count values 0, 1, . . . , 5), so roughly 90% of the fishermen in the sample caught at most 5 fish. Notice that the 640 fishermen who caught at most 2 fish caught a total of (0)(515) + (1)(65) + (2)(60) = 185 fish. A similar calculation shows that a total of 1585 fish were caught. Thus the 70% of the fishermen who caught at most two fish actually accounted for only 185/1585 = .117, or 11.7%, of the total catch. This is not as bad as 90% accounting for only 10% of the catch but will certainly not encourage the novice angler! ■

Each cumulative relative frequency is obtained by adding relative frequencies. It is also straightforward to go in the opposite direction and calculate the relative frequencies from given cumulative relative frequencies. Consider the count value 4, for which the cumulative relative frequency includes relative frequencies for 0, 1, 2, and 3 as well as for 4. To obtain the relative frequency for 4 alone, we must eliminate the unwanted relative frequencies by subtraction: Relative frequency for 4 = (cumulative relative frequency for 4) − (cumulative relative frequency for 3). In Example 8, this gives .8332 − .7750 = .0582, as expected. More generally,

$$\begin{pmatrix}\text{relative frequency}\\ \text{for a specified}\\ \text{count value}\end{pmatrix} = \begin{pmatrix}\text{cumulative relative}\\ \text{frequency for that}\\ \text{count value}\end{pmatrix} - \begin{pmatrix}\text{cumulative relative}\\ \text{frequency for next}\\ \text{smallest count value}\end{pmatrix}$$

Thus in Example 8, (relative frequency for 6) = (cumulative relative frequency for 6) − (cumulative relative frequency for 5) = .9232 − .8936 = .0296. Both the set of relative frequencies and the set of cumulative relative frequencies contain the same information, since either set can be obtained from the other.

A frequency distribution for continuous data depends on a choice of class intervals, and each relative frequency is the proportion of observations in the data set that fall in the associated class interval. The **cumulative relative frequency** for a particular class interval is the proportion of all observations that are less than the upper limit of the class interval. If 75 of the 100 observations in the data set fall either in the interval 19–<20 or intervals below this one, then the cumulative relative frequency for this interval is 75/100 = .75. That is, 75% of the observations have values less than 20. Observations that have value 20 are not included in the 75%, since such observations fall in the class interval 20–<21. The distinction between *less than 20* and *at most 20* is important if there are observations at the boundary value 20. Each cumulative relative frequency in a particular row is again the sum of the relative frequency in that row and in all rows above it in the frequency distribution.

EXAMPLE 9

The concentration of suspended solids in river water is an important environmental characteristic. The paper "Water Quality in Agricultural Watershed: Impact of Riparian Vegetation During Base Flow" (*Water Resources Bulletin* (1981): 233–39) reported on concentration (in parts per million, or ppm) for several different rivers. A frequency distribution of concentration for water samples from one such river appears in Table 7. The first two class intervals have zero frequency, but this was not true for other rivers studied.

The cumulative relative frequency for the interval 40– <50 is the sum of the relative frequencies for that interval and all four intervals below it on the measurement scale: .00 + .00 + .02 + .16 + .16 = .34. Thus 34% of the values in the data set are less than 50. This also implies that the proportion of values that are at least 50 is 1.00 − .34 = .66. Similarly, the cumulative relative frequency for 80– < 90 is .96, so only 4% of the values are at least 90.

As with count data, each relative frequency is the difference between the cumulative relative frequency for that interval and the one below it on the measurement axis. For example, the cumulative relative frequencies for 50– < 60 and 60– < 70 are .46 and .78, respectively, so the relative frequency for 60– < 70 is .78 − .46 = .32.

TABLE 7 FREQUENCY DISTRIBUTION FOR SUSPENDED SOLID CONCENTRATION IN RIVER WATER (ppm)

Class Interval	Frequency	Relative Frequency	Cumulative Relative Frequency
0–< 10	0	.00	.00
10–< 20	0	.00	.00
20–< 30	1	.02	.02
30–< 40	8	.16	.18
40–< 50	8	.16	.34
50–< 60	6	.12	.46
60–< 70	16	.32	.78
70–< 80	7	.14	.92
80–< 90	2	.04	.96
90–<100	2	.04	1.00
	50	1.00	

EXERCISES

2.10 The United States Commerce Department compiles information on per capita income by state. The following values are for the year 1982.

Per Capita Income

New England		Mideast	
Connecticut	$13,687	New Jersey	$13,027
Massachusetts	11,921	New York	12,328
Rhode Island	10,730	Maryland	12,194
New Hampshire	10,710	Delaware	11,796
Vermont	9,446	Pennsylvania	10,943
Maine	9,033		

Great Lakes		Plains	
Illinois	12,162	Kansas	11,448
Michigan	11,052	Minnesota	11,082
Ohio	10,783	North Dakota	10,476
Wisconsin	10,497	Iowa	10,532
Indiana	10,109	Nebraska	10,489
		Missouri	10,175
		South Dakota	9,506

Southeast		Southwest	
Virginia	11,003	Texas	11,352
Florida	10,875	Oklahoma	10,776
Louisiana	10,083	Arizona	10,201
Georgia	9,514	New Mexico	8,997

Southeast (continued)		Rocky Mountain	
North Carolina	9,032		
Kentucky	8,861		
West Virginia	8,856		
Tennessee	8,849	Wyoming	11,970
Alabama	8,581	Colorado	11,776
South Carolina	8,468	Montana	9,750
Arkansas	8,332	Idaho	9,259
Mississippi	7,792	Utah	8,733

Per Capita Income

Far West		Others	
California	12,543	Alaska	15,200
Nevada	11,748	Hawaii	11,602
Washington	11,635	District of Columbia	14,347
Oregon	10,392		

a. Construct a frequency distribution for per capita income.

b. The 1982 per capita income for the United States was reported as $11,506. Use your frequency distribution to approximate the proportion of states (treat the District of Columbia as a state) that have a per capita income above the national value. Then check the original data to see how far off your approximation is.

2.11 The *Journal of Marketing Research* (Feb. 1975) published the results of a study in which 22 consumers reported the number of times that they had purchased a particular brand of a product during the previous 48-week period. The results are shown below.

0	2	5	0	3	1	8	0	3	1	1
9	2	4	0	2	9	3	0	1	9	8

a. Construct a relative frequency distribution for the number of purchases.

b. What proportion of the shoppers in this study never bought the brand under investigation?

c. Suppose that each of the 22 shoppers in this study had made exactly nine purchases of the product during the previous 48 weeks. What proportion of the shoppers purchased the brand under investigation more than half of the time? All of the time?

2.12 In the paper "Reproduction in Laboratory Colonies of Bank Vole" (*Oikos* (1983): 184), the authors presented the results of a study on litter size. (According to Webster's, a vole is a small rodent with a stout body, blunt nose, and short ears!) As each new litter was born, the number of babies was recorded, and the following results were obtained.

Size of Litter

3	6	5	6	5	7	5	7	6	6	6	4	6
5	6	4	3	5	6	4	5	9	6	5	6	1
9	7	8	3	7	4	5	5	6	7	3	6	6
9	4	5	7	5	6	8	6	4	7	5	7	4
5	8	6	7	2	7	7	3	3	5	4	6	4
6	3	7	8	5	7	7	7	7	9	8	7	6
7	6	4	7	10	5	2	3	6	6	4	7	6
7	5	5	5	7	5	8	8	4	9	7	5	4
6	5	8	4	5	6	6	3	6	8	6	8	6
5	8	6	11	4	7	6	8	9	7	3	8	3
4	6	4	5	7	5	6	5	7	6	9	3	5
9	7	5	6	7	5	8	6	8	8	6	5	7
4	8	7	7	7	5	3	8	6	10	4	5	5
5												

a. Construct a relative frequency distribution for this data.

b. What proportion of the litters was greater than 6 in size? Between 3 and 8 (inclusive)?

c. Is it easier to answer questions like those posed in (b) using the relative frequency distribution than it would be using the raw data given above? Explain.

2.13 The results of the 1980 census included a state-by-state listing of population density. The values given below are the number of people per square mile for each of the 50 states.

New Jersey	986.2	Rhode Island	897.8	Massachusetts	733.3
Connecticut	637.8	Maryland	428.7	New York	370.6
Delaware	307.6	Pennsylvania	264.3	Ohio	263.3
Illinois	205.3	Florida	180.0	Michigan	162.6
Indiana	152.8	California	151.4	Hawaii	150.1
Virginia	134.7	North Carolina	120.4	Tennessee	111.6
South Carolina	103.4	New Hampshire	102.4	Louisiana	94.5
Georgia	94.1	Kentucky	92.3	Wisconsin	86.5
West Virginia	80.8	Alabama	76.7	Missouri	71.3
Washington	62.1	Vermont	55.2	Texas	54.3
Mississippi	53.4	Iowa	52.1	Minnesota	51.2
Oklahoma	44.1	Arkansas	43.9	Maine	36.3
Kansas	28.9	Colorado	27.9	Oregon	27.4
Arizona	23.9	Nebraska	20.5	Utah	17.8
Idaho	11.5	New Mexico	10.7	North Dakota	9.4
South Dakota	9.1	Nevada	7.3	Montana	5.4
Wyoming	4.8	Alaska	0.7		

a. Construct a relative frequency distribution for state population density.

b. In your relative frequency distribution of (a), did you use class intervals of equal widths? Why or why not?

c. Use the relative frequency distribution to give an approximate value for the proportion of states that have a population density of more than 100 people per square mile. Does the approximate value agree with the correct value?

2.14 In a study of warp breakage during the weaving of fabric (*Technometrics* (1982):63), 100 pieces of yarn were tested. The number of cycles of strain to breakage was recorded for each yarn sample. The resulting data is given below.

86	146	251	653	98	249	400	292	131	169	175	176
76	264	15	364	195	262	88	264	157	220	42	321
180	198	38	20	61	121	282	224	149	180	325	250
196	90	229	166	38	337	65	151	341	40	40	135
597	246	211	180	93	315	353	571	124	279	81	186
497	182	423	185	229	400	338	290	398	71	246	185
188	568	55	55	61	244	20	284	393	396	203	829
239	236	286	194	277	143	198	264	105	203	124	137
135	350	193	188								

a. Using class intervals 0–<100, 100–<200, and so on, construct a relative frequency distribution for breaking strength.

b. If weaving specifications require a breaking strength of at least 110 cycles, what proportion of the yarn samples would be considered unsatisfactory? Answer using your relative frequency distribution.

2.15 Each of 25 students was asked to identify the dictionary he or she used. The resulting responses were as follows (with A = American Heritage, F = Funk and Wagnalls, M = Macmillan, R = Random House, W = Webster's).

A	R	A	W	W	M	W	R	R	F	A	W	R
R	R	M	W	A	W	R	R	F	W	W	A	

a. Construct a relative frequency distribution for the type of dictionary used.

b. Suppose that the 25 students polled constitute a sample selected from all students at a certain state univeristy. Use the frequency distribution to estimate the proportion of all students at the university who use a Webster's dictionary.

2.16 Compute the cumulative relative frequencies for the data of Exercise 2.12 and use them to answer the following questions.

a. What proportion of observations are at most 8? At least 8?

b. What proportion of litters contain between 5 and 10 (inclusive) offspring?

2.17 The paper "Lessons from Pacemaker Implantations" (*J. Am. Med. Assoc.* (1965):231–32) gave the results of a study that followed 89 heart patients who had received electronic pacemakers. The time (in months) to the first electrical malfunction of the pacemaker was recorded:

24	20	16	32	14	22	2	12	24	6	10	20	8
16	12	24	14	20	18	14	16	18	20	22	24	26
28	18	14	10	12	24	6	12	18	16	34	18	20
22	24	26	18	2	18	12	12	8	24	10	14	16
22	24	22	20	24	28	20	22	26	20	6	14	16
18	24	18	16	6	16	10	14	18	24	22	28	24
30	34	26	24	22	28	30	22	24	22	32		

a. Summarize this data in the form of a frequency distribution using class intervals of 0–<6, 6–<12, and so on.

b. Compute the relative frequencies and cumulative relative frequencies for each class interval of the frequency distribution of part (a).

c. Show how the relative frequency for the class interval 12–<18 could be obtained from the cumulative relative frequencies.

Use the cumulative relative frequencies to give approximate answers to the following questions.

d. What proportion of those who participated in the study had pacemakers that did not malfunction within the first year?

e. If the pacemaker must be replaced as soon as the first electrical malfunction occurs, approximately what proportion required replacement between 1 and 2 years after implantation?

f. Estimate the time at which about 50% of the pacemakers had failed.

g. Estimate the time at which only about 10% of the pacemakers initially implanted were still functioning.

2.18 Birth weights for 302 eighth-born Chinese males born in Singapore are summarized in the accompanying frequency distribution (*Ann. Human Genetics* (1954):58–73).

Weight (in ounces)	Frequency
72–< 80	4
80–< 88	5
88–< 96	19
96–<104	52
104–<112	55
112–<120	61
120–<128	48
128–<136	39
136–<144	19

a. Construct the cumulative relative frequency distribution for this data.

b. What proportion of observed birth weights are less than 96? At least 96? Can you use the given information to determine what proportion of birth weights are at most 96? Explain.

c. Roughly what proportion of birth weights are less than 100? In answering this question, what assumption are you making about the 52 observations in the 96–<104 class interval?

d. Approximately what birth weight is such that 50% of the observed weights are less than that weight value?

2.4 Histograms

The proliferation of numbers in a frequency distribution can sometimes be overwhelming, particularly when the number of categories, possible count values, or class intervals is large. A **histogram** is a pictorial representation of a frequency distribution. Pictures often have more impact and stay with us longer than would tabulated numerical information. We need numbers for the more traditional confirmatory methods of inference, but histograms are great for exploratory purposes.

The general idea is to represent each relative frequency by a rectangle whose area is proportional to the relative frequency. This means, for example, that if the relative frequency for one category is three times that for a second category, then the area of the first rectangle should be three times that of the second rectangle. Similarly, if the relative frequency for one class interval is .16 and for a second is .40—2.5 times the first—then the area of the rectangle for the second interval should be 2.5 times that for the first interval. Once the rectangles have been drawn, we can see at a glance which categories or values occur relatively frequently, which occur relatively infrequently, and how values are distributed throughout the range of possible values.

Histograms for Categorical Data

To construct a histogram for categorical data, we first need a horizontal line on which the rectangles will sit. The base width of each rectangle is the same. We then draw a vertical line and mark off relative frequency values. Because the base widths are the same, areas will be proportional to relative frequencies if the height of each rectangle matches the relative frequency of the corresponding category. Using a relative frequency scale on the vertical axis makes it easy to draw the heights correctly.

EXAMPLE 10

Example 4 presented two frequency distributions for categories of the labor force encountered in a sample. The distribution in the sample resulting from 10 attempts to contact individuals had relative frequencies .524, .108, .048, .132, and .189 for the five categories. The corresponding histogram is given in Figure 10. ■

Let's agree to call the base width of each rectangle 1 (the unit of measurement doesn't matter here). Then

$$\begin{pmatrix}\text{area of a}\\\text{rectangle}\end{pmatrix} = \begin{pmatrix}\text{base}\\\text{width}\end{pmatrix}\cdot(\text{height}) = 1\cdot\begin{pmatrix}\text{relative}\\\text{frequency}\end{pmatrix} = \begin{pmatrix}\text{relative}\\\text{frequency}\end{pmatrix}$$

That is, the area of each rectangle is identical to its relative frequency. Adding relative frequencies of several categories is the same as adding the areas of the

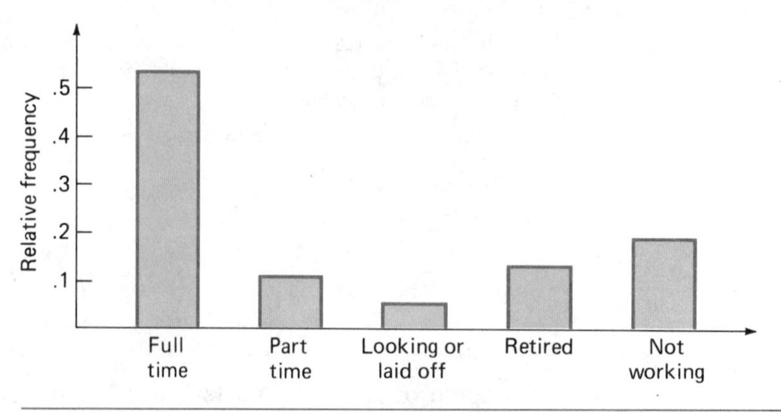

FIGURE 10 HISTOGRAM FOR THE LABOR FORCE FREQUENCY DISTRIBUTION

corresponding rectangles. In particular, the sum of all relative frequencies is 1 and so is the total area of all rectangles.

Histograms for Count Data

After marking a horizontal line with possible count values, each separated by one unit, the histogram for count data is obtained by centering each rectangle at the corresponding count value. For example, the rectangle for the count value 12 would be centered at 12, and its base would extend from 11.5 on the left to 12.5 on the right. Again, it is convenient to draw the rectangles so that their heights match the corresponding relative frequencies. Because the base width of each rectangle is 1, the area of each rectangle is (1) · (relative frequency) = relative frequency. Adding relative frequencies for a group of count values (e.g., for 2, 3, and 4) is equivalent to adding the areas of the corresponding rectangles. The total area of all rectangles in the histogram is 1, the same as the total of all relative frequencies. We shall meet these important properties again for another type of histogram when we discuss probability distributions in Chapter 5.

EXAMPLE 11

Table 2 shows a frequency distribution for the number of accidents in which bus drivers were involved over a 4-year period. The corresponding histogram appears in Figure 11. The histogram shows a peak at the values 1 and 2 and then a smooth decline in relative frequencies as the count value increases. Because the base width of each rectangle is 1, the relative frequency for any particular count value is the area of the rectangle above the count value. The proportion of drivers in the sample with at most two accidents is the sum of the areas of the rectangles above 0, 1, and 2, which is .610. The total area of the rectangles centered at 8, 9, 10, and 11 is .014, so only 1.4% of the drivers had eight or more accidents.

Figure 12 shows a computer-generated histogram (using SPSS). It is not as attractive as a hand-drawn histogram, and this is typical of histograms generated by commercial statistical computer packages. Note that frequencies rather than relative frequencies are displayed. ■

Histograms for Continuous Data

Remember that for continuous data the first step in constructing a frequency distribution was to divide the measurement axis into a number of class inter-

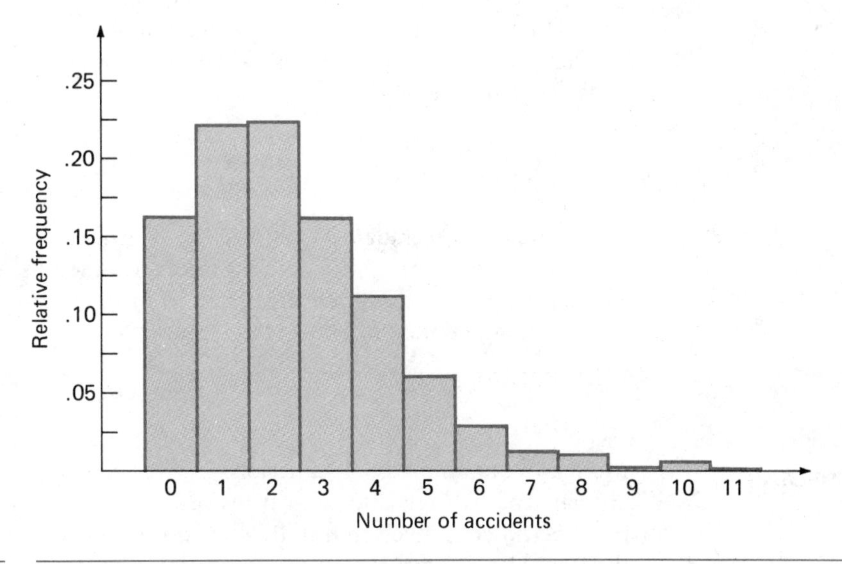

FIGURE 11 HISTOGRAM FOR NUMBER OF ACCIDENTS BY BUS DRIVERS

```
CODE
       I
   0   ****************************** (    117)
       I
   1   *************************************** (    157)
       I
   2   **************************************** (    158)
       I
   3   ****************************** (    115)
       I
   4   ********************* (     78)
       I
   5   ************ (     44)
       I
   6   ****** (     21)
       I
   7   *** (      7)
       I
   8   *** (      6)
       I
   9   * (      1)
       I
  10   ** (      3)
       I
  11   * (      1)
       I
       I.........I.........I.........I.........I.........I
       0        40        80       120       160       200
       FREQUENCY
```

FIGURE 12 HISTOGRAM FROM SPSS FOR NUMBER OF ACCIDENTS BY BUS DRIVERS

vals. The histogram results from drawing a rectangle above each class interval. The easiest case is when each class interval has the same width, so let's consider that case first.

The *area* of a rectangle in a histogram is proportional to the relative frequency of the corresponding class interval. When the class intervals have equal width, we can draw the height of each rectangle to match the corresponding relative frequency.

EXAMPLE 12

Mercury contamination is a serious environmental concern. Mercury levels are particularly high in certain types of fish, so some countries specify maximum permissible levels in fish designated for human consumption. Citizens of the Republic of Seychelles, a group of islands in the Indian Ocean, are among those who consume the most fish in the world, so it is important to gain information about mercury contamination there. The paper "Mercury Content of Commercially Important Fish of the Seychelles, and Hair Mercury Levels of a Selected Part of the Population" (*Environ. Research* (1983): 305–12) reported the following observations on mercury content (parts per million) in the hair of 40 fishermen:

13.26	32.43	18.10	58.23	64.00	68.20	35.35	10.64
33.92	23.94	18.28	22.05	39.14	31.43	18.51	29.56
21.03	5.50	6.96	5.19	28.66	26.29	13.89	40.69
25.87	9.84	26.88	16.81	37.65	19.63	21.82	12.86
31.58	30.13	42.42	16.51	21.16	32.97	9.84	13.80

A reasonable choice for class intervals is to start the first interval at zero and let each one have width 10. The resulting frequency distribution is displayed in Table 8, and the corresponding histogram appears in Figure 13.

TABLE 8

FREQUENCY DISTRIBUTION FOR HAIR MERCURY CONTENT OF SEYCHELLES FISHERMEN (ppm)

Class Interval	Frequency	Relative Frequency
0–<10	5	.125
10–<20	11	.275
20–<30	10	.250
30–<40	9	.225
40–<50	2	.050
50–<60	1	.025
60–<70	2	.050
	40	1.000

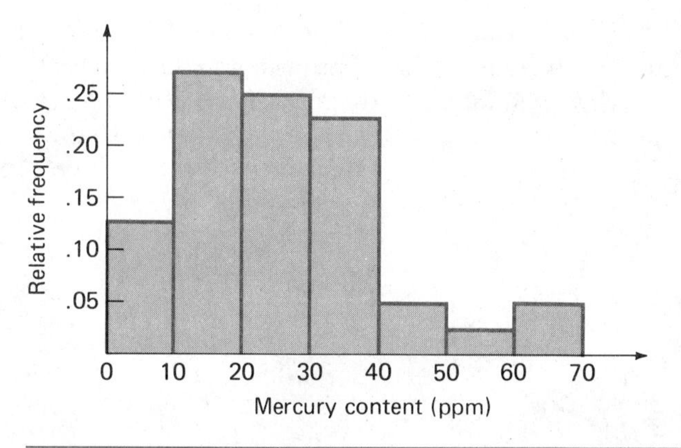

FIGURE 13 HISTOGRAM FOR HAIR MERCURY CONTENT OF SEYCHELLES FISHERMEN

The area of each rectangle in the histogram of Figure 13 is proportional to relative frequency, but the area is not equal to relative frequency. This is because the base width is 10 rather than 1. For example, the area of the rectangle above 20 – <30 is $10 \cdot (.250) = 2.50$, whereas the relative frequency is .250. It is easy to adjust heights so that area equals relative frequency.

> The area of each rectangle in the histogram will equal the corresponding relative frequency of the interval if we let
>
> $$\text{height of rectangle} = \frac{\text{relative frequency}}{\text{width of interval}}$$

This works for class intervals of unequal width also but is especially easy for equal width because each relative frequency is divided by the same number.

When height is determined according to the formula in the box, we label the scale on the vertical axis as the **density scale.** If the histogram is for task completion times in seconds, density is expressed in relative frequency per second, whereas if the histogram is for weights in grams, density is in relative frequency per gram. In general, whatever the unit of measurement for the horizontal axis, the density scale will be in relative frequency per horizontal unit. Then for any class interval, relative frequency = (interval width) · (density).

EXAMPLE 13 (*Example* 12 *continued*) The relative frequencies for the seven class intervals are .125, .275, .250, .225, .050, .025, and .050, respectively. Dividing each by the base width 10 gives the rectangle heights .0125, .0275, .0250, .0225, .0050, .0025, and .0050. The histogram of Figure 14 differs from that of Figure 13 only in that the vertical scale has been changed by a factor of 10. Now the area of the rectangle above the interval from 20 to 30 is $10 \cdot (.0250) = .25$, which is the relative frequency for that interval.

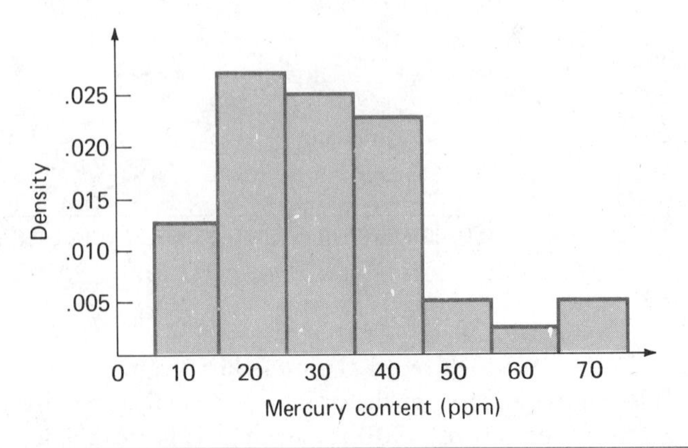

FIGURE 14 HISTOGRAM FOR HAIR MERCURY CONTENT WITH VERTICAL SCALE ADJUSTED SO THAT AREA EQUALS RELATIVE FREQUENCY

Unequal Class Widths (Optional)

When class widths are different, special care is required in drawing a histogram correctly because we must pay attention to area rather than just height. Suppose that two intervals have identical relative frequencies but that the second interval is twice as wide as the first one. Then to make the areas equal, the second rectangle should be only half as high as the first one (see Figure 15).

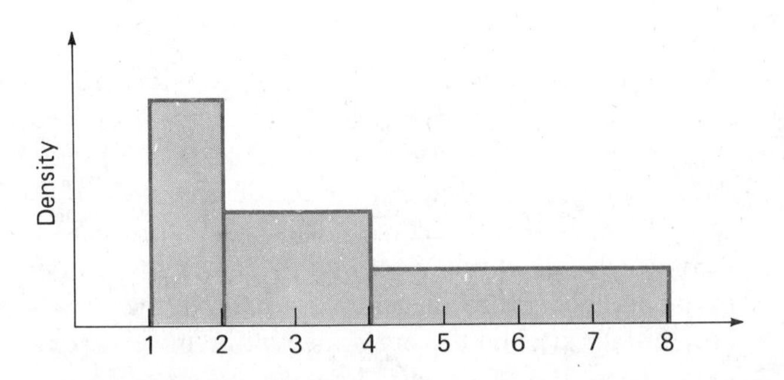

FIGURE 15 RECTANGLES FOR THREE CLASS INTERVALS WITH EQUAL FREQUENCIES AND UNEQUAL WIDTHS

To ensure that the picture is drawn correctly and that area equals relative frequency for each rectangle (so that the total area is 1), compute heights as previously suggested:

$$\text{height of rectangle} = \frac{\text{relative frequency}}{\text{width of interval}}$$

The calculations can be a bit tedious, but it's worth the price to get a correct picture. Remember, the vertical scale is chosen so that area equals relative frequency.

EXAMPLE 14

Example 7 presents a frequency distribution for error in reported grade-point average. In Table 9 we give a frequency distribution for the same data for a different choice of class intervals. The corresponding histogram appears in Figure 16. Although intervals 7, 8, and 9 have roughly the same relative fre-

TABLE 9

FREQUENCY DISTRIBUTION FOR ERRORS IN REPORTING GRADE-POINT AVERAGE

Class	Class Interval	Width	Relative Frequency	Height (Rel. Frequency/Width)
1	−1.3−<−.7	.6	.0038	.0063
2	−.7−<−.4	.3	.0190	.0633
3	−.4−<−.2	.2	.0552	.2760
4	−.2−<−.1	.1	.0971	.9710
5	−.1−<0	.1	.2095	2.0950
6	0−<.1	.1	.1886	1.8860
7	.1−<.2	.1	.1390	1.3900
8	.2−<.4	.2	.1162	.5810
9	.4−<.7	.3	.1200	.4000
10	.7−<1.3	.6	.0343	.0572
11	1.3−<1.9	.6	.0171	.0285

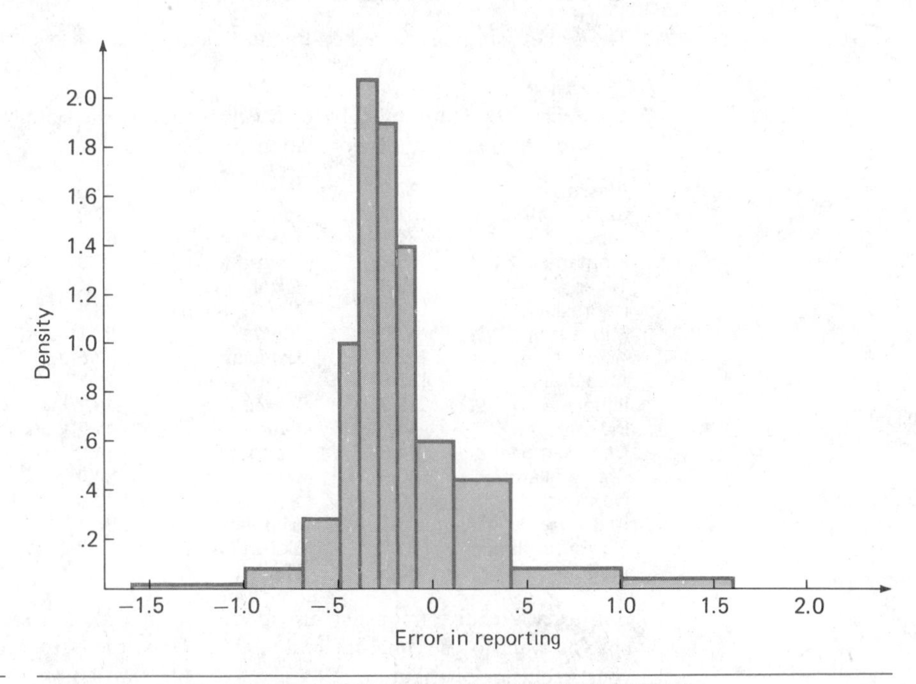

FIGURE 16

HISTOGRAM FOR ERRORS IN REPORTING GRADE-POINT AVERAGE

quencies, the heights of the three rectangles are quite different because the three base widths all differ, and we want the areas to be roughly the same.

The histogram reinforces some impressions gleaned from the frequency distribution. A large proportion of reported values were reasonably close to the true values, but large errors involved overreporting more frequently than underreporting. ■

EXERCISES

2.19 In the article "Associations Between Violent and Nonviolent Criminality" (*Multivariate Behavioral Research* (1981):237–42), the number of previous convictions for 283 adult males arrested for felony offenses was reported. The following frequency distribution is a summary of the data given in the paper.

Number of Previous Convictions	Frequency
0	0
1	16
2	27
3	37
4	46
5	36
6	40
7	31
8	27
9	13
10	8
11	2

Draw the histogram corresponding to this frequency distribution.

2.20 The 1974 edition of *Accident Facts* reported the state-by-state accidental death rates (deaths per 100,000 population) for 1973 as shown.

Alaska	137.8	Washington	51.5	Oregon	59.1
California	57.2	Idaho	77.9	Nevada	88.5
Arizona	84.3	Utah	64.7	Wyoming	106.2
Montana	85.9	Colorado	57.9	New Mexico	93.0
Texas	55.8	Oklahoma	61.4	Kansas	56.2
Nebraska	58.7	South Dakota	78.1	North Dakota	69.8
Minnesota	53.1	Iowa	56.6	Missouri	59.3
Arkansas	102.8	Louisiana	67.6	Mississippi	81.9
Alabama	73.6	Tennessee	70.1	Kentucky	64.0
Illinois	39.3	Wisconsin	49.2	Michigan	50.3
Indiana	54.2	Ohio	44.2	Maine	57.8
New Hampshire	44.8	Vermont	64.2	Massachusetts	45.6
Rhode Island	40.8	Connecticut	34.4	New York	35.5
New Jersey	40.0	Pennsylvania	42.0	West Virginia	64.2
Delaware	49.3	Maryland	36.1	Virginia	57.7
North Carolina	74.2	South Carolina	77.8	Georgia	59.7
Florida	68.5	Hawaii	37.7		

Construct a relative frequency distribution for this data. (Think about whether or not you want to use equal interval widths.) Draw the histogram corresponding to your frequency distribution.

2.21 The paper "I Hear You Knocking But You Can't Come In" (*Soc. Methods and Research* (1982):3–32) on reluctant respondents was discussed in Section 2.3. In addition to the responses to questions on status in the labor force, responses to questions about age, marital status, and dwelling type were also recorded. Frequency distributions were constructed after one attempt to interview those in the selected sample and after each of the subsequent attempts to contact the selected individuals. The two given frequency distributions show the changes in the age distribution observed after one attempt and after nine attempts to survey the people in the selected sample.

	Relative Frequency (n = 235)	Relative Frequency (n = 1002)
Age	After 1 Attempt	After 9 Attempts
18–<22	.068	.071
22–<30	.213	.242
30–<40	.187	.230
40–<50	.111	.130
50–<60	.123	.133
60–<90	.297	.196

a. Draw the histograms corresponding to these frequency distributions. (Note that the class-interval widths are not all the same).

b. Compare the two histograms. In what ways do they differ?

2.22 The paper mentioned in Exercise 2.21 also reported the following distribution of dwelling types for the 1049 people surveyed.

Dwelling Type	Relative Frequency
Single family	.606
Duplex or townhouse	.108
Low-rise apartment	.187
High-rise apartment	.068
Other	.031

Draw a histogram for this set of data.

2.23 The paper "Paraquat and Marijuana Risk Assessment" (*Am. J. Public Health* (1983):784–88) reported the results of a 1978 telephone survey on marijuana usage. The accompanying frequency distribution gives the amount of marijuana (in grams) smoked per week for those respondents who indicated that they did use the drug.

Grams Smoked per Week	Frequency
0–<3	94
3–<11	269
11–<18	70
18–<25	48
25–<32	31
32–<39	10
39–<46	5
46–<53	0
53–<60	1
60–<67	0
67–<74	1

a. Display the information given in the frequency distribution in the form of a histogram.

b. What proportion of respondents smoked 25 or more grams per week?

c. Use the histogram to estimate the proportion of respondents who smoked more than 15 grams per week.

2.24 In a study of author productivity ("Lotka's Test," *Collection Mgmt.* (1982):111–18) a large number of authors were classified according to the number of papers they had written and the results were presented in the frequency distribution given below.

Number of Papers	Number of Authors (Frequency)
1	784
2	204
3	127
4	50
5	33
6	28
7	19
8	19
9	6
10	7
11	6
12	7
13	4
14	4
15	5
16	3
17	3

Construct a histogram for this frequency distribution.

2.25 The accompanying histogram, based on data in the paper "Service Frequency, Schedule Reliability, and Passenger Wait Times at Transit Stops" (*J. of Trans. Research* (1981):465–71) shows the time (in minutes) that people had to wait for the next scheduled bus when buses were running on a 20-minute schedule. Suppose that the histogram is based on a sample of 300 waiting times. Construct the corresponding frequency distribution.

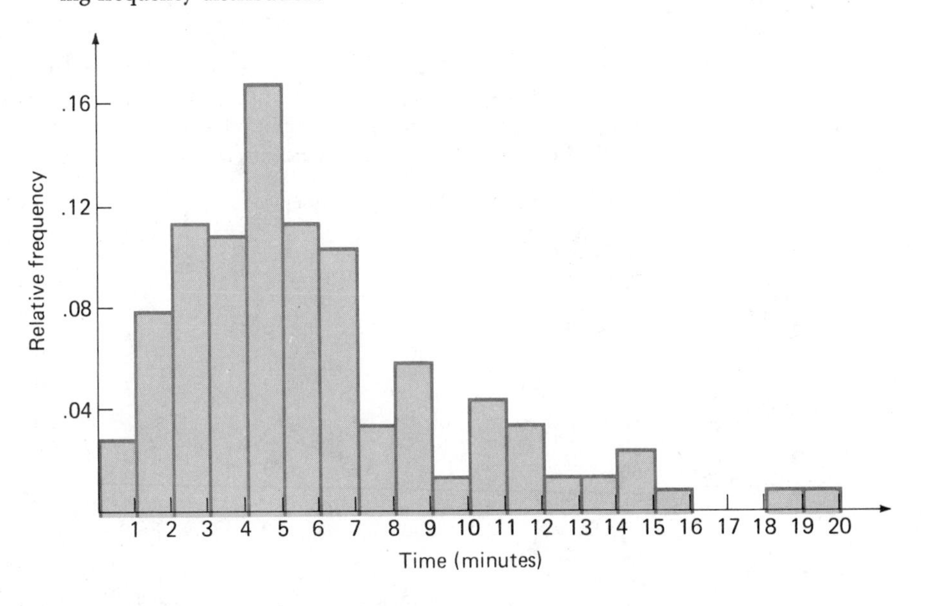

2.26 A common problem facing researchers who rely on mail questionnaires is that of nonresponse. In the paper "Reasons for Nonresponse on the Physicians' Practice Survey" (*Proc. Social Statistics Section Am. Statistical Assoc.* (1980):202), 811 doctors who did not respond to the AMA Survey of Physicians were contacted about the reason for their nonparticipation. The results are summarized in the relative frequency distribution given below.

Reason	Relative Frequency
No time to participate	.264
Not interested	.300
Don't like surveys in general	.145
Don't like this particular survey	.025
Hostility toward the government	.054
Desire to protect privacy	.056
Other reason for refusal	.053
No reason given	.103

Draw the histogram corresponding to the frequency distribution.

2.27 Many researchers have speculated that a relationship exists between birth order and vocational preferences. This theory is investigated in the paper "Birth Order and Vocational Preference" (*J. Exp. Educ.* (1980):15–18). In this study, 244 New York City high school students were given the Self-Directed Search Test. This test is designed to identify occupational preferences by classifying the student into one of six categories: realistic, investigative, artistic, social, enterprising, and conventional. The results are summarized in the two accompanying frequency distributions.

Vocational Class	Firstborn	Later Born
Conventional	38	9
Realistic	26	19
Enterprising	24	15
Social	12	15
Artistic	12	21
Investigative	10	43

a. Construct two histograms to represent occupational preference—one for firstborns and one for those born later.

b. What inferences would you make based on a comparison of the two histograms?

2.28 The mileage traveled before the first major motor failure for each of 191 buses was reported in an article that appeared in *Technometrics* ((Nov. 1980):588). The frequency distribution appearing in that paper is given below.

Distance Traveled (Thousands of Miles)	Frequency
0–<20	6
20–<40	11
40–<60	16
60–<80	25
80–<100	34
100–<120	46
120–<140	33
140–<160	16
160–<180	2
180–<200	2

a. Draw the histogram corresponding to this frequency distribution.

b. Use the histogram to estimate the proportion of all buses of this type that operate for more than 100,000 miles before the first major motor failure.

c. Use the histogram to estimate the proportion of all buses that have the first major motor failure after operating for between 50,000 and 125,000 miles.

2.29 The numbers of deaths by month for the United States in 1966 are given in the accompanying table.

Month	Number of Deaths
January	166,761
February	151,296
March	164,804
April	158,973
May	156,455
June	149,251
July	159,924
August	145,184
September	141,164
October	154,777
November	150,678
December	163,882

a. Draw the histogram corresponding to this frequency distribution.

b. Based on the histogram, do you think that the death rate remains constant throughout the year? Why or why not? What factor(s) might cause a variation in rates?

2.30 In the May 1976 issue of *Fortune* magazine, the results of a survey involving 500 chief executives were published. One of the questions asked of these people was the age at which they were appointed to their current job. The responses are summarized in the accompanying relative frequency distribution.

Age	Relative Frequency
30–<35	.021
35–<40	.067
40–<45	.158
45–<50	.248
50–<55	.263
55–<60	.210
60–<65	.031
65–<70	.002

a. Construct a histogram to represent the distribution of ages given.

b. What is the relative frequency of executives who were appointed to their jobs before reaching the age of 50?

c. What proportion of executives was appointed at 60 years of age or older?

d. Roughly what proportion was appointed while between the ages of 42 and 52?

2.5 From Samples to Populations (Optional)

Frequency distributions and histograms can be constructed when the data set consists of all population values, but it is much more frequently the case that the observed data has been sampled from a population under study. When the data consists of observations on a continuous variable, the appearance of a histogram will be determined at least in part by how the class intervals are chosen. With a small sample size, an informative histogram must necessarily be based on a small number of relatively wide class intervals. As the sample size increases, we acquire more detailed information about the distribution of population values along the measurement scale, and it is natural to use a relatively large number of narrow class intervals in constructing the histogram. Figure 17 illustrates possible histograms for data drawn from the same population when the sample size is relatively small, moderate, or large. In every histogram the area of each rectangle equals the relative frequency for the class interval, so the total area of all rectangles in the histogram is 1.

The histogram based on a relatively large sample size is much smoother than the other two histograms. This suggests that a good way of visualizing the population histogram is to use a smooth curve, as pictured in Figure 18, rather than a collection of extremely narrow rectangles. The total area under the curve should be 1, and the proportion of values in the population falling in any interval will be the area under the curve and above the interval. In later chapters, models for the population distribution of continuous variables will be based on exactly this type of smooth curve.

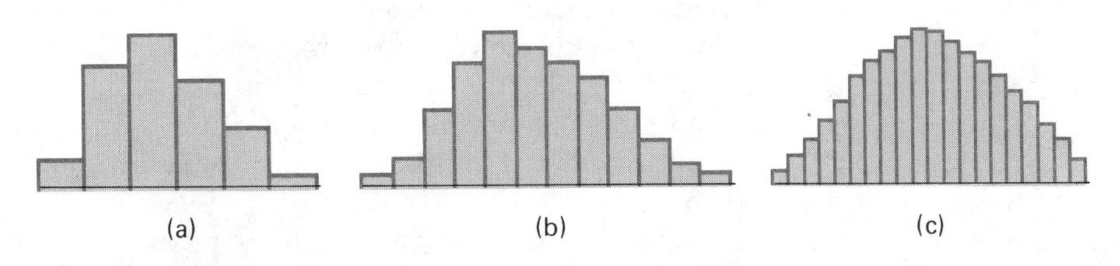

FIGURE 17 SAMPLE HISTOGRAMS BASED ON THREE DIFFERENT SAMPLE SIZES
(a) Small Sample
(b) Moderate Sample
(c) Large Sample

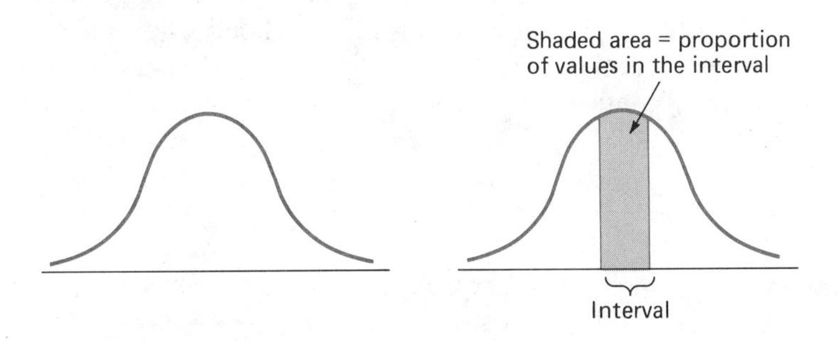

FIGURE 18 SMOOTHED POPULATION HISTOGRAMS

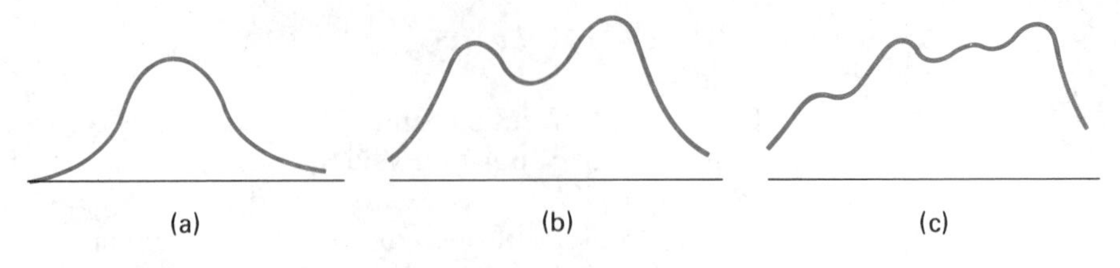

FIGURE 19 SMOOTHED HISTOGRAMS WITH VARIOUS NUMBERS OF MODES
(a) Unimodal
(b) Bimodal
(c) Multimodal

Distributional Shapes

An important aspect of a histogram is its general shape. One characterization of general shape relates to the number of peaks, or **modes.** A histogram is said to be **unimodal** if it has a single peak, **bimodal** if it has two peaks, and **multimodal** if it has more than two peaks. These shapes are illustrated in Figure 19. Many numerical variables, whether continuous or count variables, give rise to a unimodal histogram. Occasionally one encounters a bimodal histogram—an example would be a histogram of adult heights, with one peak at roughly 5 feet 6 inches for women and another peak at roughly 5 feet 9 inches for men—but very rarely does a histogram with more than two peaks occur.

Within the class of unimodal histograms, there are still several important distinctive shapes. A unimodal histogram is **symmetric** if there is a vertical *line of symmetry* such that the part of the histogram to the left of the line is a mirror image of the part to the right (bimodal and multimodal histograms can also be symmetric in this way). Several different symmetric smoothed histograms are pictured in Figure 20.

Proceeding to the right from the peak of a unimodal histogram, we move into what is called the **upper tail** of the distribution. Going in the opposite direction moves us into the **lower tail** of the distribution. For symmetric histograms the rate of decrease of the curve is the same as we move into either tail.

A unimodal histogram that is not symmetric is said to be **skewed.** If the upper tail of the histogram stretches out farther than the lower tail, then the distribution of values is **positively skewed.** If, on the other hand, the lower tail is longer than the upper tail, the distribution is **negatively skewed.** These two types of skewness are illustrated in Figure 21. Positive skewness is much more frequently encountered than is negative skewness. One example of a positively skewed distribution is the distribution of single-family home prices in Los Angeles County, where most homes are moderately priced (at least for California), whereas the relatively few homes in Beverly Hills and Malibu have much higher price tags. Another example of the same sort would be a

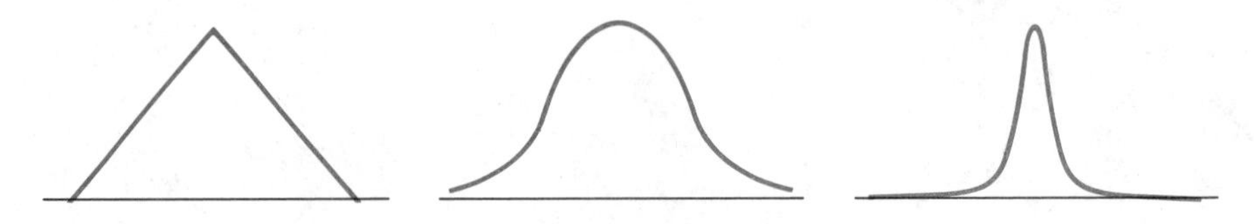

FIGURE 20 SEVERAL SYMMETRIC UNIMODAL SMOOTHED HISTOGRAMS

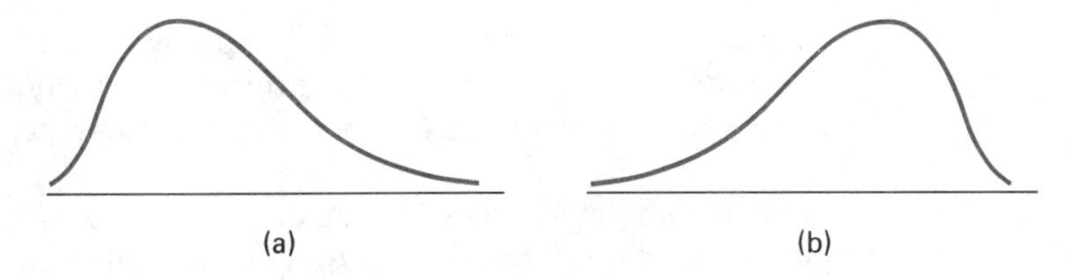

FIGURE 21 TWO EXAMPLES OF SKEWED SMOOTHED HISTOGRAMS
(a) Positive Skew
(b) Negative Skew

distribution of salaries for a large population, with the high salaries of doctors and corporate executives (but certainly not professors) strung out in the long upper tail.

There is one rather specific distributional shape that arises more frequently than any other in statistical applications, that which is described by a **normal curve.** Many variables have been found to have distributions whose shapes can be well approximated by a normal curve—for example, characteristics such as blood pressure, brain weight, adult male heights, adult female heights, and IQ scores. Here we mention briefly several of the most important qualitative properties of such a curve, postponing a more detailed discussion until Chapter 5. A normal curve is not only symmetric but also bell-shaped and looks like the middle curve in Figure 20. However, not all bell-shaped curves are normal. Starting from the top of the bell, the height of the curve decreases at a well-defined rate as one proceeds out into either tail (this rate of decrease is specified by a certain mathematical function).

A curve whose tails do not decline as rapidly as do the tails of a normal curve is said to specify a **heavy-tailed** distribution (compared to the normal curve). Similarly, a curve whose tails decrease more rapidly than the normal tail is called **light-tailed.** Figure 22 illustrates these possibilities. Many inferential procedures which work well (result in accurate conclusions) when the population distribution is approximately normal tend to break down when the population distribution is heavy-tailed, prompting much recent interest in alternative methods of analysis that are not so sensitive to heavy-tailedness. Such methods are referred to as being **robust,** and we shall give several examples of robust methods in later chapters.

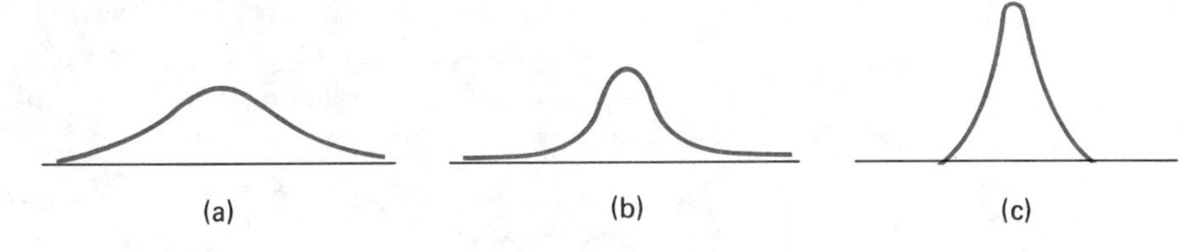

FIGURE 22 THREE EXAMPLES OF BELL-SHAPED HISTOGRAMS
(a) Normal
(b) Heavy-Tailed
(c) Light-Tailed

Do Sample Histograms Resemble the Population Histogram?

Statistical inference involves using information contained in a sample to draw conclusions about a population. The extent to which this endeavor will be successful depends on how closely various characteristics of the sample resemble the analogous population charactertistics. When we form a sample histogram, is it centered at roughly the same place as is the population histogram, does it spread out to the same extent as does the population histogram, do the two histograms have the same number of peaks, and do the peaks occur at roughly the same place or places?

An issue intimately related to the one just raised is the extent to which histograms based on different samples from the same population resemble one another. If two different sample histograms can be expected to differ from one another in obvious ways, then at least one of them will differ substantially from the population histogram, resulting in unreliable inferences. **Sampling variability**—the extent to which samples differ from one another and from the population—is a central idea in statistics. In later chapters we develop quantitative measures for assessing sampling variability, and these play a key role in our inferential methods. We next present two small examples to suggest how sample histograms resemble one another and the population histogram. The first example involves count data and the second involves sampling from a normal population.

EXAMPLE 15

Example 5 gives data on the number of accidents in which each of 708 bus drivers was involved over a certain time period. The corresponding histogram is given in Figure 11. Although the 708 observations actually constituted a sample from the population of all bus drivers, here we will regard the 708 observations as constituting the entire population. Figure 11, which is the first histogram in Figure 23, then represents the population histogram. The other four histograms in Figure 23 are based on four different samples of 50 observations each from this population.

The five histograms certainly resemble one another in a general way, but there are also some obvious dissimilarities. The population histogram rises to a peak and then declines smoothly, whereas the sample histograms tend to have more peaks, valleys, and gaps. Although the population data set contained an observation of 11, none of the four samples did. In fact, in the first two samples the largest observations were 7 and 8, respectively. Fortunately, when we study numerical characterisitics of samples and populations—most importantly, measures of center and spread—we will see that sample characterisitics tend to resemble population characteristics more closely than the pictures in Figure 23 might suggest. Although sampling variability is a problem that must be confronted, reliable methods of inference will shortly be within our grasp. ■

EXAMPLE 16

Here we took as our population curve the normal curve pictured in Figure 24. We then used MINITAB to generate five different samples of size 50 each, and the resulting sample histograms are also presented in Figure 24. As in the previous example, the general shapes and other characteristics of the sample histograms are similar to those of the normal population curve, although there are some differences in detail. Methods for making inferences about characteristics of a normal population occupy a central position in statistical inference, as we shall see beginning in Chapter 6. ■

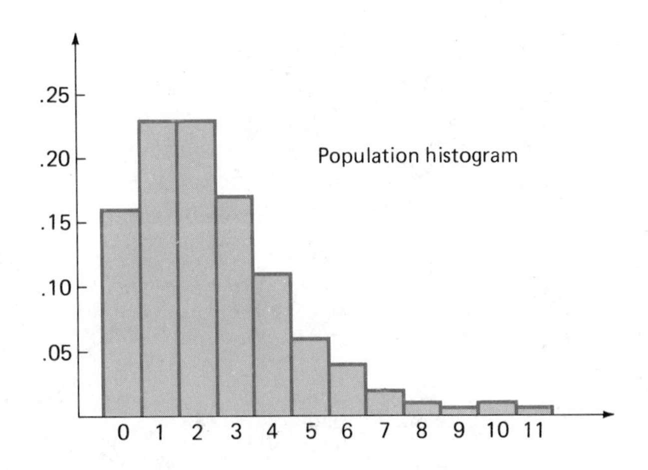

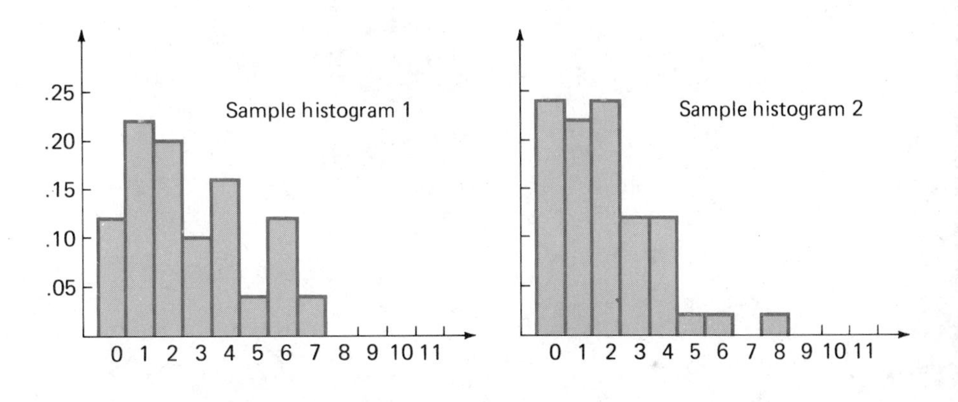

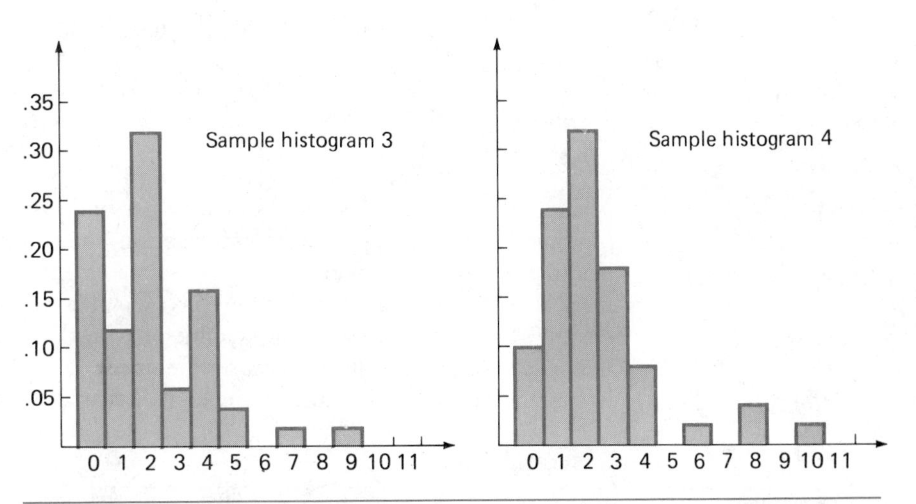

FIGURE 23 A COMPARISON OF POPULATION AND SAMPLE HISTOGRAMS FOR NUMBER OF ACCIDENTS

MIDDLE OF INTERVAL	NUMBER OF OBSERVATIONS	
70.00	0	
75.00	1	*
80.00	1	*
85.00	2	**
90.00	4	****
95.00	6	******
100.00	10	**********
105.00	10	**********
110.00	11	***********
115.00	4	****
120.00	1	*

MIDDLE OF INTERVAL	NUMBER OF OBSERVATIONS	
70.00	0	
75.00	1	*
80.00	0	
85.00	4	****
90.00	6	******
95.00	10	**********
100.00	8	********
105.00	14	**************
110.00	5	*****
115.00	2	**

MIDDLE OF INTERVAL	NUMBER OF OBSERVATIONS	
70.00	0	
75.00	1	*
80.00	0	
85.00	2	**
90.00	5	*****
95.00	10	**********
100.00	12	************
105.00	8	********
110.00	9	*********
115.00	1	*
120.00	2	**

MIDDLE OF INTERVAL	NUMBER OF OBSERVATIONS	
70.00	1	*
75.00	0	
80.00	1	*
85.00	3	***
90.00	6	******
95.00	10	**********
100.00	10	**********
105.00	8	********
110.00	8	********
115.00	3	***

MIDDLE OF INTERVAL	NUMBER OF OBSERVATIONS	
70.00	1	*
75.00	2	**
80.00	1	*
85.00	3	***
90.00	4	****
95.00	10	**********
100.00	12	************
105.00	6	******
110.00	5	*****
115.00	3	***
120.00	2	**
125.00	1	*

80 90 100 110 120

FIGURE 24 COMPARISON OF POPULATION AND SAMPLE HISTOGRAMS FOR A NORMAL POPULATION

The last two examples should help to persuade you that while sample histograms can provide qualitative information about the distribution of values in the population, to make more precise statements we need to develop numerical methods of analysis. In the next chapter we begin this task. In later chapters these methods not only enable us to make inferences but also to attach measures of reliability to those inferences. It will then be more apparent how sampling variability affects the conclusions that we might wish to draw.

EXERCISES

2.31 Construct a histogram corresponding to each of the frequency distributions given below, and state whether each histogram is symmetric, bimodal, positively skewed or negatively skewed.

	Frequency				
Class Interval	I	II	III	IV	V
0–<10	5	40	30	5	6
10–<20	10	25	10	25	5
20–<30	20	10	8	8	6
30–<40	30	8	7	7	9
40–<50	20	7	7	20	9
50–<60	10	5	8	25	23
60–<70	5	5	30	10	42

2.32 Using the following class intervals, devise a frequency distribution based on 70 observations whose histogram could be described as

a. Symmetric
b. Negatively skewed
c. Positively skewed
d. Bimodal

Class Interval
100–<120
120–<140
140–<160
160–<180
180–<200

2.33 Data on engine emissions for 46 vehicles is given (*Technometrics* (Nov. 1980):487).

Vehicle	HC	CO	Vehicle	HC	CO
1	.50	5.01	2	.65	14.67
3	.46	8.60	4	.41	4.42
5	.41	4.95	6	.39	7.24
7	.44	7.51	8	.55	12.30
9	.72	14.59	10	.64	7.98
11	.83	11.53	12	.38	4.10
13	.38	5.21	14	.50	12.10
15	.60	9.62	16	.73	14.97
17	.83	15.13	18	.57	5.04
19	.34	3.95	20	.41	3.38
21	.37	4.12	22	1.02	23.53
23	.87	19.00	24	1.10	22.92
25	.65	11.20	26	.43	3.81
27	.48	3.45	28	.41	1.85
29	.51	4.10	30	.41	2.26
31	.47	4.74	32	.52	4.29
33	.56	5.36	34	.70	14.83
35	.51	5.69	36	.52	6.35
37	.57	6.02	38	.51	5.79
39	.36	2.03	40	.49	4.62
41	.52	6.78	42	.61	8.43
43	.58	6.02	44	.46	3.99
45	.47	5.22	46	.55	7.47

a. Construct a frequency distribution and histogram for the hydrocarbon (HC) emissions data.

b. Construct a frequency distribution and histogram for the carbon monoxide (CO) data.

c. Are the HC and CO histograms symmetric or skewed? If they are both skewed, is the direction of the skew the same?

2.6 Transforming Data for Ease of Description (Optional)

A primary objective of data analysis is to find ways of describing and summarizing data that are both simple and insightful. A stem-and-leaf display or histogram of a data set may initially suggest a rather complicated distribution. In such cases, a transformation or reexpression of the data values may help to simplify the description process and increase understanding of the variable under investigation. By transforming data, we mean using some specified mathematical operation (e.g. square root, logarithm, or reciprocal) on each data value to produce a set of **transformed data.** We can then study and summarize the distribution of these transformed values. With a single data set, a transformation is usually chosen to yield a distribution of transformed values which is more symmetric (or, even better, more closely approximated by a normal curve) than was the original distribution. In later chapters we consider applying transformations to achieve other objectives.

EXAMPLE 17

A data set that has been used by several authors to introduce the concept of transformation (see "Exploratory Methods for Choosing Power Transformations," *J. Amer. Stat. Assoc.* (1982):103–08, for example) consists of values of March precipitation for Minneapolis–St. Paul over a period of 30 years. These values are displayed in Table 10, along with the square root of each value, and histograms of both the original and transformed data appear in Figure 25. The distribution of the original data is clearly skewed, with a long upper tail. The square-root transformation has resulted in a substantially

TABLE 10

ORIGINAL AND SQUARE-ROOT TRANSFORMED VALUES OF MARCH PRECIPITATION IN MINNEAPOLIS–ST. PAUL OVER A 30-YEAR PERIOD

Year	Precipitation	$\sqrt{\text{Precipitation}}$	Year	Precipitation	$\sqrt{\text{Precipitation}}$
1	.77	.88	16	1.62	1.27
2	1.74	1.32	17	1.31	1.14
3	.81	.90	18	.32	.57
4	1.20	1.10	19	.59	.77
5	1.95	1.40	20	.81	.90
6	1.20	1.10	21	2.81	1.68
7	.47	.69	22	1.87	1.37
8	1.43	1.20	23	1.18	1.09
9	3.37	1.84	24	1.35	1.16
10	2.20	1.48	25	4.75	2.18
11	3.00	1.73	26	2.48	1.57
12	3.09	1.76	27	.96	.98
13	1.51	1.23	28	1.89	1.37
14	2.10	1.45	29	.90	.95
15	.52	.72	30	2.05	1.43

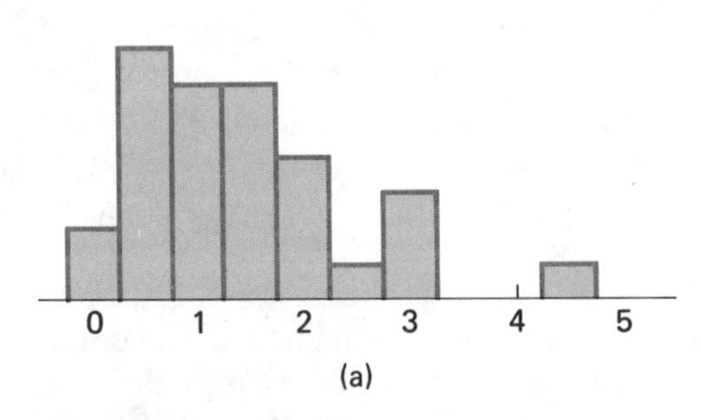

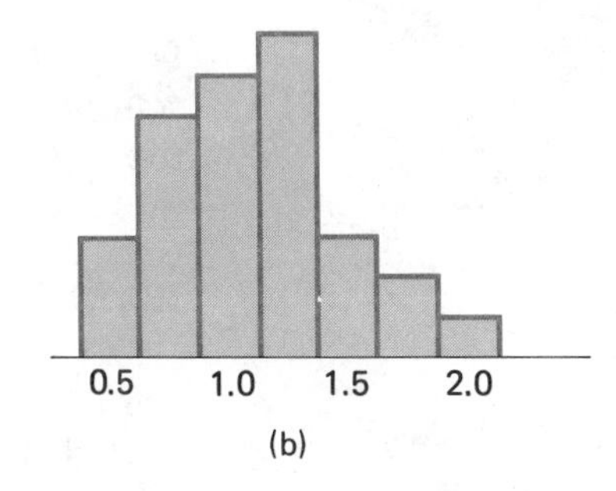

FIGURE 25

HISTOGRAMS OF PRECIPITATION DATA
(a) Untransformed Data
(b) Square-Root Transformed Data

more symmetric distribution, with a typical (i.e., central) value near the 1.25 boundary between the third and fourth class intervals. In Excercise 2.34 you are asked to consider the cube-root transformation (transformed value = (original value)$^{1/3}$), which is favored by some meteorologists. ■

Square-root and reciprocal transformations are particular cases of a general type of transformation called a **power transformation.** A transformation of this type is defined by first specifying an exponent or power, p, which could be either a positive or negative number, and then computing transformed values according to

$$\text{transformed value} = (\text{original value})^p$$

The value $p = \frac{1}{2}$ gives the square-root transformation and $p = -1$ gives the reciprocal transformation (reciprocal = 1/value). The book *Applications, Basics, and Computing of Exploratory Data Analysis,* listed among the chapter references, gives a good elementary discussion of power transformations, along with suggested quantitative methods for assessing the extent to which a transformation achieves symmetry.

Taking $p = 0$ results in every transformed value equaling one, which is not at all informative. Instead, it is customary to think of $p = 0$ as corresponding to a logarithmic transformation—taking the logarithm of each data value. This is the most frequently encountered transformation in statistics, so it is worth considering in some detail. We begin with a brief discussion of logarithms.

Logarithmic Transformations

The number 100 can be expressed as $(10)^2$, so the base 10 logarithm of 100 is 2. Similarly, $1000 = (10)^3$, so the base 10 logarithm of 1000 is 3, and $3.1623 = (10)^{.5}$, so .5 is the base 10 logarithm of 3.1623. In general, any positive number can be expressed as 10 raised to some power, and the power is called the base 10 logarithm of the number.

DEFINITION

Given a particular number, if

$$\text{number} = (10)^n$$

for a specified value n, then n is called the **base 10 logarithm** of that number. This is usually written as $\log_{10}(\text{number}) = n$.

For most numbers, computing $\log_{10}$ requires substantial calculation, so many statistics books used to include a table of such logarithms. Now many relatively inexpensive calculators have a $\log_{10}$ key, so entering a number and pressing this key immediately yields the desired logarithm. We have not included a table of logarithms, so we'll give you $\log_{10}$ values whenever we want you to work with them.

Base 10 is not the only useful base for logarithms. Because of the way in which computers work, base 2 is important in computer science. In scientific work the base 2.718281828 . . . , denoted by the letter *e*, is very useful; logarithms to this base are called **natural logarithms** and are denoted by ℓn. Many calculators have both $\log_{10}$ and ℓn keys. For simplicity, let's concern ourselves for the moment only with $\log_{10}$.

A log transformation is usually applied to data that is positively skewed (a long upper tail). This affects values in the upper tail substantially more than values in the lower tail, yielding a more symmetric—and often more normal—distribution.

EXAMPLE 18

Exposure to beryllium is known to produce adverse effects on lungs as well as on other tissues and organs in both laboratory animals and humans. The paper "Time Lapse Cinematographic Analysis of Beryllium—Lung Fibroblast Interactions" (*Envir. Research* (1983):34–43) reported the results of experiments designed to study the behavior of certain individual cells that had been exposed to beryllium. An important characteristic of such an individual cell is its interdivision time (IDT). IDT's were determined for a large number of cells both in exposed (treatment) and unexposed (control) conditions. The authors of the paper state that "The IDT distributions are seen to be skewed, but the natural logs do have an approximate normal distribution." The same property holds for $\log_{10}$ transformed data. We give representative IDT data and the resulting histograms in Figure 26, which are in agreement with the authors' statement.

IDT	$\log_{10}$(IDT)	IDT	$\log_{10}$(IDT)	IDT	$\log_{10}$(IDT)
1. 28.1	1.45	15. 60.1	1.78	29. 21.0	1.32
2. 31.2	1.49	16. 23.7	1.37	30. 22.3	1.35
3. 13.7	1.14	17. 18.6	1.27	31. 15.5	1.19
4. 46.0	1.66	18. 21.4	1.33	32. 36.3	1.56
5. 25.8	1.41	19. 26.6	1.42	33. 19.1	1.28
6. 16.8	1.23	20. 26.2	1.42	34. 38.4	1.58
7. 34.8	1.54	21. 32.0	1.51	35. 72.8	1.86
8. 62.3	1.79	22. 43.5	1.64	36. 48.9	1.69
9. 28.0	1.45	23. 17.4	1.24	37. 21.4	1.33
10. 17.9	1.25	24. 38.8	1.59	38. 20.7	1.32
11. 19.5	1.29	25. 30.6	1.49	39. 57.3	1.76
12. 21.1	1.32	26. 55.6	1.75	40. 40.9	1.61
13. 31.9	1.50	27. 25.5	1.41		
14. 28.9	1.46	28. 52.1	1.72		

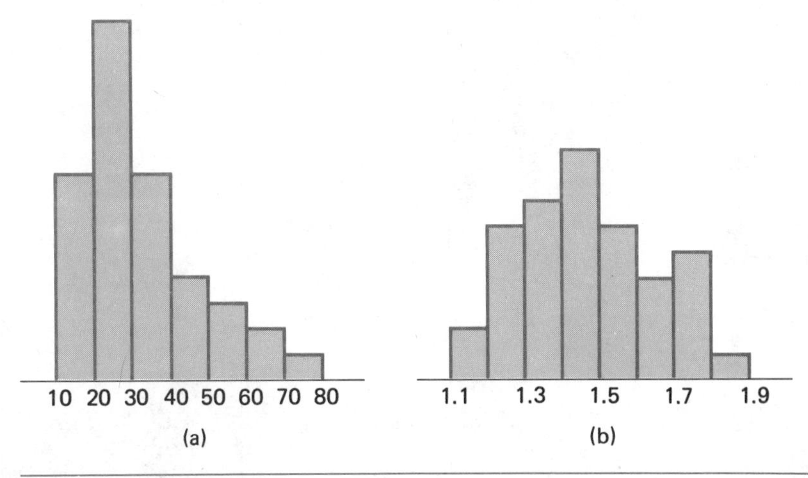

FIGURE 26

HISTOGRAMS OF IDT DATA
(a) Untransformed Data
(b) Log_{10} Transformed Data

Selecting a Transformation

A particular transformation may on occasion be dictated by some theoretical argument, but often this is not the case. Then one may wish to try several different transformations in order to find one which is satisfactory. Figure 27, taken from the paper "Distribution of Sperm Counts in Suspected Infertile Men" (*J. of Reproduction and Fertility* (1983):91–96), illustrates what can result from such a search. Other workers in this field had previously used all three of the transformations illustrated, but these investigators strongly favored the square-root transformation for their data. Information about other graphical techniques and quantitative methods for selecting a transformation can be found in several of the chapter references.

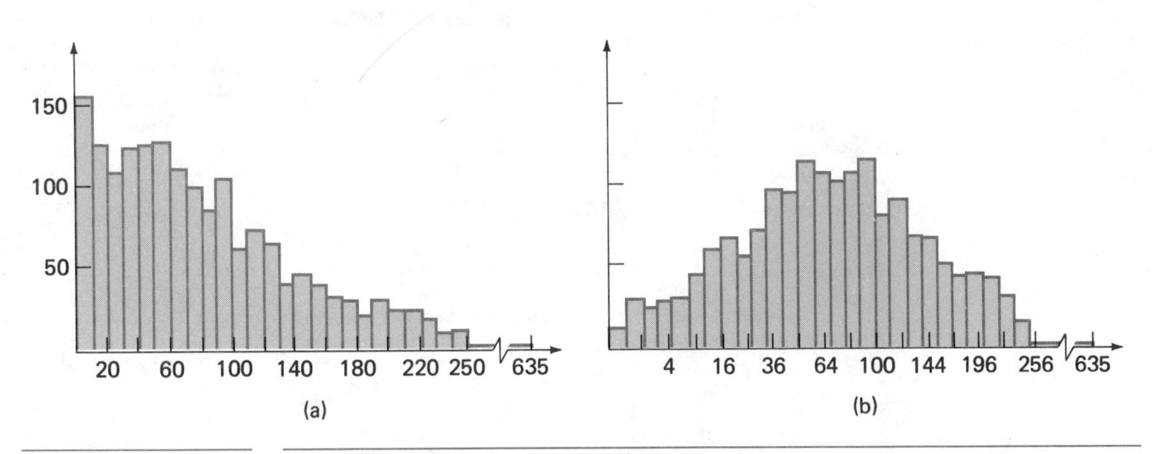

FIGURE 27

HISTOGRAMS OF SPERM CONCENTRATIONS FOR 1711 SUSPECTED INFERTILE MEN
(a) Untransformed
(b) Log Transformed
(Source: Mortimer and Lenton (1983). Reproduced by permission.)

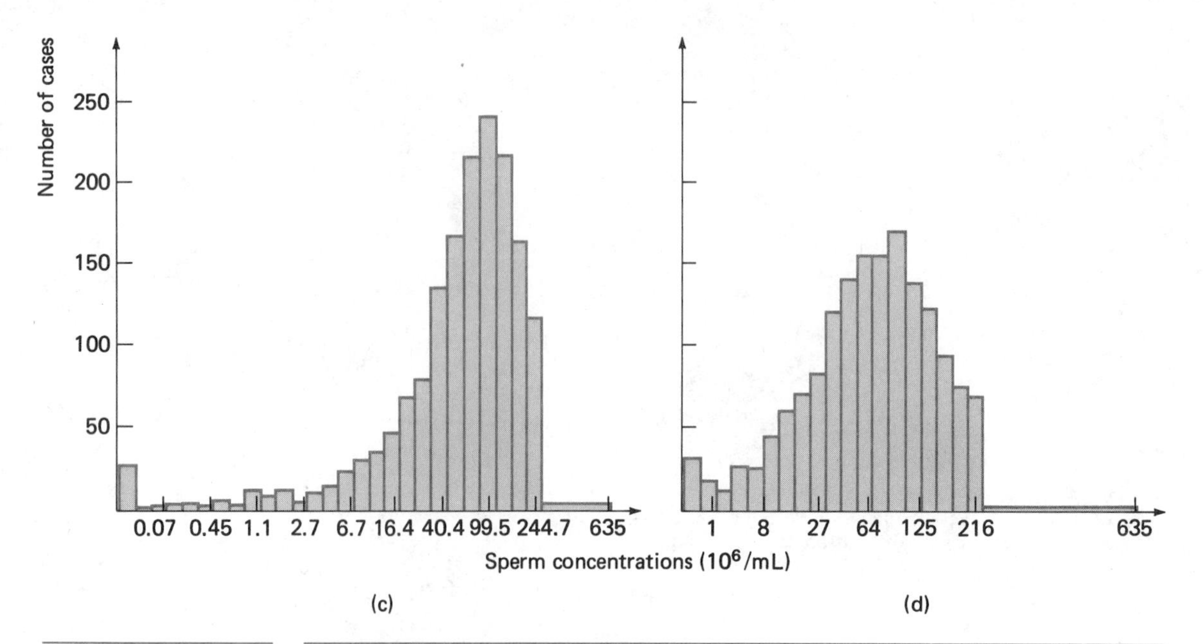

FIGURE 27 (continued)

HISTOGRAMS OF SPERM CONCENTRATIONS FOR 1711 SUSPECTED INFERTILE MEN
(c) Square-Root Transformed
(d) Cube-Root Transformed

EXERCISES

2.34 The first example in this section examined rainfall data for Minneapolis–St. Paul. The square-root transformation was used to obtain a distribution of values that was more symmetric than the distribution of the original data. Another power transformation that has been suggested by meteorologists is the cube root: Transformed value = (original value)$^{1/3}$. The original values and their cube roots (the transformed values) are given. Construct a histogram of the transformed data. Compare your histogram to those given in Figure 25. Which of the cube-root and the square-root transformations appears to result in the most symmetric histogram?

Original	.32	.47	.52	.59	.77	.81	.81	.90
Transformed	.68	.78	.80	.84	.92	.93	.93	.97
Original	.96	1.18	1.20	1.20	1.31	1.35	1.43	1.51
Transformed	.99	1.06	1.06	1.06	1.09	1.11	1.13	1.15
Original	1.62	1.74	1.87	1.89	1.95	2.05	2.10	2.20
Transformed	1.17	1.20	1.23	1.24	1.25	1.27	1.28	1.30
Original	2.48	2.81	3.00	3.09	3.37	4.75		
Transformed	1.35	1.41	1.44	1.46	1.50	1.68		

2.35 The given values represent age at onset of Parkinson's disease for 40 males suffering from the disease.

64	68	60	60	64	63	64	59	64	54	58
63	59	62	64	63	59	59	58	61	63	67
63	63	61	63	64	60	64	62	61	63	64
63	66	66	68	61	61	63				

a. Construct a relative frequency distribution for this data set and draw the corresponding histogram.

b. Would you describe this histogram as having a positive or negative skew?

c. Would you recommend transforming the data? Explain.

2.36 One hundred observations on the breaking strength of yarn were given in Exercise 2.14. Use the data given there to complete the following.

a. Construct a frequency distribution using the class intervals 0–<100, 100–<200, and so on.

b. Draw the histogram corresponding to the frequency distribution in part (a). How would you describe the shape of this histogram?

c. Find a transformation for this data that yields a histogram which is more symmetric than the original data.

2.37 The article "The Distribution of Buying Frequency Rates" (*J. of Marketing Research* (1980):210–16) reported the results of a $3\frac{1}{2}$-year study of dentifrice purchases. The authors conducted their research using a national sample of 2071 households and recorded the number of toothpaste purchases for each household participating in the study. The results are given in the accompanying frequency distribution.

Number of Purchases	Number of Households (Frequency)
10–<20	904
20–<30	500
30–<40	258
40–<50	167
50–<60	94
60–<70	56
70–<80	26
80–<90	20
90–<100	13
100–<110	9
110–<120	7
120–<130	6
130–<140	6
140–<150	3
150–<160	0
160–<170	2

a. Draw a histogram for this frequency distribution. Would you describe the histogram as positively or negatively skewed?

b. Does the square-root transformation result in a histogram which is more symmetric than that of the original data? (Be careful—this one is a bit tricky, since you don't have the raw data; transforming the endpoints of the class intervals will result in class intervals that are not necessarily of equal widths, so the histogram of the transformed values will have to be drawn with this in mind.)

KEY CONCEPTS

Types of variables and data Categorical and numerical. (p. 13)

Types of numerical variables Discrete (including counting variables as a special case) and continuous. (p. 13)

Types of data sets Univariate (observations on a single variable), bivariate, and multivariate. (p. 14)

Stem-and-leaf display A compact way of displaying numerical data. Observations are expressed using leading digits (the stems) and trailing digits (the leaves). (p. 16)

Frequency distribution A tabular summary of data. For categorical or count data, each possible category or count value is displayed, along with the number of times it occurred in the data set (its frequency). For observations on a continuous variable, the measurement scale is first subdivided into class intervals; these are then listed along with the corresponding frequencies. (p. 22)

Relative frequency The proportion of observations falling in a category, count value, or class interval; relative frequency = frequency/(number of observations in the data set). (p. 22)

Cumulative relative frequency The sum of relative frequencies from the lowest count value or class interval to a specified intermediate value or class. (p. 28)

Histogram A pictorial representation of a frequency distribution in which rectangles are drawn above the category labels, count values, or class intervals. The area of each rectangle is proportional to the corresponding frequency. (p. 35)

Density scale A vertical scale that makes the area of each rectangle in a histogram equal to the corresponding relative frequency. (p. 39)

Histogram shapes Unimodal (a single peak), bimodal, multimodal, symmetric, positively skewed, negatively skewed, normal. (p. 48)

Transformed data The result of applying a specified mathematical operation (e.g., square-root or logarithm) to each observation in the original data set. A transformation is applied in order to make description and summarization easier. (p. 54)

SUPPLEMENTARY EXERCISES

2.38 The accompanying frequency distribution of number of years of continuous service at the time of resignation from a job with an oil company appeared in the paper "The Role of Performance in the Turnover Process" (*Academy of Management J.* (1982):137–47). Construct the histogram corresponding to this frequency distribution. Which terms introduced in this chapter (symmetric, skewed, etc.) would you use to describe the histogram?

Years of Service	Frequency
0–<1	4
1–<2	41
2–<3	67
3–<4	82
4–<5	28
5–<6	43
6–<7	14
7–<8	17
8–<9	11
9–<10	7
10–<11	14
11–<12	6
12–<13	14
13–<14	5
14–<15	2

2.39 A random sample of 60 preschool-age children who were participating in a day-care program was used to obtain information about parental work status. The two given frequency distributions appeared in the paper "Nutritional Understanding of Preschool Children Taught in the Home of a Child Development Laboratory" (*Home Economics Research J.* (1984):52–60).

Work Status of Mother	Frequency
Unemployed	41
Employed part time	13
Employed full time	6

Father's Occupation	Frequency
Professional	21
Craftsman	19
Manager	6
Proprietor	7
Other	7

a. Construct a relative frequency distribution for the mothers' work statuses. What proportion of the mothers work outside the home?

b. Draw the histogram corresponding to the frequency distribution for the fathers' occupations.

2.40 Suppose that the accompanying observations are heating costs for a sample of two-bedroom apartments in Southern California for the month of January.

Heated by Gas					
25.42	26.12	25.22	23.60	27.77	28.52
21.60	29.49	26.22	25.52	20.19	23.99
26.32	23.38	26.77	31.56	25.54	22.72
27.58	29.96	26.20	23.97	28.17	18.01
22.98					

Heated with Electricity					
33.52	51.01	41.99	33.80	25.93	30.32
32.06	39.86	24.62	31.80	48.58	44.65
31.30	35.04	19.24	40.78	43.39	34.78
25.43	33.82	26.47	34.62	32.02	27.98
30.92					

Construct a comparative stem-and-leaf display contrasting heating costs for gas and electricity. How do the two sides of the display compare?

2.41 The two accompanying frequency distributions appeared in the paper "Aqueous Humour Glucose Concentration in Cataract Patients and its Effect on the Lens" (*Exp. Eye Research* (1984):605–09). The first is a frequency distribution of lens sodium concentration for nondiabetic cataract patients, while the second is for diabetic cataract patients. Draw the histogram corresponding to each frequency distribution. Do you think that the distributions for the population of all diabetic and for the population of nondiabetic cataract patients are similar? Explain.

Sodium Concentration (mM)	Nondiabetic Frequency	Diabetic Frequency
0–<20	7	0
20–<40	12	0
40–<60	5	1
60–<80	1	2
80–<100	0	3
100–<120	1	2
120–<140	1	1
140–<160	4	0
160–<180	8	1
180–<200	3	0
200–<220	2	0
220–<240	1	0

2.42 Referring to Exercise 2.41, construct the cumulative relative frequencies corresponding to the frequency distribution of lens sodium concentration (mM) of nondiabetic cataract patients. Use them to answer the following questions.

a. What proportion of the nondiabetic cataract patients had a lens sodium concentration below 100 mM?

b. What proportion of the nondiabetic cataract patients had a lens sodium concentration between 100 and 200 mM?

c. What proportion of the nondiabetic cataract patients had a lens sodium concentration that exceeded 140 mM?

d. Find a sodium concentration value for which approximately half of the observed sodium concentrations are smaller than this value.

2.43 The soil stability index (SSI) of eroded topsoil was recorded for 41 randomly selected sites under dry conditions and for 39 randomly selected sites under green conditions ("Use of Landsat Radiance Parameters to Distinguish Soil Erosion, Stablility, and Deposition in Arid Central Australia," *Remote Sensing of Environment* (1984):195–209). Construct and interpret a comparative stem-and-leaf display.

Soil Stability Index

Dry Conditions						Green Conditions					
31	44	44	44	36	36	20	20	20	20	21	21
36	45	37	45	45	38	21	24	24	24	24	25
39	39	39	39	39	39	25	25	25	25	27	27
39	39	39	39	40	40	27	27	27	28	28	28
40	40	40	40	40	40	30	30	30	41	41	41
41	41	41	41	42	42	42	42	50	50	50	50
42	42	43				50	50	59			

2.44 Americium 241 (^{241}Am) is a radioactive material used in the manufacture of smoke detectors. The article "Retention and Dosimetry of Injected ^{241}Am in Beagles" (*Radiation Res.* (1984):564–75) described a study in which 55 beagles were injected with a dose of ^{241}Am (proportional to the animals' weights). Skeletal retention of ^{241}Am (μCi/kg) was recorded for each beagle, resulting in the given data.

.196	.451	.498	.411	.324	.190	.489	.300	.346	.448
.188	.399	.305	.304	.287	.243	.334	.299	.292	.419
.236	.315	.447	.585	.291	.186	.393	.419	.335	.332
.292	.375	.349	.324	.301	.333	.408	.399	.303	.318
.468	.441	.306	.367	.345	.428	.345	.412	.337	.353
.357	.320	.354	.361	.329					

a. Construct a frequency distribution for this data and draw the corresponding histogram.

b. Write a short description of the important features of the shape of the histogram.

2.45 Construct a stem-and-leaf display using six stems (8*h*, 8*l*, 7*h*, 7*l*, 6*h*, and 6*l*) for the given data on soil pH ("Sodium-Calcium Exchange Equilibria in Soils as Affected by Calcium Carbonate and Organic Matter," *Soil Science* (1984):109).

8.53	8.52	8.01	7.99	7.93	7.89
7.85	7.82	7.80	7.72	7.85	7.73
7.58	7.40	7.35	7.30	7.27	7.27
7.23	6.84	6.50	6.40	6.36	6.19
6.16	6.17	6.15	6.09		

2.46 The two given frequency distributions of storm duration (in minutes) are based on data appearing in the article "Lightning Phenomenology in the Tampa Bay Area" (*J. of Geophysical Res.* (1984):11,789–05). Construct a histogram for each of the frequency distributions and discuss the similarities and differences between the two with respect to shape.

	Single-Peak Storms	Multiple-Peak Storms
Storm Duration	Frequency	Frequency
0–<25	1	0
25–<50	17	1
50–<75	14	1
75–<100	11	3
100–<125	8	2
125–<150	8	2
150–<175	5	1
175–<200	4	3
200–<225	3	1
225–<250	2	6
250–<275	0	4
275–<300	1	2

2.47 The paper "The Acid Rain Controversy: The Limits of Confidence" (*Am. Statistician* (1983):385–94) gave the accompanying data on average SO_2 (sulfur dioxide) emission rates from utility and industrial boilers (lb/million Btu) for 47 states (data from Idaho, Alaska, and Hawaii was not given).

2.3	2.7	1.5	1.5	0.3	0.6	4.2	1.3	1.2	0.4
0.5	2.2	4.5	3.8	1.2	0.2	1.0	0.7	0.2	1.4
0.7	3.6	1.0	0.7	1.7	0.5	0.1	0.6	2.5	2.7
1.5	1.4	2.9	1.0	3.4	2.1	0.9	1.9	1.0	1.7
1.8	0.6	1.7	2.9	1.8	1.4	3.7			

a. Summarize this set of data by constructing a relative frequency distribution.

b. Draw the histogram corresponding to the frequency distribution in part (a). Would you describe the histogram as symmetric or skewed?

c. Use the relative frequency distribution of part (a) to compute the cumulative relative frequencies.

d. Use the cumulative relative frequencies to give the approximate proportion of states with SO_2 emission rates that

i. were below 1.0 lb/million Btu;

ii. were between 1.0 and 2.0 lb/million Btu;

iii. exceeded 2.0 lb/million Btu.

2.48 The accompanying figure appeared in the paper "EDTA-Extractable Copper, Zinc and Manganese in Soils of the Canterbury Plains" (*New Zealand J. of Ag. Res.* (1984):207–17, reprinted with permission). A large number of topsoil samples were analyzed for manganese (Mn), zinc (Zn), and copper (Cu), and the resulting data was summarized using histograms. The authors transformed each data set using logarithms in an effort to obtain more symmetric distributions of values. Do you think the transformations were successful? Explain.

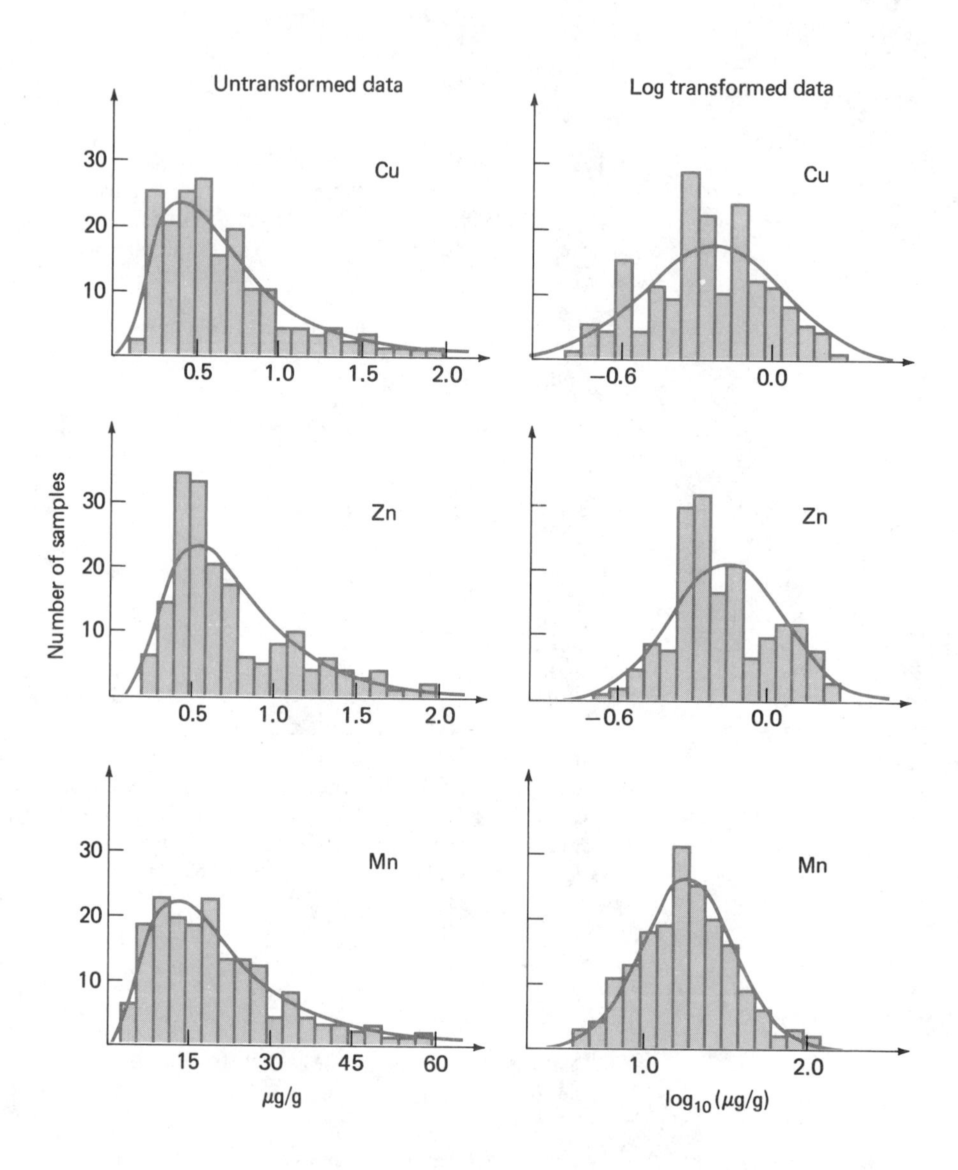

2.49 The Los Angeles Unified School District includes 49 public high schools. The *Los Angeles Times* (January 20, 1985) published the average math SAT exam score for each of the 49 schools. Use several of the methods described in this chapter to summarize this data.

Average SAT Math Score for 49 Public High Schools

341	477	461	456	349	481	499
328	471	436	440	414	448	503
399	335	332	422	356	375	488
406	375	458	341	468	404	482
464	475	398	317	466	470	463
409	478	469	404	487	439	459
464	502	472	480	481	402	339

REFERENCES

Koopmans, Lambert H. *An Introduction to Contemporary Statistics.* Boston: Duxbury, 1981. (The first part of this book contains an interesting presentation of both traditional descriptive methods and the more recently developed exploratory techniques.)

McNeil, Donald R. *Interactive Data Analysis.* New York: John Wiley, 1977. (A very informal brief introduction to some of the most useful recently developed methods for doing exploratory data analysis.)

Vellman, Paul, and David Hoaglin. *Applications, Basics, and Computing of Exploratory Data Analysis.* Boston: Duxbury, 1981. (Subtitled "ABC's of EDA," this book contains a good treatment of exploratory methods; it is a bit more comprehensive than the McNeil book.)

3 Numerical Summary Measures

INTRODUCTION

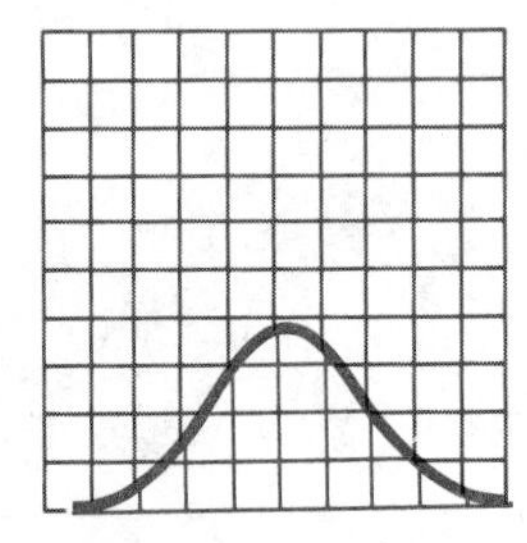

STEM-AND-LEAF displays, frequency distributions, and histograms are effective in conveying general impressions about the distribution of values in a data set. To gain deeper insights and develop methods for further analysis, however, we need more compact and precise ways of describing and characterizing data. Here we will see how some important features of a data set can be summarized and conveyed by using just a few numerical summary quantities computed from the data. What features of a data set should we try to describe in this way? The tabular and pictorial summaries of Chapter 2 point to two important features:

1. The location of a data set, and, in particular, specification of a central or representative value
2. The extent to which the values in a data set differ from one another—the amount of variability or dispersion in the data

The most useful numerical measures of center and variability are introduced in the first two sections of the chapter. The last section then illustrates how these summary measures can be used to convey salient features of the data.

The process of summarization carries with it a danger: In going from the raw data to a summary, important information may be lost. The tabular and pictorial methods discussed earlier usually involve relatively little information loss. But replacing an entire data set by just a few numerical summaries may hide some of the data's most important features. For example, suppose we have a collection of 100 observations, each one the yield of a particular stock (as a percentage of its price) listed on the New York Stock Exchange. If some of these 100 observations are for common stocks with relatively low yields, whereas others are for preferred stocks with relatively high yields, reporting only a measure of center and of spread will obscure this important feature of bimodality. On the other hand, if we are considering only yields of common stocks and their distribution can be well approximated by a normal curve, virtually all aspects of the distribution can be reconstructed using only appropriate measures of center and spread. The moral is that even the seemingly straightforward task of summarization requires forethought and intelligence on the part of the analyst.

3.1 Measuring the Center of a Data Set

An informative way to describe the location of a numerical data set is to report a central value, one that is typical of the observations in the data set. If the distribution is perfectly symmetric about some value, that value is a natural choice for the center of the data set; however, in other cases a choice of central value is not so readily apparent. The two most frequently used measures of center are the *mean* and the *median.*

The Mean

The **mean** of a set of numerical observations is just the familiar arithmetic average, the sum of the observations in the set divided by the number of observations.

DEFINITION

$$\text{mean} = \frac{\text{sum of all the observations in the data set}}{\text{number of observations in the data set}}$$

EXAMPLE 1

Ever since the Fontechevade fossils were discovered in Southern France in 1947, physical anthropologists have argued over whether or not they were definitely distinct from Neanderthals and therefore suggested the presence of a pre-Sapiens line in Europe during the Pleistocene age. The paper "A Reconsideration of the Fontechevade Fossils" (*J. of Physical Anthropology* (1973): 25–36) compared various sinus measurements from the Fontechevade I fossil to corresponding measurements on a sample of fossils known to be Neanderthal. The height of the left frontal sinus of Fontechevade I is 31 millimeters, and Table 1 gives these heights for 14 Neanderthal fossils. If the Fontechevade value is quite far from a typical Neanderthal value, that would suggest support for the separate line theory.

TABLE 1 HEIGHTS OF LEFT FRONTAL SINUSES FOR FOURTEEN NEANDERTHAL FOSSILS (mm)

Fossil	Height (mm)	Fossil	Height (mm)
1. Ehringsdorf H	42	8. La Quina H-5	34
2. Forbes' Quarry	27	9. Monte Circeo	35
3. Jebel Qafzeh	25	10. Neander Valley	25
4. Krapina C	40	11. Saccopastore I	29
5. Krapina E	33	12. Saccopastore II	30
6. La Chapelle-aux-Saints	31	13. Sala	29
7. La Ferrassie I	42	14. Spy II	35

As a typical value for Neanderthals, let's compute the mean:

$$\text{mean} = \frac{42 + 27 + 25 + 40 + 33 + 31 + 42 + 34 + 35 + 25 + 29 + 30 + 29 + 35}{14}$$

$$= \frac{457}{14} = 32.6$$

The Fontechevade value of 31 seems reasonably close to the mean of 32.6. However, if each and every one of the 14 Neanderthal measurements were 32.6, there would be no variability in the Neanderthal values with the mean still equal to 32.6. In this case the Fontechevade value would strongly suggest a non-Neanderthal fossil. We shall shortly see how the standard deviation, a measure of variability, can help us with this problem. Unfortunately for anthropologists, the data analysis in the paper was not conclusive enough to resolve the issue. ■

The measurements in Example 1 were all integers, yet we computed and reported the mean as 32.6. It is common practice to use one extra digit of decimal accuracy for the mean because the value of the mean can fall between possible observable values (e.g., the average number of children per family can be 1.8).

Most frequently the observations in a data set constitute a sample from some population rather than the collection of all population values. In this case there is some standard notation for the sample observations and sample mean that helps simplify certain formulas.

Notation: Let

x denote the numerical variable on which observations are made.
n denote the number of observations in the sample.
x_1 denote the first observation in the sample.
x_2 denote the second observation in the sample.
⋮
x_n denote the nth (last) observation in the sample.
$\overline{x}$ denote the sample mean.

Using this notation, a more compact way of writing the formula for the sample mean is

$$\text{sample mean} = \bar{x} = \frac{x_1 + x_2 + \cdots + x_n}{n}$$

EXAMPLE 2

(*Example* 1 *continued*) The variable of interest is x = the height of the left frontal sinus for a Neanderthal skull, and the number of observations is $n = 14$. The observations themselves are $x_1 = 42$, $x_2 = 27$, $x_3 = 25$, . . . , $x_{13} = 29$, and $x_{14} = 35$. The computed value of the sample mean is then $\bar{x} = 32.6$. ■

The symbol x_1 does *not* refer to the smallest observation in the sample but only the first observation obtained or the first one appearing in the list. The smallest observations in Example 1 were $x_3 = 25$ and $x_{10} = 25$, while $x_1 = 42$ was tied for the largest observation. Similarly, x_n is not usually the largest observation but only the last observation in the list of sample values. In general, *the subscript on a particular x observation is not related to the size of the observed value* but simply identifies the position of that value in the listing of sample observations. If the observations were listed in a different order, the same value of $\bar{x}$ would result, since the sum does not depend on the order in which the individual numbers are added. Also, the choice of the letter x is convenient when focusing on a single variable, but at times we may make another choice. If we want to compare Neanderthal skulls to those of modern Europeans, we might let y = frontal sinus height for a modern European, denote the observations on y by $y_1, y_2, \ldots$, and denote their mean by $\bar{y}$.

We often need to add either $x_1, x_2, \ldots$, and x_n or even more complicated expressions involving these quantities. The expression $x_1 + x_2 + \cdots + x_n$ is cumbersome to write, so mathematicians use alternative notation for such a sum: Σ (uppercase Greek sigma), called the *summation symbol.*

Summation Notation

Σx denotes the sum of all x values in the data set under consideration. When the data set is a sample, the sample mean is written as

$$\bar{x} = \frac{\Sigma x}{n}$$

The sample mean $\bar{x}$ is computed from sample values, so it is a characteristic of the particular sample in hand. It is customary to use a Roman letter to denote such a sample characteristic, as we have done in using $\bar{x}$. The **population mean** (average of all population values) is a characteristic of the entire population. Statisticians commonly use lowercase Greek letters to denote population characteristics; in particular, *let μ denote the population mean.*

EXAMPLE 3

The 50 states plus the District of Columbia contain a total of 3137 counties. Let x denote the number of residents of a county. Then there are 3137 x values in the population. The sum of these 3137 x values is 226,504,825 (1980 census), so the population average value of x is

$$\mu = \frac{226{,}504{,}825}{3137} = 72{,}204.3 \quad \text{residents per county.}$$

We used *The World Almanac and Book of Facts* to randomly select three different samples at random from this population of counties, each sample consisting of five counties. The results appear in Table 2, along with the sample mean for each sample.

TABLE 2

THREE SAMPLES FROM THE POPULATION OF ALL U.S. COUNTIES
(x = number of residents)

Sample 1		Sample 2		Sample 3	
County	x Value	County	x Value	County	x Value
Fayette, TX	18,832	Stoddard, MO	29,009	Chattanochee, GA	21,732
Monroe, IN	96,387	Johnston, OK	10,356	Petroleum, MT	655
Greene, NC	16,117	Sumter, AL	16,908	Armstrong, PA	77,768
Shoshone, ID	19,226	Milwaukee, WI	964,988	Smith, MI	15,077
Jaspar, IN	26,138	Albany, WY	29,062	Benton, MO	12,183
$\Sigma x =$	176,700	$\Sigma x =$	1,050,323	$\Sigma x =$	127,415
$\bar{x} =$	35,340.0	$\bar{x} =$	210,064.6	$\bar{x} =$	25,483.0

Not only are the three $\bar{x}$ values different from one another—because they are based on three different samples and the value of $\bar{x}$ depends on the x values in the sample—but none of the three comes close to the value of the population mean, μ. If we did not know the value of μ but had only sample 1 available, we might use $\bar{x} = 35{,}340.0$ as an *estimate* of μ, but our estimate would be far off the mark. Alternatively, we could combine the three samples into a single sample with $n = 15$ observations, denoting them by $x_1, x_2, \ldots, x_{15}$. Then $\Sigma x = 1{,}354{,}438$, so $\bar{x} = 1{,}354{,}438/15 = 90{,}295.9$. This is closer to the true value of μ but still not very satisfactory as an estimate. The problem here is that there is so much variability in the population of x values (the largest is $x = 7{,}477{,}657$ for Los Angeles County and the smallest is $x = 91$ for Loving County, Texas, which few people evidently love) that it is difficult for a sample of 15 observations, let alone just 5, to be reasonably representative of the population. But don't lose hope! Once we learn how to measure variability, we'll also see how to take it into account in making accurate inferences. ■

The mean does have one property that renders it an inappropriate measure of center in some situations. The value of the mean can be greatly influenced by the presence of a single outlying value, one which is far above or below the remaining data values. Look back at sample 2 in Table 2. The first, second, third, and fifth values are less than 30,000, but $x_4 = 964{,}988$ resulted in an extremely large value of $\bar{x}$. Imagine what would have happened if Los Angeles County had appeared in the sample! Statisticians say that *$\bar{x}$ is not a resistant measure of center.* An alternative measure of center, which is extremely resistant to outliers, is the median.

The Median

The median strip of a highway divides the highway in half, and the median of a numerical data set performs an analogous function. Once the numbers in the set have been listed in order from smallest to largest, the **median** is the middle value in the list and divides the list into two equal parts. Let's first consider the case of a sample containing n observations. When n is an odd number (e.g., 5), the sample median is the single middle value. But when n is even (e.g., 6), there are two middle values in the ordered list, so we average them to obtain the sample median.

DEFINITION

To compute the **sample median,** the n observations are first listed from smallest to largest (with any repeated values included, so that every sample observation appears in the ordered list). Then

$$\text{sample median} = \begin{cases} \text{the single middle value when } n \text{ is odd} \\ \text{the average of the two middle values when } n \text{ is even} \end{cases}$$

EXAMPLE 4

Because copper is known to be intimately involved in certain brain functions and is linked to the mechanism of action of certain antipsychotic drugs, several studies concerning copper concentration in humans have recently appeared in psychiatric and other medical journals. The paper "CSF Copper Concentration in Chronic Schizophrenia" (*Amer. J. of Psychiatry* (1983): 754–57) reported on a comparison of copper concentrations of a certain type (in ppb) for members of four groups: former heroin addicts, unmedicated and medicated schizophrenics, and normal controls. Sample values for males from two of the groups are given in the accompanying table along with the median for each sample. The authors used the analysis of variance, which is introduced in Chapter 13, to conclude that the four groups did not differ substantially from one another with respect to copper concentration.

Group	Sample Size	Sample	Ordered Values	Median
Normal controls	n = 9 (odd)	7, 6, 7, 7, 17, 4, 7, 7, 5	4, 5, 6, 7, 7, 7, 7, 7, 17	7
Former addicts	n = 8 (even)	6, 8, 8, 5, 10, 6, 6, 8	5, 6, 6, 6, 8, 8, 8, 10	$\frac{6+8}{2} = 7$

■

The **population median** plays the same role for the population as the sample median plays for the sample; it is the middle value in the ordered list consisting of all population observations. Most frequently we don't have all the population values available (otherwise there wouldn't be much business for statisticians), so we won't know the value of the population median. The sample median can then be used to draw some type of conclusion (make a statistical inference) about the population median. Just as with the sample mean, there is the problem of sampling variability—different samples from the same population yield different values of the sample median. We present a systematic discussion of such sampling variability in Chapter 6.

We previously noted that the mean—population or sample—is very sensitive to even a single value that lies far above or below the rest of the data. The value of the mean is pulled out toward such an outlying value or values. The median, on the other hand, is insensitive to such outlying values. For example, in the normal controls sample of Example 4, the largest sample value, 17, can be increased by an arbitrarily large amount without changing the value of the median. In fact, every value above the middle value(s) in the ordered list can be increased without changing the median, and every value below the middle value(s) can be decreased without causing the median to change. This stability of the median is what sometimes justifies its use as a measure of center. Income distributions and housing price distributions are commonly summarized by reporting the median rather than the mean, since otherwise those few $100,000 salaries or $1,000,000 houses would distort the resulting typical salary or price.

Comparing the Mean and the Median

Figure 1 presents several smoothed histograms that might represent either a distribution of sample values or a population distribution. Pictorially, the median is the value on the measurement axis that separates the histogram into two parts with .5 (50%) of the area under each part of the curve. The mean is a bit harder to visualize. If the histogram were placed on a fulcrum with a sharp point, it would tilt unless the fulcrum were positioned exactly at the mean. The mean is the balance point for the distribution.

When the histogram is symmetric, the point of symmetry is both the dividing point for equal areas and the balance point, so the mean and median are identical. However, when the histogram is unimodal (a single peak) with a long upper tail, the relatively few outlying values in the upper tail pull the mean up so that it lies above the median. The reverse situation occurs for a unimodal histogram with a long lower tail (see Figure 2).

Categorical Data and Proportions

Investigators frequently wish to study a population that consists of two types of individuals or objects, those with some property of interest and those without the property. Examples include high school graduates who took or didn't take a computer literacy course, automobiles that use or don't use unleaded gasoline, smoke detectors that activate or do not activate under certain condi-

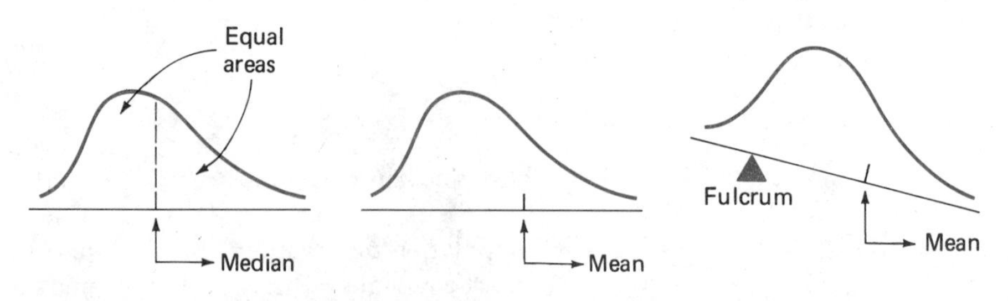

FIGURE 1 PICTURING THE MEAN AND THE MEDIAN

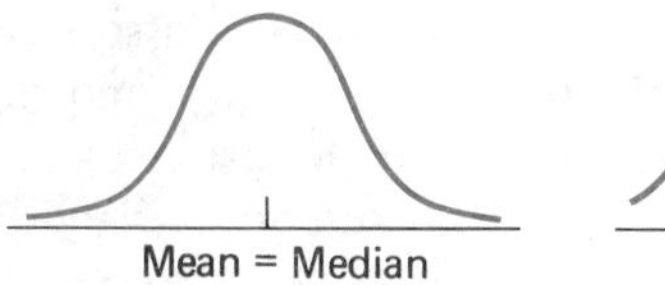

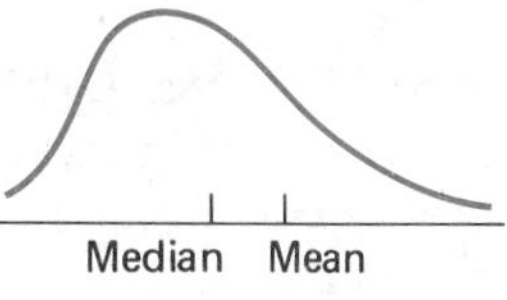

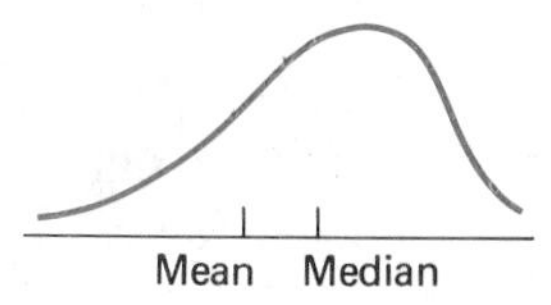

FIGURE 2 RELATIONSHIP BETWEEN THE MEAN AND THE MEDIAN

tions, households headed or not headed by a single parent, and so on. In a sample from such a population, some number of the individuals or objects would have the property and the remainder of those in the sample would not. Suppose, for example, that we denote a smoke detector that properly activates by S (for success) and one which does not activate properly by F (for failure). A sample of 10 detectors might yield (in the order in which they were tried) $S, S, S, F, S, F, S, S, S, S$—eight successes and two failures. The proportion of successes in this sample is $\frac{8}{10} = .8$. The sample proportion of successes will always be a number between 0 and 1, and the same will be true of the proportion of successes in the entire population.

In general it will be convenient to label each individual or object a success (S) or failure (F) according to whether or not it has the property of interest. Then a sample will consist of a sequence of S's and F's, and the **sample proportion of successes** is the number of S's divided by the sample size. With n again denoting the sample size and p *denoting the sample proportion of successes* (a lowercase Roman letter, as this is a sample characteristic), we have

DEFINITION

$$p = \begin{matrix}\text{sample proportion}\\ \text{of successes}\end{matrix} = \frac{\text{number of successes in the sample}}{n}$$

The Greek letter π *denotes the population proportion of successes*. The relationship between π and p is the same as that between μ and $\overline{x}$—one is a population characteristic, the other is a sample characteristic— and the latter will be used to make some type of inference about the former.

The formula for p looks somewhat like the formula for $\overline{x}$, and they can be made to look even more alike. Suppose that we code each S as a 1 and each F as a 0. Then the sample $S, S, S, F, S, F, S, S, S, S$ becomes 1, 1, 1, 0, 1, 0, 1, 1, 1, 1, and the sample mean of this numerical sample is

$$\overline{x} = \frac{1 + 1 + 1 + 0 + 1 + 0 + 1 + 1 + 1 + 1}{10} = \frac{8}{10} = .8 = p$$

If each S is coded as a 1 and each F is coded as a 0, the sample mean of the coded values is exactly the sample proportion of S's.

Because the sample proportion p is really a sample mean in disguise, whenever we present a result concerning the behavior of $\overline{x}$ (such as a measure of sampling variation), there is a parallel result for p.

Trimmed Means: A Compromise Between Mean and Median (Optional)

The extreme sensitivity of the mean to even a single outlier and the extreme insensitivity of the median to a substantial proportion of outliers can make both suspect as a measure of center. Statisticians have proposed a *trimmed mean* as a compromise between these two extremes. Before giving a general description, consider the following example.

EXAMPLE 5

The paper "Snow Cover and Temperature Relationships in North America and Eurasia" (*J. of Climate and Applied Meteorology* (1983):460–69) used statistical techniques to relate amount of snow cover on each continent to average continental temperature. Data presented there included the following 10 observations on October snow cover for Eurasia during the years 1970–1979 (in million km^2): 6.5, 12.0, 14.9, 10.0, 10.7, 7.9, 21.9, 12.5, 14.5, 9.2. The values in increasing order are 6.5, 7.9, 9.2, 10.0, 10.7, 12.0, 12.5, 14.5, 14.9, 21.9. To compute the median, we delete *all* values on both ends of the ordered list except the middle values and average what remains: median = (10.7 + 12.0)/2 = 11.35. The mean is computed by deleting *nothing* from the ordered list before averaging: $\bar{x} = (6.5 + 7.9 + \cdots + 21.9)/10 = 12.01$. A compromise between the extremes of deleting nothing and deleting virtually everything is to delete just a few observations from each end of the ordered list before averaging. Consider deleting a single observation—10% of the sample, since $n = 10$—from each end:

$$\text{10\% trimmed mean} = \frac{7.9 + 9.2 + 10.0 + 10.7 + 12.0 + 12.5 + 14.5 + 14.9}{8}$$

$$= 11.46$$

Similarly, the 20% trimmed mean results from deleting two observations from each end of the ordered list before averaging:

$$\text{20\% trimmed mean} = \frac{9.2 + 10.0 + 10.7 + 12.0 + 12.5 + 14.5}{6}$$

$$= 11.48$$

Both of these two trimmed means fall between the mean and median, and either would be a good candidate for a typical October snow-cover value. ■

DEFINITION	A **trimmed mean** is computed by first ordering the data values from smallest to largest, deleting a seleted number of values from each end of the ordered list, and finally averaging the values not deleted. The **trimming percentage** is the percentage of values deleted from each end of the ordered list.

If $n = 20$, deleting only the smallest and largest observations yields a 5% trimmed mean ($\frac{1}{20} = .05$), while if $n = 15$ and we delete two from each end, the trimming percentage is 13.33%. For a small trimming percentage, a trimmed mean is less affected by outliers than the mean and yet doesn't have the insensitivity of the median. For this reason, trimmed means are being used with increasing frequency in data exploration and analysis.

EXERCISES

3.1 The accompanying data on concentration of lead (ppm) in core samples taken at 17 Texaco drilling sites appeared in the paper "Statistical Comparison of Heavy Metal Concentrations in Various Louisiana Sediments" (*Environmental Monitoring and Assessment* (1984):163–70).

55	53	55	59	58	50	63	50	50
48	56	63	54	53	56	50	55	

a. Compute the sample mean.

b. Compute the sample median. How do the values of the sample mean and median compare?

c. The paper also gave data on zinc concentration (ppm). Compute and interpret the value of the sample mean for the data below.

86	77	91	86	81	87	94	90	70
92	90	108	112	101	88	99	98	

3.2 The paper "Penicillin in the Treatment of Meningitis" (*J. Amer. Medical Assn.* (1984):1870–73) reported the body temperatures (°F) of patients hospitalized with meningitis. Ten of the observations were as follows.

104.0	104.8	101.6	108.0	103.8	100.8	104.2	100.2	102.4	101.4

a. Compute the sample mean.

b. Do you think the 10% trimmed mean would differ much from the sample mean computed in part (a)? Why? Answer without actually computing the trimmed mean.

3.3 Suppose that the 10 patients whose temperatures were given in Exercise 3.2 received treatment with large doses of penicillin. Three days later, temperatures were again recorded, and the treatment is considered successful if there has been a reduction in a patient's temperature. Denoting success by S and failure by F, the 10 observations are:

$S \quad S \quad F \quad S \quad S \quad S \quad F \quad F \quad S \quad S$

Numerically code each success as a 1 and each failure as a 0 and compute the value of the sample mean. Note that the resulting mean is equal to the proportion of successes in the sample. If you had coded success as 0 and failure as 1 before computing the sample mean, would the interpretation of the sample mean change? Explain.

3.4 In anticipation of the 1984 Olympics, the *Los Angeles Times* (August 15, 1983) reported the ozone levels at several sites that were to be used for Olympic events the following summer. Listed below are the ozone readings (in ppm) taken at noon from July 28 to August 12 at East Los Angeles College.

10	14	13	18	12	22	14	19
22	13	14	16	3	6	7	19

a. Compute the sample mean.

b. Compute the sample median.

c. Compute the 6.25% trimmed mean (by deleting the smallest and largest values before averaging). Is this a more representative measure than the sample mean or median? Explain.

d. Consider now the ozone levels for a second Olympic site, the Coliseum.

8	13	10	9	16	12	13	14
17	13	9	16	5	9	8	12

Describe the center of this data set using the mean and the median. How do the values of the mean and median for the Coliseum compare to those for East Los Angeles College?

3.5 One of the problems with which health service administrators must deal is patient dissatisfaction. One common complaint centers on the amount of time that a patient must wait in order to see a doctor. In a survey to investigate waiting times, medical-clinic secretaries were asked to record the waiting times (times from arrival at the clinic until seeing the doctors) for a sample of patients. The data (from *Statistics and Public Policy,* cited in Chapter 1) for 1 day is given in the table.

Waiting Time (min)

40	30	40	55	30	60	35	55	40
35	5	10	65	35	35	30	30	60
35	25	65	30	30	45	85	25	25
10	10	15						

a. Describe a typical waiting time by using the mean and median. Which do you think is the most representative measure of the center of the data set? Why?

b. Compute a 10% trimmed mean. How does the trimmed mean compare in value to the mean and median calculated in part (a)?

3.6 The article "Drug Screening: The Never-Ending Search for New and Better Drugs" (from *Statistics: A Guide to the Unknown,* cited in Chapter 1) reports actual tumor weights (in grams) observed in three animals treated with a test drug and in six untreated control animals:

Treated	0.96	1.59	1.14			
Untreated	1.29	1.60	2.27	1.31	1.88	2.21

Calculate the mean and the median for each of the two samples. How do these values compare?

3.7 Consider the following statement: Over 65% of the residents of Los Angeles earn less than the average wage for that city. Could this statement be correct? If so, how? If not, why not?

3.8 The *Los Angeles Times* (December 22, 1983) reported that the average cost per day for a semiprivate hospital room for California, Oregon, and Washington was \$268, \$210, and \$220, respectively. Suppose that you were interested in the average cost of a hospital stay for the hospitals in these three West Coast states. If each state contained the same number of hospital rooms, we could average the three values given. However, this is not the case, so the three averages must be weighted in proportion to the respective number of hospital rooms in each state. In general, if $\bar{x}_1, \bar{x}_2, \ldots, \bar{x}_n$ are n group means and $w_1, w_2, \ldots, w_n$ are the respective weights, the weighted mean $\bar{x}_w$ is defined as

$$\frac{w_1\bar{x}_1 + w_2\bar{x}_2 + \cdots + w_n\bar{x}_n}{w_1 + w_2 + \cdots + w_n}$$

The number of hospital rooms in California, Oregon, and Washington are 120,000, 30,000, and 45,000, respectively. Use the formula above with $w_1 = 120{,}000$, $w_2 = 30{,}000$, and $w_3 = 45{,}000$ to find the average cost of a hospital room on the West Coast.

3.9 Suppose that an auto dealership employs clerical workers, salespeople, and mechanics. The average monthly salaries of the clerical employees, salespeople, and mechanics are \$1100, \$1800, and \$1900, respectively. If the dealership has 3 clerical employees, 10 salespeople, and 8 mechanics, find the average monthly wage for all employees of the dealership. (*Hint:* Refer to Exercise 3.8.)

3.10 a. Suppose that n, the number of sample observations, is an odd number. Under what conditions on the sample will $\bar{x}$ and the sample median be identical?

b. If $n = 10$, when will $\bar{x}$ and the 10% trimmed mean be identical?

3.11 a. A statistics instructor informed her class that the median exam score was 78 and the mean was 70. Sketch a picture of what the smoothed histogram of scores might look like.

b. Suppose that the majority of students in the class studied for the exam, but a few students had not. How might this be reflected in a smoothed histogram of exam scores?

3.2 Measuring Variability in a Data Set

A measure of the center of the data set by itself can provide only a partial summary of the data. We also want a measure of the extent to which values are spread out about the center. Figure 3 pictures three different samples, all of which have the same mean and median. Reporting only the central value would not enable anyone to differentiate between the three data sets, yet they differ in obvious ways. All six observations in the first sample are quite spread out, suggesting a great deal of variability. The variability in the second sample, while rather substantial, is due primarily to the two extreme values being quite far from the center. In the third sample, all six observations are quite close to the center, implying little variability. The simplest measure of variability is the **range**, or difference between the largest and smallest value. For sample 1 in Figure 3, the range is $70 - 20 = 50$, whereas for sample 3 it is only $50 - 40 = 10$. Generally speaking, the larger the range, the more variability there is in the data set. However, for sample 2 the range is $70 - 20 = 50$, identical to the range in the first sample, yet there is clearly more variability in sample 1 than in sample 2. Because the range depends only on the two extreme values, it is not an effective measure of variability. We need a measure that takes into account the location of *all* observations relative to the center.

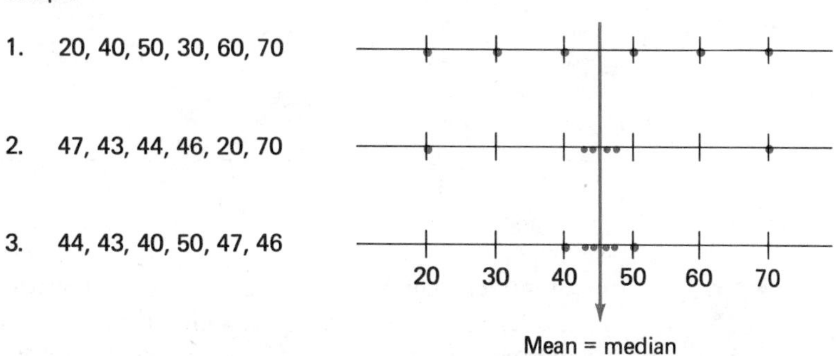

FIGURE 3 THREE SAMPLES WITH THE SAME CENTER AND DIFFERENT AMOUNTS OF VARIABILITY

Deviations from the Mean

Let's focus on measuring variability in a sample. As before, the n sample observations are denoted by $x_1, x_2, \ldots, x_n$. The sample mean, $\bar{x}$, will be our measure of center. If the first observation, x_1, lies either far above or far below $\bar{x}$, it contributes a substantial amount to variation in the sample, while if x_1 is close to $\bar{x}$, then there is little variability due to the first observation. Another way to look at the same thing is to examine $x_1 - \bar{x}$, the deviation of the first observation from the mean. This deviation could be a positive number (e.g., $47 - 45 = 2$ for sample 2, since $x_1 = 47$ is above $\bar{x} = 45$) or a negative number (e.g., $20 - 45 = -25$ for sample 1, since $x_1 = 20$ is below the mean). Deviations of the other observations from the mean are defined analogously. It is the *magnitude* of each deviation rather than its sign that gives us information about variability. A deviation will be large in magnitude if the observation lies far from the mean and small otherwise. Figure 4 illustrates this.

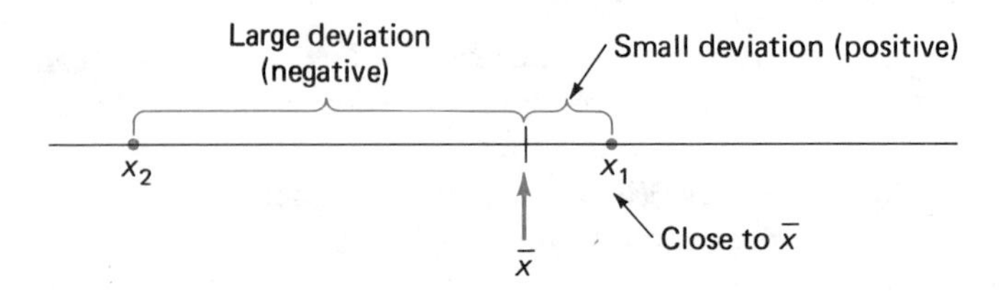

FIGURE 4 LARGE AND SMALL DEVIATIONS FROM THE MEAN

EXAMPLE 6

Example 1 gave the 14 observations in Table 3 on sinus height (mm.) of Neanderthal skulls. The sample mean was $\bar{x} = \frac{457}{14} = 32.6$, which is subtracted from each observation to obtain the 14 deviations.

TABLE 3 DEVIATIONS OF NEANDERTHAL SINUS HEIGHTS FROM THE MEAN ($\bar{x} = 32.6$)

Observation Number	Observation (x)	Deviation ($x - \bar{x}$)	Observation Number	Observation (x)	Deviation ($x - \bar{x}$)
1	42	42 − 32.6 = 9.4	8	34	1.4
2	27	27 − 32.6 = −5.6	9	35	2.4
3	25	−7.6	10	25	−7.6
4	40	7.4	11	29	−3.6
5	33	.4	12	30	−2.6
6	31	−1.6	13	29	−3.6
7	42	9.4	14	35	2.4

DEFINITION

In a numerical sample $x_1, x_2, \ldots, x_n$, the n **deviations from the mean** are $x_1 - \bar{x}, x_2 - \bar{x}, \ldots, x_n - \bar{x}$. Once these deviations have been computed, they can be combined in an appropriate fashion to yield a numerical measure of variability.

A first thought might be to average the deviations. The sum of the deviations is $(x_1 - \bar{x}) + (x_2 - \bar{x}) + \cdots + (x_n - \bar{x})$, which can be denoted more compactly by $\Sigma(x - \bar{x})$. Then

$$\text{average deviation} = \frac{\text{sum of } n \text{ deviations}}{n} = \frac{\Sigma(x - \bar{x})}{n}$$

The difficulty with this proposal is that when the deviations are added, negative and positive deviations will counteract one another.

EXAMPLE 7

(*Example 6 continued*) The sum of the 14 sinus height deviations is, from Table 3, $9.4 + (-5.6) + (-7.6) + \cdots + (-3.6) + 2.4 = .60$, so the average deviation is $.60/14 = .043$. If we use $\bar{x} = 32.64$ rather than 32.6 to compute the deviations, the sum is $9.36 + (-5.64) + (-7.64) + \cdots + (-3.64) + 2.36 = .04$, and the average deviation is $.04/14 = .0029$. Using more decimal accuracy in $\bar{x}$ when computing the deviations will yield both a sum of deviations and average deviation even closer to zero. ■

Except for effects in computing the deviations due to rounding in the value of $\bar{x}$, it is always true that $\Sigma(x - \bar{x}) = 0$.

The Variance and Standard Deviation

Because $\Sigma(x - \bar{x}) = 0$, the average deviation will equal zero, so the average deviation certainly does not measure variability. To obtain a sensible measure, we must prevent negative and positive deviations from counteracting one another. One way to do this is to square the deviations. Then deviations of -10 and 10 both lead to squared deviations of 100, so both values contribute equally to variability. The sum of the squared deviations is $(x_1 - \bar{x})^2 + (x_2 - \bar{x})^2 + \cdots + (x_n - \bar{x})^2$. Using summation notation, this is written as $\Sigma(x - \bar{x})^2$. You might think it natural to divide this sum by n to obtain the average squared deviation as a measure of variability, but there is a technical reason for doing something slightly different.

DEFINITION

The **sample variance**, denoted by s^2, is the sum of squared deviations divided by $n - 1$. That is,

$$s^2 = \frac{\Sigma(x - \bar{x})^2}{n - 1}$$

The **sample standard deviation** is the square root of the variance and is denoted by s.

A large amount of variability in the sample is indicated by a relatively large value of s^2 or of s, while a small value of s^2 or of s goes along with a small amount of variability. For most statistical purposes, s is the desired quantity, but s^2 must be computed first. Let's first look at a computational example, and then we'll discuss some properties of s^2 and s.

EXAMPLE 8

(*Example 6 continued*) Table 4 presents squared deviations for the Neanderthal sinus height data using $\overline{x} = 32.64$.

TABLE 4

DEVIATIONS AND SQUARED DEVIATIONS FOR THE NEANDERTHAL SINUS-HEIGHT DATA

Observation	$(x - \overline{x})$	$(x - \overline{x})^2$
42	9.36	87.6096
27	−5.64	31.8096
25	−7.64	58.3696
40	7.36	54.1696
33	.36	.1296
31	−1.64	2.6896
42	9.36	87.6096
34	1.36	1.8496
35	2.36	5.5696
25	−7.64	58.3696
29	−3.64	13.2496
30	−2.64	6.9696
29	−3.64	13.2496
35	2.36	5.5696
	Sum:	427.2144

With $\Sigma(x - \overline{x})^2 = 427.2144$, $s^2 = 427.2144/13 = 32.8626$ and $s = \sqrt{32.8626} = 5.73$. The same value of s would have resulted had $\overline{x} = 32.6$ been used, but for some data sets the computations are affected by the way in which $\overline{x}$ is rounded. ■

A Note Concerning Computation

The computation of s^2 using the defining formula can be a bit tedious. An alternative expression for the numerator of s^2 simplifies the arithmetic by eliminating the need to first calculate the deviations.

$$s^2 = \frac{\Sigma(x - \overline{x})^2}{n - 1} = \frac{\Sigma x^2 - n\overline{x}^2}{n - 1}$$

According to this formula, after squaring each x value and adding these to obtain Σx^2, the single quantity $n\overline{x}^2$ (i.e., $n \cdot \overline{x} \cdot \overline{x}$) is subtracted from the result. Instead of the n subtractions required to obtain the deviations, just one subtraction now suffices.

EXAMPLE 9

(*Example 6 continued*) For efficient computation it is convenient to place the x values in a single column (or row) and the x^2 values just beside (or below) them. Adding the numbers in these two columns (or rows) then gives Σx and Σx^2, respectively.

Observation Number	x	x^2
1	42	1,764
2	27	729
3	25	625
4	40	1,600
5	33	1,089
6	31	961
7	42	1,764
8	34	1,156
9	35	1,225
10	25	625
11	29	841
12	30	900
13	29	841
14	35	1,225
	$\Sigma x = 457$	$\Sigma x^2 = 15{,}345$

Thus $\bar{x} = 457/14 = 32.64$ and $n\bar{x}^2 = (14)(32.64)(32.64) = 14{,}915.17$, so

$$s^2 = \frac{15{,}345 - 14{,}915.17}{14 - 1} = \frac{429.83}{13} = 33.06$$

and $s = 5.75$. Rounding resulted in a value of s that differs slightly from the earlier value of 5.73. For practical purposes, either value can be used. ■

The computed value of s^2, whether obtained from the defining formula or the computational formula, can sometimes be greatly affected by the way in which $\bar{x}$ is rounded. Protection against adverse rounding effects can virtually always be achieved by using four or five digits of decimal accuracy beyond the decimal accuracy of the data values themselves. With this rule of thumb, $\bar{x} = 32.64286$ would have been used in place of 32.6 or 32.64 in Example 6, resulting in $s = 5.73$ when s^2 is computed using either formula.

The expression $\Sigma(x - \bar{x})^2$ and similar quantities appear repeatedly in the remainder of the text, not only in s^2 but also in various other formulas. Though we won't always remind you that the computational formula can be used, please bear it in mind.

Properties of Variance and Standard Deviation

Natural questions concerning the variance and standard deviation include the following:

1. Why might s be preferred to s^2 as a measure of variability?
2. How can we tell whether the value of s suggests a large or small amount of variability?
3. If s measures variability in a sample, is there an analogous quantity that measures variability in the population?
4. Why use the divisor $n - 1$ in s^2 rather than just n?

One reason for preferring s to s^2 is that *s is a measure of variability expressed in the same units of measurement as are the data values themselves.* The x's in Example 6 were heights measured in millimeters, so the deviations are in millimeters. Squaring each deviation, adding, and dividing by $n - 1$ gives a measure expressed in mm^2. Taking the square root to obtain s gets us back to millimeters. The standard deviation may be informally interpreted as the typical distance from the mean to an observation in the data set. Thus in Example 6 a typical distance from $\overline{x} = 32.6$ is about 5.7. Recall that for the Fontechevade fossil discussed in Example 1, sinus height was 31 mm. This value is within 1 standard deviation of the mean, so the value is certainly consistent with the hypothesis that Fontechevade is a Neanderthal fossil.

We computed $s = 5.73$ in Example 8 without saying whether this value indicated a large or small amount of variability. At this point it is easier to use s for comparative purposes than for an absolute assessment of variability. If we obtained a sample of skulls of a second type and computed $s = 2.1$ for those skulls, then we would conclude that there is more variability in our original sample than in this second sample. A particular value of s can be judged large or small only in comparison to something else.

There are measures of variability for the entire population that are analogous to s^2 and s for a sample. These measures are called the **population variance** and **population standard deviation** and are *denoted by* σ^2 *and* σ, respectively (again, a lowercase Greek letter for a population characteristic). The population standard deviation σ is expressed in the same units of measurement as are the values in the population and can easily be used to compare variability in several different populations. Given the value of σ for two different populations, the larger σ indicates a greater amount of variability.

In many statistical procedures, we would like to use the value of σ, but unfortunately it isn't available. We therefore have to use in its place a value computed from the sample that we hope is close to σ (that is, a good *estimate* of σ). For this reason the divisor $n - 1$ is used in s^2—the value of s^2 as defined tends to be a bit closer to σ^2 than if s^2 were defined using divisor n. (This is a result from statistical theory.)

Both s^2 and s are computed by starting with the deviations $x - \overline{x}$. We have already noted that the deviations have a sum of zero. Suppose, for example, that $n = 5$ and we are told that four of the deviations are -3, 5, 2, and -10. Then because $(-3) + 5 + 2 + (-10) + (\text{remaining deviation}) = 0$, or $-6 + (\text{remaining deviation}) = 0$, the remaining deviation is 6. That is, although there are five deviations, any one of the five can be determined once the other four are known—among the five deviations, only four are freely determined.

> There are n deviations $x_1 - \overline{x}, x_2 - \overline{x}, \ldots, x_n - \overline{x}$, but the relation $\Sigma(x - \overline{x}) = 0$ implies that only $n - 1$ of these are freely determined. Statisticians then say that s^2 and s are based on only **$n - 1$ degrees of freedom** (abbreviated df).

In Example 6, n is 14, so the number of degrees of freedom is 13 (the divisor in s^2).

The Interquartile Range: A Resistant Measure of Variability (Optional)

We have already seen how the value of $\overline{x}$ can be drastically affected by the presence of a single outlying value in the sample. The same thing is true of the sample standard deviation s; it is not resistant to the presence of even a single outlier. Especially for exploratory purposes, it is desirable to have a measure of variability that is resistant to the presence of a few outliers.

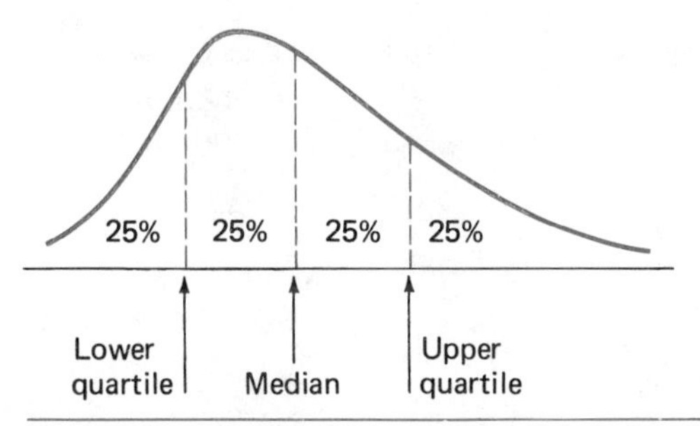

FIGURE 5 THE QUARTILES FOR A SMOOTHED HISTOGRAM

To develop one such resistant measure, the *interquartile range,* we first need to know about quartiles. The lower quartile separates the bottom 25% of the data set from the upper 75%, and the upper quartile separates the top 25% from the bottom 75%. The middle quartile is the median. Figure 5 illustrates the locations of these quartiles for a smoothed histogram.

The definition of the sample median depended on whether the sample size n was an odd or an even number. Similar care must be exercised in defining the sample quartiles. If $n = 8$, for example, the lower half of the sample consists of the four smallest sample observations and the upper half consists of the four largest observations. If $n = 9$, the median is included in both the lower and the upper half of the sample, so that each half contains five observations. The lower and upper quartiles are then the medians of the two halves of the sample (see Figure 6).

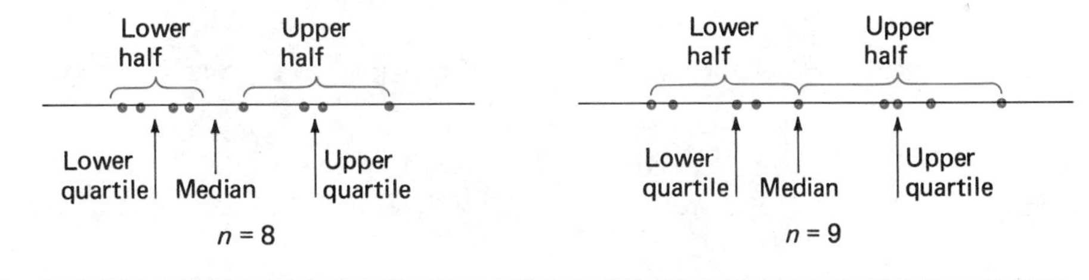

FIGURE 6 SAMPLE QUARTILES WHEN $n = 8$ AND WHEN $n = 9$

DEFINITION

Lower quartile = median of the lower half of the sample
Upper quartile = median of the upper half of the sample

(If n is odd, the median is included in both halves.)

EXAMPLE 10

Cardiac output and maximal oxygen uptake typically decrease with age in sedentary individuals, but these decreases are at least partially arrested in middle-aged individuals who engage in a substantial amount of physical exercise. To understand better the effects of exercise and aging on various circulatory functions, the paper "Cardiac Output in Male Middle-Aged Runners"

(*J. of Sports Medicine* (1982):17–22) presented data from a study of 21 male middle-aged runners. Figure 7 is a stem-and-leaf display of oxygen uptake values (mL/kg · min) while pedaling at 100 watts on a bicycle ergometer.

```
12 || 81
13 ||
14 || 95
15 || 97,83
16 ||
17 || 90
18 || 34,27
19 || 94,82
20 || 99,93,98,62,88
21 || 15
22 || 16,24              Stems: Ones
23 || 16,56              Leaves: Hundreths
HI:   35.78,36.73
```

FIGURE 7 STEM-AND-LEAF DISPLAY OF OXYGEN UPTAKE (mL/kg · min)

The sample size $n = 21$ is an odd number, so the median 20.88 is included in both halves of the sample:

Lower half	12.81	14.95	15.83	15.97	17.90	18.27
	18.34	19.82	19.94	20.62	20.88	
Upper half	20.88	20.93	20.98	20.99	21.15	22.16
	22.24	23.16	23.56	35.78	36.73	

Each half of the sample contains 11 observations, so each quartile is the sixth value in from either end of the corresponding half. This gives lower quartile = 18.27 and upper quartile = 22.16. ■

DEFINITION

The **interquartile range** (iqr), a resistant measure of variability, is given by

$$\text{iqr} = \text{upper quartile} - \text{lower quartile}$$

EXAMPLE 11

(*Example 10 continued*) The interquartile range for the sample of oxygen-uptake values is iqr = 22.16 − 18.27 = 3.89. The sample mean and standard deviation are 21.10 and 5.75, respectively. If we change the two largest values from 35.78 and 36.73 to 25.78 and 26.73 (so that they are still the two largest values), the median and the interquartile range are not affected, whereas the mean and standard deviation change to 20.14 and 3.44, respectively. ■

In general, up to 25% of the values on either end of the ordered sample can be made more extreme without affecting the interquartile range, so it is quite resistant to outliers.

The **population interquartile range** is the difference between the upper and lower population quartiles. If a histogram of the data set under consideration (whether a population or a sample) can be reasonably well approximated by a normal curve, then the relationship between the standard deviation (sd) and interquartile range is roughly sd = iqr/1.35. A value of the standard deviation much larger than iqr/1.35 suggests a histogram with heavier (or longer) tails than a normal curve. For the data of Example 10, $s = 5.75$, while iqr/1.35 = 3.89/1.35 = 2.88. This suggests that the distribution of sample values is indeed heavy-tailed compared to a normal curve, as could also be seen by trying to superimpose a normal curve on the stem-and-leaf display of Figure 7.

In the next section, we show how the median and interquartile range can be combined to yield a nice summarizing picture called a *box plot.*

EXERCISES

3.12 The paper "Improving Fermentation Productivity with Reverse Osmosis" (*Food Technology* (1984):92–96) summarized the results of an investigation into glucose concentration (g/l) for a particular blend of malt liquor. Eight batches were analyzed, resulting in the given glucose concentrations.

74 54 52 51 52 53 58 71

Describe the variability in this data set by computing the sample variance and the sample standard deviation.

3.13 The interest rates on a certificate of deposit for six randomly selected saving and loan associations (*Los Angeles Times,* September 15, 1984) were as follows.

13.43 13.11 12.95 13.10 12.74 12.25

a. Calculate the value of the sample mean.

b. Compute the six deviations from the mean, and verify that they have a sum of zero.

c. Use the deviations from part (b) to compute the value of the sample variance and of the sample standard deviation.

3.14 a. Find two sets of five numbers that have the same mean but different standard deviations.

b. Find two sets of five numbers that have the same standard deviation but different means.

3.15 Although bats are not known for their eyesight, they are able to locate prey (mainly insects) by emitting high-pitched sounds and listening for echoes. A paper appearing in *Animal Behavior* ("The Echolation of Flying Insects by Bats" (1960): 141–54) gave the following distances (in cm) at which a bat first detected a nearby insect.

62 23 27 56 52 34 42 40 68 45 83

a. Compute the sample mean distance at which a bat first detects an insect.

b. Compute the sample variance and standard deviation for this data set. How would you interpret these values?

3.16 For the data of Exercise 3.15, find the lower and upper quartiles. Use the quartiles to compute the interquartile range.

3.17 The given numbers are salinity values for water specimens taken from North Carolina's Pamlico Sound.

7.6	7.7	4.3	5.9	5.0	10.5	7.7	9.5	12.0	12.6
6.5	8.3	8.2	13.2	12.6	13.6	14.1	13.5	11.5	12.0
10.4	10.8	13.1	12.3	10.4	13.0	14.1	15.1		

a. Describe the variability of this data set by computing the sample variance and the sample standard deviation.

b. Calculate the upper and lower quartiles for this data set.

3.18 Give an example of a data set for which the interquartile range might be preferred to the sample variance as a measure of variability. Explain your choice.

3.19 The paper "Evaluating Variability of Filling Operations" (*Food Technology* (1984):51–55) gave data on the actual amount of fluid dispersed by a machine designed to disperse 10 ounces. Ten observed values were used to compute $\Sigma x = 100.2$ and $\Sigma x^2 = 1004.4$. Compute the sample variance and standard deviation.

3.20 For the data in Exercise 3.15, add -10 to each sample observation. (This is the same as subtracting 10). For the new set of values, compute the mean and the deviations from the mean. How do these deviations compare to the deviations from the mean for the original sample? How does s^2 for the new values compare to s^2 for the old values? In general, what effect does adding the same number to each observation have on s^2 and on s? Explain.

3.21 For the data of Exercise 3.15, multiply each data value by 10. How does s for the new values compare to s for the original values?

3.22 The first four deviations from the mean in a sample of $n = 5$ reaction times were .3, .9, 1.0, and 1.3. What is the fifth deviation from the mean? Give a sample for which these are the five deviations from the mean.

3.23 The standard deviation alone doesn't measure relative variation. For example, a standard deviation of $1 would be considered large if it is describing the variability from store to store in the price of an ice cube tray. On the other hand, a standard deviation of $1 would be considered small if it is describing store-to-store variability in the price of a particular brand of freezer. A quantity designed to give a relative measure of variability is the *coefficient of variation*. Denoted by CV, the coefficient of variation expresses the standard deviation as a percent of the mean. It is defined by

$$\text{CV} = \left(\frac{s}{\bar{x}}\right) \cdot 100$$

Consider the two given samples. Sample 1 gives the actual weight (in ounces) of the contents of cans of pet food labeled as having a net weight of 8 oz. Sample 2 gives the actual weight (in pounds) of the contents of bags of dry pet food labeled as having a net weight of 50 lbs.

Sample 1	8.3	7.1	7.6	8.1	7.6	8.3	8.2	7.7	7.7	7.5
Sample 2	52.3	50.6	52.1	48.4	48.8	47.0	50.4	50.3	48.7	48.2

a. For each of the given samples, calculate the mean and the standard deviation.

b. Compute the coefficient of variation for each sample. Do the results surprise you? Why or why not?

3.3 Describing and Summarizing a Distribution

A measure of center can be combined with a measure of variability to obtain informative statements about how values in a data set are distributed. Consider, for example, a data set consisting of IQ scores for which the mean and standard deviation are 100 and 15, respectively. The score 85 is 1 standard deviation below the mean (100 − 15 = 85), while 115 is 1 standard deviation above the mean. Therefore, scores within 1 standard deviation of the mean include all those between 85 and 115 (85, 86, 87, . . . , 114, 115). Similarly, scores within 2 standard deviations of the mean include all those from 70 to 130. A score that is at least 3 standard deviations above the mean is any score that is at least 100 + 3(15) = 145—i.e., 145, 146, 147, ... Just knowing the mean and standard deviation will enable us to say something about the proportion of scores that are within 2 standard deviations of the mean, the proportion of scores that exceed the mean by at least 3 standard deviations, and so on.

Chebyshev's Rule

Without knowing anything more about the data set than just the mean and standard deviation, **Chebyshev's rule** gives information about the proportion of observations that fall within a specified number (such as 2 or 3) of standard deviations of the mean. The rule applies to both a sample and a population.

Consider any number k which is at least one. Then *the proportion of observations that are within k standard deviations of the mean is at least* $1 - 1/k^2$. Substituting selected values of k gives the following table.

Number of Standard Deviations, k	*Proportion within k Standard Deviations*
2	At least $1 - \frac{1}{4} = .75$
3	At least $1 - \frac{1}{9} = .89$
4	At least $1 - \frac{1}{16} = .94$
4.472	At least $1 - \frac{1}{20} = .95$
5	At least $1 - \frac{1}{25} = .96$
10	At least $1 - \frac{1}{100} = .99$

EXAMPLE 12

Tensile strength is one of the most important properties of various wire products used for industrial purposes. The article "Fluidized Bed Patenting of Wire Rods" (*Wire J.* (June 1977):56–61) reported that for a sample of 129 specimens of a certain type of wire rod, the sample average tensile strength (kg/mm^2) was $\overline{x} = 123.6$, and the sample standard deviation was $s = 2.0$. The tensile-strength value 119.6 is 2 standard deviations below the mean (123.6 − 2(2.0)) and the value 127.6 is 2 standard deviations above the mean. According to Chebyshev's rule, the proportion of tensile-strength observations between 119.6 and 127.6 (within two standard deviations of the mean) is at least .75. Since 75% of 129 is 96, at least 96 of the 129 measurements were between 119.6 and 127.6. Similarly, the values 117.6 and 129.6 are 3 standard deviations below and above the mean, respectively, so at least 89% (at least 114) of the observations were between 117.6 and 129.6.

To determine the proportion of measurements that fell between 116.6 and 130. 6, we observe that the value 116.6 is 7 kg/mm^2 below the mean. Since 1 standard deviation is 2.0, 7 kg/mm^2 is 3.5 standard deviations below the mean (7/2.0 = 3.5). Similarly, 130.6 is 3.5 standard deviations above the mean. Therefore, the interval from 116.6 to 130.6 includes all observations within 3.5 standard deviations of the mean. Using $k = 3.5$, $1 - 1/k^2 = 1 - 1/(3.5)^2 = 1 - .082 = .918$, so that at least 91.8% (118) of the observations were between 116.6 and 130.6.

To find the proportion of the observations that are between 115 and 130, we must be careful, since these values are not symmetrically placed about the mean. The value 130 is $(130 - 123.6)/2.0 = 3.2$ standard deviations above the mean, while 115 is 4.3 standard deviations below the mean. Chebyshev's rule can be used but with the smaller number of standard deviations (to be conservative): $1 - 1/(3.2)^2 = .902$, so the proportion of observations between 115 and 130 was at least .902.

Because at least 89% of the observations were between 117.6 and 129.6, at most 11% were either below 117.6 or above 129.6. It is tempting to divide 11% in half and conclude that at most 5.5% of the observations were above 129.6, but unfortunately this conclusion cannot be justified without more information about the distribution of tensile-strength observations. The reason is that the distribution may have been very skewed, in which case the (at most) 11% outside the interval from 117.6 to 129.6 would not have been equally divided on the two ends. All that we can really say now is that at most 11% were above 129.6. ■

Example 12 shows that Chebyshev's rule must be used with caution. Because it is applicable to any data set (distribution), whether symmetric or skewed, one must be careful in making statements about the proportion above a particular value, below a particular value, or inside or outside an interval that is not centered at the mean. The rule must be used in a conservative fashion. There is another side to this conservatism. Whereas the rule states that at least 75% of the observations are within 2 standard deviations of the mean, in many data sets substantially more than 75% satisfy this condition. The same sort of understatement is frequently encountered for other values of k (numbers of standard deviations). The next example illustrates the conservative nature of Chebyshev's rule.

EXAMPLE 13

Figure 8 gives a stem-and-leaf display of IQ scores of 112 school children in one of the early studies that used the Stanford revision of the Binet-Simon intelligence scale. (See the well-known book *The Intelligence of School Children*, by L. M. Terman. Boston: Houghton Mifflin Company, 1919.) The sample average and standard deviation are $\bar{x} = 104.5$ and $s = 16.3$, respectively. Since $2s = 32.6$, the values 71.9 (from $104.5 - 32.6$) and 137.1 are 2 standard deviations below and above the mean, respectively. Thus all scores

Stem	Leaves
6	1
7	2, 5, 6, 7, 9
8	0, 0, 0, 0, 1, 2, 4, 5, 5, 5, 6, 6, 8
9	0, 0, 0, 0, 1, 1, 2, 3, 3, 3, 4, 4, 6, 6, 6, 6, 7, 7, 8, 8, 8, 9
10	0, 0, 0, 1, 1, 2, 2, 2, 2, 2, 3, 3, 3, 5, 6, 6, 6, 7, 7, 7, 7, 8, 8, 9, 9, 9, 9, 9
11	0, 0, 0, 0, 1, 1, 2, 2, 3, 3, 3, 3, 4, 4, 4, 4, 4, 4, 7, 7, 8, 9, 9
12	0, 1, 1, 1, 1, 1, 2, 3, 4, 4, 5, 6, 6, 9
13	0, 0, 6
14	2, 6
15	2

Stems: Tens
Leaves: Ones

FIGURE 8 STEM-AND-LEAF DISPLAY OF IQ SCORES

from 72 to 137 are within 2 standard deviations of the mean. Chebyshev's rule says that at least 75%, or 84, of the 112 scores should be between 72 and 137, whereas the actual number is 108 (96%). Within 3 standard deviations of the mean translates to between 56 and 153; Chebyshev's rule says that at least 89% should be between these two values, whereas 100% of the observations are actually in this interval. ■

The Empirical Rule

The fact that statements deriving from Chebyshev's rule are frequently very conservative suggests that we should look for a rule that is less conservative and more precise. The most useful such rule is the **empirical rule,** which can be applied whenever the distribution of data values can be reasonably well described by a normal curve. The word *empirical* means deriving from practical experience, and practical experience has shown that a normal curve gives a reasonable fit to many data sets.

> If the histogram of values in a data set can be reasonably well approximated by a normal curve, then
>
> roughly 68% of the observations are within 1 standard deviation of the mean.
>
> roughly 95% of the observations are within 2 standard deviations of the mean.
>
> roughly 99.7% of the observations are within 3 standard deviations of the mean.

The empirical rule makes specific instead of "at least" statements, and the percentages for k = 1, 2, and 3 standard deviations are much higher than those allowed by Chebyshev's rule.

EXAMPLE 14

One of the earliest papers to argue for the wide applicability of the normal distribution was "On the Laws of Inheritance in Man. I. Inheritance of Physical Characters" (*Biometrika* (1903):375–462). Among the data sets discussed in the paper was one consisting of 1052 measurements of mothers' stature. The mean and standard deviation were 62.484 inches and 2.390 inches, respectively, and a normal curve did provide a good fit to the data. Since $62.484 - 2.390 = 60.094$ and $62.484 + 2.390 = 64.874$, roughly 68% (715) of the observations should be between 60.094 and 64.874. The actual percentage of observations between 60 and 65 inches was 72.1%. Similarly, roughly 95% of the observed heights should be between 57.704 and 67.264, whereas the actual percentage was about 96.2%. Between 55.314 and 69.554, we should find roughly 99.7% of the height observations, and the actual percentage was 99.2%. Clearly, the empirical rule here is much more successful and informative than Chebyshev's rule would have been. ■

Our detailed study of the normal distribution and normal curve areas in Chapter 5 will enable us to make statements analogous to those of the empirical rule for values other than k = 1, 2, or 3 standard deviations. For now, note that it is rather rare to see an observation from a normally distributed population that is further than 2 standard deviations from the mean (only 5%) and shocking to see one more than 3 standard deviations away. If you encountered a mother whose stature was 72 inches, you would probably conclude that she was not part of the population described by the data set in Example 14.

Measures of Relative Standing

When you obtain your score after taking an achievement test, you will probably want to know how it compares to scores of others who have taken the test. Is your score above or below the mean and by how much? Does your score place you in the top 5% of those who took the test, or only among the top 25%? Questions of this sort are answered by finding ways to measure the position of a particular value in a data set relative to all values in the set.

Suppose that your score is 600, the mean score is 500, and the standard deviation of all scores is 62.5. Then your score is 100 points above the mean. Since 1 standard deviation is 62.5 points, 100 points is 100/62.5 = 1.6 standard deviations. Thus your score is 1.6 standard deviations above the mean. If a friend had scored 450 on the same test, that score is 50/62.5 = .8 standard deviations below the mean. Statisticians would say that your z score is 1.6 and your friend's z score is $-.8$ (negative because the original score was below the mean).

DEFINITION

The ***z* score** corresponding to a particular observation in a data set is

$$z \text{ score} = \frac{\text{observation} - \text{mean}}{\text{standard deviation}}$$

The z score gives the distance in standard deviations between the observation and the mean. It will be positive or negative according to whether the observation lies above or below the mean.

EXAMPLE 15

Suppose that two graduating seniors, one a marketing major and the other an accounting major, are comparing job offers. The marketing student has an offer for \$18,000 per year, and the accounting major has one for \$20,000 per year. However, the mean and standard deviation of accounting offers that year are \$21,000 and \$1500, respectively. The corresponding quantities for marketing offers are \$17,500 and \$1000. Thus

$$\text{accounting } z \text{ score} = \frac{20{,}000 - 21{,}000}{1500} = -.67$$

(so 20,000 is .67 standard deviations below the mean), whereas

$$\text{marketing } z \text{ score} = \frac{18{,}000 - 17{,}500}{1000} = .5$$

Relative to the appropriate data sets, the marketing offer is actually more attractive than is the accounting offer (though this may not offer much solace to the marketing major). ■

The z score is particularly useful when the distribution of observations is approximately normal. In this case, by the empirical rule, a z score outside the interval from −2 to +2 will occur in only about 5% of all cases, while a z score outside the interval from −3 to +3 will occur only about .3% of the time.

A particular observation can be located even more precisely by giving the percent of observations that fall at or below that observation. If, for example, 95% of all test scores were at or below 650, whereas only 5% were above 650, then 650 would be called the 95th percentile of the data set (or of the distribution of scores). Similarly, if 10% of all scores were at or below 400 and 90% were above 400, the value 400 would be the 10th percentile.

DEFINITION

For any particular number r between 0 and 100, the **rth percentile** is the value such that r percent of the observations in the data set fall at or below that value.

Recall that our histograms were drawn to have a total area of 1. Figure 9 shows a smoothed histogram, and the 90th percentile has been located on the measurement axis as the value having area .9 (90%) to the left and area .1 (10%) to the right. We have already met several percentiles in disguise. The median is the 50th percentile, and the lower and upper quartiles are the 25th and 75th percentiles, respectively.

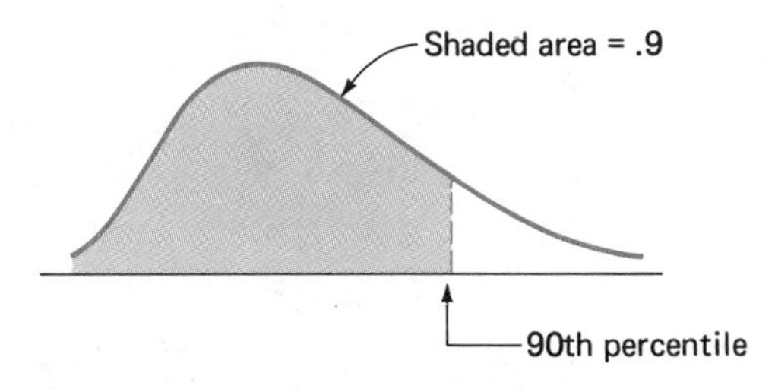

FIGURE 9 90th PERCENTILE FROM A SMOOTHED HISTOGRAM

EXAMPLE 16

The frequency distribution and histogram displayed in Figure 10 were constructed from information in the paper "An Alternative Procedure for Estimating Speed Distribution Parameters on Motorways and Similar Roads" (*Transportation Research* (1976):25–29). The data values are speeds (miles per hour) of randomly selected vehicles traveling on a major British highway. The 90th percentile is the speed value for which the proportion of vehicles moving at or below that speed is .9. In terms of the histogram, it is the value on the horizontal axis that captures 90% of the area to its left and 10% to its right. From the cumulative relative frequencies, 82.4% of the area lies to the left of 70 and 92.0% lies to the left of 75. The 90th percentile is, therefore, between 70 and 75. Since .90 − .824 = .076, we must move to the right from 70 until .076 of the area in the 70–75 rectangle is captured. This rectangle has height .096/5 = .0192, so the base of the added rectangle must satisfy (base)(.0192) = .076. Thus base = .076/.0192 = 3.96 ≈ 4, and the

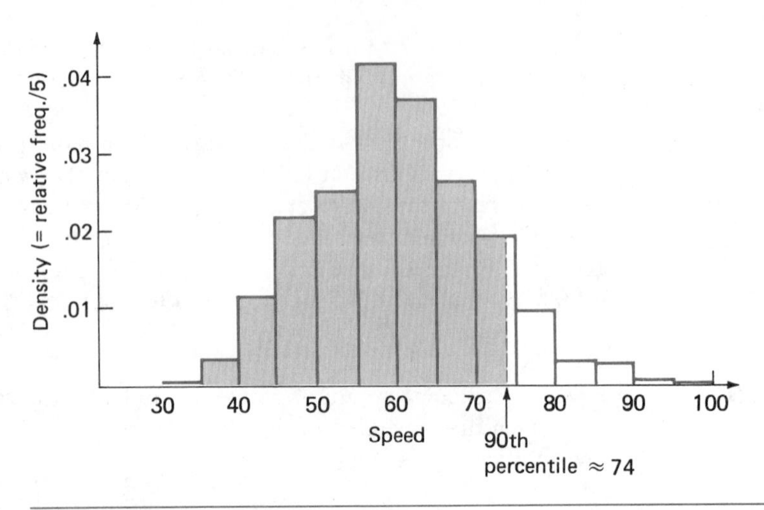

FIGURE 10 RELATIVE FREQUENCIES, CUMULATIVE RELATIVE FREQUENCIES, AND HISTOGRAM FOR THE VEHICLE-SPEED DATA

90th percentile is (approximately) $70 + 4 = 74$. Similarly, the 95th percentile is $85 + 3.19 \approx 88$. If a ticket had been issued to only the fastest 5% of all drivers, anyone moving at a speed not exceeding 88 mph would have avoided a ticket! ■

Class	30–35	35–40	40–45	45–50	50–55	55–60	60–65
Relative frequency	.003	.017	.055	.105	.124	.206	.185
Cumulative relative frequency	.003	.020	.075	.180	.304	.510	.695

Class	65–70	70–75	75–80	80–85	85–90	90–95	95–100
Relative frequency	.129	.096	.047	.015	.013	.004	.001
Cumulative relative frequency	.824	.920	.967	.982	.995	.999	1.000

Box Plots (Optional)

It would be nice to have a method of summarizing data that gives more detail than just a measure of center and spread yet less detail than a stem-and-leaf display or histogram. A box plot is one such technique. It is compact, yet provides information about center, spread, symmetry versus skewness of the data, and the presence of outlying values. To make the plot resistant to the presence of outliers, it is based on the median and interquartile range rather than the mean and standard deviation. There are several versions of box plots in current use. We'll first describe the simplest, a skeletal box plot, and then embellish it to make it more informative.

A **skeletal box plot** is constructed by first drawing a rectangular box above a horizontal measurement axis. The left edge of the box is located at the lower quartile and the right edge is at the upper quartile (so the width of the box is one interquartile range). Then a vertical line is drawn inside the box to locate the median. Finally, a horizontal line (a *whisker*) is drawn from each end of the box out to the most extreme data value on that end.

EXAMPLE 17

Example 10 of the previous section presented 21 observations on oxygen uptake of middle-aged male runners. Since $n = 21$ is odd, the median (20.88) must be included in both the lower and upper halves of the data to obtain the quartiles:

Lower half	12.81	14.95	15.83	15.97	17.90	18.27
	18.34	19.82	19.94	20.62	20.88	
Upper half	20.88	20.93	20.98	20.99	21.15	22.16
	22.24	23.16	23.56	35.78	36.73	

Then the lower and upper quartiles are 18.27 and 22.16, and the width of the box is iqr = 22.16 − 18.27 = 3.89. The lower whisker extends out to the smallest observation, 12.81, and the upper whisker extends to 36.73, the largest observation (see Figure 11). If the middle 50% of the data values (those between the lower and upper quartiles) were symmetrically distributed, the median line would be in the middle of the box, whereas here it is somewhat to the right of the middle. The upper whisker is especially long, suggesting a long upper tail in the data and the possible presence of at least one outlier at the upper end.

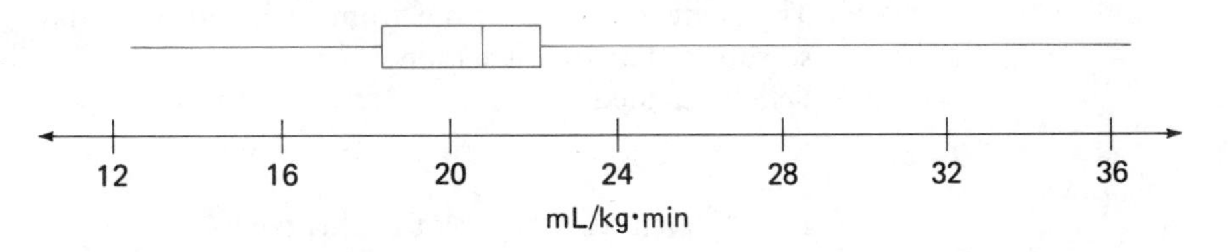

FIGURE 11 SKELETAL BOX PLOT FOR OXYGEN UPTAKE DATA ■

A more informative box plot is obtained by modifying the whiskers so that outliers can be explicitly identified. There are two types of outliers, mild and extreme. They are identified by first computing the quantities 1.5 iqr and 3 iqr. For the data of Example 17, 1.5 iqr = (1.5)(3.89) = 5.84 and 3 iqr = 11.67.

DEFINITION

An observation is a **mild outlier** if it either lies above the upper quartile by between 1.5 iqr and 3 iqr or if it lies below the lower quartile by between 1.5 iqr and 3 iqr. An **extreme outlier** is an observation that lies either more than 3 iqr above the upper quartile or more than 3 iqr below the lower quartile.

The regions associated with mild and extreme outliers are illustrated in Figure 12. The presence of such outliers, especially the extreme ones, can cause trouble for many standard inferential procedures. Such observations should be investigated to see if they resulted from errors or exceptional behavior of some sort (e.g., a bus breakdown or traffic-accident delay in the situation of Example 16). In a large data set whose histogram is well approximated by a normal curve, only .7% of the observations (7 out of every 1000) would be outliers, and only .0002% (2 out of every 1,000,000) would be extreme outliers. So, if a small sample from some population contained an extreme outlier, one would suspect that a normal curve would not give a good approximation to the population histogram.

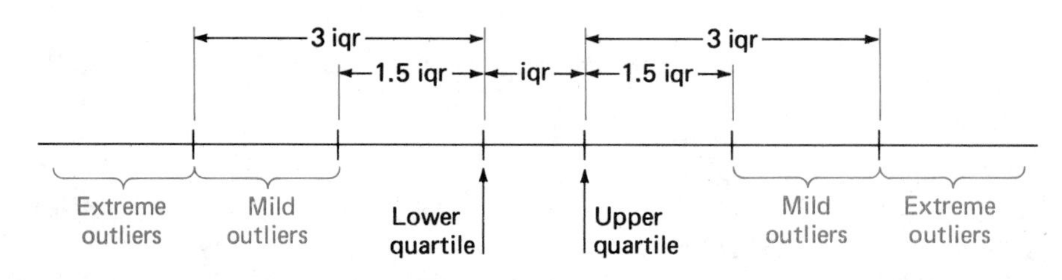

FIGURE 12 REGIONS FOR MILD AND EXTREME OUTLIERS

The more informative **box plot** starts with the box drawn as before. Then extend a whisker out from each end of the box to the furthest observation that is still within 1.5 iqr of the end. Finally, show mild outliers by shaded circles and extreme outliers by open circles.

Figure 13 pictures a box plot of this type.

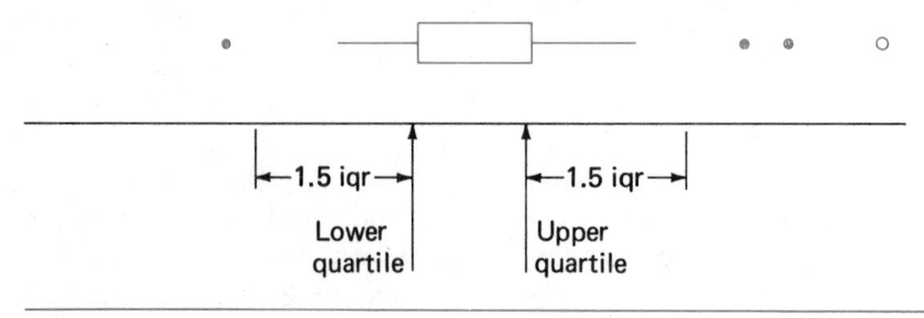

FIGURE 13 A PROTOTYPE BOX PLOT

EXAMPLE 18

(*Example 17 continued*) The box extends from the lower quartile at 18.27 to the upper quartile at 22.16. With 22.16 + 1.5 iqr = 22.16 + 5.84 = 28.00 and 22.16 + 3 iqr = 33.83, the upper whisker extends to 23.56 (the observation farthest from the box and still below 28.00), mild outliers are observations between 28.00 and 33.83, and extreme outliers are observations falling above 33.83. Since 18.27 − 5.84 = 12.43, the lower whisker extends to the smallest sample observation, 12.81, and there are no outliers on the lower end. The plot in Figure 14 shows more clearly than the skeletal box plot how the middle part of the data set behaves and what happens in the tails. Notice that the only two outliers are extreme and in the upper tail.

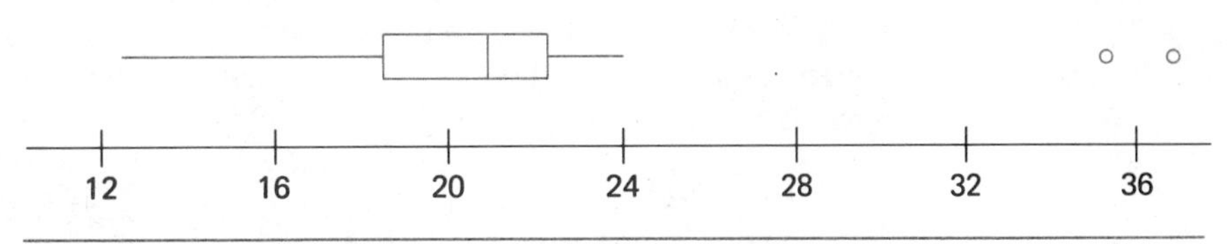

FIGURE 14 BOX PLOT FOR OXYGEN UPTAKE DATA ■

With two or more data sets consisting of observations on the same variable (e.g., miles per gallon for two types of car, weight gains for a control group and a treatment group, etc.), side-by-side box plots convey initial impressions concerning similarities and differences.

EXAMPLE 19

Two biological disturbances that are closely associated in adults suffering from endogenous depression (depression with no obvious external cause) are cortisol hypersecretion and shortened REM period latency (the elapsed time from sleep onset to the first rapid eye movement period). The paper "Plasma Cortisol Secretion and REM Period Latency in Adult Endogenous Depression" (*Amer. J. of Psychiatry* (1983):750–53) reported on a comparison of REM period latency for patients with hypersecretion and patients with normal secretion. The data values (as read from graphs) appear below:

Hypersecretion sample (n = 8): .5, 1, 2.4, 5, 15, 19, 48, 83
median = 10, lower quartile = 1.7, upper quartile = 33.5,
iqr = 31.8, 1.5 iqr = 47.7
Normal secretion sample (n = 17): 5, 5.5, 6.7, 13.5, 31, 40,
47, 47, 59, 62, 68, 72, 78, 84, 89, 105, 180
median = 59, lower quartile = 31, upper quartile = 78,
iqr = 47, 1.5 iqr = 70.5

Figure 15 displays a comparative box plot based on this data. Each sample has a mild outlier and an upper tail rather longer than the corresponding lower tail. This skewness led the investigators to transform the data by logarithms prior to a formal comparison. The impression from the box plot that normal secretion REM period latency values are substantially above those for hypersecretion was then confirmed by a formal analysis.

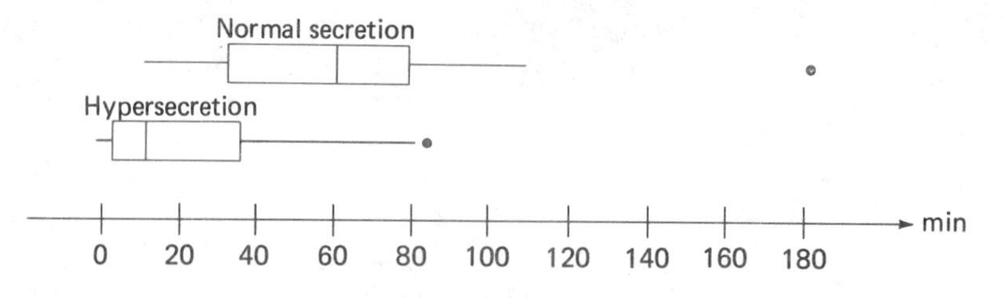

FIGURE 15 COMPARATIVE BOX PLOT FOR REM PERIOD LATENCY DATA

Figure 16 shows a computer-generated comparative box plot using MINITAB. Notice how computer keyboard characters replace hand-drawn characters—the median location is identified by the + symbol and a mild outlier by an asterisk. The first mark (tick) on the measurement axis is at zero. The symbol .3E + 1 represents $(.3)(10^1)$, or 3, so successive dashes are 3 units long. There is no lower whisker emanating from the bottom box because the whisker is so short relative to the scaling that even a single dash would be too long.

```
BOX PLOTS OF 'MINUTES' BY 'LEVEL'

                    ----------------
  1   ---------I          +       I---------                          *
                    ----------------

          ------------
  2 I  +          I---------------*
          ------------

      +---------+---------+---------+---------+---------+---------+

ONE HORIZONTAL SPACE = 0.30E+01
FIRST TICK AT        0.000
```

FIGURE 16 COMPUTER-GENERATED COMPARATIVE BOX PLOT USING MINITAB ■

EXERCISES

3.24 The average playing time of records in a large collection is 35 minutes and the standard deviation is 5 minutes.

a. Without assuming anything about the distribution of times, at least what percentage of the times are between 25 and 45 minutes?

b. Without assuming anything about the distribution of times, what can be said about the percentage of times which are either less than 20 minutes or greater than 50 minutes?

c. Assuming that the distribution of times is normal, approximately what percentage of times are between 25 and 45 minutes? Less than 20 minutes or greater than 50 minutes? Less than 20 minutes?

3.25 In a study to investigate the effect of car speed on accident severity, 5000 accident reports for fatal automobile accidents were examined and the vehicle speed at impact was recorded for each one. It was determined that the average speed was 42 miles per hour and that the standard deviation was 15 mph. In addition, a histogram revealed that vehicle speed at impact could be described by a normal curve.

Use this information to obtain an estimate of the proportion of fatal automobile accidents that occurred at speeds of over 57 mph.

3.26 The journal *Transportation Quarterly* (Jan. 1983) reported the current amount that states with a gasoline tax charge per gallon of gasoline. These values (in cents per gallon) are given below.

13	9	9.8	8	9	11	13	9	7	9
12	14	8	10	9	12	8	11.7	6.5	9
11	10	13	13	9	5	11	11	11	13.5
10.5	13	8	13.7						

a. Describe this data set using the mean and standard deviation.

b. Construct a stem-and-leaf display for the data.

c. Based on the stem-and-leaf display, would you describe this data set as being normal in shape?

d. Construct a skeletal box plot.

3.27 The accompanying 16 ozone readings (ppm) were taken at noon at East Los Angeles College (*Los Angeles Times,* Aug. 18, 1983).

10	14	13	18	12	22	14	19
22	13	14	16	3	6	7	19

a. Compute the upper quartile, the lower quartile, and the interquartile range.

b. Are there any observations that are mild outliers? Extreme outliers?

c. Construct a box plot for the ozone values.

3.28 Suppose that four additional ozone readings of 8, 13, 17, and 28 (ppm) are added to those of Exercise 3.27.

a. Does inclusion of these four values change the values of the quartiles or the interquartile range?

b. Are any of the observations mild or extreme outliers?

c. Construct a box plot for this new set of ozone values (use all 20 observations). In what way does this box plot differ from the plot of Exercise 3.27 (c)?

3.29 Suppose that your younger sister is applying for entrance to college so has taken the SAT exams. She scored at the 83rd percentile on the verbal section of the test and at the 94th percentile on the math section of the test. Since you have been studying statistics, she asks you for an interpretation of these values. What would you tell her?

3.30 The paper "Modeling and Measurements of Bus Service Reliability" (*Transportation Research* (1978):253–56) studied various aspects of bus service and presented data on travel times from several different routes. A frequency distribution for bus travel times from origin to destination on one particular route in Chicago during morning peak traffic periods is given.

Class Interval	Frequency	Relative Frequency
15–<16	4	.02
16–<17	0	.00
17–<18	26	.13
18–<19	99	.49
19–<20	36	.18
20–<21	8	.04
21–<22	12	.06
22–<23	0	.00
23–<24	0	.00
24–<25	0	.00
25–<26	16	.08

a. Construct the corresponding histogram.

b. Use the histogram to compute the following (approximate) percentiles.

i. 86th ii. 15th iii. 90th iv. 95th v. 10th

3.31 The *Los Angeles Times* (Oct. 30, 1983) reported that a typical customer of the 7-Eleven convenience stores spends $3.24. Suppose that the average amount spent by customers of 7-Eleven stores is the reported value of $3.24 and that the standard deviation for amount of sale is $8.88.

a. Based on the given mean and standard deviation, do you think that the distribution of the variable *amount of sale* could have been normal in shape? Why or why not?

b. Using Chebyshev's rule, what can be said about the proportion of all customers that spend over $20.00 on a purchase at a 7-Eleven store?

3.32 An advertisement for the "30-inch wonder" that appeared in the September 1983 issue of the journal *Packaging* claimed that the 30-inch wonder weighs cases and bags up to 110 lbs. and provides accuracy down to $\frac{1}{4}$ oz. Suppose that a 50-oz. weight was repeatedly weighed on this scale and the weight readings recorded. The mean value was 49.5 oz. and the standard deviation was .1. What can be said about the proportion of the time that the scale actually showed a weight that was within $\frac{1}{4}$ oz. of the true value of 50 oz. (*Hint:* Try to make use of Chebyshev's rule.)

3.33 Suppose your statistics professor returned your first midterm exam with only a z score written on it. She also tells you that a histogram of the scores was closely described by a normal curve. How would you interpret each of the following z scores?

a. 2.2 **b.** −.4 **c.** −1.8 **d.** 1.0 **e.** 0

3.34 The paper "Answer Changing on Multiple-Choice Tests" (*J. of Experimental Education* (1980):18–21) reported that for a group of 162 college students, the average number of responses changed from the correct answer to an incorrect answer on a test containing 80 multiple choice items was 1.4. The corresponding standard deviation was reported to be 1.5. Based on this mean and standard deviation, what can you tell about the shape of the distribution of the variable *number of answers changed from right to wrong*? What can you say about the number of students that changed at least 6 from correct to incorrect?

3.35 The article "Does Air Pollution Shorten Lives?" (from the book *Statistics and Public Policy* cited in Chapter 1) states that when the sulfate level for 117 standard metropolitan statistical areas was recorded, the resulting mean and standard deviation were 47.2 mg/m^3 and 31.3 mg/m^3, respectively. Based on this information, use Chebyshev's rule to make a statement about the proportion of metropolitan areas that have sulfate levels below 109.8 mg/m^3.

3.36 Exercise 3.17 gave salinity data for water samples from the Pamlico Sound. Construct a box plot for the salinity values given in that exercise. Does the box plot indicate that there are any extreme or unusual observations in the data set?

3.37 The average reading speed of students completing a speed reading course is 450 words per minute (wpm). If the standard deviation is 70 wpm, find the z score associated with each reading speed.

a. 320 wpm **b.** 475 wpm **c.** 420 wpm **d.** 610 wpm

3.38 The 1974 edition of *Accident Facts* reported the motor vehicle death rate for each of the fifty states. These rates, expressed in deaths per 100,000 population, are given.

23.0	38.7	31.2	17.4	41.8	25.4	30.2	54.4	18.4	30.8
45.3	18.7	32.5	26.6	28.0	48.7	18.3	27.3	35.5	34.6

57.7	20.2	26.3	21.8	24.5	44.8	16.8	33.0	16.3	16.9
31.3	34.6	21.1	28.5	24.0	28.1	35.8	33.4	27.6	22.4
25.3	22.6	34.9	46.6	13.5	30.4	23.8	33.2	29.9	39.8

a. Summarize this data set with a frequency distribution. Construct the corresponding histogram.

b. Use the histogram in part (a) to find approximate values of the following percentiles

i. 50th ii. 70th iii. 10th iv. 90th v. 40th

3.39 A quantity that has been proposed to measure the skewness of a distribution is the Pearson coefficient of skewness. Denoted by CS, the coefficient of skewness is defined as

$$CS = \frac{3(\text{mean} - \text{median})}{\text{standard deviation}}$$

A distribution that is approximately symmetric will have CS $\approx$ 0. For a skewed distribution, the sign of CS will correspond to the direction of skew in the distribution.

Calculate the coefficient of skewness for the tax-rate data of Exercise 3.26. Does this value confirm your observations about the shape of the distribution of values shown in the stem-and-leaf display constructed in that exercise?

KEY CONCEPTS

Mean The arithmetic average of the observations in a numerical data set. It is the most widely used measure of center. The sample mean is denoted by $\bar{x}$ (when data consists of observations on x), and μ is used to represent the population mean. (p. 67)

Median Another measure of center for a numerical data set. It is the middle value in the ordered list of observations. (p. 71)

Deviations from the mean The quantities obtained by subtracting the sample mean from each sample observation. (p. 78)

Sample variance A measure of how much the observations in the sample spread out about the mean. Denoted by s^2, it is the sum of the squared deviations divided by $n - 1$, where n is the sample size. (p. 79)

Sample standard deviation The square root of the sample variance, denoted by s. It is the most useful measure of sample dispersion. (p. 79)

Population variance and standard deviation Denoted by σ^2 and σ, respectively, these are population measures of spread analogous to s^2 and s. (p. 82)

Quartiles The three quartiles—lower, middle, and upper—divide a data set into four equal parts. (p. 83)

Interquartile range The difference between the third and first quartiles. It is another measure of dispersion. (p. 84)

Chebyshev's rule A rule that gives information about the proportion of data values within a specified number of standard deviations of the mean. (p. 87)

Empirical rule A rule that gives the proportion of observations within 1, 2, or 3 standard deviations of the mean when the histogram looks like a normal curve. (p. 89)

z score The distance from an observation to the mean expressed as a number of standard deviations. (p. 90)

Percentile A value such that a specified percent of the data falls at or below that value. (p. 91)

Box plot A picture based on the interquartile range that gives summary information about the distribution in the middle part of the data and identifies any extreme values (outliers). (p. 93)

SUPPLEMENTARY EXERCISES

3.40 Five randomly selected normal rats were treated with an injection of HRP (horseradish peroxidase). The total number of injured neurons in the fourth nerve nucleus was recorded. The resulting data was: 209, 187, 123, 184, and 194. (Source: "The Injury Response of Nerve Fibres in the Anterior Medullary Velum of the Adult Rat," *Brain Research* (1984):257–68). Compute and interpret the values of the sample mean and standard deviation.

3.41 Strength is an important characteristic of materials used in prefabricated housing. Each of 11 prefabricated plate elements was subjected to a severe stress test and the maximum width (in millimeters) of the resulting cracks was recorded. The given data appeared in the paper "Prefabricated Ferrocement Ribbed Elements for Low-Cost Housing" (*J. Ferrocement* (1984): 347–64). Compute the sample mean, median, and standard deviation for this data set.

Maximum Crack Width (mm)

.684	.598	.924	.483	3.520	3.130
2.650	2.540	1.4969	1.285	1.038	

3.42 Eleven sediment samples from Gannoway Lake in Texas were analyzed for concentration of iron (μg/g) and zinc (μg/g). The given data appeared in the paper "The Analysis of Aqueous Sediments for Heavy Metals" (*J. Environmental Sci. and Health* (1984):911–24).

a. Calculate the sample mean and median for the iron-concentration data. Are the numerical values of the mean and median roughly equal?

b. Calculate the sample mean and median for the zinc-concentration data. Which of the two would you recommend as a measure of location? Explain.

c. Which of the samples (iron or zinc) has a larger variance? Answer without actually computing the two sample variances. Explain the reason for your selection.

Concentration

Iron	2.5	4.5	1.5	3.2	3.3	1.8	3.4	3.4	4.0	3.9	2.9
Zinc	62	66	39	67	50	220	89	110	68	66	69

3.43 Age at death (days) for each of 12 infants who died of sudden infant death syndrome was given in the paper "Post-Mortem Analysis of Neuropeptides in Brains from Sudden Infant Death Victims" (*Brain Research* (1984):277–85). The resulting observations were 55, 56, 60, 60, 60, 105, 120, 135, 140, 154, 247, and 54.

a. Compute and interpret the values of the sample mean, median, variance, and standard deviation.

b. Calculate the upper and lower quartiles and the interquartile range. Is the value 247 a mild or extreme outlier?

c. Construct a box plot for this data set.

3.44 The air quality in major cities is monitored on a regular basis. The *Los Angeles Times* (October 25, 1984) reported that a first-stage smog alert occurs when the index of pollutants in the air is between 200 and 275. A second-stage smog alert occurs when the index exceeds 275. Suppose that the index of pollutants has a distribution with mean 125 and standard deviation 75. Without assuming anything about the shape of the distribution, what can be said about the proportion of days on which a first-stage smog alert is declared? What can be said about the proportion of days on which a second-stage smog alert is declared?

3.45 The paper "Sodium-Calcium Exchange Equilibria in Soils as Affected by Calcium Carbonate and Organic Matter" (*Soil Science* (1984):109) gave 10 observations on soil pH. The data resulted from analysis of 10 samples of soil from the Central Soil Salinity Research Institute experimental farm.

Soil pH	8.53	8.52	8.01	7.99	7.93	7.89	7.85	7.82	7.80	7.72

a. Calculate and interpret the values of the sample mean, variance, and standard deviation.

b. Compute the 10% trimmed mean and the sample median. Do either of these values differ much from the value of the sample mean?

c. Find the upper quartile, the lower quartile, and the interquartile range.

d. Illustrate the location and spread of this sample using a box plot.

3.46 The *New York Times* News Service reported that the average price of a home in the United States in 1984 was $101,000, whereas the median price was $80,900. What do the relative sizes of the mean and median imply about the shape of the distribution of home prices?

3.47 Age at diagnosis for each of 20 patients under treatment for meningitis was given in the paper "Penicillin in the Treatment of Meningitis" (*J. Amer. Medical Assn.* (1984):1870–74). The ages (in years) were as follows.

18	18	25	19	23	20	69	18	21	18
18	20	18	18	20	18	19	28	17	18

a. Calculate the values of the sample mean and standard deviation.

b. Calculate the 10% trimmed mean. How does the value of the trimmed mean compare to that of the sample mean? Which would you recommend as a measure of location? Explain.

c. Compute the upper quartile, the lower quartile, and the interquartile range.

d. Are there any mild or extreme outliers present in this data set?

e. Construct the box plot for this data set.

3.48 Suppose that the distribution of scores on an exam is closely described by a normal curve with mean 100. The 16th percentile of this distribution is 80.

a. What is the 84th percentile?

b. What is the approximate value of the standard deviation of exam scores?

c. What z score would be associated with an exam score of 90?

d. What percentile would correspond to an exam score of 140?

e. Do you think there were many scores below 40? Explain.

REFERENCES

See the references at the end of Chapter 2.

Summarizing Bivariate Data

INTRODUCTION

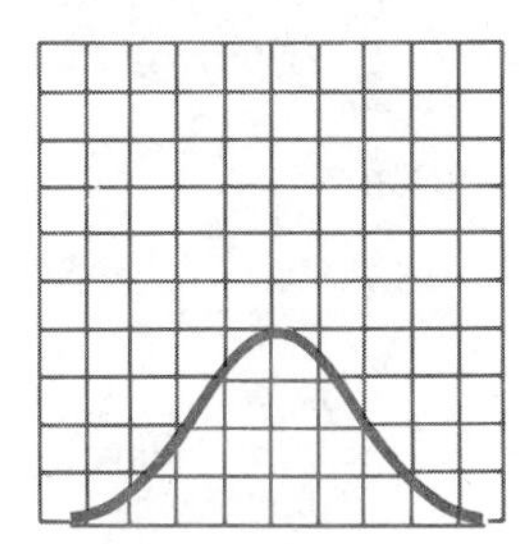

WHEN A DATA SET consists of observations on a single variable, the data set is said to be **univariate.** The methods of Chapters 2 and 3 are appropriate for describing and summarizing univariate data. Frequently an investigator is interested not in just a single attribute of individuals or objects in a population but in two or more attributes and their relationship to one another. A forester might study the growth characteristics of a certain type of tree with special attention to the relationship between age and height. An environmental researcher might wish to know how lead content of soil varies with distance from a major highway. The extent to which attitude regarding corporal punishment of school children is affected by parent's age and social class may be the subject of an investigation by sociologists. A model relating first-year college grade-point average to high school grades, SAT or ACT scores, and various family characteristics would be useful to admissions officers trying to predict whether or not an applicant would be a successful student. Many other similar examples should occur to you.

A **multivariate** data set is one that consists of measurements or observations on each of two or more variables. The most important special case involves just two variables, x and y, and the resulting data is usually called **bivariate.** Examples include $x =$ age of a tree and $y =$ height of the tree; $x =$ distance from a highway and $y =$ lead content of soil at that distance; and $x =$ religious preference and $y =$ political party affiliation. Each observation in a bivariate data set consists of a pair of values. The first element in the pair is the value of x and the second is the value of y.

The focus in this chapter is on bivariate numerical data, i.e., data for which both x and y are numerical variables. This type of data arises very frequently in applied work. In addition, a thorough grasp of this case is an important stepping stone to multivariate data analysis. As in Chapters 2 and 3, our concern here is with data description and summarization. Inferential methods for bivariate numerical data as well as for data sets where x and y are categorical variables are presented in later chapters.

4.1 Scatter Plots

When data is bivariate with both x and y numerical variables, each observation consists of a pair of numbers, such as (14, 5.2) or (27.63, 18.9), where the x value is the first number in the pair and the y value is the second number. An unorganized list of observed pairs yields little information about the distribution of x values and y values separately and even less information about how strongly and in what manner the two variables are related to one another. In Chapter 2 we saw how pictures could help make sense of univariate data. The most informative picture based on bivariate numerical data is called a scatter plot.

A **scatter plot** is a picture of bivariate numerical data in which each pair of values (each observation) is represented by a point located on a rectangular coordinate system as pictured in Figure 1(a). The horizontal axis is identified with values of x and is marked so that any x value can be easily located. Similarly, the vertical axis—the y axis—is also marked for easy location of y values. Figure 1(b) pictures the point corresponding to (4.5, 15). This point is located above the value $x = 4.5$ on the horizontal axis and to the right of the value $y = 15$ on the vertical axis.

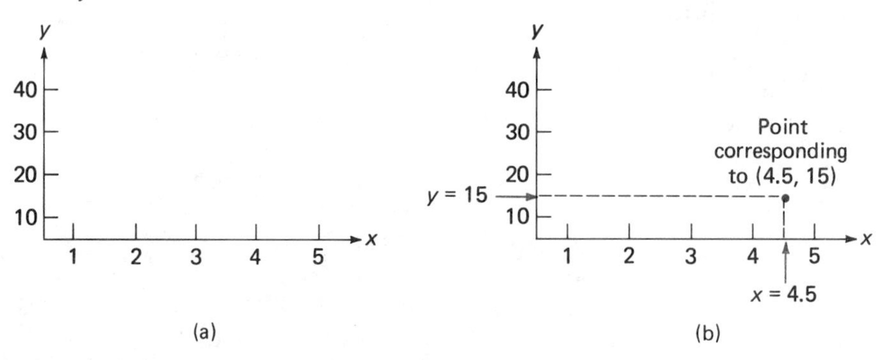

FIGURE 1 CONSTRUCTING A SCATTER PLOT
(a) Rectangular Coordinate System for a Scatter Plot of Bivariate Data
(b) The Point in the Plot Corresponding to the Observation (4.5, 15)

EXAMPLE 1 The presence of substantial quantities of lead in the blood of humans has long been recognized as a serious health hazard. The paper "A Restrospective Analysis of Blood-Lead in Mentally Retarded Children" (*Lancet* 1 (1977): 717–19) reported on a study in which both water-lead concentration in the maternal home (x) and blood-lead level (y) shortly after birth were determined for a sample of mentally retarded children for which the cause of retardation was unknown. We give 10 observations from the study in tabular format. Each x value is listed in the first row of the table and the corresponding y value appears directly below it in the second row. Thus the first pair is

(.7, 1.22), the second is (1.1, .46), and so on. The observations are listed in increasing order of the x value, which makes it somewhat easier to plot points, but this is not necessarily the order in which observations were obtained. The scatter plot of this data is shown in Figure 2. The plot shows that larger values of x tend to be paired with larger values of y. There does not appear to be any evidence of curvature in the plot, nor does any point seem to be wildly inconsistent with the other points.

Observation	1	2	3	4	5	6	7	8	9	10
x = water-lead (μmol/L)	.7	1.1	2.3	2.8	3.6	4.0	5.1	5.9	6.7	7.9
y = blood-lead (μmol/L)	1.22	.46	1.64	.76	.73	1.80	1.20	1.78	1.92	1.38

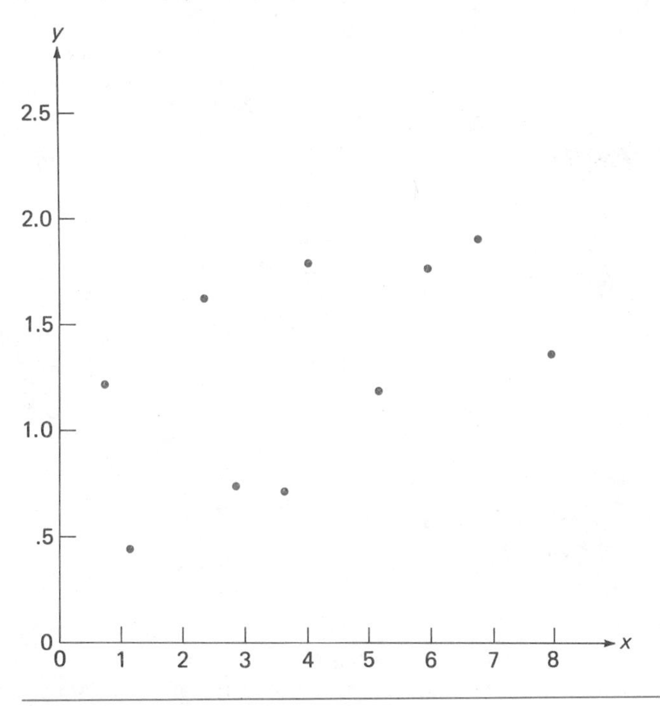

FIGURE 2 SCATTER PLOT OF DATA FROM EXAMPLE 1

The horizontal and vertical axes in the scatter plot of Figure 2 intersect at the point (0, 0). In many data sets either the values of x or of y or of both variables differ considerably from zero relative to the ranges of the values. For example, a study of how air conditioner efficiency is related to maximum daily outdoor temperature might involve observations for temperatures 80°, 82°, . . . , 98°, 100°. When this is the case, a more informative plot may result from intersecting the axes at some point other than (0, 0) and marking the axes accordingly.

EXAMPLE 2

Bicyclists are well aware that, even when a street has a bike lane, riding may pose considerable risks if the street and lane are poorly designed. The paper "Effects of Bike Lanes on Driver and Bicyclist Behavior" (*ASCE Trans. Eng. J.* (1977):243–56) reported the accompanying data on x = available travel space (the distance between a cyclist and the roadway center line) and y = separation distance between a bike and a passing car (determined by photography). Figure 3 displays two scatter plots of the data, one with the axes intersecting at (0, 0) and the other with the axes intersecting at (5, 12).

The points in the first plot are crowded together in the upper right-hand corner, making it difficult to see any patterns. At first glance there appears to be no strong evidence of curvature in this plot. The second plot is less crowded and consequently more revealing. In particular, except for the point (15.1, 7.1), the plot exhibits some curvature. If the objective were to develop a model for predicting y from x, Figure 3(b) suggests that some curve fit to the data might yield better predictions than would result from fitting a straight line to the data.

x	12.8	12.9	12.9	13.6	14.5	14.6	15.1	17.5	19.5	20.8
y	5.5	6.2	6.3	7.0	7.8	8.3	7.1	10.0	10.8	11.0

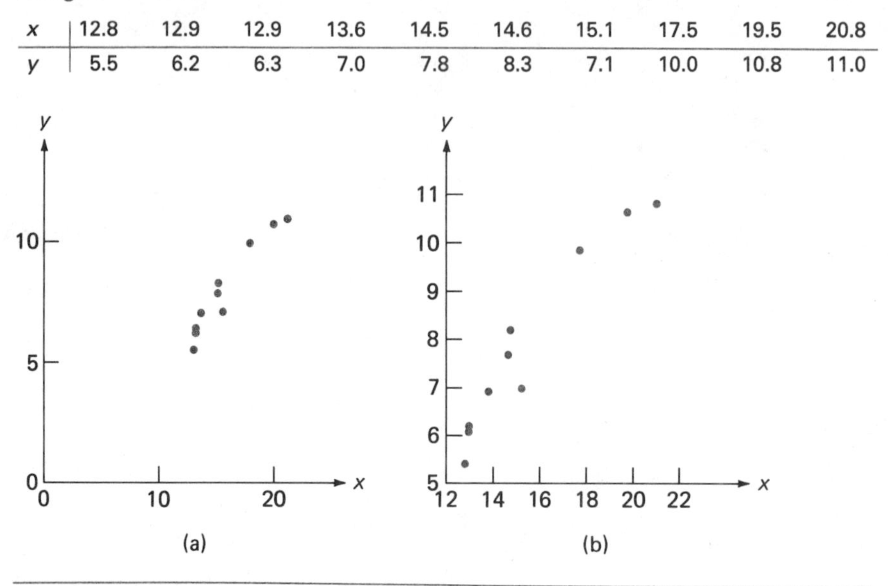

FIGURE 3 SCATTER PLOTS FOR THE DATA OF EXAMPLE 2
(a) Axes Intersecting at (0, 0)
(b) Axes Intersecting at (12, 5)

Computer-Generated Scatter Plots

One of the most attractive features of the standard statistical computer program packages is their ability to produce scatter plots and various other plots. Figure 4 shows a MINITAB scatter plot of data for which x = number of bills introduced in Congress and y = the number of bills passed. The 16 observations are for the 80th Congress (1947–48), 81st Congress, . . . , and 95th Congress (1977–78). Notice that the axes do not intersect at (0, 0). We let the computer choose the point of intersection and markings on the axes for this plot, although there is an option that allows the user to make his or her own specifications. The most interesting thing about this plot is that there is no pattern. There appears to be no strong relationship between the number of bills introduced and the number passed. This would provide support for those who have proposed a limitation on the number of bills that a senator or representative could introduce.

Sometimes two or more points are located so closely to one another that the computer cannot print separate asterisks for each point. The appearance of the number 2 on a scatter plot indicates that there are 2 points at or near that location; 3, 4, and other numbers are also used to indicate multiple observations at the same location. An illustration appears in Figure 5, where x refers to hair mercury content of a newborn child, y refers to hair mercury content of the child's mother, and 35 points are represented in each plot. The data was taken from the article mentioned in Example 12 of Chapter 2.

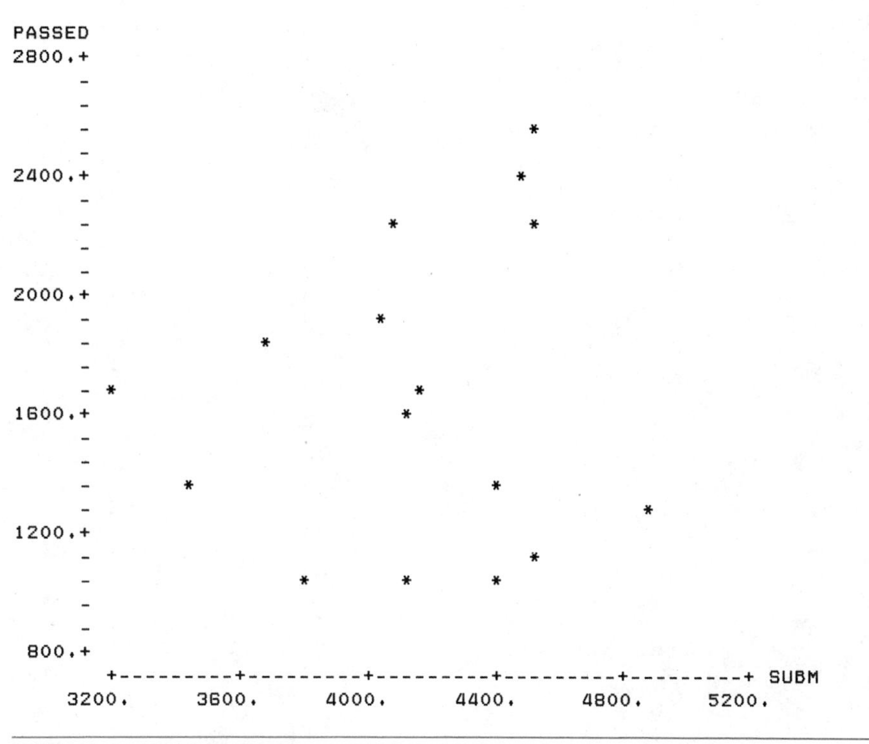

FIGURE 4 MINITAB SCATTER PLOT OF y = BILLS PASSED VERSUS x = BILLS SUBMITTED

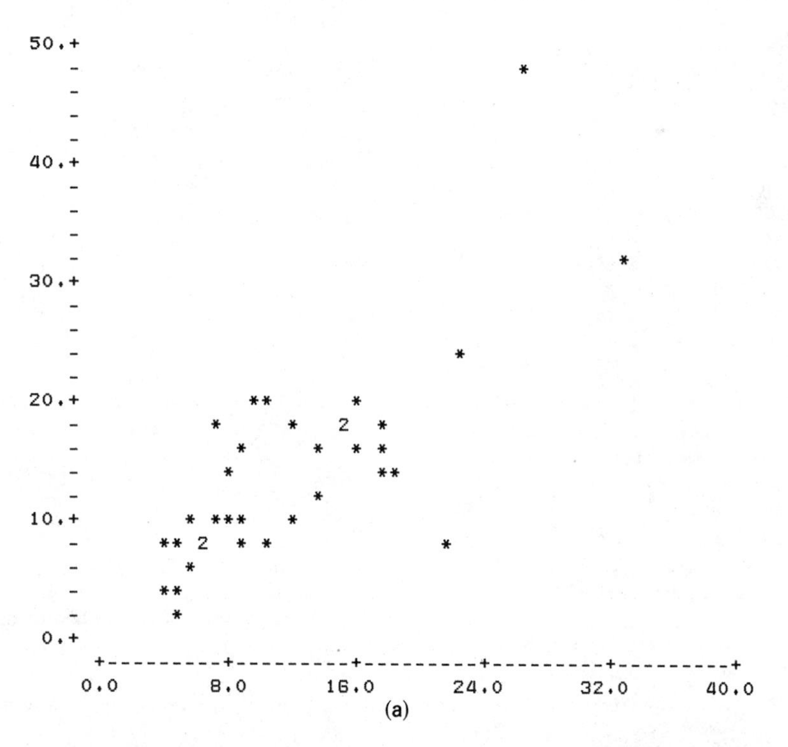

FIGURE 5 SCATTER PLOTS IN WHICH THE SYMBOL 2 REPRESENTS TWO POINTS AT THE SAME LOCATION
(a) MINITAB
(b) SPSS

(continued)

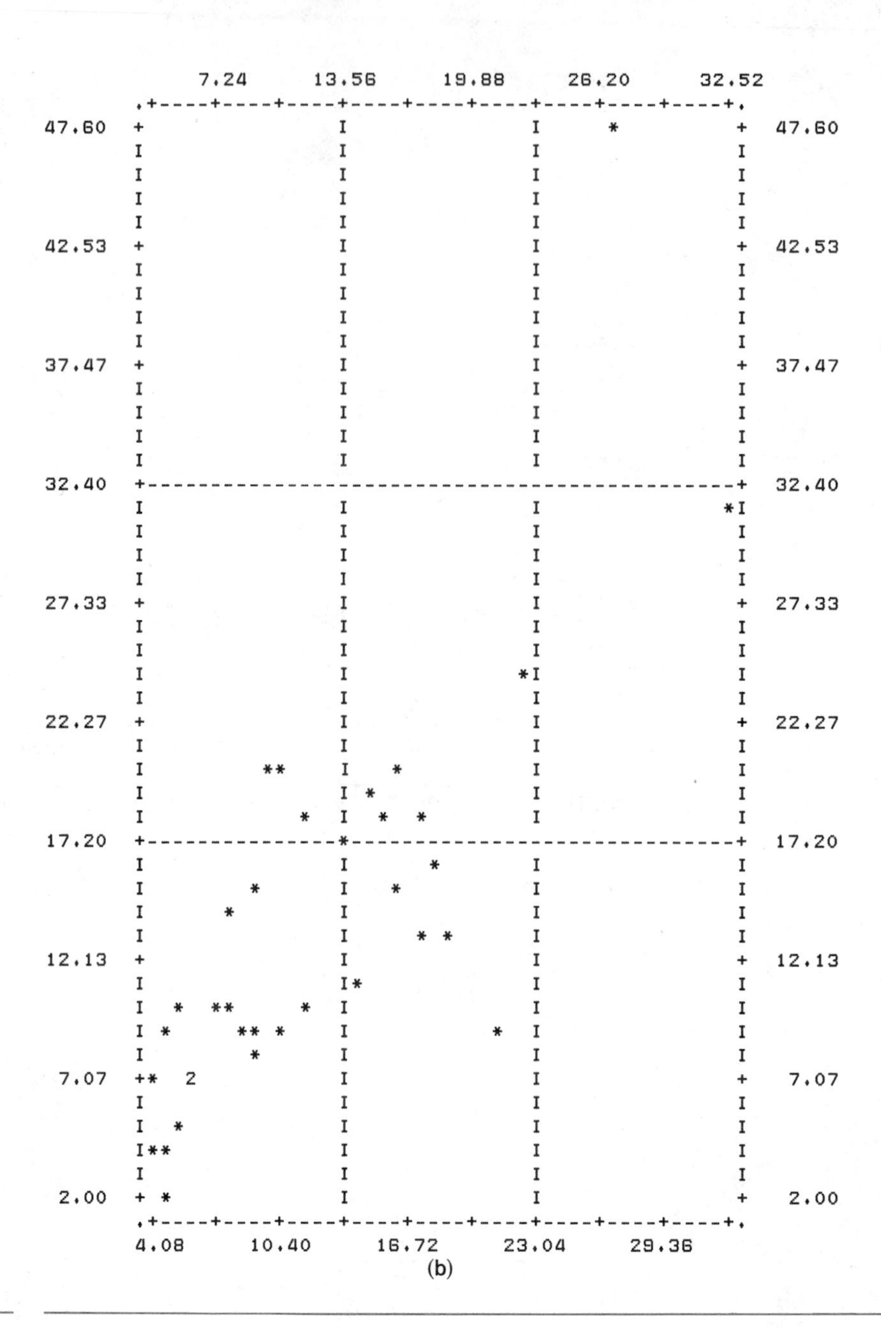

(b)

FIGURE 5
(continued)

EXERCISES

4.1 Manganese (Mn) is thought to be critical to the health of newborn infants. The paper "Manganese Intake and Serum Manganese Concentration of Human Milk-fed and Formula-fed Infants" (*Amer. J. of Clinical Nutrition* (1984):872–78) gave the data below on Mn intake and serum Mn level for eight human milk-fed infants. Use these data to construct a scatter plot. Does the plot suggest that there is a relationship between Mn intake and serum Mn level?

Intake (μg/kg/day)	.34	.35	.39	.39	.41	.41	.49	.68
Serum Mn (μg/L)	2.8	1.9	3.3	5.6	4.2	5.6	4.2	7.9

4.2 The metabolic effect of cross-country skiing was the subject of the research study described in the paper "Metabolic Modifications Caused by Sport Activity: Effect in Leisure-Time Cross-Country Skiers" (*J. Sports Med.* (1983):385–92). Subjects were participants in a 24-h cross-country relay. Age and blood CPK concentration 12 h into the relay were recorded. Use the given data to construct a scatter plot. Would you describe this plot as linear?

Skier	1	2	3	4	5	6	7	8	9
Age (x)	33	21	19	24	25	32	36	35	36
CPK (y)	180	300	520	480	580	440	380	480	520

Skier	10	11	12	13	14	15	16	17	18
Age (x)	24	25	44	51	50	52	55	62	57
CPK (y)	1040	1360	640	260	360	400	280	300	400

4.3 The paper "Effects of Enhanced UV-B Radiation on Ribulose-1, 5-Biphosphate, Carboxylase in Pea and Soybean" (*Environ. and Exper. Botany* (1984): 131–43) included the accompanying data on distance from an ultraviolet light source and an index of sunburn in pea plants. Use the data to construct a scatter plot. How would you describe the relationship between these two variables?

Distance (cm)	18	21	25	26	30	32	36	40
Sunburn units	4.0	3.7	3.0	2.9	2.6	2.5	2.2	2.0

Distance (cm)	40	50	51	54	61	62	63
Sunburn units	2.1	1.5	1.5	1.5	1.3	1.2	1.1

4.4 The given data on fish survival and ammonia concentration was taken from the paper "Effects of Ammonia on Growth and Survival of Rainbow Trout in Intensive Static-Water Culture" (*Trans. of the Amer. Fisheries Soc.* (1983):448–54). Display this data graphically using a scatter plot.

Ammonia exposure (mg/L)	10	10	20	20	25	27	27	31	50
Percent survival	85	92	85	96	87	80	90	59	62

4.5 A number of research studies have looked at the relationship between water stress and plant productivity. The paper "Water Stress Affecting Nitrate Reduction and Leaf Diffusive Resistance in Coffea Arabica L. Cultivars" (*J. of Horticultural Sci.* (1983):147–52) examined water availability and nitrate activity in five different types of coffee plants. Data on water potential (x) and nitrate activity (y) for the Angustifolia and Nacional varieties are given.

Angustifolia

x	−10	−11	−11	−14	−15	−15	−16	−16	−16	−17	−18
y	3.2	3.0	5.4	3.5	3.0	3.6	4.8	3.2	2.4	4.6	1.6

x	−19	−20	−21	−22	−23	−23	−23	−24
y	3.4	4.0	1.4	.4	.8	1.8	2.0	.2

Nacional

x	−10	−12	−12	−13	−14	−14	−14	−14	−14	−15	−15	−15
y	9.8	8.0	13.0	6.0	5.0	6.0	7.2	10.4	10.8	7.8	9.0	11.0

x	−15	−16	−16	−17	−18	−18	−18	−18	−18	−18	−19	−20
y	14.0	6.6	14.2	3.4	8.0	8.2	8.6	9.2	11.6	12.0	5.2	9.4

a. Draw a scatter plot for the Angustifolia data. Does the plot look linear?

b. Construct a scatter plot for the Nacional data. Would you describe the relationship between nitrate activity and water potential exhibited by this data set as linear?

c. Discuss the similarities and differences between the scatter plots of nitrate activity versus water potential for Angustifolia and Nacional coffee plants.

4.2 Correlation

A scatter plot of bivariate numerical data gives a visual impression of how strongly the values of x are related to the values of y with which they are paired. But to make precise statements and draw conclusions from data, we need to go beyond pictures. Our objective now is to develop a numerical measure of how strongly the x and y values in such a sample are related. It is customary to call such a measure a **correlation coefficient** (from co-relation).

Figure 6 displays several scatter plots that indicate different types of relationships between the x and y values. The plot in Figure 6(a) suggests a very strong *positive relationship* between x and y; for every pair of points in the plot, the one with the larger x value also has the larger y value. That is, an increase in x is inevitably paired with an increase in y. The plot in Figure 6(b) shows a strong *tendency* for y to increase as x does, but there are a few ex-

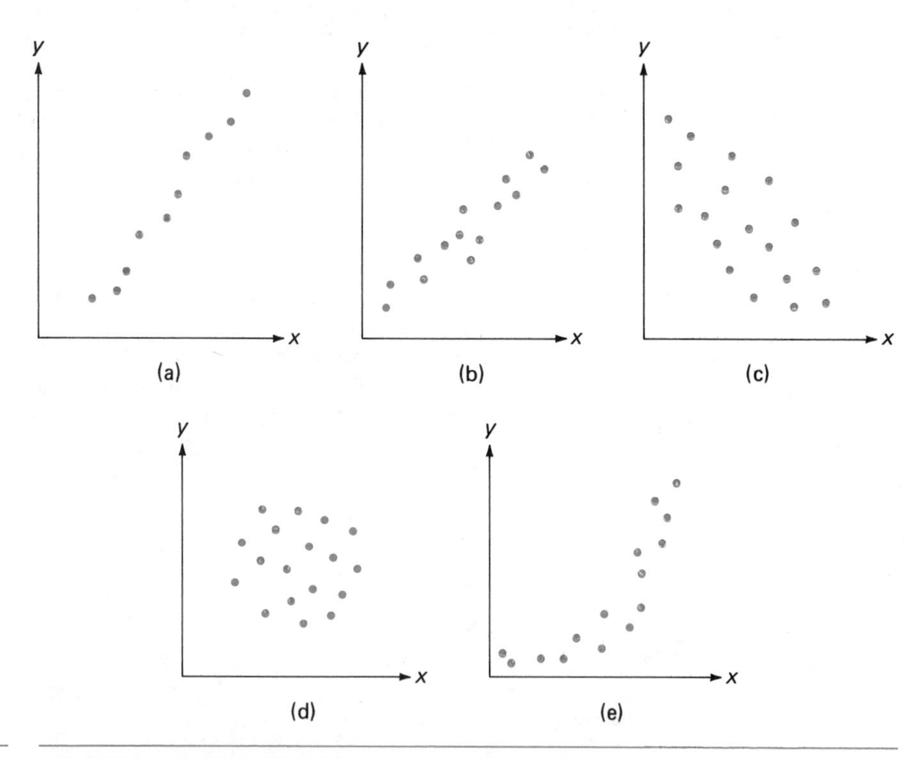

FIGURE 6

SCATTER PLOTS ILLUSTRATING VARIOUS TYPES OF RELATIONSHIPS
(a) and (b) Positive Relationship, Linear Pattern
(c) Negative Relationship, Linear Pattern
(d) No Relationship or Pattern
(e) Positive Relationship, Curved Pattern

ceptions. For example, the x and y values of the two points in the extreme upper right-hand corner of the plot go in opposite directions (x increases but y decreases). Nevertheless, a plot such as this would again indicate a rather strong positive relationship. Figure 6(c) suggests that x and y are *negatively related*—as x increases, there is a tendency for y to decrease. Obviously the negative relationship in this plot is not as strong as the positive relationship in Figure 6(b), although both plots show a well-defined linear pattern. The plot of Figure 6(d) is indicative of no strong relationship between x and y; there is no tendency for y either to increase or decrease as x increases. Finally, as illustrated in Figure 6(e), a scatter plot can show evidence of a strong positive (or negative) relationship through a pattern that is curved rather than linear in character.

Pearson's Correlation Coefficient

Let's suppose that our bivariate data set consists of n pairs of numbers selected from an entire population of pairs. As a natural extension of the notation $x_1, x_2, \ldots, x_n$ for a univariate sample of x values, we let (x_1, y_1) denote the first pair in our bivariate sample, (x_2, y_2) denote the second pair, . . . , and (x_n, y_n) denote the nth (last) pair. To illustrate, Example 1 gave $n = 10$ observations on $x =$ water-lead concentration and $y =$ blood-lead concentration. The first observation yielded an x value of .7 and a y value of 1.22, so $(x_1, y_1) = (.7, 1.22)$, while $(x_{10}, y_{10}) = (7.9, 1.38)$. As before, $\bar{x}$ denotes the mean of $x_1, \ldots, x_n$, whereas $\bar{y}$ denotes the mean of $y_1, \ldots, y_n$. Again for Example 1, $\Sigma x = 40.1$ and $\Sigma y = 12.89$, so $\bar{x} = 4.01$ and $\bar{y} = 1.289$.

Figure 7(a) shows a scatter plot that indicates a rather strong positive relationship between x and y. We marked the position of $\bar{x}$ on the x axis and drew a vertical line through it. Similarly, we marked the position of $\bar{y}$ on the y axis and drew a horizontal line through it. These horizontal and vertical lines create four regions, which we have marked I, II, III, and IV. Every point in region I has its x value above $\bar{x}$ *and* its y value above $\bar{y}$. Every point in region III has its x value below $\bar{x}$ *and* its y value below $\bar{y}$. For each point in region II and IV, the x and y values of the point are on opposite sides of their means (x above $\bar{x}$ and y below $\bar{y}$, or vice-versa). But most of the points in the plot are in regions I or III. That is, in a scatter plot consistent with a strong positive relation, both $x - \bar{x}$ and $y - \bar{y}$ are of the same sign for most points—either both positive (as in region I) or both negative (as in region III).

The situation is quite different in the scatter plot of Figure 7(b), which indicates a negative relationship. Here most of the points are in regions II and IV, in which $x - \bar{x}$ and $y - \bar{y}$ are opposite in sign (one negative and the other positive). The plot of Figure 7(c) suggests no strong relationship, and the points in the two regions for which $x - \bar{x}$ and $y - \bar{y}$ have the same sign are spread in roughly the same way that they are spread in the two regions with opposite signs.

Suppose now that for each observation (or point), we compute the product $(x - \bar{x})(y - \bar{y})$. When $x - \bar{x}$ and $y - \bar{y}$ have the same sign (positive or negative), the product is positive. When they have different signs, the product is negative. For the data of Figure 7(a), almost all products $(x - \bar{x})(y - \bar{y})$ are positive, for Figure 7(b), almost all products are negative, and in Figure 7(c), the number of negative products is roughly the same as the number of positive products.

If x and y have a strong positive relationship, almost all of the products

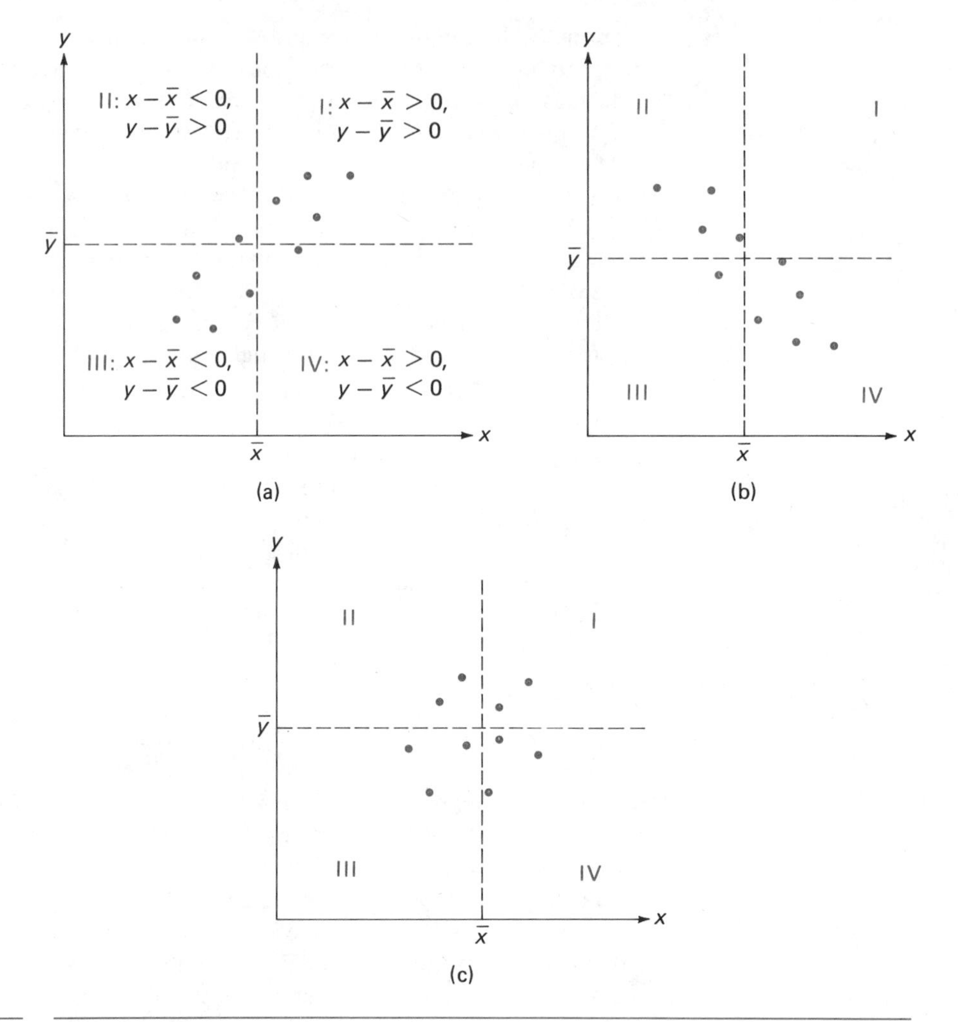

FIGURE 7 BREAKING UP A SCATTER PLOT ACCORDING TO THE SIGNS OF $x - \bar{x}$ and $y - \bar{y}$
(a) A Positive Relation
(b) A Negative Relation
(c) No Strong Relation

$(x - \bar{x})(y - \bar{y})$ are positive, so the sum of products $\Sigma(x - \bar{x})(y - \bar{y})$ is a large positive number. If the relationship is strongly negative, most of the products are negative, and the sum $\Sigma(x - \bar{x})(y - \bar{y})$ is a large negative number. If there is no strong relationship, positive and negative products will tend to offset one another and yield a sum $\Sigma(x - \bar{x})(y - \bar{y})$ close to zero.

A measure of how strongly x and y are related should not depend on the unit of measurement for either variable. If this were not the case, simply changing the unit (e.g., from feet to inches or centimeters, from degrees Centigrade to degrees Fahrenheit or degrees Kelvin, etc.) could make the measure more extreme (suggesting a stronger relationship) or closer to zero (suggesting a weaker relationship). Dependence on the units of measurement would make it very difficult to interpret a particular value. Unfortunately, the value of the sum $\Sigma(x - \bar{x})(y - \bar{y})$ is affected by a change in units. For example, if x = height, each x value when height is expressed in inches will be

12 times its measurement in feet, and this is also true of $\bar{x}$. Thus the value of each $x - \bar{x}$ changes by a factor of 12 in passing from feet to inches, and the sum itself changes by a factor of 12. To obtain a measure that is independent of the units, we make use of the sample standard deviations of the x and y values.

DEFINITION

Let $\bar{x}$ and s_x denote the sample mean and standard deviation for the x values in the pairs (x_1, y_1), (x_2, y_2), . . . , (x_n, y_n) and let $\bar{y}$ and s_y denote the sample mean and standard deviation of the y values. Then **Pearson's sample correlation coefficient** r is given by

$$r = \frac{\Sigma(x - \bar{x})(y - \bar{y})}{(n - 1)s_x s_y}$$

Before discussing various properties of r, let's look at a computational example.

EXAMPLE 3

An accurate assessment of soil productivity is an essential input to rational land-use planning. Unfortunately, as the author of the article "Productivity Ratings Based on Soil Series" (*Prof. Geographer* (1980):158–63) argues, an acceptable soil productivity index is not so easy to come by. One difficulty is that productivity is determined partly by which crop is planted, and the relationship between yield of two different crops planted in the same soil may not be very strong. To illustrate, the paper presents the accompanying data on corn yield x and peanut yield y (mT/Ha) for eight different types of soil:

Observation	1	2	3	4	5	6	7	8
x	2.4	3.4	4.6	3.7	2.2	3.3	4.0	2.1
y	1.33	2.12	1.80	1.65	2.00	1.76	2.11	1.63

The calculations leading to r are most easily carried out using a tabular format as in Table 1, which contains a row for each observation and five columns—the first two for $x - \bar{x}$ and $y - \bar{y}$, the next two for $(x - \bar{x})^2$ and $(y - \bar{y})^2$, and the last for the product $(x - \bar{x})(y - \bar{y})$. To protect against

TABLE 1

THE RECOMMENDED FORMAT FOR CALCULATIONS LEADING TO r

Observation	$x - \bar{x}$	$y - \bar{y}$	$(x - \bar{x})^2$	$(y - \bar{y})^2$	$(x - \bar{x})(y - \bar{y})$
1	−.8125	−.4700	.6602	.2209	.3819
2	.1875	.3200	.0352	.1024	.0600
3	1.3875	.0000	1.9252	.0000	.0000
4	.4875	−.1500	.2377	.0225	−.0731
5	−1.0125	.2000	1.0252	.0400	−.2025
6	.0875	−.0400	.0077	.0016	−.0035
7	.7875	.3100	.6202	.0961	.2441
8	−1.1125	−.1700	1.2377	.0289	.1891
Sum	.0000	.0000	5.7491	.5124	.5960

computational inaccuracies due to rounding, it is recommended that several extra digits be used in both $\overline{x}$ and $\overline{y}$. Here, with $\Sigma x = 25.7$ and $\Sigma y = 14.4$, the values $\overline{x} = 25.7/8 = 3.2125$ and $\overline{y} = 1.8000$ were used. From the table, $s_x^2 = 5.7491/7 = .8213$, so $s_x = .9063$; $s_y^2 = .5124/7 = .0732$, so $s_y = .2706$. This gives

$$r = \frac{\Sigma(x - \overline{x})(y - \overline{y})}{(n - 1)s_x s_y} = \frac{.5960}{7(.9063)(.2706)} = .347$$

The accompanying scatter plot (Figure 8) suggests that $r = .347$ is indicative of only a weak positive relationship.

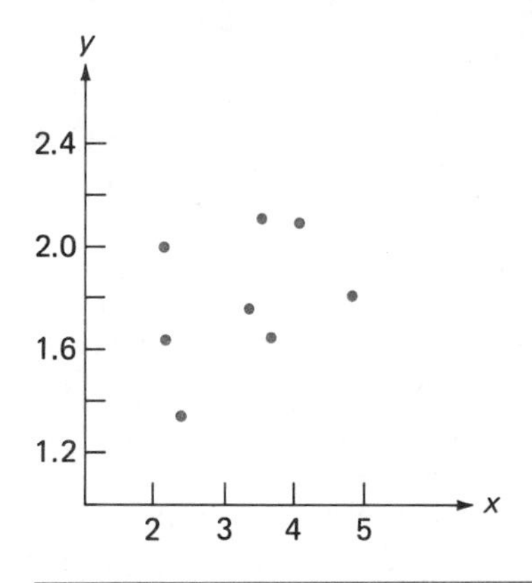

FIGURE 8 SCATTER PLOT OF y = PEANUT YIELD VERSUS x = CORN YIELD ■

A Note Concerning Computation

In Section 3.2, the expression $\Sigma x^2 - n\overline{x}^2$ was used in place of $\Sigma(x - \overline{x})^2$ to facilitate efficient computation of s^2. Replacing x^2 by y^2 and $\overline{x}^2$ by $\overline{y}^2$ yields an analagous computing formula for $\Sigma(y - \overline{y})^2$. There is a similar computing formula for the numerator of r. Use of these formulas circumvents the need to compute either the x or y deviations.

Defining Formula	*Computing Formula*
$\Sigma(x - \overline{x})^2$	$\Sigma x^2 - n\overline{x}^2$
$\Sigma(y - \overline{y})^2$	$\Sigma y^2 - n\overline{y}^2$
$\Sigma(x - \overline{x})(y - \overline{y})$	$\Sigma xy - n\overline{x}\,\overline{y}$

In particular, the numerator of r results from first multiplying each observed x value by its associated y value, summing these products, and subtracting the single quantity $n\overline{x}\,\overline{y}$ from the result. In addition to n, these formulas require the five sums Σx (for $\overline{x}$), Σy, Σx^2, Σy^2, and Σxy. A tabular format containing five columns for the x, y, x^2, y^2, and xy values helps in organizing the computations, with each derived sum recorded below the corresponding column.

EXAMPLE 4

(*Example* 3 *continued*) Here again are the $n = 8$ observations on corn yield (x) and peanut yield (y) presented earlier.

x	y	x^2	y^2	xy
2.4	1.33	5.76	1.7689	3.192
3.4	2.12	11.56	4.4944	7.208
4.6	1.80	21.16	3.2400	8.280
3.7	1.65	13.69	2.7225	6.105
2.2	2.00	4.84	4.0000	4.400
3.3	1.76	10.89	3.0976	5.808
4.0	2.11	16.00	4.4521	8.440
2.1	1.63	4.41	2.6569	3.423
25.7	14.40	88.31	26.4324	46.856
↑ Σx	↑ Σy	↑ Σx^2	↑ Σy^2	↑ Σxy

Then

$$\bar{x} = 3.2125 \qquad \bar{y} = 1.8000$$
$$n\bar{x}^2 = 8(3.2125)(3.2125) = 82.56125$$
$$n\bar{y}^2 = 8(1.8000)(1.8000) = 25.92000$$
$$n\bar{x}\bar{y} = 8(3.2125)(1.8000) = 46.26000$$

Substitution into the computational formulas yields

$$\Sigma(x - \bar{x})^2 = \Sigma x^2 - n\bar{x}^2 = 88.31 - 82.56125 = 5.74875$$
$$\Sigma(y - \bar{y})^2 = \Sigma y^2 - n\bar{y}^2 = 26.4324 - 25.92000 = .51240$$
$$\Sigma(x - \bar{x})(y - \bar{y}) = \Sigma xy - n\bar{x}\bar{y} = 46.856 - 46.26000 = .5960$$

Thus,

$$s_x = \sqrt{5.74875/7} = .90623$$
$$s_y = \sqrt{.51240/7} = .27055$$
$$r = \frac{.59600}{7(.90623)(.27055)} = \frac{.59600}{1.71626} = .347$$

This is exactly the value calculated earlier. ■

For many data sets, the numerator and denominator of r can be computed almost as quickly from the defining formulas as from the computing formulas. The choice is a matter of personal taste, so use the one with which you feel most comfortable. Even better, if you have easy access to a statistical computer package, use it—every standard package will compute and print the value of r when requested to do so.

Properties of r

A correct interpretation of r for any given data set requires an appreciation of some general properties.

1. The value of r does not depend on the unit of measurement for either variable, nor does it depend on which variable is labeled x and which is labeled y.

2. The value of r is between -1 and 1. A positive value of r indicates a positive relationship between the variables, whereas a negative value of r corresponds to a negative relationship.

3. The value $r = 1$, which indicates the strongest possible positive relationship between x and y, results only when all points in the scatter plot lie exactly on a straight line that slopes upward. The value $r = -1$, which indicates the strongest possible negative relationship, results only when all points in the scatter plot lie exactly on a straight line that slopes downward.

4. The value of r is a measure of the extent to which x and y are *linearly* related—i.e., the extent to which the points in the scatter plot fall close to a straight line. A value of r close to zero does not rule out *any* strong relationship between x and y; there could still be a strong relationship but one that is not linear.

A value of r close to $+1$ suggests a strong positive linear relationship, a value of r close to -1 suggests a strong negative linear relationship, and a value close to zero suggests the absence of any strong linear relationship. Don't assume that there is no relationship just because r is close to zero—look at a scatter plot to see if the points fall close to a well-behaved curve. A value reasonably close to $+1$ or -1 can occur even though the scatter plot has a much more pronounced curved than linear pattern, because a curve can often be well approximated by a straight line over a wide range of values. So don't stop with r—look at a scatter plot!

Frequently we wish to characterize a relationship between x and y in terms of everyday language. Here is an informal rule of thumb that you might find useful: Call the relationship strong if r is greater than $+.8$ or less than (more negative than) $-.8$, weak if r is between $-.5$ and $.5$, and moderate otherwise. This rule of thumb is summarized in Figure 9.

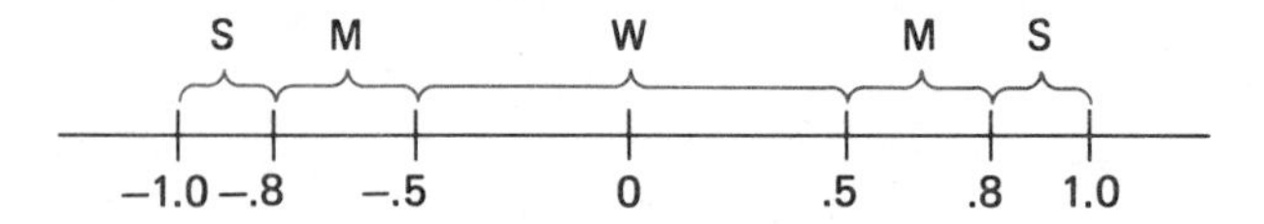

FIGURE 9 CHARACTERIZATION OF r INDICATING A STRONG (S), MODERATE (M), OR WEAK (W) RELATIONSHIP

According to this characterization, the linear relationship between yields of the two types of crops in Example 3 is weak because $r = .347$. Unfortunately, the scatter plot of Figure 8 does not suggest a strong nonlinear relationship either. It may surprise you that a value of r as large as $.5$ or $-.5$ is in the weak category. The reason for this will be apparent from the relationship between r and another quantity called the *coefficient of determination,* which is introduced later in the chapter.

The sample correlation coefficient r measures how strongly the x and y values in a *sample* of pairs are related to one another. There is an analogous measure of how strongly x and y are related in the entire population of pairs from which the sample $(x_1, y_1), \ldots, (x_n, y_n)$ was obtained. It is called the **population correlation coefficient** and is denoted by ρ (notice again the use of a Greek letter for a population characteristic and Roman letter for a sample characteristic). We'll never have to calculate ρ from the entire population of pairs, but it is important to know that ρ satisfies properties paralleling those of r:

1. ρ is a number between -1 and $+1$ that does not depend on the unit of measurement for either x or y, or on which variable is labeled x and which is labeled y.

2. $\rho = +1$ or -1 if and only if all (x, y) pairs in the population lie exactly on a straight line, so ρ measures the extent to which there is a linear relationship in the population.

Later on we show how the sample characteristic r can be used to make an inference concerning the population characteristic ρ. In particular, r can be used to decide whether or not $\rho = 0$ (no linear relationship in the population).

A value of r close to 1 indicates that relatively large values of one variable tend to be associated with relatively large values of the other variable. This is far from saying that a large value of one variable *causes* the value of the other variable to be large. Correlation (Pearson's or any other) measures the extent of association, but *association does not imply causation.* It frequently happens that two variables are highly correlated not because one is causally related to the other but because they are both strongly related to a third variable. Among elementary-school children, there is a strong positive relationship between number of cavities in a child's teeth and the size of his or her vocabulary. Yet no one advocates eating foods that result in more cavities in order to increase vocabulary size (or working to decrease vocabulary size in order to protect against cavities). Number of cavities and vocabulary size are both strongly related to age, so older children tend to have higher values of both variables than do younger ones. Among children of any fixed age, there would undoubtedly be little relationship between number of cavities and vocabulary size.

Scientific experiments can frequently make a strong case for causality by carefully controlling the values of all variables that might be related to the ones under study. Then, if y is observed to change in a "smooth" way as the experimenter changes the value of x, the most plausible explanation would be a causal relationship between x and y. In the absence of such control and ability to manipulate values of one variable, we must admit the possibility that an unidentified underlying third variable is driving both the variables under investigation. A high correlation in many uncontrolled studies carried out in different settings can marshal support for causality—as in the case of cigarette smoking and cancer—but proving causality is often a very elusive task.

Spearman's Rank Correlation Coefficient (Optional)

Pearson's correlation coefficient r identifies a strong linear relationship between x and y but may miss a strong relationship that is not linear. In addition, the value of r can be greatly affected by the presence of even one or two outlying (x, y) pairs that are far from the main part of the scatter plot. Spearman's correlation coefficient r_s is a measure that is not as sensitive as r to outlying points and identifies both linear and nonlinear relationships. It does so by using the *ranks* of the x and y observations rather than the observations themselves.

To compute r_s, first replace the smallest x value by its rank, 1, the second-smallest x value by its rank, 2, and so on. Similarly, the smallest y value is replaced by its rank, 1, the second smallest by its rank, 2, etc. Suppose, for example, that $n = 4$ and the data pairs are (110, 24.7), (125, 24.2), (116, 22.6), and (95, 23.5). Then the x values 110, 125, 116, 95 have ranks 2, 4, 3, 1, whereas the y ranks for 24.7, 24.2, 22.6, and 23.5 are 4, 3, 1, and 2, respectively. The corresponding pairs of ranks are then (2, 4) (replacing 110 by its rank and 24.7 by its rank), (4, 3), (3, 1), and (1, 2).

If there is a strong positive relationship between x and y, the x observations with small ranks tend to be paired with y observations having small ranks. An extreme case of this, the rank pairs (1, 1), (2, 2), (3, 3), (4, 4), occurs whenever a larger x value is always associated with a larger y value. Alternatively, large x ranks paired with small y ranks indicate a negative relationship, the most extreme case being (1, 4), (2, 3), (3, 2), and (4, 1) for $n = 4$ observations.

Once the rank pairs have been determined, Spearman's r_s is just Pearson's coefficient r applied to these rank pairs. To simplify the formula for r_s, note that the ranks for n observations (x values or y values) are the n integers $1, 2, 3, \ldots, n$. The average and standard deviation of this set of integers are $(n + 1)/2$ and $\sqrt{n(n + 1)/12}$, respectively. Since the denominator of r is $(n - 1)s_x s_y$, the denominator of r_s becomes

$$(n - 1)\sqrt{\frac{n(n + 1)}{12}}\sqrt{\frac{n(n + 1)}{12}} = \frac{n(n - 1)(n + 1)}{12}$$

DEFINITION

Spearman's rank correlation coefficient r_s is Pearson's correlation coefficient applied to the rank pairs obtained by replacing each x value by its rank and each y value by its rank:

$$r_s = \frac{\Sigma\left[x \text{ rank} - \left(\frac{n + 1}{2}\right)\right]\left[y \text{ rank} - \left(\frac{n + 1}{2}\right)\right]}{n(n - 1)(n + 1)/12}$$

As with r, the value of r_s is between -1 and 1. A value close to 1 or -1 indicates a strong relationship, whereas a value close to zero indicates a weak relationship.

EXAMPLE 5

The paper "The Relation Between Freely Chosen Meals and Body Habits" (*Amer. J. Clinical Nutrition* (1983):32–40) reported results of an investigation into the relationship between body build and energy intake of an individual's diet. A measure of body build is the Quetelet index (x), with a high value of x indicating a thickset individual. The variable reflecting energy intake is y = dietary energy density. There were nine subjects in the investigation, and the resulting (x, y) pairs are given. With $n = 9$, $(n + 1)/2 = 5$ and $n(n - 1)(n + 1)/12 = 9(8)(10)/12 = 60$. From the data table, the numerator of r_s is 57, so $r_s = 57/60 = .95$. This indicates a very strong positive relationship between the x and y values in the sample.

Subject	x	y	x rank	y rank	$\left(x \text{ rank} - \frac{n+1}{2}\right)\left(y \text{ rank} - \frac{n+1}{2}\right)$
1	221	.67	3	3	$(-2)(-2) = 4$
2	228	.86	6	5	$(1)(0) = 0$
3	223	.78	4	4	$(-1)(-1) = 1$
4	211	.54	1	2	$(-4)(-3) = 12$
5	231	.91	7	7	$(2)(2) = 4$
6	215	.44	2	1	$(-3)(-4) = 12$
7	224	.90	5	6	$(0)(1) = 0$
8	233	.94	8	9	$(3)(4) = 12$
9	268	.93	9	8	$(4)(3) = 12$
					Sum = 57

Figure 10 displays a scatter plot of the data. The point (268, .93) is a clear outlier and considerably distorts the linear pattern in the plot. The value of Pearson's r is only .658. Changing $x = 268$ to $x = 234$ results in the point (234, .93), which is no longer an outlier. The value of r_s remains at .95, since 234 has the same x rank as 268, but the value of r for the altered data set rises dramatically to .92.

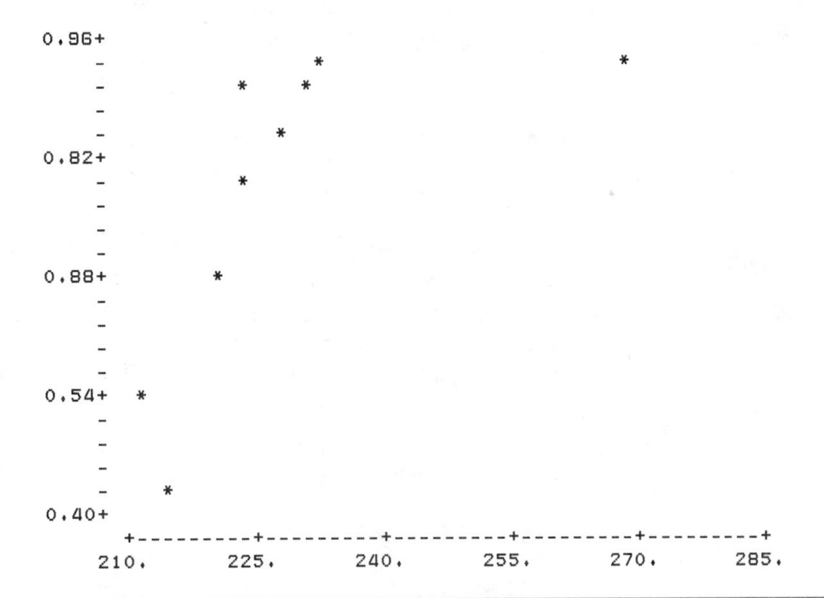

FIGURE 10 A MINITAB SCATTER PLOT FOR THE DATA OF EXAMPLE 5

The objective of some experiments is to make comparisons by having individuals rank the objects or entities in a specified group. Thus some television viewers might be asked to rank a group of programs in order of decreasing preference, or some wine connoisseurs might convene for the purpose of ranking a group of wines in order of descending quality. In such cases the data values consist of sets of rankings rather than observations on numerical variables. Spearman's r_s can be applied directly to two sets of such rankings to measure the extent to which those doing the rating agree in their judgments. The calculations are exactly like those done in Example 5 once the x and y ranks have been determined. Several examples of this type are given in the exercises.

The procedure for calculating r_s must be modified somewhat if there are ties in either the x values or the y values (e.g., three pairs with $x = 27.5$). One of the chapter references can be consulted to find out how this is done.

EXERCISES

4.6 For each of the following pairs of variables, indicate whether you would expect a positive correlation, a negative correlation, or no correlation. Explain your choice.

a. Maximum daily temperature and cooling costs

b. Interest rate and number of loan applications

c. Incomes of husbands and wives when both have full-time jobs

d. Height and IQ

e. Height and shoe size

f. Score on the math section of the SAT exam and score on the verbal section of the same test

g. Time spent on homework and time spent watching television during the same day by elementary-school children

h. Amount of fertilizer used per acre and crop yield (*Hint:* As the amount of fertilizer is increased, yield tends to increase for awhile but then tends to start decreasing.)

4.7 Is the following statement correct? Explain why or why not.

A correlation coefficient of zero implies that no relationship exists between the two variables under study.

4.8 Draw two scatter plots, one for which $r = 1$ and a second for which $r = -1$.

4.9 Data on per capita disposable personal income (x) and per capita food expenditure (y) in the United States was obtained for a 24-year period. Summary quantities are $n = 24$, $s_x = 806.54$, $s_y = 89.12$, and $\Sigma(x - \bar{x})(y - \bar{y}) = 1{,}647{,}030.264$. Compute r for this data. What does the value of r indicate about personal income and food expenditures?

4.10 An investigation of the relationship between water temperature x and calling rate y for a particular type of hybrid toad ("The Mating Call of Hybrids of the Fire-Bellied Toad and Yellow-Bellied Toad," *Oecologia* (1974):61–71) yielded the following summary quantities: $n = 17$, $s_x = 3.86$, $s_y = 8.77$, and $\Sigma(x - \bar{x})(y - \bar{y}) = 499.08$. Does this data indicate that there is a strong positive correlation between water temperature and calling rate? If water temperature decreases, what can be said about the calling rate?

4.11 A number of different methods for measuring growth rate in lobsters are discussed in the paper "A Comparison of Techniques for the Measurement of Growth in Adult Lobsters" (*Aquaculture* (1984):195–99). Twenty-three adult female lobsters were included in the study and both dry weight and volume were recorded for each one. Letting x denote dry weight (in gm) and y denote volume (in ml), summary quantities for the data given in the paper are $n = 23$, $\bar{x} = 142.5$, $\Sigma x^2 = 1{,}024{,}625.83$, $\bar{y} = 473.0$, $\Sigma y^2 = 11{,}492{,}957.01$, and $\Sigma xy = 2{,}961{,}190.58$. Use this information to calculate the correlation coefficient. Would you describe the correlation between dry weight and volume as positive or negative? As weak, moderate, or strong?

4.12 The paper "Bumblebee Response to Variation in Nectar Availability" (*Ecology* (1981):1648–61) reported a positive correlation between a measure of bumblebee abundance and amount of nectar available. Representative data taken from this paper is given. Use this data to calculate r. Interpret the value of r in terms of this study.

Resource abundance	3	8	11	10	23	23	30	35
Bumblebee abundance	4	6	12	18	11	24	22	37

4.13 The accompanying data on growth rate and research and development (R&D) expenditures for eight different industries appeared in the paper "Technology, Productivity, and Industry Structure" (*Technological Forecasting and Social Change* (1983):1–13). Calculate r for this data set. Would you describe the correlation between growth rate and R&D expenditures as weak, moderate, or strong?

Growth rate	1.90	3.96	2.44	.88	.37	−.90	.49	1.01
R&D expenditures	\$2024	\$5038	\$905	\$3572	\$1157	\$327	\$378	\$191

4.14 Data on per capita availability of crude and dietary fiber was given in the paper "Estimation of Per Capita Crude and Dietary Fiber Supply in 38 Countries" (*Amer. J. Clinical Nutrition* (1984):821–29). Let x and y denote crude and dietary fiber availability, respectively (g/day). The following summary quantities were computed from the 38 pairs of values given in the paper: $\Sigma x = 313.3$, $\Sigma x^2 = 3065.83$, $\Sigma y = 1514.1$, $\Sigma y^2 = 72334.63$, $\Sigma xy = 14702.85$. Use this data to calculate the correlation coefficient.

4.15 Sixteen different air samples were obtained at Herald Square in New York City, and both the carbon monoxide concentration x (ppm) and benzo(a) pyrene concentration y ($\mu g/10^3 m^3$) were measured for each sample ("Carcinogenic Air Pollutants in Relation to Automobile Traffic in New York City," *Environmental Science and Technology* (1971):145–50).

x	2.8	15.5	19.0	6.8	5.5	5.6	9.6	13.3
y	.5	.1	.8	.9	1.0	1.1	3.9	4.0

x	5.5	12.0	5.6	19.5	11.0	12.8	5.5	10.5
y	1.3	5.7	1.5	6.0	7.3	8.1	2.2	9.5

Compute the sample correlation coefficient for this data.

4.16 Doctors have always cautioned against extreme low-calorie weight-loss diets (under 500 calories per day). The paper "Cardiac Dysfunction in Obese Dieters: A Potentially Lethal Complication of Rapid Massive Weight Loss" (*Amer. J. Clinical Nutrition* (1984):695–702) summarized the findings of a study of healthy adults who died suddenly either during or shortly after having been on a very low-calorie diet. The data in the paper was used to rank the 16 female subjects on the basis of initial body mass (kg/m²) and length of time (months) on a low-calorie diet, and these ranks are given. Use this data to compute Spearman's rank correlation coefficient. How does the value of r_s compare to the value of Pearson's correlation coefficient, $r = .824$, given in the paper?

Ranks								
Body mass	1	2	3	4	5	6	7	8
Time on diet	1	3	4	11	8	2	7	9

Ranks								
Body mass	9	10	11	12	13	14	15	16
Time on diet	5	12	6	15	10	16	13	14

4.17 The fuel price index (FPI) and consumer price index (CPI) in the United States for the years 1970–1981 are given (*Transportation Quarterly* (1983):28).

Year	1970	1971	1972	1973	1974	1975
FPI	51.7	53.7	53.8	55.7	79.2	89.6
CPI	64.0	66.8	69.0	73.3	81.4	88.8

Year	1976	1977	1978	1979	1980	1981
FPI	97.6	100.0	106.6	153.3	206.2	229.9
CPI	93.9	100.0	107.7	119.8	136.0	150.1

a. Draw a scatter plot with x denoting FPI and y denoting CPI. Does the plot exhibit a weak, moderate, strong, or perfect linear relationship?

b. Calculate the value of r for this data set.

c. Compute Spearman's rank correlation coefficient, r_s.

d. How do the values of r and r_s compare? Does it surprise you that $r_s = 1$ but $r \neq 1$? Explain why or why not.

4.18 In a study of variables thought to be related to urban gang activity, seven Chicago-area communities were ranked according to prevalence of ganging, percent of residents who are black, median family income, and percent of families below poverty level ("Youth Gangs", *Pacific Sociol. Rev.* (1981):366). The authors of this paper analyzed these data using Spearman's rank correlation coefficient.

Community	Prevalence of Ganging	Percent Black	Median Income	Percent below Poverty Level
A	2	3	7	1
B	4	6	6	4
C	7	7	1	7
D	5	5	2	5
E	6	4	3	6
F	3	2	5	3
G	1	1	4	2

a. Calculate r_s for the following pairs of variables.
  i. Ganging prevalence and percent Black
  ii. Ganging prevalence and median income
  iii. Ganging prevalence and percent below poverty level

b. Based on the coefficients in (a), which of the variables seems to exhibit the strongest relationship with ganging prevalence?

4.19 The accompanying data (taken from "Rating the Risks" *Environment* (1979): 19) shows how three groups of people ranked the relative riskiness of 30 activities and technologies. A rank of 1 represents the most risky activity.

a. Use Spearman's rank correlation coefficient to assess the extent to which the following pairs of groups agree in their rankings of relative risks.
  i. Students and nonstudents
  ii. Students and experts
  iii. Nonstudents and experts

b. Which two groups exhibit the most agreement in ranking?

c. Where in the ranking would you put taking a statistics course?

	Ranks		
	Group 1	Group 2	Group 3
Activity	Nonstudents	College Students	Experts
Nuclear power	1	1	20
Motor vehicles	2	5	1
Handguns	3	2	4
Smoking	4	3	2
Motorcycles	5	6	6
Alcoholic beverages	6	7	3
Private aviation	7	15	12
Police work	8	8	17
Pesticides	9	4	8
Surgery	10	11	5
Fire fighting	11	10	18
Large construction	12	14	13
Hunting	13	18	23
Spray cans	14	13	26
Mountain climbing	15	22	29
Bicycles	16	24	15
Commercial aviation	17	16	16
Electric power	18	19	9
Swimming	19	30	10
Contraceptives	20	9	11
Skiing	21	25	30
X rays	22	17	7
High school/college football	23	26	27
Railroads	24	23	19
Food preservatives	25	12	14
Food coloring	26	20	21
Power mowers	27	28	28
Prescription antibiotics	28	21	24
Home appliances	29	27	22
Vaccinations	30	29	25

4.20 Ranking wines is a common practice at wine tastings. Suppose that after tasting nine wines, two judges rank the wines as follows:

Wine	A	B	C	D	E	F	G	H	I
Judge 1	7	1	3	2	8	5	9	6	4
Judge 2	9	4	1	3	7	5	6	8	2

Compute r_s as a measure of agreement between the two judges.

4.21 Suppose that sample x and y values are first expressed in standard units by means of

$$x' = \frac{(x - \bar{x})}{s_x} \quad \text{and} \quad y' = \frac{(y - \bar{y})}{s_y}$$

How does the value of $\Sigma x'y'$, the sum of products of these standardized values, relate to r?

4.22 **a.** Suppose that x and y are positive variables and that a sample of n pairs results in $r \approx 1$. If the sample correlation coefficient is computed for the n pairs $(x_1, y_1^2), (x_2, y_2^2), \ldots, (x_n, y_n^2)$—i.e., for the (x, y^2) pairs—will the resulting value also be approximately 1? Explain.

b. If x and y are positive variables, how does r_s for the (x, y) pairs compare with r_s for the (x, y^2) pairs? Explain.

4.23 Let d_1 denote the difference (x rank $-$ y rank) for the first (x, y) pair, d_2 denote this difference for the second (x, y) pair, and so on. It can be shown that

$$r_s = 1 - \frac{6\Sigma d^2}{n(n^2 - 1)}$$

a. Use this formula to compute r_s for the data of Example 5.

b. What does this formula imply about the value of r_s when there is a perfect positive relationship? Explain.

4.3 Fitting a Line to Bivariate Data

It is very rare that the points in a scatter plot fall exactly either on some straight line or along some smooth curve. However, the plot often shows a distinct linear or curved pattern. If we can find a line or curve that fits the observed points closely, this can then be used to describe approximately how a change in the value of one variable might produce or be associated with a change in the value of the other variable. In particular, we can predict the value of one variable that will be paired with a specified value of the other variable.

The correlation coefficient did not depend on which variable was labeled x and which variable was labeled y. However, in fitting a line or curve to data and using the result to make a prediction, we must distinguish between the variable whose value is to be predicted and the variable whose value is used to make the prediction. In many contexts it is obvious which variable is which. An educator might want to predict the size of a child's vocabulary from the age of the child but would not usually want to predict age from vocabulary size. Similarly, a chemist might wish to predict yield in a certain chemical reaction for a specified reaction temperature rather than the other way around. We identify y with the variable whose value is to be predicted and call it the **dependent variable.** The x variable is the one that will be used to make the prediction, so it is called the **predictor** or **independent variable.** Thus we think of the value of y as depending on the value of the independent variable x (even if changing x does not actually *cause* a change in y).

A Linear Relationship Between x and y

The most frequently occurring pattern in a scatter plot is a linear pattern. Fitting a line to such data requires that we first review some elementary facts about lines and linear relationships. Suppose that a car dealership advertises that a particular model of car can be rented for a flat fee of \$25 plus an additional \$.20 per mile. If this type of car is rented and driven for 100 miles, the dealer's revenue is $25 + (.20)(100) = 25 + 20 = 45$. If the car is driven 150 miles, the resulting revenue is $25 + (.20)(150) = 25 + 30 = 55$. With x = the number of miles driven and y = the resulting revenue, the relationship between x and y can be summarized by the equation $y = 25 + .20x$. That is, x and y are linearly related.

The general form of a linear relation between x and y is $y = a + bx$. A particular relation is specified by choosing values of a and b. Thus one such relationship is $y = 10 + 2x$, while another is $y = 100 - 5x$. If we choose some x values and compute $y = a + bx$ for each value, the points in the scatter plot of the resulting (x, y) pairs fall exactly on a straight line.

DEFINITION	The relationship $y = a + bx$ is the equation of a straight line. The value of b, called the **slope** of the line, is the amount by which y increases when x increases by 1 unit. The value of a, called the **y intercept** of the line, is the height of the line above the value $x = 0$.

The equation $y = 10 + 2x$ has slope $b = 2$, so each 1 unit increase in x results in an increase of 2 in y. When $x = 0$, $y = 10$, and the height at which the line crosses the vertical axis (where $x = 0$) is 10. This is illustrated in Figure 11(a). The slope of the line determined by $y = 100 - 5x$ is -5, so y increases by -5 (i.e., decreases by 5) when x increases by 1. The height of the line above $x = 0$ is $a = 100$. The resulting line is pictured in Figure 11(b).

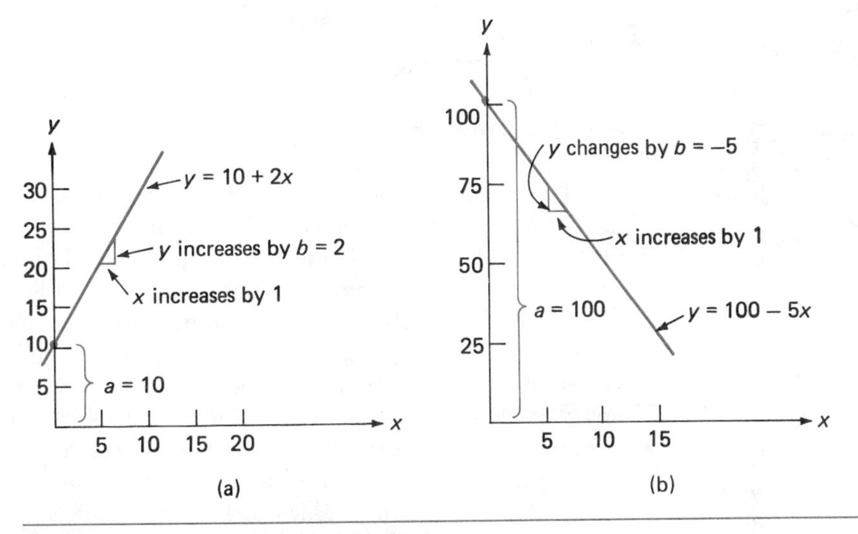

FIGURE 11 THE GRAPHS OF TWO STRAIGHT LINES
(a) A Line with Slope $b = 2$ and y Intercept $a = 10$
(b) A Line with Slope $b = -5$ and y Intercept $a = 100$

It is easy to draw the line corresponding to any particular linear equation. First choose any two x values and substitute them into the equation to obtain the corresponding y values. Then plot the resulting two (x, y) pairs as two points. The desired line is the one passing through these points. For the equation $y = 10 + 2x$, substituting $x = 5$ yields $y = 20$, while using $x = 10$ gives $y = 30$. The two points are then $(5, 20)$ and $(10, 30)$. The line in Figure 11(a) does indeed pass through these points.

Fitting a Straight Line: The Principle of Least Squares

When the points on a scatter plot do not fall on a line, an approximate relationship is specified by finding a line to which all (or at least most) of the points are close. This requires a quantitative measure of how closely any particular line fits the points in a scatter plot. Figure 12 shows a scatter plot consisting of $n = 5$ points with two different lines superimposed on the plot. Line I obviously gives a poor fit to the data, since the points in the plot deviate a great deal from this line. The fit of Line II to the data is much better; the observed points deviate by a relatively small amount from this second line.

The quantity that we shall use to measure how well a line fits the data is

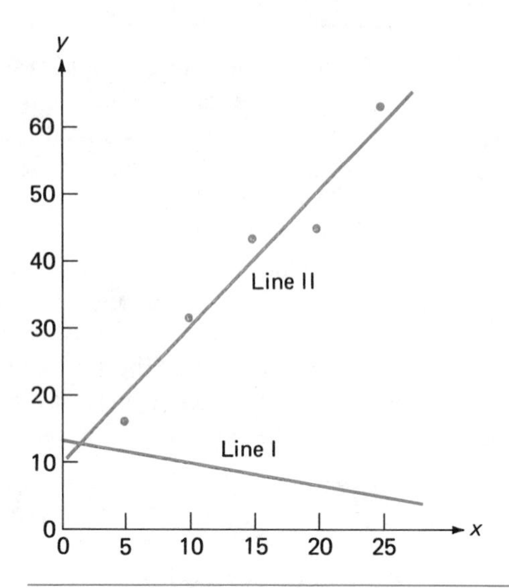

FIGURE 12 LINES I AND II GIVE POOR AND GOOD FITS, RESPECTIVELY, TO THE DATA

based on the vertical deviations of the points in the plot from the line. A vertical deviation is the difference between the height of a point (its y value) and the height of the line at the same value of x. For example, if the line is $y = 10 + 2x$, the height of the line when $x = 10$ is $y = 10 + 2(10) = 30$. The vertical deviation of the point $(10, 32)$ from the line is then $32 - 30 = 2$. The deviation of $(20, 45)$ from the same line is $45 -$ (the height of the line at $x = 20$) $= 45 - [10 + 2(20)] = 45 - 50 = -5$. The first deviation is positive because $(10, 32)$ lies above the line (see Figure 13), whereas the second deviation is negative because $(20, 45)$ lies below the line.

The vertical deviations are used because for any particular x value, the height of the line at that value gives a prediction for y. The vertical deviation (*observed* y) $-$ (*predicted* y) is then the prediction error. Figure 13 shows the vertical deviations of the indicated five points from the line $y = 10 + 2x$. Two of the points are below the line and have negative deviations, while the other three are above the line and have positive deviations. To measure the goodness-of-fit of a line to the data, the deviations must now be combined. To prevent negative and positive deviations from offsetting one another, the deviations are squared and then added together.

DEFINITION

The Principle of Least Squares

The criterion for measuring the goodness-of-fit of a line $y = a + bx$ to bivariate data $(x_1, y_1), \ldots, (x_n, y_n)$ is the sum of squared deviations about the line:

$$\Sigma[y - (a + bx)]^2 =$$
$$[y_1 - (a + bx_1)]^2 + [y_2 - (a + bx_2)]^2 + \cdots + [y_n - (a + bx_n)]^2$$

The line that gives the best fit to the data is the one that minimizes this sum; it is called the **least squares line.**

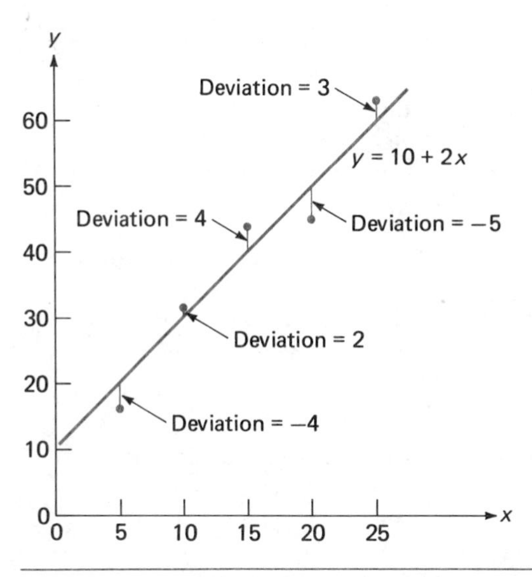

FIGURE 13 VERTICAL DEVIATIONS FROM THE LINE $y = 10 + 2x$

EXAMPLE 6

A company has tentatively selected a new plant site next to a freeway, but management is concerned about commuting time to the site during the morning rush hour. Let $x =$ the distance (km) from a point on the freeway to the site and $y =$ the time (min) necessary to travel that distance. A sample of $n = 5$ observations is obtained, resulting in the following pairs: $(5, 16)$, $(10, 32)$, $(15, 44)$, $(20, 45)$, and $(25, 63)$.

These are the pairs pictured in the scatter plot of Figure 13. The line $y = 10 + 2x$ pictured there looks as though it gives a good fit to the data. We use the accompanying tabular format to compute the sum of squared deviations for both this line and for the line $y = 7.9 + 2.14x$. This second line has a smaller y intercept than the first line but rises a bit more rapidly. The sum of squared deviations for these two lines are 70 and 65.10, respectively, so the second line gives a better fit according to the least squares criterion. In fact, as we show shortly, no other line gives a smaller sum of squared deviations than 65.10, so $y = 7.9 + 2.14x$ is the least squares line.

		$y = 10 + 2x$			$y = 7.9 + 2.14x$		
x	y	Height of Line at x	Deviation	$(\text{Deviation})^2$	Height of Line at x	Deviation	$(\text{Deviation})^2$
5	16	20	−4	16	18.6	−2.6	6.76
10	32	30	2	4	29.3	2.7	7.29
15	44	40	4	16	40.0	4.0	16.00
20	45	50	−5	25	50.7	−5.7	32.49
25	63	60	3	9	61.4	1.6	2.56
				70			65.10

■

How can we know whether a given line is the least squares line without comparing the sum of squared deviations for that line to the sum of squared deviations for every other line? In Example 6, the sum of squared deviations for the line $y = 10 + 2x$ was 70, whereas that for the line $y = 7.9 + 2.14x$ was 65.10. This implies that $y = 10 + 2x$ is not the least squares line, but maybe there is another line for which the sum of squared deviations is less than 65.10. Fortunately, statisticians have obtained formulas for the slope b and y intercept a of the least squares line, so that the sum of squared deviations for any other line need not be computed.

The **slope of the least squares line** is given by formula

$$b = \frac{\Sigma(x - \overline{x})(y - \overline{y})}{\Sigma(x - \overline{x})^2}$$

The **y intercept of the least squares** is then

$$a = \overline{y} - b\overline{x}$$

Alternative formulas for computing the numerator and denominator of b are $\Sigma xy - n\overline{x}\overline{y}$ and $\Sigma x^2 - n\overline{x}^2$, respectively.

The formula for b involves $\Sigma(x - \overline{x})(y - \overline{y})$ and $\Sigma(x - \overline{x})^2$, both of which appeared in the formula for Pearson's r. We shall shortly need $\Sigma(y - \overline{y})^2$, so we compute it along with the first two sums.

EXAMPLE 7

(*Example 6 continued*) From the commute distance–commute time data with $n = 5$, $\Sigma x = 75$, and $\Sigma y = 200$, we calculate $\overline{x} = 15.0$ and $\overline{y} = 40.0$. The same tabular format used in the calculation of r is recommended here: In addition to columns for $x - \overline{x}$ and $y - \overline{y}$, three further columns for $(x - \overline{x})^2$, $(y - \overline{y})^2$, and $(x - \overline{x})(y - \overline{y})$ are used.

x	y	$x - \overline{x}$	$y - \overline{y}$	$(x - \overline{x})^2$	$(y - \overline{y})^2$	$(x - \overline{x})(y - \overline{y})$
5	16	−10	−24	100	576	240
10	32	−5	−8	25	64	40
15	44	0	4	0	16	0
20	45	5	5	25	25	25
25	63	10	23	100	529	230
				250	1210	535

The slope of the least squares line is

$$b = \frac{\Sigma(x - \overline{x})(y - \overline{y})}{\Sigma(x - \overline{x})^2} = \frac{535}{250} = 2.14$$

and the y intercept is

$$a = \overline{y} - b\overline{x} = 40.0 - (2.14)(15.0) = 7.90$$

Computations can be streamlined a bit by using the alternative computing formulas. For this purpose, a tabular format with columns for x, y, x^2, y^2 (to obtain $\Sigma(y - \overline{y})^2$) and xy is recommended.

x	y	x^2	y^2	xy
5	16	25	256	80
10	32	100	1024	320
15	44	225	1936	660
20	45	400	2025	900
25	63	625	3969	1575
75	200	1375	9210	3535
↑	↑	↑	↑	↑
Σx	Σy	Σx^2	Σy^2	Σxy

Then

$$\Sigma(x - \bar{x})(y - \bar{y}) = \Sigma xy - n\bar{x}\bar{y} = 3535 - 5(15)(40) = 535$$
$$\Sigma(x - \bar{x})^2 = \Sigma x^2 - n\bar{x}^2 = 1375 - 5(15)(15) = 250$$
$$\Sigma(y - \bar{y})^2 = \Sigma y^2 - n\bar{y}^2 = 9210 - 5(40)(40) = 1210$$

These values are identical to those obtained from the defining formulas.

Recalling that the slope of a line is the amount by which y increases when x increases by 1, we can say that each 1-km increase in distance results in roughly 2.14 min of increased commuting time. The equation of the line can now be used to predict y for a given distance x. When $x = 17.5$, we predict commuting time to be $7.90 + 2.14(17.5) = 7.90 + 37.45 = 45.35$. The predicted commuting time when $x = 15$ is $7.9 + 2.14(15) = 40$, while the observed commuting time for this distance is 44. The prediction error for this observation is (observed y) $-$ (predicted y) $= 44 - 40 = 4$.

Suppose that a prediction of commuting time when $x = 40$ is desired. Using the least squares line gives $7.9 + 2.14(40) = 93.5$ as a predicted time. But there is a danger here: The value $x = 40$ is well outside the range of x values for which data was collected. The use of the least squares line to make this prediction amounts to assuming that the approximate relationship between x and y is described by the line not only between $x = 5$ and $x = 25$ but also for x values well above this interval. Perhaps this is indeed the case, but we have no data to support it! This is **the danger of extrapolation**—using the approximate relationship to make predictions for x values far outside the range of the data. As a more extreme example, substituting $x = 0$ gives a predicted time of 7.9 at zero distance, which is patently ridiculous even on the freeways of Los Angeles! Our data suggests an approximate linear relationship between $x = 5$ and $x = 25$, but to extrapolate much beyond this range can lead to very erroneous or even ridiculous answers. ■

We purposely chose numbers for Example 6 that make the arithmetic straightforward. The numbers in actual data sets are often not so nice. In such cases, using rounded values in intermediate computations may result in various computed values being quite different from what they should be. If, for example, $\bar{y} = 15{,}500$ and $\bar{x} = 1500$, the y intercept of the least squares line is $a = 15{,}500 - b(1500)$. Depending on whether $b = 10$, $b = 10.3$, or $b = 10.27$ is used, the value of a will be 500, 50, or 95. Using the resulting equations $y = 500 + 10x$, $y = 50 + 10.3x$, and $y = 95 + 10.27x$ to predict y when $x = 1200$ results in predictions of 12,500, 12,410, and 12,419, which differ substantially from one another.

A general rule of thumb for avoiding computational errors due to rounding is difficult to come by because there are always pathological examples that can cause special problems. Since you should be using a handheld calculator at this point, a safe approach is to use all the digits displayed by the calculator for a particular quantity in any subsequent computations involving that quantity. With more experience, you will develop an intuitive feel for deciding how many digits are important for a particular calculation.

EXAMPLE 8

The decline of visual acuity and sensitivity in adults as age increases is a well-known phenomenon. The paper "Neuron Loss in the Aging Visual Cortex of Man" (*J. of Gerontology* (1980):836–41) presented the following data on neurons per gram of tissue (y) for $n = 16$ males of various ages (x, in years). A scatter plot of the data appears in Figure 14. The plot shows a linear pattern, so we use the principle of least squares to find a line that specifies an approximate linear relationship between x and y.

x	20	24	25	29	35	41	50	51
y	4.0	4.7	4.4	4.2	5.6	3.6	2.9	2.3

x	64	68	69	69	73	76	81	83
y	3.1	2.1	2.0	3.2	2.9	3.0	2.9	2.4

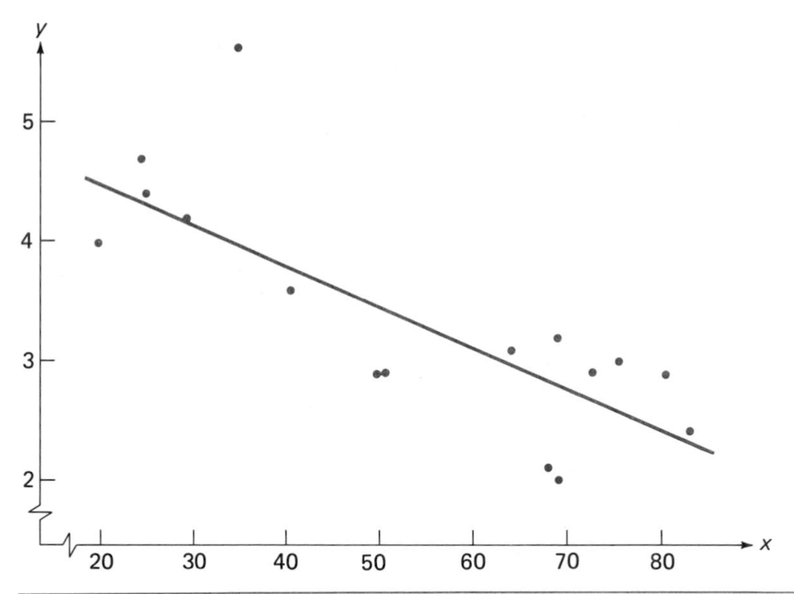

FIGURE 14

SCATTER PLOT OF THE DATA IN EXAMPLE 8 WITH THE LEAST SQUARES LINE SUPERIMPOSED

With $\Sigma x = 858$ and $\Sigma y = 53.3$, $\bar{x} = 53.6250$ and $\bar{y} = 3.331250$. These values make the calculations of the $x - \bar{x}$ and $y - \bar{y}$ values somewhat tedious, so we dispense with a display of intermediate calculations.

The results are $\Sigma(x - \bar{x})^2 = 7235.750$, $\Sigma(y - \bar{y})^2 = 15.394375$, and $\Sigma(x - \bar{x})(y - \bar{y}) = -253.91250$. Then

$$b = \frac{-253.91250}{7235.750} = -.035091$$

$$a = 3.33125 - (-.035091)(53.6250) = 3.33125 + 1.88175 = 5.2130$$

This gives $y = 5.2130 - .035091x$ as the equation of the least squares line. For most purposes the rounded equation $y = 5.21 - .0351x$ could be used. The least squares line appears superimposed on the scatter plot of Figure 14. The y (vertical) axis there does not cross the x axis at $x = 0$, so the height at which the least squares line crosses the y axis *is not* the y intercept, 5.21. We graphed the line by substituting $x = 20$ and $x = 80$ to find two points, (20, 4.51) and (80, 2.40), on the line and then connecting these points. The slope of $-.0351$ implies that for two men who are 1 year apart in age, the neuron density for the older man would be roughly .0351 smaller than that for the younger man. For a 40-year-old man, the predicted neuron density is $5.2130 - (.035091)(40) = 3.81$. This same prediction results from using the rounded equation. ■

Calculations involving the least squares line can obviously be quite tedious, and fitting a function to multivariate data is even worse. This is where the computer comes to our rescue. All the standard statistical packages fit a straight line to bivariate data when asked to do so. To accomplish this, a regression command or option is used. (This terminology is explained shortly.)

EXAMPLE 9

Figure 15 displays a small portion of the MINITAB output resulting from a regression command applied to the age-neuron density data introduced in Example 8. Instead of x and y, the variable labels *age* and *neurons* are used. The equation at the top is exactly the rounded equation whose use was suggested earlier. In the rectangular table just below the equation, the first row gives information about the y intercept, a, and the second row gives information concerning the slope, b. In particular, the "coefficient" column contains the values of a and b using more significant figures than appear in the rounded equation. (We explain the contents of the last two columns in Chapter 11.)

```
THE REGRESSION EQUATION IS
NEURONS = 5.21 - 0.0351 AGE

                                 ST. DEV.        T-RATIO =
COLUMN          COEFFICIENT      OF COEF.        COEF/S.D.
                     5.2130       0.4615            11.29
AGE               -0.035091     0.008001            -4.39
```

FIGURE 15 MINITAB OUTPUT FOR THE DATA OF EXAMPLE 8

■

Computational accuracy can be a problem even for computers. Statisticians have developed computational methods to help ensure accuracy and protect against rounding. These techniques have been implemented in the most widely used packages but not in some programs designed for use with microcomputers.

Regression

The least squares line is often called the **sample regression line.** This terminology comes from the relationship between the least squares line and Pearson's correlation coefficient. To understand this relationship, we first need alternative expressions for the slope b and equation of the line itself. A bit of algebraic manipulation yields the results

$$b = r \cdot \frac{s_y}{s_x}$$

Equation of the least squares line: $$y = \bar{y} + r \cdot \frac{s_y}{s_x}(x - \bar{x}).$$

You do not need to use these formulas in any computations, but several of their implications are important for appreciating what the least squares line does.

1. When $x = \bar{x}$ is substituted in the equation of the line, $y = \bar{y}$ results. That is, the least squares line passes through the *point of averages* $(\bar{x}, \bar{y})$.

2. Suppose for the moment that $r = 1$, so that all points lie exactly on the line whose equation is $y = \bar{y} + (s_y/s_x)(x - \bar{x})$. Consider an x value that is one standard deviation above $\bar{x}$, i.e., $x = \bar{x} + s_x$. Substituting this x into the equation yields $y = \bar{y} + s_y$. That is, with $r = 1$, when x is 1 standard deviation above its mean, we predict that the associated y value will be 1 standard deviation above its mean. Similarly, if $x = \bar{x} - 2s_x$ (2 standard deviations below its mean), then $y = \bar{y} - 2s_y$ (2 standard deviations below its mean). If $r = -1$, then $x = \bar{x} + s_x$ results in $y = \bar{y} - s_y$, so the predicted y is also 1 standard deviation from its mean but on the opposite side of $\bar{y}$ from where x is relative to $\bar{x}$. In general, if x and y are perfectly correlated, the predicted y value associated with a given x value will be the same number of standard deviations (of y) from its mean, $\bar{y}$, as x is from its mean, $\bar{x}$.

3. Suppose that x and y are not perfectly correlated. To be specific, take $r = .5$. Then the least squares line has equation $y = \bar{y} + .5 \cdot (s_y/s_x) \cdot (x - \bar{x})$. When $x = \bar{x} + s_x$ (1 standard deviation above its mean), $y = \bar{y} + .5s_y$, so we predict that y will be only $\frac{1}{2}$ standard deviation above its mean. If $x = \bar{x} - 2s_x$, a value 2 standard deviations below the mean, then $y = \bar{y} - s_y$, so the predicted y is only 1 standard deviation below its mean.

There is similar behavior in predicting y when r is negative. If $r = -.5$, then the predicted y value will be only half the number of standard deviations from $\bar{y}$ that x is from $\bar{x}$, but x and the predicted y will now be on opposite sides of their respective means.

> Consider using the least squares line to predict the value of y associated with an x value some specified number of standard deviations away from $\bar{x}$. Then the predicted y value will be only r times this number of standard deviations from $\bar{y}$. In terms of standard deviations, except when $r = 1$ or -1, the predicted y will always be closer to $\bar{y}$ than x is to $\bar{x}$.

Using the least squares line for prediction results in a predicted y which is pulled back in, or regressed, toward its mean compared to where x is relative to its mean. This regression effect was first noticed by Sir Francis Galton (1822–1911), a famous biologist, while studying the relationship between the heights of fathers and their sons. He found that the predicted height of a son whose father was above average in height would also be above average (because r is positive here) but not by as much as the father; he found a similar relationship for a father whose height was below average. This regression effect has led to the term **regression analysis** for the collection of methods involving the fitting of lines, curves, and more complicated functions to bivariate and multivariate data.

The alternate form of the regression (least squares) line emphasizes that predicting y from knowledge of x is not the same problem as predicting x from knowledge of y. The slope of the least squares line for predicting x is $r \cdot (s_x/s_y)$ rather than $r \cdot (s_y/s_x)$, and the intercepts of the lines are also usually different. It makes a difference whether y is regressed on x, as we have done, or whether x is regressed on y. The regression line of y on x should not be used to predict x, since it is not the line which minimizes the sum of squared x deviations.

EXERCISES

4.24 In addition to being a popular sport fish, largemouth bass are also considered to be the dominant carnivore in many North American lakes. The paper "Piscivorous Feeding Behavior of Largemouth Bass: An Experimental Analysis" (*Trans. Amer. Fisheries Soc.* (1983):508–16) reported a linear relationship between the distance at which a bass first reacted to a bluegill (apparently a favorite of the largemouth bass!) and the length of the bluegill. Letting x represent bluegill length (cm) and y represent reaction distance (cm), the paper gave $y = 3.98 + 22.7x$ as the least squares line. Plot this line. By how much does the height of the line increase when x increases from 3 to 4? From 5 to 6? From 5 to 15? Do you think there was a sample observation with $x = 0$?

4.25 The paper "Species and Age Differences in Accumulation of CI-DDT by Voles and Shrews in the Field" (*Environmental Pollution* (1984):327–34) used the least squares line to describe the relationship between DDT concentration in stomach contents (x) and DDT concentration in the whole body (y). For shrews, the line given is $y = 1.082 + .694x$. Plot this line. A correlation coefficient of $r = .82$ was also reported for the data analyzed in the paper. How would you interpret this value?

4.26 The accompanying data on the percent of red pine scale nymphs (a forest insect) in early and middle stages (x) and overwintering mortality rate (y) appeared in the paper "Population Dynamics of a Pernicious Parasite: Density-Dependent Vitality of Red Pine Scale" (*Ecology* (1983):710–18).

x	20	32	36	42	43	46	48	51	60
y	81	81.5	83	87	84	84.5	86.5	89	86.5

a. Construct a scatter plot for this data set.

b. The least squares line relating x and y given in the paper was $y = 76.2 + .2x$. Draw this line on your scatter diagram.

c. What would you predict for overwintering mortality rate if 50% of the scale nymphs were in early or middle substages?

d. Would it be reasonable to use this line to predict mortality rate associated with 80% of the nymphs being in early or middle substages? Explain.

4.27 The given summary quantities were obtained from a study that used regression analysis to investigate the relationship between pavement deflection and surface temperature of the pavement at various locations on a state highway. Let x = temperature (°F) and y = deflection adjustment factor. (A scatter plot in the paper

"Flexible Pavement Evaluation and Rehabilitation," *Trans. Eng. J.* (1977):75–85) displayed many more than 15 observations.)

$$n = 15 \qquad \Sigma x = 1425 \qquad \Sigma y = 10.68$$
$$\Sigma xy = 987.65 \qquad \Sigma x^2 = 139{,}037.25$$

a. Calculate the least squares line for this data.

b. What change in deflection adjustment factor would you expect to see when temperature is increased by 1°F? By 10°F?

4.28 Milk samples were obtained from 14 Holstein-Friesian cows, and each was analyzed to determine uric acid concentration (μmol/L). In addition to acid concentration, the total milk production (kg/day) was recorded for each cow, as shown in the accompanying table ("Metabolites of Nucleic Acids in Bovine Milk," *J. Dairy Sci.* (1983):723–28).

Cow	1	2	3	4	5	6	7
Milk production	42.7	40.2	38.2	37.6	32.2	32.2	28.0
Acid concentration	92	120	128	110	153	162	202

Cow	8	9	10	11	12	13	14
Milk production	27.2	26.6	23.0	22.7	21.8	21.3	20.2
Acid concentration	140	218	195	180	193	238	213

Let x denote milk production and y denote uric acid concentration.

a. Draw a scatter plot for this data.

b. Using the summary quantities below, compute the least squares line.

$$n = 14 \qquad \Sigma(x - \bar{x})^2 = 762.012 \qquad \Sigma(x - \bar{x})(y - \bar{y}) = -3964.486$$

Draw the least squares line on your scatter plot.

c. What uric acid concentration would you predict for a cow whose total milk production was 30 kg/day?

d. Would you feel comfortable using the least squares line to make a prediction for a cow whose total milk production was 10 kg/day? Explain your answer.

4.29 Data on manganese (Mn) intake x (μg/kg/day) and blood serum Mn level y (μg/L) for eight infants was given in Exercise 4.1.

a. Draw a scatter plot for this data set.

b. Will the slope of the calculated least squares line be positive or negative? Why?

c. Summary quantities for this data set include $\Sigma x = 3.46$, $\Sigma x^2 = 1.581$, $\Sigma y = 35.5$, $\Sigma y^2 = 182.75$, and $\Sigma xy = 16.536$. Compute the least squares line.

d. What would you predict blood serum Mn concentration to be for an infant whose Mn intake was .51 μg/kg/day?

4.30 The paper "Modeling Advertising Sales Relationships Involving Feedback" (*J. Marketing Research* (1982):116–25) gave the accompanying data on advertising expenditures (in thousands of dollars) and sales (in millions of dollars) for six brands of cereals.

a. Draw the scatter plot for this data using advertising as x and sales as y.

b. Calculate the correlation coefficient for this data set. Would you characterize the correlation between advertising expenditure and sales as weak, moderate, or strong?

c. Find the least squares line and draw it on your scatter plot.

d. What would you predict sales to be for a brand that spends $150,000 on advertising?

e. If the point (x, y) corresponding to a particular cereal fell exactly on the least squares line you calculated in (c), what advertising expenditure would be required to yield sales of $2 million?

f. What would happen to the equation of the least squares line if the (x, y) value for Corn Flakes were omitted?

Brand	Advertising	Sales
Corn Flakes	$501	$3.41
Special K	321	2.05
Life	137	.96
Rice Krispies	387	2.89
Alpha Bits	136	.74
Sugar Frosted Flakes	303	2.21

4.31 The March 29, 1975 issue of *Lancet* reported on the relationship between ages of a number of children who had high levels of lead absorption and a measure of wrist flexor and extensor muscle function for these children ("Neuropsychological Dysfunction in Children with Chronic Low-Level Lead Absorption"). The measure involved the number of taps with a stylus on a single metal plate during a 10-s period. Representative data is given, with x = age in months and y = taps/10 s.

x	73	84	98	112	116	132	150	160	164	180
y	35	40	50	42	46	41	52	52	51	66

a. Compute the slope and intercept of the least squares line.

b. What change in the measure of muscle function would be expected for a 1-month increase in age?

4.32 The following data is representative of that reported in the article "An Experimental Correlation of Oxides of Nitrogen Emissions from Power Boilers Based on Field Data" (*J. Eng. for Power* (1973):165–70), with x = burner area liberation rate (MBtu/h-ft^2) and y = nitrogen oxide (NO_x) emission rate (ppm).

x	100	125	125	150	150	200	200	250	250	300	300	350	400	400
y	150	140	180	210	190	320	280	400	430	440	390	600	610	670

a. Use the above data to obtain the least squares line.

b. What is the predicted NO_x emission rate when burner area liberation rate is 225?

c. By how much would you expect NO_x emission rate to change when burner area liberation rate is decreased by 50?

4.33 Athletes competing in a triathalon participated in a study described in the paper "Myoglobinemia and Endurance Exercise" (*Am. J. Sports Med.* (1984):113–18). The following data on finishing time x (h) and myoglobin level y (ng/mL) was read from a scatter plot in the paper.

x	4.90	4.70	5.35	5.22	5.20	5.40	5.70	6.00
y	1590	1550	1360	895	865	905	895	910

x	6.20	6.10	5.60	5.35	5.75	5.35	6.00
y	700	675	540	540	440	380	300

a. Compute the least squares line for this data.

b. If you were to compute the correlation coefficient, would it be positive or negative? Why?

4.34 The concentration of Microcoleus (a type of bacteria) populations found on the surface of various thicknesses of mud slurry overlays was determined to obtain the data below (Source: "Motility of the Cyanobacterium Microcoleus Chthonoplastes in Mud," *Brit. Phycological J.* (1984):117–23). Let x denote the depth of mud overlay (mm^{-1}) and y denote the concentration of Microcoleus (mm^{-2}).

x	1.6	1.7	1.8	2.2	2.5	3.0	3.6	4.5
y	12.0	9.0	7.0	7.2	5.0	3.8	4.1	4.0

x	5.0	5.0	5.0	5.5	5.5	6.0	6.0	4.1
y	3.2	3.1	2.6	1.0	0.8	0.8	0.6	0.1

a. Draw a scatter plot showing the relationship between x and y.

b. Find the least squares line and draw it on your scatter plot.

c. What concentration would you predict for each of the following values of mud overlay thickness?

i. 4.0 mm^{-1} ii. 3.1 mm^{-1} iii. 5.9 mm^{-1}

4.35 Explain why it can be dangerous to use the least squares line to obtain predictions for x values that are either substantially larger or smaller than those contained in the sample.

4.36 The sales manager of a large company selected a random sample of $n = 10$ sales people and determined for each one the values of $x =$ years of sales experience and $y =$ annual sales (in thousands of dollars). A scatter plot of the resulting (x, y) pairs showed a marked linear pattern.

a. Suppose that the sample correlation coefficient is $r = .75$ and average annual sales is $\bar{y} = 100$. If a particular salesperson is 2 standard deviations above the mean in terms of experience, what would you predict for that person's annual sales?

b. If a particular person whose sales experience is 1.5 standard deviations below average experience is predicted to have an annual sales value that is 1 standard deviation below average annual sales, what is the value of r?

4.37 In some situations y must be zero when x is zero (for example, when $x =$ number of errors and $y =$ cost of correcting errors), and a scatter plot suggests a linear pattern for all $x \geq 0$. Such data can be summarized by finding the best-fitting line which passes through the point (0, 0), i.e., a line with equation $y = bx$, where b is the slope and the y intercept is zero. Applying the principle of least squares, b is chosen to make $\Sigma(y - bx)^2$ as small as possible. It can be shown that the minimizing b value is $b = \Sigma xy/\Sigma x^2$. Suppose that a random sample of $n = 8$ rats is selected, a particular insulin dosage x is given to each one, and the amount of reduction in blood sugar y is observed, resulting in the accompanying data.

x	.1	.2	.2	.3	.3	.4	.5	.5
y	8	25	21	30	38	42	55	60

a. Construct a scatter plot of the data. Does it appear as though a line through (0, 0) would provide a reasonable summary of the relationship?

b. Determine the equation of the best-fitting line that passes through (0, 0) and use it to predict the amount of reduction in blood sugar when dosage is .25.

c. Does the line through (0, 0) provide a better fit to the data (in terms of sum of squared deviations) than the least squares line $y = a + bx$ discussed in this chapter? (*Hint:* You should be able to answer this and give an explanation without doing any computation.)

4.38 Explain why the slope b of the least squares line always has the same sign (positive or negative) as does the sample correlation coefficient r.

4.39 The accompanying data resulted from an experiment in which weld diameter, x, and shear strength, y (lb), were determined for five different spot welds on steel. A scatter plot shows a pronounced linear pattern. With $\Sigma(x - \bar{x})^2 = 1000$ and $\Sigma(x - \bar{x})(y - \bar{y}) = 8577$, the least squares line is $y = -936.22 + 8.577x$.

x	200	210	220	230	240
y	813.7	785.3	960.4	1118.0	1076.2

a. Since 1 lb = .4536 kg, strength observations can be reexpressed in kilograms through multiplication by this conversion factor: new y = .4536(old y). What is the equation of the least squares line when y is expressed in kilograms?

b. More generally, suppose that each y value in a data set consisting of n (x, y) pairs is multiplied by a conversion factor c (which changes the units of measurement for y). What effect does this have on the slope b (i.e., how does the new value of b compare to the value before conversion), on the y intercept a, and on the equation of the least squares line? Verify your conjectures by using the given formulas for b and a. (*Hint:* Replace y by cy and see what happens—and remember, this conversion will affect $\bar{y}$.)

4.4 Assessing the Fit of a Line

When a scatter plot exhibits a pronounced linear pattern, the relationship between the variables x and y can be described approximately by finding a straight line that fits as closely as possible the points in the plot. The most frequently used method for obtaining such a line involves using the principle of least squares as the criterion for measuring fit. The least squares line $y = a + bx$ is the line that results in the smallest possible sum of squared vertical deviations for the (x, y) pairs in the sample.

Once a best-fit line has been obtained, it is natural to ask how effectively the line summarizes the relationship between x and y. In particular, how much of the sample variation in y can be attributed to the combination of an approximate linear relationship and variation in x? Additionally, does the data suggest that the selected line is not the most effective summary of the relationship and that some type of curve might yield a better summary? Are there any "unusual" observations in the sample, observations that either lie very far from the selected line or else have had great influence in determining the line? The most informative quantities for purposes of answering questions such as these and assessing a linear fit are the residuals from the chosen line.

Predicted Values and Residuals

A particular line $y = a + bx$ gives a good fit to the points $(x_1, y_1), \ldots, (x_n, y_n)$ in a scatter plot if the vertical deviations from the line are small. Consider substituting each x value at which an observation was made into the equation to obtain a predicted y value. The resulting n predicted values are just the heights of the line above the x values $x_1, x_2, \ldots, x_n$. Then the n differences between observed and predicted y values are the vertical deviations. The difference is positive when the corresponding point lies above the line and negative when the point lies below the line. The magnitudes of these differences indicate how well the line fits the data. Most or all of the differences are small (negative or positive) when the line provides a good fit, whereas a line that fits poorly results in many large differences.

DEFINITION

The **predicted values** corresponding to sample observations at $x_1, x_2, \ldots, x_n$ based on the line $y = a + bx$ are the n quantities $\hat{y}_1 = a + bx_1, \hat{y}_2 = a + bx_2, \ldots, \hat{y}_n = a + bx_n$. The n **residuals** are the differences $y_1 - \hat{y}_1, y_2 - \hat{y}_2, \ldots, y_n - \hat{y}_n$ between observed and predicted values.

The predicted values and residuals are included in regression output from the standard statistical computer packages.

EXAMPLE 10

The ability of proteins to bind fat is important in improving flavor and texture of meat. The paper "A Simple Turbidimetric Method for Determining the Fat Binding Capacity of Proteins" (*J. Ag. and Food Chem.* (1983): 58–63) proposed the use of regression methods for predicting fat binding capacity (FBC) y from other protein characteristics. The first predictor used in the study was x = surface hydrophobicity, a measure of the extent to which water is not attracted to the protein's surface.

Observations on x and y for $n = 8$ proteins and a scatter plot are given in Figure 16. The predominant pattern in the plot is linear, but it is clear that for any line, some of the residuals will be large. It is easy to check that

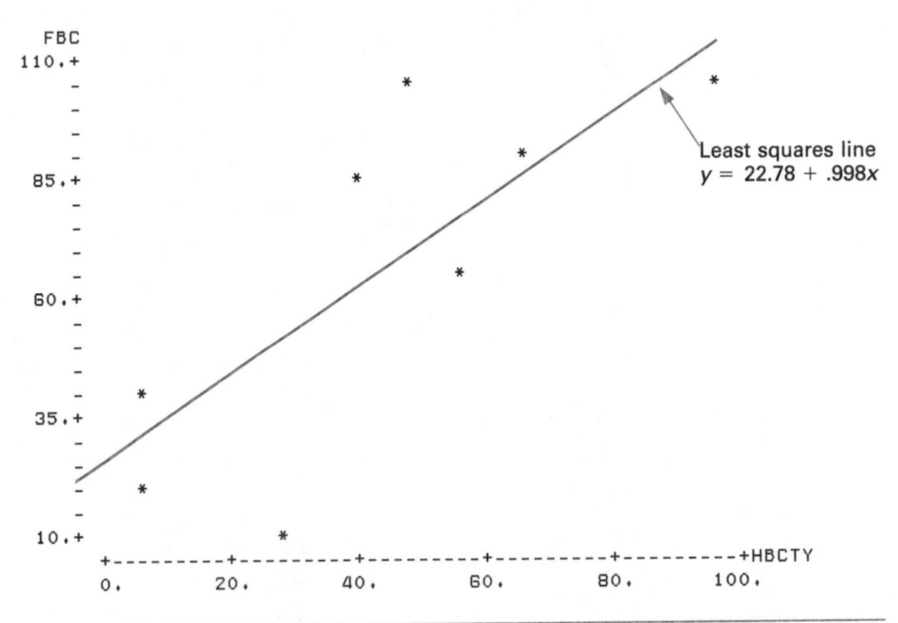

FIGURE 16 DATA, SCATTER PLOT, PREDICTED VALUES, AND RESIDUALS FROM MINITAB FOR EXAMPLE 10

ROW	HBCTY	Y FBC	PRED. Y VALUE	RESIDUAL
1	6.0	37.70	28.77	8.93
2	28.0	10.10	50.71	-40.61
3	95.0	105.90	117.54	-11.64
4	39.0	85.30	61.68	23.62
5	66.0	92.30	88.62	3.68
6	5.0	19.10	27.77	-8.67
7	47.0	105.80	69.66	36.14
8	55.0	66.20	77.64	-11.44

FIGURE 16
(continued)

$\overline{x} = 42.625$, $\overline{y} = 65.3$, $\Sigma(x - \overline{x})(y - \overline{y}) = 6429.80$, and $\Sigma(x - \overline{x})^2 = 6445.875$. The coefficients of the least squares line are then $b = 6429.80/6445.875 = .99750616$ and $a = 65.3 - (.99750616)(42.625) = 22.78129993$, giving $y = 22.78129993 + .99750616x$ as the least squares line. The first predicted value is (since $x_1 = 6.0$) $\hat{y}_1 = 22.78129993 + .99750616(6.0) \approx 28.77$, and the corresponding residual is $y_1 - \hat{y}_1 = 37.70 - 28.77 = 8.93$. The remaining $\hat{y}$'s and residuals appear in the columns headed PRED. Y VALUE and RESIDUAL in Figure 16. There is one very large negative residual, and two positive residuals are quite large. ■

The residuals from the least squares line satisfy one nice property that is useful for checking hand computations: Their sum is (except for rounding effects) 0. The residuals displayed in Figure 16 have a sum of .01. For other methods of finding a best-fit line, this property will not generally hold.

The Coefficient of Determination

Because the least squares line is chosen to minimize the sum of squared residuals, it is natural to use this sum as an indicator of fit.

DEFINITION

The **residual sum of squares,*** denoted by SSResid, is given by

$$\text{SSResid} = \Sigma(y - \hat{y})^2 = (y_1 - \hat{y}_1)^2 + (y_2 - \hat{y}_2)^2 + \cdots + (y_n - \hat{y}_n)^2$$

When the residuals are computed from the least squares line, an alternative formula for residual sum of squares is

$$\text{SSResid} = \Sigma(y - \overline{y})^2 - b\Sigma(x - \overline{x})(y - \overline{y})$$

Clearly, the residual sum of squares, as with any sum of squares, cannot be negative. The smallest possible value is zero, obtained when every point in the sample falls exactly on the line. SSResid will be small when the points fall close to the chosen line and large when they fall far from the line. The computing formula for SSResid is very sensitive to rounding. You should, therefore, use as many digits as possible in the value of b to assure an accurate result.

EXAMPLE 11

The residuals for the surface hydrophobicity–fat binding capacity data of Example 10 were 8.93, −40.61, −11.64, 23.62, 3.68, −8.67, 36.14, and −11.44. The residual sum of squares is then

$$\text{SSResid} = (8.93)^2 + (-40.61)^2 + \cdots + (-11.44)^2 = 3947.996.$$

*Some sources refer to this as *error sum of squares* and denote it by SSE.

Using four digits beyond the decimal point in each residual (e.g., 8.9337 instead of 8.93) yields SSResid = 3947.898. The alternative formula requires $\Sigma(y - \bar{y})^2$, which is 10,361.66. Then

$$\text{SSResid} = 10{,}361.66 - .99750616(6429.80) = 3947.895.$$

Use of MINITAB gives SSResid = 3947.9. From a practical viewpoint, there is little difference between these values. However, if $b = 1.00$ (rather than .99750616) is used, SSResid = 3931.86 from the computational formula. With other data, rounding may produce an even greater discrepancy. ■

The residual sum of squares is sometimes referred to as a measure of **unexplained variation**—the amount of variation in y that cannot be attributed to a linear relationship between x and y (in the sample). Figure 17 shows three different scatter plots with the least squares line superimposed on each one. To emphasize the substantial variability in y in each plot, we have drawn dashed lines from the points over to the corresponding y values on the vertical axis. The line fits all points perfectly in the first plot and SSResid = 0. All y variability in the data set (100%) can be attributed to the fact that x is also varying and that the two variables are linearly related.

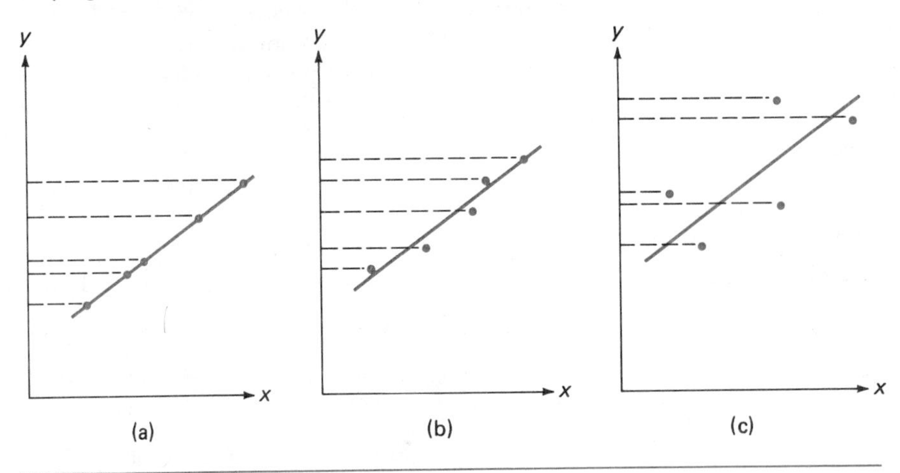

FIGURE 17

EXPLAINING VARIATION IN y BY A LINEAR RELATIONSHIP TO x
(a) SSResid = 0, No Unexplained Variation
(b) SSResid Relatively Small, Little Unexplained Variation
(c) SSResid Relatively Large, Much Unexplained Variation

In Figure 17(b) the least squares line gives a very good—though not perfect—fit, and SSResid is a relatively small number. The variation in x in concert with the approximate linear relationship explains most of the variation in y. On the other hand, SSResid is relatively large for the data pictured in Figure 17(c). There is much variability in y here that cannot be attributed to variation in x and an approximate linear relationship.

The standard measure of how much variation in y is attributable to an approximate linear relationship involves comparing SSResid, the unexplained variation, to a measure of total variability in the sample y's.

Total variation, or **total sum of squares,** denoted by SSTo, is defined by SSTo = $\Sigma(y - \bar{y})^2$. Notice the distinction between SSResid and SSTo. In the former, each predicted value, $\hat{y}$ is subtracted from the corresponding y before squaring and summing, and if all x values are different, all $\hat{y}$ values are also

However, in SSTo the same number ($\bar{y}$) is subtracted from each y before squaring and adding. A very important fact is that for residuals computed from the least squares line, SSResid $\leq$ SSTo. (Were this not the case, unexplained variation could exceed total variation, which would certainly run counter to the intuitive use of this terminology.) This is easily justified by representing SSTo pictorially. Figure 18(a) displays a scatter plot with the least squares line superimposed, and Figure 18(b) shows the horizontal line at height $\bar{y}$ superimposed on the plot. SSResid is the sum of squared vertical deviations about the least squares line, while SSTo is the sum of squared deviations $(y_1 - \bar{y})^2 + \cdots + (y_n - \bar{y})^2$ about the pictured horizontal line. By definition, the sum of squared deviations about the least squares line is smaller than the sum of squared deviations about any other line. Since SSTo is the sum of squared deviations about another line, SSResid $<$ SSTo unless the horizontal line itself is the least squares line, in which case the two values are identical.

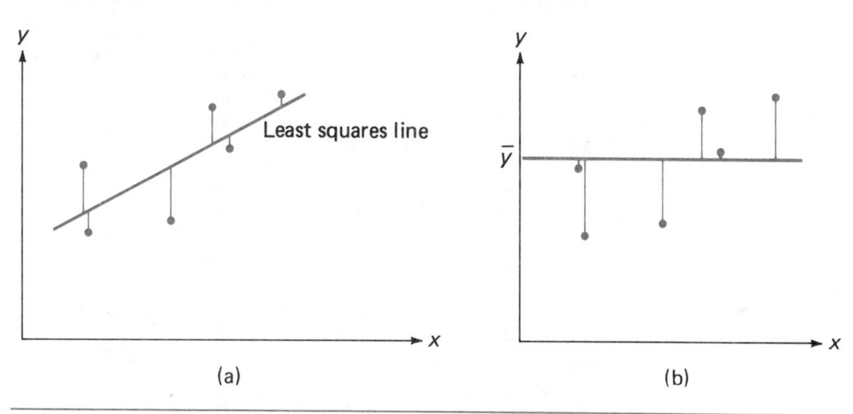

FIGURE 18

GEOMETRIC INTERPRETATION OF SSResid AND SSTo
(a) SSResid = Sum of Squared Deviations from Least Squares Line
(b) SSTo = Sum of Squared Deviations from Horizontal Line at Height $\bar{y}$

The difference SSTo − SSResid can be called **explained variation.** Dividing this by SSTo yields the proportion of total variation explained by or attributable to an approximate linear relationship.

DEFINITION

The **coefficient of determination,** denoted by r^2, is given by

$$r^2 = \frac{\text{SSTo} - \text{SSResid}}{\text{SSTo}} = 1 - \frac{\text{SSResid}}{\text{SSTo}}$$

It is the proportion of variation in y that can be attributed to a linear relationship between x and y in the sample.

When r^2 is multiplied by 100, the result is the percentage of y variation attributable to a linear relationship. Either r^2 or $100r^2$ is an integral part of regression output from any of the standard statistical packages.

EXAMPLE 12

We found for the hydrophobicity–fat binding capacity data of Examples 10 and 11 that SSResid = 3947.9 and SSTo = $\Sigma(y - \bar{y})^2$ = 10,361.66. Then $r^2 = 1 - 3947.9/10{,}361.66 = 1 - .381 = .619$, so 61.9% of the sample variation in FBC is attributable to an approximate linear relationship with hydrophobicity. In many situations 61.9% of variation explained would be quite respectable, but in this particular instance the investigators were not satisfied, so they proceeded to an analysis involving more than a single predictor variable. ■

EXAMPLE 13

Forest managers are increasingly concerned about the damage done to natural populations when forests are clear-cut. Woodpeckers are a valuable forest asset both because they provide nest and roost holes for other animals and birds and because they prey on many forest insect pests. The paper "Artificial Trees as a Cavity Substrate for Woodpeckers" (*J. Wildlife Management* (1983):790–98) reported on a study of woodpecker behavior when provided with polystyrene cylinders as an alternative roost and nest cavity substrate. We give selected values of x = ambient temperature (°C) and y = cavity depth (cm); these values were read from a scatter plot that appeared in the paper. Our plot (Figure 19) as well as the original plot give evidence of a strong negative linear relationship between x and y.

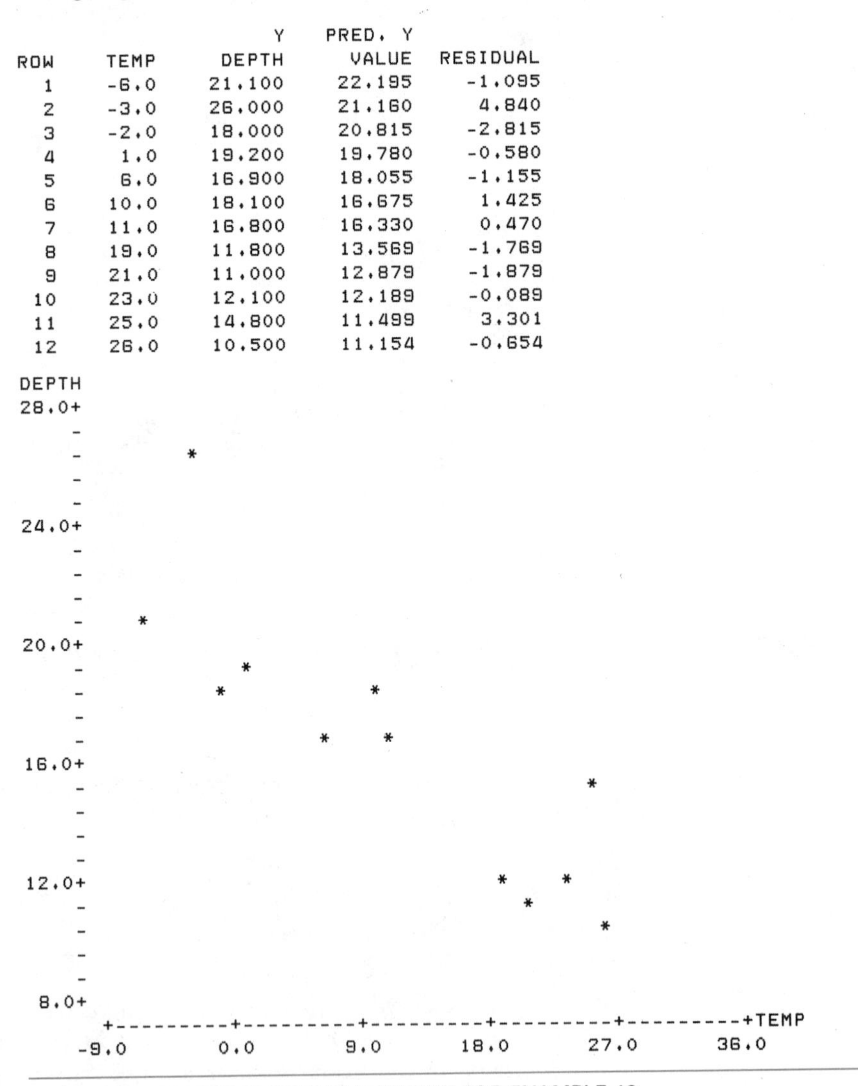

ROW	TEMP	Y DEPTH	PRED. Y VALUE	RESIDUAL
1	-6.0	21.100	22.195	-1.095
2	-3.0	26.000	21.160	4.840
3	-2.0	18.000	20.815	-2.815
4	1.0	19.200	19.780	-0.580
5	6.0	16.900	18.055	-1.155
6	10.0	18.100	16.675	1.425
7	11.0	16.800	16.330	0.470
8	19.0	11.800	13.569	-1.769
9	21.0	11.000	12.879	-1.879
10	23.0	12.100	12.189	-0.089
11	25.0	14.800	11.499	3.301
12	26.0	10.500	11.154	-0.654

FIGURE 19 DATA AND SCATTER PLOT FROM MINITAB FOR EXAMPLE 13

Summary values include $\Sigma x = 131$ and $\Sigma y = 196.3$, so $\bar{x} = 10.916667$ and $\bar{y} = 16.358333$. Then using the computational formulas, $\Sigma(x - \bar{x})^2 = 1508.916579$, $\Sigma(x - \bar{x})(y - \bar{y}) = -520.641688$, and $\Sigma(y - \bar{y})^2 = 234.109298$. The least squares coefficients are

$$b = \frac{-520.641688}{1508.916579} = -.345043$$

$$a = 16.358333 - (-.345043)(10.916667) = 16.358333 + 3.766720$$
$$= 20.125053$$

The least squares line has equation $y = 20.125 - .345x$. In addition,

$$\begin{aligned}\text{SSResid} &= \Sigma(y - \bar{y})^2 - b\Sigma(x - \bar{x})(y - \bar{y}) \\ &= 234.109298 - (-.345043)(-520.641688) \\ &= 234.109298 - 179.643770 \\ &= 54.465528\end{aligned}$$

so

$$r^2 = 1 - \frac{54.465}{234.109} = 1 - .233 = .767$$

Roughly 76.7% of the variation in observed cavity depth can be explained by an approximate linear relation between cavity depth and ambient temperature. ■

The use of r to denote Pearson's correlation coefficient and r^2 to denote the coefficient of determination suggests that the two quantities are related in the obvious way. If, for example, $r = .9$ for a particular data set, then a line fit using the principle of least squares would explain $100r^2 = 100(.9)^2 = 81\%$ of the variation in y through an approximate linear relation with x. If $r^2 = .64$, this implies that either $r = .8$ or $r = -.8$—the former if the scatter plot slopes upward (b positive) and the latter if the plot tilts down (b negative). This relationship between r and r^2 is what motivated our earlier rule of thumb for describing a linear relationship as strong, moderate, or weak. The value $r = .5$ may seem substantial, but only 25% of y variation is attributable to an approximate linear relationship. In Example 13 we calculated $r^2 = .767$ and b was negative, so $r = -\sqrt{.767} = -.876$. This is indeed indicative of a strong negative linear relationship between temperature and cavity depth.

The coefficient of determination measures the extent of variation about the best fit line *relative* to overall variation in y. A high value of r^2 does not by itself promise that the deviations from the line are small in an absolute sense. A typical observation could deviate from the line by quite a bit, yet these deviations might still be small relative to overall y variation. Recall that in Chapter 3 the sample standard deviation $s = \sqrt{\Sigma(x - \bar{x})^2/(n - 1)}$ was used as a measure of variability in a single sample; roughly speaking, s is the typical amount by which a sample observation deviates from the mean. There is an analogous measure of variability when a line is fit by least squares.

DEFINITION	The **standard deviation about the least squares line** is given by $s_e = \sqrt{\text{SSResid}/(n - 2)}$.

Roughly speaking, s_e is the typical amount by which an observation deviates from the least squares line. The subscript e serves to distinguish s_e from our earlier s based on a sample of x observations. Notice that SSResid $= \Sigma(y - \hat{y})^2$ plays the same role here that the sum of squared deviations $\Sigma(x - \bar{x})^2$ played in defining s. We shall see in Chapter 11 that the choice of e as a subscript is a natural outgrowth of notation used in the simple linear regression model, and a justification is given there for dividing SSResid by $n - 2$ rather than $n - 1$.

EXAMPLE 14

Recall the context of Example 6, in which the variables under study were x = commuting distance and y = commuting time. We give three different data sets, each of which might arise during such an investigation. The corresponding scatter plots appear in Figure 20.

Sample	1		2		3	
	x	y	x	y	x	y
	15	42	5	16	5	8
	16	35	10	32	10	16
	17	45	15	44	15	22
	18	42	20	45	20	23
	19	49	25	63	25	31
	20	46	50	115	50	60
$\Sigma(x - \bar{x})^2$	17.50		1270.8333		1270.8333	
$\Sigma(x - \bar{x})(y - \bar{y})$	29.50		2722.5		1431.6667	
b	1.685714		2.142295		1.126557	
a	13.666672		7.868852		3.196729	
SSTo $= \Sigma(y - \bar{y})^2$	114.83		5897.5		1627.33	
SSResid	65.10		65.10		14.48	
r^2	.433		.989		.991	
s_e	4.03		4.03		1.90	

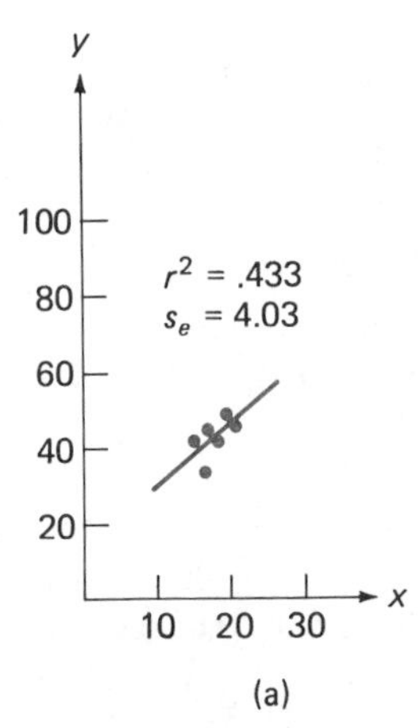

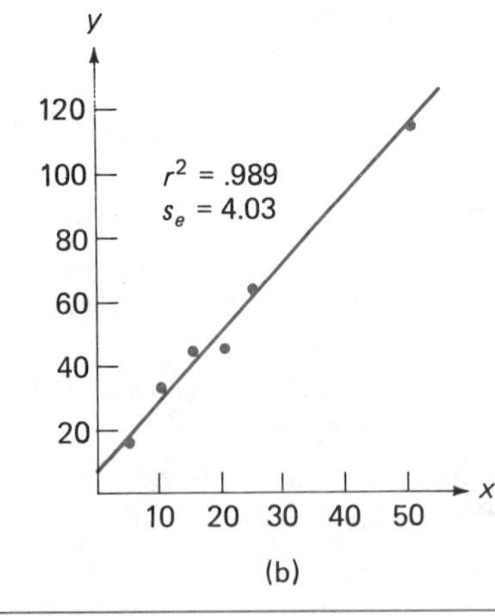

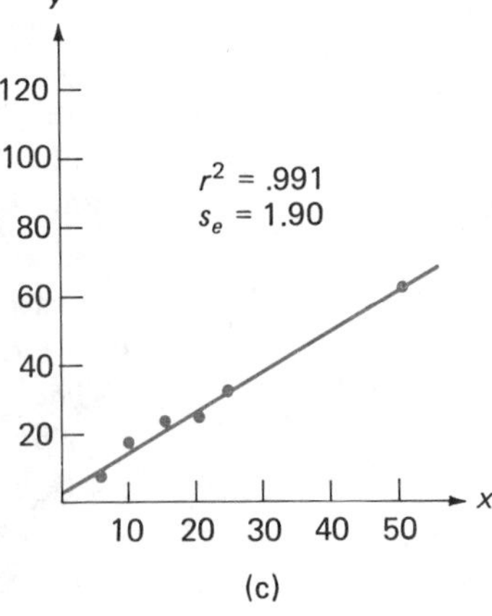

FIGURE 20

SCATTER PLOTS FOR EXAMPLE 14
(a) Sample 1
(b) Sample 2
(c) Sample 3

For sample 1, a rather small proportion of variation in y can be attributed to an approximate linear relationship, and a typical deviation from the least squares line is roughly 4. The amount of variability about the line for sample 2 is the same as for sample 1 but the value of r^2 is much much higher because y variation is much greater overall in sample 2 than in sample 1. Sample 3 yields roughly the same high value of r^2 as does sample 2, but the typical deviation from the line for sample 3 is only half that for sample 2. A complete picture of variation requires that both r^2 and s_e be computed. ■

Plotting the Residuals (Optional)

A plot of the (x, residual) pairs, called a **residual plot,** is often helpful in identifying unusual or highly influential observations and in revealing patterns in the data that may suggest how an improved fit can be achieved. The points in this plot are $(x_1, y_1 - \hat{y}_1), (x_2, y_2 - \hat{y}_2), \ldots, (x_n, y_n - \hat{y}_n)$. A desirable plot is one that exhibits no particular pattern (such as curvature or much greater spread in one part of the plot than in another part) and has no point that is far removed from all others. A point falling far above or below the horizontal line at height zero corresponds to a large residual, which may indicate some type of unusual behavior, such as a recording error, nonstandard experimental condition, or atypical experimental subject. A point whose x value differs greatly from others in the data set may have exerted excessive influence in determining the fitted line. One method for assessing the impact of such an isolated point on the fit is to delete it from the data set and then recompute the best-fit line and various other quantities. Substantial changes in the equation, predicted values, r^2, and s_e warn of instability in the data. More information may then be needed before reliable conclusions can be drawn.

EXAMPLE 15

In Example 5 we presented $n = 9$ observations on x = Quetelet index (a measure of body build) and y = dietary energy density. The least squares line is $y = -.898457 + .00733014x$, $r^2 = .433$, and $s_e = .148$. The predicted values and residuals are displayed in the accompanying table, and both a scatter plot and residual plot appear in Figure 21.

x	y	$\hat{y} = -.898 + .00733x$	Residual $= y - \hat{y}$	Points in the Residual Plot
221	.67	.722	−.052	(221, −.052)
228	.86	.773	.087	(228, .087)
223	.78	.736	.044	(223, .044)
211	.54	.648	−.108	(211, −.108)
231	.91	.795	.115	(231, .115)
215	.44	.678	−.238	(215, −.238)
224	.90	.744	.156	(224, .156)
233	.94	.810	.130	(233, .130)
268	.93	1.066	−.136	(268, −.136)

There is an obvious pattern in the residual plot—the points on the extremes lie below the zero line, whereas the points in the middle are above this line. The primary reason for this is the presence of the observation (268, .93). Deleting it and using the remaining eight observations results in least squares line $y = -4.226 + .02249x$ with $r^2 = .834$ and $s_e = .0822$. Clearly, a, b, r^2, and s_e all change substantially as a result of this deletion. Notice, though, that the residual −.136 corresponding to this influential observation is not the

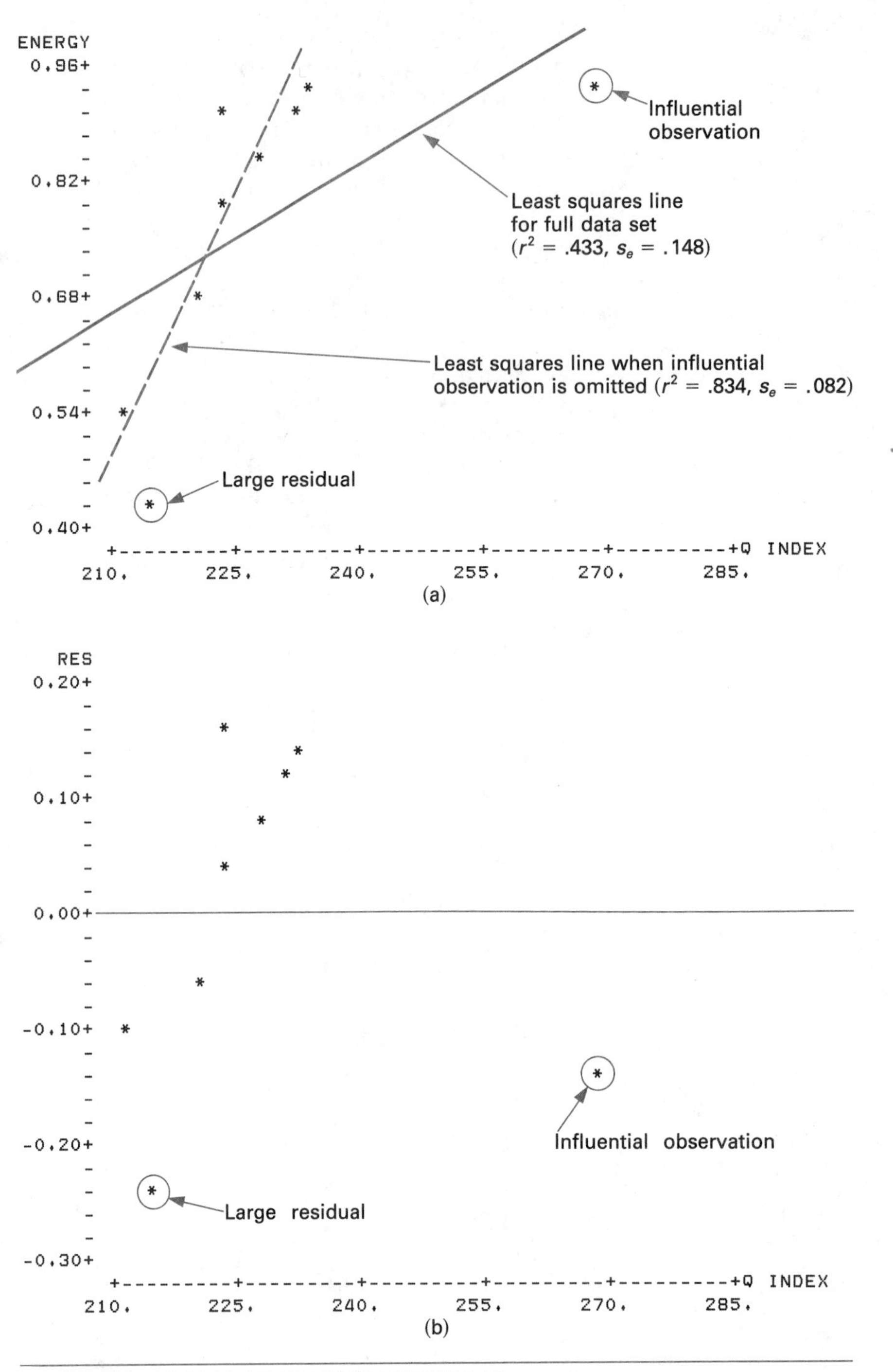

FIGURE 21

PLOTS FROM MINITAB FOR DATA IN EXAMPLE 15
(a) Scatter Plot
(b) Residual Plot

largest residual in the data set. In trying to make the squared deviations small, the least squares line is pulled very far toward the discrepant point. Either this point represents a very unusual departure from a strong approximate linear relationship, or else it suggests the possibility that y may begin to decrease as x continues to increase (a curved relationship). More data is required before a judgment can be made.

Deletion of the observation corresponding to the largest residual (−.238) also changes the value of the important summary quantities, but these changes are not profound. For example, the new equation is $y = -.460 + .00555x$, and the predicted values do not change dramatically. This observation does not appear to be all that unusual. ■

Looking at a residual plot after fitting a line amounts to examining y after removing any linear dependence on x. This can sometimes more clearly bring out the existence of a nonlinear relationship.

EXAMPLE 16

Consider the accompanying data on x = height (in) and y = average weight (lb) for American females aged 30–39 (taken from the 1984 edition of *The World Almanac and Book of Facts*). The scatter plot displayed in Figure 22(a) appears quite straight. However, when the residuals from the least squares line ($y = -98.23 + 3.596x$) are plotted, substantial curvature is apparent (even though $r^2 \approx .99$). It is not accurate to say that weight increases in direct proportion to height (linearly with height). Instead, average weight increases somewhat more rapidly in the range of relatively large heights than it does for relatively small heights.

x	58	59	60	61	62	63	64	65
y	113	115	118	121	124	128	131	134

x	66	67	68	69	70	71	72
y	137	141	145	150	153	159	164

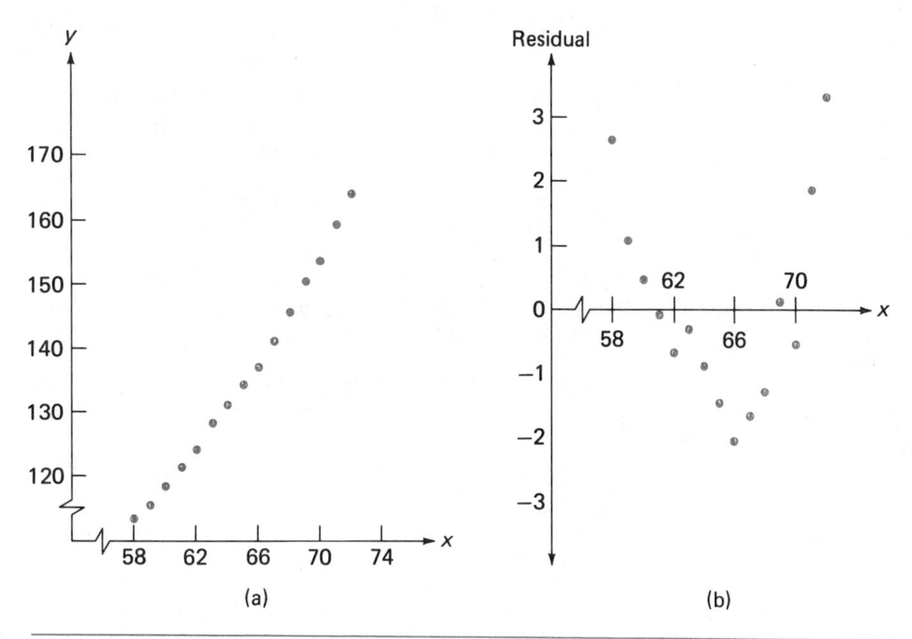

FIGURE 22

PLOTS FOR DATA FROM EXAMPLE 16
(a) Scatter Plot
(b) Residual Plot

■

Resistant Lines

The least squares line can be greatly affected by the presence of even a single observation that shows large discrepancy in the x or y direction from the rest of the data. When the data set contains such unusual observations, it is desirable to have a method for obtaining a summarizing line that is resistant to the influence of these stray values. In recent years many methods for obtaining a resistant (or robust) line have been proposed. You should consult a statistician or a book on exploratory data analysis to obtain information about such methods.

EXERCISES

4.40 The article "Effects of Gamma Radiation on Juvenile and Mature Cuttings of Quaking Aspen" (*Forest Sci.* (1967):240–45) reported the accompanying data on x = exposure time to radiation (kR/16h) and y = dry weight of roots (mg $\times 10^{-1}$). The least squares line $y = 126.6 - 6.65x$ was used to obtain the predicted values ($\hat{y}$) given. Compute the residuals for this data set.

x	0	2	4	6	8
y	110	123	119	86	62
Predicted y	127	113.7	100.4	87.1	73.8

4.41 The accompanying data on fish survival and ammonia concentration was first given in Exercise 4.4. With x = ammonia exposure and y = percent survival, the least squares line is $y = 100.79 - .78x$.

x	10	10	20	20	25	27	27	31	50
y	85	92	85	96	87	80	90	59	62

a. Compute the predicted values and the residuals. Are there any unusually large residuals? What is the sum of the residuals?

b. Use the residuals in (a) to compute SSResid.

c. Would the line $y = 100 - 1x$ have a larger or smaller SSResid than the line $y = 100.79 - .78x$? Explain your answer.

4.42 The accompanying data on x = R&D expenditures and y = growth rate was first presented in Exercise 4.13. The least squares line for this data is $y = .29 + .0006x$. Compute the residuals and SSResid.

x	2024	5038	905	3572	1157	327	378	191
y	1.90	3.96	2.44	.88	.37	−.90	.49	1.01

4.43 Data on x = soil pH and y = Cl^- ion retention (mL/100 g) appeared in the paper "Single Equilibration Method for Determination of Cation and Anion Retention by Variable Charge Soils" (*Soil Sci. Plant Nutr.* (1984):71-76).

x	6.15	6.11	5.88	6.45	5.80	6.06	5.83	6.33	7.35
y	0.14	0.37	1.47	1.12	2.08	1.79	3.18	2.15	0.51

x	8.18	7.69	7.29	6.53	5.01	5.34	6.19	5.81
y	0.32	0.76	2.13	2.75	6.69	5.59	2.87	4.22

Summary quantities are:

$$n = 17 \quad \Sigma(x - \bar{x})^2 = 11.179 \quad \Sigma(x - \bar{x})(y - \bar{y}) = -16.905$$

a. Determine the least squares line.

b. Use the least squares line to compute the residuals and SSResid. Are there any unusually large residuals?

c. Plot the residuals versus the observed x values. Does the residual plot suggest that the relationship between x and y is not linear?

4.44 Anthropologists often study soil composition for clues as to how the land was used during different time periods. The accompanying data on x = soil depth (cm) and y = percent montmorillonite in the soil was taken from a scatter plot in the paper "Ancient Maya Drained Field Agriculture: Its Possible Application Today in the New River Floodplain, Belize, C.A." (*Ag. Ecosystems and Environment* (1984): 67–84).

x	40	50	60	70	80	90	100
y	58	34	32	30	28	27	22

a. Draw a scatter plot of y versus x.

b. Find the least squares line for this data set and draw it on your scatter plot.

c. Based on the scatter plot and the least squares line, do you think that there are any large residuals? Explain.

d. Compute the residuals and SSResid for this data set.

4.45 There have been numerous studies on the effects of radiation. Data on the relationship between degree of exposure to ^{242}Cm alpha particles (x) and the percentage of exposed cells without aberrations (y) appeared in the paper "Chromosome Aberrations Induced in Human Lymphocytes by D-T Neutrons" (*Radiation Research* (1984):561–73). Find the least squares line for this data and use it to compute the residuals and SSResid.

x	.106	.193	.511	.527	1.08	1.62	1.73	2.36	2.72	3.12	3.88	4.18
y	98	95	87	85	75	72	64	55	44	41	37	40

4.46 Data on advertising expenditures and sales for six brands of cereal was given in Exercise 4.30 of the previous section. The computed values for SSResid and SSTo for this data set are 1.178 and 5.485, respectively. Use this information to compute r^2. How would you interpret this value?

4.47 The accompanying data is representative of that appearing in the paper "Predicting the Optimum Harvest Dates for Apples Using Temperature and Full Bloom Records" (*J. Horticultural Sci.* (1983):37–44), with x representing the number of days from March 31 to full bloom (a measure of the length of the growing season) and y representing the number of days from full bloom to harvest for Orange Pippin apples. What proportion of the variation in the number of days from full bloom to harvest can be attributed to a linear relationship between the number of days from full bloom to harvest and the length of the growing season?

x	31	39	39	47	44	44	40	45	45	40	57
y	142	142	139	138	133	134	135	129	132	131	127

4.48 Compute r^2 and s_e for the pH-ion retention data given in Exercise 4.43. How would you interpret these values?

4.49 Data on x = soil depth and y = percent montmorillonite in the soil was given in Exercise 4.44. What are the values of r^2 and s_e for this data set? How would you interpret these values? Do they suggest that a straight line provides a good description of the relationship between x and y?

4.50 Use the radiation data given in Exercise 4.45 to determine the proportion of the variability in the percentage of cells without aberrations that can be attributed to a linear relationship between the degree of exposure and the percent of cells without aberrations.

4.51 **a.** Is it possible that both r^2 and s_e could be large for a bivariate data set? Explain. (A picture might be helpful.)

b. Is it possible that a bivariate data set could yield values of r^2 and s_e that are both small? Explain. (Again, a picture might be helpful.)

c. Explain why it is desirable to have r^2 large and s_e small if the relationship between two variables x and y is to be described using a straight line.

4.52 Construct a residual plot for the soil data of Exercise 4.44. Does the residual plot suggest that the relationship between x and y is nonlinear? Do you think there are any particularly influential observations? Explain.

4.53 Use the radiation data and computed residuals of Exercise 4.45 to construct a residual plot. How would you interpret this plot? Does it provide any useful information about the relationship between dose and the percent of cells without aberrations?

4.54 A scatter plot appearing in the article "Thermal Conductivity of Polyethylene: The Effects of Crystal Size, Density, and Orientation on the Thermal Conductivity" (*Polymer Eng. and Sci.* (1972):204–08) suggests the existence of a relationship between y = thermal conductivity and x, a measure of lamellar thickness. In the accompanying data, there is an x value that is much larger than the other x values.

x	240	410	460	490	520	590	745	8300
y	12.0	14.7	14.7	15.2	15.2	15.6	16.0	18.1

a. The least squares line for the given data set is $y = 14.5 + .0004x$ and the computed values of r^2 and s_e are .533 and 1.246, respectively. Construct a residual plot.

b. The (x, y) pair (8300, 18.1) looks like it might be an influential observation. Try omitting it from the data set and computing the least squares line for the remaining data. Do the values of the intercept and slope differ much from those computed using the full data set?

c. Construct a residual plot using the least squares line in (b) and the data set that does not include (8300, 18.1) How does its appearance compare with the residual plot in (a)?

d. Again using the data set and line resulting from omission of (8300, 18.1), compute r^2 and s_e and compare these to the values reported in (a) for the full data set.

4.55 **a.** Show that the sum of the residuals from the least squares line, $\Sigma[y - (a + bx)]$, is zero. (*Hint:* Substitute $a = \bar{y} - b\bar{x}$ and then use facts about $\Sigma(x - \bar{x})$ and $\Sigma(y - \bar{y})$.)

b. Let the residuals be denoted by $e_1, e_2, \ldots, e_n$ and consider fitting a line to the (x, e) pairs—i.e., to the points in the residual plot—using the method of least squares. Then

$$\text{slope of the least squares line} = \frac{\Sigma(x - \bar{x})(e - \bar{e})}{\Sigma(x - \bar{x})^2}$$

From (a), $\bar{e} = 0$, so $e - \bar{e} = e = y - (a + bx) = y - \bar{y} - b(x - \bar{x})$. Use this to show that the slope in the given expression is zero, so that the residual plot has no tilt (though it may still have a nonlinear pattern if a straight-line fit is not appropriate).

4.56 Some straightforward but slightly tedious algebra shows that SSResid = $(1 - r^2)\Sigma(y - \bar{y})^2$, from which it follows that

$$s_e = \sqrt{\frac{(n-1)}{(n-2)}}\sqrt{1 - r^2}\, s_y$$

Unless n is quite small, $(n - 1)/(n - 2) \approx 1$, so $s_e \approx \sqrt{1 - r^2}\, s_y$.

a. For what value of r is s_e as large as s_y? What is the least squares line in this case?

b. For what values of r will s_e be much smaller than s_y?

c. A study by the Berkeley Institute of Human Development (see the book *Statistics* by Freedman, Pisani, and Purves listed in the Chapter 1 references) reported the following summary data for a sample of $n = 66$ California boys:

   $r \approx .80$
   At age 6, average height $\approx$ 46 in., standard deviation $\approx$ 1.7 in.
   At age 18, average height $\approx$ 70 in., standard deviation $\approx$ 2.5 in.

   What would s_e be for the least squares line used to predict 18-year-old height from 6-year-old height?

d. Referring to (c), suppose you wanted to predict the past value of 6-year-old height from knowledge of 18-year-old height. What is the equation for the appropriate least squares line, and what is the corresponding value of s_e?

4.5 Transforming Data to Straighten a Plot and Fit a Curve (Optional)

When the points in a scatter plot exhibit a linear pattern, it is relatively easy to find a line that gives a good fit to the points in the plot. Such a line then describes an approximate relationship between x and y. A linear relationship is easy to interpret (for example, the impact on y of a specified change in x is easily assessed), departures from the line are easily detected, and using the line to predict y from knowledge of x is straightforward. Often, though, a scatter plot shows a strong curved pattern rather than a straight-line pattern. In this case, finding a "nice" curve that fits the observed data well may not be very easy, and departures from the selected curve may not be as apparent as departures from a straight line would be.

An alternative to fitting a curve is to find a way to transform x values and/or y values so that a scatter plot of the transformed data has a linear appearance. Sometimes a transformation is suggested by a theoretical model that relates y to x, but frequently the investigator would like the data to suggest an appropriate transformation. In this latter case, several transformations may have to be tried to find one which works. A type of transformation that statisticians have found useful for straightening a plot is a **power transformation.** A power (exponent) is first selected, and each original value is raised to that power to obtain the corresponding transformed value. Table 2 displays a ladder of the most frequently used power tranformations. The power 1 corresponds to no transformation at all. Using the power 0 would transform every value to 1, which is certainly not informative, so statisticians use the logarithmic transformation in its place. Other powers intermediate to or more extreme than those listed can be used, of course, but they are less frequently needed than those on the ladder. Notice that these are the same transformations suggested in Chapter 2 for transforming a single data set to obtain a more symmetric distribution.

Figure 23 is designed to suggest where on the ladder we should go to find an appropriate transformation. The four curved segments labeled 1, 2, 3, and

TABLE 2

POWER TRANSFORMATION LADDER: TRANSFORMED VALUE = (ORIGINAL VALUE)^{POWER}

Power	Transformed Value	Name
3	(Original value)^3	Cube
2	(Original value)^2	Square
1	Original value	No transformation
$\frac{1}{2}$	$\sqrt{\text{Original value}}$	Square root
$\frac{1}{3}$	$\sqrt[3]{\text{Original value}}$	Cube root
0	Log (original value)	Logarithm
−1	1/(original value)	Reciprocal

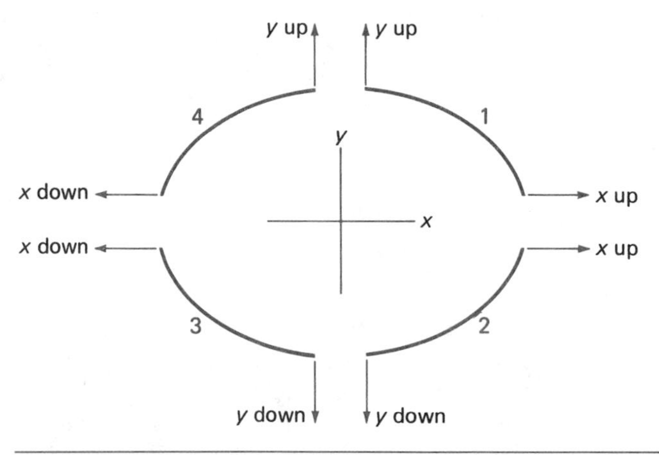

FIGURE 23

SCATTER PLOT SHAPES AND WHERE TO GO ON THE TRANSFORMATION LADDER TO STRAIGHTEN THE PLOT

4 represent shapes of curved scatter plots that are commonly encountered. Suppose that a scatter plot looks like the curve labeled 1. Then to straighten the plot, we should use a power on x that is up the ladder from the no-transformation row (x^2 or x^3) and/or a power on y that is also up the ladder from the power 1. Thus we might be led to squaring each x value, cubing each y, and plotting the transformed pairs. If the curvature looks like curved segment 2, a power up the ladder from no transformation for x and/or a power down the ladder for y (e.g., $\sqrt{y}$ or log (y)) should be used.

EXAMPLE 17

In many parts of the world a typical diet consists mainly of cereals and grains, and many individuals suffer from a substantial iron deficiency. The paper "The Effects of Organic Acids, Phytates, and Polyphenols on the Absorption of Iron from Vegetables" (*British J. Nutrition* (1983): 331–42) reported the accompanying data on x = proportion of iron absorbed when a particular vegetable was consumed and y = polyphenol content of the vegetable (mg/g). The scatter plot of the data in Figure 24(a) shows a clear curved pattern, which resembles the curved segment 3 in Figure 23. This suggests that x and/or y should be transformed by a power down the ladder from 1. The authors of the paper applied a square-root transformation to each variable. The resulting scatter plot in Figure 24(b) is reasonably straight.

Vegetable	x	y	$\sqrt{x}$	$\sqrt{y}$
Wheat germ	.007	6.4	.084	2.53
Aubergine	.007	3.0	.084	1.73
Butter beans	.012	2.9	.110	1.70
Spinach	.014	5.8	.118	2.41
Brown lentils	.024	5.0	.155	2.24
Beetroot greens	.024	4.3	.155	2.07
Green lentils	.032	3.4	.179	1.84
Carrot	.096	.7	.310	.84
Potato	.115	.2	.339	.45
Beetroot	.185	1.5	.430	1.22
Pumpkin	.206	.1	.454	.32
Tomato	.224	.3	.473	.55
Broccoli	.260	.4	.510	.63
Cauliflower	.263	.7	.513	.84
Cabbage	.320	.1	.566	.32
Turnip	.327	.3	.572	.55
Saurkraut	.327	.2	.572	.45

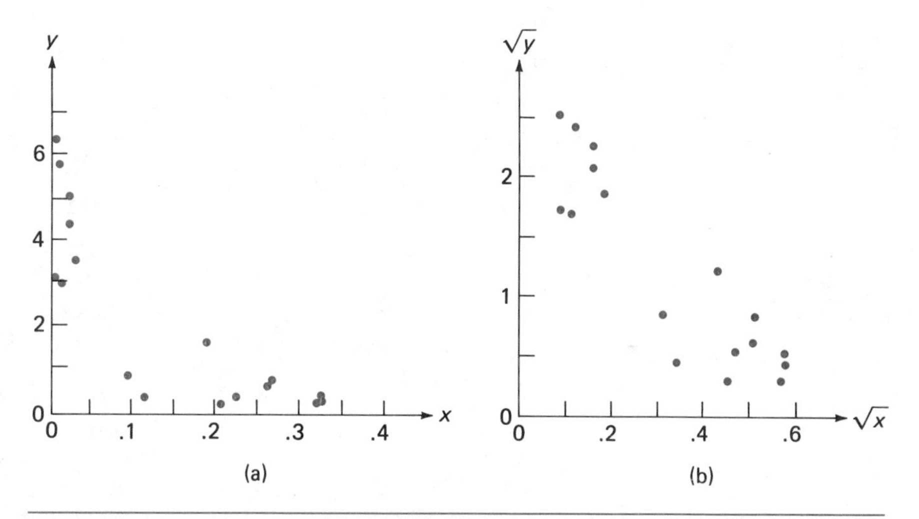

FIGURE 24 SCATTER PLOT OF DATA FROM EXAMPLE 17
(a) Original Data
(b) Square-Root Transformed Data

EXAMPLE 18

The article "Reduction in Soluble Protein and Chlorophyll Contents in a Few Plants as Indicators of Automobile Exhaust Pollution" (*Intl. J. of Environ. Studies* (1983):239–44) reported the accompanying data on x = distance from a highway (m) and y = lead content of soil at that distance (ppm).

x	.3	1	5	10	20	25
y	62.75	37.51	29.70	20.71	17.65	15.41

x	25	30	40	50	75	100
y	14.15	13.50	12.11	11.40	10.85	10.85

Figure 25(a) displays a scatter plot of the data. The curvature in the plot is very pronounced. The authors of the paper did not suggest a transformation for straightening the plot. Since the curvature is like that of the curved segment labeled 3 in Figure 23, we tried a number of transformations down the ladder for both x and y. The plot in Figure 25(b) is the result of transforming x by logarithms and leaving y untransformed. Figure 25(c) displays the plot resulting from a log transformation of both x and y. These latter two plots are quite straight, and there is little reason for preferring one to the other.

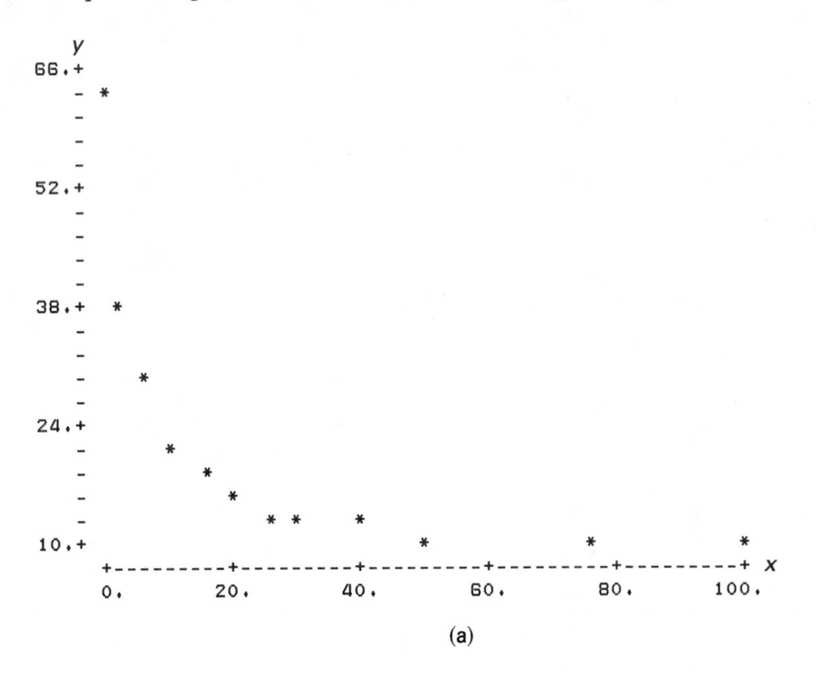

(a)

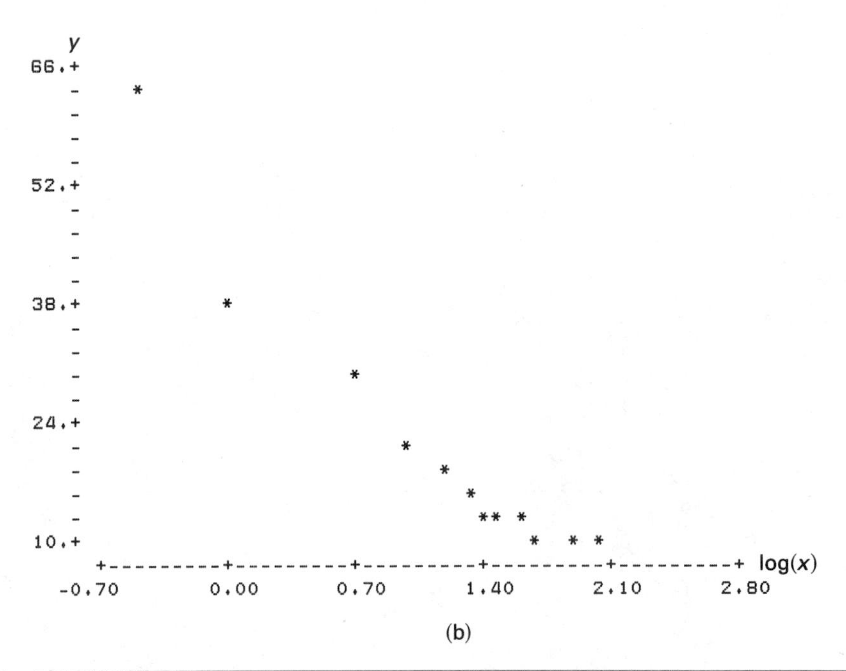

(b)

FIGURE 25 MINITAB SCATTER PLOTS FOR ORIGINAL AND TRANSFORMED DATA FROM EXAMPLE 18

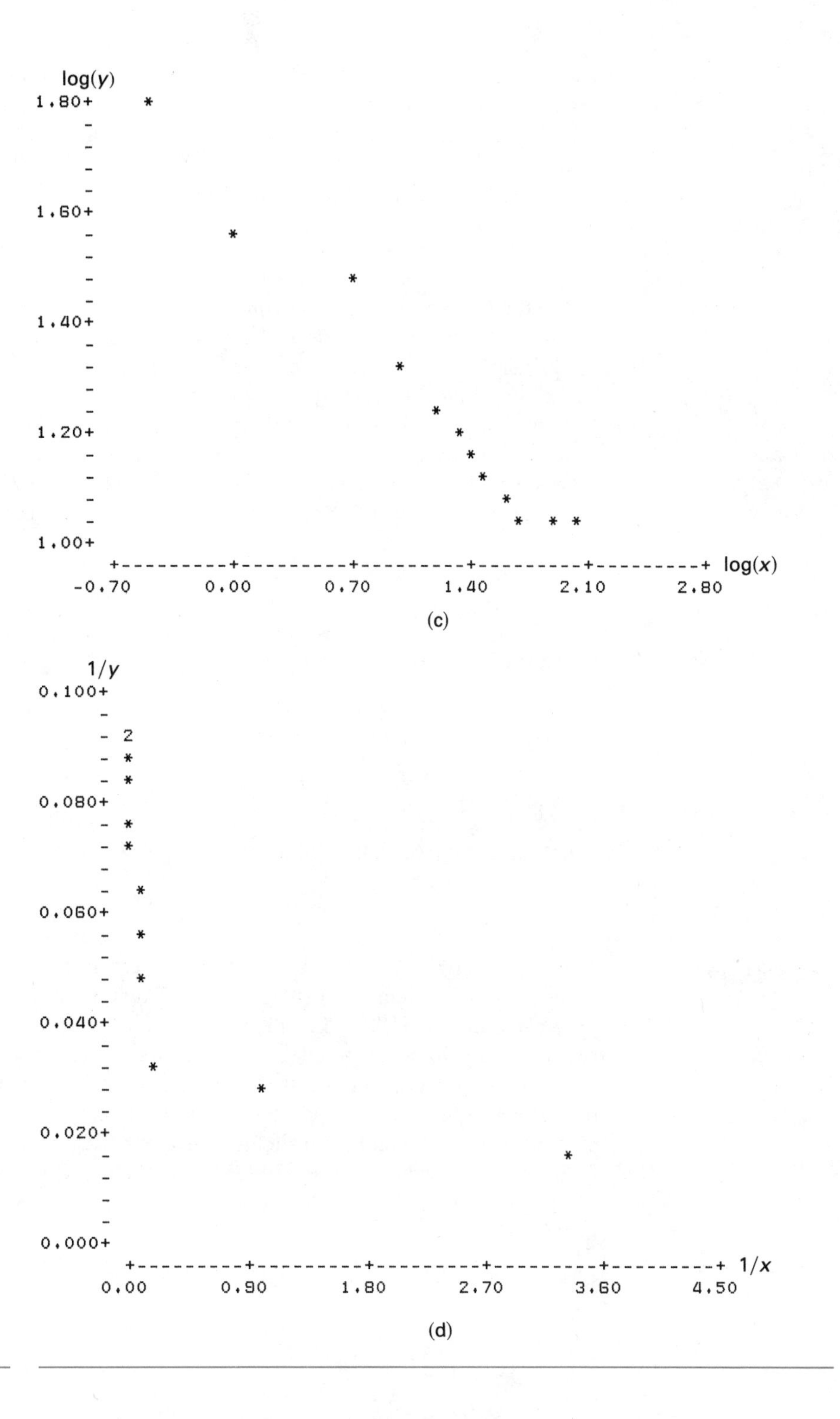

FIGURE 25
(continued)

We also went further down the ladder and tried a reciprocal transformation for both x and y ($1/y$ versus $1/x$). The resulting plot, shown in Figure 25(d), is quite curved, indicating that we have gone too far down the ladder. You might think that going too far would produce the curvature of segment

1 in Figure 23, which is opposite what we started with (segment 3). However, the plot of $1/y$ versus $1/x$ actually looks like segment 3 because taking reciprocals reverses the ordering of numbers: $2 < 4$, but $\frac{1}{2} > \frac{1}{4}$. Some authors have suggested using $-1/x$ and $-1/y$, which preserves order, and a plot of $-1/y$ versus $-1/x$ does indeed look like segment 1. ■

Fitting a Curve

When the scatter plot shows a curved pattern, a transformation of x and/or y can often straighten the plot. Once such a transformation has been identified, the principle of least squares can be used to fit a line to the transformed data. After obtaining the least squares line, the transformation can be reversed to yield a curved relationship between the original variables. For example, suppose that a plot of $\log(y)$ versus $\log(x)$ is reasonably straight and that the least squares line for the transformed data has y intercept 2 and slope -3. Then $\log(y)$ and $\log(x)$ are approximately related by the equation $\log(y) = 2 - 3\log(x)$. To reverse this transformation, we now take the antilog of each side of the equation:

$$(10)^{\log(y)} = (10)^{2 - 3[\log(x)]}$$

Using the properties

$$(10)^{\log(y)} = y$$
$$(10)^{2 - 3[\log(x)]} = (10)^2(10)^{-3[\log(x)]}$$
$$(10)^{-3[\log(x)]} = x^{-3}$$

the resulting equation is $y = 100x^{-3}$ (i.e., $y = 100/x^3$). The graph of this equation is a curve shaped like the arc labeled 3 in Figure 23.

EXAMPLE 19

The problem of soil erosion is faced by farmers all over the world. The paper "Soil Erosion by Wind from Bare Sandy Plains in Western Rajasthan, India" (*J. Arid Environ.* (1981):15–20) reported on a study of the relationship between wind velocity x (km/h) and soil erosion y (kg/day) in a very dry environment, where erosion control is especially important. We present selected data extracted from the paper. Figure 26(a) displays a scatter plot of the data. Comparison of the plot with the arcs of Figure 23 suggests moving down the ladder in y and/or up the ladder in x. Figure 26(b) displays a plot of $y' = \log(y)$ versus x, which is quite straight (so transformation of x is not necessary).

Observation	x	y	$y' = \log(y)$	Observation	x	y	$y' = \log(y)$
1	13.5	5	.6990	8	21	140	2.1461
2	13.5	15	1.1761	9	22	75	1.8751
3	14	35	1.5441	10	23	125	2.0969
4	15	25	1.3979	11	25	190	2.2788
5	17.5	25	1.3979	12	25	300	2.4771
6	19	70	1.8451	13	26	240	2.3802
7	20	80	1.9031	14	27	315	2.4983

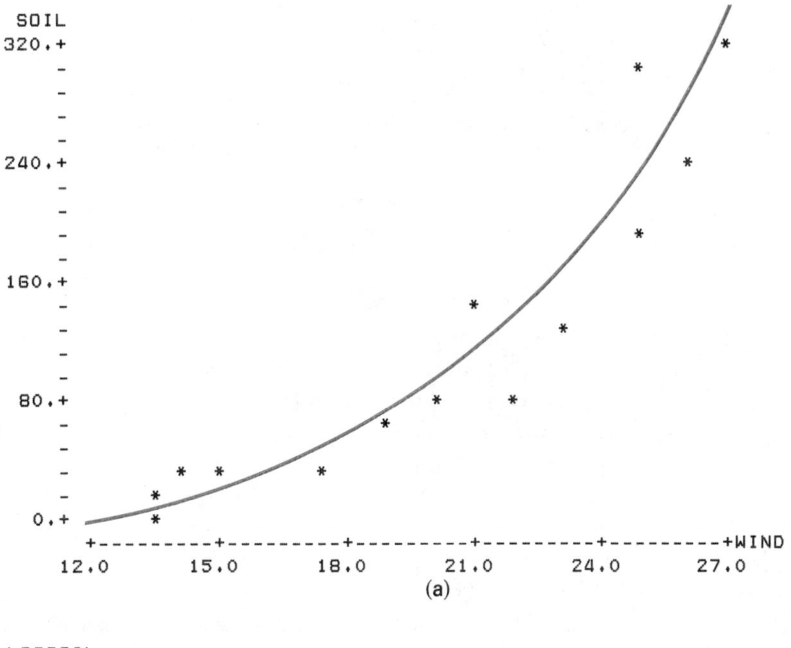

(a)

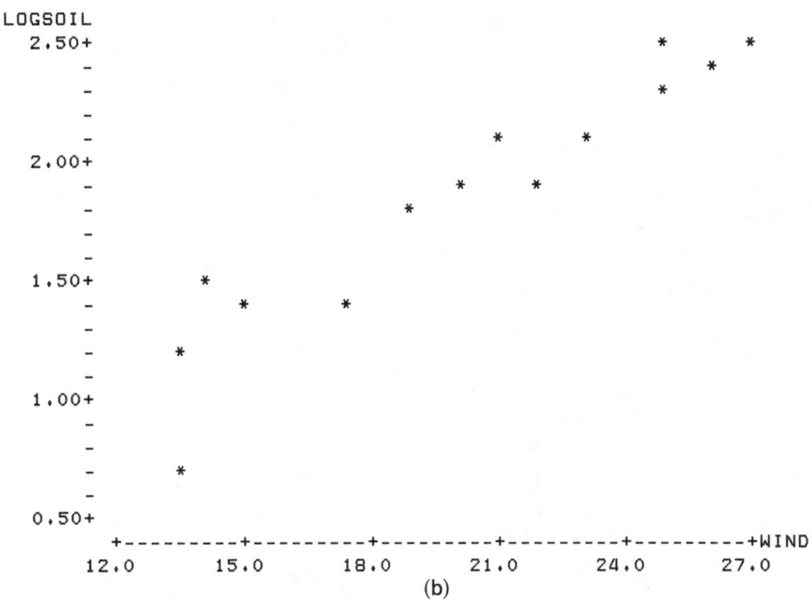

(b)

FIGURE 26

MINITAB PLOTS FOR THE DATA OF EXAMPLE 19
(a) y versus x
(b) $\log(y)$ versus x

Finding the least squares line for the 14 (x, y') pairs requires that we first compute $\overline{x} = \Sigma x/14 = 20.107143$ and $\overline{y}' = \Sigma y'/14 = 1.836836$. These values are then used to obtain $x - \overline{x}$, $y' - \overline{y}'$, $(x - \overline{x})^2$, and $(x - \overline{x})(y' - \overline{y}')$ in the recommended tabular format (so the y' values are used exactly as the y values were in the previous examples). Doing this and summing results in $\Sigma(x - \overline{x})^2 = 301.589286$ and $\Sigma(x - \overline{x})(y' - \overline{y}') = 31.198362$, so

$$b = \frac{31.198362}{301.589286} = .103447,$$

$$a = 1.836836 - (.103447)(20.107143) = -.243188$$

The approximate linear relationship between x and $y' = \log(y)$ is $\log(y) = -.243188 + .103447x$. The antilog of $\log(y)$ is y itself; equating this to the antilog of the right-hand side of the equation gives

$$y = 10^{-.243188+.103447x} = (10^{-.243188})(10^{.103447})^x$$
$$= .5712(1.2690)^x$$

The resulting equation $y = .5712(1.2690)^x$ is called an **exponential function** (of x), since x appears as an exponent on the right-hand side. By substituting various x values and using a calculator which can raise a number to any desired power (e.g., $(1.2690)^{15}$, $(1.2690)^{17.5}$, $(1.2690)^{20}$), points on the graph of this function can be identified. They can then be connected in a smooth fashion to obtain a prediction curve, which we have superimposed on the scatter plot of Figure 26(a). For any x, the corresponding predicted y value is the height of the curve above that x value. The predicted y value when $x = 15$ is $(.5712)(1.2690)^{15} = 20.357$ (the observed y for this value of x is 25). The predicted y value when $x = 16$ is $(.5712)(1.2690)^{16} = 25.83$ (there is no y observed for this value of x). It would be dangerous to make a prediction for x much larger than 27, the largest x value in the sample, since the amount of soil erosion presumably levels off at some point rather than continuing to increase explosively. ■

EXERCISES

4.57 The accompanying data on leaf diffusive resistance (y) and water potential (x) for one variety of coffee plant has been read from a scatter plot that appeared in the paper mentioned in Exercise 4.5.

x	−10	−10	−11	−11	−12	−12	−13	−13	−14	−16	−17
y	14	16	13	15	10	14	8	11	13	10	11

x	−17	−18	−22	−22	−23	−24	−24	−24	−25	−25	−26
y	18	12	17	32	20	24	26	44	36	54	52

a. Construct a scatter plot for the given data set.

b. Can you suggest a transformation from Table 2 that might help straighten this plot? Explain the rationale for your suggestion (you need not actually construct the plot).

4.58 The scatter plot of the data given in Exercise 4.57 shows quite a bit of curvature. Examination of Figure 23 suggests that two possible transformations for straightening this plot are log y and $1/y$. Use the given transformed values to construct a scatter plot using log y and x and a scatter plot using $1/y$ and x. Which plot appears to be the straightest?

y	14	16	13	15	10	14	8	11	13	10	11
log y	1.15	1.20	1.11	1.18	1.00	1.15	.90	1.04	1.11	1.00	1.04
1/y	.071	.063	.077	.067	.100	.071	.125	.091	.077	.100	.091

y	18	12	17	32	20	24	26	44	36	54	52
log y	1.26	1.08	1.23	1.51	1.30	1.38	1.41	1.64	1.56	1.73	1.72
1/y	.056	.083	.059	.031	.050	.042	.038	.023	.028	.019	.019

4.59 The paper "Aspects of Food Finding by Wintering Bald Eagles" (*The Auk* (1983):477–84) examined the relationship between the time that eagles spend aerially searching for food (indicated by the percentage of eagles soaring) and relative food availability. The data below is taken from a scatter plot that appeared in this paper. Let x denote salmon availability and y denote the percent of eagles in the air.

x	0.0	0.0	0.2	0.5	0.5	1.0
y	28.2	69.0	27.0	38.5	48.4	31.1

x	1.2	1.9	2.6	3.3	4.7	6.5
y	26.9	8.2	4.6	7.4	7.0	6.8

a. Draw a scatter plot for this data set. Would you describe the plot as linear or curvilinear?

b. One possible transformation that might lead to a straighter plot involves taking the square root of both the x and y values. Use Figure 23 to explain why this might be a reasonable choice of a transformation.

c. Construct a scatter plot using the variables $\sqrt{x}$ and $\sqrt{y}$. Is this scatter plot straighter than the plot in (a)?

d. Using Table 2 of this section, can you suggest another transformation that might be used to straighten the original plot?

4.60 Data on salmon availability (x) and the percent of eagles in the air (y) is given in the previous exercise.

a. Calculate the correlation coefficient for this data.

b. Since the scatter plot of the original data appeared curved, transforming both the x and y values by taking square roots was suggested. Calculate the correlation coefficient for the variables $\sqrt{x}$ and $\sqrt{y}$. How does this value compare with that of (a)? Does this indicate that the transformation was successful in straightening the plot?

4.61 Penicillin was administered orally to five horses and the concentration of penicillin in the blood was determined after five different lengths of time (a different horse was used each time). The accompanying data appeared in the paper "Absorption and Distribution Patterns of Oral Phenoxymethyl Penicillin in the Horse" (*Cornell Veterinarian* (1983):314–23).

(x) Time elapsed (h)	1	2	3	6	8
(y) Penicillin concentration (mg/mL)	1.8	1.0	.5	.1	.1

Construct scatter plots using the following variables. Which transformation, if any, would you recommend?

a. x and y

b. $\sqrt{x}$ and y

c. x and $\sqrt{y}$

d. $\sqrt{x}$ and $\sqrt{y}$

e. x and $\log(y)$ (the values of $\log(y)$ are .26, 0, −.30, −1, −1)

4.62 The paper "Population Pressure and Agricultural Intensity" (*Annals of the Assoc. of Amer. Geog.* (1977):384–96) reported a positive association between population density and agricultural intensity. The given data consists of measures of population density (x) and agricultural intensity (y) for 18 different subtropical locations.

x	1.0	26.0	1.1	101.0	14.9	134.7	3.0	5.7	7.6
y	9	7	6	50	5	100	7	14	14

x	25.0	143.0	27.5	103.0	180.0	49.6	140.6	140.0	233.0
y	10	50	14	50	150	10	67	100	100

a. Construct a scatter plot of agricultural intensity versus population density. Is the scatter plot compatible with the statement of positive association made in the paper?

b. The scatter plot in (a) is curved upward like segment 2 in Figure 23, suggesting a transformation that is up the ladder for x or down the ladder for y. Try a scatter plot that uses y and x^2. Does this transformation straighten the plot?

c. Try drawing a scatter plot that uses log y and x. The log y values, given in order corresponding to the y values, are .95, .85, .78, 1.70, .70, 2.00, .85, 1.15, 1.15, 1.00, 1.70, 1.15, 1.70, 2.18, 1.00, 1.83, 2.00, and 2.00. How does this scatter plot compare with that of (b)?

d. Next consider a scatter plot that uses both a transformation on x and a transformation on y: log y and x^2. Is this effective in straightening the plot? Explain.

4.63 The growth rate of lichen was the subject of a study reported in the paper "Lichen Growth Responses to Stress Induced by Automobile Exhaust Pollution" (*Science* (1979):423–24). The accompanying data is taken from a scatter plot illustrating the relationship between initial population size (area in mm^2) and percent area increase over a 6-month period. Construct a scatter plot for this data. If the scatter plot exhibits curvature, try to find a transformation that straightens the plot.

Population size (x)	.02	.02	.03	.05	.06	.06	.08	.09	.14
Percent increase (y)	200	190	150	155	150	130	90	140	50

Population size (x)	.14	.15	.22	.22	.55	.61	1.49	1.65
Percent increase (y)	80	40	40	30	30	35	50	55

4.64 Determining the age of an animal can sometimes be a difficult task. One method of estimating the age of harp seals is based on the width of the pulp canal in the seal's canine teeth. To investigate the relationship between age and the width of the pulp canal, age and canal width were measured for seals of known age. The accompanying data is a portion of a larger data set that appeared in the paper "Validation of Age Estimation in the Harp Seal Using Dentinal Annuli" (*Canadian J. of Fisheries and Aquatic Sci.* (1983):1430–41). Let x denote age (years) and y denote canal length (mm).

x	.25	.25	.50	.50	.50	.75	.75	1.00	1.00	1.00
y	700	675	525	500	400	350	300	300	250	230

x	1.00	1.00	1.25	1.25	1.50	1.50	2.00	2.00	2.50	2.75
y	150	100	200	100	100	125	60	140	60	50

x	3.00	4.00	4.00	5.00	5.00	5.00	5.00	6.00	6.00
y	10	10	10	10	15	10	10	15	10

Construct a scatter plot for this data set. Would you describe the relationship between age and canal length as linear? If not, suggest a transformation that might straighten the plot.

4.65 A frequently encountered problem in crop planting situations involves deciding when to harvest in order to maximize yield. The accompanying data on x = date of harvesting (number of days after flowering) and y = yield (kg/ha) of paddy, a grain farmed in India, appeared in the paper "Determination of Biological Maturity and Effect of Harvesting and Drying Conditions on Milling Quality of Paddy" (*J. of Ag. Eng.* (1975):353–61). Construct a scatter plot of this data. Can the methodology discussed in this section be used to select a straightening transformation? Why or why not? What kind of a curve might provide a reasonable fit to the plot?

x	16	18	20	22	24	26	28	30
y	2508	2518	3304	3423	3057	3190	3500	3883

x	32	34	36	38	40	42	44	46
y	3823	3646	3708	3333	3517	3241	3103	2776

4.66 An investigation of the influence of sodium benzoate concentration on the critical pH necessary for the inhibition of iron (Fe) yielded the accompanying data ("Mechanism of the Corrosion Inhibition of Fe by Sodium Benzoate" *Corrosion Sci.* (1971):675–82).

Concentration (x)	.01	.025	.1	.95
pH (y)	5.1	5.5	6.1	7.3

a. Draw a scatter plot of this data.

b. One transformation that might straighten this plot involves taking the logarithm of the x values. Use the given logarithms to draw a scatter plot using y and $\log x$. Does this plot look straighter than the plot of the untransformed data?

x	.01	.025	.1	.95
$\log x$	−2	−1.6	−1	−.02

c. Find the least squares line using y and the transformed x values.

d. Use the equation in (c) as a basis for predicting the pH when sodium concentration is .5.

4.67 In the article "Ethylene Synthesis in Lettuce Seeds: Its Physiological Significance" (*Plant Physiology* (1972):719–22), ethylene content of lettuce seeds y (nL/g dry wt) was studied as a function of exposure time x (min) to an ethylene absorbant.

x	2	10	20	30	40	50	60	70	80	90	100
y	408	274	196	137	90	78	51	40	30	22	15

a. Find the least squares line for this data set.

b. Use the least squares line to compute predicted values and residuals. Also compute r^2, and say whether its value supports the use of a straight line summary.

c. Construct a residual plot. Does the plot indicate that the relationship between x and y is not linear? Explain.

d. Since the residual plot is quite curved, the authors of this paper used a transformation on the data. The selected transformation involved taking the logarithm of each y value and leaving the x values unchanged. Use the given logarithms to find the least squares line for the transformed data.

y	408	274	196	137	90	78	51	40	30	22	15
$\log y$	2.61	2.44	2.29	2.13	1.95	1.89	1.71	1.60	1.48	1.34	1.18

e. Compute the residuals and construct a residual plot for the transformed data. Does this residual plot suggest that a straight line provides a good description of the relationship between x and $\log y$?

KEY CONCEPTS

Bivariate data Data for which each observation consists of a value of one variable, x, paired with a value of a second variable, y. (p. 103)

Scatter plot A picture of bivariate numerical data in which each observation (x, y) is represented as a point located with respect to a horizontal x axis and a vertical y axis. (p. 104)

Pearson's correlation coefficient A quantitative measure of the extent to which x and y values in a bivariate numerical data set are linearly related. It is denoted by r if the data set is a sample of pairs and ρ if the data set consists of all pairs in a population. (p. 113)

Spearman's correlation coefficient Another measure of how strongly x and y values in a bivariate numerical data set are related. It is based on the ranks of the x and y observations and will identify strong nonlinear as well as linear relationships. (p. 118)

Principle of least squares A general principle used to select a line that summarizes an approximate linear relationship between a dependent variable y and an independent (or predictor) variable x. The sum of squared vertical deviations from a line to the points in the scatter plot is used to measure how well the line fits the data. The least squares line is the line with the smallest sum of squared deviations. (p. 126)

Residual sum of squares (SSResid) The sum of squares of residuals (vertical deviations from the least squares line), which measures the variation in observed y values that cannot be attributed to an approximate linear relationship between x and y. (p. 139)

Total sum of squares (SSTo) A measure of the total amount of variation in observed y values. (p. 140)

Coefficient of determination (r^2) The proportion of total variation in y that can be attributed to an approximate linear relationship; $r^2 = 1 - \text{SSResid/SSTo}$. (p. 141)

Standard deviation about the least squares line (s_e) The typical amount by which an (x, y) point deviates from the least squares line, given by $s_e = \sqrt{\text{SSResid}/(n - 2)}$. (p. 144)

Power transformation An exponent, or power, p is first specified, and then new (transformed) data values are calculated as *transformed value* = (original value)^p. A logarithmic transformation is identified with $p = 0$. When the scatter plot of original data exhibits curvature, a power transformation of x and/or y will often result in a scatter plot that has a linear appearance. (p. 151)

SUPPLEMENTARY EXERCISES

4.68 The accompanying data on x = blood glucose level (mM) and y = glucose level (mM) in the aqueous humour of the eye for eight diabetic cataract patients was read from a scatter plot in the paper "Aqueous Humour Glucose Concentration in Cataract Patients and its Effect on the Lens" (*Exper. Eye Research* (1984): 605–09).

x	8.0	8.0	10.0	12.0	13.0	14.5	21.0	22.0
y	4.8	5.4	5.8	7.0	10.2	6.9	13.0	6.0

a. Construct a scatter plot. Does the relationship between y and x look linear?
b. Compute the correlation coefficient. Would you characterize the strength of the linear relationship as weak, moderate, or strong?
c. Calculate the slope and intercept of the least squares line.

4.69 Residual concentration of ClO_2 (mg/L) in water disinfected with ClO_2 was measured for various concentrations of applied ClO_2 (mg/L). The accompanying data appeared in the paper "Determination of Oxidants Formed Upon the Disinfection of Drinking Water with Chlorine Dioxide" (*J. Environ. Sci. Health* (1984): 943–57).

Applied ClO_2	.90	.80	.80	.80	.60	.60
Residual ClO_2	.60	.10	.20	.29	.15	.12

a. Describe the strength of linear relationship between applied and residual ClO_2 using the correlation coefficient.

b. Compute Spearman's rank correlation coefficient.

4.70 Two methods (HPLC and GLC) for determining acid in bile were compared in the article "Development and Validation of a Method for Measuring the Glycine in Bile by High-Performance Liquid Chromatography" (*J. Chromatography* (1984): 249–57). Twelve volunteers supplied bile samples. Each sample was analyzed for taurodeoxycholic acid concentration using both the HPLC and the GLC methods. Calculate the value of r, the sample correlation coefficient. Would you say that there is a strong correlation between the acid concentration determinations using the two methods?

	Concentration (mol)											
HPLC	0	6	10	14	17	17	19	21	24	28	30
GLC	4	10	9	13	16	18	20	22	23	29	30

4.71 In an Italian study of dietary factors influencing the concentration of tritium in urine, two different prediction models were proposed ("Validation of a Metabolic Model for Tritium" *Radiation Research* (1984):503–09). The actual tritium concentration (pCi/mL) for seven subjects is given, along with the corresponding predicted values for the two models under consideration. Calculate the residuals and SSResid for each of the two models. Based on SSResid, would you recommend one of the models over the other?

Subject	Actual Concentration	Model A Predicted Value	Model B Predicted Value
1	1.4	1.4	1.4
2	.7	.6	.5
3	1.8	1.5	1.5
4	.7	2.7	2.6
5	1.9	.3	1.3
6	.5	.5	.5
7	.9	1.0	1.0

4.72 The article "Capital Expenditures Report" (*Food Eng.* (1984):93–101) ranked various food and beverage companies with respect to capital spending. Rankings for 1979 and 1983 are given. Use Spearman's rank correlation coefficient to describe the association between 1979 and 1983 expenditures. Interpret the value of r_s.

	Rank	
Company	1983	1979
Anheuser-Busch	1	1
Coca-Cola	2	3
Pepsico	3	2
General Foods	4	7
Nabisco Brands	5	9
R. J. Reynolds	6	10
CPC International	7	6
Campbell Soup	8	5
Philip Morris	9	4
Kellogg's	10	8

4.73 The accompanying values of x = interest rate on a certificate of deposit and y = mortgage rate resulted from a sample of savings and loan associations (*Los Angeles Times*, September 15, 1984).

Observation	1	2	3	4	5	6
x	13.43	13.11	12.95	12.25	13.10	12.74
y	13.46	15.00	13.32	13.92	13.80	13.24

Observation	7	8	9	10	11	12
x	12.82	12.60	13.00	12.82	12.68	13.08
y	13.70	13.83	13.48	13.09	13.05	13.33

a. Does a scatter plot of the data suggest any relationship between x and y? Does the data set contain any "unusual" observations—in particular, an observation whose deletion would yield a data set for which x and y appeared to be somewhat related?

b. Compute the sample correlation coefficient r. Does the value of r suggest any linear relationship? (*Hint:* $\Sigma(x - \bar{x})^2 = 1.0052$, $\Sigma(y - \bar{y})^2 = 3.0428$, and $\Sigma(x - \bar{x})(y - \bar{y}) = .1200$.)

c. Suppose you subtracted 12 from each sample x value and 13 from each sample y value and then computed r. What would the result be? (Answer without computation.) Explain.

d. Delete the observation (13.11, 15.00) and recompute r. What does this result suggest? (*Hint:* For the remaining 11 observations, $\bar{x} = 12.8609$, $\bar{y} = 13.4745$, $\Sigma(x - \bar{x})^2 = .9483$, $\Sigma(y - \bar{y})^2 = .9097$, and $\Sigma(x - \bar{x})(y - \bar{y}) = -.2283$.)

4.74 **a.** Referring back to Exercise 4.73, compute the value of Spearman's rank correlation coefficient for the given data. Does this value suggest any strong relationship (linear or nonlinear) between x and y?

b. Suppose you computed $\log(x)$ and $\log(y)$ and then computed r_s for the $(\log(x), \log(y))$ pairs. What would the result be? Explain. (*Hint:* If c and d are both positive numbers with $c < d$, then $\log(c) < \log(d)$.)

4.75 The paper "Environmental Significance of Trace Elements in Human Hair—A Case Study from Sri Lanka" (*Inter. J. Environ. Studies* (1984):41–48) gave the accompanying observations on x = lead concentration (ppb) and y = cadmium concentration (ppb) for $n = 11$ college students in Sri Lanka. Compute and interpret the value of Spearman's rank correlation coefficient.

x	2,000	12,000	2,300	3,100	17,000	1,100	2,200	10,000	2,300	1,000	14,500
y	3,000	3,000	10,000	2,000	19,000	0	2,000	12,000	5,000	0	14,000

4.76 The paper "Biomechanical Characteristics of the Final Approach Step, Hurdle, and Take-Off of Elite American Springboard Divers" (*J. Human Movement Studies* (1984):189–212) gave the data below on y = judge's score and x = length of final step (m) for a sample of seven divers performing a forward pike with a single somersault.

y	7.40	9.10	7.20	7.00	7.30	7.30	7.90
x	1.17	1.17	.93	.89	.68	.74	.95

a. Construct a scatter plot.

b. Calculate the slope and intercept of the least squares line. Draw this line on your scatter plot.

c. Calculate and interpret the value of the correlation coefficient.

d. Compute the value of Spearman's rank correlation coefficient, r_s. (For tied scores, assign the average of the ranks that would have been assigned had they differed slightly; for example, if the values of 7.3 in the score row were slightly different, they would be ranked 4 and 5, so each is assigned rank $(4 + 5)/2 = 4.5$). How does the value of r_s compare to that of r from (c)?

4.77 The paper cited in Exercise 4.76 also gave the accompanying data on score and flight time for seven divers completing a reverse half-somersault. Compute the value of the correlation coefficient. Would you characterize the relationship between these two variables as a strong positive linear relationship? Explain.

Score	7.5	8.3	7.6	6.8	7.4	7.5	6.5
Time (s)	.45	.45	.45	.40	.41	.40	.46

4.78 The accompanying data on movie production costs, promotion costs, and worldwide ticket sales (all in millions of dollars) appeared in an article on "Dumb Movies" in the *Los Angeles Times* (January 20, 1985).

Movie	Production Costs	Promotion Costs	Ticket Sales
Animal House	\$ 2.9	\$3	\$150
Meatballs	1.4	2	70
Caddyshack	4.8	4	60
Stripes	10.5	4.5	85
Spring Break	4.5	5	24
Porky's	4.8	9	160
Fast Times at Ridgemont High	5.0	4.9	50
Porky's II	7.0	7.5	55
Hot Dog—The Movie	2.0	4	22
Bachelor Party	7.0	7.5	38
Revenge of the Nerds	7.0	7.5	42
Police Academy	4.5	4	150

a. Compute and interpret the value of the correlation coefficient for production costs and ticket sales.

b. Compute and interpret the value of the correlation coefficient for promotion costs and ticket sales.

4.79 Consider the four (x, y) pairs $(0, 0)$, $(1, 1)$, $(1, -1)$, and $(2, 0)$.

a. What is the value of the sample correlation coefficient r?

b. If a fifth observation is made at the value $x = 6$, find a value of y for which $r > .5$.

c. If a fifth observation is made at the value $x = 6$, find a value of y for which $r < -.5$.

4.80 The accompanying data on y = concentration of penicillin-G in pig's blood plasma (units/mL) and x = time (min) from administration of a dose of penicillin (22 mg/kg body weight) appeared in the paper "Calculation of Dosage Regimens of Antimicrobial Drugs for Surgical Prophylaxis" (*J. Amer. Vet. Med. Assoc.* (1984): 1083–87).

x	5	15	45	90	180	240	360	480	1440
y	32.6	43.3	23.1	16.7	5.7	6.4	9.2	.4	.2

a. Construct a scatter plot for this data.

b. Using the ladder of transformations of Section 4.5, suggest a transformation that might straighten the plot. Give reasons for your choice of transformation.

4.81 The accompanying data resulted from an experiment in which x was the amount of catalyst added to accelerate a chemical reaction and y was the resulting reaction time.

x	1	2	3	4	5
y	49	46	41	34	25

a. Calculate r. Does the value of r suggest a strong linear relationship?
b. Construct a scatter plot. On the basis of this plot, does the word *linear* really provide the most effective description of the relationship between x and y? Explain.

REFERENCES

The book by Freedman et al. (Chapter 1) as well as the three books listed in Chapter 2 all contain interesting material relating to the summarizing of bivariate data.

Probability and Probability Distributions

INTRODUCTION

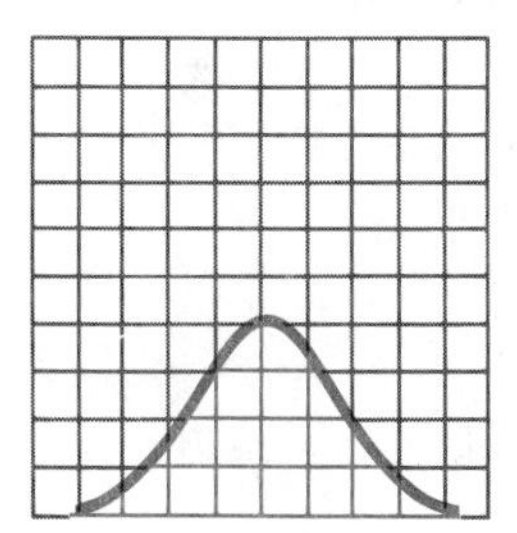

ALMOST ALL SITUATIONS that we confront in our everyday activities involve some aspects of uncertainty. There is uncertainty concerning the number of cars queued up at a bank's drive-up window, concerning whether or not an appliance will need repair while still under warranty, concerning the amount of weight one might lose on a particular diet, and so on. **Probability** is the scientific discipline whose objective is to study uncertainty in a systematic fashion.

The first growth spurt of probability occurred during the seventeenth century in attempts to answer questions concerning games of chance. Even today games of chance suggest many interesting questions that can be answered using methods from probability. For example, it used to be thought that the odds in blackjack virtually always favored the house (the dealer's employer), but in the 1960s probability methods were used to discover many situations (involving cards not yet dealt) in which the advantage lay with the individual bettor. In the twentieth century the scope of probability has enlarged considerably as investigators have attempted to come to grips with the pervasiveness of uncertainty in both scientific contexts and in everyday life. Probability methods have recently been used to increase understanding of such diverse phenomena as the spread of an epidemic through a population, the mechanism of memory recall, the operating characteristics of various computer time-sharing systems, the diffusion of particles through a membrane, changes in consumers' brand preferences over time, social class mobility through succeeding generations, and (of course) what tomorrow's weather might be like. Our goal in this chapter is to introduce you to just a few of the most important concepts and methods of probability. To study probability in much depth requires the use of some rather sophisticated mathematical tools, but our discussion is informal and intuitive. It focuses on those aspects of probability that bear most directly on the logic and methods of inference.

Frequently the objective of inference is to estimate the value of some population characteristic by using sample information. Probability can be used to indicate the reliability of an estimate by saying how close it might be to what is being estimated. Still another type of inference problem involves trying to ascertain the plausibility of a claim put forth by someone. Perhaps the claim is based on a certain genetic theory and involves the expected percentage of a particular genotype resulting from a large plant breeding experiment. Such a claim can be checked by comparing the percentage predicted by a theory to what was actually observed. Sample results usually deviate somewhat from what is predicted, but a very large deviation would suggest that alternative theories (perhaps not yet developed) are more plausible than the current one. In this setting probability would be used to help decide whether a deviation is so large as to cast substantial doubt on the claim.

5.1 Probability

The basic ideas and terminology of probability are easiest to introduce in situations that are both familiar to you and reasonably simple. We first address uncertainties associated with such experiments as tossing a coin once or several times, selecting one or more cards from a deck, and rolling a single die or several dice. These activities are not always very interesting in and of themselves, so after putting the basics in place we move on to more realistic problem situations.

Chance Experiments and Relative Frequency

When a single coin is tossed, it can land with its head side up or its tail side up. When a single card is selected from a well-mixed deck, it can be the ace of spades, the five of diamonds, or any one of the 50 other possibilities. When a red die is rolled and then a green die is rolled, possible outcomes include the following:

1. Four on the red die and one on the green die, denoted by (4, 1)
2. Six on each die, denoted by (6, 6)
3. One on the red die and four on the green die, denoted by (1, 4)

There are 33 other possibilities. (To see this, start listing pairs in the order (1, 1), (1, 2), . . . , (1, 6), (2, 1), etc.) We use the phrase *chance experiment* to refer to any activitity or situation in which the outcome is uncertain. Rolling a die might not strike you as much of an experiment in the usual sense of the word, but our usage is broader than what is usually implied. A chance experiment could also refer to ascertaining whether or not each person in a sample supports the death penalty (an opinion poll or survey) or to an investigation carried out in a laboratory to study how varying the amount of a certain chemical input affects the yield of a product.

EXAMPLE 1

One of the simplest chance experiments involves tossing a coin just once. Frequently we hear a coin described as fair, or we are told that there is a 50% chance of the coin landing with its head side up. Such a description cannot refer to the result of a single toss, since a single toss cannot result in both a head and a tail. Might fairness and 50% refer to 10 successive tosses yielding exactly 5 heads and 5 tails? Not really, since it is easy to imagine a coin characterized as fair landing heads up on only 3 or 4 of the 10 tosses.

Suppose that we take such a coin and begin to toss it over and over. After each toss, we compute the relative frequency of heads observed so far, i.e., (number of heads)/(number of tosses). The results for the first 10 tosses are as follows:

Toss Number	1	2	3	4	5	6	7	8	9	10
Outcome	T	H	H	H	T	T	H	H	T	T
Cum. Number of Heads	0	1	2	3	3	3	4	5	5	5
Relative Frequency	0	.5	.667	.75	.6	.5	.571	.625	.556	.5

Figure 1 illustrates how the relative frequency of heads fluctuates during the first 50 tosses. Much empirical evidence suggests that *as the number of tosses increases, the relative frequency of heads does not continue to fluctuate wildly but instead stabilizes and approaches some fixed number*. This stabilization is illustrated for a sequence of 1000 tosses in Figure 2.

Because each relative frequency is between 0 and 1, the limiting value is also. It is then natural to call the coin *fair* if the limiting value is .5. The 50%

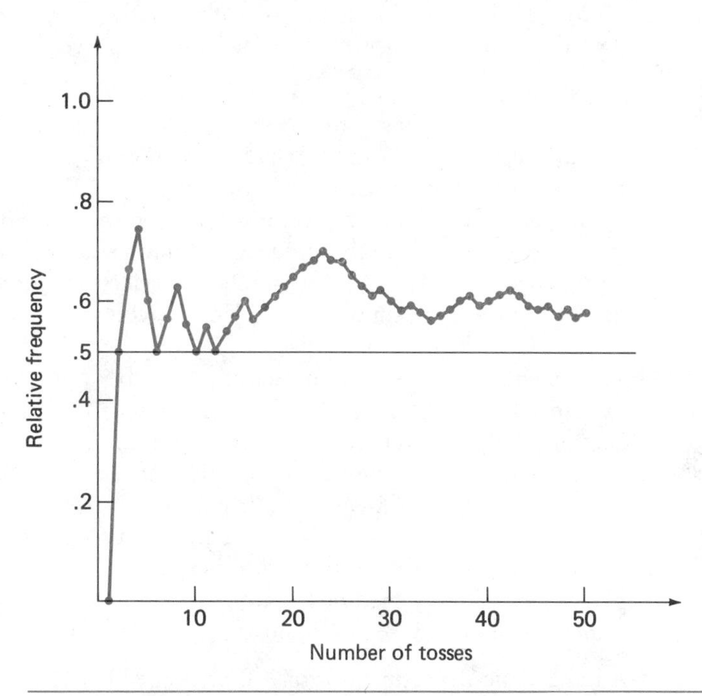

FIGURE 1 RELATIVE FREQUENCY OF HEADS IN THE FIRST 50 OF A LONG SERIES OF TOSSES

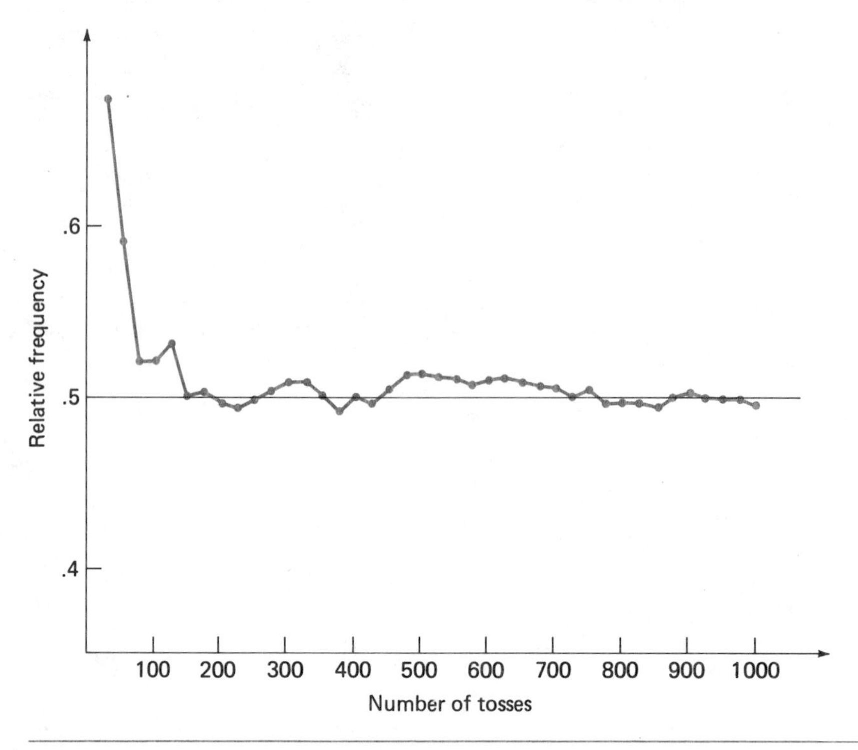

FIGURE 2 STABILIZATION OF THE RELATIVE FREQUENCY OF HEADS IN COIN TOSSING

chance doesn't refer to exact results in some fixed number of tosses, such as 10 or 100, but to what happens to the relative frequency of heads as we repeat the chance experiment over and over and over again. ■

EXAMPLE 2

Consider selecting a single card from a well-mixed deck of 52 cards. Card players would surely say that there is a 25% chance of selecting a card whose suit is hearts (as opposed to spades, diamonds, or clubs). To interpret this, think of performing this chance experiment over and over again: Select a first card, replace and shuffle, select a second card, replace and shuffle, and so on. If we examine the relative frequency (number of times a heart is selected)/(number of selections), as the number of selections increases the relative frequency stabilizes at the value .25. This is why we say there is a 25% chance of selecting a heart.

Let's complicate matters by selecting 5 cards (without replacement) rather than just 1 (some people call this a poker hand). The resulting 5 cards may all be from the same suit (called a flush in poker) or they may not be. Consider performing this experiment repeatedly and tracking the relative frequency (number of times all 5 cards are from the same suit)/(number of replications of the experiment). This relative frequency also stabilizes as the number of replications increases, but the limiting value is not obvious even to poker players. Probability methods can be used to show that it is .00198. Thus the chance of obtaining such a hand is much less than 1%. ■

EXAMPLE 3

An academic department with 8 faculty members has voted by secret ballot to elect from its ranks either candidate A or candidate B to serve on a grievance panel. Suppose that the votes are on identical slips of paper in a box and that A has actually received 5 votes. The slips are removed from the box one by one and a cumulative tally is kept. How likely is it that A remains ahead of B throughout the vote count?

Consider the following four replications of the experiment (the process of removing and counting ballots).

Vote Number	1	2	3	4	5	6	7	8
Replication 1	A	A	A	B	A	B	B	A
Replication 2	A	A	B	B	A	A	B	A
Replication 3	B	A	A	B	A	A	B	A
Replication 4	A	A	B	A	B	A	A	B

On the first and fourth replications, candidate A remains ahead all the way. On the second replication B catches up to A when the fourth vote is recorded, whereas on the third replication B goes ahead when the first vote is recorded. Suppose now that we continue to replicate this experiment. It turns out that the relative frequency

$$\frac{\text{number of replications in which A is ahead of B throughout the counting}}{\text{number of replications of the experiment}}$$

stabilizes at the value .25. We can say that there is a 25% chance that A will stay ahead of B throughout the counting. ■

To proceed further, we need some terminology.

DEFINITION	An **event** is any collection of possible outcomes of a chance experiment.

EXAMPLE 4

For the experiment in which first a red die and then a green die are rolled, some events include the following:

1. The single outcome (1, 4) (1 on red and 4 on green)
2. The three outcomes (1, 3), (2, 2), and (3, 1) (the only outcomes for which the sum of the two numbers is four)
3. The 11 outcomes (1, 6), (2, 6), (3, 6), (4, 6), (5, 6), (6, 6), (6, 5), (6, 4), (6, 3), (6, 2), and (6, 1) (all outcomes in which at least one 6 is rolled)

Many more events can be obtained by forming different groupings consisting of one or more of the 36 individual outcomes. ■

EXAMPLE 5

Here are some events for the ballot problem described in Example 3:

1. AABAABBA, an event consisting of a single outcome
2. The event consisting of the six outcomes BBBAAAAA, ABBBAAAA, AABBBAAA, AAABBBAA, AAAABBBA, and AAAAABBB (all outcomes in which the three B votes are selected one after the other)
3. The event consisting of the four outcomes BAABAABA, BAABAAAB,

BAAABAAB, and ABAABAAB (all outcomes in which at least two A votes separate every B vote) ■

Probability as Limiting Relative Frequency

When any given chance experiment is performed, some events are relatively likely to occur, whereas others are not so likely to occur. For a specified event E, we want to assign a number to this event that gives a precise indication of how likely it is that E will occur. This number is called *the probability of the event* E and is denoted by $P(E)$. For chance experiments that can be replicated as we've discussed, $P(E)$ is the limiting value of E's relative frequency of occurrence. $P(E)$ is then a number between 0 and 1; E can be judged relatively unlikely or likely according to where the value of $P(E)$ lies between the extremes of 0 and 1.

DEFINITION	The **probability of an event E,** denoted by $P(E)$, is the value approached by the relative frequency of occurrence of E in a very long series of replications of a chance experiment.

For a single toss of a fair coin, $P(\text{head side up}) = .5$. When a single card is selected from a well-mixed deck, $P(\text{selected card is a heart}) = .25$. In the ballot problem of Example 3, $P(\text{A remains ahead of B throughout the vote count}) = .25$. When we informally speak of a 10% chance of occurrence, we mean a probability of .10.

Our definition of probability depends on being able to perform a chance experiment repeatedly under identical conditions. Often probability language and concepts are used in situations in which replication is not feasible. When a new product is introduced, the marketing manager might state that the probability of its being successful is .3. Or, a public utility executive may testify that the probability of a nuclear plant meltdown during the next decade is .00000001. The most common alternative to the definition of probability based on relative frequency is a subjective, or personal, interpretation. Here probability is a measure of how strongly a person believes that an event will occur. This interpretation permits two people with different opinions to assign different probabilities to the same event (reflecting differing strengths of belief). However, both subjective probabilities and those based on limiting relative frequencies do satisfy the same general rules of probability, so probabilities of complex events can be calculated once probabilities of simple events have been specified. We do not pursue subjective probabilities any further. The relative frequency definition is intuitive, very widely used, and most relevant for the inferential procedures that we present. Whenever you want to interpret a probability, we recommend that you think in terms of limiting relative frequency.

There would seem to be a practical difficulty at this point: How can we find $P(E)$ without performing a long series of chance experiments? Fortunately, there are many situations in which experience suggests appropriate probabilities for individual outcomes. The probability of any event E is then easily obtained from the probabilities of all outcomes contained in E.

> The probability of an event E is the sum of the probabilities of all outcomes contained in E.
>
> When the probabilities of all possible experimental outcomes are added together, the result must be 1 (this is analogous to the sum of all the relative frequencies being 1).

EXAMPLE 6

A fair die is one for which each of the six possible outcomes 1, 2, 3, 4, 5, and 6 has the same long-run frequency of occurrence. Thus $P(1) = P(2) = \cdots = P(6)$ and, because their sum is 1, it follows that each outcome has probability $\frac{1}{6}$. Let E denote the event that a single toss results in an odd-numbered outcome. This event consists of the outcomes 1, 3, and 5, so $P(E) = P(1) + P(3) + P(5) = \frac{1}{6} + \frac{1}{6} + \frac{1}{6} = \frac{3}{6} = .5$. That is, if the die is tossed repeatedly, in the long run 50% of the tosses result in an odd number.

Now suppose that the die is loaded so that in the long run a 2 occurs twice as frequently as 1, a 3 occurs three times as frequently as a 1, and so on. The only probability assignment consistent with this loading is $P(1) = \frac{1}{21}$, $P(2) = \frac{2}{21}, \ldots, P(6) = \frac{6}{21}$. With E again denoting the event that an odd number results, now $P(E) = P(1) + P(3) + P(5) = \frac{1}{21} + \frac{3}{21} + \frac{5}{21} = \frac{9}{21} = .429$. In the long run, this die will land with an odd number showing only 42.9% of the time. ■

The fair die experiment just described is one of many in which outcomes are equally likely. The probability that the toss of a fair die results in an odd outcome is the ratio of the number of outcomes in this event (3) to the number of possible experimental outcomes (6). When outcomes are equally likely, the probability of any event is given by such a ratio.

> Consider an experiment that can result in any one of N equally likely outcomes. Then the probability of any particular outcome is $\frac{1}{N}$, and the probability of any specified event E is
>
> $$P(E) = \frac{\text{number of outcomes in } E}{N}$$

EXAMPLE 7

Table 1 displays all 56 possible outcomes for the ballot problem described in Example 3. If the slips have been well mixed before selection, no one of these outcomes is any more or less likely than any other. Thus the probability of each outcome is $\frac{1}{56}$. If the experiment is repeated over and over, any particular outcome occurs roughly $\frac{1}{56}$, or 1.79%, of the time. Among the 56 outcomes, there are 14 in which A remains ahead of B throughout the ballot count. In replications of the experiment, one of these 14 outcomes would occur roughly $\frac{14}{56}$, or 25%, of the time, so $P(\text{A remains ahead of B}) = .25$. Similarly, $P(\text{B's votes all occur in succession}) = \frac{6}{56} = .107$.

TABLE 1 POSSIBLE OUTCOMES FOR THE BALLOT COUNTING PROBLEM

BBBAAAAA	BAABAAAB	ABABAABA	*AABABAAB
BBABAAAA	BAAABBAA	ABABAAAB	*AABAABBA
BBAABAAA	BAAABABA	ABAABBAA	*AABAABAB
BBAAABAA	BAAABAAB	ABAABABA	*AABAAABB
BBAAAABA	BAAAABBA	ABAABAAB	AAABBBAA
BBAAAAAB	BAAAABAB	ABAAABBA	*AAABBABA
BABBAAAA	BAAAAABB	ABAAABAB	*AAABBAAB
BABABAAA	ABBBAAAA	ABAAAABB	*AAABABBA
BABAABAA	ABBABAAA	AABBBAAA	*AAABABAB
BABAAABA	ABBAABAA	AABBABAA	*AAABAABB
BABAAAAB	ABBAAABA	AABBAABA	*AAAABBBA
BAABBAAA	ABBAAAAB	AABBAAAB	*AAAABBAB
BAABABAA	ABABBAAA	AABABBAA	*AAAABABB
BAABAABA	ABABABAA	*AABABABA	*AAAAABBB

* Denotes an outcome for which A remains ahead of B throughout the counting.

In order to compute probabilities of complex events, such as P(odd number) or P(A remains ahead of B), it is necessary to have probabilities for simpler events. In general, the probabilities of some events must be known before the rules of probability can be used to compute the probabilities of other events. The use of probability methods requires that we draw on personal knowledge, opinion, and past experience to specify at least partially the probability structure for the chance experiment. Once that is done, other probabilities can be computed using various rules.

Often an investigator has determined the probability of an event E and of another event F and then wishes to obtain the probability that at least one of these two events—E or F—will occur. This is easily done when E and F cannot occur together, i.e., when no outcome is contained in both E and F. For example, let E be the event that the next customer to make a purchase at a certain store pays with a Visa credit card, and let F denote the event that a Mastercard is used. If 30% of all customers pay with Visa and 25% pay by Mastercard, then 30 + 25 = 55% pay with Visa or Mastercard. That is $P(E \text{ or } F) = P(E) + P(F) = .3 + .25 = .55$.

Let E and F be two events that have no outcomes in common. Then

$$P(E \text{ or } F) = P(E) + P(F)$$

More generally, if events $E_1, E_2, \ldots, E_k$ have no outcomes in common, then

$$P(E_1 \text{ or } E_2 \text{ or } \ldots \text{ or } E_k) = P(E_1) + P(E_2) + \cdots + P(E_k)$$

In words, the probability that one of these k events occurs is the sum of the probabilities of the individual events.

EXAMPLE 8

A computer center employs graduate students to answer programming questions. Let E_1 denote the event that the next query concerns a Fortran program, and define E_2 and E_3 analogously for BASIC and Pascal programs. If $P(E_1) = .30$, $P(E_2) = .20$, and $P(E_3) = .25$, then the probability that the next query concerns a program written in one of these three languages is

$$P(E_1 \text{ or } E_2 \text{ or } E_3) = P(E_1) + P(E_2) + P(E_3) = .75$$

■

If the events under consideration have outcomes in common, more than one of these events can occur when the experiment is performed. For example, let E be the event that a randomly selected automobile has defective tires and let F be the event that the selected car has defective headlights. Then the probability of any outcome (car) for which both E and F occur is included once in $P(E)$ and a second time in $P(F)$, so $P(E \text{ or } F)$ is not simply the sum of $P(E)$ and $P(F)$. The general rule for calculating $P(E_1 \text{ or } E_2 \text{ or } \ldots \text{ or } E_k)$ when these events have outcomes in common is rather complicated and is not needed for our purposes.

Independence

Sometimes the knowledge that one event has occurred considerably changes the likelihood that another event will occur. Consider a disease whose current incidence rate in a particular population is .1%, so the probability that a randomly selected individual has the disease is only .001. The presence of the disease cannot be discerned from outward appearances, but there is a diagnostic test available. Unfortunately, this test is not completely reliable—80% of those with positive test results actually have the disease and the other 20% of those with positive results are *false positives.* The experiment consists of selecting an individual and performing the diagnostic test. The two events of interest are that the result of the diagnostic test is positive (event F) and that the individual has the disease (event E). Before the diagnostic test is administered, $P(E) = .001$. However, once it is known that F occurred, the probability of E rises to .8 (the 80% true positives).

As another example, suppose that cars are being inspected for unsafe tires. Let E be the event that the right front tire is unsafe and F be the event that the left front tire on the same car is unsafe. The percentage of cars with unsafe tires may not be very high. But once it is known that F has occurred, it is highly likely that E will also occur, since we would expect a car's two front tires to be in similar condition. On the other hand, if event G is that the car's color is blue, being told that G occurred should not affect the likelihood of E, since car color and tire condition would seem to have nothing to do with one another.

DEFINITION

Two events are said to be **independent** if the chance that one event occurs is not affected by knowledge of whether or not the other occurred. If the occurrence of one event changes the probability that the other event occurs, the events are **dependent.** Similarly, if there are more than two events under consideration, they are independent if knowledge that some of the events have occurred doesn't change the probabilities that any of the other events occur.

Suppose a chance experiment consists of both tossing a fair coin and of selecting one card from a well-mixed deck. Because we can't tell anything about the card selected from knowing how the coin landed, the events *head side up* and *spade selected* are independent. If this experiment is repeated many times, on roughly 50% of the replications the head side of the coin lands up, and a spade is selected roughly 25% of the time. On what percentage of the replications do both a head and a spade result? Focus only on those replications on which a head occurs. By independence, a spade will occur on approximately 25% of these replications. Therefore, the two should occur together on about 25% of 50%, or 12.5%, of all replications. Using probabilities in place of percentages, (.25)(.5) = .125, so multiplying the individual event probabilities gives the probability that both events occur.

> If two events E and F are independent, then
>
> $$P(E \text{ and } F) = P(E) \cdot P(F)$$
>
> More generally, let $E_1, E_2, \ldots, E_k$ be k independent events. Then
>
> $$P(E_1 \text{ and } E_2 \text{ and } \ldots \text{ and } E_k) = P(E_1) \cdot P(E_2) \ldots P(E_k)$$
>
> In words, the probability that all the events occur together is the product of the probabilities of the individual events.

EXAMPLE 9

Let E be the event that your statistics professor begins class on time and let F be the event that your philosophy professor does likewise. With $P(E) = .9$ and $P(F) = .6$, assuming independence of these two events gives the probability that both classes begin on time as (.9)(.6) = .54. ■

EXAMPLE 10

A microcomputer system consists of a monitor, a disk drive, and the computer itself. Let E_1 be the event that a newly purchased monitor operates properly, and let events E_2 and E_3 be defined analogously for the disk drive and computer. All three components must work properly for the system to function. Assuming that the events E_1, E_2, and E_3 are independent with $P(E_1) = .99$, $P(E_2) = .9$, and $P(E_3) = .95$, the probability that the system functions is $P(E_1 \text{ and } E_2 \text{ and } E_3) = (.99)(.9)(.95) = .85$. ■

Two events are dependent when knowledge that one event has occurred changes the likelihood that the other occurs. The situation involving the diagnostic test for a rare disease illustrates this. A .1% incidence rate for the disease implies that a randomly selected individual has the disease with probability .001. However, given that a person tests positive for the disease, the probability that he or she has the disease is .8. Thus the events *individual has the disease* (E) and *positive test result* (F) are dependent. Whereas the original probability of E is .001, this probability changes to .8 given that F occurred. The probability .8 is a **conditional probability** and is denoted by $P(E|F)$, read "the probability that E occurs conditional on the event F having occurred."

As another example, suppose that 5% of all outstanding automobile loans go into default. Then with E denoting the event that a randomly selected loan goes into default, $P(E) = .05$. When a person applies for an automobile loan, most banks routinely perform a credit check and might rate an applicant as a poor, average, or good risk. Let F be the event that the selected person was classified as a good risk. Then if only 1% of all good risks default, $P(E|F) = .01$. The events E and F are dependent because $P(E|F) \neq P(E)$.

In some situations conditional probabilities can easily be reasoned out. For example, suppose that two copies of a book are to be randomly selected from among 20 copies, of which 8 are first printings and 12 are second printings. Let F denote the event that the first book selected is a first printing, and let E represent the event that the second copy selected is a second printing. If F occurs, there are 7 first printings and 12 second printings among the 19 remaining copies, so $P(E|F) = \frac{12}{19} = .632$.

There are general rules for determining conditional probabilities when they cannot be reasoned out. In addition, when events $E_1, E_2, \ldots, E_k$ are dependent, calculation of $P(E_1 \text{ and } E_2 \text{ and } \ldots \text{ and } E_k)$ involves certain conditional probabilities. For a more comprehensive discussion of these topics, please consult one of the chapter references.

Sampling with and without Replacement

Many statistical problems involve repeated sampling from a single population. Here is a simple example that introduces an important distinction.

EXAMPLE 11

Consider selecting three cards from a deck. This selection can be made either **without replacement** (dealing three cards off the top) or **with replacement** (replacing each card and shuffling before selecting the next card). You can probably already guess that one of these selection methods gives independence of successive selections, whereas the other does not. To see this more clearly, define these events:

E_1 = the first card is a spade

E_2 = the second card is a spade

E_3 = the third card is a spade

For sampling with replacement, the probability of E_3 is .25 regardless of whether or not either E_1 or E_2 occur, since replacing selected cards gives the same deck for the third selection as for the first two selections. Whether or not either of the first two cards is a spade has no bearing on the third card selected, and the three events E_1, E_2, and E_3 are independent.

When sampling is without replacement, the chance of a spade on the third draw very definitely depends on the results of the first two draws. If both E_1 and E_2 occur, only 11 of the 50 cards remaining are spades. Since any one of these 50 has the same chance of being selected, the probability of E_3 is $\frac{11}{50} = .22$. On the other hand, if neither E_1 nor E_2 occurs, all 13 spades remain in the deck, so the probability of E_3 is $\frac{13}{50} = .26$. Information about the occurrence of E_1 and E_2 affects the chance that E_3 will occur, so for sampling without replacement the events are not independent. ■

In opinion polls and other types of surveys, sampling is virtually always done without replacement. For this method of sampling, the results of successive selections are not independent of one another. This is too bad, because many results from probability and statistics are much easier to state and use when independence can be assumed. The next example suggests that under certain circumstances the selections in sampling without replacement are approximately independent.

EXAMPLE 12

A lot of 10,000 industrial components consists of 2500 manufactured by one firm and 7500 manufactured by a second firm, all mixed together. Three components are to be randomly selected without replacement. Let E_1, E_2, and E_3 denote the events that the first, second, and third component selected were made by the first firm. Reasoning as in the card selection example, if E_1 and E_2 both ocur, the probability of E_3 is $2498/9998 = .24985$. If neither E_1 nor E_2 occurs, E_3 has probability $2500/9998 = .25005$. While these two probabilities differ slightly, to three decimal places they are both .250. We conclude that the occurrence or nonoccurrence of E_1 or E_2 has virtually no effect on the chance that E_3 will occur. For practical purposes, the three events can be considered independent. ■

The essential difference between the situations of Example 11 and Example 12 is the size of the sample relative to the size of the population. In the former example, a relatively large proportion of the population was sampled (3 out of 52), whereas in the latter example the proportion of the population sampled was quite small (only 3 out of 10,000).

> If the individuals or objects in a sample are selected without replacement from a population but the sample size is small relative to the population size, the successive selections are approximately independent. As a reasonable rule of thumb, independence can be assumed if at most 5% of the population is sampled.

This result justifies the assumption of independence in many statistical problems, yielding formulas and statistical methods that are relatively simple.

EXERCISES

5.1 The probability of an event can be estimated empirically by observing the result of a long sequence of trials and then computing a relative frequency. Use a standard deck of playing cards to estimate empirically the probability that a randomly selected card is a heart. Each trial should consist of shuffling the cards, choosing a card at random, and noting whether or not it is a heart. After 50 trials, compute the relative frequency of hearts. How close is your relative frequency to the theoretical probability of .25? Construct a graph like that of Figure 1 to illustrate your results.

5.2 **a.** Consider the following events, which might occur when a card is selected at random from a deck of playing cards. Determine the probability of occurrence for each of these events. (*Hint:* The 52 outcomes are equally likely.)

i. Drawing a face card (jack, queen, or king)
ii. Drawing a red card
iii. Drawing a card whose face value is 6, 7, 8, or 9
iv. Drawing a 2 or 3
v. Drawing a 2
vi. Drawing a 2 or a red card

b. Use the technique described in Exercise 5.1 to estimate the probability of each event of (a) empirically (based on the same 50 trials).

c. Compare your empirical estimates with the true probabilities of (a). Do all your estimates appear to be of about the same accuracy? What does this suggest about the number of trials necessary to obtain relatively accurate empirical estimates of probabilities?

5.3 Suppose the accompanying information on births in the United States over a given period of time is available to you.

Type of Birth	Number of Births
Single birth	41,500,000
Twins	500,000
Triplets	5,000
Quadruplets	100

Use this information to approximate the probability that a randomly selected pregnant woman who reaches full term:

a. Delivers twins

b. Delivers quadruplets

c. Gives birth to more than a single child

5.4 Consider the event E and suppose that $P(E) = .6$, so that in the long run E will occur 60% of the time when the experiment is performed repeatedly. Let *not* E denote the event that E does not occur. What is $P(\textit{not } E)$? What is the general relationship between $P(E)$ and $P(\textit{not } E)$?

5.5 Suppose that the probability that a particular type of smoke detector will function properly and sound an alarm in the presence of smoke is .7.

a. If you have two such alarms in your home, what is the probability that at least one functions when a fire occurs? Assume that the smoke detectors operate independently of one another. (*Hint:* Let E be the event that at least one functions. What is *not* E? Now use the result of Exercise 5.4.)

b. How many smoke detectors would you need so that the probability of at least one functioning is .99 or greater?

5.6 A men's tennis tournament involves participants in four different divisions (A, B, C, and D). In each division, the resulting title match is between the first-seeded player and the second-seeded player. Suppose that the outcomes of the four matches are independent of one another and that in each one, $P(\textit{number 1 seed beats number 2 seed}) = .6$.

a. What is the probability that all four number 1 seeds are victorious?

b. Using the result of Exercise 5.4, what is the probability that at least one of the underdogs is victorious?

c. What is the probability that all four number 1 seeds lose their title matches?

5.7 After all students have left the classroom, a statistics professor notices that four copies of the text were left under desks. At the beginning of the next lecture, the professor distributes the four books in a completely random fashion to each of the four students (1, 2, 3, and 4) who claim to have left books. One possible outcome is that 1 receives 2's book, 2 receives 4's book, 3 receives his or her own book, and 4 receives 1's book. This outcome can be abbreviated as (2, 4, 3, 1).

a. List the other 23 possible outcomes.

b. Which outcomes are contained in the event that exactly two of the books are returned to their correct owners? Assuming equally likely outcomes, what is the probability of this event?

c. What is the probability that exactly one of the four students receives his or her book?

d. What is the probability that exactly three receive their own books?

e. What is the probability that at least two of the four students receive their own books?

5.8 Four potential donors—A, B, C, and D—are waiting at a blood bank, which needs a type O donor immediately. A and B have type O blood, while C and D do not (but the bank and donors don't know this). The donors are to be selected for blood typing in random order (think of drawing slips of paper one by one from a box), with typing terminating as soon as a type O donor is identified. One possible outcome is CDB (C is selected first but doesn't have type O blood, D is then selected and is also not a type O donor, and finally B is selected and is found to have type O blood); another outcome is A (A is selected first and found to be type O, so that no further typing is necessary).

a. List the remaining 8 possible outcomes.

b. What outcomes are contained in the event that at most one typing is necessary?

c. What is the probability that the first individual selected is A? B? What is the probability that at most one typing is necessary?

d. Using the result of (c), are the 10 possible outcomes equally likely?

5.9 After mixing a deck of 52 cards very well, 5 cards are dealt out.

a. It can be shown that (disregarding the order in which the cards are dealt) there are 2,598,960 possible hands, of which only 1287 consist entirely of spades. What is the probability that a hand will consist entirely of spades? What is the probability that a hand will consist entirely of a single suit?

b. It can be shown that exactly 63,206 hands contain only spades and clubs with both suits represented. What is the probability that a hand consists entirely of spades and clubs with both suits represented?

c. What is the probability that a hand contains no red cards?

d. Using the results of (b), what is the probability that a hand contains cards from exactly two suits?

5.10 This case study is reported in the article "Parking Tickets and Missing Women," which appears in the book *Statistics: A Guide to the Unknown.* In a Swedish trial on a charge of overtime parking, a policeman testified that he had noted the position of the two air valves on the tires of a parked car—to the closest hour one was at the one o'clock position and the other at the six o'clock position. After the allowable time for parking in that zone had passed, the policeman returned, noted that the valves were in the same position, and ticketed the car. The owner of the car claimed that he had left the parking place in time and had returned later. The valves just happened by chance to be in the same positions. An "expert" witness computed the probability of this occurring as $(\frac{1}{12})(\frac{1}{12}) = \frac{1}{144}$.

a. What reasoning did the expert use to arrive at the probability of $\frac{1}{144}$?

b. Can you spot the error in the reasoning that leads to the stated probability of $\frac{1}{144}$? What effect does this error have on the probability of occurrence? Do you think that $\frac{1}{144}$ is larger or smaller than the correct probability of occurrence?

5.11 A particular airline has 10 A.M. flights from Chicago to New York, Atlanta, and Los Angeles. Let A denote the event that the New York flight is full, and define events B and C analogously for the other two flights. Suppose that $P(A) = .6$, $P(B) = .5$, $P(C) = .4$, and that the three events are independent.

a. What is the probability that all three flights are full? That at least one flight is not full?

b. What is the probability that only the New York flight is full? That exactly one of the three flights is full?

5.12 A shipment of 5000 printed circuit boards contains 40 that are defective. Two boards are chosen at random, without replacement. Consider the two events

E_1: first board selected is defective

E_2: second board selected is defective

a. Are E_1 and E_2 dependent events? Explain in words.

b. Let *not* E_1 be the event that the first board selected is not defective (the event that E_1 does not occur). What is $P(not\ E_1)$?

c. How do the two probabilities $P(E_2|E_1)$ and $P(E_2|not\ E_1)$ compare?

d. Based on your answer to (c), would it be reasonable to view E_1 and E_2 as approximately independent? Explain your answer.

5.2 Random Variables and Probability Distributions

When a chance experiment is performed, the focus is usually on some numerical characteristic of the resulting outcome. For example, after selecting a single individual from a population, we may inquire about the individual's blood pressure, annual salary, IQ, or number of traffic citations received during the last 3 years. The experiment may consist of tossing both a red die and a green die and then noting the sum of the two resulting numbers, the larger of the two numbers, or the difference between the larger and smaller number. After selecting and testing four batteries intended for use in a portable cassette recorder, one might focus on the number of defective batteries among the four. An experiment might involve opening sealed bids for oil drilling rights in an offshore tract, and interest would naturally center on the amount of the highest bid.

In each of these examples, once the numerical characteristic or variable of interest is specified, there is a single value of the variable associated with each possible experimental outcome. For each individual selected, there is a single value for the variable *blood pressure*. Each possible outcome of tossing a red and a green die has a particular sum (e.g., (2, 4) yields a sum of 6). Each selection of four batteries leads to some specified number—either 0, 1, 2, 3, or 4—that are defective. Before the battery experiment is performed, we don't know which of the five possible values of the variable *number of defectives* will result. There is uncertainty concerning the resulting value, so it is natural to speak of the variable as being random. Similarly, the sum of the numbers resulting from tossing both a red and a green die could be any of the values 2, 3, 4, . . . , 11, 12. Before the experiment is performed, there is uncertainty as to which value of the sum will be observed, so we call the sum a random variable.

DEFINITION	A **random variable** associates a single number with each outcome of an experiment.

We denote random variables by lowercase letters from the end of the alphabet, such as x, y, or z.

EXAMPLE 13

Consider the experiment in which a family with three children is selected and the sex of each child (oldest, middle, and youngest) is noted. Let the random variable x denote the number of female children in the selected family. Possible values of x are 0, 1, 2, and 3. Before the family is selected, there is uncertainty as to which value of x will be observed, so x is a variable quantity whose value is determined by a chance experiment. We list the eight possible experimental outcomes (where BGG denotes the outcome in which the oldest child is a boy and the other two children are girls). Clearly, for each outcome there is a single value of x, so x satisfies the condition necessary for it to be a random variable.

Outcome	BBB	GBB	BGB	BBG	GGB	GBG	BGG	GGG
Value of x	0	1	1	1	2	2	2	3

■

EXAMPLE 14

Reconsider the ballot problem introduced in Example 3. There, candidate A had 5 votes to 3 votes for B, each vote was written on a slip of paper, and the slips were selected one by one in random order. Table 1 lists all 56 possible outcomes. Define a random variable x as the vote number on which A goes ahead of B for the remainder of the vote count. For the outcome ABBAAABA, the two candidates are tied after the fourth vote is counted, but A moves ahead on the fifth vote and stays ahead, so $x = 5$ for this outcome. The value of x for the outcome ABABABAA is 7, while $x = 1$ for the outcome AAABAABB. One of the possible x values 1, 3, 5, or 7 is associated with each of the other 53 outcomes. Before the experiment is performed, there is uncertainty about the outcome and the associated x value, so x is a random variable. ■

EXAMPLE 15

A 2-ft-long metal rod, as pictured, will be cut at a randomly located point along the length of the rod. Let the random variable x denote the length of the shorter piece. If the cut point is .5 ft from the left end of the rod, the two pieces are .5 ft and 1.5 ft long, so $x = .5$. If the cut point is 1.25 ft from the left end, then $x = .75$. In a similar manner, any other cut point (outcome) results in a unique value of x. Any number between 0 and 1 is a possible value of x, so the set of possible values is an entire interval on the number line.

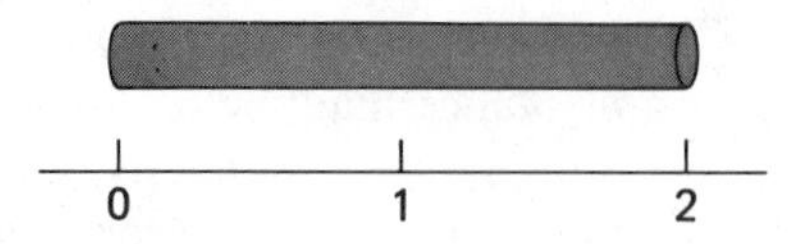

■

In Section 2.1 we distinguished between two types of numerical variables. A discrete variable is one whose set of possible values is a set of isolated points on the number line. The most frequently occurring variable of this type is one whose possible values are whole numbers (integers), such as the variables of Example 13 and Example 14. The variable of Example 15 is continuous rather than discrete because its possible values comprise an entire interval on the number line.

DEFINITION	A random variable is **discrete** if its set of possible values is a set of isolated points on the number line and **continuous** if its set of possible values consists of an entire interval on the number line.

Examples of discrete random variables include the number of defective transistors in a lot and the number of directory assistance calls during a 24-h period. Examples of continuous random variables include the tensile strength of a piece of wire, the gasoline efficiency (mi/gal) of an automobile, and the lifetime of a lightbulb.

Probability Distributions for Discrete Random Variables

The outcome that actually occurs when an experiment is performed determines which of the possible values of a discrete random variable x is observed. Some x values may have a very small chance of occurrence, whereas others might be relatively likely to be observed. The total probability associated with all outcomes of a chance experiment is 1. The chance of observing any particular x value is determined once we know how the total probability of 1 is distributed among the possible values of x.

The concept of a probability distribution follows directly from our work with relative frequency distributions and our interpretation of probability. Suppose, for example, that components of a certain type are packed four to a box for shipment to distributors. Let the random variable x denote the number of defective components in a randomly selected box. Possible x values are 0, 1, 2, 3, and 4. If we sample 100 such boxes, we might find 46 with $x = 0$, 38 with $x = 1$, 13 with $x = 2$, 3 with $x = 3$, and none with $x = 4$. Dividing these observed frequencies by the sample size 100 yields this relative frequency distribution:

x value	0	1	2	3	4
Relative frequency	.46	.38	.13	.03	.00

Continuing to sample boxes, the relative frequency distribution after 1000 boxes are examined might be

x value	0	1	2	3	4
Relative frequency	.417	.403	.151	.027	.002

As more and more boxes are examined, the relative frequency for each x value stabilizes and approaches a limit. The limiting value of each relative frequency is just the probability associated with the corresponding x value. For example, if the limiting relative frequency of the x value 2 is .1536, then the probability of the event $x = 2$ is .1536. This can be written as $P(x = 2) = .1536$. Often $P(x = 2)$ is written more compactly as $p(2)$, so we write $p(2) = .1536$. Similarly, if the limiting relative frequency of the event $x = 0$ is .4096, we write $p(0) = .4096$.

DEFINITION	The **probability distribution of a discrete random variable** x gives the probability associated with each possible x value. Each probability is the limiting relative frequency of occurrence of the corresponding x value when the experiment is repeatedly performed.

Often it is not necessary to replicate the experiment to obtain the probability distribution. Assumptions about probabilities of outcomes and rules of probability can be used instead.

EXAMPLE 16

(*Example 7 continued*) In the ballot problem, the random variable x = the vote number on which candidate A forges ahead for good has possible values 1, 3, 5, and 7. Table 1 lists all possible outcomes; these are equally likely, so each has probability $\frac{1}{56}$. The variable x takes the value 1 for the 14 outcomes for which A is ahead from the start. If the experiment is repeated over and over, in the long run the event $x = 1$ occurs on $\frac{14}{56}$, or 25%, of the replications. Thus $p(1) = P(x = 1) = .25$. Again from Table 1, it can be verified that 10 outcomes have $x = 3$, 12 outcomes have $x = 5$, and 20 outcomes have $x = 7$, so $p(3) = \frac{10}{56} = .179$, $p(5) = \frac{12}{56} = .214$, and $p(7) = \frac{20}{56} = .357$. The distribution is often displayed in a tabular format.

x value	1	3	5	7
Probability $p(x)$	.250	.179	.214	.357

■

EXAMPLE 17

Suppose that a family with three children is selected at random. Assuming that each child is equally likely to be a boy or a girl (probability .5 for each of the two possibilities), the eight outcomes *BBB, GBB, BGB, BBG, BGG, GBG, GGB,* and *GGG* are equally likely, so each has probability $\frac{1}{8} = .125$. Let x be the number of girls in the selected family. Since there is just one outcome with $x = 0$, $p(0) = .125$. There are three outcomes with $x = 1$, so $p(1) = \frac{3}{8} = .375$. Similarly, $p(2) = .375$ and $p(3) = .125$. ■

EXAMPLE 18

Beer manufacturers have advertised heavily in an attempt to persuade us that premium beers (such as Lowenbrau and Michelob) taste better than regular beers and are worth the extra cost. Suppose that two glasses are filled with regular beer and one is filled with premium beer. Each of three individuals is asked to sip from all three glasses and identify the premium beer. The random variable of interest is x = the number of correct identifications. With S (success) denoting an individual who successfully identifies the premium glass and F one who does not, the outcomes and associated x values are as follows:

Outcome	FFF	SFF	FSF	FFS	FSS	SFS	SSF	SSS
x value	0	1	1	1	2	2	2	3

Suppose that there is really no difference whatsoever between the two types of beer (other than the packaging). Then any of the three glasses is equally likely to be designated the premium glass, so the probability of an individual being a success is $\frac{1}{3}$ and the probability of being a failure is $\frac{2}{3}$. Since judgments of different individuals are independent, individual S and F probabilities can be multiplied to obtain probabilities of the outcomes. For example, $P(FFF) = (\frac{2}{3})(\frac{2}{3})(\frac{2}{3}) = \frac{8}{27} = .296$. Similarly, $P(SFF) = (\frac{1}{3})(\frac{2}{3})(\frac{2}{3}) = .148$, and the two other outcomes with $x = 1$ also have probabilities .148. Thus $P(x = 0) = p(0) = .296$ and $p(1) = 3(.148) = .444$. A bit more calculation yields $p(2) = .222$ and $p(3) = .037$. ■

The experiments and random variables of these last two examples are quite similar. Each experiment consisted of three trials (first, second, and third child or first, second, and third individual), each trial could result in one of two outcomes (a dichotomy), the trials were independent, and the probabilities of the two outcomes were the same on each trial. The random variable of interest counted the number of trials on which a specified one of the two possibilities occurred. More generally, a **binomial experiment** consists of some fixed number of trials (not necessarily three) in which the other conditions just described are satisfied. The random variable of interest is then called a **binomial random variable.** The probability distribution of a binomial random variable is discussed in more detail in Section 5.6.

EXAMPLE 19

(*Example 18 continued*) The April 1978 issue of *Consumer Reports* reported the result of such a beer taste-testing experiment to compare Lowenbrau to Miller High Life. The number of participating individuals was 24 (24 trials of a binomial experiment). Under the hypothesis of no discernible difference, the success and failure probabilities are again $\frac{1}{3}$ and $\frac{2}{3}$. It is much more tedious to list outcomes here than in the case of just 3 individuals, but the general binomial distribution formula can be used to obtain the probability distribution of x = the number of correct identifications. The distribution, assuming no discernible difference, is given in the table.

x	0	1	2	3	4	5	6	7	8	9
$p(x)$	.000	.000	.004	.015	.039	.079	.125	.161	.171	.152

x	10	11	12	13	14	15	16	17	. . .	24
$p(x)$	.114	.072	.039	.018	.007	.002	.001	.000	. . .	.000

If this experiment were performed over and over again, each time with 24 tasters, in the long run the proportion of replications with 12 correct identifications (exactly half) would be .039. The proportion of replications with at least 11 correct identifications would be $.072 + .039 + .018 + .007 + .002 + .001 = .139$. Thus if there is really no discernible difference between the premium and regular beers, it would not be terribly surprising to see 11 or more correct identifications. The *Consumer Reports* article stated that there were indeed 11 correct identifications, so the experimental evidence is reasonably consistent with the no-difference hypothesis. If $x = 15$ had been observed, that would provide strong evidence for a discernible difference, since with no difference the probability that x is at least 15 is $P(15 \leq x) = .002 + .001 = .003$. This probability is so small that either something very unusual has been observed (something that would occur only about three times in a thousand), or else the probability assignment resulting from the hypothesis of no discernible difference is incorrect. We opt for the latter judgment. ■

In tabular form, the probability distribution for a discrete random variable looks exactly like a relative frequency distribution of the sort discussed in Chapter 2. There we introduced a histogram as a pictorial representation of a relative frequency distribution. An analogous picture for a discrete probability distribution is called a **probability histogram.** The picture has a rectangle centered above each possible value of x, and the area of each rectangle is the probability of the corresponding value. Figure 3 displays the probability histogram for the probability distribution of Example 18.

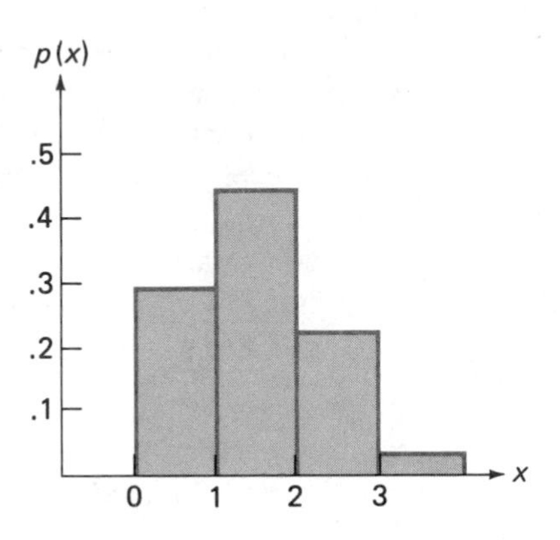

FIGURE 3 PROBABILITY HISTOGRAM FOR THE DISTRIBUTION OF EXAMPLE 18

Probability Distributions for Continuous Random Variables

To see how a probability distribution for a continuous random variable x is specified, recall how a histogram for continuous data was constructed. First, class intervals were selected, and then a rectangle was drawn above each interval. The area of each rectangle was the proportion (i.e., relative frequency) of observations falling in the corresponding interval. To complete the picture, the vertical axis was labeled the density scale, so that (density)·(interval width) = area = relative frequency. One implication of this was that the total area of all rectangles was 1. Also, adding areas of adjacent rectangles gave the proportion of observations falling in an interval wider than the class intervals themselves. Now suppose that by repeating the experiment a very large number of times, we obtain a very large number of observed x values. In this situation,

1. The histogram can be based on very narrow class intervals.
2. The area of each rectangle is close to the probability that x falls in the corresponding class interval (by the limiting relative frequency definition of probability).

Thus the more observations we have on x, the more closely the histogram can be made to look like a smooth curve (see Figures 18 and 19 in Chapter 2). This suggests that we specify a continuous probability distribution by selecting an appropriate smooth curve such that the total area under the curve is 1. Then the probability that x falls in a particular interval would be identified with the area under the curve and above the interval.

DEFINITION	A **probability distribution for a continuous random variable** x is specified by selecting a smooth curve, called the **density curve,** such that the total area under the curve is one. Then the probability that x falls in any particular interval is the area under the density curve and above the interval.

For any two numbers a and b with $a < b$, let $a < x < b$ denote the event that x is observed to fall in the interval between a and b. Similarly, let $x < a$ denote the event that x is observed to be less than a, and let $b < x$ denote the event that x exceeds b. Figure 4 illustrates how the probabilities of these events are identified with areas under a particular density curve.

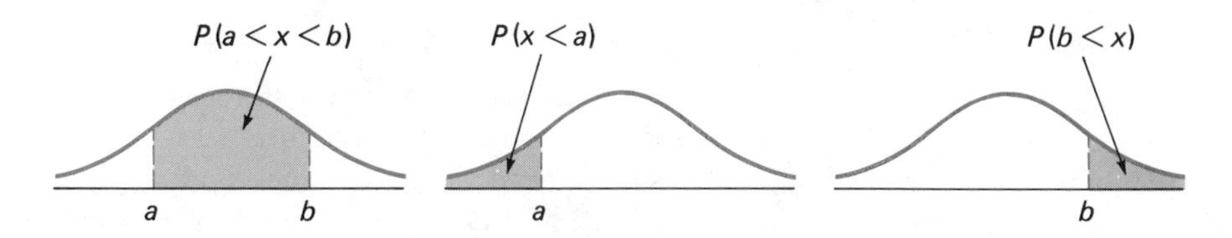

FIGURE 4 PROBABILITIES AS AREAS UNDER A PROBABILITY DENSITY CURVE

EXAMPLE 20

Let x be the amount of time it takes for a particular clerk to process a certain type of application form. Suppose that the density curve is as pictured in Figure 5. This distribution, often referred to as *uniform,* is especially easy to use since calculating probabilities requires only finding areas of rectangles using the formula $(base)(height) = area$.

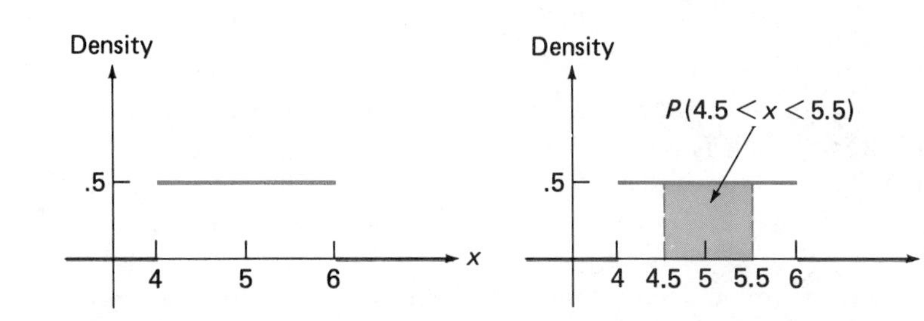

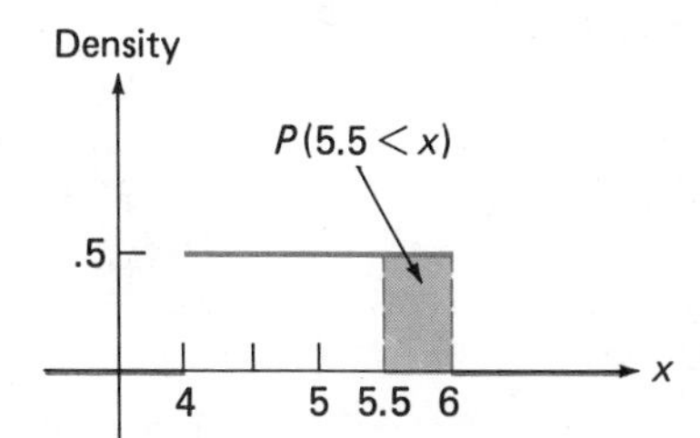

FIGURE 5 THE UNIFORM DISTRIBUTION FOR EXAMPLE 20

The curve has positive height only between $x = 4$ and $x = 6$, so according to this model the smallest possible x value is 4 and the largest value is 6. The total area under the curve is just the area of the rectangle whose base extends from 4 to 6 and whose height is .5. This gives area $= (6 - 4)(.5) = 1$ as required.

The probability that x is between 4.5 and 5.5 is the area under the curve and above this interval, which is again the area of a rectangle: $P(4.5 < x < 5.5) = (5.5 - 4.5)(.5) = .5$. Similarly, x is greater than 5.5 if and only if it is between 5.5 and 6, so $P(5.5 < x) = (6 - 5.5)(.5) = .25$.

Thus, according to this uniform model, in the long run 25% of all forms processed will result in an x value of more than 5.5 min.

Suppose the clerk's supervisor suspects that processing times tend to take longer than those that would be expected with this model. The supervisor surreptitiously monitors processing time on 10 forms and finds that each of the 10 times exceeds 5.5 min. Assuming independent processing times for the 10 forms, the uniform distribution model implies that

$$\begin{aligned} P(\text{all 10 times exceed 5.5}) &= P(\text{1st exceeds 5.5}) \\ &\quad \cdot P(\text{2nd exceeds 5.5}) \cdots P(\text{10th exceeds 5.5}) \\ &= (.25)(.25) \cdots (.25) = .00000095 \end{aligned}$$

If a very large number of batches of 10 forms were each monitored, all 10 times in a batch would exceed 5.5 min roughly once in a million batches! Either something very unusual has occurred, or (more plausibly) this uniform distribution model was not the correct probability distribution for x. ■

Suppose that the discrete random variable x represents the number of left-handed writers among 10 randomly selected adults. Then the event $3 \le x \le 7$ is comprised of the x values 3, 4, 5, 6, and 7 (the interval's end values 3 and 7 are included), whereas $3 < x < 7$ includes only the values 4, 5, and 6. Since $P(x = 3)$ and $P(x = 7)$ are both greater than zero, $P(3 \le x \le 7)$ differs from (is larger than) $P(3 < x < 7)$. If, however, x is a continuous random variable such as task completion time, then the probability of the event $3 \le x \le 7$ is the same as the probability of the event $3 < x < 7$ because the area under the density curve above a single value, such as 3, is zero. Thus the area under the density curve is not affected by the inclusion of either end value. Similarly, $P(x < 7) = P(x \le 7)$ and $P(3 < x) = P(3 \le x)$ when x is a continuous random variable.

> When x is a continuous random variable, the probability that x falls in some interval is not affected by the inclusion or exclusion of an end value of the interval. This is not usually the case when x is a discrete random variable.

A comprehensive treatment of probability distributions for continuous random variables is beyond our scope because most of the density curves that are useful in applied work are specified by complicated mathematical functions. Also, finding areas under such curves typically involves using calculus. Fortunately, for our purposes a comprehensive treatment is unnecessary. The most important continuous distribution is specified by a normal density curve, and we see in Section 5.4 how to use an appropriate table to obtain areas under a normal curve. Later on we encounter other continuous distributions in connection with statistical inference, and again the use of appropriate tables yields all information necessary for drawing conclusions.

EXERCISES

5.13 Airlines sometimes overbook flights. Suppose that for a plane with 100 seats, an airline takes 110 reservations. Define the variable x as the number of people who actually show up for a sold-out flight. From past experience, the probability distribution of x is:

x	95	96	97	98	99	100	101	102	103	104	105	106	107	108	109	110
$p(x)$	.05	.10	.12	.14	.24	.17	.06	.04	.03	.02	.01	.005	.005	.005	.0037	.0013

a. What is the probability that the airline can accommodate everyone who shows up for the flight?
b. What is the probability that not all passengers can be accommodated?
c. If you are trying to get a seat on such a flight and you are number 1 on the standby list, what is the probability that you will be able to take the flight? What if you are number 3?

5.14 Let x denote the number of courses for which a randomly selected student at a particular university is registered. Consider the accompanying probability distribution of x.

x	1	2	3	4	5	6	7
$p(x)$	.02	.03	.11	.32	.38	.13	.01

a. Draw a probability histogram for this probability distribution.
b. What is $P(x \leq 3)$?
c. What is $P(x \leq 6)$?
d. What is $P(3 \leq x \leq 6)$?

5.15 Many manufacturers have quality control programs that include inspection of incoming materials for defects. Suppose that a computer manufacturer receives computer boards in lots of 5. Two boards are selected from each lot for inspection. We can represent possible outcomes of the selection process by pairs. For example, the pair (1, 2) represents selection of boards 1 and 2 for inspection.

a. List the 10 different possible outcomes.
b. Suppose that boards 1 and 2 are the only defective boards in a lot of 5. Two boards are to be chosen at random. Define x to be the number of defective boards observed among those inspected. Find the probability distribution of x.

5.16 Simulate the experiment described in Exercise 5.15 using five slips of paper with two marked defective and three marked nondefective. Place the slips in a box, mix them well and draw out two. Record the number of defectives. Replace the slips and repeat until you have 50 observations on the variable x. Construct a relative frequency distribution for the 50 observations and compare this with the probability distribution obtained in Exercise 5.15.

5.17 Suppose that on any given day a particular stock has probability .6 of going up in price. If the stock is observed on three randomly chosen days and x is defined as the number of days on which the stock went up, find the probability distribution of x. (*Hint:* One outcome is *UDD*, with $x = 1$; assuming independence, this outcome has probability $(.6)(.4)(.4) = .096$.)

5.18 Refer back to Exercise 5.7 and let the random variable x be the number of students who receive their own books. What are the possible values of x? Assuming that outcomes are equally likely, obtain the probability distribution of x.

5.19 Some parts of California are particularly earthquake-prone. Suppose that in one such area, 40% of all homeowners are insured against earthquake damage. Four homeowners are to be selected at random. Let x denote the number among the four who have earthquake insurance.

a. Find the probability distribution of x. (*Hint:* Let S denote a homeowner who has insurance and F denote one who doesn't. Then one possible outcome is

SFSS, with probability (.4)(.6)(.4)(.4) and associated x value 3. There are 15 other outcomes.)

b. What is the most likely value for x?

5.20 Let x be the amount of time (min) that a particular San Francisco commuter must wait for a BART train. Suppose that the density curve is as pictured (a uniform distribution).

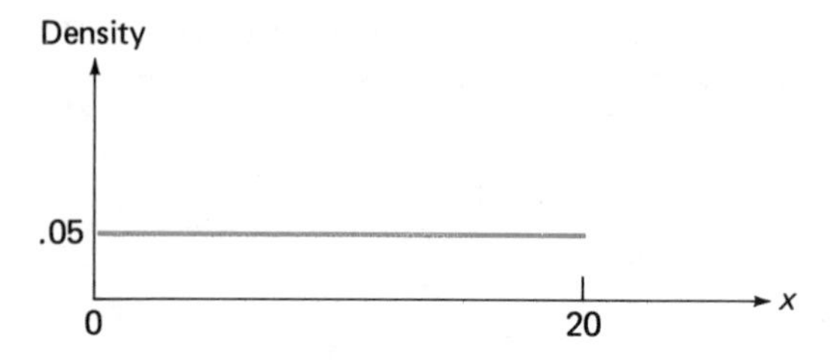

a. What is the probability that x is less than 10 min? More than 15 min?

b. What is the probability that x is between 7 and 12 min?

c. Find the value c for which $P(x < c) = .9$.

5.21 Referring to Exercise 5.20, let x and y be waiting times on two independently selected days. Define a new random variable w by $w = x + y$, the sum of the two waiting times. The set of possible values for w is the interval from 0 to 40 (since both x and y can range from 0 to 20). It can be shown that the density curve of w is as pictured. (It is called a *triangular distribution* for obvious reasons!)

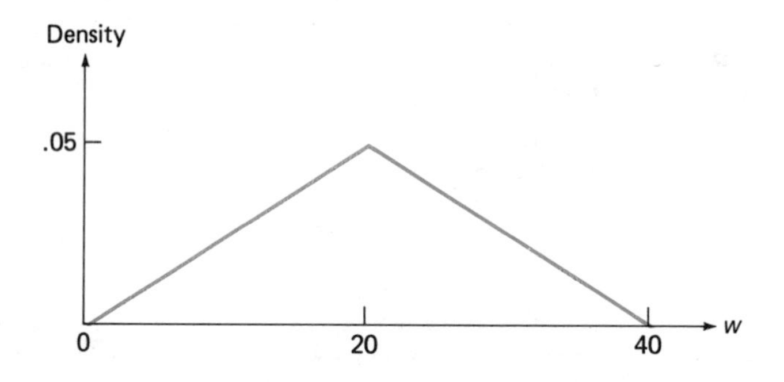

a. Verify that the total area under the density curve is equal to 1. (*Hint:* The area of a triangle is $(\frac{1}{2})$(base)(height).)

b. What is the probability that w is less than 20? Less than 10? More than 30?

c. What is the probability that w is between 10 and 30? (*Hint:* It might be easier first to find the probability that w is *not* between 10 and 30.)

5.3 The Mean Value and Standard Deviation of a Random Variable

The purpose in studying some random variable x, such as the number of insurance claims made by a homeowner (a discrete variable) or the tensile strength of a certain type of wire (a continuous variable) is to learn something about how its values are distributed along the measurement scale.

When a sample of observations on x is available, a sample histogram gives a picture of how these sample values are distributed. The sample mean $\overline{x}$ is a summary measure describing where the sample is centered, and the sample standard deviation s describes the amount of spread, or variability, in the sample.

The long-run behavior of x when many, many observations are made is described by its probability distribution. The probability distribution can be pictured by drawing a probability histogram (in the discrete case) or density curve (in the continuous case). The mean value of the random variable x is a measure of where the probability distribution is centered. It describes the long-run average behavior of x when the experiment is repeatedly performed. Similarly, the standard deviation of x describes the extent to which the probability distribution spreads out about the mean value. If the probability histogram or density curve is quite spread out, the standard deviation will be large, suggesting much long-run variability in observed x values. Conversely, a small standard deviation suggests that there will be relatively little variability in a long sequence of observed x values, so that most observations will be close to the mean value.

The Mean Value of a Discrete Random Variable

Consider the experiment in which an automobile licensed in a particular state is randomly selected. Let the discrete random variable x be the number of low-beam headlights on the selected car that need adjustment. Possible x values are 0, 1, and 2, and the probability distribution of x and corresponding probability histogram might look as follows:

x value	0	1	2
Probability	.5	.3	.2

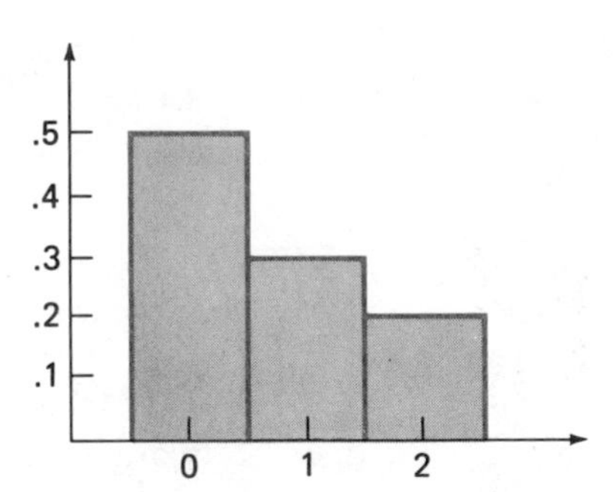

In a sample of 100 such cars, the sample relative frequencies might differ somewhat from these probabilities (limiting relative frequencies). We might see 46 cars with $x = 0$ (relative frequency .46), 33 cars with $x = 1$, and 21 cars with $x = 2$. The sample average value of x for these 100 observations is then the sum of 46 zeroes, 33 ones, and 21 twos, all divided by 100. This can be written as

$$\overline{x} = \frac{(46)(0) + (33)(1) + (21)(2)}{100} = \left(\frac{46}{100}\right)(0) + \left(\frac{33}{100}\right)(1) + \left(\frac{21}{100}\right)(2)$$

$$= \begin{pmatrix}\text{relative}\\\text{frequency}\\\text{of } 0\end{pmatrix}(0) + \begin{pmatrix}\text{relative}\\\text{frequency}\\\text{of } 1\end{pmatrix}(1) + \begin{pmatrix}\text{relative}\\\text{frequency}\\\text{of } 2\end{pmatrix}(2) = .75$$

As the sample size increases, each relative frequency approaches the corresponding probability. In a very long sequence of experiments, the value of $\overline{x}$ will approach

$$\begin{pmatrix}\text{probability}\\ \text{that}\\ x = 0\end{pmatrix}(0) + \begin{pmatrix}\text{probability}\\ \text{that}\\ x = 1\end{pmatrix}(1) + \begin{pmatrix}\text{probability}\\ \text{that}\\ x = 2\end{pmatrix}(2)$$

$$= (.5)(0) + (.3)(1) + (.2)(2) = .70 = \text{mean value of } x$$

Notice that $\overline{x}$ appears above as a *weighted average* of possible x values; the weight for each value is the observed relative frequency. Similarly, the mean value of x is a weighted average, but now the weights are the probabilities from the probability distribution. The mean value of x is thus a characteristic of the probability distribution rather than any particular sample.

DEFINITION

The **mean value of a discrete random variable** x, denoted by μ_x (or just μ when the identity of x is obvious), is computed by first multiplying each possible x value by the probability of observing that value and then adding the resulting quantities. Symbolically,

$$\mu_x = \sum_{\substack{\text{all possible}\\ x \text{ values}}} x \cdot (\text{probability of } x)$$

It is no accident that the symbol μ for the mean value is the same symbol used earlier for a population mean. When the probability distribution describes how x values are distributed among the members of a population (so probabilities are population relative frequencies), the mean value of x is exactly the average value of x in the population.

EXAMPLE 21

At 1 min after birth and again at 5 min, each newborn child is given a numerical rating called an *Apgar score.* Possible values of this score are 0, 1, 2, . . . , 9, and 10. A child's score is determined by five factors: muscle tone, skin color, respiratory effort, strength of heartbeat, and reflex, with a high score indicating a healthy infant. Let the random variable x denote the Apgar score of a randomly selected newborn infant at a particular hospital, and suppose that x has the given probability distribution.

x	0	1	2	3	4	5	6	7	8	9	10
Probability	.002	.001	.002	.005	.02	.04	.17	.38	.25	.12	.01

The mean value of x is

$$\mu = (0)\cdot P(x = 0) + (1)\cdot P(x = 1) + \cdots + (9)\cdot P(x = 9) + (10)\cdot P(x = 10)$$
$$= (0)(.002) + (1)(.001) + \cdots + (9)(.12) + (10)(.01) = 7.16$$

The sample average Apgar score for a sample of newborn children born at this hospital may be $\overline{x} = 7.05$, $\overline{x} = 8.30$, or any other number between 0 and 10. However, as child after child is born and rated, the average Apgar score will approach the value 7.16.

Suppose that the given distribution refers just to those children born at the hospital within the last year, so that each probability is a population relative frequency. Then the average value of x in this particular population of children is $\mu = 7.16$, the population mean Apgar score.

A particular pediatrician presently sees 36 children who were born at this hospital. Looking back through birth records, the pediatrician discovers that the sample average Apgar score for these 36 children is $\bar{x} = 7.76$. This appears to be substantially higher than the long-run average value 7.16 for all children born at the hospital. Assuming that the distribution of scores for this pediatrician's patients is identical to the one displayed above, it can be shown that the chance of obtaining a sample of 36 scores with an $\bar{x}$ value at least as large as the observed value 7.76 is only roughly .001 (we do such calculations in the next chapter). Either something very unlikely has occurred, or—more plausibly—the value of μ for this pediatrician's patients is higher than the hospital average of 7.16. ■

The Standard Deviation of a Discrete Random Variable

The mean value μ provides only a partial summary of a probability distribution. Two different distributions may both have the same value of μ, yet a long sequence of sample values from one distribution may exhibit considerably more variability than a long sequence of values from the other distribution.

EXAMPLE 22

A television manufacturer receives certain components in lots of four from two different suppliers. Let x and y denote the number of defective components in randomly selected lots from the first and second suppliers, respectively. The probability distributions and associated probability histograms for x and y are given in Figure 6.

x	0	1	2	3	4
Probability	.4	.3	.2	.1	0

y	0	1	2	3	4
Probability	.2	.6	.2	0	0

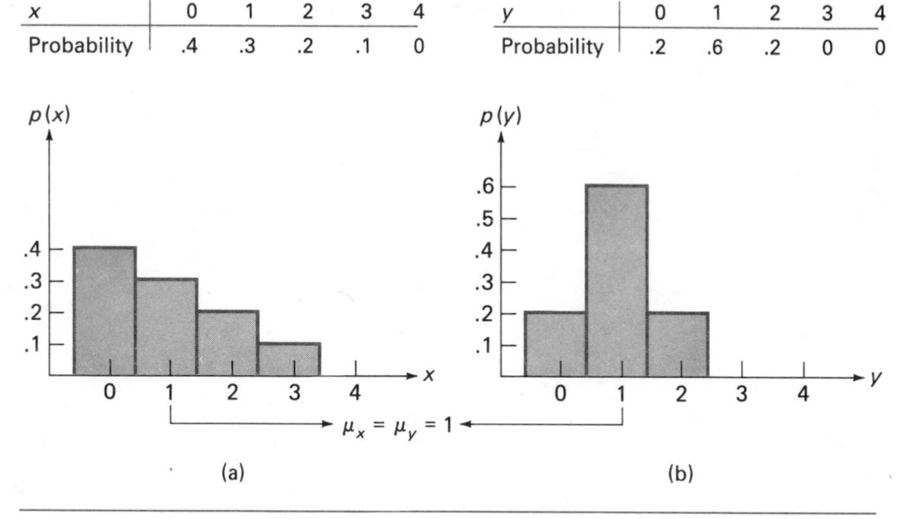

FIGURE 6

PROBABILITY DISTRIBUTION FOR THE NUMBER OF DEFECTIVE COMPONENTS
(a) In a Lot from Supplier 1
(b) In a Lot from Supplier 2

The mean values of both x and y are 1, so for either supplier the long-run average number of defectives per lot is 1. However, the two probability histograms show that the probability distribution for the second supplier is concentrated nearer the mean value than is the first supplier's distribution. The greater spread of the first distribution implies that there will be more variability in a long sequence of observed x values than in an observed sequence of y values. For example, the y sequence will contain no 3's, whereas in the long run, 10% of the observed x values would be 3. ■

As with the sample variance and standard deviation, the variance and standard deviation of x involve squared deviations from the mean. A value far from the mean results in a large squared deviation. However, such a value contributes substantially to variability in x only if the probability associated with that value is not too small. For example, if $\mu_x = 1$ and $x = 25$ is a possible value, the squared deviation is $(25 - 1)^2 = 576$. If, however, $P(x = 25) = .000001$, the value 25 will hardly ever be observed, so it won't contribute much to variability in a long sequence of observations. This is why each squared deviation is multiplied by the probability associated with the value (and thus weighted) to obtain a measure of variability.

DEFINITION

The **variance of a discrete random variable** x, denoted by σ_x^2 or just σ^2, is computed by first subtracting the mean from each possible x value to obtain the deviations, squaring each deviation and multiplying the result by the probability of the corresponding x value, and finally adding these quantities. Symbolically,

$$\sigma^2 = \sum_{\substack{\text{all possible} \\ x \text{ values}}} (x - \mu)^2 \cdot (\text{probability of } x)$$

The **standard deviation** of x, denoted by σ_x or just σ, is the square root of the variance.

When the probability distribution describes how x values are distributed among members of a population (so probabilities are population relative frequencies), σ^2 and σ are the population variance and standard deviation (of x), respectively.

EXAMPLE 23

(*Example 22 continued*) For $x =$ the number of defectives in a lot from the first supplier,

$$\begin{aligned}\sigma_x^2 &= (0 - 1)^2 \cdot P(x = 0) + (1 - 1)^2 \cdot P(x = 1) \\ &\quad + (2 - 1)^2 \cdot P(x = 2) + (3 - 1)^2 \cdot P(x = 3) \\ &= (1)(.4) + (0)(.3) + (1)(.2) + (4)(.1) = 1.0\end{aligned}$$

so $\sigma_x = 1.0$. For $y =$ the number of defectives in a lot from the second supplier,

$$\sigma_y^2 = (0 - 1)^2(.2) + (1 - 1)^2(.6) + (2 - 1)^2(.2) = .4$$

Then $\sigma_y = \sqrt{.4} = .632$. The fact that $\sigma_x > \sigma_y$ confirms what we already guessed from Figure 6: There is more variability in the x distribution than in the y distribution. ■

The Mean and Standard Deviation when x Is Continuous

The mean value of a discrete random variable involves adding the quantities (x value) · (probability of x value), whereas the variance involves adding the quantities (x value $- \mu)^2$ · (probability of x value). Figure 7 illustrates how the density curve for a continuous random variable can be approximated by a probability histogram of a discrete random variable. Thus a continuous probability distribution can be approximated by the probability distribution of a discrete random variable. Computing the mean value and standard deviation using this discrete approximation gives an approximation to μ and σ for the continuous random variable x. If an even more accurate approximating probability histogram is used (narrower rectangles), better approximations to μ and σ result.

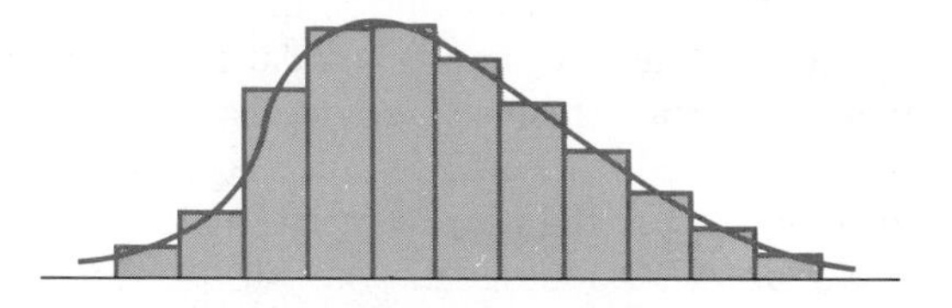

FIGURE 7 APPROXIMATING A DENSITY CURVE BY A PROBABILITY HISTOGRAM

In practice, it is usually not necessary to compute μ and σ by using such a sequence of approximations. Instead, in the continuous case μ and σ can be formally defined by using methods from calculus. The details of such a definition need not concern us. What is important is that μ and σ play exactly the same role here as they did in the discrete case. The mean value μ measures the center of the continuous distribution and gives the approximate long-run average of many observed x values. The standard deviation σ measures to what extent the continuous distribution (density curve) spreads out about μ and gives information about the amount of variability that can be expected in a long sequence of observed x values.

For example, suppose that a company receives concrete of a certain type from two different suppliers. Let x, a continuous random variable, denote the compressive strength (psi) of a randomly selected batch of concrete from the first supplier, and let y be defined similarly for the second supplier. If $\mu_x = 4680$ psi and $\mu_y = 4500$ psi, then the long-run average compressive strength per batch for many, many batches from supplier 1 is roughly 4680, which is 180 psi greater than the long-run average for batches from supplier 2. If, in addition, $\sigma_x = 200$ and $\sigma_y = 265$, a long sequence of batches from supplier 1 would exhibit substantially less variability in compressive strength values than would a similar sequence from supplier 2. Both in terms of average value and variability, the first supplier would be preferred to the second one.

More on Interpreting μ and σ

When the standard deviation was first introduced in Chapter 3, we commented that it was a less natural summary measure than the mean—more difficult to interpret and less obvious in its use. We did discuss Chebyshev's rule and the empirical rule as examples of how knowledge of the mean and standard deviation can lead to useful statements about the distribution of values in a data set. These two rules can also be used with μ_x and σ_x to make useful statements about the probability distribution of x.

EXAMPLE 24

In Example 21 we computed $\mu = 7.16$ from the probability distribution of Apgar score x given there. Applying the definition of σ to this distribution yields $\sigma = 1.25$. Suppose for the moment that the entire distribution is unknown to us but that the values $\mu = 7.16$ and $\sigma = 1.25$ are given. Adding 2 standard deviations to the mean gives $7.16 + 2(1.25) = 9.66$, so the only possible score that is more than 2 standard deviations above the mean is 10. Similarly, $\mu - 2\sigma = 4.66$, so the scores 0, 1, 2, 3, and 4 are all more than 2 standard deviations below the mean.

Chebyshev's rule can be reworded to state in part that whatever the distribution of x, the chance of observing a value more than 2 standard deviations from the mean is at most .25. That is, in the long run at most 25% of observed x values fall further than 2 standard deviations from the mean. For our example involving Apgar scores, we can say that at most 25% of the scores from a long sequence of newborns will be 0, 1, 2, 3, 4, or 10 (those scores which are further than 2σ from μ). At least 75% of the scores will be 5, 6, 7, 8, or 9. Looking at the distribution previously displayed, $P(x = 0) + \cdots + P(x = 4) + P(x = 10) = .04$, so Chebyshev's rule is actually quite conservative. This is the price that must be paid for a rule that applies to all distributions.

If the probability distribution (probability histogram or density curve) were approximately bell-shaped, then the empirical rule could be applied to conclude that the chance of observing an x value more than 2 standard deviations away from the mean is only .05, much smaller than what Chebyshev says. However, the probability histogram for the Apgar score distribution is quite skewed, so the empirical rule should not be applied (even though .05 is much closer to the actual value .04 than is .25). ■

As we proceed, the usefulness of the standard deviation will become more apparent. In particular, this measure of variability is indispensable when using methods of statistical inference to draw conclusions about a population mean μ.

The Mean and Standard Deviation of a 0-1 Variable

Many statistical problems involve drawing a conclusion about the proportion of individuals in a population who possess some characteristic—have a certain chromosomal mutation, own a foreign car, have completed high school, etc. Let the Greek letter π denote the proportion of the population possessing the characteristic of interest, so π is a number between 0 and 1 (e.g., $\pi = .25$ means that 25% of the population members have the characteristic). Suppose that we now attach a numerical label to each population member, a 1 if the individual has the characteristic and a 0 otherwise. Then π is the proportion of the population having label 1.

Let x be the label on an individual selected at random from the population. Then x is a discrete random variable whose only possible values are 0 and 1, with $P(x = 1) = \pi$ and $P(x = 0) = 1 - \pi$. The mean value of x is

$$\mu_x = \sum_{\substack{\text{all possible} \\ x \text{ values}}} x \cdot (\text{probability of } x)$$

$$= (0) \cdot P(x = 0) + (1) \cdot P(x = 1) = (0)(1 - \pi) + (1)(\pi)$$
$$= \pi$$

That is, the mean value of x is just the proportion of 1 labels in the population.

The variance of x is

$$\sigma_x^2 = \sum_{\substack{\text{all possible} \\ x \text{ values}}} (x - \text{mean value})^2 \cdot (\text{probability of } x \text{ value})$$

$$= (0 - \pi)^2 \cdot P(x = 0) + (1 - \pi)^2 \cdot P(x = 1)$$
$$= (-\pi)^2(1 - \pi) + (1 - \pi)^2(\pi) = \pi(1 - \pi)$$

The standard deviation of x is then $\sigma_x = \sqrt{\pi(1 - \pi)}$.

> Consider a population in which each individual has either the label 1 or the label 0, and let x denote the label on a single randomly selected individual. With π denoting the population proportion of 1's, $\mu_x = \pi$ and $\sigma_x = \sqrt{\pi(1 - \pi)}$.

The result $\mu_x = \pi$ makes intuitive sense. It says that the proportion of 1's observed in a long sequence of observations is exactly the proportion of 1 labels in the population. The expression for σ_x is less obvious. Notice that if $\pi = 0$, $\sigma_x = 0$. In this case, every label is zero, so there is no long-run variability in the observed sequence. Similarly, $\sigma_x = 0$ if $\pi = 1$ (since then every label is 1). The largest value of $\sqrt{\pi(1 - \pi)}$ occurs when $\pi = .5$, so there will be more long-run variability for a 50-50 split between 0's and 1's than for any other split. In general, there is more variability when π is close to .5 than when π is near to either 0 or 1.

EXERCISES

5.22 A personal computer salesperson working on a commission receives a fixed amount for each system sold. Suppose that for a given month, the probability distribution of x = number of systems sold is given in the accompanying table.

x	1	2	3	4	5	6	7	8
$p(x)$	.05	.10	.12	.30	.30	.11	.01	.01

a. Find the mean value of x (the mean number of systems sold).

b. What is the probability that x is within 2 of its mean value?

c. Find the variance and standard deviation of x. How would you interpret these values?

5.23 A local television station sells 15-s, 30-s, and 60-s advertising spots. Let x denote the length of a randomly selected commercial appearing on this station, and suppose that the probability distribution of x is given in the table.

x	15	30	60
$p(x)$	.1	.3	.6

a. Find the average length for commercials appearing on this station.

b. If a 15-s spot sells for \$500, a 30-s spot for \$800, and a 60-s spot for \$1000, find the average amount paid for commercials appearing on this station. (*Hint:* Consider a new variable $y = cost$ and then find the probability distribution and mean value of y.)

5.24 An author has written a book and submitted it to a publisher. The publisher offers to print the book and gives the author the choice between a flat payment of \$10,000 or a royalty plan. Under the royalty plan, the author would receive \$1 for each copy of the book sold. The author thinks that the accompanying table gives the probability distribution of the variable x = number of books that will be sold. Which payment plan should the author choose? Why?

x	1,000	5,000	10,000	20,000
$p(x)$	.05	.3	.4	.25

5.25 A grocery store has an express line for customers purchasing at most five items. Let x be the number of items purchased by a randomly selected customer using this line. Give examples of two different assignments of probabilities such that the resulting distributions have the same mean but have standard deviations that are quite different.

5.26 Refer to Exercise 5.13 and compute the mean value of the number of people holding reservations who show up for the flight.

5.27 After shuffling a deck of 52 cards, 5 are dealt out. Let x denote the number of suits represented in the 5-card hand (the four suits are spades, hearts, diamonds, and clubs). Using some counting rules, it can be shown that the probability distribution of x is as given in the accompanying table. Compute the mean value and the standard deviation of x.

Value of x	1	2	3	4
Probability of value	.002	.146	.588	.264

5.28 When an attorney takes a case, she must decide whether to charge a flat fee or to take a fixed percent of any settlement. Suppose the attorney has a case for which she could either charge a flat fee of \$5000 or take 20% of any settlement. The attorney feels that the probability of losing the case is .4. If the case is won, the judge will award either \$50,000 or \$100,000. The probability of winning the case and receiving a judgment of \$50,000 is thought to be .5, whereas the probability of winning with a judgment of \$100,000 is thought to be .1. Does the attorney expect to make more money by charging the flat fee or by taking 20% of any settlement?

5.29 Suppose that the number of calls per day to an ambulance service has the given probability distribution.

x	0	1	2	3	4
$p(x)$	.2	.2	.3	.2	.1

a. What is the long-run average number of calls per day for this ambulance service?

b. Find the variance and standard deviation of x.

5.30 Let x denote the outcome when a fair die is rolled once. Then it is easily seen that the mean value of x is 3.5. Suppose that someone offers to give you either \$10 or else the amount $35/x$ after rolling a fair die. Which would you choose? (*Hint:* If you gamble, what are the possible values of your payoff and the associated probabilities? What is the mean value of the payoff?)

5.4 The Normal Distribution

A normal curve was first introduced in Section 2.6 as one whose shape gives a very good approximation to sample histograms for many different data sets. There are actually many different normal curves rather than just one. Every one is bell-shaped, and a particular normal curve results from specifying where the curve is centered and how much it spreads out about its center. Each such normal curve can be formally (mathematically) defined by specifying a *height function,* which gives the height of the curve above each point x on the measurement axis. The height function is rather complicated, but fortunately we needn't deal with it explicitly. For our purposes an acquaintance with some general properties and a table that gives certain normal curve areas will suffice.

DEFINITION

A continuous random variable x is said to have a **normal distribution** if the density curve of x is a normal curve. The mean value μ determines where the curve is centered on the measurement axis, and the standard deviation σ determines the extent to which the curve spreads out about μ.

Figure 8 illustrates normal density curves for several different values of μ and σ. As with all density curves, the total area under each curve is 1.

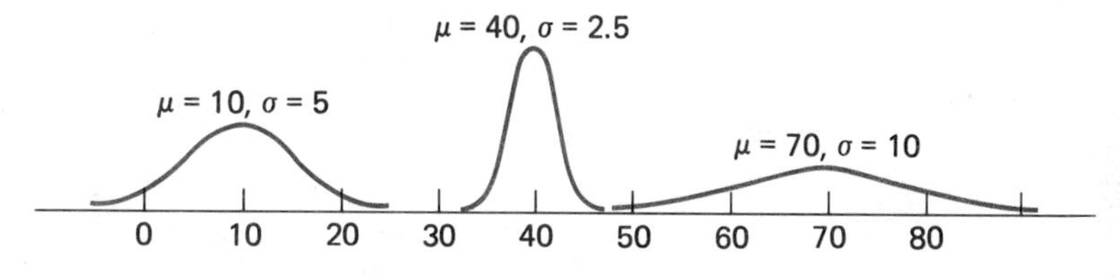

FIGURE 8 SEVERAL NORMAL DENSITY CURVES

The value of μ is the number on the measurement axis lying directly below the top of the bell. The value of σ can also be ascertained from a picture of the curve. Consider the normal curve pictured in Figure 9. Starting at the top of the bell (above 100) and moving to the right, the curve turns downward

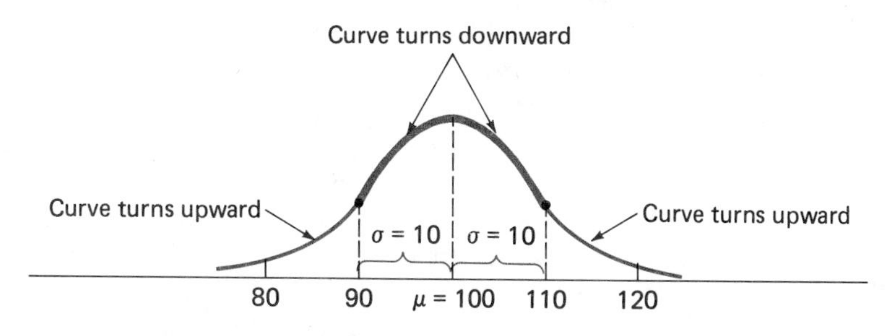

FIGURE 9 PICTORIAL IDENTIFICATION OF μ AND σ

until it is above the value 110. After that point it continues to decrease in height but is turning up rather than down. Similarly, to the left of 100 it turns down until it reaches 90 and then begins to turn up. The curve changes from turning down to turning up at a distance 10 on either side of μ, and thus $\sigma = 10$. In general, σ is the distance to either side of μ at which a normal curve changes from turning downward to turning upward.

Once it is assumed that a random variable x has a normal distribution with specified values of μ and σ, the probability that an observed x value falls in some interval is the area under the corresponding normal density curve and above the interval. For example, let x denote the number of miles per gallon achieved by a particular type of car in a fuel efficiency test. Assuming that x is a normally distributed variable with $\mu = 27.0$ and $\sigma = 1.5$, Figure 10 illustrates the curve areas that correspond to the probabilities $P(25.5 < x < 30)$, $P(x < 25.0)$, and $P(30.0 < x)$. The shaded areas in Figure 10 cannot, unfortunately, be determined using simple geometric arguments (as was done for the uniform distribution example at the end of Section 5.2). It turns out, though, that a single table of certain normal curve areas can be used to compute probabilities for any specified values of μ and σ.

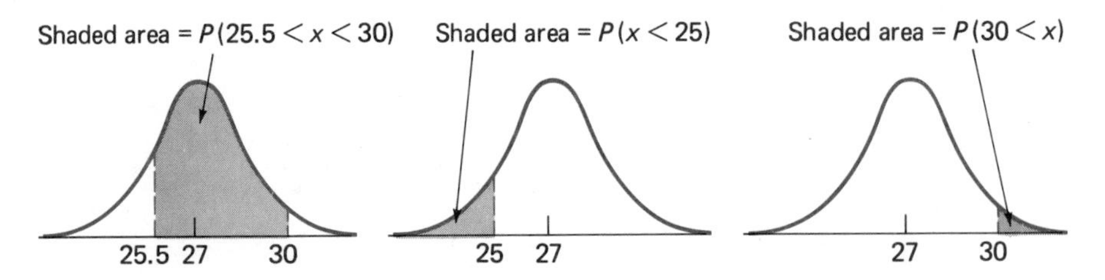

FIGURE 10 PROBABILITIES AS CURVE AREAS WHEN x HAS A NORMAL DISTRIBUTION WITH $\mu = 27.0$ AND $\sigma = 1.5$

The Standard Normal Curve

Rather than tabulate normal curve areas separately for each different combination of values of μ and σ, statisticians have chosen a particular normal curve as a reference curve. Once we learn to use the table containing areas for this reference curve, we shall see how areas under other normal curves (i.e., probabilities) are easily obtained.

DEFINITION

The normal curve with $\mu = 0$ and $\sigma = 1$ is called the **standard normal, or *z*, curve.** Table I in the appendices gives the area under the z curve both to the left of and to the right of different values along the measurement axis.

Figure 11 illustrates the types of z curve areas contained in Table I. Because the total area under the z curve is 1, either shaded area in Figure 11 is easily calculated once the other is known just by subtracting the known area from 1. We have chosen to tabulate both areas to avoid extra arithmetic.

The area under the z curve and above an interval is just the difference between two tabulated areas: the area to the left of the upper endpoint minus the area to the left of the lower endpoint. This is pictured in Figure 12.

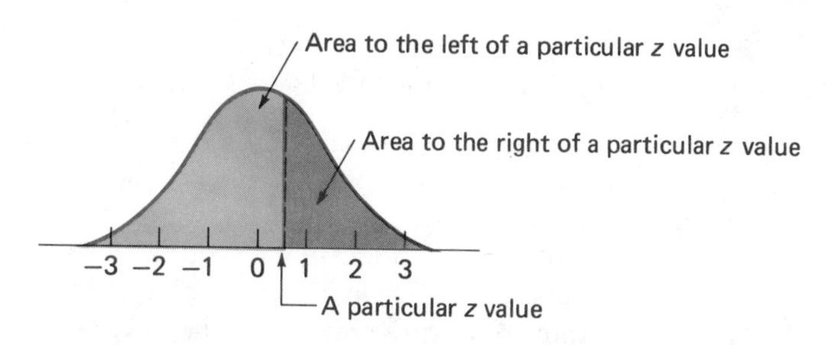

FIGURE 11

STANDARD NORMAL CURVE AREAS OF THE TYPE GIVEN IN TABLE I OF THE APPENDICES

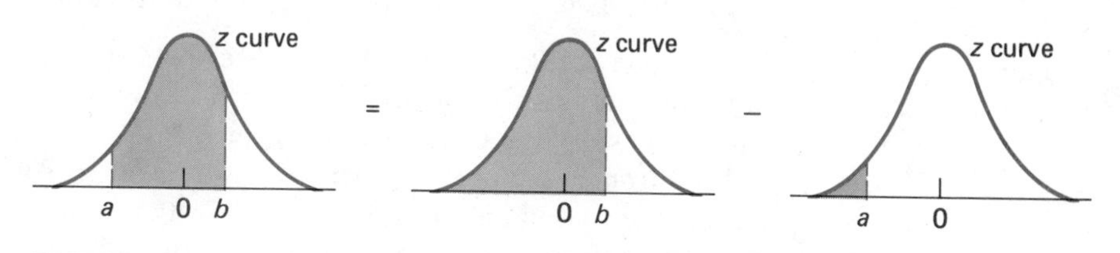

FIGURE 12

THE AREA ABOVE AN INTERVAL IS THE DIFFERENCE BETWEEN TWO TABULATED AREAS

EXAMPLE 25

To find the area under the z curve to the left of 1.2, locate 1.2 in the z column of Table I. The number next to 1.2, .8849, is the desired area. Right next to .8849 is .1151, the area under the z curve to the right of the value 1.2. Similarly, looking down the z column to −2.6 and then across, the area to the left of −2.6 is .0047 and the area to the right of −2.6 is .9953. The area under the z curve between −2.6 and 1.2 is .8849 − .0047 = .8802, as illustrated.

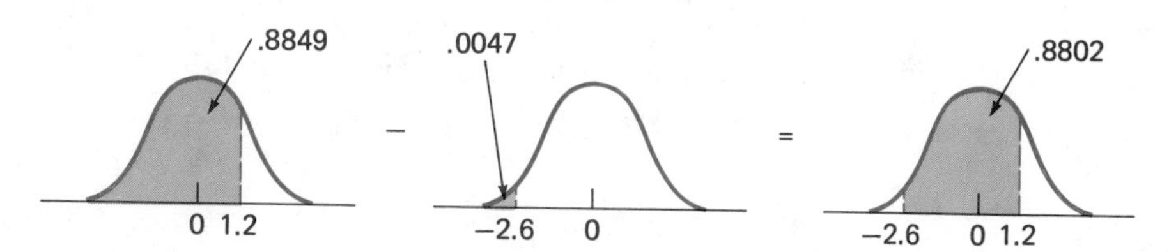

The area to the left of −2.0 is .0228, and the area to the right of 2.0 is also .0228; that these two are equal is a consequence of the symmetry of the z curve about zero. Thus the area under the z curve *outside* the interval between −2.0 and 2.0 is .0456. Since $\mu = 0$ and $\sigma = 1$ for the z curve, the value 2 (−2) is 2 standard deviations to the right (left) of the mean. Thus the total area above all values that are at least 2 standard deviations away from the mean is only .0456. The total area above all values that are within 2 standard deviations of the mean $(-2.0 < z < 2.0)$ is $1 - .0456 = .9544$ (roughly 95%). Similarly, the area above the interval $-3.0 < z < 3.0$ (within 3 standard deviations of the mean) is $.9987 - .0013 = .9974$. There is very little area under the z curve above values that are more than 3 standard deviations from the mean. ■

Example 25 illustrates how a z curve area can be obtained once the endpoint(s) of an interval are specified. In many applications, it is a curve area that is specified—often an upper tail area or lower tail area such as .05 or .01—and the corresponding z value(s) are desired.

EXAMPLE 26

What value on the z measurement axis is such that the area to the right of that value under the curve is only .01 (1% of the total area)? Since the area to the right of the desired value is specified, we look down the "area to the right of z" column of Table I to locate the specified area. The area .01 appears in this column opposite the value $z = 2.33$, so 2.33 is the desired z value. Similarly, if we want to capture area .01 in the lower tail, we look for .01 in the "area to the left of z" column. This appears opposite $z = -2.33$, so 1% of the area under the z curve lies to the left of -2.33. Notice that the desired upper and lower tail z values are opposite in sign but equal in magnitude. ■

EXAMPLE 27

Rather than a tail area, suppose that we want an interval extending an equal distance to either side of 0, above which lies .95 (95%) of the area. That is, with z a positive number, we want the value of z for which the area above the interval from $-z$ to z is .95. This situation is pictured; the area left over for the two tails combined is $1 - .95 = .05$. By symmetry, half of this area, or .025, must be contained in each tail. Now looking down the "area to the right of z" column, .025 appears opposite $z = 1.96$, so the desired interval extends from -1.96 on the left to 1.96 on the right.

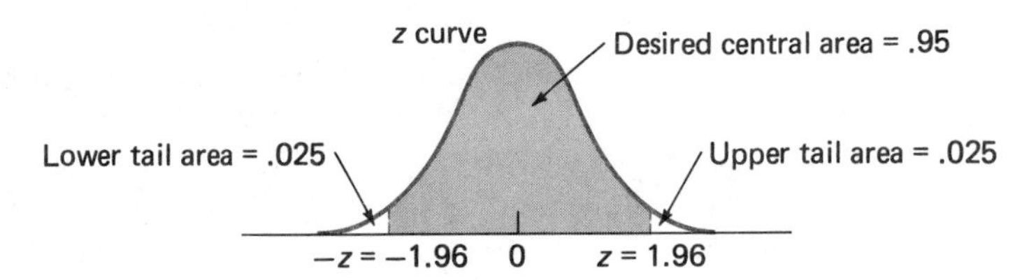

■

Many large-sample statistical inference procedures involve finding a z value that determines either a particular tail area or a particular central area. Such a value is often called a **z critical value.** Table 2 displays the most frequently used z critical values.

TABLE 2

THE MOST USEFUL z CRITICAL VALUES

Critical Value, z	Area to the Right of z	Area to the Left of $-z$	Area Between $-z$ and z
1.28	.10	.10	.80
1.645	.05	.05	.90
1.96	.025	.025	.95
2.33	.01	.01	.98
2.58	.005	.005	.99
3.10	.001	.001	.998
3.30	.0005	.0005	.999

Computing Probabilities for Any Normal Distribution

Suppose that x has a normal distribution with mean value μ and standard deviation σ. The method for computing probabilities such as $P(a < x < b)$, $P(x < a)$, and $P(b < x)$ is to **standardize** the values a and b by subtracting μ and dividing by σ and then refer to the z table.

To compute $P(a < x < b)$, first compute the two quantities

$$\text{standardized lower end value} = \frac{a - \mu}{\sigma}$$

$$\text{standardized upper end value} = \frac{b - \mu}{\sigma}$$

The desired probability is then the area under the z curve between the two standardized values. Similarly, $P(x < a)$ is the area under the z curve to the left of the standardized value $(a - \mu)/\sigma$, and $P(b < x)$ is the area under the z curve to the right of the standardized value $(b - \mu)/\sigma$.

EXAMPLE 28

Let x denote the systolic blood pressure (mm) of an individual selected at random from a certain population. Suppose that x has a normal distribution with mean $\mu = 120$ and standard deviation $\sigma = 10$. (The article "Oral Contraceptives, Pregnancy, and Blood Pressure" (*J. Amer. Med. Assoc.* 222 (1972):1507–10) reported on the results of a large study in which the distribution of blood pressure among women of similar ages was found to be well approximated by a normal curve.) To find the probability that x is between 110 and 140, we first standardize these values:

$$\text{standardized lower end value} = \frac{110 - 120}{10} = -1.0$$

$$\text{standardized upper end value} = \frac{140 - 120}{10} = 2.0$$

Then

$$\begin{aligned} P(110 < x < 140) &= \begin{pmatrix}\text{area under the } z \text{ curve} \\ \text{between } -1.0 \text{ and } 2.0\end{pmatrix} \\ &= \begin{pmatrix}\text{area to the} \\ \text{left of } 2.0\end{pmatrix} - \begin{pmatrix}\text{area to the} \\ \text{left of } -1.0\end{pmatrix} \\ &= .9772 - .1587 = .8185 \end{aligned}$$

Thus if blood pressure values for many, many women from this population were determined, roughly 82% of the values would fall between 110 and 140. Similarly, standardizing 125 yields $(125 - 120)/10 = .5$, so

$$P(x < 125) = \begin{pmatrix}\text{area under the } z \text{ curve} \\ \text{to the left of } .5\end{pmatrix} = .6915$$

Since x is a continuous random variable, $P(x \leq 125) = .6915$ also, and $P(110 \leq x \leq 140) = P(110 < x < 140) = .8185$.

Suppose that an individual's blood pressure is reported to exceed 160. Is it plausible that this individual was selected from the population under discussion? If this were the case, then because $(160 - 120)/10 = 4.0$,

$$P(160 < x) = \begin{pmatrix}\text{area under the } z \text{ curve} \\ \text{to the right of 4.0}\end{pmatrix} = .000033$$

Either something *very* unusual has occurred (a blood pressure value very far out in the upper tail of the distribution) or, more plausibly, the individual came from a population whose blood pressure distribution was something other than normal with $\mu = 120$ and $\sigma = 10$ (e.g., one with $\mu > 120$). ■

EXAMPLE 29

The Environmental Protection Agency has in recent years developed a testing program to monitor vehicle emission levels of several pollutants. The article "Determining Statistical Characteristics of a Vehicle Emissions Audit Procedure" (*Technometrics* (1980):483–93) describes the program, which involves using different vehicle configurations (combinations of weight, engine, transmission, and axle ratios) and a fixed driving schedule (including cold and hot start phases, idling, accelerating, and decelerating). Data presented in the paper suggest that the normal distribution is a plausible model for the amount of oxides of nitrogen (g/mi) emitted. Let x denote the amount of this pollutant emitted by a randomly selected vehicle with a particular configuration. Suppose that x has a normal distribution with $\mu = 1.6$ and $\sigma = .4$ and we wish to calculate the probability that x exceeds 2.0. Since $(2.0 - 1.6)/.4 = 1.0$,

$$P(2.0 < x) = \begin{pmatrix}\text{area under the } z \text{ curve} \\ \text{to the right of 1.0}\end{pmatrix} = .1587$$

Similarly, $(2.2 - 1.6)/.4 = 1.5$, and the $z = 1.5$ row of Table I in the appendices gives $P(2.2 < x) = .0668$. Thus roughly 16% of all vehicles with this configuration have emission levels exceeding 2.0, whereas only 6.7% of all such vehicles have emission levels exceeding 2.2.

What emission level is exceeded by only 1% of all vehicles? Denoting this level by c, the value of c should satisfy the condition $P(c < x) = .01$. The probability $P(c < x)$ is the area under the z curve to the right of $(c - 1.6)/.4$, and we want this area to be .01. Looking down the "area to the right of z" column of Table I for .01, this amount of area lies to the right of 2.33. Thus we must have $(c - 1.6)/.4 = 2.33$. Solving this equation for c gives $c = (.4)(2.33) + 1.6 = 2.53$. Any vehicle emitting more than 2.53 g/mi of this pollutant is in the worst 1% of all such vehicles. ■

The process of standardizing—subtracting μ and dividing by σ—is just a way of reexpressing distances. Suppose, for example, that $\mu = 20$ and $\sigma = 2$. The value 24 is 4 units above (to the right of) the mean. Since one standard deviation is 2 units, 24 is 2 standard deviations above the mean. When 24 is standardized, the result is $(24 - 20)/2 = 2$. Similarly, 15 is 2.5 standard deviations below the mean (since $20 - 15 = 5$ and $(2.5)(2) = 5$), and standardizing 15 gives $(15 - 20)/2 = -2.5$. The minus sign here indicates that we started with a value below the mean. In fact, the operation of standardizing was used previously in Section 3.3 to compute the z score of an observation in a data set. The resulting z score was the distance of an observation from the mean expressed as some number of standard deviations.

Standardizing a value gives the distance from that value to the mean in units of standard deviation. A value above the mean yields a positive standardized value, while a value below the mean yields a negative standardized value.

EXAMPLE 30

Suppose that reaction time x to a certain stimulus has a normal distribution with mean μ and standard deviation σ. What is the probability that a particular reaction time will fall within 1 standard deviation of the mean? Since 1 standard deviation below the mean corresponds to $\mu - \sigma$ and 1 standard deviation above the mean corresponds to $\mu + \sigma$, an x value within 1 standard deviation of μ is one in the interval from $\mu - \sigma$ to $\mu + \sigma$. The desired probability is then $P(\mu - \sigma < x < \mu + \sigma)$. You might think that numerical values of μ and σ would have to be supplied before this probability could be computed, but this is not the case. Standardizing the upper limit $\mu + \sigma$ gives $[(\mu + \sigma) - \mu]/\sigma = \sigma/\sigma = 1$, no surprise since $\mu + \sigma$ is one standard deviation above μ. Similarly, standardizing $\mu - \sigma$ gives -1. Finally,

$$P(\mu - \sigma < x < \mu + \sigma) = \begin{pmatrix}\text{area under the } z \text{ curve} \\ \text{between } -1 \text{ and } +1\end{pmatrix}$$
$$= .8413 - .1587 = .6826$$

Roughly 68% of all reaction times are within 1 standard deviation of the mean. ■

Although x in Example 30 denoted reaction time, the resulting probability, .6826, could equally well have applied to any other context as long as x is normally distributed. Furthermore, the empirical rule discussed in Chapter 3 stated that if a histogram could be closely approximated by a normal curve, roughly 68% of the values would lie within 1 standard deviation of the mean. The result of Example 30 is just a restatement of this fact using the language of probability and the normal distribution. Repeating the calculation of the example for values within 2 standard deviations of the mean (from $\mu - 2\sigma$ to $\mu + 2\sigma$) and for values within 3 standard deviations of the mean leads to a restatement of the empirical rule.

The probability that a normal random variable x assumes a value lying within some specified number of standard deviations of the mean is the same for all values of μ and σ. The probabilities corresponding to within 1, 2, and 3 standard deviations of the mean are .6826, .9544, and .9974, respectively.

Percentiles

The statement that 120 s is the 90th percentile of reaction time means that 90% of all times are at most 120, whereas 10% of all times exceed 120. If x is the random variable *reaction time,* an alternative way to state this is to write $P(x \leq 120) = .90$. Similarly, to say that 4.8 lb is the 5th percentile of birthweight is to say that only 5% of all weights are at most 4.8. Equivalently, we can write $P(y \leq 4.8) = .05$, where y denotes the random variable *birthweight*.

In general, a value c is the 90th percentile of the distribution of a random variable x if $P(x \leq c) = .9$. Other percentiles result from replacing .9 by the corresponding probability—.75 for the 75th percentile, .10 for the 10th percentile, and so on. When x has a normal distribution, it is easy to obtain percentiles.

EXAMPLE 31

In Example 29, the amount of pollutant emitted was assumed to be normally distributed with $\mu = 1.6$ and $\sigma = .4$. We then wanted the value c such that only 1% of all vehicles had emission levels exceeding c. This means that 99% of all vehicles have x values that are at most c, or $P(x \leq c) = .99$, so c is the 99th percentile. To obtain c, we found the area .01 in the "area to the right of z" column of Table I. This is equivalent to finding .99 in the "area to the left of z" column. These areas appear in the "$z = 2.33$" row of the table. The desired value of c was then computed as $c = \mu + 2.33\sigma = 1.6 + (2.33)(.4) = 2.53$.

Suppose that the 5th percentile has been requested (area .05 to the left and .95 to the right). Looking in the "area to the left of z" column, .05 is found in the row for -1.645. The 5th percentile is then $\mu + (-1.645)\sigma = 1.6 - 1.645(.4) = .94$. Only 5% of all vehicles have emission levels of at most .94. ■

The general procedure for finding percentiles of a normal distribution parallels what is done in Example 31. For a specified percentile, we first find the value of z that captures the corresponding area to the left of this value. Then

$$\text{percentile} = \mu + (z \text{ value})\sigma$$

If, for example, the 90th percentile is desired, the corresponding z value is 1.28 (since the area under the z curve to the left of 1.28 is .9). Assuming that reaction time is normally distributed with $\mu = .5$ sec and $\sigma = .08$ sec, the 90th percentile is $.5 + (1.28)(.08) = .6024$, or roughly .60. The 10th percentile (found by using -1.28) is roughly .40.

The Normal Curve and Discrete Variables

The probability distribution of a discrete random variable x is represented pictorially by a probability histogram in which the probability of a particular value is the area of the rectangle centered at that value. The possible values of x are isolated points on the number line, usually whole numbers. For example, if x = IQ of a randomly selected 8-year-old child, then x is a discrete random variable, since an IQ score must be a whole number.

Often, however, a probability histogram can be very well approximated by a normal curve, as illustrated in Figure 13. When this is the case, instead of using the discrete probability distribution and probability histogram to compute probabilities, it is customary to say that x has *approximately* a normal distribution. The approximating normal curve is then used to calculate approximate probabilities of events involving x. For example, x = IQ score might be regarded as having approximately a normal distribution with $\mu = 100$ and $\sigma = 15$. Probabilities such as $P(90 \leq x \leq 110)$ and $P(140 < x)$ could then be calculated by standardizing and using the z table.

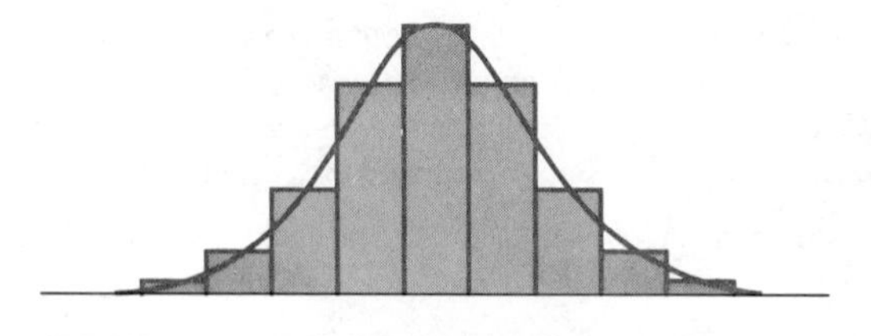

FIGURE 13 A NORMAL CURVE APPROXIMATION TO A PROBABILITY HISTOGRAM

The important point here is that many applied studies use the normal distribution model for a discrete random variable x even though technically a normal random variable is continuous. As long as the appropriate normal curve (the one with μ and σ selected to give the best approximation) provides the sort of fit illustrated in Figure 13, probabilities calculated from the normal model will be reasonably accurate approximations to the actual probabilities.

EXERCISES

5.31 Give two examples of variables whose probability distributions might be well approximated by a normal curve. Give two variables whose distributions would not resemble a normal curve. Explain the reason for your choices.

5.32 Let z be a variable whose distribution is given by the standard normal curve. Calculate the following probabilities, drawing pictures wherever appropriate.

a. $P(0 \leq z \leq 2.2)$
b. $P(0 \leq z \leq 1)$
c. $P(-2.5 \leq z \leq 0)$
d. $P(-2.5 \leq z \leq 2.5)$
e. $P(z \leq 1.37)$
f. $P(-1.8 \leq z)$
g. $P(-1.5 \leq z \leq 2.0)$
h. $P(1.4 \leq z \leq 2.5)$
i. $P(1.5 \leq z)$

5.33 Find the approximate values of the following percentiles of the standard normal distribution.

a. 91st **b.** 9th **c.** 75th **d.** 25th **e.** 95th

5.34 Let x denote the duration of a randomly selected pregnancy (the time elapsed between conception and birth). Accepted values for the mean value and standard deviation of x are 266 days and 16 days, respectively. Suppose that the probability distribution of x is (approximately) normal.

a. What is the probability that the duration of pregnancy is between 250 and 300 days?

b. What is the probability that the duration of pregnancy is at most 240 days?

c. What is the probability that the duration of pregnancy is within 16 days of the mean duration?

d. A January 20, 1973, *Dear Abby* column contained a letter from a woman who stated that her duration of pregnancy was exactly 310 days (she wrote that the last visit with her husband, who was in the navy, occurred 310 days prior to birth). What is the probability that the duration of pregnancy is at least 310 days? Does this probability make you a bit skeptical of the claim?

e. Some insurance companies will pay the medical expenses associated with childbirth only if the insurance has been in effect for more than 9 months (275 days). This restriction is designed to ensure that the insurance company has to pay benefits only for those pregnancies where conception occurred during coverage. Suppose that conception were to occur 2 weeks after coverage began. What is the probability that the insurance company will refuse to pay benefits because of the 275-day insurance requirement?

5.35 If a normal distribution has mean 25 and standard deviation 5, what is the 91st percentile of the distribution? The 6th percentile?

5.36 A machine producing vitamin E capsules operates so that the actual amount of vitamin E in each capsule is normally distributed with mean 5 mg and standard deviation .05. What is the probability that a randomly selected capsule contains less than 4.9 mg of vitamin E? At least 5.2 mg?

5.37 Accurate labeling of packaged meat is difficult because of weight decrease due to moisture loss (defined as a percentage of the package's original net weight). Suppose that moisture loss for a package of chicken breasts is normally distributed with mean value 4.0% and standard deviation 1.0% (this model is suggested in the paper "Drained Weight Labeling for Meat and Poultry: An Economic Analysis of a Regulatory Proposal" (*J. of Consumer Affairs* (1980):307–25)). Let x denote the moisture loss for a randomly selected package.

a. What is the probability that x is between 3.0 and 5.0?

b. What is the probability that x is at most 4.0?

c. What is the probability that x is at least 7.0?

d. What value is such that 90% of all packages have moisture losses below that value?

e. What is the probability that moisture loss differs from the mean value by at least 1%?

5.38 The *Wall Street Journal* (February 15, 1972) reported that General Electric was being sued in Texas for sex discrimination over a minimum height requirement of 5 ft 7 in. The suit claimed that this restriction eliminated more than 94% of adult females from consideration. Let x represent the height of a randomly selected adult woman. Suppose that x is approximately normally distributed with mean 66 in. (5 ft 6 in.) and standard deviation 2 in.

a. Is the claim that 94% of all women are shorter than 5 ft 7 in. correct?

b. What proportion of adult women would be excluded from employment due to the height restriction?

5.39 Suppose that your statistics professor tells you that the scores on a midterm exam were approximately normally distributed with a mean of 78 and a standard deviation of 7. The top 15% of all scores have been designated as A's. Your score is 89. Did you receive an A?

5.40 Suppose that the pH of soil samples taken from a certain geographic region is normally distributed with mean pH 6.00 and standard deviation .10. If the pH of a randomly selected soil sample from this region is determined, answer the following questions.

a. What is the probability that the resulting pH is between 5.90 and 6.15?

b. What is the probability that the resulting pH exceeds 6.10?

c. What is the probability that the resulting pH is at most 5.95?

d. What value will be exceeded by only 5% of all such pH values?

5.41 The light bulbs used to provide exterior lighting for a large office building have an average lifetime of 700 h. If length of life is approximately normally distributed with a standard deviation of 50 h, how often should all of the bulbs be replaced so that no more than 20% of the bulbs will have already burned out?

5.42 A machine that cuts corks for wine bottles operates so that the diameter of the corks produced is approximately normally distributed with mean 3 cm and standard deviation .1 cm. The specifications call for corks whose diameters are between 2.9 and 3.1 cm. A cork not meeting the specifications is considered defective (a cork that is too small leaks and causes the wine to deteriorate, while a cork that is too large doesn't fit in the bottle). What proportion of corks produced by this machine are defective?

5.43 Refer to Exercise 5.42. Suppose that there are two machines available for cutting corks. One, described above, produces corks whose diameters are normally distributed with mean 3 cm and standard deviation .1 cm. The second machine produces corks whose diameters are normally distributed with mean 3.05 cm and standard deviation .01 cm. Which machine would you recommend? (*Hint:* Which machine would produce the fewest defective corks?)

5.44 Suppose that SAT math and verbal scores are approximately normally distributed. Both the math and verbal sections of the test have a distribution of scores with average 500 and standard deviation 100.

a. If a student scored 620 on the verbal section, what percentile is associated with his or her score?

b. If the same student scored 710 on the math section, what is the corresponding percentile?

c. What score does a student have to achieve in order to be at the 90th percentile?

5.5 Checking for Normality (Optional)

Sometimes the main purpose of a statistical investigation is simply to describe how the values of some variable x are distributed in the population. More frequently, investigators wish to use sample data to draw conclusions of some sort about the population distribution. Often a particular inferential procedure yields reliable conclusions if the population distribution is of a certain specified type (e.g., normal) but cannot be trusted to do so if the distribution differs markedly from the specified type. That is, the validity of a statistical procedure is often based on some specific assumption(s) about the nature of the population distribution. If the assumption is grossly violated, use of the procedure can lead to grave errors in conclusions concerning the population.

It is, therefore, useful to have methods, based on sample data, which can indicate whether or not the assumption of some specific distribution is plausible. Because many statistical procedures are based on an assumption of normality, our discussion focuses on assessing the plausibility of this particular distribution. Analogous methods are routinely used by data analysts to check the plausibility of other distributions.

Statisticians have developed a number of different methods for checking normality. In recent years methods involving various plots have gained prominence, in part because such plots can be quickly and easily constructed using a statistical computer package once the data has been properly entered. While there are variations in the details of such plots, the basic idea is always the same. The points in the plot should fall reasonably close to a straight line if the underlying distribution is normal but should deviate substantially from a linear pattern if the underlying distribution is very unlike the normal distribution (e.g., very skewed, heavy-tailed, or light-tailed).

Expected Normal Scores

Some plots for checking normality require the use of special graph paper, called *normal probability paper*. An alternate plot, which we shall describe here, does not require the use of special paper but instead involves some quantities called *expected normal scores.* To introduce these quantities in a specific context, suppose the diameter of a certain type of bolt is a continuous random variable x having a normal distribution with $\mu = .500$ and $\sigma = .010$. Consider obtaining a sample of five such bolts and determining the x value for each one. The resulting values might be .502, .489, .506, .514, and .498. The smallest of the five is then .489, the second smallest is .498, etc.

If a different sample is selected, the smallest value may be something other than .489, the second smallest may not be .498, and so on. Think of performing this experiment over and over again, each time with five different bolts. In the long run, what will be the average value of the smallest observation (that is, the mean value of the smallest observation)? Similarly, what will be the long-run average (i.e., mean value) of the second smallest observation, third smallest observation, and so on? Statisticians have shown that

$$\begin{aligned}
\text{mean value of the smallest } x &= \mu + (-1.163)\sigma = .488 \\
\text{mean value of the second smallest } x &= \mu + (-.495)\sigma = .495 \\
\text{mean value of the third smallest } x &= \mu + (0)\sigma = .500 \\
\text{mean value of the fourth smallest } x &= \mu + (.495)\sigma = .505 \\
\text{mean value of the largest } x &= \mu + (1.163)\sigma = .512
\end{aligned}$$

To see where the numbers -1.163, $-.495$, 0, .495, and 1.163 come from, let's change the scenario. Instead of taking a sample of five observations from a normal distribution with $\mu = .500$ and $\sigma = .010$, consider taking a sample of five observations from a normal distribution with $\mu = 0$ and $\sigma = 1$, i.e., from a standard normal distribution. Then the mean value of the smallest among these five observations is -1.163. That is, in a very long sequence of such experiments, the average value of the smallest observation is roughly -1.163. Similarly, the mean value of the second smallest observation is $-.495$, the mean value of the third smallest (middle) observation is 0, and so on.

DEFINITION

Consider obtaining a sample of n observations from a normal distribution with mean value μ and standard deviation σ and then ordering the observations from smallest to largest. The mean value of the smallest observation is called the smallest **expected normal score,** the mean value of the second smallest observation is called the second smallest expected normal score, and so on. Expected normal scores for selected values of n when $\mu = 0$, $\sigma = 1$ are given in Table II of the appendices. Then for any other values of μ and σ,

$$\text{expected normal score} = \mu + \begin{pmatrix}\text{corresponding expected} \\ \text{score for } \mu = 0, \sigma = 1\end{pmatrix} \cdot \sigma$$

The values -1.163, $-.495$, 0, .495, and 1.163 are called expected normal scores for a sample of size five from a standard normal distribution. The values .488, .495, .500, .505, and .512 are the expected normal scores for a sample of size five from a normal distribution with $\mu = .500$ and $\sigma = .010$. In general, whatever the values of μ and σ, the smallest expected normal score in a sample of size five is $\mu + (-1.163)\sigma$, the second smallest expected normal score is $\mu + (-.495)\sigma$, etc. Once the expected normal scores for $\mu = 0$, $\sigma = 1$ are available, those for any other values of μ and σ are easily calculated.

For example, Table II shows that when $n = 25$, the third smallest expected normal score when $\mu = 0$, $\sigma = 1$ is -1.263. If observations are taken in groups of 25, the long-run average of the third smallest one will be roughly -1.263. If height is normally distributed with $\mu = 69$ in., $\sigma = 3$ in., and height values are obtained in groups of 25, the long-run average value of the third smallest height value would be $69 + (-1.263)(3) = 65.211$.

A Plot for Checking Normality

The expected normal scores for a sample of size $n = 5$ when $\mu = 50$, $\sigma = 10$ are $50 + (-1.163)(10) = 38.37$, 45.05, 50, 54.95, and 61.63. Suppose we obtain a sample of five x values from a particular population under study. If the population distribution is actually normal with $\mu = 50$, $\sigma = 10$, then the five observed scores should be reasonably close to the expected normal scores. Consider the pairs (smallest expected normal score, smallest observed value), (second smallest expected normal score, second smallest observed value), . . . , (largest expected normal score, largest observed value). If every observed value coincides exactly with the corresponding expected normal score, the resulting pairs are (38.37, 38.37), (45.05, 45.05), (50, 50), (54.95, 54.95), and (61.63, 61.63). Figure 14(a) shows a plot of these points with the horizontal axis identified with expected normal score and the vertical axis identified with observed value. The five points fall exactly on a 45° line passing through the point (0, 0).

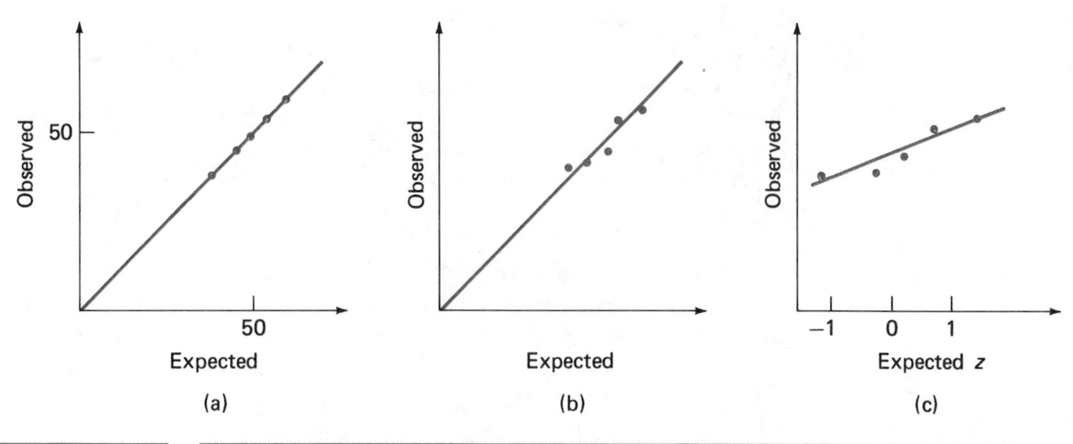

FIGURE 14 NORMAL PROBABILITY PLOTS
(a) When Observed Values = Expected Scores
(b) When Observed Values Are Close to Expected Scores
(c) When Expected Standard Normal Scores Are Used

In practice, even when the underlying distribution is identical to the normal distribution for which the expected normal scores are computed (here $\mu = 50$, $\sigma = 10$), observed values usually do not coincide exactly with the expected normal scores. The observed scores (ordered) might be 39.2, 43.5, 49.1, 55.4, and 55.9, yielding the (expected, observed) pairs (38.37, 39.2), (45.05, 43.5), (50, 49.1), (54.95, 55.4), and (61.63, 59.9). Figure 14(b) displays a plot of these pairs. The points fall reasonably close to the 45° line.

Suppose that instead of plotting observed versus expected normal scores when $\mu = 50$, $\sigma = 10$, we plot observed versus expected normal scores when $\mu = 0$, $\sigma = 1$ (the standard normal expected scores). For our example, this involves plotting (−1.163, 39.2), (−.495, 43.5), (0, 49.1), (.495, 55.4), and (1.163, 59.9). Figure 14(c) shows the resulting plot. It has exactly the same general pattern as the plot of Figure 14(b); the points fall close to a straight line, but it is no longer the 45° line of the two earlier plots. This suggests that plotting observed values versus expected scores when $\mu = 0$, $\sigma = 1$ will yield a straight line pattern if the underlying distribution is normal irrespective of the actual values of μ and σ, and this is indeed the case.

To construct a plot for checking normality, order the n sample observations from smallest to largest, obtain the expected normal scores when $\mu = 0$, $\sigma = 1$ from Table II (or some other more complete source), and form the n (expected score, observed value) pairs. If the distribution from which the sample was obtained is normal (at least approximately), a plot of these pairs should show a reasonably strong linear pattern. If the underlying distribution is distinctly nonnormal, the plot should show a rather pronounced departure from linearity. The plot is usually called a **normal probability plot.**

Figure 15 contains four such plots for samples of $n = 25$ observations each. The first plot shows the expected straight-line pattern consistent with a normal distribution. The second plot is the sort of picture that typically results from sampling a distribution that is symmetric but has heavier tails than the normal curve. The middle part of the plot is reasonably linear, but on the left end the points fall below a straight line through the middle part and on the right end the points fall above such a straight line. This difference occurs since, because of the heavy tails, observations at the upper end tend to be larger than what is expected from a normal distribution. Corresponding points at the upper end of the plot then tend to be higher than those resulting from a normal distribution. Figure 15(c) illustrates what would occur if the underlying distribution had lighter tails than a normal distribution. The nature of the departure from linearity at extreme ends of the plot is the mirror image of what happens in the heavy-tailed case. The plot in Figure 15(d) is typical of what results from sampling a distribution that is quite skewed—a rather strong curved pattern in the plot.

In the best of all possible worlds, the points in a normal probability plot fall exactly on a straight line, but in practice, sampling variability precludes such an ideal picture. How far can the pattern in the plot deviate from linearity before the assumption of normality should be judged implausible? This is

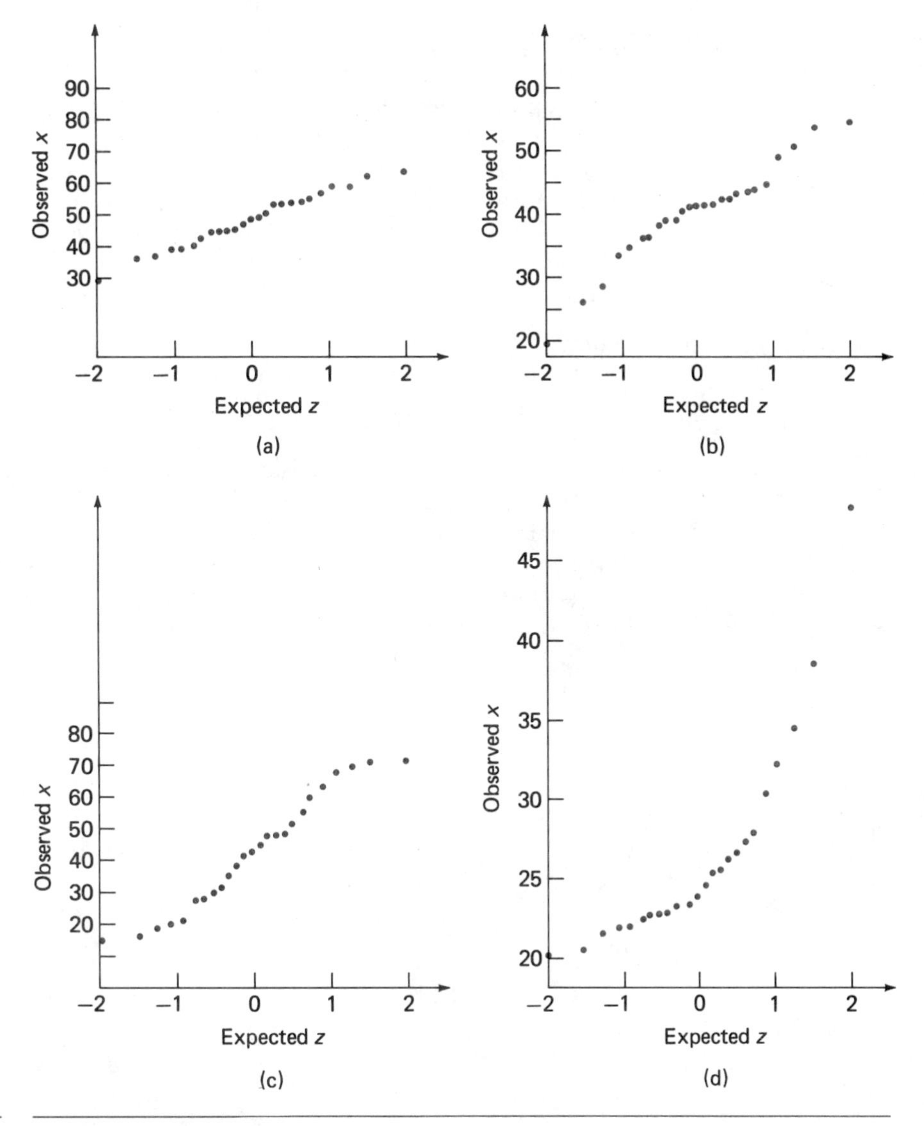

FIGURE 15

NORMAL PROBABILITY PLOTS
(a) Observations from a Normal Distribution
(b) Observations from a Heavy-Tailed Distribution
(c) Observations from a Light-Tailed Distribution
(d) Observations from a Positively Skewed Distribution

not an easy question to answer. To get a feeling for what the plot might look like when the distribution is normal, for each of a number of different values of the sample size *n*, we could generate a number of samples of *n* normally distributed observations and study the resulting plots. This obviously requires a large investment of space (for us) and time (for you), both of which are in short supply. The book *Fitting Equations to Data* by Cuthbert Daniel and Fred Wood (New York: John Wiley, 1980) presents a number of such plots on pages 33–43. These plots suggest that with small sample sizes (e.g., $n < 20$), there can be so much sampling variability that substantial departures from linearity can result even when the distribution is normal. So be careful about deciding against the plausibility of a normal distribution based on a normal probability plot when n is small.

As previously mentioned, the computer is very good at constructing normal probability plots. In particular, all the most frequently used packages of statistical computer programs have a command that will produce such a plot. The details of the plot vary somewhat from package to package—for example, MINITAB and BMDP use (different) internally calculated approximations to the expected normal scores rather than tabulated values—but the key idea is always to look for linearity.

EXAMPLE 32

Example 29 of the previous section referred to an article that contained $n = 46$ observations on the amount of oxides of nitrogen (NO_x) emitted by a particular type of automobile. Figure 16 shows a normal probability plot of the data produced by MINITAB. Where two points in the plot fell so close to one another that separate asterisks would not show, the character 2 was used instead, and three close points were replaced by the character 3. The plot has a reasonably well defined straight-line character, though there is a bit of wobbliness in the two tails. With only a moderately large sample size, such wobbling is quite common. We conclude that an assumption of normality for the distribution of NO_x emissions is quite plausible.

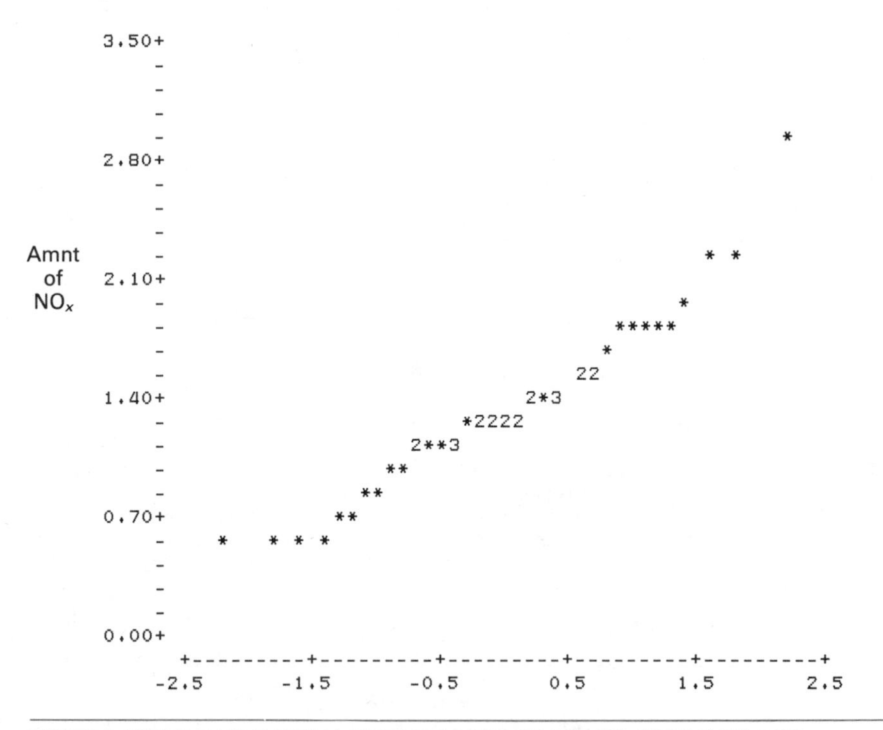

FIGURE 16 MINITAB GENERATED NORMAL PROBABILITY PLOT FOR NO_x EMISSION DATA

■

EXAMPLE 33

Example 18 of Chapter 2 presented data on cell interdivision times (IDT's). A sample histogram of the original data was quite skewed, but transforming by logarithms yielded a reasonably bell-shaped histogram. Figure 17 displays

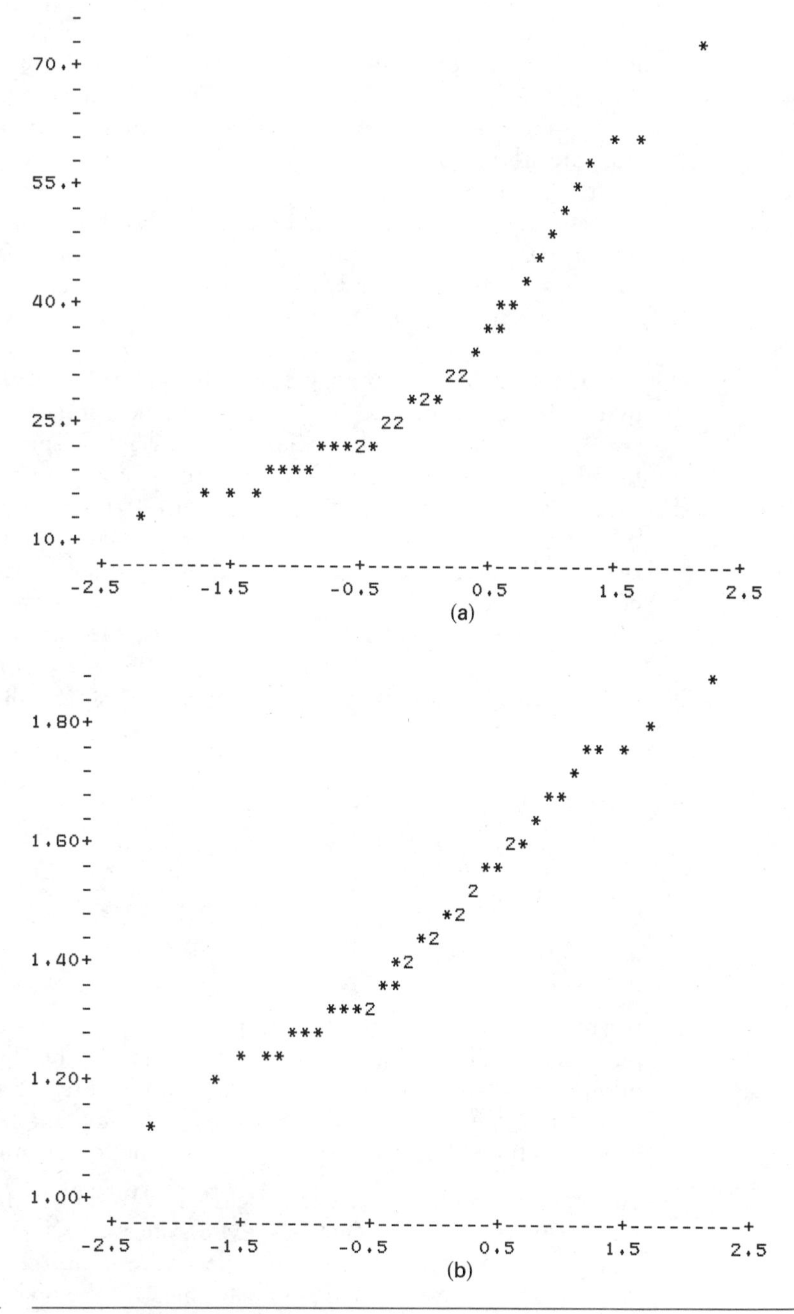

FIGURE 17

MINITAB GENERATED NORMAL PROBABILITY PLOT
(a) Original IDT Data
(b) Log-Transformed IDT Data

normal probability plots for the original data and for the log-transformed data. The latter plot is clearly much more linear in appearance than is the former. Not only has a log transformation resulted in a symmetric histogram but also in one that is very well fit by a normal curve. ■

Using the Correlation Coefficient to Check Normality

The correlation coefficient r was introduced in Chapter 4 as a quantitative measure of the extent to which the points in a scatter plot fall close to a straight line. Suppose that we use the label y to denote an observed value of the variable under study and the label x to denote the corresponding expected standard normal score. Then the correlation coefficient can be computed using the defining equation for r given in Chapter 4. The normal probability plot always slopes upward (since it is based on values ordered from smallest to largest), so r will be a positive number. A value of r quite close to 1 indicates a very strong linear relationship in the normal probability plot. If r is too much smaller than 1, normality of the underlying distribution is questionable.

How far below 1 does r have to be before we begin to doubt seriously the plausibility of normality? The answer depends on the sample size n. If n is small, an r value somewhat below 1 would not be surprising even when the distribution is normal, but if n is large, only an r value very close to 1 would support the assumption of normality. For selected values of n, Table 3 gives critical values to which r can be compared in checking for normality. If the computed value of r is less than the tabulated value for the given sample size, considerable doubt is cast on the assumption of normality. If your sample size is in between two tabled values of n, use the critical value for the larger sample size (e.g., if $n = 46$, use the value .966 for sample size 50). In addition to providing a normal probability plot, MINITAB will automatically compute r upon request.

TABLE 3

VALUES TO WHICH r CAN BE COMPARED TO CHECK FOR NORMALITY

n	5	10	15	20	25	30	40	50	60	75
Critical r	.832	.880	.911	.929	.941	.949	.960	.966	.971	.976

Source: MINITAB User's Manual

EXAMPLE 34

For the NO_x emission data of Example 32, MINITAB yielded $r = .973$ (to compute this by hand would require the original data and the expected standard normal scores for $n = 46$, but these latter values are not contained in our abbreviated Table II). The critical r value is .966 for sample size 50. Since .973 is not less than .966, the assumption of normality remains plausible. ■

EXAMPLE 35

For the untransformed IDT data discussed in Example 33, $r = .950$, whereas for the log-transformed data, $r = .988$. Since $n = 40$ for this data set, the critical r value is .960. Since .950 is less than .960, the assumption that the distribution of IDT is normal is implausible. But the normality of $\log_{10}$(IDT) is strongly supported by the very high value of r, much larger than the critical value. ■

How were the critical values in Table 3 obtained? Consider the critical value .941 for $n = 25$. Suppose the underlying distribution is actually normal. Consider obtaining a large number of different samples, each one consisting of 25 observations, and computing the value of r for each one. Then it can be shown that only 1% of the samples result in an r value less than the critical value .941. That is, .941 was chosen to guarantee a 1% error rate—in only 1% of all cases will we judge normality implausible when the distribution really is normal. The other critical values are also chosen to yield a 1% error rate for the corresponding sample sizes. It might have occurred to you that another type of error is possible—obtaining a large value of r and concluding that normality is a reasonable assumption when the distribution is actually quite nonnormal. This type of error is more difficult to control than the type mentioned above, but the procedure we have described does a good job of controlling this type as well.

EXERCISES

5.45 Ten measurements of steam rate (lb/h) of a distillation tower were used to construct the given normal probability plot (Source: "A Self-Descaling Distillation Tower" *Chem. Eng. Process* (1968):79–84). Based on this plot, do you think it is reasonable to assume that the normal distribution provides an adequate description of the steam rate distribution? Explain.

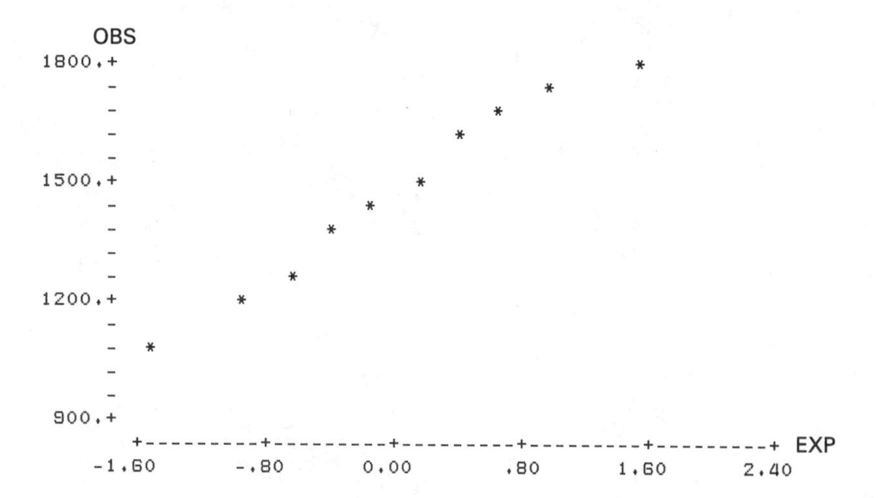

5.46 The accompanying normal probability plot was constructed using part of the data appearing in the paper "Trace Metals in Sea Scallops" (*Environ. Concentration and Toxicology* 19:326–1334). The variable under study was the amount of cadmium in North Atlantic sea scallops. Does the sample data suggest that the cadmium concentration distribution is not normal? Explain.

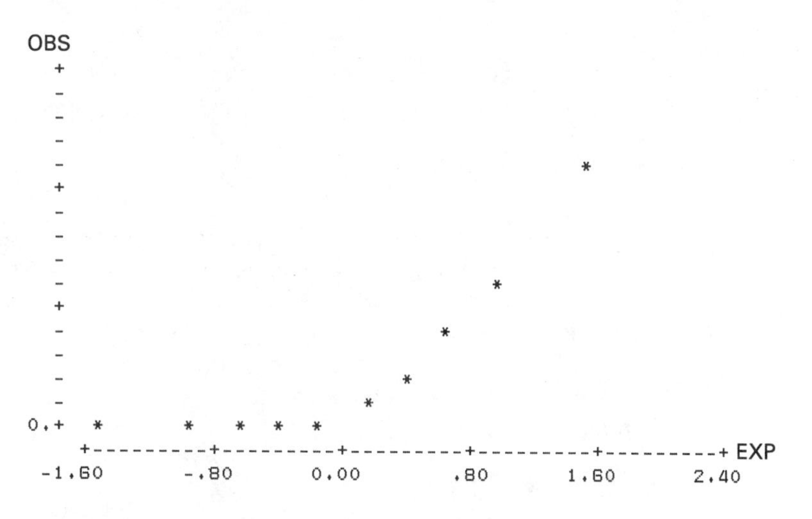

In Exercises 5.47–5.50, use the methods described in this section to determine whether the normal distribution provides an adequate description of the population distribution from which the sample was selected.

5.47 DDT concentrations in the blood of 20 people (from *Statistics from Scratch*, 1977).

24	26	30	35	35	38	39	40	40	41
42	42	52	56	58	61	75	79	88	102

5.48 Twenty-five observations of the coefficient of friction for a metal (*Technometrics*, May 1979). (*Note:* multiplying each observation by 1000 gives a set of values that is equivalent to the original set for purposes of assessing normality and is easier to work with.)

.0175	.0375	.0225	.0325	.0225	.0375	.0225	.0375
.0275	.0425	.0275	.0375	.0275	.0375	.0275	.0425
.0325	.0525	.0325	.0425	.0325	.0475	.0325	.0475
.0325							

5.49 Sodium content (mg/tsp) of 25 margarines (*Consumer Reports*, January 1983)

40	32	31	32	35	34	34	44	31	33	33	35	31
38	35	34	30	35	40	37	38	35	34	33	33	

5.50 Miles per gallon achieved by 25 subcompact cars (*Consumer Reports*, December 1982)

21	21	21	17	39	38	34	31	24	24	20	25	27
19	18	24	26	23	16	31	28	42	37	27	22	

5.51 In Section 2.6 we considered the Minneapolis–St. Paul rainfall data given here.

.77	1.20	3.00	1.62	2.81	2.48
1.74	.47	3.09	1.31	1.87	.96
.81	1.43	1.51	.32	1.18	1.89
1.20	3.37	2.10	.59	1.35	.90
1.95	2.20	.52	.81	4.75	2.05

a. Construct and interpret a normal probability plot for this data set.

b. In Chapter 2, we determined that the square-root transformation resulted in a histogram that was more symmetric than that of the untransformed data. Construct a normal probability plot for the transformed values. Compare the plots for the transformed and untransformed values.

5.6 The Binomial Distribution (Optional)

Suppose we decide to record the sex of each of the next 25 newborn children at a particular hospital. What is the chance that at least 15 are female? What is the chance that between 10 and 15 are female? Or suppose that the same coin is to be tossed 20 times. What is the chance that at least half the tosses result in heads? How many heads can we expect to see? These and other similar questions can be answered by studying the binomial probability distribution.

The Binomial Experiment and Random Variable

The binomial probability distribution arises when the experiment of interest has the following properties:

1. The experiment consists of a fixed number of smaller experiments, each one referred to as a *trial.*
2. Each trial can result in one of only two possible outcomes; the possibilities on each trial are denoted by S (for success) and F (for failure).
3. Outcomes of different trials are independent of one another (knowing the outcomes of some trials gives no information concerning outcomes on the remaining trials).
4. The probability that a trial results in S is the same for each trial in the experiment.

Any experiment satisfying 1–4 is called a **binomial experiment.** The assignment of the S-F labels in any particular problem context is arbitrary; in coin tossing, for example, S can be identified either with a head or with a tail, as long as future calculations are then consistent with the assignment. The binomial probability distribution depends both on the number of trials and on the probability of success. A statement of general results is easier with some notation for these quantities.

> Let n denote the number of trials in a binomial experiment and let π denote the probability of S on any particular trial.

The experiment in which a fair coin is tossed $n = 20$ times has, with $S =$ head, $\pi = P(S) = .5$ and $1 - \pi = P(F) = .5$. If S denotes a car that turns left at a dead-end intersection, the value of π might be .6 for one intersection (in the long run, 60% of all cars turn left) and only .3 for another intersection.

The result of performing a binomial experiment is an observed sequence of S's and F's. The ordering of the S's and F's is not usually of interest (unless one suspects dependence in outcomes of successive trials, in which case the experiment would not be binomial). An investigator is typically concerned with the number of S's in the n trials—number of female children born, number of heads, number of cars turning left, etc. If $n = 10$, the number of S's resulting could be $0, 1, \ldots, 10$. Before the experiment is performed, there is uncertainty concerning how many S's will result.

DEFINITION

Let x denote the number of S's in a binomial experiment consisting of n trials. Then x is a discrete random variable, called a **binomial random variable,** with possible values $0, 1, 2, \ldots, n$. The probability distribution of x is called the **binomial probability distribution.**

The Binomial Probability Distribution

The probabilities for the various possible x values depend on both the number of trials n and the success probability π. Fortunately, there is a compact general formula that can be used to compute these binomial probabilities. An understanding of the formula is facilitated by first looking at a specific example.

EXAMPLE 36

Suppose that at a particular dead-end intersection, $\pi = P(\text{left turn}) = P(S) = .6$. The turning direction of each of $n = 4$ different cars is observed. Each outcome of this binomial experiment is a sequence of four S's and/or F's. Table 4 lists all 16 possible outcomes, the probability of each one (we explain these momentarily), and the x value for each outcome.

TABLE 4

OUTCOMES AND PROBABILITIES FOR A BINOMIAL EXPERIMENT IN WHICH $n = 4$ AND $\pi = .6$

Outcome	Probability	x Value	Outcome	Probability	x Value
FFFF	(.4)(.4)(.4)(.4)	0	*FSSF*	(.4)(.6)(.6)(.4)	2
SFFF	(.6)(.4)(.4)(.4)	1	*FSFS*	(.4)(.6)(.4)(.6)	2
FSFF	(.4)(.6)(.4)(.4)	1	*FFSS*	(.4)(.4)(.6)(.6)	2
FFSF	(.4)(.4)(.6)(.4)	1	*FSSS*	(.4)(.6)(.6)(.6)	3
FFFS	(.4)(.4)(.4)(.6)	1	*SFSS*	(.6)(.4)(.6)(.6)	3
SSFF	(.6)(.6)(.4)(.4)	2	*SSFS*	(.6)(.6)(.4)(.6)	3
SFSF	(.6)(.4)(.6)(.4)	2	*SSSF*	(.6)(.6)(.6)(.4)	3
SFFS	(.6)(.4)(.4)(.6)	2	*SSSS*	(.6)(.6)(.6)(.6)	4

You can see from the table that the probability of any particular outcome involves multiplying S and F probabilities together the appropriate number of times. Multiplication is used because successive trials are independent. One of our basic results from Section 5.1 said that when events are independent, the probability that they all happen together is the product of their individual probabilities. Thus

$$\begin{aligned} P(SFSS) &= P\begin{pmatrix} S \text{ on 1st} & & F \text{ on 2nd} & & S \text{ on 3rd} & & S \text{ on 4th} \\ \text{trial} & \text{and} & \text{trial} & \text{and} & \text{trial} & \text{and} & \text{trial} \end{pmatrix} \\ &= P\begin{pmatrix} S \text{ on 1st} \\ \text{trial} \end{pmatrix} \cdot P\begin{pmatrix} F \text{ on 2nd} \\ \text{trial} \end{pmatrix} \cdot P\begin{pmatrix} S \text{ on 3rd} \\ \text{trial} \end{pmatrix} \cdot P\begin{pmatrix} S \text{ on 4th} \\ \text{trial} \end{pmatrix} \\ &= (.6)(.4)(.6)(.6) = (.6)^3(.4) = .0864 \end{aligned}$$

To interpret this probability, think of observing turning directions for a first group of four cars, then a second group, then a third, and so on. In the long run, 8.64% of the time the first car will turn left, the second car right, the third car left, and the fourth car left. Probabilities of the other outcomes are computed and interpreted in the same manner.

The five possible values of x are 0, 1, 2, 3, and 4. Let's focus on $x = 3$. From Table 4, this value results from any of the four outcomes *FSSS, SFSS, SSFS,* or *SSSF*. Applying one of our basic rules of probability, $P(x = 3)$ is obtained by adding together the probabilities of these four outcomes:

$$\begin{aligned} p(3) &= P(x = 3) = P(FSSS \text{ or } SFSS \text{ or } SSFS \text{ or } SSSF) \\ &= P(FSSS) + P(SFSS) + P(SSFS) + P(SSSF) \\ &= (.6)^3(.4) + (.6)^3(.4) + (.6)^3(.4) + (.6)^3(.4) \\ &= 4[(.6)^3(.4)] = .3456 \end{aligned}$$

Similarly, $P(x = 2)$ is the sum of the probabilities of the six outcomes for which $x = 2$. Since each such outcome probability is $(.6)^2(.4)^2$,

$$p(2) = P(x = 2) = 6[(.6)^2(.4)^2] = .3456$$

Using the same reasoning for the other three x values, the probability distribution of x is as follows.

x	0	1	2	3	4
$p(x)$	.0256	.1536	.3456	.3456	.1296

If we observe turning directions for successive groups of four cars at this intersection, in the long run 12.96% of the time all four will turn left, 34.56% of the time three out of four will turn left, and so on. ■

If n is even moderately large, listing all outcomes in order to compute probabilities becomes extremely tedious. A general formula eliminates the necessity for such a listing. To obtain such a formula, notice first that every outcome resulting in a particular x value has the same probability. For example, with $n = 4$ and $\pi = .6$ as in Example 36, each of the four outcomes with $x = 3$ had probability $(.6)^3(.4)$—a factor of .6 for each of the three S's and one of .4 for the single F. Similarly, if $n = 10$ and $\pi = .3$, each outcome with $x = 6$ has probability $(.3)^6(.7)^4$—a factor of .3 for each of the six S's and a factor of .7 for each of the four F's. In this case,

$$p(6) = P(x = 6) = \begin{pmatrix} \text{number of outcomes} \\ \text{for which } x = 6 \end{pmatrix} \cdot (.3)^6(.7)^4$$

This is because in computing $P(x = 6)$, we add $(.3)^6(.7)^4$ once for each outcome with $x = 6$. More generally, without specifying the numerical values of n and π,

$$p(x) = P(x\ S\text{'s in } n \text{ trials}) = \begin{pmatrix} \text{number of outcomes} \\ \text{with } x\ S\text{'s} \end{pmatrix} \cdot \pi^x(1 - \pi)^{n-x}$$

If an expression for the number of outcomes having a specific x value can be obtained, we would have our general formula. The development of this expression requires a digression into counting methods. Rather than impede the flow here, we discuss the appropriate counting argument at the very end of this section. We do need new notation at this point, **factorial notation.** The expression 5! stands for (5)(4)(3)(2)(1) = 120. Similarly, 12! (read "twelve factorial") = (12)(11)(10) · · · (3)(2)(1) = 479,001,600. In general, for any positive whole number k, $k! = (k)(k - 1)(k - 2) \cdots (3)(2)(1)$, a descending product of whole numbers. Notice that 1! = 1. In addition, 0! is also defined to be 1.

The box displays the general formula for the binomial probability distribution. The expression involving factorials represents the number of outcomes with a specified x value.

The **probability distribution of a binomial random variable** x based on n trials and success probability π is given by

$$p(x) = P(x\ S\text{'s in the } n \text{ trials})$$
$$= \left(\frac{n!}{x!\,(n - x)!}\right) \cdot \pi^x(1 - \pi)^{n-x} \qquad x = 0,1,2,\ldots,n$$

EXAMPLE 37

(*Example 36 continued*) Let's check to see that the formula gives the same probabilities calculated earlier using basic properties of probability. If $n = 4$, and $\pi = .6$, then

$$p(x) = \frac{4!}{x!\,(4-x)!}(.6)^x(.4)^{4-x} \qquad x = 0, 1, 2, 3, 4$$

Then, since $4! = 24$ and $3! = 6$,

$$\begin{aligned} p(3) = P(x = 3) &= \frac{4!}{3!1!}(.6)^3(.4)^1 = \frac{24}{(6)(1)}(.6)^3(.4) \\ &= 4(.0864) = .3456 \end{aligned}$$

Similarly,

$$p(4) = P(x = 4) = \frac{4!}{4!0!}(.6)^4(.4)^0 = 1(.6)^4 = .1296$$

The other three probabilities are easily checked. ■

EXAMPLE 38

Let S denote an automobile that passes a vehicle emissions inspection test in a certain state, and suppose that $\pi = .8$. The outcome of each of the next 20 inspections is recorded. What is the probability that exactly 15 are S's? That at most 15 are S's? What is the probability that at least 75% (a proportion of at least .75) are S's? With $x =$ the number of S's (vehicles that pass) among the next $n = 20$ inspected, the probability distribution of x is given by

$$p(x) = \left(\frac{20!}{x!\,(20-x)!}\right)\cdot(.8)^x(.2)^{20-x} \qquad x = 0, 1, \ldots, 20$$

Then

$$\begin{aligned} p(15) = P(x = 15) &= \frac{20!}{15!\,5!}(.8)^{15}(.2)^5 \\ &= \frac{[(20)(19)(18)(17)(16)]\cdot(\not{15!})}{\not{15!}\,5!}(.8)^{15}(.2)^5 \\ &= (15{,}504)(.8)^{15}(.2)^5 = .175 \end{aligned}$$

According to this calculation, there are 15,504 sequences of length 20 that consist of exactly 15 S's, far too many to list.

The second question asks for $P(x \le 15)$. This is the probability that $x = 0, 1, 2, \ldots, 14$, or 15. The above formula must be used for each of these x values and the resulting probabilities added together. Omitting the tedious arithmetic,

$$P(x \le 15) = p(0) + p(1) + \cdots + p(14) + p(15) = .370$$

Lastly, 75% of 20 is 15, so at least 75% of the trials will be S's when $15 \le x$. Then

$$\begin{aligned} P(\text{at least 75\% } S\text{'s}) &= P(15 \le x) = p(15) + p(16) + \cdots + p(20) \\ &= .805. \end{aligned}$$

That is, roughly 80% of all binomial experiments with $n = 20$ and $\pi = .8$ would result in at least 75% of the trials being S's. ■

Clearly, using the binomial distribution formula can be tedious unless n is very small. We have included as Table III in the appendices a tabulation of binomial probabilities for $n = 10$ and $n = 20$ in combination with various values of π. This should help you practice using the binomial distribution without getting bogged down in arithmetic. To obtain the probability of a particular x value, find the column headed by your value of π and move down to the row labeled with the desired value of x. Doing this for $n = 20$, $\pi = .8$ (a column), and $x = 15$ (a row) yields $p(15) = .175$, as above. Although $p(x)$ is positive for every possible x value, many probabilities are zero to three decimal places, so they appear as .000 in the table. There are much more extensive binomial tables available. Alternatively, it is easy to program a computer or some calculators to calculate these probabilities.

Suppose that a population consists of N individuals or objects, each one classified as an S or an F. If sampling is carried out without replacement (as it almost always is), then successive draws are dependent. However, when the sample size n is much smaller than N, the extent of this dependence is minimal. In this case, x, the number of S's in the sample, has approximately a binomial distribution. The usual rule of thumb is that the binomial distribution gives accurate results if n is at most 5% of N.

EXAMPLE 39

An article in the December 11, 1983, *Los Angeles Times* discussed the environmental problems caused by underground gasoline storage tanks that leak. There are over 2 million tanks in the United States, and several studies have suggested that roughly 25% of them leak. Suppose that a random sample of $n = 20$ tanks is selected from the population of all tanks (so n is much less than 5% of N). Let S denote a tank that leaks (not what would ordinarily be thought a success!), and suppose that $\pi = .25$. Then the probability that exactly half the tanks in the sample leak is, from Table III, $p(10) = .010$. The probability that at most 50% of the tanks sampled leak is $P(x \leq 10) = p(0) + p(1) + \cdots + p(10) = .996$ (almost a sure thing). Conversely, $P(10 < x) = .004$, so if more than half the 20 tanks sampled had leaks, this would cast substantial doubt on the validity of the value $\pi = .25$. ■

Formulas for the mean value μ and standard deviation σ of a discrete random variable x are given in Section 5.3. Substituting the binomial probabilities into these formulas results (after much algebra) in especially simple expressions for μ and σ here.

> When x is a binomial random variable based on n trials and success probability π, the mean value of x is $\mu = n\pi$ and the standard deviation of x is $\sigma = \sqrt{n\pi(1 - \pi)}$.

EXAMPLE 40

(*Example 39 continued*) The mean value of the number of tanks in a sample of 20 that leak is $\mu = (20)(.25) = 5$. The standard deviation of the number that leak is $\sigma = \sqrt{20(.25)(.75)} = \sqrt{3.75} = 1.94$. ■

The Normal Approximation

Figure 18 shows the probability histograms for two binomial distributions, one with $n = 25$, $\pi = .4$ and the other with $n = 25$, $\pi = .1$. For each distribution we computed $\mu = n\pi$ and $\sigma = \sqrt{n\pi(1 - \pi)}$ and then superim-

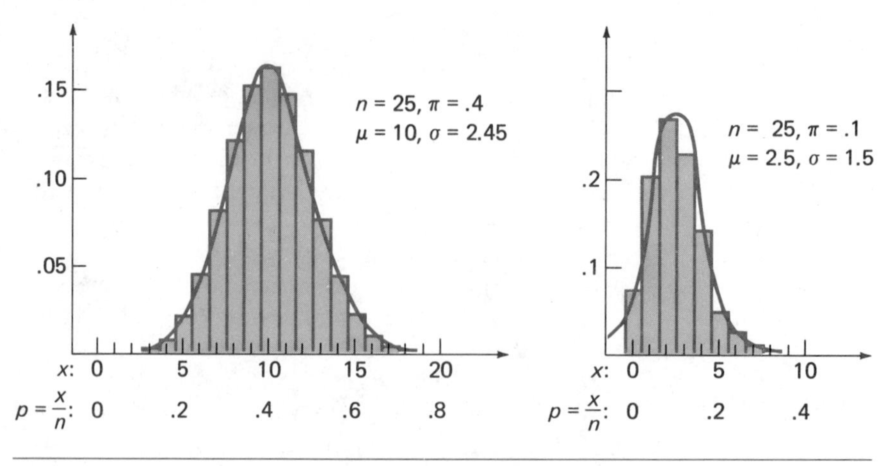

FIGURE 18 NORMAL APPROXIMATIONS TO BINOMIAL DISTRIBUTIONS

posed a normal curve with this μ and σ on the corresponding probability histogram. The normal curve fits the probability histogram very well in the first case. When this happens, binomial probabilities (areas of rectangles in the histogram) can be accurately approximated by areas under the normal curve. Because of this, statisticians say that both x (the number of S's) and x/n (the proportion of S's) are approximately normally distributed. In the second case, the normal curve does not give a good approximation because the probability histogram is quite skewed, whereas the normal curve is symmetric. A normal curve would also give a poor approximation when $n = 25$, $\pi = .9$, since then the probability histogram would be centered at $\mu = 22.5$ and skewed in the opposite direction.

> If n is even moderately large and π is not too near 0 or 1, the probability distributions of both x and x/n are approximately normal. In practice, the approximation is adequate if both $n\pi \geq 5$ and $n(1 - \pi) \geq 5$.

Counting and the Binomial Distribution

The formula for binomial probabilities requires an expression for the number of outcomes that have exactly x S's in n trials. Consider the case $n = 5$ and $x = 3$. Let's label the five trials 1, 2, 3, 4, 5. Then the number of outcomes with three S's is the number of ways of selecting three trial labels to be the S trials. Think of the five labels resting in an urn. We reach in and select a first label, then a second, and then a third. There are 5 possibilities for the first selection and 4 for the second, so there is a total of $(5)(4) = 20$ for the first two selections (5 for each way of selecting the first and 4 further possibilities for the second, so multiplication is correct here). For each of these 20 ways, there are 3 further ways to select the third label. Therefore, the total number of ways of selecting 3 from the 5 is $(5)(4)(3) = 60$.

But there is a difficulty here. Among the 60 possibilities, we've counted

1, 2, 3   1, 3, 2   2, 1, 3   2, 3, 1   3, 1, 2   3, 2, 1

That is, we counted not only the number of different sets of 3 labels but every different ordering as well. These 6 possibilities all correspond to the same set of labels 1, 2, 3. Similarly, since there are $3 \times 2 \times 1 = 6$ ways to order the labels 2, 4, 5, another 6 of our ordered sets correspond to the single set of labels 2, 4, 5. In this case ($n = 5$ and $x = 3$), there are $3 \times 2 \times 1$ ordered sets corresponding to every set of interest to us. Therefore, the number of ways of selecting 3 from the 5 labels (trials) without counting different orderings of the same set is

$$\frac{(5)(4)(3)}{(3)(2)(1)} = \frac{60}{6} = 10$$

It is easy to list these 10: (1, 2, 3), (1, 2, 4), (1, 2, 5), (1, 3, 4), (1, 3, 5), (1, 4, 5), (2, 3, 4), (2, 3, 5), (2, 4, 5), and (3, 4, 5) (and each can be ordered in 6 ways).

In the general case, the number of ways to select x from among the n trial labels when all orderings are considered is $n(n - 1)(n - 2) \cdots (n - x + 1)$ (select a first from the n possibilities, a second from the remaining $n - 1$, and so on, until finally the xth is selected from the remaining $n - x + 1$). To eliminate different orderings of the same set of labels, we divide by $(x)(x - 1)(x - 2) \cdots (2)(1) = x!$, the number of ways of ordering a set of x labels. This gives

$$\begin{pmatrix}\text{number of ways to}\\ \text{select the } x \ S \text{ labels}\end{pmatrix} = \frac{n(n - 1)(n - 2) \cdots (n - x + 1)}{x!}$$

Multiplying both the numerator and denominator by $(n - x)!$ (which doesn't change the value of the expression) yields $n!/[x!(n - x)!]$, which appears in the binomial probability distribution formula.

EXERCISES

5.52 The *Los Angeles Times* (December 13, 1983) reported that only 58% of tenth graders in Los Angeles high schools graduate from those schools 3 years later (of the 42% who did not graduate from Los Angeles schools, some moved to other school districts, but most are presumed to have dropped out). Suppose that four tenth graders are randomly selected from the Los Angeles schools. This is a binomial experiment with $\pi = .58$, $n = 4$, and $x =$ the number among the four who graduate.

- **a.** What is the probability that all four graduate 3 years later from a Los Angeles school (i.e., that $x = 4$)?
- **b.** What is the probability that exactly three of the four graduate from Los Angeles schools?
- **c.** What is the probability that at least three graduate from Los Angeles schools?
- **d.** What is the probability that none of the four graduate from Los Angeles schools?

5.53 A breeder of show dogs is interested in the number of female puppies in a litter. If a birth is equally likely to result in a male or female puppy, give the probability distribution of

$$x = \text{number of female puppies in a litter of size 5}$$

5.54 A manufacturer of camera flash bars notes that defective flash bulbs are sometimes produced. If the probability that any given bulb will be defective is .05, what is the probability that there are no defective flashes on a flash bar containing 10 bulbs? What is the probability that at most one of the 10 bulbs is defective?

5.55 Industrial quality control programs often include inspection of incoming materials from suppliers. If parts are purchased in large lots, a typical plan might be to select 20 parts at random from the lot and inspect them. The lot might be judged as being of acceptable quality if 1 or fewer defective parts are found among those inspected. Otherwise, the lot is rejected and returned to the supplier. Use Table III to find the probability of accepting lots that have each of the following.

a. 5% defective parts

b. 10% defective parts

c. 20% defective parts

(*Hint:* Identify success with a defective part.)

5.56 In an experiment to investigate whether a graphologist (handwriting analyst) could distinguish a normal person's handwriting from that of a psychotic, a well-known expert was given 10 files, each containing a handwriting sample for a normal person and a person diagnosed as psychotic. The graphologist was then asked to identify the psychotic's handwriting. The graphologist made correct identifications in 6 of the 10 trials (data taken from Larsen and Stroup, *Statistics in the Real World* (New York: MacMillan, 1976)). Does this evidence indicate that the graphologist has an ability to distinguish the handwriting of psychotics? (*Hint:* What is the probability of correctly *guessing* 6 or more times out of ten? Your answer should depend on whether this probability is relatively small or large.)

5.57 If the temperature in Florida falls below 32°F during certain periods of the year, there is a chance that the citrus crop will be damaged. Suppose that the probability is .1 that any given tree will show measurable damage when the temperature falls to 30°F. If the temperature drops to 30°F, what is the expected number of trees showing damage in an orchard of 2000 trees? What is the standard deviation of the number of trees which show damage?

5.58 You are to take a multiple choice exam consisting of 100 questions with 5 possible responses to each. Suppose that you have not studied and so must guess (select one of the 5 answers in a completely random fashion) on each question. Let x represent the number of correct responses on the test.

a. What kind of probability distribution does x have?

b. What is your expected score on the exam? (*Hint:* Your expected score is the mean value of the x distribution.)

c. Compute the variance and standard deviation of x.

d. Based on your answers to (b) and (c), is it likely that you would score over 50 on this exam? Explain the reasoning behind your answer.

5.59 There are two different types of videocassette recorders (VCR's), VHS and Beta. Past records suggest that 60% of all those purchasing a VCR at a certain store choose a VHS type.

a. What is the probability that at most 4 of the next 10 purchasers choose VHS?

b. What is the probability that more than 5 of the next 10 purchasers choose VHS?

c. The store currently has 8 VCR's of each type in stock. What is the probability that each of the next 10 customers can obtain his or her desired type from current stock?

KEY CONCEPTS

Event A collection of possible outcomes for a chance experiment. (p. 171)

Probability of an event The long-run relative frequency of occurrence of the event when a chance experiment is repeatedly performed. The probability of an event E is denoted by $P(E)$. (p. 172)

Probability rules Relationships between the probabilities of various complex events and the probabilities of simpler events. The simplest rule says that $P(E)$ is the sum of probabilities of all outcomes contained in E. A more general addition rule states that when events have no common outcomes, the probability that one of these events will occur is the sum of the individual event probabilities. (p. 173, 174)

Independent events Two events are independent if the probability that one of the events occurs is not affected by the occurrence of the other event. The concept extends to more than two events. When events are independent, the probability that they all occur is the product of probabilities of the individual events. (p. 175)

Sampling with and without replacement Successive draws from a population are independent when sampling is with replacement. While this is not true in general for sampling without replacement, it is approximately true when the population size greatly exceeds the sample size. (p. 177)

Random variable A numerical variable that associates a unique value with each experimental outcome. (p. 182)

Discrete random variable A variable whose set of possible values is a set of isolated points on the number line. (p. 183)

Continuous random variable A variable whose set of possible values is an interval on the number line. (p. 183)

Binomial experiment An experiment involving a fixed number of independent trials. Every trial results in one of two outcomes, generically labeled success and failure, and P(success) is the same on each trial. (p. 185)

Binomial random variable The number of successes in a binomial experiment. (p. 185)

Probability distribution For a discrete random variable x, the probability associated with each possible x value. For a continuous random variable x, a curve (called the *density curve*) such that the area under the curve and above any interval gives the probability that x will fall in that interval. (p. 184, 187)

Mean value of a random variable x This value describes where the probability distribution of x is centered. (p. 192)

Variance and standard deviation of a random variable Measures of the extent to which the probability distribution speads out about the mean value. (p. 194)

Normal distribution The most important continuous probability distribution. Its density curve is bell-shaped in appearance. (p. 199)

Standard normal (z) curve The normal curve corresponding to $\mu = 0$ and $\sigma = 1$. Other normal curve areas (probabilities) can be obtained from z curve areas through the operation of standardizing. (p. 200)

z critical value A number on the z measurement scale that captures a specified area under the z curve. (p. 202)

Normal probability plot A plot whose appearance will be roughly linear if the sampled population is normally distributed. (p. 212)

SUPPLEMENTARY EXERCISES

5.60 A student who has a 7 A.M. class on Monday, Wednesday, and Friday has only six socks left in his drawer on Friday morning. Two of these socks are blue, two are brown, and two are green. He reaches into his drawer and selects two socks at random. What is the probability that they are the same color?

5.61 An auto repair shop charges \$40, \$50, and \$60 for labor to tune cars with four-, six-, and eight-cylinder engines, respectively. From past records, 50% of all tuneups are four-cylinder jobs, 30% are six-cylinder jobs, and 20% are eight-cylinder jobs. On a particular day, two car owners call independently of one another to schedule tuneups. Let x = the total revenue for labor from these two tuneups.

- **a.** What is $P(x = 80)$? (*Hint:* $x = 80$ results only when the first and second tuneups are both \$40 jobs.)
- **b.** What is $P(x = 100)$? (*Hint:* One way to have $x = \$100$ is for the first job to generate \$40 and the second \$60. What are the other possibilities? What is the probability of each of these possibilities?)
- **c.** Determine the probability distribution of x, and display it in table form.

5.62 Suppose that net typing rate in words per minute for experienced electric-typewriter touch typists is approximately normally distributed with mean value 60 and standard deviation 15. (The paper "Effects of Age and Skill in Typing" (*J. of Exper. Psych.* (1984): 345–71) describes how net rate is obtained from gross rate by using a correction for errors.)

- **a.** What is the probability that a randomly selected typist's net rate is at most 60? Less than 60?
- **b.** What is the probability that a randomly selected typist's net rate is between 45 and 90?
- **c.** Would you be surprised to find a typist in this population whose net rate exceeded 105? Explain. (*Note:* The largest net rate in a sample described in the paper cited is 104.)
- **d.** Suppose that two typists are independently selected. What is the probability that both their typing rates exceed 75?

5.63 Two friends, A and B, have agreed to meet between 1 P.M. and 6 P.M. on a particular day. In fact, A is equally likely to arrive at 1 P.M., 2 P.M., 3 P.M., 4 P.M., 5 P.M., or 6 P.M. B is also equally likely to arrive at any one of these six times, and the two arrival times are independent of one another. There are thus 36 equally likely (A, B) arrival-time pairs (for example, (2, 5) or (6, 3)). If each person waits for exactly 1 hour after arriving, what is the probability that one person will arrive after the other one has left?

5.64 Only four students—A, B, C, and D—are enrolled in a seminar class. The professor has just asked a question, to which only A and B know the answer. Suppose that the students are called upon in random order (by placing four slips of paper labeled A, B, C, and D in a box and selecting them one by one) until a student who knows the answer is selected. There are 10 possible outcomes, two of which are B and CDA.

- **a.** List the remaining 8 outcomes. Are the outcomes equally likely? Explain.

b. Let x denote the number of students whose names are called. Determine the probability distribution of x.

c. Calculate the mean value, variance, and standard deviation of the random variable x described in (b).

5.65 The Rockwell hardness of a metal is determined by pressing a hardened point into the surface of the metal and then measuring the depth of penetration of the point. Suppose that the Rockwell hardness of a particular alloy is normally distributed with mean 70 and standard deviation 3 (assume that Rockwell hardness is measured on a continuous scale).

a. If a specimen is acceptable only if its hardness is between 65 and 75, what is the probability that a randomly chosen specimen has an acceptable hardness?

b. If the acceptable range is as in (a), what is the probability that at most three of four independently selected specimens are acceptable?

5.66 To determine whether A or B moves first in a game, five slips of paper numbered 1, 2, 3, 4, and 5 are placed in a box. First A selects a slip at random, and then B selects one of the four remaining slips.

a. If the person who selects the slip with the higher number moves first, what is the probability that A moves first?

b. Suppose that A will move first only if the number he or she draws exceeds that drawn by B by more than 1. What is the probability that A moves first?

5.67 Refer to Exercise 5.66. Let x denote the difference between the larger number drawn and the smaller number selected. Suppose that the person who draws the higher number gets x consecutive moves before the other person makes his or her move.

a. What is the probability distribution of x?

b. Calculate the mean value, variance, and standard deviation of x.

5.68 Two friends, Abe and Ben, are going to play a series of Trivial Pursuit games. The first person to win three games will be declared the series winner. Suppose that the probability of Abe winning any particular game is .6 (so .4 is the probability that Ben wins the game) and that outcomes of successive games are independent of one another.

a. What is the probability that Abe wins the series in three games?

b. What is the probability that the series terminates after just three games?

c. What is the probability that Abe wins the series?

5.69 Referring back to the previous problem, let y denote the number of games played until a winner can be declared.

a. What is the probability distribution of y?

b. What is the mean value of y?

c. Let w denote the number of games won by the series loser. Determine the probability distribution of w.

5.70 A particular type of gasoline tank for a compact car is designed to hold 15 gallons. Suppose that the actual capacity x of a randomly selected tank of this type is normally distributed with mean 15 and standard deviation .2.

a. What is the probability that a randomly selected tank will hold at most 14.8 gal?

b. What is the probability that a randomly selected tank will hold between 14.7 and 15.1 gal?

c. If the car on which a randomly chosen tank is mounted gets exactly 25 mi/gal, what is the probability that the car can travel 370 mi without refueling?

5.71 Suppose that 75% of all consultations handled by student consultants at a computing center involve programs with syntax errors. Let x be the number of programs with syntax errors in a sample of 10 randomly chosen consultations.

a. What is the probability that exactly 7 of the 10 programs contain syntax errors?

b. What is the probability that fewer than 4 of the 10 programs contain syntax errors?

c. What is the probability that more than half of the 10 programs contain syntax errors?

5.72 Assume that development time for a particular type of photographic printing paper when it is exposed to a light source for 5 sec is normally distributed with a mean of 25 and a standard deviation of 1.3 sec.

a. What is the probability that a particular print will require more than 26.5 sec to develop?

b. What is the probability that development time is at least 23 sec?

c. What is the probability that development time differs from the mean (25 sec) by more than 2.5 sec?

5.73 A friend of ours recently planned a camping trip. He had two flashlights, one that required a single 6-V battery and another that used two size-D batteries. He had previously packed two 6-V and four size-D batteries in his camper. Suppose the probability that any particular battery works is .5 and that batteries work or fail independently of one another. Our friend wants to take just one flashlight. Which one has the greater chance of working with the batteries he has packed?

5.74 Alvie Singer lives at the corner labeled *A* in the accompanying diagram. He wants to visit Annie Hall, who lives at the corner marked *I*, so he will always travel toward *I* (either north or east). Suppose that at each corner at which he has a choice of direction (*A*, *B*, *D*, and *E*), Alvie tosses a fair coin and moves north if the coin lands heads up and east otherwise. What is the probability that he visits the corner labeled *G*? The corner labeled *E*?

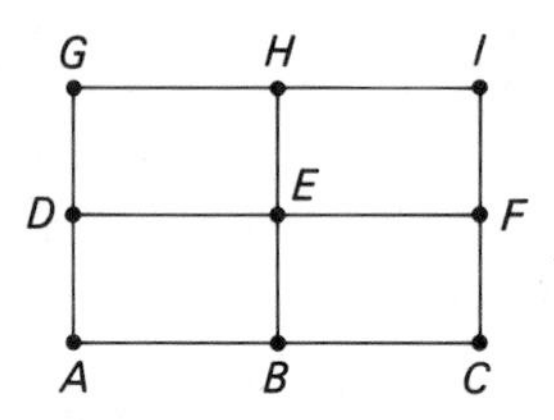

5.75 Five friends—A, B, C, D, and E—have tickets to see a play. The aisle in which they will sit has only five seats. A seating arrangement will be determined by mixing the stack of tickets and dealing one to each individual (as in cards). It can be shown that there are 120 possible arrangements, each with the same probability of occurring (equally likely outcomes).

Seat number	1	2	3	4	5
One possible arrangement	B	E	A	C	D

a. What is the probability that A and E end up in seats 1 and 5, respectively? (*Hint:* First list all the ways that this can happen.)

b. What is the probability that A and E sit at the ends of the aisle (separated by the other three people)?

c. What is the probability that A and E sit next to one another in seats 1 and 2, respectively?

d. What is the probability that A and E sit next to one another?

e. Let x = the number of seats that separate A and E, so x is a random variable. What is $p(0)$ (i.e., $P(x = 0)$)? $p(3)$?

f. It isn't difficult to show that $p(2) = \frac{24}{120}$. Use this fact to obtain the probability distribution of x.

REFERENCES

McClave, James, and Frank Dietrich. *Statistics,* 3rd ed. (New York: Dellen/Macmillan, 1985). (Includes an elementary discussion of probability that goes into more detail than our book does.)

Mosteller, Frederick, Robert Rourke, and George Thomas. *Probability with Statistical Applications*. (Reading, MA.: Addison-Wesley, 1970). (Although a bit old, there is no more recently published book that provides a better in-depth coverage of probability at a very modest mathematical level).

6 Sampling Distributions

INTRODUCTION

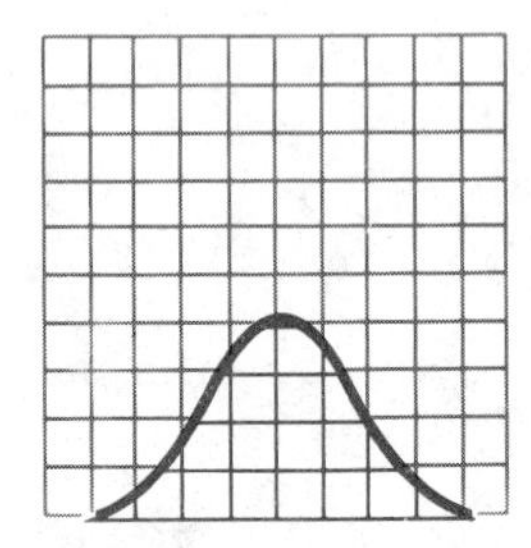

THE PROBABILITY DISTRIBUTION of a numerical variable x describes the long-run behavior when a single value of x is repeatedly observed. When the objective of a study is to draw conclusions about the population distribution of a variable x, an investigator virtually always obtains a sample consisting of more than just one x value. Various numerical quantities, such as the sample mean, $\bar{x}$, or sample standard deviation, s, can be calculated from the sample values and used to make inferences. As an example, let x denote the nicotine content of a randomly selected cigarette of a certain brand. To learn something about the mean nicotine content, μ, a sample of $n = 4$ cigarettes might be obtained, resulting in $x_1 = 1.68$, $x_2 = 1.89$, $x_3 = 1.73$, and $x_4 = 1.95$, so $\bar{x} = 1.813$. Suppose that a second sample of four cigarettes is obtained. The resulting x values might be $x_1 = 1.92$, $x_2 = 1.67$, $x_3 = 1.58$, and $x_4 = 1.87$, yielding a sample mean value of $\bar{x} = 1.760$.

Now consider repeating this experiment over and over again, each time taking a new sample of four cigarettes, determining the four x values, and computing $\bar{x}$. In the long run, what proportion of $\bar{x}$ values will be between 1.75 and 1.85? What proportion of $\bar{x}$ values will be at least 2.0? What proportion of $\bar{x}$ values will be below 1.70? These questions, as well as many others, can be answered by studying what is called the *sampling distribution* of $\bar{x}$. We shall see that a sampling distribution is just a probability distribution. The word *sampling* is used in place of *probability* to emphasize that the distribution refers to a quantity calculated from an entire sample of x values rather than just a single x value.

Two frequently encountered problems in elementary statistical inference concern drawing conclusions about a population mean μ or about the proportion π of individuals in the population who have some specified property. For this reason, after introducing some terminology and the notion of a random sample, we shall focus on the sampling distribution of a sample mean $\overline{x}$ and of a sample proportion (the fraction of individuals in a sample who have the specified property). Other sampling distributions are used in later chapters as a basis for various inferential procedures.

6.1 Statistics and Random Samples

The objective of many investigations and research projects is to draw conclusions about how the values of some numerical variable x are distributed in a population. Although many aspects of the distribution might be of interest, attention is often focused on one particular population characteristic. Examples include

1. x = fuel efficiency (mi/gal) for a 1986 Ford Escort automobile, with interest centered on the mean value μ of fuel efficiency for all such cars.
2. x = the diameter of a certain type of bearing, with the major concern being diameter variability as described by σ, the standard deviation of x.
3. x = the time until first major repair of the top-of-the-line Maytag washing machine, with attention focusing on the proportion of times that exceed the manufacturer's warranty period (it sometimes seems as though all breakdowns occur immediately after a warranty expires!).

Statistics

Statistical methods for drawing conclusions about the distribution of a variable x are based on obtaining a sample of x values. As in earlier chapters, the letter n denotes the number of observations in the sample (sample size). The observations themselves are denoted by $x_1, x_2, \ldots, x_n$, where x_1 is the first x value in the sample (the one obtained from the first individual or object selected or from making a first measurement), x_2 is the second x value in the sample, etc. Once the values of $x_1, x_2, \ldots, x_n$ are available, an investigator must decide which quantities computed from these sample values will be most informative in drawing the desired type of conclusion. For example, we shall see in the next few chapters that several standard statistical procedures for drawing conclusions about μ utilize both $\overline{x}$ and s. Statisticians use the term *statistic* for any quantity such as $\overline{x}$ or s that is computed from sample data.

DEFINITION	Any quantity computed from values in a sample is called a **statistic.**

Although we did not refer to them as such, several statistics other than $\overline{x}$ and s were introduced in Chapter 3. These include the sample median (middle value in the ordered list of sample observations), the sample range (difference between the largest and smallest sample values), a trimmed mean, and a sample percentile.

It is very important to appreciate the distinction between a population characteristic (i.e., a characteristic of the x distribution) and a statistic, which is a sample characteristic. The mean value of x, or population mean, μ, is a population characteristic. The value of μ is a fixed number, such as 1.75 mg for nicotine content or 28.5 mi/gal for fuel efficiency. However, the value of μ is generally not known, which is why we take a sample of x values. The sample mean, $\overline{x}$, is a characteristic of the sample. Its value is a fixed number for any particular sample, but different samples typically result in different $\overline{x}$ values. It would be nice if the value of $\overline{x}$ turned out to be exactly μ, but in practice this almost never happens because of sampling variability. Some samples yield x values that are not very representative of the population distribution. Thus the value of μ might be 1.75, but one sample might yield $\overline{x} = 1.63$, whereas another might result in $\overline{x} = 2.04$.

The population under study is sometimes composed of just two types of individuals, those possessing a certain property (the successes) and those without the property (the failures). Here it is common practice to introduce a numerical variable with value 1 for a success and value 0 for a failure (so each member of the population can be regarded as a 1 or a 0). The proportion π of successes in the population is then the same as the proportion of 1's. Drawing a conclusion about π amounts to saying something about how the 1's and 0's are distributed in the population.

The proportion of successes, π, in a population is a population characteristic. Its value is some number between 0 and 1. For example, the proportion of all beer drinkers who could correctly state which of two glasses contains Lowenbrau and which contains Schlitz might be $\pi = .6$. But this value would surely be unknown to an investigator, so to make an inference about π it would be necessary to perform an experiment with a sample of beer drinkers. Selecting a sample of size $n = 5$ might result in values $x_1 = 1$, $x_2 = 0$, $x_3 = 1$, $x_4 = 1$, $x_5 = 1$ (four successful identifications). The sample proportion of successes, p, is a statistic and for this sample has value $p = \frac{4}{5} = .80$. Another sample might result in only 2 out of 5 successes, whence $p = \frac{2}{5} = .40$. If the value of π really were .6, a sample consisting of three successes would yield $p = \frac{3}{5} = .6 = \pi$; in this case the value of the statistic would be exactly the value of the population proportion. However, it can be shown that the probability of this happening is only .3456, so it is more likely than not that p will differ from π. The important point is that the value of the population characteristic π is fixed, whereas the value of the statistic p is a variable quantity depending on which sample values are observed.

Sampling Distributions

Inferential methods involve using the values of various statistics to reach conclusions about one or more population characteristics. Unfortunately, the value of a statistic is subject to sampling variability. This variability may result in different samples yielding different conclusions. Information about a statistic's long-run behavior is needed in order to attach a measure of reliability to conclusions. This information is obtained by applying probability concepts to determine a statistic's sampling distribution.

DEFINITION

The value of a statistic depends on the particular sample selected from the population and changes from sample to sample. A statistic is, therefore, a random variable and as such has a probability distribution. The probability distribution of a statistic is called its **sampling distribution.** The sampling distribution of a statistic describes the long-run behavior of the statistic's values when many different samples, each of size *n*, are obtained and the value of the statistic is computed for each one.

EXAMPLE 1

The record library of a particular classical music radio station contains five different recordings of Beethoven's Fifth Symphony. Because listener polls have identified this as the most popular classical work, the station manager has insisted that three different recordings of this symphony be played each month. The recordings all differ somewhat in playing time (due to differences in interpretations by conductors and orchestras). These times are 29.6, 29.9, 30.0, 30.2, and 30.8 (min).

Suppose that the three records to be played in a given month are determined by writing each of the five playing times on a different slip of paper, mixing up the slips, and selecting three at random. Then any set of three times has the same chance of selection as any other set. Let the statistic of interest be the sample median playing time. We list each of the 10 possible samples (ordered for convenience) and the corresponding value of this statistic.

Sample	Sample Median	Sample	Sample Median
29.6, 29.9, 30.0	29.9	29.6, 30.2, 30.8	30.2
29.6, 29.9, 30.2	29.9	29.9, 30.0, 30.2	30.0
29.6, 29.9, 30.8	29.9	29.9, 30.0, 30.8	30.0
29.6, 30.0, 30.2	30.0	29.9, 30.2, 30.8	30.2
29.6, 30.0, 30.8	30.0	30.0, 30.2, 30.8	30.2

Since each sample has the same chance of occurring as any other sample, the probability that any particular one results is $\frac{1}{10}$, or .10. That is, in a very long sequence of months, each sample would occur $\frac{1}{10}$ (or 10%) of the time. The only three possible values of the sample median are 29.9, 30.0, and 30.8. Since 30.0 is the sample median for four of the possible samples, $P(\text{sample median} = 30.0) = \frac{4}{10} = .40$. Similarly, $P(\text{sample median} = 29.9) = .3$ and $P(\text{sample median} = 30.2) = .3$. The sampling distribution of the sample median can now be summarized in a probability distribution table:

Value of sample median	29.9	30.0	30.2
Probability of value	.3	.4	.3

Thus, in the long run, 40% of all months result in a sample median playing time of 30.0. Notice that the population median time is 30.0 (the middle value when all five possible times are ordered). In 40% of all months, the sample median equals the population median, but in 60% of all months the two differ. The population median remains fixed in value, but the sample median varies in value from sample to sample. ■

You should notice several things about Example 1. First of all, even though this experiment was quite simple and involved relatively few outcomes, obtaining the sampling distribution required some careful thought and calculation. Things would have been much worse if the radio station owned 10 different recordings (a larger population) and five (a larger sample) were to be played each month. Just listing the possible outcomes would be extremely tedious. Second, the same reasoning could be used to obtain the sampling distribution of any other statistic. For example, replacing the value of the sample median by the value of $\overline{x}$ (29.83 for the first ordered sample, etc.) would lead to the sampling distribution of $\overline{x}$. Third, the above sampling distribution resulted from sampling without replacement. If sampling had been with replacement, the same time might have been chosen twice or even on all three selections. Additional samples and corresponding values of the sample median would be possible (e.g., 29.6 and 30.8), resulting in a more complicated sampling distribution. The sampling distribution depends not only on which statistic is under consideration but also on the sample size and the method of sampling.

EXAMPLE 2

A company maintains three offices in a certain area, each staffed by two employees. Information concerning yearly salaries (in thousands of dollars) is given.

Office	1		2		3	
Employee	1	2	3	4	5	6
Salary	14.7	18.6	15.2	18.6	10.8	14.7

A survey is to be carried out to obtain information on average salary levels. Two of these six employees will be selected for inclusion in the survey. Suppose that six slips numbered 1, 2, . . . , 6 are placed in a box and two are drawn without replacement. This ensures that each of the 15 possible employee pairs has the same probability—namely, $\frac{1}{15}$—of being chosen. Computing the sample average salary $\overline{x}$ for each possible selection leads to the sampling distribution of $\overline{x}$ shown. For example, $\overline{x} = 16.65$ occurs for the four pairs (1, 2), (1, 4), (2, 6), and (4, 6), so $P(\overline{x} = 16.65) = \frac{4}{15} = .267$.

$\overline{x}$ value	12.75	13.00	14.70	14.95	16.65	16.90	18.60
Probability	.133	.067	.200	.133	.267	.133	.067

Thus if this experiment is performed over and over again, each time with a new selection of two employees, in the long run the value $\overline{x} = 14.70$ will occur 20% of the time, the value $\overline{x} = 18.60$ only 6.7% of the time, and so on.

Now consider what happens if the method for obtaining a sample is changed. One of the three offices is selected at random (using just three slips of paper), and both employees from that office are included. This is called a *cluster sample* because a cluster (in this case an office) is selected at random and all individuals in the cluster are included in the sample. There are only three possible employee pairs, each one having probability $\frac{1}{3}$ (or .333) of occurring. The resulting sampling distribution of $\overline{x}$ is as follows:

$\overline{x}$ value	12.75	16.65	16.90
Probability	.333	.333	.333

If this experiment is repeated many times over, each of these three $\overline{x}$ values will occur 33.33% of the time in the long run. The above two sampling distributions of $\bar{x}$ are obviously quite different, demonstrating clearly that the sampling distribution depends on the method of sampling used. ■

Random Samples

When an investigator takes a sample from a population of interest, the sampling is virtually always done without replacement. As we discussed in the section on probability at the outset of Chapter 5, the results of successive selections are dependent when sampling is without replacement. However, if the sample size is small relative to the population size, then the dependence is quite negligible. In this case, sampling without replacement is almost like sampling with replacement, where there is no dependence in successive draws. When a sample is obtained with replacement, the distribution of possible values on any particular selection is exactly the same as on any other selection.

DEFINITION

A sample $x_1, x_2, \ldots, x_n$ of values of a numerical variable x is called a **random sample** if the sampled values are selected independently from the same population distribution.

In most applications, sampling is without replacement, but the sample size is much smaller than the population size (at most 5% of the population is sampled). For practical purposes, the successive observations can be regarded as independent, so the sample can be considered a random sample as we have defined it. The x values in a random sample are usually obtained by first selecting n individuals or objects and then observing or determining the value of x for each one. It is customary to refer to the selected individuals or objects, as well as to the x values themselves, as a random sample. Thus we may speak of a random sample of students when the variable of interest is x = grade point average, or a random sample of houses when a study is concerned with x = January electricity usage.

Methods for analyzing both random samples and other types of samples are based on concepts of probability and results concerning sampling distribution of various statistics. However, the statistics and methods employed are most easily understood in the case of random samples. Throughout the remainder of this book, we deal with random samples.

EXERCISES

6.1 What is the difference between a population characteristic and a statistic?

6.2 What is the difference between $\bar{x}$ and μ? Between σ and s?

6.3 Example 1 illustrates the construction of the sampling distribution of the median of a sample of size 3 from a small population. Use the same population, sample size, and procedure to find each of the following.

a. Sampling distribution of the sample mean

b. Sampling distribution of the sample standard deviation

6.4 Explain briefly the difference between sampling with replacement and sampling without replacement.

6.5 In Example 2, suppose that a sample of size 3 is selected without replacement from the entire population of six employees. Determine the sampling distribution of $\bar{x}$. How does this distribution compare to the distribution based on $n = 2$?

6.6 Assume that the 435 members of the U.S. House of Representatives are listed in alphabetical order. Explain how you would select a random sample of 20 members.

6.7 The three offices in Example 2 can be regarded as distinct segments, or *strata*, of the population. A *stratified sample* involves selecting a specified number of individuals from each strata. When values within a strata are similar but values in some strata differ substantially from those in other strata, stratified sampling can help ensure a representative sample. In the context of Example 2, suppose that a stratified sample consists of one of the two individuals from each strata (each selected by tossing a fair coin). There are then eight possible samples. List them, and obtain the sampling distribution of $\bar{x}$.

6.8 Describe how you might go about selecting a random sample of each.

a. Doctors practicing in Los Angeles County
b. Students enrolled at a particular university
c. Boxes in a warehouse
d. Registered voters in your community
e. Subscribers to a local newspaper
f. Radios from a shipment of 100

6.2 The Sampling Distribution of $\bar{x}$: Some Empirical Results

When the objective of a statistical investigation is to draw some type of conclusion concerning μ, the mean value of a numerical variable x, it is natural to base the conclusion on the value of the sample mean, $\bar{x}$. If μ is the true average fuel efficiency (mi/gal) of a certain type of car, a sample of $n = 5$ cars might be selected. The resulting sample values might be $x_1 = 27.6$, $x_2 = 29.3$, $x_3 = 28.6$, $x_4 = 27.8$, and $x_5 = 27.2$, from which $\bar{x} = 28.10$. Does this value of $\bar{x}$ strongly contradict the automobile manufacturer's claim that μ is (at least) 28.50? Perhaps the claim is correct and the sample consisted of some unusually small x values. It is possible that another sample of size 5 would yield $x_1 = 29.5$, $x_2 = 27.8$, $x_3 = 28.6$, $x_4 = 28.2$, and $x_5 = 29.3$, giving $\bar{x} = 28.68$, which supports the company's claim.

To better understand how inferential procedures based on $\bar{x}$ work, we must first study how $\bar{x}$ varies in value from sample to sample. The behavior of $\bar{x}$ computed from different samples is described by its sampling distribution. Characteristics of the x population distribution—its shape (normal, skewed, or something else), mean value μ, and standard deviation σ—and the sample size n on which $\bar{x}$ is based are important in determining characteristics of the sampling distribution of $\bar{x}$. Starting with a specified x distribution, what can be said about the shape of the $\bar{x}$ distribution (this shape determines how the $\bar{x}$ values in a very long sequence are distributed along the measurement axis)? How is the mean value of the $\bar{x}$ distribution, which determines the long-run average value of the $\bar{x}$'s obtained from many different samples, related to the mean value (center) μ of the x distribution? And how is the standard deviation of $\bar{x}$'s sampling distribution, which determines the amount of variability in a long sequence of $\overline{x}$ values, related to the population standard deviation σ?

Answers to these questions are provided by some general rules to be stated shortly. However, to aid in developing intuition, it is helpful to look first at the results of some sampling experiments. In each example that follows, we started with a specified x population distribution, fixed a sample size n, and selected 500 different random samples of this size from the x distribution. We then computed $\overline{x}$ for each sample and constructed a sample histogram of these 500 $\overline{x}$ values. Because 500 is reasonably large (a reasonably long sequence of samples), the sample histogram should rather closely resemble the true sampling distribution of $\overline{x}$ (obtained from an unending sequence of $\overline{x}$ values). To see how the choice of sample size affects the sampling distribution, we repeated the experiment for several different values of n. Thinking hard about the characteristics of these sample histograms should better equip you for understanding and applying the general rules that follow.

EXAMPLE 3

The paper "Platelet Size in Myocardial Infarction" (*Brit. Med. J.* (1983): 449–51) presented evidence that suggests that the distribution of platelet volume was approximately normal in shape both for patients after acute myocardial infarction (a heart attack) and for control subjects who had no history of serious illness. The suggested values of μ and σ for the control-subject distribution were $\mu = 8.25$ and $\sigma = .75$. Figure 1 pictures the corresponding normal curve. The curve is centered at 8.25, the mean value of platelet volume. The value of the population standard deviation, .75, determines the extent to which the x distribution spreads out about its mean value.

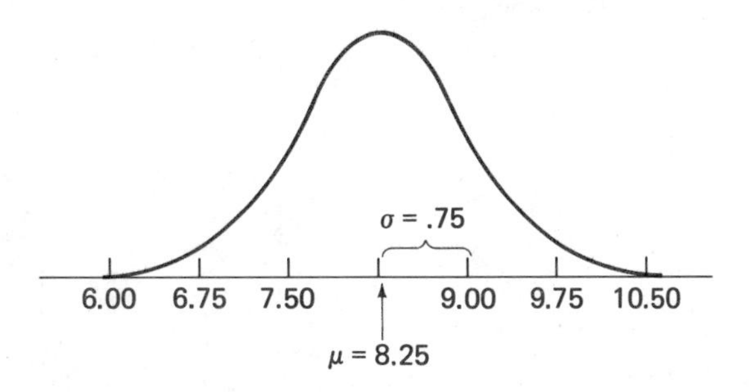

FIGURE 1 NORMAL DISTRIBUTION OF PLATELET VOLUME x WITH $\mu = 8.25$ AND $\sigma = .75$

We first used MINITAB to select 500 random samples from this normal distribution, each one consisting of $n = 5$ observations. A histogram of the resulting 500 $\overline{x}$ values appears in Figure 2(a). This procedure was repeated for samples of size $n = 10$ (again obtaining 500 samples), then again for $n = 20$, and finally for $n = 30$. The resulting sample histograms of $\overline{x}$ values are displayed in Figure 2(b), (c), and (d). Note that using relative frequencies on the vertical axis results in a histogram in which the total area of all rectangles is not equal to 1, as would be the case for a probability distribution. This can be remedied by using the density scale, as described in Chapter 2, to mark the vertical axis.

The first thing to notice about the histograms is their shape. To a reasonable approximation, each of the four has the shape of a normal curve. The deviation of each from a normal curve can be explained by the fact that each

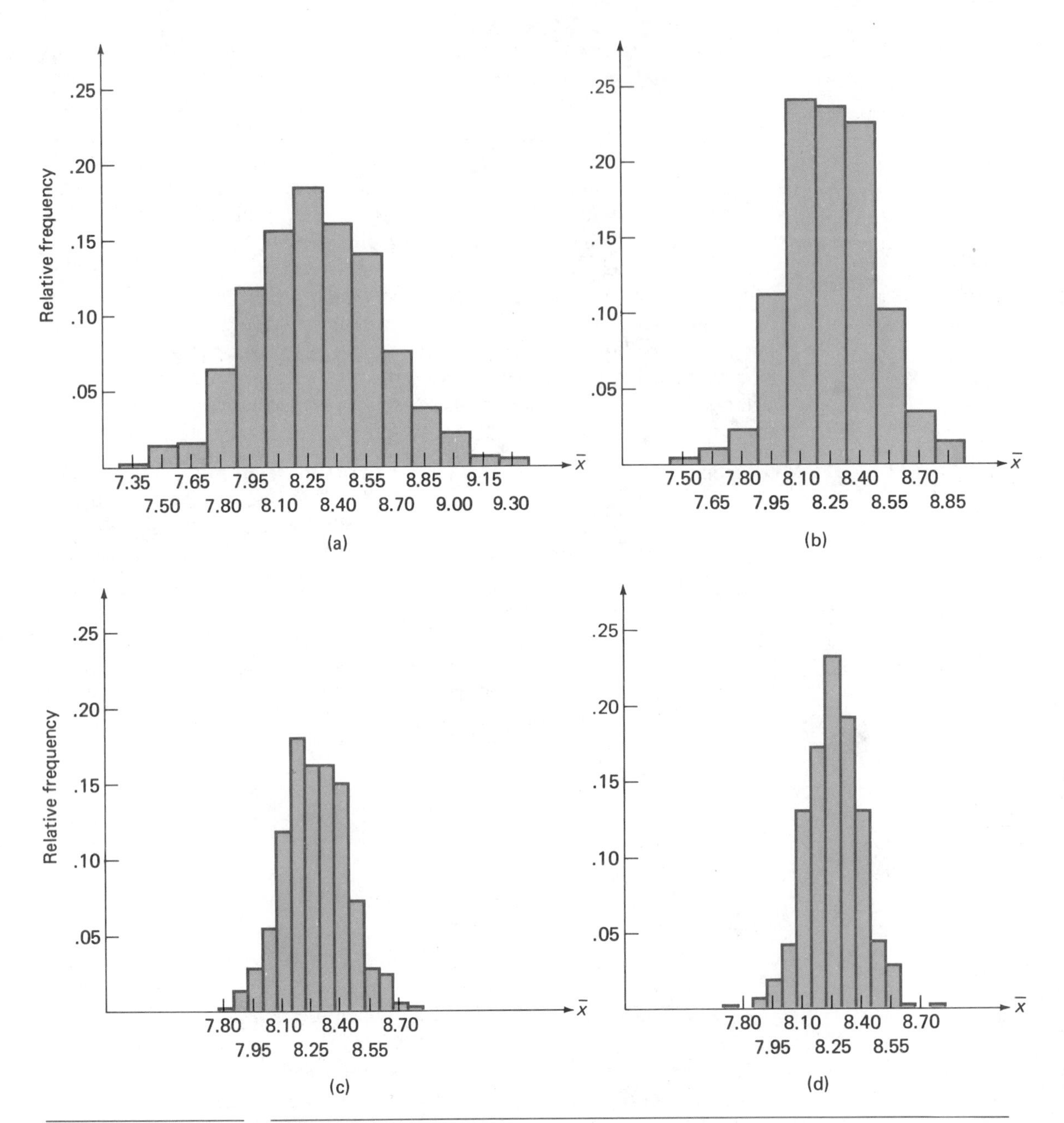

FIGURE 2

SAMPLE HISTOGRAMS FOR $\overline{x}$ BASED ON 500 SAMPLES, EACH CONSISTING OF n OBSERVATIONS
(a) $n = 5$
(b) $n = 10$
(c) $n = 20$
(d) $n = 30$

histogram is based on only 500 $\overline{x}$ values rather than on an unending sequence. One of our general rules states that when the population distribution is normal, the sampling distribution of $\overline{x}$ is always (for any n) described by a normal curve. Second, each histogram is centered approximately at 8.25, the mean of the population being sampled. Another general rule says that the true sampling-distribution histogram is centered at exactly the population mean, here 8.25, for any sample size n. Our four histograms are centered at slightly different values (mean values 8.293, 8.242, 8.264, and 8.255, respectively) because, again, each is based on only 500 $\overline{x}$ values.

The final aspect of the histograms that you should note is their spread relative to one another. The smaller the value of n, the greater the extent to which the sampling distribution spreads out about the mean value. This is why the histograms for $n = 20$ and $n = 30$ are based on narrower class intervals than those for the two smaller sample sizes. For the larger sample sizes, most of the $\overline{x}$ values are quite close to 8.25. This is the effect of averaging. When n is small, a single unusual x value can result in an $\overline{x}$ value far from the center. With a larger sample size, any unusual x values, when averaged in with the other sample values, still tends to yield an $\overline{x}$ value close to μ. Another of our general rules relates the spread of the $\overline{x}$ distribution (as described by its standard deviation) to the population standard deviation σ. A larger n always implies less spread in the $\overline{x}$ distribution. Combining these insights yields a result that should appeal to your intuition—$\overline{x}$ based on a large n tends to be closer to μ than does $\overline{x}$ based on a small n. ■

EXAMPLE 4

The symmetry of the x population distribution and resulting $\overline{x}$ sample histograms in Example 3 made comparisons and identification of salient features relatively straightforward. Now consider properties of the $\overline{x}$ sampling distribution when the population distribution is quite skewed (and thus very unlike a normal distribution). The May 8, 1983, issue of the *Los Angeles Times* contained data on the amount spent per pupil by each of 254 school districts located in Southern California. Figure 3 displays a MINITAB histogram of the data. The histogram has several peaks and a very long upper tail.

Let's now regard the 254 values as comprising a population, so that the histogram of Figure 3 shows the distribution of values in the population. Because of the skewed shape, identification of the mean value from the picture is not as easy as for a normal distribution. We found the average of these 254 values to be $\mu = 1864$, so that is the balance point for the population histogram. The median population value is 1818, less than μ and reflective of the distribution's positively skewed nature.

As we did in Example 3, for each of the sample sizes $n = 5$, 10, 20, and 30, as well as for $n = 50$, 500 random samples of size n were selected. These were selected with replacement to approximate more nearly the usual situation in which the sample size n is only a small fraction of the population size (without replacement, $n = 30$ results in more than 10% of the population being sampled; see the discussion at the end of Section 6.1). The value of $\overline{x}$ for each of the 500 samples was computed, and a histogram of these values was constructed for each of the five sample sizes. The resulting histograms are displayed in Figure 4(a)–(e).

The first thing to notice about the histograms is that, unlike the normal population case, they all differ in shape. In particular, the histograms become progressively less skewed as the sample size, n, increases. The histograms for $n = 10$ and $n = 20$ are substantially more symmetric than is the population distribution, and the shapes of the distributions for $n = 30$ and $n = 50$ are very much like a normal curve. This is the effect of averaging. Even when n is large, one of the few large x values in the population appears only infrequently in the sample. When one does appear, its contribution to $\overline{x}$ is

```
MIDDLE OF   NUMBER OF
INTERVAL    OBSERVATIONS
 1625.0       6     ******
 1675.0      49     *************************************************
 1725.0      24     ************************
 1775.0      30     ******************************
 1825.0      48     ************************************************
 1875.0      29     *****************************
 1925.0      15     ***************
 1975.0       8     ********
 2025.0      11     ***********
 2075.0       9     *********
 2125.0       6     ******
 2175.0       6     ******
 2225.0       3     ***
 2275.0       2     **
 2325.0       0
 2375.0       0
 2425.0       1     *
 2475.0       1     *
 2525.0       1     *
 2575.0       0
 2625.0       1     *
 2675.0       0
 2725.0       1     *
 2775.0       1     *
 2825.0       0
 2875.0       0
 2925.0       1     *
 2975.0       0
 3025.0       0
 3075.0       0
 3125.0       0
 3175.0       0
 3225.0       0
 3275.0       0
 3325.0       0
 3375.0       0
 3425.0       0
 3475.0       0
 3525.0       0
 3575.0       0
 3625.0       0
 3675.0       0
 3725.0       1     *
```

FIGURE 3 THE POPULATION DISTRIBUTION FOR EXAMPLE 4

swamped by the contributions of more typical sample values. The normal curve shape of the histograms for $n = 30$ and $n = 50$ are exactly what is predicted by the central limit theorem, to be introduced shortly. According to this theorem, even if the population distribution bears no resemblance whatsoever to a normal curve, the $\overline{x}$ sampling distribution is approximately normal when the sample size n is reasonably large.

The averages of the 500 $\overline{x}$ values for sample sizes $n = 5, 10, 20, 30,$ and 50 are 1863.5, 1865.4, 1865.0, 1862.1, and 1864.5, the centers (balance points) of the corresponding five histograms. These are all quite close to the population mean $\mu = 1864$. If each histogram had been based on an unending sequence of $\overline{x}$ values rather than just 500 values, each histogram would have been centered at exactly 1864. Thus different values of n change the

```
MIDDLE OF   NUMBER OF
INTERVAL    OBSERVATIONS
 1685.0       0
 1700.0       3      ***
 1715.0       5      *****
 1730.0      17      *****************
 1745.0      24      ************************
 1760.0      29      *****************************
 1775.0      31      *******************************
 1790.0      22      **********************
 1805.0      41      *****************************************
 1820.0      38      **************************************
 1835.0      27      ***************************
 1850.0      32      ********************************
 1865.0      34      **********************************
 1880.0      34      **********************************
 1895.0      28      ****************************
 1910.0      20      ********************
 1925.0      22      **********************
 1940.0      15      ***************
 1955.0      15      ***************
 1970.0      10      **********
 1985.0       8      ********
 2000.0       8      ********
 2015.0       4      ****
 2030.0       6      ******
 2045.0       4      ****
 2060.0       2      **
 2075.0       0
 2090.0       4      ****
 2105.0       2      **
 2120.0       2      **
 2135.0       4      ****
 2150.0       0
 2165.0       0
 2180.0       0
 2195.0       0
 2210.0       0
 2225.0       1      *
 2240.0       1      *
 2255.0       3      ***
 2270.0       2      **
 2285.0       0
 2300.0       0
 2315.0       1      *
 2330.0       0
```

(a)

```
EACH * REPRESENTS 2 OBSERVATIONS

MIDDLE OF   NUMBER OF
INTERVAL    OBSERVATIONS
 1685.0       0
 1700.0       1      *
 1715.0       1      *
 1730.0       3      **
 1745.0       4      **
 1760.0       8      ****
 1775.0      18      *********
 1790.0      31      ****************
 1805.0      49      *************************
 1820.0      50      *************************
 1835.0      53      ***************************
 1850.0      43      **********************
 1865.0      49      *************************
 1880.0      44      **********************
 1895.0      25      *************
 1910.0      25      *************
 1925.0      20      **********
 1940.0      20      **********
 1955.0       9      *****
 1970.0       8      ****
 1985.0       8      ****
 2000.0       7      ****
 2015.0       6      ***
 2030.0       2      *
 2045.0       6      ***
 2060.0       1      *
 2075.0       2      *
 2090.0       2      *
 2105.0       1      *
 2120.0       1      *
 2135.0       1      *
 2150.0       0
 2165.0       0
 2180.0       1      *
 2195.0       0
 2210.0       0
 2225.0       0
 2240.0       0
 2255.0       1      *
```

(b)

(continued)

FIGURE 4 MINITAB HISTOGRAMS OF 500 $\overline{x}$ VALUES, EACH BASED ON n OBSERVATIONS
(a) $n = 5$
(b) $n = 10$

```
EACH * REPRESENTS 2 OBSERVATIONS

MIDDLE OF   NUMBER OF
INTERVAL    OBSERVATIONS
 1685.0        0
 1700.0        0
 1715.0        0
 1730.0        0
 1745.0        1     *
 1760.0        3     **
 1775.0       12     ******
 1790.0       22     ***********
 1805.0       30     ***************
 1820.0       52     **************************
 1835.0       67     **********************************
 1850.0       65     *********************************
 1865.0       58     *****************************
 1880.0       38     *******************
 1895.0       34     *****************
 1910.0       44     **********************
 1925.0       20     **********
 1940.0       14     *******
 1955.0       15     ********
 1970.0       13     *******
 1985.0        3     **
 2000.0        3     **
 2015.0        2     *
 2030.0        2     *
 2045.0        1     *
 2060.0        1     *
```

(c)

```
EACH * REPRESENTS 2 OBSERVATIONS

MIDDLE OF   NUMBER OF
INTERVAL    OBSERVATIONS
 1685.0        0
 1700.0        0
 1715.0        0
 1730.0        0
 1745.0        0
 1760.0        1     *
 1775.0        1     *
 1780.0       12     ******
 1805.0       33     *****************
 1820.0       44     **********************
 1835.0       77     ***************************************
 1850.0       77     ***************************************
 1865.0       80     ****************************************
 1880.0       53     ***************************
 1895.0       45     ***********************
 1910.0       30     ***************
 1925.0       22     ***********
 1940.0       13     *******
 1955.0        3     **
 1970.0        4     **
 1985.0        4     **
 2000.0        1     *
```

(d)

(continued)

FIGURE 4
(continued)

MINITAB HISTOGRAMS OF 500 $\overline{x}$ VALUES, EACH BASED ON n OBSERVATIONS
(c) $n = 20$
(d) $n = 30$

```
EACH * REPRESENTS 2 OBSERVATIONS

MIDDLE OF   NUMBER OF
INTERVAL    OBSERVATIONS
 1775.0        0
 1790.0        4      **
 1805.0       10      *****
 1820.0       28      **************
 1835.0       72      ************************************
 1850.0       91      **********************************************
 1865.0      102      ***************************************************
 1880.0       93      ***********************************************
 1895.0       68      **********************************
 1910.0       22      ***********
 1925.0        8      ****
 1940.0        2      *
```

(e)

FIGURE 4 (continued)

MINITAB HISTOGRAMS OF 500 $\overline{x}$ VALUES, EACH BASED ON n OBSERVATIONS
(e) $n = 50$

shape but not the center of $\overline{x}$'s sampling distribution. Comparison of the five $\overline{x}$ histograms with one another and with the population distribution shows that as n increases, not only does the $\overline{x}$ histogram more nearly resemble a normal curve but the extent to which the histogram spreads out about its center decreases markedly. Increasing n both changes the shape of the distribution and squeezes it in toward the center, so that $\overline{x}$ based on a large n is a less variable quantity than is $\overline{x}$ based on a small value of n. ■

EXERCISES

6.9 An experiment consists of rolling a die three times. Upon completion of the experiment, the mean and range (largest value minus the smallest value) of the three observed values are to be computed. This is equivalent to taking a sample of size 3 from the population whose distribution is as follows:

x	1	2	3	4	5	6
$p(x)$	$\frac{1}{6}$	$\frac{1}{6}$	$\frac{1}{6}$	$\frac{1}{6}$	$\frac{1}{6}$	$\frac{1}{6}$

a. Draw a histogram to represent the population distribution.

b. Compute μ and σ, the mean and standard deviation of the population distribution. (Refer to Section 5.3.)

c. Simulate this experiment 100 times, each time recording the mean and the range. (*Hint:* You can put 6 slips of paper in a box and sample 3 *with* replacement.)

d. Draw a histogram to represent the distribution of $\overline{x}$ values. In what way is the $\overline{x}$ histogram similar to the population histogram? In what ways do they differ?

e. Draw a histogram to represent the sampling distribution of the sample range. Note that the population range is 5. Based on your histogram, do you think that the sample range tends to be smaller or larger than the population range?

6.10 Suppose that a sample of size 3 is to be drawn with replacement from a population whose distribution is as follows:

x	1	2	3	4	5
$p(x)$	.1	.2	.4	.2	.1

This experiment could be simulated by using 10 slips of paper marked as follows: 1, 2, 2, 3, 3, 3, 3, 4, 4, 5. The slips are mixed, one is selected, and its number is noted. The chosen slip is then replaced and the process is repeated two more times, resulting in a sample of size 3. Refer to Exercise 6.9 and answer (a)–(e) for the population distribution of this problem. For (c), a simulation of 50 samples is acceptable.

6.11 Describe how you would use simulation to approximate the sampling distribution of the 10% trimmed mean when the sample size is $n = 10$.

6.3 General Rules Concerning the Sampling Distribution of $\overline{x}$

This section presents some general results regarding the sampling distribution of $\overline{x}$. The first relates the center and spread of the $\overline{x}$ sampling distribution to the population characteristics μ and σ. We then turn to rules that describe the specific shape of the $\overline{x}$ distribution in certain situations.

The Mean Value and Standard Deviation of the $\overline{x}$ Distribution

The $\overline{x}$ sample histograms discussed in Examples 3 and 4 suggest that for any n, the center of the $\overline{x}$ sampling distribution (the mean value of $\overline{x}$) coincides with the mean of the population being sampled but that the spread of the $\overline{x}$ distribution decreases with increasing n. Since the spread of a distribution (amount of variability) is usually described by its standard deviation, the pictures suggest that the standard deviation of $\overline{x}$ is smaller for large n than for small n. Here are precise statements of these results.

> Let $\overline{x}$ denote the sample mean of the observations in a random sample of size n selected from a population of x values having mean μ and standard deviation σ. Denote the mean value of $\overline{x}$ (the center of $\overline{x}$'s sampling distribution) by $\mu_{\overline{x}}$ and the standard deviation of $\overline{x}$ (a measure of the spread of the $\overline{x}$ distribution about its mean value) by $\sigma_{\overline{x}}$. Then
>
> $$\mu_{\overline{x}} = \mu, \qquad \sigma_{\overline{x}} = \frac{\sigma}{\sqrt{n}}$$

The rule $\mu_{\overline{x}} = \mu$ says that the $\overline{x}$ sampling distribution is always centered at the mean of the population being sampled. The second rule, $\sigma_{\overline{x}} = \sigma/\sqrt{n}$, not only says that the spread of the $\overline{x}$ distribution decreases as n increases but gives a precise relationship between the standard deviation of $\overline{x}$ and the population standard deviation. When $n = 4$, for example (so that $\sqrt{n} = 2$), $\sigma_{\overline{x}} = \sigma/2$, so the $\overline{x}$ distribution spreads out only half as much about μ as does the population distribution. Figure 5 illustrates these rules by showing several $\overline{x}$ sampling distributions superimposed on a graph of the x population distribution.

EXAMPLE 5

Let x denote the duration of a randomly selected song for a certain type of songbird. Suppose that the mean value of song duration is $\mu = 1.5$ and that

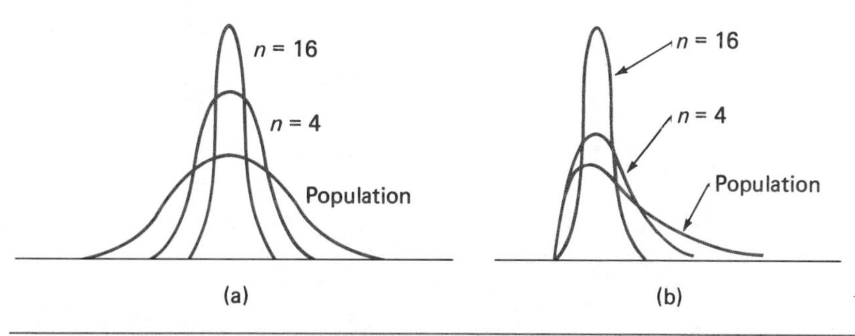

FIGURE 5 POPULATION DISTRIBUTION AND SAMPLING DISTRIBUTIONS OF $\overline{x}$
(a) A Symmetric Population Distribution
(b) A Skewed Population Distribution

the standard deviation of song duration is $\sigma = .9$ (these values are suggested by the results of a large random sample of house finch songs as reported in "Song Dialects and Colonization in the House Finch" (*Condor* (1975): 407–22)). The sampling distribution of $\overline{x}$ based on a random sample of $n = 25$ song durations then also has mean value $\mu_{\overline{x}} = 1.5$. That is, the sampling distribution of $\overline{x}$ is centered at 1.5. The standard deviation of $\overline{x}$ is $\sigma_{\overline{x}} = \sigma/\sqrt{n} = .9/\sqrt{25} = .18$, only one-fifth as large as the population standard deviation σ.

If an ornithologist reported a sample average duration of $\overline{x} = 2.65$ for a random sample of size $n = 25$, is it plausible that the songs were sung by the type of bird described above? For that type of bird, $\mu_{\overline{x}} = 1.5$ and $\sigma_{\overline{x}} = .18$. Thus $\overline{x} = 2.65$ is roughly 6 standard deviations above the mean value ($6\sigma_{\overline{x}} \approx 1.1$). Chebyshev's rule says that at least $100\,[1 - (1/6)^2] = 97.2\%$ of all values in a set or distribution are within 6 standard deviations of the mean. So either the songs came from the type of bird described above and the $\overline{x}$ value is quite unusual or, perhaps more plausibly, this sample is for a different type of bird. If the $\overline{x}$ distribution were approximately normal, there would be little doubt—we virtually never see a value 6 standard deviations above the mean of a normal distribution. ■

The Case of a Normal Population Distribution

The rules $\mu_{\overline{x}} = \mu$ and $\sigma_{\overline{x}} = \sigma/\sqrt{n}$ give summary information about the two most important characteristics of $\overline{x}$'s sampling distribution, the mean value and standard deviation. But these quantities by themselves do not allow us to calculate probabilities such as $P(27.5 \le \overline{x} \le 30.0)$ or $P(\overline{x} \le 27.0)$. For this purpose, more specific information concerning the sampling distribution of $\overline{x}$ is needed. The following result gives the exact form of the sampling distribution when the x distribution is normal.

> When the x population distribution is normal and $\overline{x}$ is based on a random sample consisting of n observations, the sampling distribution of $\overline{x}$ is also a normal distribution (with mean value $\mu_{\overline{x}} = \mu$ and standard deviation $\sigma_{\overline{x}} = \sigma/\sqrt{n}$) for any value of n. Thus probabilities involving $\overline{x}$ can be computed by standardizing (subtracting μ and dividing by $\sigma/\sqrt{n}$) and referring to the z table. That is, the standardized variable

$$z = \frac{\bar{x} - \mu_{\bar{x}}}{\sigma_{\bar{x}}} = \frac{\bar{x} - \mu}{\sigma/\sqrt{n}}$$

has a probability distribution described by the standard normal (z) curve.

EXAMPLE 6

A meat market claims that the average fat content of its ground chuck is 12%. Let x denote the fat content of a randomly selected package of ground chuck. Suppose that x is normally distributed with $\sigma = 1.6\%$. Consider selecting $n = 16$ packages and determining the fat content of each one (observations $x_1, x_2, \ldots, x_{16}$), and let $\bar{x}$ denote the resulting sample average fat content. Because the x distribution is normal, $\bar{x}$ is also normally distributed. If the market's claim is correct, the $\bar{x}$ sampling distribution has mean value $\mu_{\bar{x}} = \mu = 12$, and the standard deviation of $\bar{x}$ is $\sigma_{\bar{x}} = \sigma/\sqrt{n} = 1.6/\sqrt{16} = .40$. To calculate a probability involving $\bar{x}$, we simply standardize by subtracting the mean value, 12, and dividing by the standard deviation (of $\bar{x}$), .40. For example, the probability that the sample average fat content is between 11.6 and 12.8 is calculated by first standardizing the interval limits: $(11.6 - 12)/.40 = -1.0$ and $(12.8 - 12)/.40 = 2.0$. Then

$$P(11.6 \le \bar{x} \le 12.8) = \begin{pmatrix}\text{area under the } z \\ \text{curve between} \\ -1.0 \text{ and } 2.0\end{pmatrix}$$

$$= \begin{pmatrix}\text{area to the} \\ \text{left of } 2.0\end{pmatrix} - \begin{pmatrix}\text{area to the} \\ \text{left of } -1.0\end{pmatrix}$$

$$= .9772 - .1587 = .8185$$

The probability that the sample average fat content is at least 13% is, since $(13 - 12)/.40 = 2.5$,

$$P(13 \le \bar{x}) = \begin{pmatrix}\text{area under the } z \text{ curve} \\ \text{to the right of } 2.5\end{pmatrix} = .0062$$

Thus if the x distribution is as described and the claim is correct, a sample average fat content based on 16 observations exceeds 13 for less than 1% of all such samples. If a value of $\bar{x}$ that exceeds 13 is observed, it is reasonable to reject the market's claim and conclude that $\mu > 12$.

Suppose that we decide to test the market's claim as described above: Obtain a sample of $n = 16$ fat-content observations and reject the claim in favor of the conclusion that $\mu > 12$ if $\bar{x} > 13$. If the market's claim is incorrect and the true average fat content μ is actually 13.5, what is the chance that the claim will not be rejected (an erroneous conclusion)? We still have $\sigma_{\bar{x}} = .40$, but now $\bar{x}$ must be standardized using $\mu_{\bar{x}} = \mu = 13.5$. Since $(13 - 13.5)/.4 = -1.3$,

$$P(\text{the claim is not rejected}) = P(\bar{x} \le 13 \text{ when } \mu = 13.5)$$

$$= \begin{pmatrix}\text{area under the } z \text{ curve} \\ \text{to the left of } -1.3\end{pmatrix} = .0968$$

The chance of making an incorrect decision by not rejecting the claim when $\mu = 13.5$ and this decision rule is used is rather small. ■

Example 6 emphasizes that when standardizing $\overline{x}$, the appropriate divisor is not σ, the population standard deviation, but instead $\sigma/\sqrt{n}$, the standard deviation of the statistic being standardized. In general, standardization requires subtracting the mean value of the statistic and then dividing by its standard deviation. Such standardization appears frequently in later chapters.

The Large-Sample Case

When the x distribution is normal, the sampling distribution of $\overline{x}$ is normal for any sample size n, and the standardized variable $z = (\overline{x} - \mu) / (\sigma/\sqrt{n})$ has a standard normal distribution for any n. The $\overline{x}$ sample histograms of Example 4 suggest that these results will continue to hold when n is large even if the population distribution is very nonnormal. This insight is confirmed by the following general result.

Central Limit Theorem

Let $\overline{x}$ be the sample mean of a random sample of n observations drawn from a population distribution having mean value μ and standard deviation σ. If n is sufficiently large, then the sampling distribution of $\overline{x}$ is approximately normal (again with $\mu_{\overline{x}} = \mu$ and $\sigma_{\overline{x}} = \sigma/\sqrt{n}$). In this case, a good approximation to a probability involving $\overline{x}$ can be computed by standardizing as before and using the z table. In particular, the distribution of the standardized variable

$$z = \frac{\overline{x} - \mu}{\sigma/\sqrt{n}}$$

is described approximately by the standard normal (z) curve.

In essence, the central limit theorem says that when n is large, probabilities involving $\overline{x}$ can be computed exactly as we did in the normal case just considered—provided the qualifier *approximate* is appended to each probability so calculated.

Application of the central limit theorem in specific problem situations requires a rule of thumb for deciding whether n is indeed sufficiently large. Such a rule is not as easy to come by as you might think. Look at Figure 4, which shows the approximate sampling distribution of $\overline{x}$ for $n = 5$, 10, 20, 30, and 50 when the population distribution is quite skewed. Certainly the histogram for $n = 10$ is not well described by a normal curve, and this is still true for the histogram for $n = 20$, particularly in the tails of the histogram (far away from the mean value). Among the five, only the histograms for $n = 30$ and $n = 50$ have a reasonably normal shape.

On the other hand, when the population distribution is normal, the sampling distribution of $\overline{x}$ is normal for *any n*. If the population distribution is somewhat skewed but not to the extent of Figure 3, we might expect the $\overline{x}$ sampling distribution to be a bit skewed for $n = 5$ but quite well fit by a normal curve for n as small as 10 or 15. The value of n necessary for a normal curve to give a good approximation to $\overline{x}$'s sampling distribution depends on how much the population distribution differs from a normal distribution. The closer the population distribution is to being normal, the smaller the value of n for which the central limit theorem approximation is accurate.

The rule that many statisticians recommend is conservative.

> The central limit theorem can safely be applied if n exceeds 30.

If the population distribution is believed to be reasonably close to a normal distribution, an n of 15 or 20 is often large enough for $\bar{x}$ to have approximately a normal distribution. At the other extreme, we can imagine a distribution with a much longer tail than that of Figure 3, in which case even $n = 40$ or 50 would not suffice for approximate normality of $\bar{x}$. In practice, however, very few population distributions are likely to be this badly behaved.

EXAMPLE 7

When someone files for bankruptcy, he or she must list the debts and the time taken to acquire them on the petition. There is quite a bit of variability in the time different individuals take to acquire their debts before filing. The paper "Petitioners Under Chapter XIII of the Bankruptcy Act" (*J. Consumer Affairs* (1969):26–40) reported on a sample of 250 petitioners who had a sample average time of 35.4 months, a sample median time of 34, and a sample standard deviation of 21.3 months. The fact that the sample median is smaller than the sample mean suggests that a histogram of values would be positively skewed. Thus the distribution of x, the time for a petitioner to acquire debts, is probably somewhat nonnormal.

Suppose that the mean value of x is actually $\mu = 30$ and that the standard deviation of x is $\sigma = 20$. Let $\bar{x}$ represent the sample average time to acquire debts for a random sample of $n = 100$ petitioners. The sampling distribution of $\bar{x}$ is centered at $\mu_{\bar{x}} = \mu = 30$, and the standard deviation of the sampling distribution is $\sigma_{\bar{x}} = \sigma/\sqrt{n} = 20/\sqrt{100} = 2$. The sample size 100 is large enough so that the central limit theorem can be applied. The probability that $\bar{x}$ is between 25 and 35, $P(25 \le \bar{x} \le 35)$, is approximately the area under the z curve between $(25 - 30)/2 = -2.5$ and $(35 - 30)/2 = 2.5$:

$$P(25 \le \bar{x} \le 35) \approx \begin{pmatrix}\text{area under the } z \text{ curve}\\ \text{to the left of } 2.5\end{pmatrix} - \begin{pmatrix}\text{area under the } z \text{ curve}\\ \text{to the left of } -2.5\end{pmatrix}$$
$$= .9938 - .0062 = .9876$$

To a reasonable approximation, between 98% and 99% of all samples based on 100 observations would have $\bar{x}$ values between 25 and 35 months. Because $(35.4 - 30)/2 = 2.7$,

$$P(35.4 \le \bar{x}) \approx \begin{pmatrix}\text{area under the } z \text{ curve}\\ \text{the right of } 2.7\end{pmatrix} = .0035$$

The smallness of this probability implies that a sample average (based on $n = 100$) of at least 35.4 is very unlikely if $\mu = 30$ and $\sigma = 20$. You can verify that for $n = 250$, an $\bar{x}$ at least as large as 35.4 would be even less likely (use $\sigma_{\bar{x}} = 20/\sqrt{250} = 1.26$). A plausible conclusion based on the sample data described above is that μ exceeds 30 months. ■

EXAMPLE 8

A cigarette manufacturer claims that one of its brands of cigarettes has an average nicotine content of $\mu = 1.8$ mg per cigarette. Presumably smokers of this brand would not be disturbed to find that μ is less than 1.8 but would be unhappy if μ were actually greater than the specified value. Let x denote the nicotine content of a randomly selected cigarette, and suppose that the standard deviation of the x distribution is .4.

To test the company's claim that $\mu = 1.8$ against the alternative claim that $\mu > 1.8$, an independent testing organization is asked to analyze a random sample of $n = 36$ cigarettes. Let $\bar{x} =$ the sample average nicotine content for this sample. The sample size $n = 36$ is large enough to invoke the central limit theorem and regard the $\bar{x}$ distribution as being approximately normal. The standard deviation of $\bar{x}$ is $\sigma_{\bar{x}} = .4/\sqrt{36} = .0667$, and when the company's claim is correct, $\mu_{\bar{x}} = \mu = 1.8$.

Due to sampling variation, even when $\mu = 1.8$ it is quite likely that $\bar{x}$ will deviate from this value. A reasonable decision rule would then consist of rejecting the company's claim only if $\bar{x}$ is observed to be much larger than 1.8. This ensures that there is a small chance of (erroneously) rejecting the claim when it is correct. Rather than use $\bar{x}$ itself, consider the following decision rule based on the standardized version of $\bar{x}$:

$$\text{reject the company's claim if } \frac{\bar{x} - 1.8}{.0667} \text{ exceeds 2.33}$$

Here is the rationale for this rule: When $\mu_{\bar{x}} = \mu = 1.8$ and $\sigma_{\bar{x}} = \sigma/\sqrt{n} = .0667$, the central limit theorem states that the probability distribution of $(\bar{x} - 1.8)/.0667$ is given approximately by the standard normal (z) curve. From Table I in the appendices, the area in the upper tail of the z curve to the right of 2.33 is .01. Thus the probability of rejecting the claim when $\mu = 1.8$—i.e., of $(\bar{x} - 1.8)/.0667$ exceeding 2.33 when $\mu = 1.8$—is only .01. The choice of 2.33 in the rule fixes the probability of making this type of error at approximately .01.

Using this decision rule, if $\bar{x} = 1.92$ is observed, then the value of $(\bar{x} - 1.8)/.0667$ is $(1.92 - 1.8)/.0667 = 1.80$. Because 1.80 does not exceed 2.33, the company's claim should not be rejected. Although $\bar{x} = 1.92$ does exceed 1.8, it does not do so by enough to cast substantial doubt on the company's claim. Our conclusion is that the difference between the observed $\bar{x}$ (1.92) and the claimed value of μ (1.8) can be attributed to sampling variation. You might wonder why we didn't use a value even larger than 2.33 to obtain an error probability even less than .01. (Why tolerate a 1% chance of an erroneous conclusion?) The answer is that there is another type of error, not rejecting the company's claim when μ actually exceeds 1.8. Using a value larger than 2.33 makes this second type of error more likely. We say more about this in the first chapter on hypothesis testing. ■

Other Cases

We now know a great deal about the sampling distribution of $\bar{x}$ in two cases, that of a normal population distribution and the case of a large sample size. What about the case in which the population distribution is not normal and n is small? Unfortunately, while it is still true that $\mu_{\bar{x}} = \mu$ and $\sigma_{\bar{x}} = \sigma/\sqrt{n}$, there is no general result about the shape of the $\bar{x}$ distribution. When the objective is to make an inference about the center of such a population, one way to proceed is to replace the normality assumption with some other distributional model for the population. Statisticians have proposed and studied a number of such models. Then theoretical methods or simulation can be used to describe the $\bar{x}$ distribution corresponding to the assumed model. An alternative path is to use an inferential procedure based on statistics other than $\bar{x}$. Several procedures of this sort appear in later chapters.

EXERCISES

6.12 Let x denote the time (min) that it takes a fifth grade student to read a certain passage. Suppose that the mean value and standard deviation of x are $\mu = 2$ and $\sigma = .8$, respectively.

a. If $\overline{x}$ is the sample average time for random sample of $n = 9$ students, where is the $\overline{x}$ distribution centered and how much does it spread out about the center (as described by its standard deviation)?

b. Repeat (a) for a sample of size $n = 20$ and again for a sample of size $n = 100$. How do the centers and spreads of the three $\overline{x}$ distributions compare to one another? Which sample size would be most likely to result in an $\overline{x}$ value close to μ, and why?

6.13 In a learning experiment, untrained mice are placed in a maze and the time required for each mouse to traverse the maze and find the exit is recorded. Let the variable x represent this time. Suppose that the time required for an untrained mouse has a mean value of 50 sec and a standard deviation of 15 sec.

a. If 64 randomly selected untrained mice are placed in the maze and the time is recorded for each one, describe the sampling distribution of $\overline{x}$, the sample mean of the 64 times.

b. What is the approximate probability that the sample mean is between 50 and 53?

c. What is the approximate probability that $\overline{x}$ differs from 50 by more than 5?

6.14 Referring to Exercise 6.13, suppose 64 mice that have previously run the same maze several times are again placed in the maze. The times are recorded and the sample mean is 42 sec. Do you think that these mice have benefited from training? (*Hint:* If these mice are behaving like untrained mice, what is the probability that the sample mean would be 42 or less?)

6.15 Suppose that the mean value of interpupillary distance for all adult males is 65 mm and the population standard deviation is 5 mm.

a. If the distribution of interpupillary distance is normal and a sample of $n = 25$ adult males is selected, what is the probability that the sample average distance $\overline{x}$ for these 25 will be between 64 and 67 mm? At least 68 mm?

b. Suppose that a sample of 100 adult males is obtained. Without assuming that interpupillary distance is normally distributed, what is the approximate probability that the sample average distance is between 64 and 67 mm? At least 68 mm?

6.16 College students with a checking account typically write relatively few checks in any given month, while full-time residents typically write many more checks during a month. Suppose that 50% of a bank's accounts are held by students and 50% are held by full-time residents. Let x denote the number of checks written in a given month by a randomly selected bank customer.

a. Give a sketch of what the probability distribution of x might look like.

b. Let the mean value of x be 22.0 and the standard deviation of x be 16.5. If a random sample of $n = 100$ customers is selected and $\overline{x}$ denotes the sample average number of checks written during a particular month, where is the sampling distribution of $\overline{x}$ centered, and what is the standard deviation of the $\overline{x}$ distribution? Sketch a rough picture of the sampling distribution.

c. Refering to (b), what is the approximate probability that $\overline{x}$ is at most 20? At least 25? What result are you using to justify your computations?

6.17 Suppose that a sample of size 100 is to be drawn from a population with standard deviation 10.

a. What is the probability that the sample mean will be within 2 of the value of μ?

b. For this example ($n = 100$, $\sigma = 10$), complete each statement by computing the appropriate value.
i. Approximately 95% of the time, $\bar{x}$ is within ? of μ.
ii. Approximately .3% of the time, $\bar{x}$ is further than ? from μ.

6.18 An airplane with room for 100 passengers has a total baggage limit of 6000 lb. Suppose that the total weight of baggage checked by a passenger is a random variable with mean value 50 lb and standard deviation 20 lb. If 100 passengers board a flight, what is the approximate probability that the total weight of their baggage will exceed the limit? (*Hint:* With $n = 100$, the total weight exceeds the limit precisely when the average weight $\bar{x}$ exceeds 6000/100.)

6.19 The time that a randomly selected individual waits for an elevator in an office building has a uniform distribution over the interval from 0 to 1 min (the uniform distribution is discussed briefly in Section 5.2). It can be shown that, for this distribution, $\mu = .5$ and $\sigma = .289$.

a. Let $\bar{x}$ be the sample average waiting time for a random sample of 16 individuals. What are the mean value and standard deviation of $\bar{x}$'s sampling distribution?

b. Answer (a) for a random sample of 50 individuals. In this case, sketch a picture of a good approximation to the actual $\bar{x}$ distribution.

6.4 The Sampling Distribution of a Sample Proportion

The objective of many statistical investigations is to draw a conclusion about the proportion of individuals or objects in a population that possess a specified property—Maytag washers that don't require service during the warranty period, Europeans who favor the deployment of a certain type of missile, smokers who regularly smoke nonfilter cigarettes, etc. In such situations, any individual or object that possesses the property of interest is labeled a success (S), and one that does not possess the property is termed a failure (F). The letter π denotes the proportion of S's in the population. The value of π is a number between 0 and 1, and 100π is the percentage of S's in the population. Thus $\pi = .75$ means that 75% of the population members are S's, while $\pi = .01$ identifies a population containing only 1% S's and 99% F's.

The value of π is usually unknown to an investigator. When a random sample of size n is selected from this type of population, some of the individuals in the sample are S's and the remaining individuals in the sample are F's. The statistic that will provide a basis for making inferences about π is p, the **sample proportion of S's:**

$$p = \frac{\text{the number of } S\text{'s in the sample}}{n}$$

For example, if $n = 5$ and three S's result, then $p = \frac{3}{5} = .6$.

Just as making inferences about μ requires knowing something about the sampling distribution of the statistic $\bar{x}$, to make inferences about π we must first learn about properties of the sampling distribution of the statistic p. For example, when $n = 5$, possible values of p are 0, .2 (from $\frac{1}{5}$), .4, .6, .8, and 1.

The sampling distribution of p gives the probability of each of these six possible values (the long-run proportion of the time each value would occur if samples with $n = 5$ were selected over and over again). Similarly, when $n = 100$ the 101 possible values of p are 0, .01, .02, . . . , .98, .99, and 1. The sampling distribution of p can then be used to calculate probabilities such as $P(.3 \leq p \leq .7)$ (the long-run proportion of samples of size 100 for which the sample proportion of S's is between .3 and .7), $P(p \leq .1)$, and $P(.75 \leq p)$.

Before stating some general rules concerning p's sampling distribution, we present the results of two sampling experiments as aids to developing an intuitive understanding of the rules. In each example, we selected a population having a specified value of π and obtained 500 random samples, each of size n, from the population. We then computed the sample proportion of S's (p) for each sample and constructed a sample histogram of these 500 values of p. As with $\bar{x}$, this was repeated for several different values of n to show how the sampling distribution changes with increasing sample size. One way to carry out such an experiment is to sample slips of paper repeatedly from a box in which a proportion π of the slips are marked S (e.g., for $\pi = .75$, we could sample with replacement from a box containing 15 S and 5 F slips). However, this would obviously be time-consuming and tedious, so we instead used a computer to perform these simulation experiments.

EXAMPLE 9

The percentage of females in the labor force varies widely from country to country in Europe. The publication *European Marketing Data and Statistics* (17th edition, 1981) reports that Ireland's work force had the lowest percentage, 26.5%, while the U.S.S.R., with 50.4%, was the highest. We decided to simulate sampling from Ireland's labor force with S denoting a female worker and F a male worker, so $\pi = .265$. The computer was used to select 500 samples of size $n = 5$, then 500 samples of size $n = 10$, then 500 samples with $n = 25$, and finally 500 samples with $n = 50$. The sample histograms of the 500 values of p for the four sample sizes are displayed in Figure 6.

The most noticeable feature of the histogram shapes is the progression toward the shape of a normal curve as n increases. The histogram for $n = 5$ is definitely skewed. The two smallest possible values of p, 0 and .2, occurred 122 and 174 times, respectively, in the 500 samples, while the two largest values, .8 and 1, occurred only 3 and 1 times, respectively. The histogram for $n = 10$ is considerably more bell-shaped, although it still has a slight positive skew. The histogram for $n = 25$ exhibits very little skewness, and the histogram for $n = 50$ looks like a normal curve.

Although the skewness of the first two histograms makes the location of their centers (balance points) a bit difficult, all four histograms appear to be centered at the same place. The average of the 500 values of p from the experiment where $n = 5$ is .2596, and the averages of p for the other three experiments are .2656, .2631, and .2647. All these are very close to .265, the value of π for the population being sampled. Had the histograms been based on an unending sequence of samples (or equivalently, on theoretical calculations using the binomial probability distribution) instead of just 500 samples, each histogram would have been centered at exactly .265. Finally, as was the case with the sampling distribution of $\bar{x}$, the histograms spread out more for small n than for large n. As your intuition should suggest, the value of p

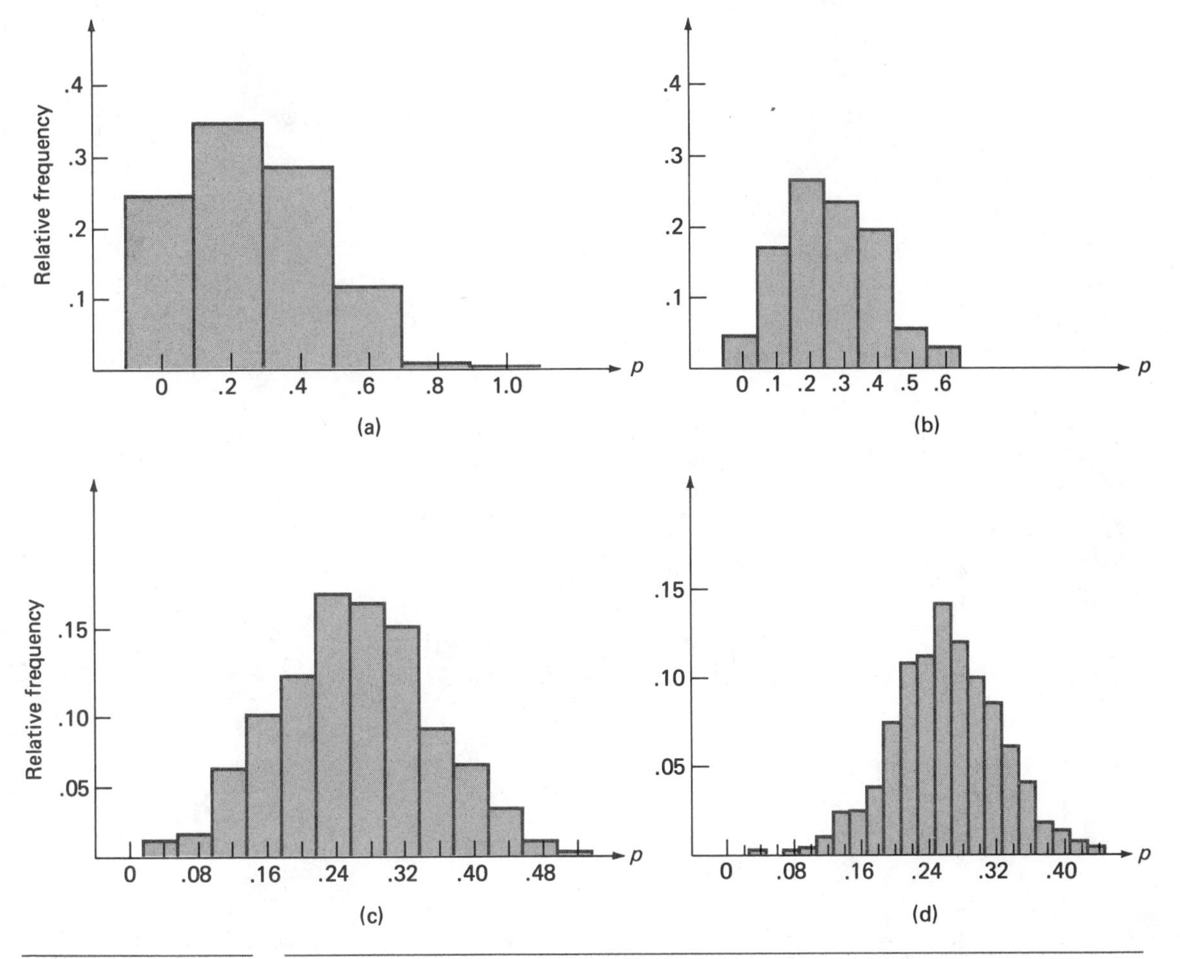

FIGURE 6

HISTOGRAMS OF 500 VALUES OF p, EACH BASED ON A RANDOM SAMPLE OF SIZE n ($\pi = .265$)
(a) $n = 5$
(b) $n = 10$
(c) $n = 25$
(d) $n = 50$

based on a large sample size tends to be closer to the population proportion of S's, π, than does p from a small sample. ■

Our next example shows what happens to the sampling distribution of p when π is either quite close to 0 or quite close to 1.

EXAMPLE 10

The development of viral hepatitis subsequent to a blood transfusion can cause serious complications for a patient. The paper "Hepatitis in Patients with Acute Nonlymphatic Leukemia" (*Amer. J. of Med.* (1983):413–21) reported that in spite of careful screening for those having a hepatitis antigen, viral hepatitis occurs in 7% of blood recipients. Here we simulate sampling from the population of blood recipients, with S denoting a recipient who contracts hepatitis (not the sort of characteristic one thinks of as identifying a success, but the S-F labeling is arbitrary), and $\pi = .07$. Figure 7 displays histograms of 500 values of p for the four sample sizes $n = 10$, 25, 50, and 100.

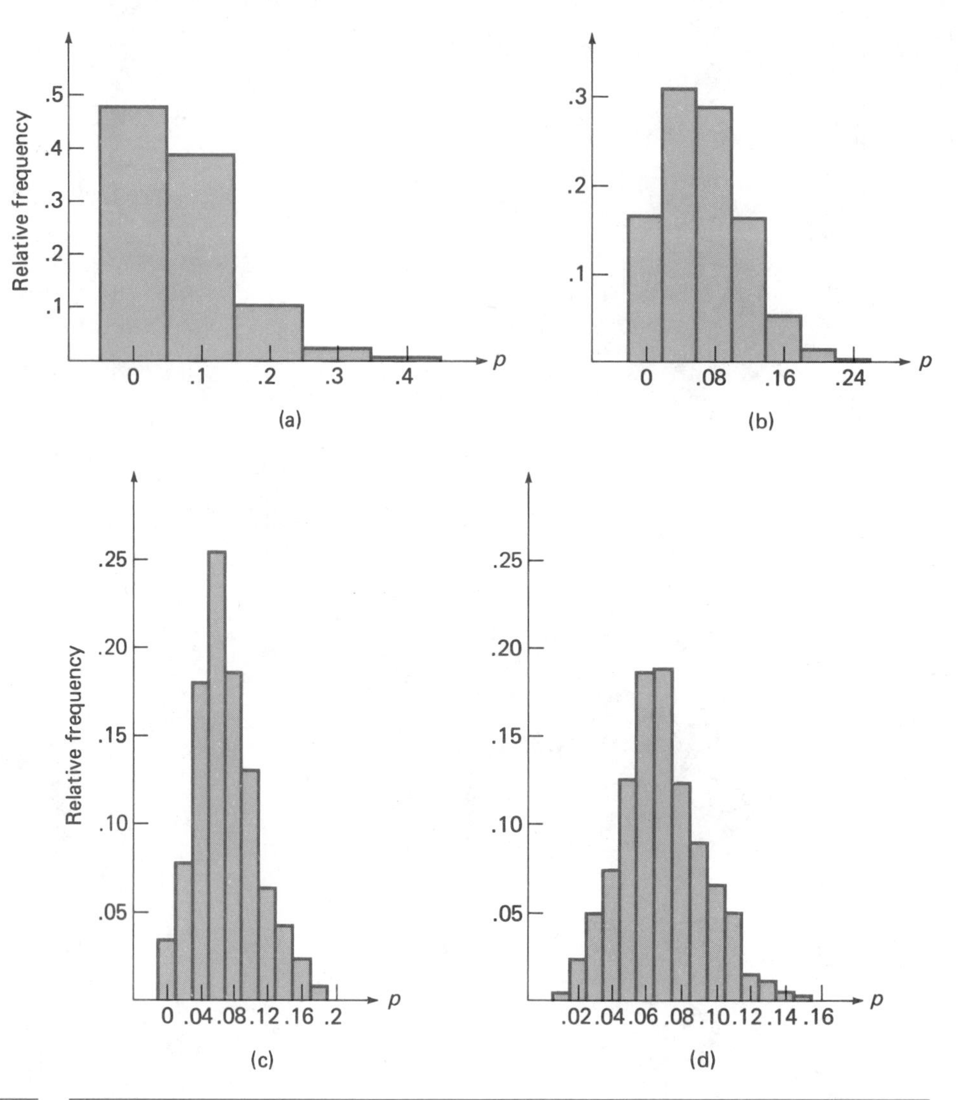

FIGURE 7 HISTOGRAMS OF 500 VALUES OF p, EACH BASED ON A RANDOM SAMPLE OF SIZE n ($\pi = .07$)
(a) $n = 10$
(b) $n = 25$
(c) $n = 50$
(d) $n = 100$

As was the case in the previous example, all four histograms are centered at approximately the value of π for the population being sampled (the average values of p are .0690, .0677, .0707, and .0694). If the histograms had been based on an unending sequence of samples, they would all have been centered at exactly $\pi = .07$. Again the spread of a histogram based on a large n is smaller than the spread of a histogram resulting from a small sample size. The larger the value of n, the closer the sample proportion, p, tends to be to the value of the population proportion, π. Furthermore, there is a progression toward the shape of a normal curve as n increases. However, the progression is much slower here than in the previous example because the value of π is so extreme (the same thing would happen for $\pi = .93$, except that the histograms would be negatively rather than positively skewed). The histograms for $n = 10$ and $n = 25$ exhibit substantial skew, the skew of the histogram for $n = 50$ is still moderate (compare Figure 7(c) to Figure 6(d)), and only the histogram for $n = 100$ is reasonably well fit by a normal curve. You can see that whether a normal curve provides a good approximation to the sampling distribution of p depends on the values of both n and π. Knowing just that $n = 50$ is not enough to guarantee that the shape of the histogram is approximately normal. ■

General Rules for the Sampling Distribution of p

At the beginning of Section 6.1 we discussed how a population of *S*'s and *F*'s can be viewed as a numerical population by introducing a variable x that has value 1 for a success and value 0 for a failure. Then the population consists of 1's and 0's rather than *S*'s and *F*'s, and the population proportion of *S*'s, π, is now the proportion of 1's in the population. Similarly, p is now the proportion of 1's in the sample. For example, the sample *SFSSS* corresponds to the sample of x values $x_1 = 1, x_2 = 0, x_3 = 1, x_4 = 1, x_5 = 1$. The resulting value of p is $\frac{4}{5} = .8$, which can be written as the sample mean of the observed x values: $p = (1 + 0 + 1 + 1 + 1)/5 = \frac{4}{5} = .8$.

That is, with this 0–1 coding, p is the mean of the x values in the sample and π is the population mean value of x. We used this coding at the end of Section 5.3 to show not only that $\mu_x = \pi$ but also that $\sigma_x = \sqrt{\pi(1 - \pi)}$. The point of this numerical coding is that having recognized that p is just a special case of a sample mean, the results concerning the sampling distribution of $\overline{x}$ carry over to results concerning the sampling distribution of p.

> Let p denote the sample proportion of successes based on a random sample of size n from a population having proportion of successes π. Then
>
> $$\mu_p = \pi \quad \text{and} \quad \sigma_p = \sqrt{\frac{\pi(1 - \pi)}{n}}$$
>
> Furthermore, if n is sufficiently large, the sampling distribution of p is approximately normal, and the distribution of the standardized variable
>
> $$z = \frac{p - \pi}{\sqrt{\dfrac{\pi(1 - \pi)}{n}}}$$
>
> is described approximately by the standard normal (z) curve.

Thus the sampling distribution of p is always centered at the value of the population success proportion π (as the $\overline{x}$ distribution is centered at μ), and the extent to which the distribution spreads out about π decreases as the sample size n increases.

EXAMPLE 11

Figure 6 (see Example 9) suggests that when $\pi = .265$, the sampling distribution of p is approximately normal when $n = 50$, so the normal approximation should be even better when $n = 100$. Suppose we plan to select a random sample of size $n = 100$ from the Irish work force, and let

p = the proportion of females in the sample. Then $\mu_p = .265$ and $\sigma_p = \sqrt{(.265)(.735)/100} = \sqrt{.001948} = .0441$. To compute $P(.20 \le p \le .30)$, the probability that between 20% and 30% of those in the sample are female, the interval limits must first be standardized:

$$\frac{.20 - .265}{.0441} = -1.5, \qquad \frac{.30 - .265}{.0441} = .8$$

Then

$$\begin{aligned} P(.20 \le p \le .30) &\approx \text{area under the } z \text{ curve between } -1.5 \text{ and } .8 \\ &= \begin{pmatrix}\text{area to the left} \\ \text{of } .8\end{pmatrix} - \begin{pmatrix}\text{area to the left} \\ \text{of } -1.5\end{pmatrix} \\ &= .7881 - .0668 = .7213 \end{aligned}$$

Thus in approximately 72% of all samples of 100 workers, the proportion of females is between .2 and .3. Similarly, since $(.35 - .265)/.0441 = 1.9$,

$$P(.35 \le p) \approx \begin{pmatrix}\text{area under the } z \text{ curve} \\ \text{to the right of } 1.9\end{pmatrix} = .0287$$

Less than 3% of all samples have a sample proportion of females that exceeds .35 (35%).

Suppose that the proportion .265 refers to a particular year. Several years later an investigator wants to know if the population proportion has changed, so the investigator decides to select a sample of size 100. Intuitively, since p usually differs somewhat from the population proportion π, a change is suggested only by a value of p that differs substantially from .265. Rather than use p itself, consider the standardized quantity

$$z = \frac{p - \pi}{\sigma_p} = \frac{p - .265}{.0441}$$

Then z gives the distance between p and .265 in standard deviations. For example, $z = 2$ corresponds to a value of p that is 2 standard deviations larger than what we expect when there is no change. An extreme positive or negative value of z suggests that the value of π has changed.

Consider the following decision rule: Conclude that π has changed if either $z < -1.96$ or if $1.96 < z$. If $p = .32$, then $z = (.32 - .265)/.0441 = 1.2$. Since 1.2 is neither greater than 1.96 nor less than -1.96, we conclude that π has not changed significantly. On the other hand, if $p = .15$, then $z = -2.61$. Because $-2.61 < -1.96$, the appropriate conclusion is that π has changed.

Why is it reasonable to use 1.96 in the decision rule? According to our general rules for the sampling distribution of p, when $\pi = .265$ the standardized variable $z = (p - .265)/.0441$ has a distribution that is well approximated by the z curve. Thus for the proposed decision rule,

$$P\begin{pmatrix}\text{concluding that } p \text{ has changed} \\ \text{when it really hasn't}\end{pmatrix} = P(z < -1.96 \quad \text{or} \quad 1.96 < z \quad \text{when } \pi = .265)$$

$$= \begin{pmatrix}\text{area under the } z \text{ curve} \\ \text{to the left of } -1.96\end{pmatrix} + \begin{pmatrix}\text{area under the } z \text{ curve} \\ \text{to the right of } 1.96\end{pmatrix}$$

$$= .025 + .025 = .05$$

The value 1.96 in the decision rule ensures that there is only a 5% chance of concluding that a change has occurred when it really hasn't. By replacing 1.96 with a larger number, this probability could be made smaller, but that results in an increased chance of concluding that π has not changed when it really has. ■

Examples 9 and 10 indicate that both π and n must be considered in judging whether p's sampling distribution is approximately normal.

> The further the value of π is from .5, the larger must be the value of n before the normal approximation to the sampling distribution of p is accurate. A conservative rule of thumb is that if both $n\pi \geq 5$ and $n(1 - \pi) \geq 5$, then it is safe to use the normal approximation.

A sample size of $n = 100$ is not by itself sufficient to justify the use of the normal approximation. If $\pi = .01$, the distribution of p is very positively skewed, so a bell-shaped curve does not give a good approximation. Similarly, if $n = 100$ and $\pi = .99$ (so $n(1 - \pi) = 1 < 5$), the distribution of p has a substantial negative skew. The conditions $n\pi \geq 5$ and $n(1 - \pi) \geq 5$ are designed to ensure that the sampling distribution of p is not too skewed. If $\pi = .5$, the normal approximation can be used for n as small as 10, while for $\pi = .05$ or .95, n should be at least 100.

EXAMPLE 12

The proportion of all blood recipients stricken with viral hepatitis was assumed in Example 10 to be $\pi = .07$. For a sample of size $n = 225$, $n\pi = 15.75$ and $n(1 - \pi) = 209.25$, both of which are at least 5. Determination of $P(p \leq .05)$ requires $\sigma_p = \sqrt{(.07)(.93)/225} = .017$ and the standardized value $(.05 - .07)/.017 = -1.2$. Then

$$P(p \leq .05) \approx \begin{pmatrix}\text{area under the } z \text{ curve} \\ \text{to the left of } -1.2\end{pmatrix} = .1151$$

This says that about 12% of all samples with $n = 225$ will result in a value of p that is at most .05.

Suppose a new treatment has been developed that is believed will reduce the incidence rate of viral hepatitis. This treatment is given to 225 blood recipients. Let $z = (p - .07)/.017$ (the standardized value of p assuming no reduction) and consider the following decision rule: Conclude that the new treatment reduces the incidence rate if $z < -2.33$. When there is actually no reduction in incidence, $\pi = .07$ and $\sigma_p = .017$, so z has approximately a standard normal distribution. Since the area under the standard normal curve to the left of -2.33 is .01, using -2.33 in the decision rule ensures that the chance of falsely concluding that the new treatment reduces viral hepatitis is only .01. If only 8 of the 225 recipients subsequently contract viral hepatitis,

$p = .036$ and $z = (.036 - .07)/.017 = -2.00$. Since -2.00 is not less than -2.33, we conclude that the new treatment does not reduce the incidence rate. ■

EXERCISES

6.20 Explain how the quantities p and π are different.

6.21 Describe how you might use simulation to approximate the sampling distribution of a sample proportion p computed from a sample of size 10 drawn from a population where $\pi = .1$. Be specific: Discuss how you would represent the population, how you would select your samples, how many samples you would take, and how you would summarize your results.

6.22 A certain chromosome defect occurs in only 1 out of 200 Caucasian adult males. A random sample of $n = 100$ males is obtained.

a. What is the mean value of the sample proportion p (number of defects divided by 100) and what is the standard deviation of the sample proportion?

b. Does p have approximately a normal distribution in this case? Explain.

c. What is the smallest value of n for which the sampling distribution of p is approximately normal?

6.23 In Exercise 6.22, the population consists only of Caucasian males. Now let p be the sample proportion of defects for a random sample of 2000 adult males from another racial group.

a. If the defect rate for this group is identical to that for Caucasians, does p have an approximately normal distribution?

b. Suppose that in the sample there are 25 individuals with defects. Does this suggest that the defect rate for this type of individual exceeds 1 in 200? (*Hint:* The computed value of p is how many standard deviations above what you would expect if the 1 in 200 rate were correct?)

6.24 A column in the April 22, 1985, issue of *Newsweek* reported that 55% of all women in the American work force regard themselves as underpaid. Although this conclusion was based on sample data, suppose that in fact $\pi = .55$, where π represents the true proportion of working women who believe that they are underpaid.

a. Would p based on a random sample of only 10 working women have approximately a normal distribution? Explain why or why not.

b. What are the mean value and standard deviation of p based on a random sample of size 400?

c. When $n = 400$, what is $P(.5 \leq p \leq .6)$?

d. Suppose now that $\pi = .4$. For a random sample of $n = 400$ working women, what is $P(.5 < p)$?

6.25 The article "Thrillers" (*Newsweek,* April 22, 1985) states "Surveys tell us that more than half of America's college graduates are avid readers of mystery novels." Let π denote the actual proportion of college graduates who are avid readers of mystery novels. Consider p based on a random sample of 225 college graduates.

a. If $\pi = .5$, what are the mean value and standard deviation of p? Answer this question when $\pi = .6$. Does p have approximately a normal distribution in both cases? Explain.

b. Calculate $P(p \geq .6)$ both when $\pi = .5$ and when $\pi = .6$.

c. Without doing any calculation, how do you think the probabilities in (b) would change if n is 400 rather than 225?

6.5 Estimating a Population Characteristic

The objective of inferential statistics is to draw some type of conclusion about a population by using sample data. Usually an investigator is interested in some specific characteristic of the population, such as the population mean, μ, population standard deviation, σ, or population proportion of successes, π. The simplest type of conclusion involving such a characteristic is to compute a single number that can be regarded as the most plausible value for that characteristic. Thus sample data might suggest that 1.1 mg is the most plausible value for true average nicotine content of Players 100-mm cigarettes (this is the value stated on the package). As another example, a survey carried out in a university library might result in .03 being declared the most plausible value for $\pi =$ the true proportion of misshelved books.

DEFINITION	A **point estimate** of a population characteristic is a single number that is based on sample data and represents the most plausible value of the characteristic.

In the examples just given, 1.1 is the point estimate of μ and .03 is the point estimate of π. Similarly, based on a random sample of houses for sale in a metropolitan area, we might settle on \$76,500 as our point estimate for the median price in the population of all homes for sale in the area. The adjective *point* reflects the fact that the estimate is a single point on the number line.

A sensible way to obtain a point estimate of a particular population characteristic is first to select an appropriate statistic. The estimate is then the computed value of the statistic for the given sample.

EXAMPLE 13

The sale of human organs for transplantation raises some difficult ethical issues both inside and outside the medical community. Let π denote the proportion of all U.S. doctors who oppose such activity (with success identified as opposition, π is the population proportion of successes). The very large number of doctors makes it impractical to determine the exact value of π by soliciting an opinion from each and every doctor. Suppose that a random sample of n doctors is obtained. Then the statistic $p =$ (number of successes in the sample)/n, the sample proportion of successes, is an obvious candidate for use in obtaining a point estimate of π. An article in the November 24, 1978, *Los Angeles Times* reported that in a sample of $n = 244$ doctors, the number opposed to the sale of organs was 184. Thus the point estimate of π based on this sample is $p = 184/244 = .754$. ■

For purposes of estimating a population proportion, π, you would be hard put to suggest a statistic other than p that could reasonably be used to obtain an estimate. However, there are other situations in which each of several different statistics might be used to compute an estimate. For example, an investigator might wish to obtain a point estimate of μ, the population mean value of some variable x, by using the results of a random sample $x_1, x_2, \ldots, x_n$ of x values. The investigator believes that the distribution of x can be described by a bell-shaped curve. The curve might actually be a normal curve, or it might be one with heavier tails or lighter tails than a normal curve.

An obvious statistic for use in estimating μ is the sample mean $\overline{x}$. However, because a bell-shaped curve is symmetric, the population median is also equal to μ. This suggests that a point estimate for μ can be obtained by using as a statistic the sample median (middle value in the ordered list of sample values). Yet another possible statistic for computing a point estimate of μ is a trimmed mean, introduced in Chapter 3 as a compromise between $\overline{x}$ and the sample median.

EXAMPLE 14

The paper "Transport of Oxygen to the Brain in Patients with Elevated Haematocrit Values Before and After Venesection" (*Brain* (1983):513–23) gave $n = 20$ observations on blood viscosity for a sample of patients, each of whom had an abnormally high red-blood-cell level. These observations are summarized in Figure 8. The largest sample value, 8.82, is clearly an outlier. A normal probability plot is reasonably straight except for the point corresponding to this outlier. This is the sort of sample that might result from a distribution that has heavier tails than the normal distribution. Compared to $\overline{x}$, both the sample median and a trimmed mean reduce the influence of an outlying observation.

Stem	Leaves
4h	.96
5l	.30
5h	.62, .73, .87
6l	.02, .02, .09, .13, .24, .30, .48
6h	.65, .66, .73, .77, .87
7l	.38, .45
7h	
8l	
8h	.82

FIGURE 8 STEM-AND-LEAF DISPLAY OF OBSERVATIONS ON BLOOD VISCOSITY

Three statistics and the corresponding estimates of μ are

$$\text{sample mean } \overline{x} = \frac{\Sigma x}{20} = \frac{128.09}{20} = 6.405$$

$$10\% \text{ trimmed mean} = \text{average of middle 16 values} = \frac{101.56}{16} = 6.348$$

$$\text{sample median} = \text{average of two middle values} = \frac{6.24 + 6.30}{2} = 6.270$$

The estimates differ somewhat from one another. The one that should be reported depends on which statistic tends to produce an estimate closest to the true value. We consider this issue shortly. Here, we think the estimate 6.348 is a good choice. ■

EXAMPLE 15

A statistics professor has an apartment that overlooks part of a marathon race route. Let θ denote the number of runners in the race (a Greek letter for a population characteristic, the population size). Without regard to running ability, the runners have been assigned the numbers $1, 2, \ldots, \theta$. Each runner's number is displayed on his or her back. During the race, the professor glances out the window and records the numbers on a sample of $n = 5$ runners. The sampled values are x_1, x_2, x_3, x_4, x_5. On the basis of this sample, the professor wants to estimate θ.

Because θ is the largest x value in the population, a reasonable approach here is to estimate θ by the largest x value in the sample. That is, the statistic used to compute an estimate is max $\{x_1, x_2, x_3, x_4, x_5\}$. If $x_1 = 217$, $x_2 = 29$, $x_3 = 326$, $x_4 = 475$, and $x_5 = 188$, then the point estimate is max $\{217, 29, 326, 475, 188\} = 475$. The problem with using this statistic is that unless the largest x value in the population appears in the sample, the estimate is smaller than θ. For $\theta = 500$ and $n = 5$, it can be shown that the probability that the runner with number θ appears in the sample is only .01. Thus with probability .01, the estimate equals θ, whereas with probability .99 it is less than θ. The use of the suggested statistic results in a systematic tendency to underestimate the true value.

There are several ways to adjust for this systematic underestimation by the above statistic. The average x value in the population is $(\theta + 1)/2$ (the average of the numbers $1, 2, \ldots, \theta$). Thus the sample mean $\bar{x}$ estimates $(\theta + 1)/2$, from which it follows that $2\bar{x} - 1$ estimates θ. For the above sample, $\Sigma x = 1235$, so $2\bar{x} - 1 = 2(1235/5) - 1 = 493$. Another estimate is based on the idea that the largest sample value is less than θ by roughly the same amount that the smallest sample value exceeds the smallest population number, 1. This suggests using the statistic $\max(x) + [\min(x) - 1]$. For the above sample, $\min(x) = 29$, so the estimate is $475 + [29 - 1] = 503$. Many statisticians adjust $\max(x)$ upward and use $[(n + 1)/n] \cdot \max(x) - 1$. When applied to the given sample, this statistic has value$(6/5)(475) - 1 = 569$. If this statistic were used, the point estimate of θ would be 569.

This estimation problem arose in a more realistic context during World War II. The Allies had captured a sample of German tanks. The tanks had been serially numbered starting with 1, and the sample of captured tanks was used to estimate the number of tanks that had been manufactured. ■

Choosing a Statistic for Computing an Estimate

A main point of the previous two examples is that there may be more than one statistic that can reasonably be used to obtain a point estimate of a specified population characteristic. Loosely speaking, the statistic used should be one that tends to yield an accurate estimate, i.e., an estimate close to the value of the population characteristic. Information on the accuracy of estimation when a particular statistic is used is provided by the statistic's sampling distribution. Figure 9 pictures the sampling distributions of three different statistics. The value of the population characteristic, which we refer to as the true value, is marked on the measurement axis.

The statistic whose distribution is pictured in Figure 9(a) is unlikely to yield an estimate close to the true value. The distribution is centered quite a bit to the right of the true value, making it very likely that an estimate (value of the statistic for a particular sample) will be substantially larger than the true value. If this statistic is used to compute an estimate based on a first sample, then another estimate based on a second sample, then another estimate based on a third sample, and so on, the long-run average value of the estimates will considerably exceed the true value. For example, if the mean value of the sampling distribution is 110, whereas the true value is 100, there will be a long-run tendency to overestimate the true value by an amount

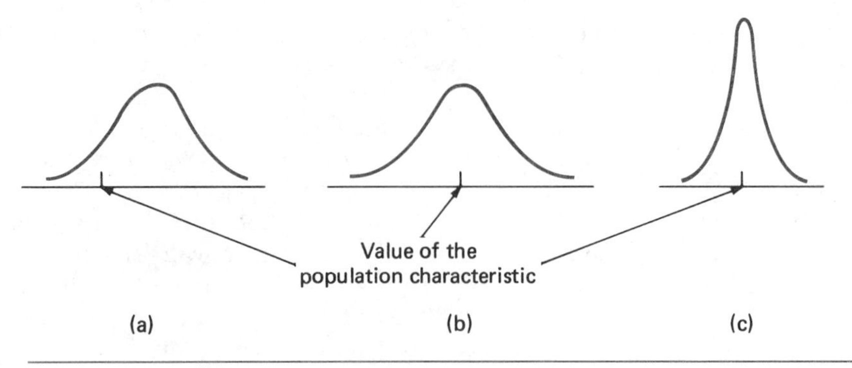

FIGURE 9 SAMPLING DISTRIBUTIONS OF THREE DIFFERENT STATISTICS

110 − 100 = 10. Using such a statistic would be like purchasing a scale from a manufacturer whose average scale overweighs by 10 lb.

The sampling distribution of the statistic of Figure 9(b) is centered at the true value. Thus while one estimate may be smaller than the true value and another may be larger, when this statistic is used many times over with different samples, there will be no long run tendency to overestimate or underestimate the true value. However, notice that while this sampling distribution is correctly centered, it spreads out quite a bit about the true value. That is, the statistic's standard deviation is relatively large. Because of this, some estimates resulting from the use of this statistic will be either a fair distance to the right of or to the left of the true value—even though there is no systematic tendency to underestimate or overestimate the true value. In contrast, the mean value of the statistic whose distribution appears in Figure 9(c) is exactly the true value of the population characteristic (implying no systematic estimation error) and the statistic's standard deviation is relatively small. Estimates resulting from the use of this third statistic will almost always be quite close to the true value, certainly more so than estimates resulting from using the second statistic.

DEFINITION	A statistic whose mean value is always equal to the value of the population characteristic being estimated is said to be an **unbiased statistic.** A statistic that is not unbiased is said to be **biased,** and the **bias** is the difference between the statistic's mean value and the value of the population characteristic.

If a statistic has a positive bias—for example, a mean value 110 when the value of the population characteristic is 100 (bias 110 − 100 = 10)—then using the statistic over and over with different samples results in a long-run tendency to overestimate the true value. Similarly, using a statistic whose bias is negative results in estimates that tend to be smaller than the true value by the amount of the bias. Many statisticians see unbiasedness as an attractive property for a statistic to possess and recommend the use of an unbiased—as opposed to a biased—statistic for the purpose of obtaining a point estimate.

EXAMPLE 16

Consider estimating the proportion of successes π in a population based on a random sample of size n. We previously suggested using the statistic p, the sample proportion of successes, to obtain a point estimate. A rule from Section 6.4 states that $\mu_p = \pi$ whatever the value of π. This says that p is an unbiased statistic for estimating π—the sampling distribution of p is always centered at the value of π. Even though we have not suggested any other statistic as an alternative to p, it is nice to know that the repeated use of p entails no systematic tendency to underestimate or overestimate π. In addition, $\sigma_p = \sqrt{\pi(1 - \pi)/n}$, which will be very small when the sample size n is large. The value of p based on a large sample will almost always be very close to the true value of π. ■

EXAMPLE 17

The fact that we define the sample variance as $s^2 = \Sigma(x - \overline{x})^2/(n - 1)$ suggests that this statistic is a good choice for obtaining a point estimate of the population variance σ^2. When s^2 was introduced in Chapter 3, we noted that dividing by $n - 1$ rather than n seemed unnatural. An alternative statistic for estimating σ^2 is $\Sigma(x - \overline{x})^2/n$, the average squared deviation. We have focused attention in this chapter on the sampling distribution of $\overline{x}$ and p and have said nothing about the sampling distribution of s^2. It can be shown, though, that s^2 is an unbiased statistic for estimating σ^2. That is, whatever the value of σ^2, the sampling distribution of s^2 is centered at that value. It is for precisely this reason—to obtain an unbiased statistic—that the divisor $n - 1$ is used. Using the statistic with divisor n results in estimates that tend to be somewhat smaller than the value of σ^2; the bias of this statistic is $-\sigma^2/n$, which can be large if n is small.

As an example, consider the following sample consisting of $n = 10$ observations on breaking strength of linen thread: $x_1 = 32.5$, $x_2 = 21.2$, $x_3 = 35.4$, $x_4 = 21.3$, $x_5 = 28.4$, $x_6 = 26.9$, $x_7 = 34.6$, $x_8 = 29.3$, $x_9 = 24.5$, $x_{10} = 31.0$. The sample mean for this sample is $\overline{x} = 28.51$, and $\Sigma(x - \overline{x})^2 = (32.5 - 28.51)^2 + \cdots + (31.0 - 28.51)^2 = 231.41$. Using the statistic s^2 to estimate the population variance σ^2 yields the estimate $231.41/9 = 25.71$. Using the average squared deviation (with divisor $n = 10$), the resulting estimate is $231.41/10 = 23.14$. Because s^2 is an unbiased statistic, most statisticians would recommend using the estimate 25.71.

The obvious choice of a statistic for estimating the population standard deviation σ is the sample standard deviation, s. For the above breaking-strength data, $s = 5.07$, so our point estimate for the population standard deviation σ is 5.07. Unfortunately, the fact that s^2 is an unbiased statistic for estimating σ^2 does not imply that s is an unbiased statistic for estimating σ. The sample standard deviation tends to underestimate slightly the true value of σ. However, unbiasedness is not the only criterion by which a statistic can be judged, and there are other good reasons for using s to estimate σ. In what follows, whenever we need to estimate σ based on a single random sample, we use the statistic s to obtain a point estimate. ■

EXAMPLE 18

(*Example 15 continued*) For estimating θ, the number of runners in a marathon race, based on a random sample of size n, we suggested four statistics: $\max(x)$, $2\overline{x} - 1$, $\max(x) + [\min(x) - 1]$, and $[(n + 1)/n]\cdot\max(x) - 1$. Intuitively, the statistic $\max(x)$ is biased, because it never overestimates θ and usually underestimates it. The mean value of this statistic (the place at which its sampling distribution is centered) is smaller than the value of θ. It is for this reason that we introduced the three other statistics. Using some advanced methods, it can be shown that each of these three other statistics is unbiased. Here is a case in which there are several unbiased statistics, so a criterion other than unbiasedness must be used to decide between them. The criterion is suggested by Figure 9(b) and (c); use the unbiased statistic whose standard deviation is smallest. The standard deviation of the statistic $[(n + 1)/n]\max(x) - 1$ is smaller than the standard deviations of the other two unbiased statistics, which is why most statisticians recommend its use. ■

Using an unbiased statistic whose standard deviation is small guarantees not only that there will be no systematic tendency to underestimate or overestimate the value of the population characteristic but also that estimates will almost always be relatively close to the true value.

> Given a choice between several unbiased statistics that could be used for estimating a population characteristic, the best statistic to use is the one with the smallest standard deviation.

Let's now return to the problem of estimating a population mean μ. The obvious choice of statistic for obtaining a point estimate of μ is the sample mean $\bar{x}$. One of the basic rules of Section 6.3 stated that $\mu_{\bar{x}} = \mu$. That is, whatever the value of μ, the sampling distribution of $\bar{x}$ is centered at that value. This is exactly what it means to say that $\bar{x}$ is an unbiased statistic for estimating μ—and this is true whatever the shape of the population distribution.

However, when the population distribution is symmetric, $\bar{x}$ is not the only unbiased statistic. Other unbiased statistics for estimating μ in this case include the sample median and any trimmed mean (with the same number of observations trimmed from each end of the ordered sample). To select a statistic, we should look for the unbiased statistic with the smallest standard deviation. We know from Section 6.3 that the standard deviation of $\bar{x}$ is $\sigma_{\bar{x}} = \sigma/\sqrt{n}$. Unfortunately, the formulas for the standard deviations of the other unbiased statistics are quite complicated. They depend not only on the sample size n and population standard deviation σ but also on other aspects of the population distribution. This makes a choice of statistic difficult (if everything in statistics were easy, there wouldn't be much demand for statisticians). The following proposition partially resolves this dilemma.

> If the population distribution is normal, then $\bar{x}$ has smaller standard deviation than any other unbiased statistic for estimating μ. However, in this case a trimmed mean with a small trimming percentage (such as 10%) performs almost as well as $\bar{x}$. When the population distribution is symmetric with heavy tails compared to the normal curve, a trimmed mean is a better statistic than $\bar{x}$ for estimating μ.

When the population distribution is unquestionably normal, the choice is clear—use $\overline{x}$ to estimate μ. But with a heavy-tailed distribution, a trimmed mean gives protection against one or two outliers in the sample that might otherwise drastically affect the value of the estimate. In Example 14, we would use 6.348, the 10% trimmed mean, as a point estimate of true average blood viscosity, μ.

Even though we have suggested caution in using $\overline{x}$ to estimate μ, many of the standard inferential procedures presented in later chapters are based on $\overline{x}$. This is largely because at the moment statisticians know a great deal about properties of $\overline{x}$ and somewhat less about other statistics like trimmed means. We think this will change as statisticians develop other procedures that are resistant to the effects of outliers.

EXERCISES

6.26 Twelve specimens of a certain brand of white bread were analyzed to determine the carbohydrate content (percentage of nitrogen-free extract) of each one. The resulting observations were as follows.

76.93	76.88	77.07	76.68	76.39	75.09
76.88	77.67	78.15	76.50	77.16	76.42

a. Assuming that the distribution of carbohydrate content is normal, give a point estimate for the average carbohydrate content for all specimens of this brand of bread (i.e., for the population mean value).

b. Fourteen specimens of a second brand of bread were analyzed, yielding a sample average carbohydrate content of 74.28. Let μ_1 denote the population mean value for the first brand and μ_2 denote the population mean value for the second brand. Then $\mu_1 - \mu_2$ is the difference between the two population mean values. Compute a point estimate for $\mu_1 - \mu_2$. If $\overline{x}$ denotes the sample average for the first brand and $\overline{y}$ denotes the sample average for the second brand, what statistic (based on $\overline{x}$ and $\overline{y}$) did you use to obtain your estimate?

6.27 A random sample of $n = 10$ 4-year-old red pine trees was selected, and the diameter (in.) of each tree's main stem was measured. The resulting observations were as follows.

11.3	10.7	12.4	15.2	10.1	12.1	16.2	10.5	11.4	11.0

a. Compute a point estimate of σ, the population standard deviation of main stem diameter. What statistic did you use to obtain your estimate?

b. Making no assumption whatsoever about the shape of the population distribution of diameter, give a point estimate for the population median diameter (that is, for the middle diameter value in the entire population of 4-year-old red pine trees). What statistic did you use to obtain the estimate?

c. Suppose that the population distribution of diameter is symmetric but with heavier tails than the normal distribution. Give a point estimate of the population mean diameter based on a statistic that gives some protection against the presence of outliers in the sample. What statistic did you use?

d. Suppose that the diameter distribution is normal. Then the 90th percentile of the diameter distribution is $\mu + 1.28\sigma$ (90% of all trees have diameters below this value). Compute a point estimate for this percentile. (*Hint:* First compute an estimate of μ in this case; then use it along with your estimate of σ from (a).)

6.28 Referring to Exercise 6.26, suppose that the population standard deviation of carbohydrate content of the first brand of bread is identical to that for the second brand. Let σ denote this population standard deviation, so that σ^2 is the population variance (for both brands). Let s_1^2 denote the sample variance for the brand #1 sample and s_2^2 denote the sample variance for the brand #2 sample. (Even though σ^2 is the same for both brands, because of sampling variability s_1^2 and s_2^2 generally differ.) Each s^2 separately estimates σ^2, and it seems natural to combine them to obtain a better estimate. A first thought might be to use the simple average $(s_1^2 + s_2^2)/2$. But if the two sample sizes are different, the s^2 based on the larger sample size tends to be more accurate than the other s^2. This suggests computing a weighted average of s_1^2 and s_2^2. It can be shown that the following weighted average is an unbiased statistic for estimating σ^2:

$$\frac{(\text{first sample size} - 1)s_1^2 + (\text{second sample size} - 1)s_2^2}{\text{total of two sample sizes} - 2}$$

If 12 observations on brand 1 yield $s_1 = .75$ and 14 observations on brand 2 yield $s_2 = .86$, use this statistic to compute an estimate of σ^2 and then of σ.

6.29 An investigator wishes to estimate the true average lifetime of a certain type of light bulb. One possibility is to select n bulbs, start them all burning at the same time, keep the experiment going until every one has burned out, and use $\bar{x}$ (the sample average lifetime) to estimate μ. However, if μ is a very large number, it will take a great deal of time to run this experiment because the lifetime of the longest-lasting bulb is typically very large. An alternative is to put n bulbs on test, but specify some number $r < n$ and let the experiment continue until r bulbs have burned out (this is called a *censored experiment*). Let y_1 = the time until the first failure among the n bulbs, y_2 = the time until the second failure, . . . , and y_r = the time until the rth failure (so the experiment terminates at time y_r with $n - r$ of the bulbs still working). It can be shown that a reasonable statistic for estimating μ is

$$\frac{y_1 + y_2 + y_3 + \cdots + y_r + (n - r)y_r}{r} = \frac{\begin{pmatrix}\text{total burn time}\\ \text{for all } n \text{ bulbs}\end{pmatrix}}{r}$$

Suppose that $n = 20$ bulbs are tested, that $r = 10$, and that the resulting y's are 11, 15, 29, 33, 35, 40, 47, 55, 58, and 72 (so that the experiment terminates after 72 hours with 10 bulbs still burning). Use the above statistic to compute an estimate of μ. (Note: The disadvantage of this censored experiment is that the standard deviation of the above statistic is larger than the standard deviation of $\bar{x}$ based on n uncensored observations.)

6.30 A sample of n captured Pandamonian jet fighters results in serial numbers x_1, $x_2, x_3, \ldots, x_n$. The CIA knows that the aircraft were numbered consecutively at the factory starting with α and ending with β, so that the total number of planes manufactured is $\beta - \alpha + 1$ (e.g., if $\alpha = 17$ and $\beta = 29$, then $29 - 17 + 1 = 13$ planes having serial numbers 17, 18, 19, . . . , 28, 29 were manufactured). However, the CIA does not know the values of α or β. A CIA statistician suggests using the statistic $\max(x) - \min(x) + 1$ to estimate the total number of planes manufactured.

a. If $n = 5$, $x_1 = 237$, $x_2 = 375$, $x_3 = 202$, $x_4 = 525$, and $x_5 = 418$, what is the estimate using the above statistic?

b. Under what conditions on the sample will the value of the estimate be exactly equal to the true total number of planes? Will the estimate ever be smaller than the true total? Do you think the statistic is unbiased for estimating $\beta - \alpha + 1$? Explain in one or two sentences.

KEY CONCEPTS

Statistic Any quantity whose value is computed from sample data. Important statistics include the sample mean, $\overline{x}$, sample standard deviation, s, and sample proportion of successes, p. (p. 233)

Sampling distribution The probability distribution of a statistic, which describes the long-run behavior of the statistic's values resulting from an unending sequence of samples. (p. 235)

Random sample A sample for which the successively sampled values are independent of one another and each value is selected from the same population distribution. (p. 237)

Sampling distribution of $\overline{x}$ The probability distribution of the sample mean based on a random sample of n observations. The mean value and standard deviation of this sampling distribution are $\mu_{\overline{x}} = \mu$ and $\sigma_{\overline{x}} = \sigma/\sqrt{n}$, respectively, where μ is the population mean and σ is the population standard deviation. (p. 238, 246)

Sampling distribution of p The probability distribution of the sample proportion of successes based on a random sample of size n. The mean value and standard deviation of this sampling distribution are $\mu_p = \pi$ and $\sigma_p = \sqrt{\pi(1 - \pi)/n}$, where π is the population proportion of successes. (p. 253)

Central limit theorem For sufficiently large n, the sampling distribution of $\overline{x}$ is approximately normal irrespective of the population distribution's shape. In this case, the sampling distribution of the standardized variable $z = (\overline{x} - \mu) / (\sigma/\sqrt{n})$ is well approximated by the z curve. Since p can also be regarded as the mean of a sample of 0's and 1's, this theorem also implies that when n is large, the sampling distribution of p is approximately normal. (p. 249)

Point estimate of a population characteristic A single number, based on sample data, that represents the most plausible value of the population characteristic. To obtain a point estimate, select a suitable statistic and compute its value for the given sample. (p. 261)

Unbiased statistic A statistic is unbiased for estimating a particular population characteristic if the sampling distribution of the statistic is always centered at the value of the population characteristic, i.e., if the mean value of the statistic is equal to the value of the population characteristic. The statistic $\overline{x}$ is unbiased for estimating μ, p is unbiased for estimating π, and s^2 is unbiased for estimating σ^2. (p. 264)

SUPPLEMENTARY EXERCISES

6.31 A local clothing store carries three different brands of blue oxford cloth shirts, priced at $12, $15, and $20, respectively. Based on past experience, the probability distribution of the price x paid by a randomly selected customer purchasing such a shirt is known to be the following:

x	12	15	20
$p(x)$	.5	.2	.3

Let x_1 and x_2 denote the prices paid by two randomly selected customers. Then x_1 and x_2 constitute a random sample of size 2 from the above population distribution.

a. List all possible pairs of values for (x_1, x_2). (*Hint:* There are nine such pairs.)

b. The pairs listed in (a) are not all equally likely. For example, the probability of observing the pair (12, 12) is (.5)(.5) = .25. Compute the probability of observing each one of the other eight pairs.

c. Display the sampling distribution of $\bar{x}$, the mean of a random sample of size 2 from the population distribution given above, in table form.

6.32 Water permeability of concrete is an important characteristic in assessing suitability for various applications. Permeability can be measured by letting water flow across the surface and determining the amount lost (in./h). Suppose that the permeability index x for a randomly selected concrete specimen of a particular type is normally distributed with mean value 1000 and standard deviation 150.

a. How likely is it that a single specimen will have a permeability index between 850 and 1300?

b. If the permeability index is determined for each specimen in a random sample of size 10, how likely is it that the sample average permeability index will be between 950 and 1100? Between 850 and 1300?

c. Would a sample average permeability index that exceeded 1150 for a random sample of 12 specimens strongly suggest that $\mu > 1000$? Explain.

6.33 A sample of 100 washing machines of a certain type is selected, and it is determined that 40 of these needed service during the warranty period.

a. Give a point estimate for the proportion π of all machines of this type which needed or will need service during the warranty period.

b. An apartment owner purchases two of these machines. Suppose they operate independently of one another. Give a point estimate for the probability that both machines need service during the warranty period. (*Hint:* If the value of π were known, what would this probability be?)

6.34 Although a lecture period at a certain university lasts exactly 50 min, the actual lecture time of a statistics instructor on any particular day is a random variable (what else?) with mean value 52 min and standard deviation 2 min. Suppose that times of different lectures are independent of one another and the instructor gives 36 lectures in a particular course. Let $\bar{x}$ represent the mean of the 36 lecture times.

a. What are the mean value and standard deviation of the sampling distribution of $\bar{x}$?

b. What is the probability that the sample mean exceeds 50 min? 55 min?

6.35 Suppose that you want to estimate the number of pages in the shortest adult fiction book in a certain library. You randomly select 10 such books and determine the number of pages in each one.

a. What statistic (sample based quantity) might you naturally employ to obtain a point estimate?

b. If the numbers of pages for books in the sample are 314, 209, 427, 862, 223, 179, 245, 298, 152, and 256, what would your point estimate be?

c. Does your statistic of (a) ever lead to an underestimate? Do you think your statistic is unbiased? Explain.

6.36 An ecologist has a square sampling frame, called a *quadrat,* which is 2 m on each side. She randomly selects 10 locations in a particular geographical region, drops the sampling frame on the ground at each location, and counts the number of plants of a certain species that fall within the frame. Suppose that the resulting numbers are 3, 12, 1, 5, 8, 10, 4, 1, 0, and 9.

a. Let λ denote the density of plants (plants per square meter) in this region. Use the given data to compute a point estimate of λ. (*Hint:* What is the density of plants for the sample?)

b. If the region of interest extends over 100 km^2, how would you use the above data to estimate the total number of plants in the region, and what is your point estimate? Assume that the above counts came from nonoverlapping plots.

6.37 Let $x_1, x_2, \ldots, x_{100}$ denote the actual net weights of 100 randomly selected bags of fertilizer. Suppose that the weight of a randomly selected bag is a random variable with mean 50 and variance 1. Let $\bar{x}$ be the sample mean weight ($n = 100$).

a. Describe the sampling distribution of $\bar{x}$.

b. What is the probability that the sample mean is between 49.75 and 50.25?

c. What is the probability that the sample mean is less than 50?

6.38 Suppose that 60% of all students taking elementary statistics write their names in their textbooks. A random sample of 100 students is to be selected.

a. What is the mean value of the proportion among the 100 sampled who have their names in their texts? What is the standard deviation of the sample proportion?

b. What is the chance that at most 50% of those sampled have their names in their texts?

c. Answer (b) if there are 400 students in the random sample.

6.39 The nicotine content in a single cigarette of a particular brand is a random variable with mean .8 mg and standard deviation .1 mg. If 100 of these cigarettes are analyzed, what is the probability that the resulting sample mean nicotine content will be less than .79? Less than .77?

6.40 a. A random sample of 10 houses in a particular area, each of which is heated with natural gas, is selected and the amount of gas (therms) used during the month of January is determined for each house. The resulting observations are 103, 156, 118, 89, 125, 147, 122, 109, 138, and 99. Let μ_J denote the average gas usage during January by all houses in this area. Compute a point estimate of μ_J.

b. Suppose that there are 10,000 houses in this area that use natural gas for heating. Let τ denote the total amount of gas used by all these houses during January. Estimate τ using the data of (a). What statistic (sample based quantity) did you use in computing your estimate?

c. Use the data in (a) to estimate π, the proportion of all houses that use at least 100 therms.

d. Give a point estimate of the population median usage (middle value in the population of all houses) based on the sample of (a). What statistic did you use?

6.41 Referring to Exercise 6.40, suppose that August gas usage is determined for these same 10 houses, yielding observations (in the same order) 42, 57, 50, 26, 43, 62, 68, 50, 47, and 29. Let μ_d denote the average difference between January and August usage for all houses in this area. Use this data along with that of 6.40(a) to compute a point estimate of μ_d. What statistic did you use?

REFERENCES

The books by Freedman et al. and by Moore, both listed in the Chapter 1 references, give excellent informal discussions of sampling distributions at a very elementary level.

7 Interval Estimation Using a Single Sample

INTRODUCTION

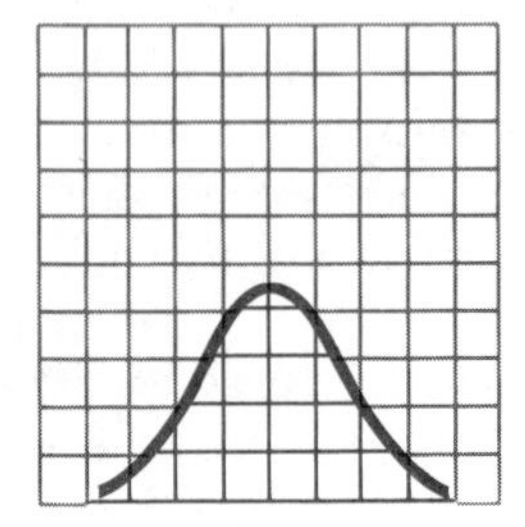

THE LAST SECTION of Chapter 6 discussed estimating a population characteristic by a single number, called a point estimate. Recall that a point estimate results from selecting an appropriate statistic and computing its value for the given sample. It would be nice if a statistic could be found for which the resulting point estimate was exactly the value of the characteristic being estimated. However, the estimate (value of the statistic) depends on which sample is selected. Different samples generally yield different estimates, due to sampling variability. In practice, only rarely is a sample selected for which the estimate is exactly equal to the value of the population characteristic. We can only hope that the chosen statistic produces an estimate *close* to the value of the population characteristic. These considerations suggest the desirability of indicating how precisely the population characteristic has been estimated. The point estimate by itself conveys no information about its closeness to the value of the population characteristic. While an estimate of 157.6 may represent our best guess for the value of μ, it is not the only plausible value.

Suppose that instead of reporting a point estimate as the single most credible value for the population characteristic, we report an entire interval of reasonable values based on the sample data. We might, for example, be highly confident that for some population the value of the average serum cholesterol level μ is in the interval from 156.4 to 158.8. The narrowness of this interval implies that we have rather precise information about the value of μ. If, with the same high degree of confidence, we could state only that μ was between 145.3 and 169.9, it would be clear that our knowledge concerning the value of μ is relatively imprecise.

A **confidence interval,** or interval estimate, for a population characteristic is an interval of plausible values for the characteristic. It is constructed so that with a suitably high degree of confidence, the value of the characteristic will be captured inside the interval. The width of the interval conveys information about how precisely the value of the characteristic is known. In this chapter, confidence intervals for μ and for π, each based on a random sample from the population under investigation, are presented. Many other confidence intervals are introduced in later chapters.

7.1 A Large-Sample Confidence Interval for a Population Mean

Let $x_1, x_2, \ldots, x_n$ denote a random sample of size n from a population with mean μ and standard deviation σ. Confidence intervals for μ presented in this chapter are based on the sample mean $\overline{x}$ and its sampling distribution. Recall that whatever the value of n, $\mu_{\overline{x}} = \mu$ and $\sigma_{\overline{x}} = \sigma/\sqrt{n}$. In this section we assume that n is large enough for the central limit theorem to apply, so the sampling distribution of $\overline{x}$ is described approximately by a normal curve. When this is the case, a confidence interval for μ is possible without making *any* specific assumptions about the population distribution. A confidence interval when n is small and the population distribution is normal (a specific assumption) is discussed in the next section.

The Case in Which σ Is Known

The development of a confidence interval for μ and various properties of the interval are most easily understood when the value of σ is known. Knowledge of σ is not, however, realistic in most applied situations. If the value of any population characteristic is known, it is much more likely to be μ than σ. Fortunately, the large-sample interval when σ is unknown requires only a minor modification of the interval appropriate for known σ.

Table I in the appendices, the table of standard normal (z) curve areas, shows that the area under the z curve to the left of -1.96 is .025; this is also the area under the curve to the right of 1.96. The area under the z curve above the interval from -1.96 to 1.96 is then $1 - .025 - .025 = .95$. This information also appears in Table 2 of Section 5.4, a table of the most useful z critical values (values that capture specified tail or central areas under the z curve). Figure 1 illustrates these areas.

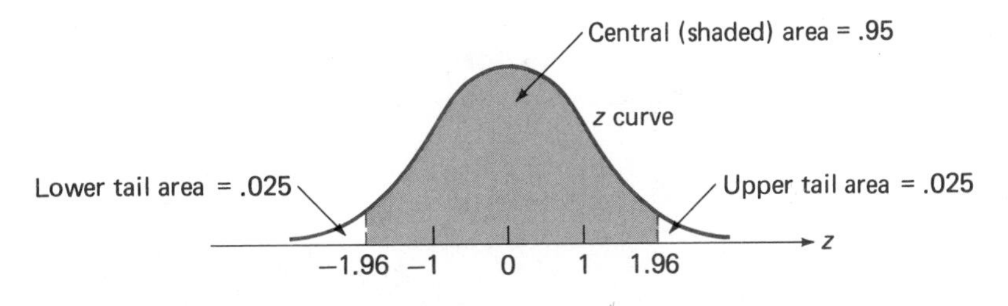

FIGURE 1 z CURVE AREAS FOR CRITICAL VALUES 1.96 AND -1.96

The discussion in Section 5.4 then implies that the area under *any* normal curve and above the interval of values within 1.96 standard deviations of the mean is .95 (using 2 instead of 1.96, as in the empirical rule, gives area .9544, a bit larger than what we want here). The normal curve of interest is the one that approximates the sampling distribution of $\bar{x}$. It is centered at the population mean μ (because $\mu_{\bar{x}} = \mu$) and has standard deviation $\sigma/\sqrt{n}$ (*not* σ, the population standard deviation). As illustrated in Figure 2, the area under this curve and above values within $1.96\sigma/\sqrt{n}$ of μ is .95. This says that approximately 95% of all samples of size n selected from the population will result in a value of $\bar{x}$ that is within $1.96\sigma/\sqrt{n}$ of μ.

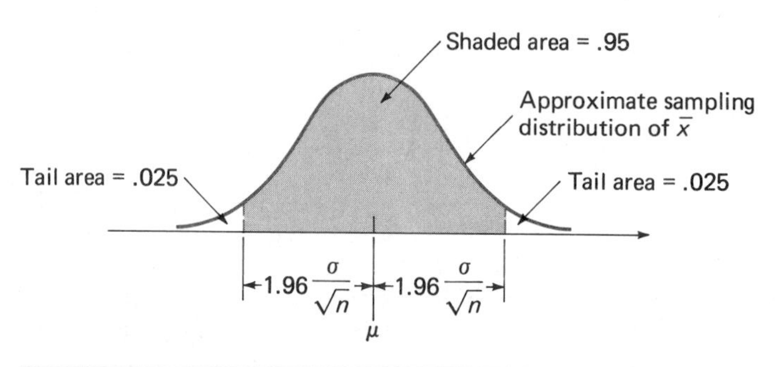

FIGURE 2

THE APPROXIMATE SAMPLING DISTRIBUTION OF $\bar{x}$ (NORMAL CURVE, MEAN VALUE μ, STANDARD DEVIATION $\sigma/\sqrt{n}$)

Now suppose that a sample has yielded a particular value of $\bar{x}$. To obtain an interval of plausible values for μ, it is natural to consider all values within a specified distance of the point estimate $\bar{x}$. In particular, consider the interval extending from $\bar{x} - 1.96\sigma/\sqrt{n}$ to $\bar{x} + 1.96\sigma/\sqrt{n}$. This is the interval of values that extends $1.96\sigma/\sqrt{n}$ (1.96 standard deviations of the statistic $\bar{x}$) to either side of $\bar{x}$. Whether or not this interval includes μ depends on the value of $\bar{x}$. Figure 3 illustrates three different cases. In the first case, $\bar{x}$ lies within $1.96\sigma/\sqrt{n}$ of μ, and μ is captured between the limits $\bar{x} - 1.96\sigma/\sqrt{n}$ and $\bar{x} + 1.96\sigma/\sqrt{n}$. In the second case, $\bar{x}$ lies more than $1.96\sigma/\sqrt{n}$ below μ, re-

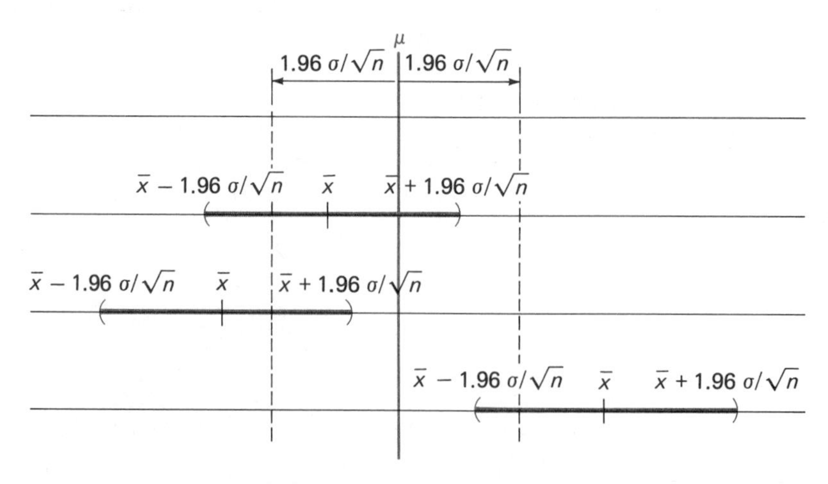

FIGURE 3

μ IS CAPTURED IN THE INTERVAL FROM $\bar{x} - 1.96\sigma/\sqrt{n}$ TO $\bar{x} + 1.96\sigma/\sqrt{n}$ WHEN $\bar{x}$ IS WITHIN $1.96\sigma/\sqrt{n}$ OF μ

sulting in an interval that falls entirely below μ. Similarly, in the third case the interval falls entirely above μ because $\overline{x}$ is more than $1.96\sigma/\sqrt{n}$ to the right of μ.

Figure 3 shows that the interval with lower limit $\overline{x} - 1.96\sigma/\sqrt{n}$ and upper limit $\overline{x} + 1.96\sigma/\sqrt{n}$ will capture μ precisely when $\overline{x}$ is within $1.96\sigma/\sqrt{n}$ of μ. From Figure 2, this will be the case for approximately 95% of all samples that might be selected, so that intervals that don't include μ, like the second and third intervals of Figure 3, are only infrequently obtained.

When n is large, approximately 95% of all samples result in

$$\overline{x} - 1.96\frac{\sigma}{\sqrt{n}} < \mu < \overline{x} + 1.96\frac{\sigma}{\sqrt{n}}$$

The interval $(\overline{x} - 1.96\sigma/\sqrt{n},\ \overline{x} + 1.96\sigma/\sqrt{n})$ obtained by substituting the value of $\overline{x}$ computed from a given sample is called a **95% confidence interval for μ.** An abbreviated formula for the interval is $\overline{x} \pm 1.96\sigma/\sqrt{n}$, where + gives the upper limit and − gives the lower limit of the interval.

The confidence interval extends $1.96\sigma/\sqrt{n}$ to either side of the point estimate $\overline{x}$, so the width of the interval is $2(1.96\sigma/\sqrt{n}) = 3.92\ \sigma/\sqrt{n}$. If the value of σ is relatively large, a small sample size will result in a wide interval (an imprecise interval estimate of μ).

EXAMPLE 1

Many of us who watch television think that far too much time is devoted to commercials. People involved in various aspects of television advertising obviously have a different perspective. The paper "The Impact of Infomercials: Perspectives of Advertisers and Advertising Agencies" (*J. Ad. Research* (1983):25–32) reported on a survey of $n = 62$ such individuals. Each person was asked what he or she believed to be the optimum amount of allocated time per hour for commercials during prime time. The resulting sample average was $\overline{x} = 8.20$ min. Let μ denote average allocation of time believed optimal for the population of all individuals involved in television advertising, and let σ denote the population standard deviation. For illustrative purposes, suppose that $\sigma = 4.5$ (the paper reported a *sample* standard deviation of 4.43). Then a 95% confidence interval for μ is $\overline{x} \pm 1.96\sigma/\sqrt{n} = 8.20 \pm (1.96)(4.5)/\sqrt{62} = 8.20 \pm 1.12 = (7.08,\ 9.32)$. That is, $7.08 < \mu < 9.32$ with 95% confidence. Obtaining a substantially more precise interval estimate (a substantially narrower interval) would require a sample size much larger than 62 because the value $\sigma = 4.5$ indicates a reasonable amount of variability in the population. ■

The 95% confidence interval for μ in Example 1 is (7.08, 9.32). One might now be tempted to say that there is a 95% chance of μ being between 7.08 and 9.32. Don't yield to this temptation! The 95% refers to the percentage of all possible samples that result in an interval that includes μ. Said another way, if we take sample after sample from the population and use each one separately to compute a 95% confidence interval, in the long run roughly 95% of these intervals will capture μ. Figure 4 illustrates this for 100 inter-

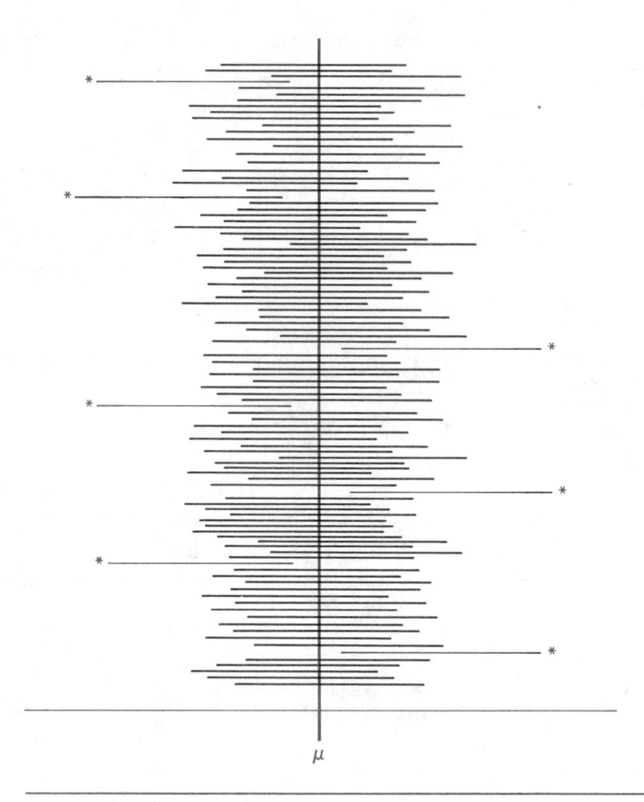

FIGURE 4 ONE HUNDRED 95% CONFIDENCE INTERVALS FOR μ COMPUTED FROM 100 DIFFERENT SAMPLES

vals—93 of the intervals include μ, whereas 7 do not. Our interval (7.08, 9.32) either includes μ or does not (remember, the value of μ is fixed but not known to us). We cannot make a chance (probability) statement concerning this particular interval. *The* **confidence level** *95% refers to the method used to construct the interval rather than to any particular interval such as the one we obtained.*

The confidence level 95% is the one most frequently used by investigators, but another confidence level can be obtained by using an appropriate z critical value in place of 1.96. For example, Table 2 of Section 5.4 shows that 99% of the area under the z curve lies between -2.58 and 2.58 (with area $.01/2 = .005$ in each tail). A 99% confidence interval for μ then results from replacing 1.96 in the formula for the 95% interval by 2.58.

The general formula for a **large-sample confidence interval for μ when the value of σ is known is**

$$\overline{x} \pm (z \text{ critical value}) \cdot \frac{\sigma}{\sqrt{n}}$$

The desired confidence level determines which critical value is used. The three most commonly used confidence levels, 90%, 95%, and 99%, require the critical values 1.645, 1.96, and 2.58, respectively.

Why settle for 95% confidence when 99% confidence is possible? The higher confidence level comes with a price tag: It is wider than the 95% interval. The width of the 95% interval is $2(1.96\sigma/\sqrt{n})$, whereas the 99% interval has width $2(2.58\sigma/\sqrt{n})$. The higher reliability of the 99% interval (where reliability is specified by the confidence level) entails a loss in precision (as indicated by the wider interval). Many investigators think that a 95% interval gives a reasonable compromise between reliability and precision. After all, how impressed would you be by the statement that with 99.999% confidence, the average IQ of students at your college was between 60 and 160? While the associated confidence level is very high, the computed interval is useless as an estimate of μ.

A Confidence Interval When σ Is Unknown

Only rarely in practice will the value of σ be known to an investigator. Fortunately, when n is large, a valid confidence interval requires only that we replace σ in our previous formula by the sample standard deviation s. This replacement is legitimate because when n is large, $s \approx \sigma$ for almost any sample that is likely to be selected. Replacing σ by s introduces very little extra variability, so no other change in the interval is required.

> A **large-sample confidence interval for a population mean μ** is given by the formula
>
> $$\bar{x} \pm (z \text{ critical value}) \cdot \frac{s}{\sqrt{n}}$$
>
> For a confidence level of approximately 95%, the z critical value 1.96 should be used; for a 90% confidence level, the value 1.645; and for a 99% confidence level, the value 2.58. As a rule of thumb, this interval is appropriate when the sample size exceeds 30.

EXAMPLE 2

(*Example 1 continued*) A 95% confidence interval for μ, the average value of commercial time believed optimal for 1 hour of prime time television by individuals in the television advertising business, was computed using the assumed value $\sigma = 4.5$. Now the sample standard deviation s, which was 4.43, can be used in place of σ. With $n = 62$ and $\bar{x} = 8.20$, the 95% interval is

$$\bar{x} \pm (1.96) \cdot \frac{s}{\sqrt{n}} = 8.20 \pm (1.96) \cdot \frac{4.43}{\sqrt{62}}$$
$$= 8.20 \pm 1.10 = (7.10, 9.30)$$

■

EXAMPLE 3

A primary cause of hearing loss in many individuals is exposure to noise. The paper "Effects of Steady State Noise Upon Human Hearing Sensitivity from 8,000 to 20,000 Hz" (*Amer. Indus. Hygiene Assoc. J.* (1980):427–32) reported on a study involving a sample of 44 individuals who had substantial exposure to industrial noise (jet engines, turbines, and the like). Each individual was asked to identify the loudness level at which signals at various frequencies became audible. The sample average loudness level for detecting a 10-kHz frequency signal was $\bar{x} = 32$ db, and the sample standard deviation was $s = 22$. A 90% confidence interval for μ, the average detection level for the population of all individuals exposed to noise of this type, is

$$\bar{x} \pm (1.645) \cdot \frac{s}{\sqrt{n}} = 32 \pm (1.645) \cdot \frac{22}{\sqrt{44}}$$
$$= 32 \pm 5.5 = (26.5, 37.5)$$

Even though the confidence level is only 90% rather than 95%, the resulting interval is rather wide because s is quite large—there is much variability in detection level—and n is not terribly large. However, individuals with minimal noise exposure have an average detection level on the order of 20. So even though our estimate of μ is relatively imprecise, it is clear that noise exposure can have negative effects. Aficionados of loud rock music (and Wagner), beware! ■

The variable of interest in each of these examples (time in Example 2 and loudness level in Example 3) is continuous. The large-sample confidence interval formulas are valid not only in this case but also when μ is the mean value of a discrete variable. (The central limit theorem applies whether the sample data consists of observations on a continuous or a discrete random variable.) Furthermore, even when the only possible values of a discrete variable x are whole numbers (a count variable), the mean value μ need not be a whole number. For example, let x denote the number of traffic citations given to a randomly selected driver during the last three years. Then μ, the mean value of x, might be .64, 1.22, 2, or any one of a number of other possibilities.

EXAMPLE 4

The behavior of bats has long been a fascinating subject to many people. One aspect of particular interest has been foraging behavior. Individuals of most species do not forage continuously at night. Instead, they commonly return to the roost for varying amounts of time between foraging periods. The paper "Night Roosting Behavior of the Little Brown Bat" (*J. Mammology* (1982): 464–74) reported that when bats of this species return to the roost, they often have trouble gaining a foothold to hang in their usual upside-down position. Let x denote the number of attempts made by a randomly selected little brown bat before gaining a foothold, and let μ denote the mean value of x (if you picture a population of bats returning to roost, μ is the mean value of x in the population). For a sample of $n = 48$ bats attempting to roost, the sample average number of attempts to succeed was $\bar{x} = 13.7$ and the sample standard deviation was $s = 14.1$. A 95% confidence interval for μ is then

$$\bar{x} \pm (1.96) \cdot \frac{s}{\sqrt{n}} = 13.7 \pm (1.96) \cdot \frac{14.1}{\sqrt{48}}$$
$$= 13.7 \pm 4.0 = (9.7, 17.7)$$

While this interval estimate is not terribly precise, one would presumably go batty acquiring enough data to substantially increase the precision. ■

Choosing the Sample Size

The 95% confidence interval for μ when σ is known is based on the fact that for approximately 95% of all random samples, $\bar{x}$ will be within $1.96\sigma/\sqrt{n}$ of μ. The quantity $1.96\sigma/\sqrt{n}$ is sometimes called the **bound on the error of estimation** associated with a 95% confidence level—with 95% confidence, the point estimate $\bar{x}$ will be no further than this amount from μ. The bound on error depends both on σ and on the sample size n. Before collecting any data, an investigator may wish to determine a sample size for which a particular value of the bound is achieved. For example, with μ representing the average fuel efficiency (mi/gal) for all cars of a certain type, the objective of an investigation may be to estimate μ to within 1 mi/gal with 95% confidence. The value of n necessary to achieve this is obtained by equating 1 with $1.96\sigma/\sqrt{n}$ and solving for n.

In general, suppose it is desired to estimate μ to within an amount B (the specified error of estimation) with 95% confidence. The necessary sample size involves solving $B = 1.96\sigma/\sqrt{n}$ for n. The result is

$$n = \left[\frac{1.96\sigma}{B}\right]^2$$

Notice that a large value of σ forces n to be large, as does a small value of B.

EXAMPLE 5

The current version of a certain bias-ply tire is known to give an average of 20,000 mi of tread wear with a standard deviation of 2000. A change in the manufacture of the tire has been proposed that, it is hoped, will increase the average tread wear without changing the standard deviation. To find how many prototype tires of the new type should be manufactured and tested in order to estimate the true average tread life μ to within 500 mi with 95% confidence, use $\sigma = 2000$ and $B = 500$ in the formula for n:

$$n = \left[\frac{(1.96)(2000)}{500}\right]^2 = [7.84]^2 = 61.5$$

A sample size of $n = 62$ will suffice to achieve the desired precision. ■

Of course, the value of σ is hardly ever known. The bound on the error of estimation then results from using s in place of σ. The necessary sample size is $n = [1.96s/B]^2$. Unfortunately, calculation of n requires the value of s, which is not known before the sample is actually obtained. One possibility is to carry out a preliminary study and use the resulting sample standard deviation (or a somewhat larger value, to be conservative) to determine n for the main part of the study. Another possibility is simply to make a rough guess about what value of s might result and use it in calculating n. For a population distribution that is not greatly skewed, dividing the range (difference between the largest and smallest value) by 4 gives a rough idea of the standard deviation. In the tire-wear problem of Example 5, suppose that σ is unknown but tread lives are believed to be between 28,000 and 36,000 mi. Then a reasonable value to use as the standard deviation in the formula for n would be $(36{,}000 - 28{,}000)/4 = 2000$.

The General Form of a Confidence Interval

Many confidence intervals have the same general form as the large-sample intervals for μ just considered. This form is more apparent when the interval is expressed in words rather than symbols. We started with a statistic $\overline{x}$ from which a point estimate for μ was obtained. The standard deviation of this statistic is $\sigma/\sqrt{n}$, which can be computed when the value of σ is known. This resulted in a confidence interval of the form

$$\left(\begin{array}{c}\text{point estimate using}\\ \text{a specified statistic}\end{array}\right) \pm \left(\begin{array}{c}\text{critical}\\ \text{value}\end{array}\right) \cdot \left(\begin{array}{c}\text{standard deviation}\\ \text{of the statistic}\end{array}\right)$$

When σ was unknown, we estimated the standard deviation of the statistic by $s/\sqrt{n}$, yielding the interval

$$\begin{pmatrix}\text{point estimate using}\\ \text{a specified statistic}\end{pmatrix} \pm \begin{pmatrix}\text{critical}\\ \text{value}\end{pmatrix} \cdot \begin{pmatrix}\text{estimated standard de-}\\ \text{viation of the statistic}\end{pmatrix}$$

For a population characteristic other than μ, a statistic for estimating the characteristic will be selected. Then (drawing on statistical theory) a formula for the standard deviation of the statistic will be given. In practice it will almost always be necessary to estimate this standard deviation (using something analogous to $s/\sqrt{n}$ rather than $\sigma/\sqrt{n}$), so the second interval will be the prototype confidence interval. The estimated standard deviation of the statistic is sometimes referred to as the **standard error** in published articles.

The appropriate critical value for an interval based on a large sample size (or large sample sizes in multisample problems) is a z critical value such as 1.96 or 2.58. However, other critical values are needed to obtain the desired confidence levels in small-sample problems. Here is a slightly different derivation of the confidence interval for μ, which should give you added insight into how other confidence intervals are constructed. We know that $\mu_{\bar{x}} = \mu$, $\sigma_{\bar{x}} = \sigma/\sqrt{n}$, and—when n is large—the sampling distribution of $\bar{x}$ is approximately normal. These facts imply that the sampling distribution of the standardized variable

$$z = \frac{\bar{x} - \mu}{\sigma/\sqrt{n}}$$

is described approximately by the z curve. In particular, since the interval from -1.96 to 1.96 captures area .95 under the z curve, approximately 95% of all samples yield an $\bar{x}$ satisfying

$$-1.96 < \frac{\bar{x} - \mu}{\sigma/\sqrt{n}} < 1.96$$

These inequalities can be manipulated to yield an equivalent set of inequalities with μ isolated in the middle:*

$$\bar{x} - 1.96\frac{\sigma}{\sqrt{n}} < \mu < \bar{x} + 1.96\frac{\sigma}{\sqrt{n}}$$

The two extreme terms in this set of inequalities are exactly the lower and upper limits, respectively, of the 95% confidence interval for μ. To achieve any other specified confidence level (e.g., 99%), replace the critical value 1.96 by an appropriate value (e.g., 2.58).

In the more frequently encountered situation in which σ is not known, it is natural to replace it by s and use the standardized variable

$$\frac{\bar{x} - \mu}{s/\sqrt{n}}$$

Statistical theory says that when n is large, this standardized variable also has a sampling distribution described approximately by the z curve. This is because when n is large, $s \approx \sigma$ in most cases, so replacing σ by s contributes lit-

*The steps necessary to achieve this are (1) multiply each term by $\sigma/\sqrt{n}$, (2) subtract $\bar{x}$ from each term, (3) multiply by -1 to change $-\mu$ to μ (which changes each $<$ to $>$), and (4) rewrite the inequalities from right to left so that $<$ appears again.

tle extra variability to the resulting standardized variable. Thus for roughly 95% of all samples, it will be the case that

$$-1.96 < \frac{\bar{x} - \mu}{s/\sqrt{n}} < 1.96$$

Manipulating this set of inequalities to isolate μ yields the large-sample interval in which s rather than σ appears in each limit. This interval is often written as $\bar{x} \pm 1.96s_{\bar{x}}$, where $s_{\bar{x}} = s/\sqrt{n}$ (the estimated standard deviation of the statistic $\bar{x}$).

EXERCISES

7.1 The formula used to compute a confidence interval for μ when n is large and σ is known is

$$\bar{x} \pm (z \text{ critical value}) \frac{\sigma}{\sqrt{n}}$$

What is the appropriate z critical value for each of the following confidence levels?

a. 95% **b.** 90% **c.** 99% **d.** 80%

e. 85% (*Hint:* What critical values capture a central area of .85 under the z curve?)

7.2 Suppose that a random sample of 50 bottles of a particular brand of cough medicine is selected and the alcohol content of each bottle is determined. Let μ denote the average alcohol content for the population of all bottles of the brand under study. Suppose that this sample resulted in a 95% confidence interval for μ of (7.8, 9.4).

a. Would a 90% confidence interval have been narrower or wider than the given interval? Explain your answer.

b. Consider the following statement: There is a 95% chance that μ is between 7.8 and 9.4. Is this statement correct? Why or why not?

c. Consider the following statement: If the process of selecting a sample of size 50 and then computing the corresponding 95% confidence interval is repeated 100 times, 95 of the resulting intervals will include μ. Is this statement correct? Why or why not?

7.3 Computer equipment can be very sensitive to high temperatures. As a result, when testing computer components, the temperature at which a malfunction occurs is of interest. Let μ denote the average temperature at which components of a certain type fail. A random sample of 49 components was tested by exposing the components to increasing temperatures and recording the temperature at which each component first failed. The resulting observations were used to compute $\bar{x} = 89°\text{F}$ and $s = 14°\text{F}$. Find a 95% confidence interval for μ.

7.4 One hundred employees at a large facility were randomly chosen and the one-way commuting distance was recorded for each of those selected. This resulted in a sample mean of 12 mi and a sample standard deviation of 2.5 mi. Use this information to construct a 95% confidence interval for μ, the mean one-way commuting distance for the population of all employees.

7.5 A manufacturer of video recorder tapes sells tapes labeled as giving 6 hours of playing time. Sixty-four of these tapes are selected and the actual playing time for each is determined. If the mean and standard deviation of the 64 observed playing times are 352 min and 8 min, respectively, construct a 99% confidence interval for μ, the true average playing time for 6-h tapes made by this manufacturer. Based on your interval, do you think that the manufacturer could be accused of false advertising? Explain.

7.6 An investigator is interested in estimating the bacteria count for unhomogenized milk. A random sample of fifty 1-pt containers is selected and a laboratory analysis performed to determine a bacteria count (bacteria/mL) for each container. Sample results were as follows: $\bar{x} = 2.3$, $s = .4$.

a. Construct a 90% confidence interval for μ, the true average bacteria count for unhomogenized milk.

b. Based on your interval, are values of μ greater than 2.5 plausible? Explain.

7.7 A manufacturer of college text books is interested in estimating the strength of the bindings produced by a particular binding machine. Strength can be measured by recording the force required to pull the pages from the binding. If this force is measured in pounds, how many books should be tested in order to estimate the average force required to break the binding to within .1 lb with 95% confidence? Assume that σ is known to be .8.

7.8 The amount of time spent on housework by women who work outside the home was examined in the paper "The Effects of Wife's Employment Time on Her Household Work Time" (*Home Ec. Research J.* (1983):260–65). Each person in a sample of 362 working women was asked to indicate how much time she spent each day on certain household activities. Some of the results are summarized in the accompanying table.

Activity	Average (min/day)	Standard Deviation
All housework	348.95	176.9
Food preparation	74.43	50.4
Cleaning	72.09	72.6

a. Construct a 90% confidence interval for the average amount of time that all working women spend on housework.

b. Construct a 95% confidence interval for the mean time that working women spend on food preparation.

c. Would a 95% confidence interval for the average amount of time spent on cleaning be wider or narrower than the 95% interval for the average amount of time spent in food preparation? Explain.

d. How would you interpret the very large standard deviation for the amount of time spent on cleaning?

7.9 The paper "The Variability of Blood Pressure Measurements in Children" (*Amer. J. of Public Health* (1983):1207–11) reported the results of a study of 99 third-grade children. The sample consisted of 53 boys and 46 girls. The values of several variables, including systolic blood pressure, diastolic blood pressure, and pulse, were recorded for each child. Means and standard deviations are given in the accompanying table.

Boys	Average	Standard Deviation
Systolic blood pressure	101.7	9.8
Diastolic blood pressure	59.2	9.8
Pulse	86.7	10.3
Girls	**Average**	**Standard Deviation**
Systolic blood pressure	101.8	9.8
Diastolic blood pressure	60.3	10.1
Pulse	86.7	10.2

a. Construct 90% confidence intervals for the mean systolic blood pressure for boys and for girls. Are the intervals similar? Based on your intervals, do you think that there is a difference in mean systolic blood pressure for boys and girls? Explain.

b. Construct a 99% confidence interval for boys' average pulse rate.

7.10 Suppose that the study discussed in Exercise 7.9 is to be considered a pilot study and that a large-scale study will follow.

a. The researchers would like to estimate the true average pulse rate for girls to within 1 with 95% confidence. How many girls should be included in the study?

b. Suppose that in addition to estimating the average pulse rate for girls to within 1, the researchers also want to be able to estimate both the average systolic blood pressure and the average diastolic blood pressure to within 1.5 with 95% confidence. How many girls should be studied in order to achieve both of these goals?

7.11 In a recent study of blood-lead levels, researchers working for the federal government sampled the blood of 27,801 people across the country. They found that the sample average blood-lead level was 9.2 μg/dL. Suppose that the corresponding sample standard deviation was 10.4 μg/dL.

a. Based on the given mean and standard deviation, is it reasonable to think that blood-lead levels are normally distributed? Why or why not?

b. Construct a 99% confidence interval for the true average blood-lead level for U.S. residents. Note that the interval is relatively narrow. To what can this be attributed?

c. According to standards set by the Center for Disease Control, a blood-lead level that exceeds 30 μg/dL is cause for concern. What can be said about the proportion of U.S. residents who have blood-lead levels in excess of 30? (*Hint:* Use Chebyshev's rule with the sample mean and standard deviation, since μ and σ are not known.)

7.12 Anyone who has owned a dog or cat knows that caring for a pet can be expensive. The paper "Veterinary Health Care Market for Dogs" (*J. of the Amer. Vet. Med. Assoc.* (1984):207–08) studied annual veterinary expenditures for households owning dogs. The average veterinary expenditure was reported as $74 per year, and the corresponding standard deviation was (approximately) $40. Suppose that these statistics had been calculated based on a random sample of size 144 (the actual sample was much larger). Construct a 90% confidence interval for the average yearly veterinary expenditure for dog-owning households.

7.13 The study on attitudes toward commercial time and length of commercials discussed in Example 1 reported the results of a survey of people working in the television advertising business. In addition to questions regarding the optimum amount of time per hour that should be allocated to commercials, the 62 respondents were also asked to indicate the optimum length of a prime-time commercial. The resulting average and standard deviation were 39.8 s and 15.66 s, respectively. Let μ denote the average length believed optimal for the population of all individuals involved in television advertising. Construct a 95% confidence interval for μ. Does the interval suggest that μ has been precisely estimated? Explain.

7.14 The paper "National Geographic, The Doomsday Machine" which appeared in the *J. of Irreproducible Results* (yes, there really is a journal by that name—it's a spoof of technical journals!) predicted dire consequences resulting from a nationwide buildup of *National Geographic*. The author's predictions are based on the observations that the number of subscriptions for *National Geographic* is on the rise and that no one ever throws away a copy of the *National Geographic*. A key variable in the analysis presented in the paper is the weight of an issue of the magazine. Suppose that you were assigned the task of estimating the average weight of a particular issue of the *National Geographic*. How many copies should you sample in order to estimate the average weight to within .1 oz with 95% confidence? Assume that σ is known to be 1 oz.

7.15 The formula for determining sample size described in this section corresponds to a confidence level of 95%. What would be the appropriate formula for determining sample size when the desired confidence level is 90%? 99%?

7.2 A Small-Sample Confidence Interval for the Mean of a Normal Population

The large-sample confidence intervals for μ discussed in Section 7.1 are appropriate whatever the shape of the population distribution. This is because they are based on the central limit theorem, according to which $\bar{x}$ has approximately a normal distribution when n is large irrespective of the population distribution. Because the standardized variable $z = (\bar{x} - \mu)/(\sigma/\sqrt{n})$ then has a probability distribution described approximately by the standard normal curve, $-1.96 < (\bar{x} - \mu)/(\sigma/\sqrt{n}) < 1.96$ for approximately 95% of all possible samples. Manipulating these inequalities to isolate μ in the middle (see the end of Section 7.1) results in

$$\bar{x} - (1.96)\frac{\sigma}{\sqrt{n}} < \mu < \bar{x} + (1.96)\frac{\sigma}{\sqrt{n}}.$$

The extreme left term is the lower limit of the 95% confidence interval for μ and the extreme right term is the upper limit.

When n is small, the central limit theorem cannot be applied. One way to proceed in this case is to make a specific assumption about the shape of the population distribution and then use a confidence interval that is valid only under this assumption. The confidence intervals to be discussed in this section are based on the assumption that the population distribution is (at least approximately) normal. If the population distribution is believed to deviate substantially from the normal (as suggested, for example, by a normal probability plot or a box plot), these intervals should not be used. A statistician should be consulted if evidence suggests that the population distribution is quite skewed.

The assumption of population normality implies that $\bar{x}$ has a normal sampling distribution for any sample size n, so $(\bar{x} - \mu)/(\sigma/\sqrt{n})$ has a standard normal (z) distribution even when n is small. The value of σ is almost never known, in which case the relevant standardized variable is $(\bar{x} - \mu)/(s/\sqrt{n})$, obtained by using the sample standard deviation s in place of σ. The substitution of s for σ has little effect when n is large, but for small n, it results in a great deal of extra variability that wasn't there before the substitution. Intuition suggests that the sampling distribution of $(\bar{x} - \mu)/(s/\sqrt{n})$ should be more spread out than the z curve. A description of this sampling distribution requires that we first learn about some new probability distributions called t distributions.

t Distributions

Recall that there are many different normal distributions, one for each different combination of μ and σ. Similarly, there are many different t distributions. Each different t distribution corresponds to a different value of a quantity called the *number of degrees of freedom* (df) of the distribution. There is a t distribution with 1 df, another with 2 df, and so on. In general, for each positive whole number, there is a corresponding t distribution with that number of degrees of freedom. The important properties of t distributions are as follows:

1. The t curve corresponding to any fixed number of degrees of freedom is bell-shaped and centered at zero, just as is the standard normal (z) curve.
2. Any t curve is more spread out than the z curve.
3. As the number of degrees of freedom increases, the spread of the corresponding t curve decreases.
4. As the number of degrees of freedom grows arbitrarily large, the corresponding t curve approaches the z curve more and more closely.

These properties are illustrated in Figure 5, which shows several t curves along with a z curve.

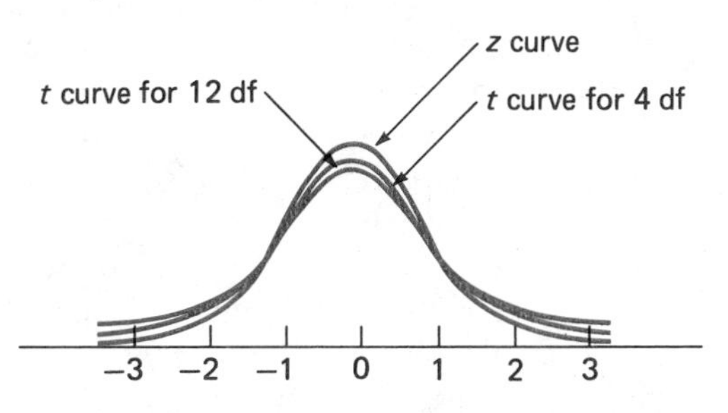

FIGURE 5 COMPARISON OF THE z CURVE AND t CURVES FOR 12 df AND 4 df

The large-sample z confidence intervals utilized z critical values such as 1.645, 1.96, and 2.58, which captured specified central areas and tail areas under the z curve. The 95% interval was based on the fact that the interval from -1.96 to 1.96 captured central area .95 under the z curve (leaving area .025 for each tail). Critical values for t distributions are defined in the same manner, but they now depend not only on the area to be captured but also on the number of degrees of freedom under consideration. For example, for the t curve with 4 df, the interval from -2.78 to 2.78 captures central area .95 (leaving area .025 for each tail). The lower- and upper-tail values are equidistant from zero because the t curve is symmetric. Furthermore, since the t curve is more spread out than the z curve, one must go further out in the tail of the t curve to capture central area .95 and tail area .025. Thus the t critical values ± 2.78 exceed the corresponding z critical values ± 1.96. For the t curve with 12 df, the critical values that capture central area .95 and area .025 in each tail are ± 2.18. These values are between those for the z curve and the t curve with 4 df because the t curve with 12 df has spread between those of the two other curves.

Table IV of the appendices gives selected upper-tail critical values for various t distributions. The central areas for which values are tabulated are .80, .90, .95, .98, .99, .998, and .999. To find a particular critical value, go down the left margin of the table to the row labeled with the desired number of degrees of freedom. Then move over in that row to the column headed by the desired central area. For example, the value in the 12-df row under the column corresponding to central area .95 is 2.18, so 95% of the area under the 12-df t curve lies between -2.18 and 2.18. Moving over two columns, the critical value for central area .99 (still with 12 df) is 3.06. Moving down the .99 column to the 20-df row, the critical value is 2.85, so area .99 lies between -2.85 and 2.85 under the t curve with 20 df. Notice that the critical values increase as one moves to the right in each row. This is necessary in order to capture a larger central area and smaller tail area. In each column, the critical values decrease as one moves downward, reflecting decreasing spread for t distributions with greater numbers of degrees of freedom.

The larger the number of degrees of freedom, the more closely the t curve resembles the z curve. To emphasize this, we have included the z critical values 1.645, 1.96, 2.58, and several others as the last row of the t table. Furthermore, once the number of degrees of freedom exceeds 30, the critical values change very little as the number of degrees of freedom increases. For this reason, our table jumps from 30 df to 40 df, then to 60 df, then to 120 df, and finally to the row of z critical values. Many tables jump directly from 30 to z. If you need a critical value for a number of degrees of freedom between those tabulated, you can either use the closest number of degrees of freedom or else just use the z critical values.

The t Confidence Interval

The fact that the sampling distribution of $(\overline{x} - \mu) / (\sigma/\sqrt{n})$ is approximately the z (standard normal) distribution when n is large led to the large-sample z intervals of Section 7.1. In the same way, the following proposition provides the key to obtaining a small-sample confidence interval when the population distribution is normal.

> Let $x_1, x_2, \ldots, x_n$ constitute a random sample from a normal population distribution. Then the sampling distribution of the standardized variable
>
> $$t = \frac{\overline{x} - \mu}{s/\sqrt{n}}$$
>
> is the t distribution with $n - 1$ degrees of freedom.

The $n - 1$ df results from having used $\overline{x}$ as an estimate of μ in computing the deviations $x_1 - \overline{x}, \ldots, x_n - \overline{x}$ on which s is based. Because these deviations have a sum of zero, they contain only $n - 1$ independent pieces of information. Statisticians customarily say that the estimation of μ has resulted in a loss of 1 df. Later we shall see situations in which having to estimate several characteristics entails losing several degrees of freedom.

To see how this result leads to the desired confidence interval, consider the case $n = 20$, so that the number of degrees of freedom is 19. According to Table IV, the interval between -2.09 and 2.09 captures area .95 under the t curve with 19 df. Thus

$$-2.09 < \frac{\overline{x} - \mu}{s/\sqrt{20}} < 2.09$$

for 95% of all samples with $n = 20$. Manipulating these inequalities to isolate μ in the middle yields $\bar{x} - 2.09 \cdot s/\sqrt{20} < \mu < \bar{x} + 2.09 \cdot s/\sqrt{20}$. The 95% confidence interval for μ in this case then extends from the lower limit $\bar{x} - (2.09)(s/\sqrt{20})$ to the upper limit $\bar{x} + (2.09)(s/\sqrt{20})$. This can be written as $\bar{x} \pm (2.09)(s/\sqrt{20})$. The major difference between this interval and our large-sample interval is the use of the t critical value 2.09 rather than the z critical value 1.96. The extra uncertainty that results from having to estimate σ using a small sample causes the t interval to be wider than the z interval.

If the sample size is something other than 20 or if the desired confidence level is something other than 95%, the critical value 2.09 should be replaced by a value from another row or column of Table IV.

Let $x_1, x_2, \ldots, x_n$ constitute a random sample from a normal population distribution with mean value μ. Then a **small-sample confidence interval for μ** has the form

$$\bar{x} \pm (t \text{ critical value}) \cdot \frac{s}{\sqrt{n}}$$

where the critical value is based on $n - 1$ df. Table IV gives critical values appropriate for each of the confidence levels 90%, 95%, and 99%, as well as several other less frequently used confidence levels.

This confidence interval is appropriate for small n only when the population distribution is approximately normal. If this is not the case, an alternative method should be used.

EXAMPLE 6

The use of synthetic male hormones (technically, anabolic steroids) is widespread in sports that require great muscular strength. The article "Side Effects of Anabolic Steroids in Weight Trained Men" (*The Physician and Sports Med.* (December 1983):87–98) reported on a study of a sample of 20 body builders who were current users of such steroids. The sample average weekly dosage for oral agents was $\bar{x} = 173$ mg, and the sample standard deviation was $s = 45$ mg. Suppose that the distribution of weekly dosage in the population of all body builders who use oral steroids is normal with mean value μ. To compute a 95% confidence interval for μ based on the given sample information, the t critical value for $n - 1 = 19$ df is needed. From Table IV, this value is 2.09. The 95% confidence interval is then

$$\bar{x} \pm (t \text{ critical value}) \cdot \frac{s}{\sqrt{n}} = 173 \pm (2.09) \cdot \frac{45}{\sqrt{20}}$$
$$= 173 \pm 21 = (152, 194)$$

We can be highly confident that the true average weekly dosage is between 152 and 194. The article reports that a manufacturer's recommended dosage for a certain oral steroid is between 35 and 70 mg per week, so our analysis suggests excessive steroid use among body builders. ■

EXAMPLE 7

Chronic exposure to asbestos fiber is a well-known health hazard. The paper "The Acute Effects of Chrysotile Asbestos Exposure on Lung Function" (*Environ. Research* (1978):360–72) reported results of a study based on a sample of construction workers who had been exposed to asbestos over a prolonged period. Among the data given in the article were the following (ordered) values of pulmonary compliance ($cm^3/cm\ H_2O$) for each of 16 subjects 8 months after the exposure period (pulmonary compliance is a measure of lung elasticity, or how effectively the lungs are able to inhale and exhale):

167.9	180.8	184.8	189.8	194.8	200.2	201.9	206.9
207.2	208.4	226.3	227.7	228.5	232.4	239.8	258.6

Construction of a box plot (Figure 6) for the data requires the first quartile, median, and third quartile, which are 192.3, 207.05, and 228.1, respectively. The interquartile range is iqr = 228.1 − 192.3 = 35.8, and 1.5iqr = 53.7. No observation lies more than 53.7 from the closest quartile, so the box plot shows no outliers or substantial skewness that might warn against using the t interval.

```
                  -------------------
------------I          +           I---------------
                  -------------------

-------+---------+---------+---------+---------
      180       200       220       240
```

	PUL COMP
N	16
MEAN	209.8
MEDIAN	207.1
TMEAN	209.3
STDEV	24.2
SEMEAN	6.0
MAX	258.6
MIN	167.9
Q3	228.3
Q1	191.1

	N	MEAN	STDEV	SE MEAN	90.0 PERCENT C.I.
PUL COMP	16	209.8	24.2	6.0	(199.2, 220.3)

FIGURE 6 MINITAB-GENERATED BOX PLOT, DESCRIPTIVE QUANTITIES, AND CONFIDENCE INTERVAL*

The sample total is $\Sigma x = 3356.0$, so $\overline{x} = 3356.0/16 = 209.75$. Using either the computational formula or the 16 deviations gives $\Sigma(x - \overline{x})^2 = 8752.82$. Then $s^2 = \Sigma(x - \overline{x})^2/(n - 1) = 8752.82/15 = 583.52$, and $s = \sqrt{583.52} = 24.16$. Let μ denote the average value of pulmonary compliance for the population of all men who have had extensive exposure to asbestos. A 90% confidence interval for μ requires use of the appropriate t critical value based on $n - 1 = 15$ df. From Table IV, this critical value is 1.75. The confidence interval is then

$$209.75 \pm (1.75) \cdot \frac{24.16}{\sqrt{16}} = 209.75 \pm (1.75)(6.04)$$

$$= 209.75 \pm 10.57 = (199.18, 220.32)$$

We can then say that with 90% confidence, $199.18 < \mu < 220.32$. This interval also appears in the MINITAB output of Figure 6. ■

In Section 7.1, we showed that $\overline{x}$ would estimate μ to within an amount B with 95% confidence provided that $n \geq [1.96\sigma/B]^2$. When σ is unknown, a guess or preliminary estimate must be used in its place. An additional

*MINITAB's method for computing the quartiles differs slightly from the method that we have been using.

difficulty is that 1.96 should be used only if it is known that n will be large. If it is suspected that a small n might work, a t critical value should be used in place of 1.96. A conservative approach is to use a value somewhat larger than 1.96 in the formula.

EXERCISES

7.16 What percentage of the time will the value of a variable that has a t distribution with the specified degrees of freedom fall in the indicated region?

a. 10 df, between -1.81 and 1.81

b. 10 df, between -2.23 and 2.23

c. 24 df, between -2.06 and 2.06

d. 24 df, between -2.80 and 2.80

e. 24 df, outside the interval from -2.80 to 2.80

f. 24 df, to the right of 2.80

g. 10 df, to the left of -1.81

7.17 The formula used to compute a confidence interval for the mean of a normal population when n is small is

$$\bar{x} \pm (t \text{ critical value}) \cdot \frac{s}{\sqrt{n}}$$

What is the appropriate t critical value for each of the following confidence levels and sample sizes?

a. 95% confidence, $n = 17$

b. 90% confidence, $n = 12$

c. 99% confidence, $n = 24$

d. 90% confidence, $n = 25$

e. 90% confidence, $n = 13$

f. 95% confidence, $n = 10$

7.18 A wine manufacturer sells a Cabernet whose label asserts an alcohol content of 11%. Sixteen bottles of this Cabernet are randomly selected and analyzed for alcohol content. The resulting sample mean and sample standard deviation are 10.2 and .8, respectively.

a. Construct a 95% confidence interval for μ, the average alcohol content of bottles of Cabernet produced by this manufacturer (assume that alcohol content is normally distributed).

b. Based on your interval in (a), do you think that the manufacturer is incorrect in its label claim? Explain.

7.19 Fifteen students at a particular university were selected at random and asked to indicate the amount of money spent on text books during the previous semester. Results were (in dollars):

90	110	40	170	80	75	100	80
60	110	95	130	100	115	130	

Assuming that amount spent is normally distributed, find a 90% confidence interval for μ, the mean amount spent on books per semester by all students at this university.

7.20 The annual yields for 10 randomly selected common stocks (in percent) during a particular year were as follows:

8.7	9.2	10.8	6.4	8.9	9.8	8.8	9.4	8.7	7.9

a. Use the data to construct a 95% confidence interval for the mean annual yield of all common stocks during the given year.

b. What assumptions about the distribution of common stock yields must be true in order for the interval obtained in (a) to be valid?

7.21 Five students visiting the student health center for a free dental examination during National Dental Hygiene Month were asked how many months had passed since their last visit to a dentist. Their responses were as follows:

6 17 11 22 29

If these five students can be considered to be a random sample of all students participating in the free check-up program, construct a 95% confidence interval for the mean number of months elapsed since the last visit to a dentist for the population of students participating in the program.

7.22 In an effort to estimate the average starting salary of students who graduate with a degree in computer science, 10 graduating computer science students who had accepted job offers were randomly selected, and the yearly starting salary was noted for each. The observed values were (in thousands of dollars) as follows:

28 31 26 32 27 28 27 30 31 29

Find a 99% confidence interval for the mean starting salary of computer science graduates.

7.23 Suppose that the body temperatures of 5 people who had contracted a particular virus were (at the time of diagnosis) as follows:

99.2 98.9 99.6 99.9 100.4

a. Construct a 90% confidence interval for μ, the mean body temperature of those who have the virus.

b. What assumptions about the body temperature of people who have this virus must be satisfied in order for your interval in (a) to be valid?

c. Based on your interval in (a), explain whether or not you would feel comfortable making the statement "Evidence suggests that the average temperature of people suffering from this virus is above normal (98.6°)."

7.24 Authors of the paper "Quality of Carrots Dehydrated by Three Home Methods" (*Home Ec. Research J.* (September 1983):81) examined the time and energy required to dehydrate carrots. Five 1-qt containers of sliced carrots were dehydrated using each method, and the accompanying data was reported.

Time Required for Dehydration (min)

Method	Mean	(Approximate) Standard Deviation
Convection oven	428	35
Food dehydrator	513	20
Microwave oven	135	4

Energy Required for Dehydration (Wh)

Method	Mean	(Approximate) Standard Deviation
Convection oven	2199	100
Food dehydrator	3920	170
Microwave oven	1431	30

a. Compute a 90% confidence interval for the average time required to dehydrate 1 qt of carrots using a convection oven.

b. Would a 90% confidence interval for the average time required for dehydration using a food dehydrator be wider or narrower in width than the 90% interval for the convection oven method computed in (a)? Explain without actually computing the interval.

c. Construct a 95% confidence interval for the average energy required to dehydrate 1 qt of carrots for each of the three methods. Based on these intervals, what conclusions can you draw about the relative efficiency of the three methods?

7.25 An appliance repair company routinely assigns seven repair jobs a day to each of its service technicians. Due to complaints that the employees are being given more work than they can complete in 8 h, the company randomly selects 25 files on previously completed repair jobs. Since the repair people record total time spent on each job, the company is able to compute an average and standard deviation for the 25 selected repair times. The resulting sample mean and sample standard deviation were 80 min and 20 min, respectively. Compute a 90% confidence interval for μ, the mean time required to complete a repair job. Does your interval support the employees claim that they are overworked?

7.26 The paper "Surgeons and Operating Rooms: Underutilized Resources" (*Amer. J. of Public Health* (1983):1361–65) investigated the number of operations per year performed by doctors in various medical specialties.

a. Nine plastic surgeons were asked to indicate the number of operations performed in the previous year. The resulting sample mean was 263.7. If the sample standard deviation was 50.4, find a 90% confidence interval for the average number of operations per year for the population of all plastic surgeons.

b. Twenty-two neurosurgeons were also surveyed concerning the number of operations performed during the previous year. The sample mean was 58.5 operations. If the sample standard deviation had been 12.1, find a 99% confidence interval for μ, the mean number of operations performed per year by a neurosurgeon.

c. What assumptions about the distribution of the number of operations performed in one year for the two populations of doctors are required in order for the intervals in (a) and (b) to be appropriate?

7.27 Family food expenditures were investigated in the paper "Household Production of Food: Expenditures, Norms, and Satisfaction" (*Home Ec. Research J.* (March 1983):27). A sample of Iowa homes resulted in an average weekly food expenditure of $164 and a standard deviation of $85. Assuming these results were based on a random sample of size 25, construct a 95% confidence interval for μ, the mean weekly food expenditure for Iowa families. Interpret your interval.

7.28 Describe how a normal probability plot could be used to decide whether the small-sample confidence interval of this section is an appropriate way to estimate μ.

7.29 A. Duncan, in his book *Quality Control and Industrial Statistics* (Homewood, Il.: Irwin, 1974), reported shear strength observations for stainless steel welds in pounds per square inch. The data for 22 welds is given along with a MINITAB histogram and a normal probability plot. Discuss whether it would be reasonable to assume that the distribution of the strengths of stainless steel welds is approximately normal. If a confidence interval for the mean strength of stainless steel welds is desired, would you recommend using the t confidence interval of this section? Why or why not?

```
2340.00   2440.00   2370.00   2340.00   2360.00
2340.00   2330.00   2380.00   2350.00   2360.00
2390.00   2360.00   2400.00   2320.00   2360.00
2350.00   2340.00   2320.00   2350.00   2330.00
2320.00   2300.00
```

```
              DATA
MIDDLE OF    NUMBER OF
INTERVAL     OBSERVATIONS
  2300.         1     *
  2320.         3     ***
  2340.         6     ******
  2360.         7     *******
  2380.         2     **
  2400.         2     **
  2420.         0
  2440.         1     *

                                HISTOGRAM
2460.+
     -
     -                                                          *
     -
     -
2400.+                                                       *
     -                                                    *
     -                                                 *  *
     -                                             4
     -                                         3
2340.+                                     4
     -                                 2
     -                            3
     -     *
     -
2280.+
      +---------+---------+---------+---------+---------+
     -2.0      -1.0       0.0       1.0       2.0       3.0
```

Normal probability plot

7.30 The U.S. Department of Agriculture's Office of Marketing released the results of a study on strength of a particular type of yarn. The following strengths (lb) were reported:

66	117	132	111	107
85	89	79	91	97
138	103	111	86	78
96	93	101	102	110
95	96	88	122	115

Construct a normal probability plot for this data. If appropriate, find a 90% confidence interval for μ, the mean strength for this type of yarn.

7.3 A Large-Sample Confidence Interval for a Population Proportion

An investigator often wishes to draw some conclusions concerning the proportion of individuals or objects in a population that possess a particular property (e.g., smoke cigarettes, file income taxes using a short form, need repair while under warranty, etc.). As before, the label "success" (S) is used for any individual or object having the property, and π denotes the proportion of S's in the population. In this section we present a confidence interval for π based on the sample proportion of successes, p = (number of S's)/n, resulting from a random sample of size n.

Although a confidence interval for π can be obtained when n is small, our focus is on the large-sample case. Then the sampling distribution of p can be well approximated by a normal curve. Most applied problems are of this latter type. To justify the formula for the large-sample interval, it is first necessary to review properties of the sampling distribution of p (turn back to Section 6.4 for a more detailed discussion):

1. The mean value of p is $\mu_p = \pi$, so p is an unbiased statistic for estimating π (the sampling distribution of p is centered at the value of π).
2. The standard deviation of p is $\sigma_p = \sqrt{\pi(1 - \pi)/n}$.
3. As long as n and π satisfy $5 \le n\pi$ and $5 \le n(1 - \pi)$, our rule of thumb for characterizing the sample size as large, the sampling distribution of p is accurately approximated by a normal curve.

Properties 1–3 imply that when n is large, the sampling distribution of the standardized variable

$$z = \frac{\text{statistic} - \text{mean value}}{\text{standard deviation}} = \frac{p - \pi}{\sqrt{\pi(1 - \pi)/n}}$$

is well approximated by the standard normal (z) curve. Furthermore, for large n, the sampling distribution is approximately the z curve even after replacing the denominator by $\sqrt{p(1 - p)/n}$, an estimate of σ_p. Since 95% of the area under the z curve lies above the interval from -1.96 to 1.96, it follows that

$$-1.96 < \frac{p - \pi}{\sqrt{p(1 - p)/n}} < 1.96$$

for approximately 95% of all samples. This statement parallels statements in the two previous sections, which led directly to the 95% z and t confidence intervals for μ. Manipulating these inequalities to isolate π in the middle results in $p - 1.96\sqrt{p(1 - p)/n} < \pi < p + 1.96\sqrt{p(1 - p)/n}$ for approximately 95% of all samples. The quantities on the far left and far right are the lower and upper limit, respectively, of the 95% confidence interval for π. Any other desired confidence level can be achieved by using a different z critical value in place of 1.96. Because the value of π is unknown, the rule of thumb for saying whether n is large must be modified, as indicated below.

The general form of the **large-sample confidence interval for π** is

$$p \pm (z \text{ critical value}) \sqrt{\frac{p(1 - p)}{n}}$$

The three most commonly used confidence levels, 90%, 95%, and 99%, are achieved by using the critical values 1.645, 1.96, and 2.58, respectively. This interval can safely be used as long as $5 \le np$ and $5 \le n(1 - p)$.

A word of warning is needed: If n is small, it is not valid to replace a z critical value by a t critical value. A completely different approach is required, and you should consult a statistician for further information.

EXAMPLE 8

The 1983 Tylenol poisoning episode focused attention on the desirability of packaging various commodities in a tamper-resistant manner. The article "Tamper-Resistant Packaging: Is It Really?" (*Package Engr.* (June 1983): 96–104) reported the results of a survey dealing with consumer attitudes toward such packaging. One question asked of the sample of 270 consumers was, "Would you be willing to pay extra for tamper-resistant packages?" The number of yes responses was 189. Let π denote the proportion of all consumers who would pay extra for such packaging. A point estimate of π is $p = 189/270 = .700$. Since $np = (270)(.700) = 189$ and $n(1-p) = (270)(.300) = 81$ are both at least 5, the large-sample interval can be used. A 95% confidence interval for π is then

$$.700 \pm (1.96) \cdot \sqrt{(.700)(.300)/270} = .700 \pm (1.96)(.028) = .700 \pm .055 = (.645, .755)$$

We can be 95% confident that π is between .645 and .755, or equivalently, that between 64.5% and 75.5% of all consumers would pay extra for tamper-resistant packaging.

This result should be encouraging to those manufacturers who want to pass on extra packaging costs to the consumer. There is a caveat, though: People often do not do what they say they will, especially when money is involved. The state of California carried out a survey asking whether people would pay an extra \$25 for reflectorized license plates. The survey results prompted the state to manufacture a large number of reflectorized plates, but sales have been quite disappointing. ■

Choosing the Sample Size

In the estimation of μ using a large sample, $1.96\sigma/\sqrt{n}$ is called the bound on the error of estimation associated with a 95% confidence level because for 95% of all samples, $\bar{x}$ is within this amount of μ. Similarly, when p is used to estimate π, $1.96\sqrt{\pi(1-\pi)/n}$ is the bound on the error of estimation—in 95% of all samples, p is within this amount of π.

Suppose now that an investigator wishes to estimate π to within an amount B (the specified error bound) with 95% confidence. Then the sample size n should be chosen to satisfy $B = 1.96\sqrt{\pi(1-\pi)/n}$. Solving this equation for n results in $n = \pi(1-\pi)(1.96/B)^2$. But there is a problem with this formula—using it requires a value of π, which is, of course, unknown. A conservative approach relies on the fact that $\pi(1-\pi)$ can be at most .25 (its value when $\pi = .5$). Using the maximum value .25 for $\pi(1-\pi)$ yields $n = .25(1.96/B)^2$. This value of n ensures that p will be within B of π for (at least) 95% of all samples irrespective of the value of π.

EXAMPLE 9

The article "What Kinds of People Do Not Use Seat Belts" (*Amer. J. Public Health* (1977):1043–49) reported on a survey whose objective was to study characteristics of drivers and seat belt usage. Let π denote the proportion of all drivers of cars with seat belts who use them. A sample of $n = 1024$ drivers using cars equipped with seat belts was selected. Of the 1024 drivers in the sample, only 264 were regular seat-belt users, so the estimate of π is $p = 264/1024 = .258$.

Suppose that at the outset of the study, the investigators wished to estimate π to within an amount .02 with 95% confidence, whatever the value of π. Then the required sample size is $n = .25(1.96/.02)^2 = .25(9604) = 2401$, so the sample size should have been more than twice what it was. On the other hand, with $B = .05$, $n = 384$, so the sample size used was more than sufficient to estimate π to within an amount .05. ■

EXERCISES

7.31 The use of the interval

$$p \pm (z \text{ critical value}) \cdot \sqrt{\frac{p(1-p)}{n}}$$

requires a large sample. For each of the following combinations of n and p, indicate whether the given interval would be appropriate.

a. $n = 50$ and $p = .30$
b. $n = 50$ and $p = .05$
c. $n = 15$ and $p = .45$
d. $n = 100$ and $p = .01$
e. $n = 100$ and $p = .70$
f. $n = 40$ and $p = .25$
g. $n = 60$ and $p = .25$
h. $n = 80$ and $p = .10$

7.32 In order to estimate the proportion of students at a particular university who favor the sale of beer on campus, a random sample of 100 students was selected. Of the selected students, 43 supported the sale of beer. Let π denote the true proportion of the university's students who favor the sale of beer on campus. Estimate π using a 90% confidence interval. Does the width of the interval suggest that precise information about π is available? Why or why not?

7.33 Studies have shown that children start giving serious consideration to career choices as early as junior high school. A survey of 150 randomly selected junior high school girls showed that 37 planned careers in a technical field. Construct a 95% confidence interval for π, the proportion of all junior high school girls that are planning careers in a technical field.

7.34 The paper "Television and Human Values: A Case for Cooperation" (*J. of Home Ec.* (Summer 1982):18–23) reported that 48% of U.S. homes had more than one television set.

a. Suppose that the statistic reported in this paper had been based on a random sample of 200 homes. Construct a 95% confidence interval for π, the proportion of all U.S. homes that have more than one television set.

b. The same paper reported that the average viewing time per household was about 6 h per day. If you wanted to construct an interval estimate for the average television viewing time for all U.S. households, would you use the interval

$$p \pm (z \text{ critical value}) \cdot \sqrt{\frac{p(1-p)}{n}}$$

Explain your answer.

7.35 In recent years a number of student deaths have been attributed to participation in high school sports. The paper "Concussion Incidences and Severity in Secondary School Varsity Football Players" (*Amer. J. of Public Health* (1983):1370–75) reported the results of a survey of 3063 high school varsity football players. Each participant was asked to provide information on injuries and illnesses incurred as a result of participation in the 1977 football season. Loss of consciousness due to concussion was reported by 528 players. Use this information to construct a 90%

confidence interval for π, the proportion of all high school football players who suffer loss of consciousness due to concussions.

7.36 The *Los Angeles Times* (January 18, 1984) reported that the number of adjustable rate mortgages is on the rise. A study of several hundred new loans made during the first 6 months of 1983 found 44% to have an adjustable interest rate. Assuming $n = 300$, construct a 90% confidence interval for the proportion of home loans made during the first half of 1983 that had adjustable interest rates.

7.37 The paper "Racial Exclusion in Juries" (*J. of Applied Behavioral Sci.*, (1982): 29–47) examined the criteria used in jury selection. A sample of district clerks and jury commissioners from seven southern states was selected and each participant was asked to indicate reasons for disqualifying a person from jury service. Among reasons mentioned were lack of intelligence, commission of a felony offense, and illiteracy.

a. Of 25 respondents from Alabama, 13 reported that they would eliminate people who were not intelligent from consideration. Use this information to construct a 90% confidence interval for π, the proportion of Alabama jury commissioners and clerks who routinely disqualify those judged as not intelligent.

b. A similar sample of 38 individuals was obtained in Arkansas. Twenty-five said they would disqualify a person who was not intelligent. Construct a 90% confidence interval for the proportion of commissioners and district clerks in Arkansas who would disqualify an unintelligent person.

c. Of the participants from Arkansas, 6 said they would disqualify a person who had been convicted of a felony offense. Let π denote the proportion of jury commissioners and district clerks who disqualify felons from jury duty. Estimate π using a 90% confidence interval.

7.38 Many universities allow courses to be taken on a credit/no credit basis. This encourages students to take courses outside their areas of emphasis without having to worry about the effect on their grade-point averages. The paper "Effects of Transition from Pass/No Credit to Traditional Letter Grade System" (*J. of Exper. Educ.* (Winter 1981/82):88–90) compared student performance under the pass/no credit and letter-grade systems. Sixty-three university professors who had taught under both grading systems participated in the study. Thirty-five responded yes when asked whether they felt that students under the letter grade system performed better on tests and assignments. Construct a 99% confidence interval for π, the proportion of university professors who have taught under both systems and feel that student performance is better under the letter grade system.

7.39 The paper "Worksite Smoking Cessation Programs: A Potential for National Impact" (*Amer. J. of Public Health* (1983):1395–96) investigated the effectiveness of smoking cessation programs that appeal to *all* smokers at a particular worksite and not just those who have expressed an interest in giving up smoking. The program tested involved group meetings and monetary incentives for attending meetings and for not smoking. Of those who chose to participate in the experiment, 91% successfully stopped smoking and were still abstinent 6 months later. Suppose that 70 people were involved in the experiment and that these 70 are considered to be a sample of all people who participate in such a program. Let π denote the success rate (proportion who are still nonsmokers 6 months after completing the program) for this program. Find a 99% confidence interval for π.

7.40 Nesting ecology and reproduction of painted turtles was the subject of the paper "Nesting Frequency and Success: Implications for the Demography of Painted Turtles" (*Ecology* (1981):1426–32). In an effort to determine the proportion of female painted turtles that reproduce in a given year, 120 adult females were observed during the nesting season. Of the 120 studied, 31 were judged to be reproductive.

Estimate π, the proportion of adult female painted turtles that reproduce during a particular nesting season, using a 95% confidence interval.

7.41 The *Los Angeles Times* reported that of 671 dog bites reported to authorities during 1983, German shepherds were responsible for 284. Fifty-one of the bites were attributed to Doberman pinschers.

a. Based on these statistics, would you conclude that German shepherds are more vicious than Dobermans? Why or why not? Would you want any additional information before drawing a conclusion?

b. Would you use this information and the methods of this section to estimate the proportion of *all* dog bites that can be attributed to German shepherds? Why or why not?

7.42 The paper "Effects of the Islamic Revolution in Iran on Medical Education: The Shiraz University School of Medicine" (*Amer. J. of Public Health* (1983):1400) reported that of the 173 full-time faculty employed at the Shiraz Medical School in 1978, 108, or 63%, had left the university by the end of 1982. Define the population of interest for this study. Is the proportion given here (.63) a statistic or a population characteristic? Explain your answer.

7.43 A survey of 1010 Californians found that 750 favored development of a satellite system to defend the United States against a nuclear missile attack (*San Luis Obispo Telegram-Tribune,* March 14, 1984). Let π denote the proportion of all Californians who favored development of a satellite defense system at the time of the survey.

a. Estimate π using a 90% confidence interval.

b. If π were to denote the proportion of all U.S. citizens who favored a satellite defense system, would you use the interval computed in (a) as an estimate of π? Explain.

7.44 In an attempt to estimate the proportion of customers who paid by credit card, a small business examined 100 transactions and found that 39 were credit card purchases. The true proportion of customers paying by credit card, π, was then estimated as .39, and a bound on the error of estimation was stated to be .049.

a. How was the stated bound on the error of estimation obtained?

b. How would you interpret the bound on the error of estimation of .049?

7.45 A consumer group is interested in estimating π, the proportion of packages of ground beef sold at a particular store that have an actual fat content exceeding the fat content stated on the label. How many packages of ground beef should be tested in order to estimate π to within .05 with 95% confidence?

7.46 A manufacturer of small appliances purchases plastic handles for coffee pots from an outside vendor. If a handle is cracked, it is considered defective and must be discarded. A very large shipment of plastic handles is received. The proportion of defective handles, π, is of interest. How many handles from the shipment should be inspected in order to estimate π to within .1 with 95% confidence?

7.47 Cornell University's Cooperative Education Department conducted a study of soda consumption among children (*Consumers' Research* (November 1983)). An estimate of the proportion of children under age 3 who drink soda at least once every 3 days was desired. How large a sample should be selected in order to obtain an estimate that is within .05 of the true value with 95% confidence?

7.48 The formula given in the text for computing the sample size necessary to estimate π to within an amount B has an associated confidence level of 95%. How would you modify the formula to obtain the sample size required for 99% confidence? Will the value of n for 99% confidence be larger than for 95% confidence (based on the same B)? Explain.

KEY CONCEPTS

Confidence interval (interval estimate) for a population characteristic   An interval of plausible values for the characteristic computed from sample data. (p. 273)

Confidence level   The percentage of samples that result in an interval that captures the value of the population characteristic. Traditional confidence levels are 95%, 99%, and 90%. (p. 276)

Large-sample confidence interval for μ $\bar{x} \pm (z \text{ critical value})(s/\sqrt{n})$. (If σ is known, it should be used in place of s.) (p. 277)

t distributions   Distribution whose density curves are bell-shaped, centered at 0, and are more spread out than the z curve. Each different t distribution corresponds to a different number of degrees of freedom. (p. 285)

Small-sample confidence interval for the mean μ of a normal population $\bar{x} \pm (t \text{ critical value})(s/\sqrt{n})$, where the t critical value is based on $n - 1$ df. (p. 287)

Large-sample confidence interval for a population proportion π $p \pm (z \text{ critical value}) \sqrt{p(1-p)/n}$. (p. 293)

SUPPLEMENTARY EXERCISES

7.49 Television advertisers are becoming concerned over the use of video cassette recorders (VCR's) to tape television shows, since many viewers fast-forward through the commercials when viewing the taped shows. A survey conducted by A. C. Nielsen Co. of 1100 VCR owners found that 715 used the fast-forward feature to avoid commercials on taped programs (*Los Angeles Times,* September 2, 1984). Construct and interpret a 95% confidence interval for the proportion of all VCR owners who use the fast-forward feature to avoid advertisements.

7.50 Stock researcher Norman Fosback published a study assessing the effect of stock tips that appeared in the *Wall Street Journal.* A sample of companies receiving favorable mention showed a mean 1-day price increase of 5.5 points (*Los Angeles Times,* April 30, 1984). Suppose the sample consisted of 100 observations and that the sample standard deviation was 3.6. Construct a 95% confidence interval for μ, the true mean 1-day increase of companies receiving positive mention.

7.51 The paper "The Market for Generic Brand Products" (*J. Marketing* (1984): 75–83) reported that in a random sample of 1442 shoppers, 727 purchased generic brands. Estimate the true proportion of all shoppers who purchase generic brands using a 99% confidence interval.

7.52 The effectiveness of various drugs used to treat horses was discussed in the paper "Factors Involved in the Choice of Routes of Administration of Antimicrobial Drugs" (*J. Amer. Vet. Assoc.* (1984):1076–82). One characteristic of interest is the half-life of a drug (the length of time until the concentration of the drug in the blood is one-half the initial value). Given below are the reported values of the sample size and the half-life sample mean and standard deviation for three drugs under study.

Drug	n	Sample Mean	Sample Standard Deviation
Gentamicin	7	1.85 h	.231 h
Trimethoprin	6	3.16 h	.845 h
Sulfadimethoxine	6	10.62 h	2.560 h

a. Construct a 90% confidence interval for the mean half-life of gentamicin. Is this confidence interval valid whatever the distribution of half-lives? Explain.

b. Construct individual 90% confidence intervals for the average half-life of trimethoprin and sulfadimethoxine.

c. Interpret each of the intervals in (a) and (b). Do any of the intervals overlap? If a shorter half-life is desirable (since it would indicate quicker absorption of the drug), based on your confidence intervals would you be able to recommend one of the three drugs over the others? Explain.

7.53 Each year as Thanksgiving draws near, inspectors from the Department of Weights and Measures weigh turkeys randomly selected from grocery store freezers to see if the marked weight is accurate. The *San Luis Obispo Telegram-Tribune* (November 22, 1984) reported that of 1000 birds weighed, 486 required remarking. Estimate π, the true proportion of all frozen turkeys whose marked weight is incorrect, using a 95% confidence interval. Based on your interval, do you think that it is plausible that more than half of all frozen turkeys are incorrectly marked? Explain.

7.54 About 14% of all dogs suffer from an infection of the urinary tract sometime during their lifetime. The preferred treatment for an infection caused by the urinary bacteria *pseudomonas* is tetracycline. In a study of healthy adult dogs who received a daily dose of tetracycline of 55 mg/kg body weight, the mean concentration of tetracycline in the urine was 138 μg/mL and the sample standard deviation was 65 μg/mL. (*Source:* "Therapeutic Strategies Involving Antimicrobial Treatment of the Canine Urinary Tract," *J. Amer. Vet. Assoc.* (1984):1162–64). Suppose the mean and standard deviation given had been computed using a sample of $n = 10$ observations. Construct and interpret a 95% confidence interval for the mean concentration of tetracycline in the urine of dogs receiving tetracycline (55 mg/kg body weight) daily.

7.55 The paper "Chlorinated Pesticide Residues in the Body Fat of People in Iran" (*Environ. Research* (1978):419–22) summarized the results of an Iranian study of a sample of $n = 170$ tissue specimens. It was found that the sample mean DDT concentration and sample standard deviation were 8.13 ppm and 8.34 ppm, respectively.

a. Construct and interpret a 95% confidence interval for μ, the true mean DDT concentration.

b. If the above summary data had resulted from a sample of only $n = 15$ specimens, do you think a t interval would have been appropriate? Explain.

7.56 In a survey of 1515 people, 606 said they thought that autoworkers were overpaid (Associated Press, August 15, 1984). Treating the 1515 people as a random sample of the American public, use a 90% confidence interval to estimate the true proportion of Americans who think autoworkers are overpaid.

7.57 The effect of anaesthetic on the flow of aqueous humour (a fluid of the eye) was investigated in the paper "A Method for Near-Continuous Determination of Aqueous Humour Flow: Effects of Anaesthetics, Temperature and Indomethacin" (*Exper. Eye Research* (1984):435–53). Summary quantities for aqueous flow rate (μL/min) observed under three different anaesthetics are given.

Anaesthetic	n	Mean Flow Rate	Standard Deviation
Pentobarbitol	191	.99	.235
Urethane	13	1.47	.314
Ketamine	16	.99	.164

a. Construct a 95% confidence interval for the true mean flow rate when under the effects of pentobarbitol.

b. Construct a 95% confidence interval for the true mean flow rate under urethane. Give two reasons why this interval is wider than that in (a).

c. Construct a 95% confidence interval for the true mean flow rate under the anaesthetic ketamine. Note that the sample mean was the same for the pentobarbitol and the ketamine samples and yet the corresponding 95% confidence intervals are different. What factors contribute to this difference? Explain.

7.58 In a random sample of 31 inmates selected from residents of the prison in Angola, Louisiana ("The Effects of Education on Self-Esteem of Male Prison Inmates," *J. Correctional Educ.* (1982):12–18), 25 were Caucasian. Use a 90% confidence interval to estimate the true proportion of inmates (at this prison) who are Caucasian.

7.59 When n is large, the statistic s is approximately unbiased for estimating σ and has approximately a normal distribution. The standard deviation of this statistic when the population distribution is normal is $\sigma_s \approx \sigma/\sqrt{2n}$, which can be estimated by $s/\sqrt{2n}$. A large-sample confidence interval for the population standard deviation σ is then

$$s \pm (z \text{ critical value}) \cdot \frac{s}{\sqrt{2n}}$$

Use the data of Exercise 7.57 to obtain a 95% confidence interval for the true standard deviation of flow rate under pentobarbitol.

7.60 The interval from -2.33 to 1.75 captures area .95 under the z curve. This implies that another large-sample 95% confidence interval for μ has lower limit $\bar{x} - (2.33)s/\sqrt{n}$ and upper limit $\bar{x} + (1.75)s/\sqrt{n}$. Would you recommend using this 95% interval over the 95% interval $\bar{x} \pm (1.96)s/\sqrt{n}$ discussed in the text? Explain. (*Hint:* Look at the width of each interval.)

7.61 Suppose that an individual's morning waiting time for a certain bus is known to have a uniform distribution on the interval from 0 min to an unknown upper limit θ min. A 95% confidence interval for θ based on a random sample of n waiting times can be shown to have lower limit $\max(x)$ and upper limit $\max(x)/(.05)^{1/n}$. If $n = 5$ and the resulting waiting times are 4.2, 3.5, 1.7, 1.2, and 2.4, obtain the confidence interval. (*Hint:* $(0.5)^{1/5} = .5493$.) Notice that the confidence interval here is not of the form (estimate) $\pm$ (critical value)(standard deviation).

REFERENCES

Again, the books by Freedman et. al. and by Moore listed in the Chapter 1 references contain very informal and lucid discussions of confidence intervals at a level comparable to that of this text.

8 Hypothesis Testing Using a Single Sample

INTRODUCTION

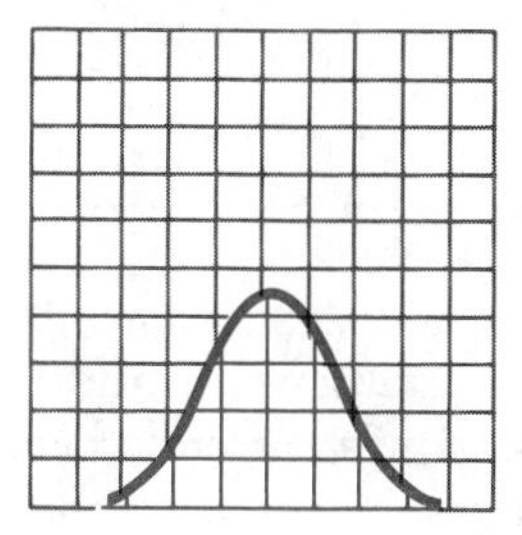

A HYPOTHESIS is a claim or statement either about the value of a single population characteristic or about the values of several characteristics. One example of a hypothesis is the claim $\mu = 100$, where μ is the average IQ for all first-born children. Another example is the statement $\pi > .1$, where π is the proportion of all television sets of a certain brand that need repair while under warranty. The statements $\bar{x} > 110$ and $p = .15$ are not hypotheses because neither $\bar{x}$ nor p is a population characteristic (each is a statistic whose value changes from sample to sample, while the values of μ and π are fixed but usually unknown). As another example of a hypothesis, let μ_1 denote the true average breaking strength of one type of twine and μ_2 denote the true average breaking strength of a second type of twine. Then the statement $\mu_1 - \mu_2 = 0$ (equivalently, $\mu_1 = \mu_2$) is one possible hypothesis, and another is that $\mu_1 - \mu_2 > 5$ (this latter claim states that μ_1 exceeds μ_2 by more than 5).

In any hypothesis-testing problem, there are two contradictory hypotheses under consideration. One hypothesis might be $\mu = 100$ and the other $\mu \neq 100$, or one might be $\pi = .1$ and the other $\pi > .1$. The objective is to decide, based on sample information, which of the two hypotheses is correct. In this chapter we focus on testing hypotheses about a population mean μ and about a population proportion π. Decision procedures are based on a random sample selected from the population of interest. Subsequent chapters present methods for testing hypotheses about characteristics of two or more populations.

8.1 Hypotheses and Test Procedures

A familiar situation in which a choice between two contradictory claims must be made is a criminal trial. The person accused of a crime must be judged either innocent or guilty. Under the American system of justice, the individual on trial is initially presumed innocent. Only strong evidence to the contrary will cause the innocence claim to be rejected in favor of a guilty verdict. The burden of proof is thus put on the prosecution to prove the guilty claim. The French perspective in criminal proceedings is the opposite of ours. There, once enough evidence has been presented to justify bringing an individual to trial, the initial assumption is that the accused is guilty. The burden of proof then falls on the accused to establish innocence.

A **test of hypotheses** is a method for deciding which of the two contradictory claims (hypotheses) is the correct one. As in a judicial proceeding, we shall initially assume that a particular one of the two hypotheses is the correct one. In carrying out a test, this claim will then not be rejected in favor of the second (alternative) claim unless sample evidence is very incompatible with the initial assumption.

DEFINITION

The **null hypothesis,** denoted by H_0, is the claim that is initially assumed to be true. The other hypothesis is referred to as the **alternative hypothesis** and is denoted by H_a. In carrying out a test of H_0 versus H_a, H_0 will be rejected in favor of H_a only if sample evidence strongly suggests that H_0 is false. If the sample does not contain such evidence, H_0 will not be rejected. The two possible conclusions are then *reject* H_0 and *fail to reject* H_0.

EXAMPLE 1

Consider a machine that produces ball bearings. Because of variation in the machining process, bearings produced by this machine do not have identical diameters. Let μ denote the true average diameter for bearings currently being produced. Suppose that the machine was initially calibrated to achieve the design specification $\mu = .5$ in. However, the manufacturer is now concerned that diameters no longer conform to this specification. That is, the hypothesis $\mu \neq .5$ must now be considered a possibility. If sample evidence suggests that $\mu \neq .5$, the production process would have to be halted while recalibration takes place. Because this is costly, the manufacturer wants to be quite sure that $\mu \neq .5$ before undertaking recalibration. Under these circumstances, it is sensible to select the null hypothesis as $H_0 : \mu = .5$ (the specification is being met, so recalibration is unnecessary) and the alternative hypothesis as $H_a : \mu \neq .5$. Only compelling sample evidence would then result in H_0 being rejected in favor of H_a. ■

EXAMPLE 2

A pack of a certain brand of cigarettes displays the statement "1.5 mg nicotine average per cigarette by FTC method." Let μ denote the average nicotine content per cigarette for all cigarettes of this brand. Then the advertised claim is that $\mu = 1.5$. People who smoke this brand would probably be unhappy if it turned out that μ exceeded the advertised value. Suppose a sample of cigarettes of this brand is selected, and the nicotine content of each cigarette is determined. The sample results can then be used to test the hypothesis $\mu = 1.5$ against the hypothesis $\mu > 1.5$. The accusation that the company is understating average nicotine content is a serious one, and it is reasonable to require compelling sample evidence before concluding that $\mu > 1.5$. This suggests that the claim $\mu = 1.5$ should be selected as the null hypothesis and $\mu > 1.5$ as the alternative hypothesis. Then $H_0 : \mu = 1.5$ would be rejected in favor of $H_a : \mu > 1.5$ only when sample evidence strongly suggests that the initial assumption $\mu = 1.5$ is no longer tenable. ■

Because the alternative of interest in Example 1 was that $\mu \neq .5$, it was natural to state H_0 as the equality claim $\mu = .5$. However, the alternative hypothesis in Example 2 was stated as $\mu > 1.5$ (true average nicotine content exceeds the advertised level), from which it might seem more reasonable to state H_0 as $\mu \leq 1.5$ rather than as $\mu = 1.5$. After all, the average level might actually be less than what the company advertises! Suppose, though, that sample evidence leads to the rejection of $\mu = 1.5$ in favor of the claim $\mu > 1.5$. Then, intuitively, the sample would offer even less support to values of μ smaller than 1.5 when compared to the claim $\mu > 1.5$. Thus explicitly testing $H_0 : \mu = 1.5$ is equivalent to implicitly testing the null hypothesis $\mu \leq 1.5$. We have chosen to state a null hypothesis as an equality claim.

The form of a null hypothesis is

$$H_0 : \text{population characteristic} = \text{hypothesized value}$$

where the hypothesized value is a specific number determined by the problem context. The alternative hypothesis then has one of the following three forms:

$$H_a : \text{population characteristic} > \text{hypothesized value}$$
$$H_a : \text{population characteristic} < \text{hypothesized value}$$
$$H_a : \text{population characteristic} \neq \text{hypothesized value}$$

Thus we might test $H_0 : \pi = .1$ versus $H_a : \pi < .1$ but not consider testing $H_0 : \mu = 50$ versus $H_a : \mu > 100$. The number appearing to the right of the inequality in the alternative hypothesis must be identical to the hypothesized value in H_0.

We previously noted that the American and French judicial systems operate from different perspectives when it comes to the initial presumption of innocence or guilt. Similarly, the selection of a null hypothesis—the claim that is initially assumed true—sometimes depends on the viewpoint of an investigator.

EXAMPLE 3

A customer is considering the purchase of many components of a certain type from a particular manufacturer and so is concerned about the long-run percentage of defective components. After reflection, the customer decides that 10% is the dividing line between acceptable and unacceptable defective rates. Let π denote the true proportion of this manufacturer's components that are defective. The manufacturer may be the only one currently making this component or may have offered the customer favorable purchase terms. In that case, the customer would want to purchase from this manufacturer unless sample evidence strongly suggested an unacceptable defective rate. It would then be sensible to test $H_0 : \pi = .1$ versus $H_a : \pi > .1$, so that the alternative hypothesis is identified with an unacceptable defective percentage.

On the other hand, the customer might wish to place the burden of proof on the manufacturer to show that its defective rate is acceptable. This would suggest testing $H_0 : \pi = .1$ versus $H_a : \pi < .1$, with the alternative now stating that the defective rate is acceptable. The purchase from this manufacturer would then not be made unless sample evidence strongly suggested rejecting H_0 in favor of H_a. ■

Test Statistics and Procedures

Once H_0 and H_a have been formulated, a **test procedure** (method for making a decision) is based on a sample selected from the population under investigation. Recall that a statistic is any quantity whose value can be computed from sample data. Carrying out a test requires that we first select a particular statistic to serve as our decision maker. The decision as to whether or not H_0 should be rejected then depends on the extent to which the value of this statistic computed from the sample is consistent with H_0. Generally speaking, if the computed value is very different from what would be expected when H_0 is true, rejection of H_0 is appropriate. However, if the computed value is one that might reasonably have resulted when H_0 is true, then there is no strong reason to reject H_0 in favor of H_a.

EXAMPLE 4

(*Example 2 continued*) With μ denoting the true average nicotine content for cigarettes of the brand being studied, the hypotheses to be tested are $H_0 : \mu = 1.5$ versus $H_a : \mu > 1.5$. For ease of exposition, let's assume that the population standard deviation of nicotine content is known from previous studies to be $\sigma = .20$ mg. (As in confidence interval problems, this assumption is almost always unrealistic; situations in which this assumption is unnecessary are considered shortly.) A random sample of $n = 36$ cigarettes is selected and the nicotine content of each cigarette is determined, resulting in sample observations $x_1, x_2, \ldots, x_{36}$. Suppose that the sample average nicotine content is then computed, yielding $\overline{x} = \Sigma x/36 = 1.65$ mg.

We already know that $\overline{x}$ is an unbiased statistic for estimating μ, so its computed value should be quite helpful in deciding whether to reject H_0. When H_0 is true, the mean, or expected, value of $\overline{x}$ is $\mu_{\overline{x}} = \mu = 1.5$, whereas when H_0 is false and H_a is true, the value of $\overline{x}$ can be expected to exceed 1.5. However, because of sampling variability, the observed value of $\overline{x}$ usually differs somewhat from its mean value μ. Thus a value of $\overline{x}$ just a bit larger than 1.5 would not be incompatible with H_0. Only an $\overline{x}$ value that greatly exceeds 1.5 would cast substantial doubt on the validity of H_0 and suggest accepting H_a instead.

The computed value of $\overline{x}$, 1.65, certainly exceeds 1.5, but does it do so by a large enough amount to justify rejecting H_0? The answer is highly dependent on the amount of variability in the nicotine content distribution. If the population standard deviation σ is large, then $\sigma_{\overline{x}} = \sigma/\sqrt{n}$ will also be large. In this case an $\overline{x}$ value rather far from μ would not be very unusual. So when H_0 is true but $\sigma_{\overline{x}}$ is large, a value as much above 1.5 as is our value of $\overline{x}$ would not raise many eyebrows. On the other hand, with $\sigma_{\overline{x}}$ small, the value

1.65 would be many standard deviations above what would be expected were H_0 true.

It is easier to take explicit account of variability in reaching a decision if a standardized statistic rather than $\bar{x}$ itself is used as a decision maker. The sample size is large here, so by the central limit theorem, $\bar{x}$ has approximately a normal distribution. The standard deviation of $\bar{x}$ is $\sigma_{\bar{x}} = .20/\sqrt{36} = .0333$, and when H_0 is true, $\mu_{\bar{x}} = 1.5$. Standardizing $\bar{x}$ assuming that H_0 is true yields the standardized statistic

$$z = \frac{\bar{x} - \text{hypothesized value}}{\text{standard deviation of } \bar{x}} = \frac{\bar{x} - 1.5}{.0333}$$

This statistic expresses the distance between $\bar{x}$ and its expected value when H_0 is true as a number of standard deviations. An $\bar{x}$ substantially larger than 1.5 yields a large positive value of z, so such values of z are more consistent with H_a than with H_0. In particular, when H_0 is true, the sampling distribution of z is approximately the standard normal curve, so a z value much above 2 would be quite unusual. The value of z for our sample is $(1.65 - 1.5)/.0333 = 4.5$; $\bar{x}$ fell 4.5 standard deviations above what would have been expected under H_0! This z value is so extreme that it argues overwhelmingly for the rejection of H_0 in favor of H_a.

Suppose that $\bar{x} = 1.54$ rather than 1.65. Then $z = (1.54 - 1.5)/.0333 = 1.2$, so the value of $\bar{x}$ is only 1.2 standard deviations ($1.2\sigma_{\bar{x}}$) above what would have been expected were H_0 true. From Table I in the appendices, a value at least this far out in the upper tail of the z curve would occur with probability .115. That is, if H_0 were true, roughly 11.5% of all samples would result in z being at least as large as 1.2. Because .115 is not very small, we feel that $z = 1.2$ is not unusual enough to be inconsistent with H_0, so we would not reject H_0 for this value of z. ■

The decisions for $z = 4.5$ and $z = 1.2$ in Example 4 seemed quite clear. But what if $\bar{x}$ is such that $z = 1.8$? With H_0 true, the approximate sampling distribution of z is again the standard normal curve. A value at least as large as 1.8 would occur with probability .036 (in 3.6% of all samples). Some might think that the smallness of .036 indicates an unusual enough occurrence to justify rejecting H_0 in favor of H_a, whereas others would not. As a decision maker, though, you must have a rule that for each possible value of z says whether or not H_0 should be rejected. We decided in Example 4 that H_0 should be rejected if z is "sufficiently large." Thus we need to specify a cutoff value c and agree to reject H_0 if z exceeds c ($z > c$) and not reject H_0 otherwise. For example, if the cutoff 2.33 is used (we see shortly why this might be reasonable), then H_0 would be rejected if $z > 2.33$ and not rejected if $z \leq 2.33$. The inequality $z > 2.33$ specifies what is called the *rejection region* for this test procedure. The rejection region consists of all values of the decision-making statistic for which H_0 would be rejected. If the rejection region $z > 2.33$ is used and $\bar{x} = 1.56$ in Example 4, the resulting value of z is $(1.56 - 1.50)/.0333 = 1.8$. Because $z = 1.8$ is not in the rejection region (1.8 does not exceed 2.33), H_0 is not rejected. If $z = 3.0$ (from $\bar{x} = 1.60$), then use of the rejection region $z > 2.33$ would result in the rejection of H_0 because 3.0 does fall in the rejection region.

DEFINITION

A test procedure is determined by both a **test statistic** and a **rejection region.** The test statistic is the sample-based quantity used in making a decision. The rejection region consists of all values of the test statistic for which H_0 would be rejected in favor of H_a.

EXAMPLE 5

In Example 1, the population characteristic of interest is μ = the current true average diameter of ball bearings. The hypotheses to be tested are $H_0 : \mu = .5$ versus $H_a : \mu \neq .5$. Intuitively, an $\overline{x}$ value either much less than or much greater than .5 would suggest rejecting H_0. Suppose that the value of σ is known to be .006, and a sample of $n = 40$ bearings is obtained. Then $\sigma_{\overline{x}} = \sigma/\sqrt{n} = .006/\sqrt{40} = .000949$ and, when H_0 is true, $\mu_{\overline{x}} = \mu = .5$. Consider the test statistic obtained by standardizing $\overline{x}$, assuming that H_0 is true:

$$z = \frac{\overline{x} - \text{hypothesized value}}{\text{standard deviation of } \overline{x}} = \frac{\overline{x} - .5}{.000949}$$

The test statistic is labeled as z because when H_0 is true, its sampling distribution is described approximately by the standard normal (z) curve (the central limit theorem again). A value of $\overline{x}$ far above or below .5 corresponds to a value of z that is extreme and positive or extreme and negative. Therefore, it makes intuitive sense to specify a positive cutoff value c and use a rejection region of the form

$$\text{reject } H_0 \quad \text{either if} \quad z > c \quad \text{or if} \quad z < -c$$

We shall shortly see that a reasonable choice for c is 1.96. This gives as the rejection region either $z > 1.96$ or $z < -1.96$. If $\overline{x} = .503$, then $z = (.503 - .500)/.000949 = 3.16$. This value of z falls in the rejection region ($3.16 > 1.96$), so H_0 would be rejected in favor of the conclusion that μ is something other than .5. Similarly, $\overline{x} = .497$ results in $z = -3.16$, which is also in the rejection region ($-3.16 < -1.96$). The value $\overline{x} = .499$ results in $z = (.499 - .500)/.000949 = -.001/.000949 = -1.05$. Because -1.05 is neither greater than 1.96 nor less than -1.96, it does not fall in the rejection region. Thus $\overline{x} = .499$ would not lead to rejection of H_0. ■

EXERCISES

8.1 Explain why the statement $\overline{x} = 17$ is not a legitimate hypothesis.

8.2 For the following pairs, indicate which don't comply with our rules for setting up hypotheses and explain why.

a. $H_0 : \mu = 15, H_a : \mu = 15$

b. $H_0 : \pi = .4, H_a : \pi > .6$

c. $H_0 : \mu = 123, H_a : \mu < 123$

d. $H_0 : \mu = 123, H_a : \mu = 125$

e. $H_0 : p = .1, H_a : p = .1$

8.3 In order to determine whether the pipe welds in a nuclear power plant meet specifications, a random sample of welds is selected, and tests are conducted on each weld in the sample. Weld strength is measured as the force required to break the weld. Suppose that the specifications state that mean strength of welds should exceed 100 lb/in.^{2}. The inspection team decides to test $H_0 : \mu = 100$ versus $H_a : \mu > 100$. Explain why it might be preferable to use this H_a rather than $\mu < 100$.

8.4 Microwave ovens work by emitting high-frequency electromagnetic waves similar to radio waves. Since a microwave oven is a small broadcasting system, the Federal Communications Commission regulates the frequency of microwaves. The average wavelength for microwaves is supposed to be 4.81 in. (radio waves are much longer). Suppose the FCC decides to randomly select 15 ovens made by a particular manufacturer and determine the frequency of the microwaves in each one. What hypotheses should be tested using the resulting data if the FCC is interested in ascertaining whether or not the manufacturer's ovens are broadcasting at the correct frequency?

8.5 A consumer advocacy group has received complaints from a number of people who felt that a particular cereal manufacturer was underfilling its boxes. The consumer group decides to investigate this complaint and will file a false advertising suit if there is sufficient evidence to suggest that the average weight of boxes claiming to contain 12 oz is in fact less than 12 oz. What hypotheses should the consumer group test? Explain your choice of the null and alternative hypotheses.

8.6 A city council member must decide how to vote on a measure to appropriate funds for road repairs. His fiscal decisions have been criticized in the past and so he decides to take a survey of constituents to find out whether they favor spending money for road repair. He will vote to appropriate funds only if he can be fairly certain that a majority of the people in his district favor the measure. What hypotheses should he test?

8.7 Many older homes have electrical systems that use fuses rather than circuit breakers. A manufacturer of 40-A fuses wants to make sure that the mean amperage at which its fuses burn out is in fact 40. If the mean amperage is lower than 40, customers will complain because the fuses require replacement too often. If the mean amperage is higher than 40, the manufacturer might be liable for damage to an electrical system due to fuse malfunction. In order to verify the amperage of the fuses, a sample of fuses is to be selected and inspected. If a hypothesis test were to be performed on the resulting data, what null and alternative hypotheses would be of interest to the manufacturer?

8.8 A radio station that plays a mix of current hits and oldies (songs from the authors' heyday!) is thinking of changing its programming to consist entirely of oldies. A survey of listeners will be conducted to determine if the change in programming would meet with a favorable response. The conversion will be made if there is evidence that more than 75% of current listeners think the change is a good idea. Give two different pairs of hypotheses that might reasonably be tested in this situation. Which pair would you recommend testing and why?

8.9 In calibrating a scale, a 1-g metal cylinder is weighed 10 times and each observed weight value is recorded.

a. The scale will be recalibrated if there is evidence that the mean value in repeated weighings of the metal cylinder is different from 1 g. What hypotheses should be tested in order to determine whether a scale needs recalibration?

b. Excessive variability in the values obtained in repeated weighings would also be cause for concern, as it would indicate that the scale is not very precise. Suppose that a standard deviation that exceeds .15 g is indicative that the scale needs repair. What hypotheses should be tested to determine if a scale should be sent for repair?

8.10 An automobile manufacturer offers a 50,000-mi extended warranty on its new cars. You plan on buying one of these cars and must decide whether to purchase the extended warranty (the ordinary warranty is for 12,000 mi). Suppose that a recent magazine article has reported the number of miles at which 30 cars made by this manufacturer first needed major repair. This information can be used to conduct a test of hypotheses that will aid you in your decision.

a. What hypotheses would you test in each case?

i. The extended warranty is very expensive.

ii. The extended warranty is not very expensive.

b. Explain your choice of hypotheses in (a).

8.11 Referring to Exercise 8.5, let μ denote the average weight of all boxes. Suppose that $\sigma = .5$ and that $n = 100$ boxes are randomly selected. Then $\sigma_{\bar{x}} = \sigma/\sqrt{n} = .05$, and the test statistic $z = (\bar{x} - 12)/.05$ is appropriate.

a. For testing the appropriate hypotheses, which rejection region would you recommend?

i. $z > 1.645$

ii. $z < -1.645$

iii. either $z > 1.96$ or $z < -1.96$

b. If $\bar{x} = 11.83$, what should be concluded based on the rejection region chosen in (a)?

c. What is the chance that H_0 will be rejected when it is actually true for the rejection region chosen in (a)?

8.12 Let μ denote the true average activation temperature of a certain device. The device has been designed to activate at 130°F. A random sample of $n = 64$ observations is to be selected and used to decide whether the true average activation temperature differs from the design value. Suppose that $\sigma = 4$, so that $\sigma_{\bar{x}} = .5$.

a. What hypotheses should be tested?

b. Consider the test statistic $z = (\bar{x} - 130)/.5$. If $\bar{x} = 128$, what is z and how would you interpret this value? Repeat for $\bar{x} = 130.5$.

c. Propose a reasonable rejection region based on z. Using this region, what would you conclude if $\bar{x} = 128$? If $\bar{x} = 130.5$?

8.13 Let π denote the probability that a toss of a particular coin results in a head (the long-run proportion of heads when the coin is tossed repeatedly).

a. What hypotheses are appropriate if the question of interest is whether or not the coin is fair?

b. Let p denote the sample proportion of heads (successes) resulting from 100 tosses. What are μ_p and σ_p when H_0, the null hypothesis specified in (a), is true?

c. Use the results of (b) to specify a standardized test statistic for testing the coin. What is the value of this statistic if $p = .54$?

d. Propose a reasonable rejection region based on the standardized statistic of (c), and use it to draw a conclusion when $p = .54$.

8.2 Errors in Hypothesis Testing

In the examples considered thus far, the test statistic and form of the rejection region were motivated by inituitive considerations, whereas the choice of the specific cutoff value for the rejection region seemed arbitrary. Understanding more about what is involved in specifying a test procedure requires considering the likelihood of drawing an erroneous conclusion when a particular procedure is employed.

In reaching a judgment on the innocence or guilt of a defendant in a criminal trial, two different types of errors must be considered. The defendant may be found guilty when in fact he or she is innocent, or the defendant may be found innocent even though guilty. Similarly, there are two different types of errors that might be made when making a decision in a hypothesis-testing problem. One type of error involves rejecting H_0 even though H_0 is true. The other type of error involves not rejecting H_0 even though H_0 is false and H_a is true.

DEFINITION

The error of rejecting H_0 when H_0 is true is called a **type I error.** The error in which H_0 is not rejected when it is false and H_a is true is called a **type II error.**

No reasonable test procedure comes with a guarantee that neither type of error will be made; this is the price paid for basing an inference on a sample rather than examining the entire population. With any procedure, there is some chance that a type I error will be made when H_0 is actually true, and also some chance that a type II error will result when H_0 is false.

EXAMPLE 6

For testing $H_0 : \mu = 1.5$ versus $H_a : \mu > 1.5$ in Example 4, we proposed using the test statistic

$$z = \frac{\overline{x} - \text{hypothesized value}}{\text{standard deviation of } \overline{x}} = \frac{\overline{x} - 1.5}{.0333}$$

Consider the rejection region $z > 2.33$. It is easy to check that z will exceed 2.33 precisely when $\overline{x}$ exceeds $1.5 + (2.33)(.0333) \approx 1.578$. Thus if $\overline{x} > 1.578$, H_0 will be rejected, but if $\overline{x} \leq 1.578$, H_0 will not be rejected.

Now suppose that H_0 is true, so that $\mu = 1.5$. Then to reject H_0 would be to commit a type I error. Yet even though $\mu = 1.5$, it is possible that the value of $\overline{x}$ will exceed 1.578, resulting in rejection of H_0 and a type I error. Figure 1(a) illustrates the approximate sampling distribution of $\overline{x}$ when H_0 is true, a normal curve (by the central limit theorem) with mean value 1.5 and standard deviation .0333. The shaded area in the upper tail (to the right of 1.578) represents the probability of a type I error for the test procedure under discussion. That this probability is .01 is a consequence of using the z critical value 2.33 to specify the rejection region. The interpretation of this error probability is that when H_0 is true here, only 1% of all samples will result in a value of z falling in the rejection region—only 1% of all samples will lead to incorrect rejection of H_0.

Let's now investigate the chance of making a type II error. Consider the case in which H_0 is false because $\mu = 1.6$ rather than 1.5. A type II error will result when $\overline{x} \leq 1.578$ (corresponding to $z \leq 2.33$). Figure 1(b) pictures the approximate sampling distribution of $\overline{x}$ in this case. It is a normal curve centered at 1.6 with standard deviation .0333. The probability of a type II error is represented by the shaded area under this curve to the left of 1.578. A straightforward normal probability calculation gives this error probability as the area under the z curve to the left of $(1.578 - 1.6)/.0333 = -.66$. From Table I, this area is approximately .25. Thus when $\mu = 1.6$, roughly 25% of all samples would result in $z \leq 2.33$ and incorrect nonrejection of H_0.

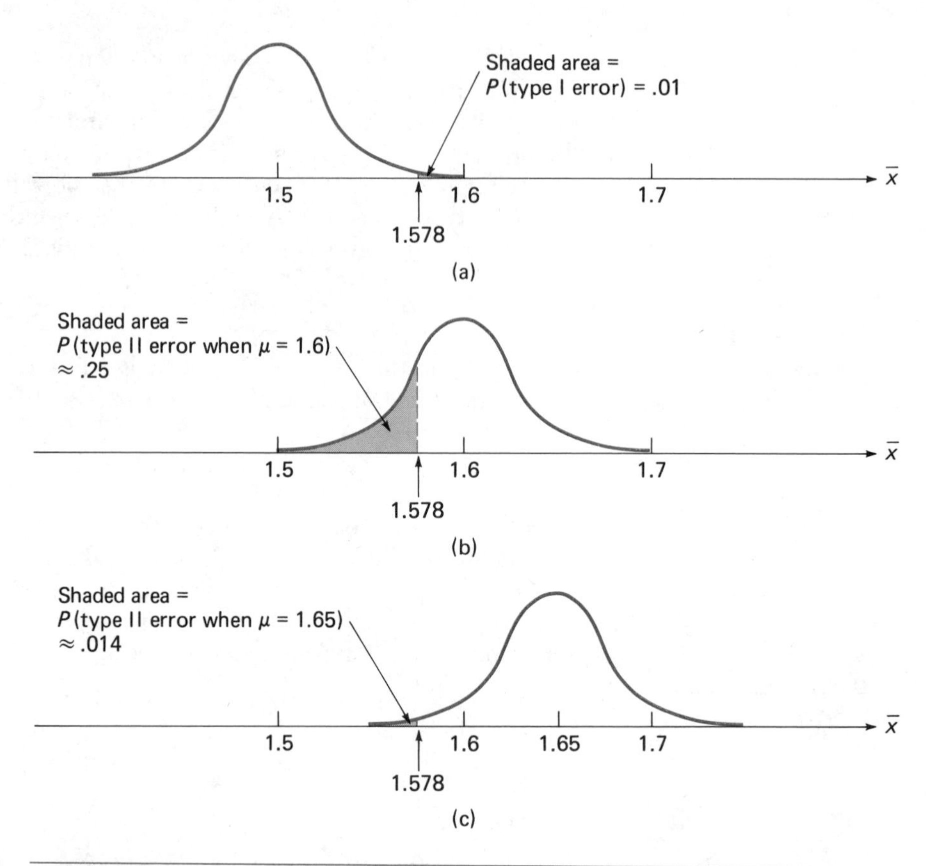

FIGURE 1

THE SAMPLING DISTRIBUTION OF $\overline{x}$ AND ERROR PROBABILITIES FOR EXAMPLE 6
(a) $\mu = 1.5$ (H_0 true)
(b) $\mu = 1.6$ (H_0 false)
(c) $\mu = 1.65$ (H_0 false)

Figure 1(c) shows the probability of a type II error when $\mu = 1.65$. This probability is smaller than for the case $\mu = 1.6$ because $\overline{x} \leq 1.578$ ($z \leq 2.33$) is less likely when $\mu = 1.65$ than when $\mu = 1.60$. Put another way, the chance of a type II error is smaller for a value of μ far from the hypothesized value 1.5 than for a value of μ closer to 1.5. Although we illustrated P(type II error) only for the alternative values $\mu = 1.6$ and $\mu = 1.65$, this probability can also be computed for any other value of μ exceeding 1.5 (i.e., any other value for which H_0 is false and H_a is true). ■

The following notation and terminology for error probabilities is commonly used.

DEFINITION	The probability of a type I error is often denoted by α and is called the **level of significance** of the test. Thus a test with $\alpha = .01$ would be said to have level of significance .01 or to be a level .01 test. The probability of a type II error is frequently denoted by β.

Because our H_0 will always be an equality claim, there will be a single α. The test procedure of Example 6, which used the z critical value 2.33, had $\alpha = .01$. However, there will not be a single β, but rather there will be one for each alternative value of the population characteristic. In Example 6 there is a β for each μ larger than 1.5. We calculated $\beta \approx .25$ when $\mu = 1.6$ and $\beta \approx .014$ when $\mu = 1.65$.

The value $\alpha = .01$ in Example 6 would be small enough to satisfy many decision makers. However, $\beta \approx .25$ when $\mu = 1.6$ might be viewed as rather large (with $\mu = 1.6$, roughly 25% of all samples would lead to a type II error). If it is considered very important to detect such a departure from H_0, this type II error probability would not be very satisfactory. Consider the effect of replacing the z critical value 2.33 in the rejection region by the value 1.645. Then rejecting H_0 if $z > 1.645$ is equivalent to rejecting H_0 if $\bar{x} > 1.5 + (1.645)(.0333) \approx 1.555$. Now look at Figure 1 and see what happens if we slide the $\bar{x}$ cutoff value from 1.578 back to 1.555. Because α is the area to the right of this value under the $\bar{x}$ curve, α increases. However, β is the area to the left of this value, so it will decrease for each μ larger than 1.5. It is easy to check that for this new rejection region, $\alpha = .05$ (which is a consequence of using 1.645), whereas $\beta \approx .09$ when $\mu = 1.6$ and $\beta \approx .002$ when $\mu = 1.65$. Clearly, the use of this new rejection region has improved β, but the price paid for this is an increase in α.

> Once a test statistic has been selected and a sample size fixed, changing the rejection region to yield a decrease in β inevitably results in an increase in α. Only by using a larger sample size can β be made smaller without increasing α.

EXAMPLE 7

The standardized test statistic proposed in Example 5 for testing $H_0 : \mu = .5$ versus $H_a : \mu \neq .5$ was

$$z = \frac{\bar{x} - \text{hypothesized value}}{\text{standard deviation of } \bar{x}} = \frac{\bar{x} - .5}{.000949}$$

We suggested rejecting H_0 either if $z > 1.96$ or if $z < -1.96$ (corresponding to an $\bar{x}$ value that deviates substantially from .5). This is equivalent to rejecting H_0 if either $\bar{x} > .5 + (1.96)(.000949) = .50186$ or $\bar{x} < .5 - (1.96)(.000949) = .49814$. Figure 2 pictures the approximate sampling distribution of $\bar{x}$ both when H_0 is true and when it is false because $\mu = .501$. The type I error probability is the sum of the two shaded tail areas (to the left of .49814 and to the right of .50186) in Figure 2(a): $\alpha = .025 + .025 = .05$. The probability of a type II error is the shaded area in Figure 2(b) between .49814 and .50186 (since that is when H_0 would not be rejected), resulting in $\beta \approx .815$. This error probability is extremely large because with $\sigma = .006$, it is difficult to distinguish between a population with $\mu = .5$ and one with $\mu = .501$ using a sample of size 40. It can also be verified by symmetry that $\beta \approx .815$ when $\mu = .499$, since this alternative value is as far to the left of .5 as .501 is to the right. As Figure 2(c) shows, $\beta \approx .25$ for $\mu = .5025$ (and also for $\mu = .4975$), which is smaller than .815 but certainly not negligible.

As long as the sample size n is fixed at 40, changing the critical value 1.96 to decrease β results in an increase in α. For example, replacing 1.96 by

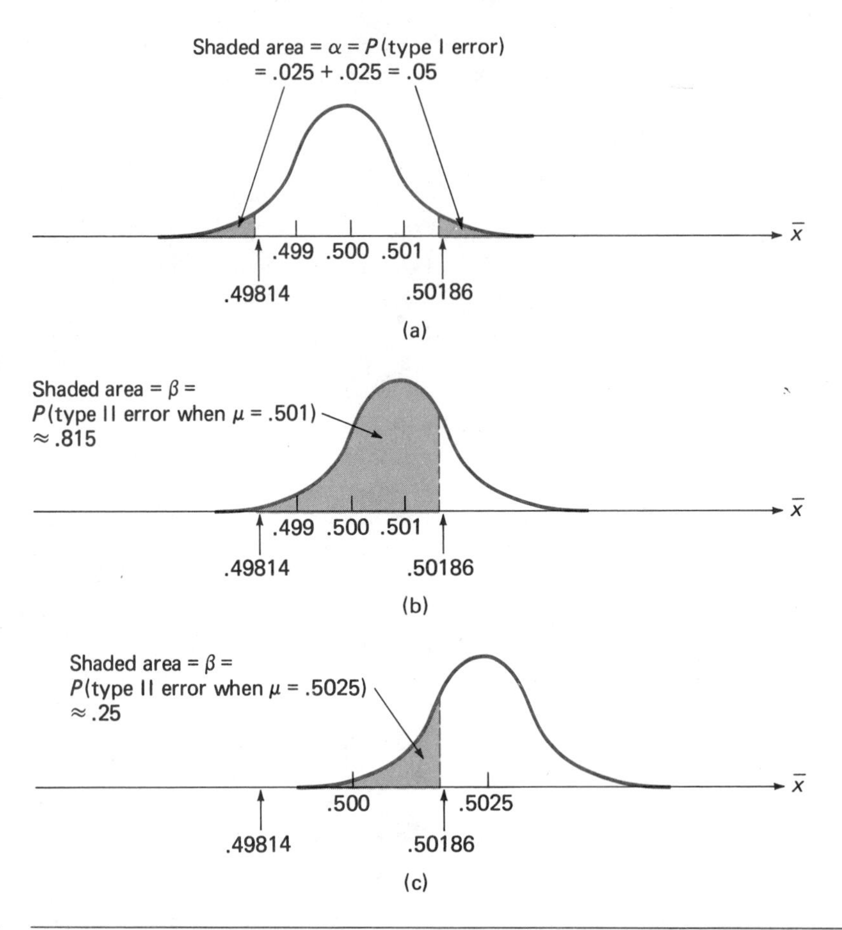

FIGURE 2 THE SAMPLING DISTRIBUTION OF $\bar{x}$ AND ERROR PROBABILITIES FOR EXAMPLE 7
(a) $\mu = .5$ (H_0 true)
(b) $\mu = .501$ (H_0 false)
(c) $\mu = .5025$ (H_0 false)

1.645 decreases β substantially for each μ different from .5. However, α for this new critical value is .10, which might be regarded as intolerably high by some people. ■

A General Principle

The ideal test procedure would have both $\alpha = 0$ and $\beta = 0$. However, once n is fixed, it is not possible to find a procedure (test statistic and rejection region) for which both α and β are arbitrarily small. One must compromise between small α and small β. This leads to the following widely accepted principle for specifying a test procedure.

> After thinking about the consequences of a type I error, identify the largest α that you think is tolerable in your problem. Then employ a test statistic and rejection region that uses this maximum acceptable value—rather than anything smaller—as the level of significance (because using a smaller α increases β).

Thus if you decide that $\alpha = .05$ is tolerable, you should not use a test with $\alpha = .01$ because the smaller α inevitably results in larger β. The values of α most frequently used in practical problems are .05 and .01 (a 1 in 20 or 1 in 100 chance of rejecting H_0 when it is actually true), but the choice in any given problem depends on the seriousness of a type I error in that context.

EXAMPLE 8

A television manufacturer claims that (at least) 90% of its sets will need no service during the first 3 years of operation. A consumer agency wishes to check this claim, so it obtains a random sample of $n = 100$ purchasers and asks each whether or not the set purchased needed repair during the first 3 years. Let $p =$ the sample proportion of responses indicating no repair (so that no repair is identified with a success). Let π denote the true proportion of successes. The agency does not want to claim false advertising unless sample evidence strongly suggests that $\pi < .9$. The appropriate hypotheses are then

$$H_0 : \pi = .9 \quad \text{versus} \quad H_a : \pi < .9$$

Intuitively, a value of p substantially smaller than .9 would suggest rejecting H_0 in favor of H_a. Rather than using p itself as a test statistic, let's standardize p assuming that H_0 is true. In this case, $\mu_p = \pi = .9$ and $\sigma_p = \sqrt{\pi(1 - \pi)/n} = \sqrt{(.9)(.1)/100} = .03$. Furthermore, because $n\pi = 100(.9) = 90$ and $n(1 - \pi) = 10$ are both at least 5, n is large enough for p to have approximately a normal sampling distribution. The standardized test statistic is then

$$z = \frac{p - \text{hypothesized value}}{\text{standard deviation of } p} = \frac{p - .9}{.03}$$

Rejecting H_0 when p is considerably below .9 is the same as rejecting H_0 when the value of z is extreme and negative. That is, for some negative cutoff value $-c$, the rejection region should have the form $z < -c$.

A type I error consists of saying that $\pi < .9$, i.e., that the manufacturer's claim is fallacious, when in fact the manufacturer is correct in its claim. After reflection, suppose it is decided that a type I error probability of .05, but no larger, can be tolerated. Then $-c$ should be chosen to yield $\alpha = .05$, a test with level of significance .05. When H_0 is true, our test statistic has approximately a standard normal sampling distribution. From Table I, the area under the standard normal curve to the left of -1.645 is .05, so the rejection region $z < -1.645$ gives a level .05 test. If 86 of the 100 sample responses are successes, then $p = 86/100 = .86$ and $z = (.86 - .9)/.03 = -1.33$. Since -1.33 is not less than -1.645, the value of z is not in the rejection region. H_0 should not be rejected at level of significance .05. The sample does not provide strong evidence against the manufacturer's claim. ■

Our general principle essentially says to use a procedure for which the type I error probability is controlled at a suitably small value. This is especially compelling when, as is frequently the case, a type I error is quite serious. In research situations, the null hypothesis is often a statement of accepted theory or behavior, while H_a postulates a new theory developed by an investigator (the *research hypothesis*). Rejecting a widely accepted theory in favor of a new explanation that is actually not correct typically has very serious consequences, so a small α is certainly in order. In addition, if the researcher uses a small α and then discovers that H_0 can be rejected at that level in favor of the research hypothesis (which is what he or she hopes will happen), the evidence for the new theory becomes very compelling.

Finding a test procedure that gives a specified α requires that the sampling distribution of the test statistic when H_0 is true be known (at least approximately). This is because the rejection region is specified under the assumption that H_0 is true; α is then the probability that the test statistic value lands in the rejection region. The type II error probability β, on the other hand, is computed by assuming that H_0 is false in some specific way. For many statistical tests the computation of β is difficult, so our focus is on presenting tests that have a specified level α. Many of these turn out to be based on standardized z or t statistics, so that a specified α is achieved by appropriate choice of a z or t critical value. Later on, we encounter other testing problems that require introducing some other sampling distributions.

Finally, the test procedures that we present are motivated on intuitive grounds rather than derived using theoretical tools. However, many of these tests can be shown to be "best" tests with the specified level of significance—that is, tests that make β as small as possible for a given choice of α.

EXERCISES

8.14 A new prescription drug to be marketed as a tablet is supposed to contain 5 mg of codeine in each tablet. The FDA must decide whether to allow sale of this medication. Letting μ denote the true mean dosage of codeine per tablet, the FDA decides to include a test of $H_0 : \mu = 5$ mg versus $H_a : \mu \neq 5$ mg as part of its investigation.

a. Why do you think the above alternative hypothesis was chosen over the other two possibilities, $H_a : \mu > 5$ and $H_a : \mu < 5$?

b. For the null and alternative hypotheses given (and in the context of this problem), describe type I and type II errors and discuss the possible consequences of making each type of error.

8.15 A manufacturer of handheld calculators receives very large shipments of printed circuits from a supplier. It is too costly and time consuming to inspect all incoming circuits, so when each shipment arrives, a sample is selected for inspection. Information from the sample is then used to test $H_0 : \pi = .05$ versus $H_a : \pi < .05$, where π is the true proportion of defectives in the shipment. If the null hypothesis is rejected, the shipment is accepted and the circuits are used in the production of calculators. If the null hypothesis cannot be rejected, the entire shipment is returned to the supplier due to inferior quality (a shipment is defined to be of inferior quality if it contains 5% or more defectives).

a. In this context, define type I and type II errors.

b. From the calculator manufacturer's point of view, which type of error would be considered more serious?

c. From the printed circuit supplier's point of view, which type of error would be considered more serious?

8.16 Occasionally, warning flares of the type contained in most automobile emergency kits fail to ignite. A consumer advocacy group is to investigate a claim against a manufacturer of flares brought by a person who claims that the proportion of defectives is much higher than the value of .1 claimed by the manufacturer. A large number of flares will be tested and the results used to decide between $H_0 : \pi = .1$ and $H_a : \pi > .1$, where π represents the true proportion of defectives for flares made by this manufacturer. If H_0 is rejected, charges of false advertising will be filed against the manufacturer.

a. Explain why the alternative hypothesis was chosen to be $H_a : \pi > .1$.

b. In this context, describe type I and type II errors and discuss the consequences of each.

8.17 Water samples are taken from water used for cooling as it is being discharged from a power plant into a river. It has been determined that as long as the mean temperature of the discharged water is at most 150°F, there will be no negative effects on the river ecosystem. To investigate whether the plant is in compliance with regulations that prohibit a mean discharge water temperature above 150°, 50 water samples will be taken at randomly selected times, and the water temperature of each sample will be recorded. A z statistic

$$z = \frac{\overline{x} - 150}{\sigma/\sqrt{n}}$$

will then be used to decide between the hypotheses

$$H_0 : \mu = 150 \quad \text{and} \quad H_a : \mu > 150$$

where μ is the true mean temperature of discharged water. Assume that σ is known to be 10.

a. Explain why use of the z statistic would be appropriate in this setting.

b. Describe type I and type II errors in this context.

c. The rejection of H_0 when $z > 1.8$ corresponds to what value of α (the probability of making a type I error)?

d. Suppose that the true value for μ is 153 and that the rejection region $z > 1.8$ is used. Draw a sketch (similar to that of Figure 1) of the sampling distribution of $\overline{x}$ and shade the region that would represent β, the probability of making a type II error.

e. For the hypotheses and rejection region described, compute the value of β when $\mu = 153$.

f. For the hypotheses and rejection region described, what value would β take if $\mu = 160$?

g. If the rejection region $z > 1.8$ is used and $\overline{x} = 152.4$, what is the appropriate conclusion? What type of error might have been made in reaching this conclusion?

8.18 A university provides enough bicycle parking to accommodate 10% of its student body. In response to complaints that the number of bike racks is inadequate, the administration has agreed to conduct a survey of 500 randomly selected students. Each student participating in the survey will be asked if he or she rides a bike to school. Letting π denote the true proportion of students who ride bicycles to campus, the hypotheses $H_0 : \pi = .1$ versus $H_a : \pi > .1$ will be tested. The sampling distribution of p, the sample proportion, is approximately normal with mean $\mu_p = \pi$ and standard deviation $\sigma_p = \sqrt{\pi(1 - \pi)/n}$.

a. If the null hypothesis is true, what are the values of μ_p and σ_p?

b. The statistic

$$z = \frac{p - .1}{\sqrt{(.1)(.9)/500}}$$

(obtained by standardizing p assuming that H_0 is true) can be used as a basis for choosing between H_0 and H_a. Suppose H_0 will be rejected in favor of H_a whenever $z > 1.645$. If H_0 is true, what is the probability of incorrectly rejecting H_0?

c. The administration has agreed to provide more bike parking only if H_0 is rejected. Suppose that in fact $\pi = .13$. Using the test procedure proposed in (b), what is the probability that the null hypothesis $H_0 : \pi = .1$ will not be rejected?

d. Suppose the test procedure described in (b) is employed. If π is really .2, would the probability of not rejecting H_0 be larger than the probability computed in (c)? Explain.

e. If only 46 of the sampled students ride bikes to school, what conclusion is appropriate using the rejection region of (b)? In what type of error might this conclusion result?

f. Answer (e) if 80 of the sampled students ride bikes to school.

8.19 A maker of aspirin claims that, when placed in water, its tablets dissolve completely in at most 2 min. Fifty randomly selected tablets are to be examined. Each will be dropped into a beaker of water and the time until the tablet completely dissolves will be recorded. The resulting data will be used to test $H_0 : \mu = 2$ versus $H_a : \mu > 2$, where μ represents the true mean dissolving time for aspirin tablets made by this particular manufacturer. Since $n = 50$ is large, the test statistic $z = (\bar{x} - 2)/\sigma_{\bar{x}}$ will have approximately a standard normal distribution when H_0 is true.

a. Consider rejecting H_0 when $z > 1.9$. What is the level of significance for this test?

b. The value 1.9 given in (a) is called the critical value for the test. If the critical value were changed to 2.2, so that H_0 is rejected if $z > 2.2$, would the significance level be larger or smaller than that corresponding to the critical value of 1.9? Explain.

c. If H_0 is rejected when $z > 1.645$, the corresponding level of significance would be .05. If a smaller significance level were desired, would a critical value greater than or less than 1.645 be required? Explain.

d. Would the type II error probability be larger for $\mu = 2.5$ or for $\mu = 3.0$ (irrespective of which α is used)? Explain.

8.20 Suppose that as an inspector for the Fish and Game Department, you are given the task of determining whether to prohibit fishing along part of the California coast. You will close an area to fishing if it is determined that fish in that region have an unacceptably high mercury content.

a. If a mercury concentration of 5 ppm is the maximum considered safe, which pair of hypotheses would you test?

$$H_0 : \mu = 5 \quad \text{versus} \quad H_a : \mu > 5$$

or

$$H_0 : \mu = 5 \quad \text{versus} \quad H_a : \mu < 5$$

Give the reasons for your choice.

b. Would you prefer a significance level of .1 or .01 for your test? Explain.

8.3 Large-Sample Hypothesis Tests for a Population Mean

The first two sections presented the most important general concepts of hypothesis testing. Now we turn our attention to a detailed development of large-sample procedures for testing hypotheses about μ. As long as n is large enough for the central limit theorem to apply, these test procedures are valid without any restrictions on the population distribution (as were the large-sample confidence intervals of Chapter 7).

The central limit theorem states that when n is large, the sampling distribution of $\overline{x}$ is approximately normal. Since $\mu_{\overline{x}} = \mu$ and $\sigma_{\overline{x}} = \sigma/\sqrt{n}$, for large n the sampling distribution of the standardized variable

$$z = \frac{\overline{x} - \mu}{\sigma / \sqrt{n}}$$

is described approximately by the standard normal (z) curve. In practice, the value of σ is rarely known, so it is natural to use s, the sample standard deviation, in its place. For n large, the resulting standardized variable

$$z = \frac{\overline{x} - \mu}{s/\sqrt{n}}$$

also has a sampling distribution that is well approximated by the z curve. The general form of a null hypothesis for tests concerning μ will be

$$H_0 : \mu = \text{hypothesized value}$$

In any given problem, the hypothesized value will be a particular number determined by the focus of the investigation. This value will usually be apparent from the problem statement.

The large-sample test statistic now results from replacing μ in the standardized variable z by its hypothesized value according to H_0:

$$\text{test statistic} = z = \frac{\overline{x} - \text{hypothesized value}}{s/\sqrt{n}}$$

When H_0 is true, $\mu_{\overline{x}} = \mu =$ hypothesized value. Thus the test statistic is formed by standardizing $\overline{x}$ (the statistic for estimating μ) under the assumption that H_0 is true. This implies that *when H_0 is true, the sampling distribution of z is approximately described by the standard normal curve.* Knowledge of the (approximate) sampling distribution of the test statistic when H_0 is true tells us how z will behave in this case—which values are consistent with H_0 and which values are quite unusual. This is precisely what is needed for selecting the rejection region to yield a specified level of significance α (i.e., type I error probability).

The appropriate rejection region depends on which of the three inequalities, $>$, $<$, or $\neq$, appears in the alternative hypothesis H_a. If the alternative is $H_a : \mu >$ hypothesized value, then H_0 should be rejected in favor of H_a if $\overline{x}$ considerably exceeds the hypothesized value. This is equivalent to rejecting H_0 if the value of z is quite large and positive (a z value far out in the upper tail of the z curve). The desired α is then achieved by using the z critical value which captures upper-tail z curve area α as the cutoff value for the rejection region. Similarly, if the inequality $<$ appears in H_a, then H_0 should be rejected if $\overline{x}$ is considerably less than the hypothesized value or, equivalently, if z is too far out in the lower tail of the z curve to be consistent with H_0. In the case $H_a : \mu \neq$ hypothesized value, rejection of H_0 is appropriate if the computed value of z is more than a specified distance out in either tail of the z curve. Figure 3 illustrates the choice of rejection region to control α in each of these three cases.

Once H_a is identified, carrying out the test requires that an appropriate z critical value be obtained. Critical values corresponding to the most frequently used significance levels—.10, .05, .01, and .001—appear in Table 2 of Section 5.4. A more convenient place to find these values, though, is in the bottom row of Table IV, the t table (remember, the bottom row contains z critical values for the large-sample z confidence intervals of Chapter 7).

Summary of Large-Sample z Tests for μ

Null hypothesis: $H_0 : \mu =$ hypothesized value

Test statistic: $$z = \frac{\overline{x} - \text{hypothesized value}}{s/\sqrt{n}}$$

Alternative hypothesis	*Rejection region*
$H_a : \mu >$ hypothesized value	Reject H_0 if $z > z$ critical value (upper-tailed test)
$H_a : \mu <$ hypothesized value	Reject H_0 if $z < -z$ critical value (lower-tailed test)
$H_a : \mu \neq$ hypothesized value	Reject H_0 either if $z > z$ critical value or if $z < -z$ critical value (two-tailed test)

The z critical value in the rejection region is determined by the desired level of significance. The bottom row of Table IV contains critical values corresponding to the most frequently used significance levels.

Suppose, for example, that the null hypothesis is $H_0 : \mu = 25$ and that $\alpha = .05$ is specified. If the alternative is $H_a : \mu > 25$, the appropriate test procedure is upper-tailed. The level $\alpha = .05$ is then located in the "level of significance for a one-tailed test" row along the bottom margin of Table IV. The desired critical value, 1.645, appears directly above $\alpha = .05$ in the "z critical value" row. In this case, H_0 will be rejected if $z > 1.645$ but not otherwise. Similarly, if the alternative is $H_a : \mu < 25$, the test is lower-tailed and, therefore, one-tailed, so for $\alpha = .05$ the same critical value 1.645 is used. Here H_0 will be rejected if $z < -1.645$. Finally, $H_a : \mu \neq 25$ requires a two-tailed test. The correct critical value for this test when $\alpha = .05$ is *not* 1.645 because, as illustrated in Figure 3(c), .05 must be divided equally between the two tails, giving .025 in each tail. Then the critical value capturing upper-tail z curve area .025 is identified. This is accomplished by entering the "level of significance for a two-tailed test" row along the bottom margin of Table IV, moving over to $\alpha = .05$, and looking directly above to the corresponding entry in the "z critical value" row. This gives z critical value $= 1.96$, so H_0 will be rejected in favor of $H_a : \mu \neq 25$ either if $z > 1.96$ or if $z < -1.96$.

EXAMPLE 9

In recent years a substantial amount of research has focused on possible relationships between chemical contamination of various sorts and mental retardation. The article "Increased Lead Burdens and Trace-Mineral Status in Mentally Retarded Children" (*J. Special Educ.* (1982):87–89) reported data

Alternative Hypothesis	Rejection Region and $\alpha = P$(type I error)
(a) $H_a : \mu >$ hypothesized value	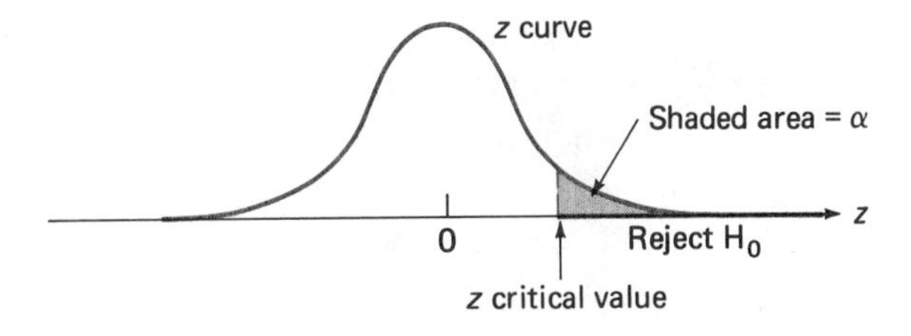
(b) $H_a : \mu <$ hypothesized value	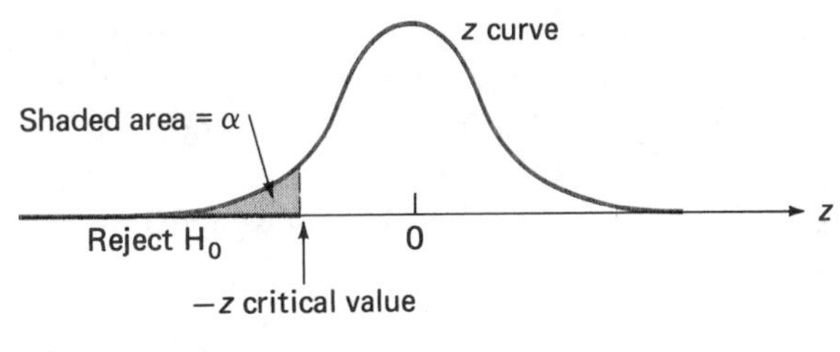
(c) $H_a : \mu \neq$ hypothesized value	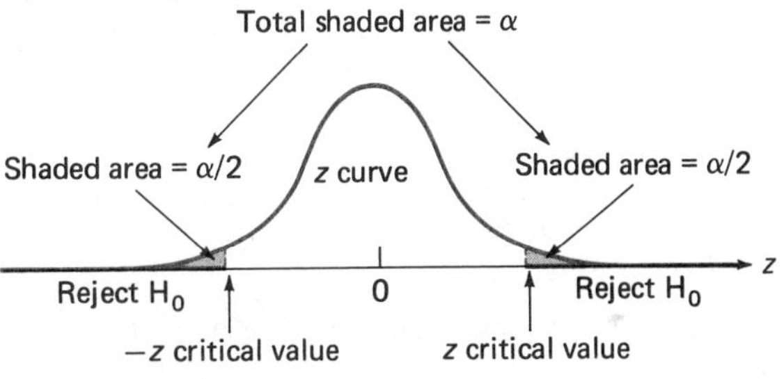

FIGURE 3 ALTERNATIVE HYPOTHESIS, REJECTION REGION, AND α FOR A LARGE-SAMPLE z TEST
(a) Upper-tailed test
(b) Lower-tailed test
(c) Two-tailed test

from a sample of $n = 40$ mentally retarded children for which the cause of retardation was unknown. The sample average hair-lead concentration was given as 15.90 ppm, and the sample standard deviation was 8.40. The paper states that 15 ppm is considered the acceptable upper limit of hair-lead concentration. Does the above sample data support the research hypothesis that true average hair-lead concentration for all such mentally retarded children exceeds the acceptable upper limit?

In answering this question, it is helpful to proceed in an organized manner by following a fixed sequence of steps.

1. *Population characteristic of interest:* $\mu =$ the average hair-lead concentration for all mentally retarded children with the cause of retardation unknown.
2. *Null hypothesis:* $H_0 : \mu = 15$.
3. *Alternative hypothesis:* $H_a : \mu > 15$.
4. *Test statistic:*

$$z = \frac{\bar{x} - 15}{s/\sqrt{n}}$$

5. *Rejection region:* The inequality in H_a is $>$, so the test is upper-tailed with rejection $z > z$ critical value. Using level of significance $\alpha = .01$, Table IV yields the critical value 2.33. H_0 will be rejected in favor of H_a if $z > 2.33$ and not rejected otherwise.

6. *Computations:* The values $n = 40$, $\overline{x} = 15.90$, and $s = 8.40$ are given. Thus

$$z = \frac{15.90 - 15}{8.40/\sqrt{40}} = \frac{.90}{1.33} = .68$$

7. *Conclusion:* Since .68 is less than 2.33, the computed value of z does not fall into the rejection region. At level of significance .01, H_0 is not rejected. The data does not provide support for concluding that true average hair-lead concentration exceeds 15. A reasonable explanation for the observed difference between $\overline{x} = 15.9$ and the hypothesized value 15 is sampling variation. ■

We recommend that the sequence of steps illustrated in Example 9 be used in any hypothesis-testing analysis.

Steps in a Hypothesis-Testing Analysis

1. Describe the population characteristic about which hypotheses are to be tested.

2. State the null hypothesis, H_0.

3. State the alternative hypothesis, H_a.

4. Display the test statistic to be used, with substitution of the hypothesized value identified in Step 2 but *without* any computation at this point.

5. Identify the rejection region. This is accomplished by first using the inequality in H_a to determine whether an upper, lower, or two-tailed test is appropriate and then going to an appropriate table to obtain the critical value corresponding to the selected level of significance α.

6. Compute all quantities appearing in the test statistic; then compute the value of the test statistic itself.

7. State the conclusion (which will be to reject H_0 if the value of the test statistic falls in the rejection region and to not reject H_0 otherwise). The conclusion should be stated in the context of the problem, and the level of significance used should be included.

Steps 1–3 constitute a statement of the problem, Steps 4 and 5 state how the conclusion will be reached, and Steps 6 and 7 give the analysis and conclusion.

EXAMPLE 10

A certain type of brick is being considered for use in a particular construction project. It is decided that the brick will be used unless sample evidence strongly suggests that the true average compressive strength is below 3200 lb/in.^{2} A random sample of 36 bricks is selected and each is subjected to a compressive strength test. The resulting sample average compressive strength and sample standard deviation of compressive strength are 3109 lb/in.^{2} and 156 lb/in.^{2}, respectively. State the relevant hypotheses and carry out a test to reach a decision using level of significance .05.

1. μ = true average compressive strength for this type of brick.

2. $H_0 : \mu = 3200$.

3. $H_a : \mu < 3200$ (so that the brick will be used unless H_0 is rejected in favor of H_a).

4. Test statistic:

$$z = \frac{\overline{x} - 3200}{s/\sqrt{n}}$$

5. Because < appears in H_a, a lower-tailed test is used. For $\alpha = .05$, the appropriate one-tailed critical value is obtained from Table IV as 1.645. H_0 will now be rejected if $z < -1.645$.

6. $n = 36$, $\overline{x} = 3109$, and $s = 156$, so

$$z = \frac{3109 - 3200}{156/\sqrt{36}} = \frac{-91}{26} = -3.50$$

That is, $\overline{x}$ has been observed to fall 3.5 estimated standard deviations (of $\overline{x}$) below what would have been expected were H_0 true.

7. Since $-3.50 < -1.645$, H_0 is rejected at the 5% level of significance. We conclude that true average compressive strength is below 3200, so the brick should not be used. In reaching this conclusion, we may have made a type I error (rejecting H_0 when it is true), but the low level of significance .05 has made that unlikely. ■

EXAMPLE 11

An automobile manufacturer recommends that any purchaser of one of its new cars bring it in to a dealer for a 3000-mi checkup. The company wishes to know whether the true average mileage at which this initial servicing is done differs from 3000. A random sample of 50 recent purchasers resulted in a sample average mileage of 3208 mi and a sample standard deviation of 273 mi. Does the data strongly suggest that true average mileage for this checkup is something other than the value recommended by the company? State and test the relevant hypotheses using level of significance .01.

1. μ = true average mileage of cars brought to the dealer for 3000-mi checkups.

2. $H_0 : \mu = 3000$.

3. $H_a : \mu \neq 3000$ (which says that μ differs from what the manufacturer recommends).

4. Test statistic:

$$z = \frac{\overline{x} - 3000}{s/\sqrt{n}}$$

5. The alternative hypothesis uses $\neq$, so a two-tailed test should be used. Table IV gives the $\alpha = .01$ critical value as 2.58 (*not* 2.33, the *one*-tailed critical value for $\alpha = .01$). H_0 will now be rejected if either $z > 2.58$ or $z < -2.58$.

6. $n = 50, \bar{x} = 3208$, and $s = 273$, which gives

$$z = \frac{3208 - 3000}{273/\sqrt{50}} = \frac{208}{38.61} = 5.39$$

7. Since 5.39 is in the upper tail of the two-tailed rejection region ($5.39 > 2.58$), H_0 is rejected using $\alpha = .01$. The data does strongly suggest that true average initial checkup mileage differs from the manufacturer's recommended value. ■

Statistical versus Practical Significance

Carrying out a test amounts to deciding whether the value obtained for the test statistic could plausibly have resulted when H_0 is true. If the value doesn't deviate too much from what is expected when H_0 is true, there is no compelling reason for rejecting H_0 in favor of H_a. But suppose that the observed value is quite far out in the appropriate tail of the test statistic's sampling distribution when H_0 is true (e.g., a large positive value of z when H_a contains the inequality $>$). One could continue to believe that H_0 is true and that such a value arose just through chance variation (a very unusual and "unrepresentative" sample). However, in this case a more plausible explanation for what was observed is that H_0 is false and H_a is true.

When the value of the test statistic falls in the rejection region, it is customary to say that the result is **statistically significant** at the chosen level α. The finding of statistical significance means that, in the investigator's opinion, the observed deviation from what was expected under H_0 cannot plausibly be attributed just to chance variation. Unfortunately, though, statistical significance cannot be equated with the conclusion that the true situation differs from what H_0 states in any practical sense. That is, even after H_0 has been rejected, the data may suggest that there is no practical difference between the true value of the population characteristic and what the null hypothesis states that value to be.

EXAMPLE 12

Parents in the state of Euphoria believe that their children are on average smarter than children in all other states. To substantiate this belief, 2500 Euphorian children are randomly selected and an IQ test is administered to each one. The resulting sample average IQ and sample standard deviation are $\bar{x} = 101$ and $s = 15$. Let μ denote the average IQ for all children in Euphoria. The average IQ for all children across the nation is 100, so the appropriate hypotheses are $H_0 : \mu = 100$ versus $H_a : \mu > 100$. The form of the inequality in H_a implies an upper-tailed test, and the value of z is $(101 - 100)/(15/\sqrt{2500}) = 1/.3 = 3.33$. From Table IV, the critical value for $\alpha = .001$ is 3.09 and $3.33 > 3.09$. Even at the very small level of significance .001, H_0 is rejected. The evidence is very strong for the conclusion that $\mu > 100$, so euphoria reigns supreme among Euphorian parents—but unjustifiably so.

Admittedly, when $\mu = 100$ and σ is on the order of 15, it is very unlikely to observe a value of z anywhere near as large as what was observed. A much more plausible explanation for what was observed is that $\mu > 100$. However, with $n = 2500$, the point estimate $\bar{x} = 101$ is almost surely very close to the true value of μ (a 99% confidence interval has lower limit 100.4 and upper limit 101.6). So it looks as though H_0 was rejected because $\mu \approx 101$

rather than 100. And from a practical point of view, a 1-point IQ difference has no significance! So the statistically significant result does not have any practical consequences. ■

Although the context for this example was somewhat whimsical, the moral is important. A statistically significant result can be obtained from data in which a large sample size magnifies a departure from H_0 that has no practical import. One must look at more than just the computed value of a test statistic to get an assessment of practical significance.

EXERCISES

8.21 Let μ denote the true average amount of surface area covered by 1 gal of a particular oil-based paint. A researcher wishes to test the hypotheses $H_0 : \mu = 400$ versus $H_a : \mu > 400$ using a sample size of 50. Give the appropriate test statistic and rejection region for each of the given significance levels.
a. .01 **b.** .05 **c.** .10 **d.** .13

8.22 A sample of size 75 is to be used to decide between the hypotheses $H_0 : \mu = 14$ and $H_a : \mu \neq 14$, where μ is the true average filled weight for containers coming off a certain production line. Give the test statistic and rejection region associated with each of the given significance levels.
a. .05 **b.** .01 **c.** .10 **d.** .24

8.23 A scale for rating a politician's image solicits responses from -5 (very negative) to 5 (very positive). Let μ denote the true average image rating of a particular politician. A large-sample z statistic is to be used to test the hypotheses $H_0 : \mu = 0$ versus $H_a : \mu < 0$. Determine the appropriate rejection region for the given significance levels.
a. .01 **b.** .05 **c.** .10

8.24 A television manufacturer claims that a current of at most 250 μA is needed to attain a certain brightness level with a particular type of set. A sample of 40 sets yields a sample average current of 257.3 and a sample standard deviation of 15. Let μ denote the true average current necessary to achieve the desired brightness with sets of this type. Test at level .05 the null hypothesis that μ is (at most) 250 against the appropriate alternative.

8.25 The Food and Nutrition Board of the National Academy of Sciences reports that mean daily sodium intake should be at least 1100 mg and should not exceed 3300 mg. In a study of sodium intake (*Consumer Reports* (1984):17–22), a sample of U.S. residents was found to have a mean daily sodium intake of approximately 4600. Suppose that this statistic was based on a sample of size 100 and that the sample standard deviation was 1100. Does this data suggest that mean daily sodium intake for U.S. residents exceeds the maximum recommended level? Use a level .05 test.

8.26 Minor surgery on horses under field conditions requires a reliable short-term anesthetic producing good muscle relaxation, minimal cardiovascular and respiratory changes, and a quick, smooth recovery with minimal aftereffects so that horses can be left unattended. The article "A Field Trial of Ketamine Anesthesia in the Horse" (*Equine Vet. J.* (1984):176–79) reported that for a sample of $n = 73$ horses to which ketamine was administered under certain conditions, the sample average lateral recumbency (lying-down) time was 18.86 min and the standard deviation was 8.6 min. Does this data suggest that true average lateral recumbency time under these conditions is less than 20 min? Use the seven-step procedure to test the appropriate hypotheses at level of significance .10.

8.27 One of the biggest problems facing researchers who must solicit survey data by mail is that of nonresponse. One method that has been proposed as a way of increasing both the response rate and the quality of responses is to offer a monetary incentive for returning a questionnaire. The paper "The Effect of Monetary Inducement on Mailed Questionnaire Response Quality" (*J. of Marketing Research* (1980): 265–68) examines some of these issues. One hypothesis of interest to the researchers was that providing a cash incentive for completing a survey would lead to fewer questions left unanswered. An appliance warranty questionnaire containing 17 questions that had been used extensively was mailed to people who had purchased a major appliance during the previous year. A $.25 payment was included with each questionnaire. In the past, when no money was included with the survey, the mean number of questions left unanswered on returned forms was 1.38. In this experiment, 174 surveys were returned and the number of unanswered questions was determined for each one. The resulting sample mean was .81. Suppose that the sample standard deviation was 1.8. Using a level .01 test, can you conclude that including $.25 with each survey results in a mean number of unanswered questions that is smaller than 1.38?

8.28 The owner of a gas station that makes headlight inspections for the state is trying to decide whether or not to discontinue the service. To make the operation profitable, the station must average in excess of 15 inspections per week. Unless data indicates strongly that this is the case, inspections will be discontinued. Data for a random sample of 36 weeks yields a sample average of 16.7 inspections per week and a sample standard deviation of 4.5. Is this strong enough evidence to cause the owner to retain the inspection service? Use a level .05 test.

8.29 The Environmental Protection Agency sets limits on the maximum allowable concentration of certain chemicals in water. For the substance PCB, the limit has been set at 5 ppm. A random sample of 36 water specimens from a well results in a sample mean PCB concentration of 4.82 and a standard deviation of .6.

a. Is there sufficient evidence to substantiate the claim that the well water is safe? Use a .01 level of significance.

b. Would you recommend using a significance level greater than .01? Why or why not?

8.30 Vomitoxin is a poison produced by the fungus *F. graminearum*. In cool wet years vomitoxin contamination of corn and cereal grains used for animal feed can sometimes be a problem ("Survey of Vomitoxin-contaminated Feed Grains in Midwestern United States, and Associated Problems in Swine" *J. Amer. Vet. Med. Assoc.* (1984):189–94). A concentration of vomitoxin exceeding 15 ppm in feed causes severe vomiting (thus the name vomitoxin). Even at lower concentrations, loss of weight and food refusal have been observed. After the 1981 corn harvest, 274 feed samples from a particular supplier were analyzed for vomitoxin contamination. The mean concentration of the toxin was found to be 3.14 ppm and the sample standard deviation was (approximately) 10. Noticeable symptoms occur when contamination exceeds 1 ppm. Is there sufficient evidence to indicate that the mean concentration of vomitoxin in corn from this supplier exceeds 1 ppm? Use a level .10 test.

8.31 To check a manufacturer's claim that its audio tapes have an average playing time of at least 90 min, 900 tapes are randomly selected and timed. These yield a sample average playing time of 89.95 min and a sample standard deviation of .3 min. Does the data refute the manufacturer's claim? Comment on the statistical and practical significance of this result.

8.4 P-Values

One way to report the result of a hypothesis-testing analysis is simply to say whether or not the null hypothesis was rejected at a specified level of significance. Thus an investigator might state that H_0 was rejected at level of significance .05 or that use of a level .01 test resulted in nonrejection of H_0. This type of statement is somewhat inadequate because nothing is said about whether the computed value of the test statistic just barely fell into the rejection region or whether it exceeded the critical value by a very large amount. A related difficulty is that such a report imposes the specified significance level on other decision makers. There are many decision situations in which individuals might have different views concerning the consequences of type I and type II errors. Each individual would then want to select his or her own personal significance level—some selecting $\alpha = .05$, others .01, and so on—and reach a conclusion accordingly. This could result in some individuals rejecting H_0, whereas others might conclude that the data does not show a strong enough contradiction of H_0 to justify its rejection.

EXAMPLE 13

The true average time to initial relief of pain for the current best-selling pain reliever is known to be 10 min. Let μ denote the true average time to relief for a company's newly developed pain reliever. The company wishes to produce and market this product only if it provides quicker relief than does the current best seller, so it wishes to test $H_0 : \mu = 10$ versus $H_a : \mu < 10$. Only if experimental evidence leads to rejection of H_0 will the new pain reliever be introduced. After weighing the relative seriousness of the two types of errors, a single level of significance must be agreed upon and a decision—to reject H_0 and introduce the pain reliever or not to do so—must be made at that level.

Now suppose that the new product has been introduced. The company supports its claim of quicker relief by stating that, based on an analysis of experimental data, $H_0 : \mu = 10$ was rejected in favor of $H_a : \mu < 10$ using level of significance $\alpha = .10$. Any particular individual contemplating a switch to this new pain reliever would naturally want to reach his or her own conclusion concerning the validity of the claim. Individuals who are satisfied with the current best seller would view a type I error (concluding that the new product provides quicker relief when it actually doesn't) as serious, so they might wish to use $\alpha = .05$, .01, or an even smaller level. Unfortunately the nature of the company's statement prevents an individual decision maker from reaching a conclusion at such a level. The company has imposed its own choice of significance level on others. The report could have been done in a manner that allowed each individual flexibility in drawing a conclusion at a personally selected α. ■

A P-value conveys a great deal of information about the strength of evidence against H_0 and allows an individual decision maker to draw a conclusion at any specified level α. Before we give a general definition, consider how the conclusion in a hypothesis-testing problem depends on the selected level α.

EXAMPLE 14

The problem involving nicotine content discussed in Example 6 involved testing $H_0 : \mu = 1.5$ versus $H_a : \mu > 1.5$. Because of the inequality in H_a, the rejection region was upper-tailed, with H_0 rejected if $z >$ z critical value. Suppose that $z = 2.10$. The accompanying table displays the rejection region for each of four different values of α along with the resulting conclusion.

Level of Significance α	Rejection Region	Conclusion
.05	$z > 1.645$	Reject H_0
.025	$z > 1.96$	Reject H_0
.01	$z > 2.33$	Don't reject H_0
.005	$z > 2.58$	Don't reject H_0

For α relatively large, the z critical value is not very far out in the upper tail; 2.10 exceeds the critical value, and so H_0 is rejected. However, as α decreases, the critical value increases. For small α the z critical value is large; 2.10 does not exceed it, and H_0 is not rejected.

Recall that for an upper-tailed z test, α is just the area under the z curve to the right of the critical value. That is, once α is specified, the critical value is chosen to capture upper-tail area α. Table I shows that the area to the right of 2.10 is .0179. Using an α larger than .0179 corresponds to z critical value < 2.10. An α less than .0179 necessitates using a z critical value that exceeds 2.10. The decision at a particular level α thus depends on how the selected α compares to the tail area captured by the computed z. This is illustrated in Figure 4. Notice in particular that .0179, the captured tail area, is the smallest level α at which H_0 would be rejected, because using any smaller α results in a z critical value that exceeds 2.10, so that 2.10 is not in the rejection region. ■

In general, suppose that the sampling distribution of a test statistic when H_0 is true has been determined. Then for specified α, the rejection region is determined by finding a critical value or values that capture tail area α (upper-, lower-, or two-tailed, whichever is appropriate) under the sampling distribution curve. The smallest α for which H_0 is rejected is the tail area captured by the computed value of the test statistic. This smallest α is the P-value.

DEFINITION

The ***P*-value** is the smallest level of significance at which H_0 can be rejected. Once the P-value has been determined, the conclusion at any particular level α results from comparing the P-value to α:

1. P-value $\leq \alpha \Rightarrow$ reject H_0 at level α
2. P-value $> \alpha \Rightarrow$ do not reject H_0 at level α

An easy way to visualize the comparison of α and the P-value is to draw a picture like that of Figure 5. The calculation of the P-value depends on whether the test is upper-, lower-, or two-tailed. But once the P-value has been calculated, the comparison with α does not depend on which type of test was used.

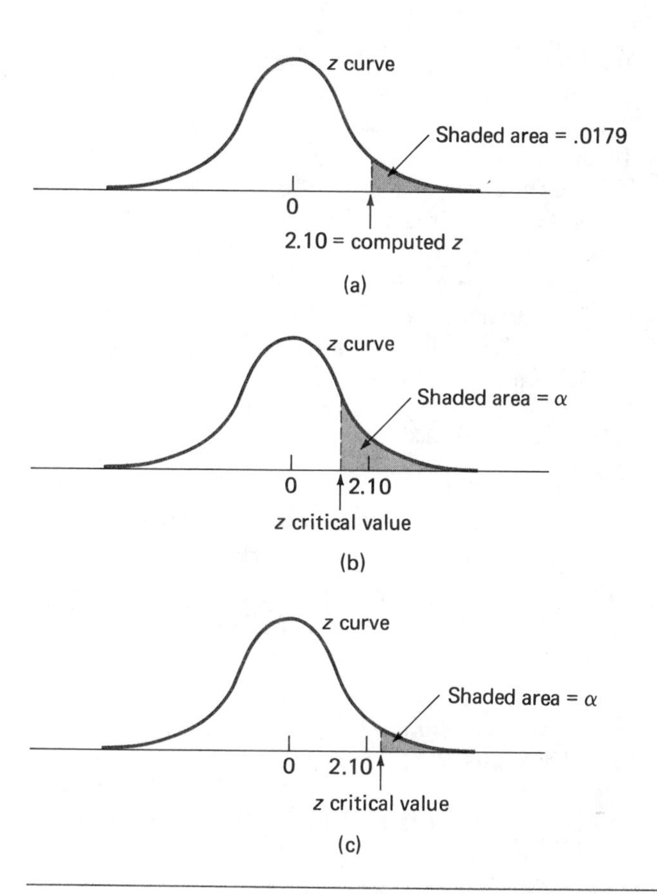

FIGURE 4

RELATIONSHIP BETWEEN α AND TAIL AREA CAPTURED BY COMPUTED z
(a) Tail area captured by computed z
(b) When $\alpha > .0179$, z critical value < 2.10 and H_0 is rejected
(c) When $\alpha < .0179$, z critical value > 2.10 and H_0 is not rejected

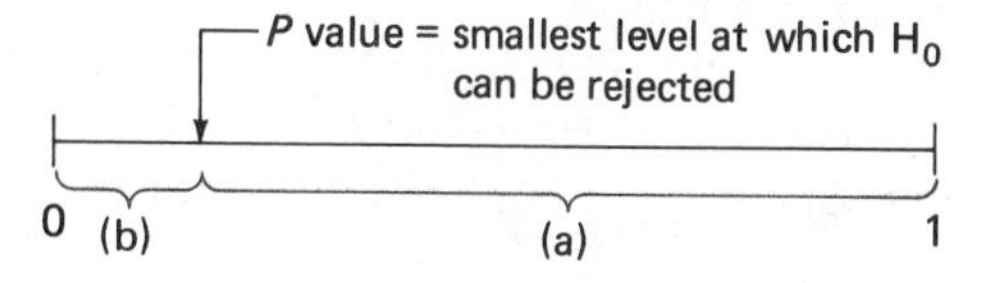

FIGURE 5

COMPARING α AND THE *P*-VALUE
(a) Reject H_0 when α lies here
(b) Do not reject H_0 when α lies here

EXAMPLE 15

(*Example 13 continued*) Suppose that when data from an experiment involving the new pain reliever was analyzed, the *P*-value for testing $H_0 : \mu = 10$ versus $H_a : \mu < 10$ was calculated as .0384. Since $\alpha = .05$ is larger than the *P*-value (.05 lies in interval (a) of Figure 5), H_0 would be rejected by anyone carrying out the test at level .05. However, at level .01, H_0 would not be rejected because .01 is smaller than the smallest level (.0384) at which H_0 can be rejected. ■

Most standard statistical computer packages automatically calculate and print out a *P*-value when a hypothesis-testing analysis is performed. We give some typical output in the next section after discussing small-sample tests.

Calculating the P-Value for a z Test

With the aid of Table I the (approximate) P-value can be calculated for any z test (not only the large-sample z test concerning μ discussed in Section 8.3 but also others presented in later sections and chapters). The test of Example 14 is upper-tailed, so the P-value is the area captured under the z curve to the right of the computed value of z. If the test had been lower-tailed, the P-value would have been the area under the z curve to the left of the computed z (the lower-tail area). For a two-tailed test, finding the area captured in the tail in which z falls—e.g., the upper tail when $z = 2.43$ or the lower tail when $z = -.92$—determines $\alpha/2$ (because α is the sum of the upper- and lower-tail areas). In this case the P-value is twice the captured area. This is summarized in the accompanying table.

Type of z Test	*P-Value*
Upper-tailed	Area under the z curve to the right of computed z
Lower-tailed	Area under the z curve to the left of computed z
Two-tailed	Twice the area captured in the tail in which computed z falls

EXAMPLE 16

(*Example 15 continued*) Suppose a z test for testing $H_0 : \mu = 10$ versus $H_a : \mu < 10$ results in $z = -1.49$. The closest entry in Table I is for $z = -1.5$, and the area in the lower tail to the left of this value is .0668. Thus P-value $\approx .0668$. At level .10, H_0 would be rejected, but rejection of H_0 is not justified at level .05. ■

EXAMPLE 17

(*Example 11 continued*) With μ denoting the true average mileage at the first checkup, the hypotheses of interest were $H_0 : \mu = 3000$ versus $H_a : \mu \neq 3000$. Suppose that $z = -2.7$, which falls in the lower tail of the z curve. The area to the left of this value is .0035, so the P-value is $2(.0035) = .007$. If we had instead computed $z = 2.7$, the area captured in the upper tail (to the right of 2.7) would have been .0035, yielding the same P-value, .007. In either case H_0 would be rejected at level .01 (α exceeds the P-value, as in (a) of Figure 5) but not at level .005 or .001. ■

Bounds on the P-Value for a z Test

Rather than computing an exact P-value, it is usually sufficient to say how the P-value compares to the most frequently quoted levels of significance. This is easily done using the z critical values in the bottom row of Table IV. First considering an upper-tailed test, there are three possible cases:

1. If the computed z falls between two consecutive tabulated critical values, the P-value lies between the corresponding one-tailed levels of significance.
2. If the computed z exceeds the largest critical value 3.29, then P-value $< .0005$ (the corresponding one-tailed level of significance).
3. If the computed z is smaller than the smallest z critical value 1.28, then P-value $> .10$.

EXAMPLE 18

(*Example 14 continued*) With the hypotheses of interest as $H_0 : \mu = 1.5$ versus $H_a : \mu > 1.5$, suppose that $n = 36$, $\bar{x} = 1.58$, and $s = .26$. This gives $z = (1.58 - 1.5)/(.26/\sqrt{36}) = 1.85$. Since $1.645 < 1.85 < 1.96$, the P-value satisfies $.025 < P$-value $< .05$ (which simply says that H_0 would be rejected at level .05, since $1.85 > 1.645$ but not at level .025 because 1.85 is not greater than 1.96). According to these bounds, H_0 would be rejected if the selected α were at least .05 and would not be rejected if α were at most .025. If the selected level were between .025 and .05 (an unusual choice), a conclusion could not be reached just from these bounds.

If, instead, $z = 3.62$, then P-value $< .0005$, so H_0 would be rejected for $\alpha = .10, .05, .01$, or .001 (since each of these is larger than the smallest level at which H_0 could be rejected). Similarly, if $z = .93$, then P-value $> .10$, so even at the rather large level .10, H_0 could not be rejected (.10 is smaller than the smallest α at which rejection of H_0 is allowed). ■

When the test is lower-tailed, bounds are obtained by changing the sign on z and proceeding exactly as described earlier (using significance levels for a one-tailed test). Thus if $z = -2.16$, because $1.96 < 2.16 < 2.33$, $.01 < P$-value $< .025$. In this case H_0 would be rejected at level .05 or .10 but not at level .01 or .001. For a two-tailed test, bounds are obtained by proceeding exactly as above except that significance levels for a two-tailed rather than a one-tailed test are used. As an example, suppose that $z = 2.08$, so that $1.96 < 2.08 < 2.33$. Since the two-tailed significance levels corresponding to 1.96 and 2.33 are .05 and .02, respectively, $.02 < P$-value $< .05$. If $z = -2.08$, the minus sign is disregarded and exactly the same bounds result. If $z = 3.54$ or $z = -3.54$, P-value $< .001$.

EXERCISES

8.32 For which of the given P-values would the null hypothesis be rejected when performing a level .05 test?
a. .001 **b.** .021 **c.** .078 **d.** .047 **e.** .148

8.33 Pairs of P-values and significance levels, α, are given. For each pair, state whether the observed P-value would lead to rejection of H_0 at the given significance level.

a. P-value $= .084$, $\alpha = .05$
b. P-value $= .003$, $\alpha = .001$
c. P-value $= .498$, $\alpha = .05$
d. P-value $= .084$, $\alpha = .10$
e. P-value $= .039$, $\alpha = .01$
f. P-value $= .218$, $\alpha = .10$

8.34 Let μ denote the true average reaction time to a certain stimulus. For a large-sample z test of $H_0 : \mu = 5$ versus $H_a : \mu > 5$, find the P-value associated with each of the given values of the test statistic.

a. 1.4 **b.** .9 **c.** 1.9 **d.** 2.4 **e.** $-.1$

8.35 Newly purchased automobile tires of a certain type are supposed to be filled to a pressure of 30 lb/in.^{2} Let μ denote the true average pressure. Find the P-value associated with each given z statistic value for testing $H_0 : \mu = 30$ versus $H_a : \mu \neq 30$.

a. 2.1 **b.** -1.7 **c.** $-.5$ **d.** 1.4 **e.** -5

8.36 Use the z critical values from the last row of the t table (Table IV) to find an upper and/or lower bound for the P-value for each value of z in Exercise 8.34.

8.37 Use the z critical values from the last row of Table IV to find an upper and/or lower bound for the P-value for each value of z given in Exercise 8.35.

8.38 A soda manufacturer is interested in determining whether its bottling machine tends to overfill. Each bottle is supposed to contain 12 oz of fluid. A random sample of size 36 is taken from bottles coming off the production line, and the contents of each bottle are carefully measured. It is found that the mean amount of soda for the sample of bottles is 12.2 oz and that the sample standard deviation is .4 oz. The manufacturer will use this information to test $H_0 : \mu = 12$ versus $H_a : \mu > 12$.

a. What value does the z test statistic take for this data?

b. Find the P-value associated with the value of z computed in (a).

c. If the manufacturer had decided on a level .05 test, should H_0 be rejected in favor of the conclusion that the machine is overfilling?

8.39 A sample of 40 speedometers of a particular brand is obtained and each is checked for accuracy at 55 mi/h. The resulting sample average and sample standard deviation are 53.8 and 1.3, respectively. Let μ denote the true average reading when the actual speed is 55 mi/h. Compute a P-value and use it and a significance level of .01 to decide whether the sample evidence strongly suggests that μ is not 55.

8.40 The drying time of a particular brand and type of paint is known to have a mean of 75 min. In an attempt to improve drying time, a new additive has been developed. Use of the additive in 100 test samples of the paint yields an observed mean drying time of 68.5 min and a standard deviation of 9.4 min. Using a significance level of .01, does the experimental evidence indicate that the additive improves (shortens) drying time? Use a P-value to conduct your test.

8.41 The national mean cholesterol level is approximately 210 (*Science 84* (April 1984):16). Each person in a group of men with unusually high cholesterol levels (over 265) was treated with a new drug, cholestyramine. After taking the drug for a given length of time, cholesterol determinations were made. Suppose that 100 men participated in the study. After treatment with the drug, the mean cholesterol level for the 100 men was 228 and the sample standard deviation was 12. Let μ denote the average cholesterol level for all men taking this drug. One question of interest is whether men taking the drug still have a mean cholesterol level that exceeds the national average. Compute the P-value for this data. If a .05 significance level is chosen, what conclusion would you draw?

8.42 The mean systolic blood pressure for white males aged 35–44 in the United States is 127.2. The paper "Blood Pressure in a Population of Diabetic Persons Diagnosed After 30 Years of Age" (*Amer. J. of Public Health* (1984):336–39) reports that the mean blood pressure and standard deviation of a sample of 101 diabetic males aged 35–44 are 130 and 8, respectively.

a. Use a level .05 test to determine if there is sufficient evidence to indicate that mean systolic blood pressure of 35–44-year-old male diabetics differs from that of 35–44-year-old males in the general population.

b. If you were to report the results of your analysis using a P-value, what value should be given?

8.5 Small-Sample Hypothesis Tests for the Mean of a Normal Population

The large-sample hypothesis testing procedures for μ discussed in Section 8.3 can be used without having to make any specific assumptions about the population distribution. The justification for these procedures is invalid, though, when n is small because the central limit theorem can no longer be used. As with confidence intervals, one way to proceed is to make a specific assumption about the nature of the population distribution and develop testing procedures that are valid in this more specialized situation. Here we shall restrict consideration to the case of a normal population distribution. The result on which a test procedure is based is then the same one used in Chapter 7 to obtain a t confidence interval.

When $x_1, x_2, \ldots, x_n$ constitute a random sample of size n from a normal distribution, the sampling distribution of the standardized variable

$$t = \frac{\bar{x} - \mu}{s/\sqrt{n}}$$

is the t distribution with $n - 1$ df.

The null hypothesis is stated just as it was for the large-sample test concerning μ:

$$H_0 : \mu = \text{hypothesized value}$$

When H_0 is true, replacing μ in t by the hypothesized value gives a test statistic whose sampling distribution is known (t with $n - 1$ df). A rejection region giving the desired significance level is then obtained by using the appropriate t critical value from the $n - 1$ df row of Table IV.

Summary of t Tests for the Mean of a Normal Population

Null hypothesis: $H_0 : \mu = \text{hypothesized value}$

Test statistic: $t = \dfrac{\bar{x} - \text{hypothesized value}}{s/\sqrt{n}}$

Alternative hypothesis	*Rejection region*
$H_a : \mu >$ hypothesized value	Reject H_0 if $t > t$ critical value (upper-tailed test)
$H_a : \mu <$ hypothesized value	Reject H_0 if $t < -t$ critical (lower-tailed test)
$H_a : \mu \neq$ hypothesized value	Reject H_0 if either $t > t$ critical value or $t < -t$ critical value (two-tailed test)

The t critical value in the rejection region is based on $n - 1$ df and is determined by the desired level of significance. Table IV contains one-tailed and two-tailed critical values corresponding to the most frequently used significance levels.

The test statistic here is the same as the large-sample z statistic of Section 8.3. It is labeled t to emphasize that it has a t sampling distribution when H_0 is true rather than the z distribution.

EXAMPLE 19

Extensive data collected during the first half of this century showed clearly that in those years American-born Japanese children grew faster than did Japanese-born Japanese children. A recent paper ("Do American Born Japanese Children Still Grow Faster than Native Japanese?" *Amer. J. Phys. Anthropology* (1975):187–94) conjectured that improved economic and environmental conditions in postwar Japan had greatly narrowed this gap. To investigate the validity of this conjecture, a large sample of Hawaiian-born Japanese children was obtained, and the children were categorized with respect to age. There were thirteen 11-year-old boys in the sample (most of the children sampled were older). The sample average height of these 13 was 146.3 cm and the sample standard deviation was 6.92 cm. The average height of native-born 11-year-old Japanese children at that time was known to be 139.7 cm. Does this data suggest that the true average height for Hawaiian-born male 11-year-olds exceeds that for their native-born counterparts? The investigators were willing to assume that the population height distribution was normal. The relevant hypotheses can then be tested using a t test with level of significance .01. The steps in the analysis are as follows.

1. μ = true average height for all Hawaiian-born 11-year-old male Japanese children.
2. $H_0 : \mu = 139.7$.
3. $H_a : \mu > 139.7$.
4. Test statistic:

$$t = \frac{\bar{x} - 139.7}{s/\sqrt{n}}$$

5. Rejection region: Because the inequality in H_a is $>$, an upper-tailed test is appropriate. H_0 should be rejected in favor of H_a if $t > t$ critical value. With df $= n - 1 = 13 - 1 = 12$, moving along the bottom margin of Table IV to one-tailed level of significance .01 and up to the 12 df row gives a t critical value of 2.68. The rejection region is $t > 2.68$.
6. With $n = 13$, $\bar{x} = 146.3$, and $s = 6.92$,

$$t = \frac{146.3 - 139.7}{6.92/\sqrt{13}} = \frac{6.60}{1.92} = 3.44$$

7. Since 3.44 is in the rejection region ($3.44 > 2.68$), at level of significance .01 H_0 is rejected in favor of H_a. It seems clear that at this age the average height of Hawaiian-born Japanese children exceeds that of their native-born counterparts.

While the growth differential at age 11 appears to be substantial, analysis of data on older children suggests that the previous gap has narrowed considerably. ■

Strictly speaking, the validity of the one-sample t test requires that the population distribution be normal. In practice, the test can be used even when the population distribution is somewhat nonnormal as long as n is not too small. Statisticians say that the test is robust to mild departures from normality.

EXAMPLE 20

The low population density of the Amazon region has long puzzled geographers and other social scientists. Some have suggested that environmental conditions are inimical to support of a large population. The paper "Anthrosols and Human Carrying Capacity in Amazonia" (*Annals of the Assoc. of Amer. Geog.* (1980):553–66) suggests otherwise. The author's case for this viewpoint rests largely on an analysis of black-earth soil samples, which gives evidence for the presence of large and sedentary Indian populations prior to the European influx.

4*l*	.2, .3
4*h*	.5, .6, .6, .6, .6, .6, .6, .9
5*l*	.3, .3, .3, .4, .4, .4
5*h*	.5, .5, .5, .6, .6, .6, .7, .9, .9
6*l*	.1, .2
6*h*	
7*l*	.0
7*h*	.9

The accompanying stem-and-leaf display gives pH values for the 29 black-earth soil samples discussed in the paper (pH is a numerical measure of acidity and is related to availability of soil nutrients). Does this data indicate that true average black-earth pH differs from 5.0, the pH value for many other types of soil in the region? The display gives evidence of a somewhat skewed distribution (confirmed by a normal probability plot) and the largest value, 7.9, is a mild outlier. However, $n = 29$ is close to the sample size required for the large-sample z test, which requires no specific assumption about the pH distribution, so it seems safe to use the t test here. Let's state and test the relevant hypotheses at level of significance .05.

1. $\mu =$ the true average black-earth soil pH in Amazonia.
2. $H_0 : \mu = 5.0$.
3. $H_a : \mu \neq 5.0$ ($\neq$ because of the phrase *differs from,* which indicates a departure from H_0 in either direction).
4. Test statistic:

$$t = \frac{\overline{x} - 5.0}{s/\sqrt{n}}$$

5. Rejection region: Reject H_0 either if $t >$ t critical value or if $t < -t$ critical value. Because the test is two-tailed, first locate .05 in the two-tailed significance level row along the bottom margin of Table IV. Moving up that column to the $n - 1 = 28$ df row gives critical value 2.05, so H_0 will be rejected if either $t > 2.05$ or $t < -2.05$.
6. $\Sigma x = 155.6$ and $\Sigma(x - \overline{x})^2 = 18.23$ (either from the deviations or by using the computational formula $\Sigma x^2 - n\overline{x}^2$). Thus $\overline{x} = 155.6/29 = 5.37$ and $s^2 = 18.23/28 = .651$, so $s = .807$. The computed value of t is

$$t = \frac{5.37 - 5.0}{.807/\sqrt{29}} = \frac{.37}{.150} = 2.47$$

7. $t = 2.47$ is in the upper tail of the rejection region ($2.47 > 2.05$), so H_0 is rejected at level .05. The true average pH of black-earth soil does appear to be something other than 5.0. ■

P-Values for t Tests

In the previous section we used tail areas for the z distribution given in Table I to compute P-values for z tests. Because of the limited information about t distributions in Table IV—only seven critical values are given for each one—the best that we can do is establish an upper and/or lower bound for a t test P-value. The method parallels what was done for a z test, except that appropriate t critical values (df $= n - 1$) rather than z critical values are used.

EXAMPLE 21

(*Example 19 continued*) The test was upper-tailed, based on 12 df, and t was computed as 3.44. This falls between the critical values 3.06 ($\alpha = .005$ for a one-tailed test) and 3.93 ($\alpha = .001$), so $.001 < P\text{-value} < .005$. Thus H_0 would be rejected at levels .10, .05, and .01 but not at level .001 or anything smaller. ■

EXAMPLE 22

(*Example 20 continued*) The computed value of t was 2.47, and the 28-df row of Table IV shows the $\alpha = .02$ two-tailed critical value to be 2.47 also. Because the computed value coincides with a critical value, we have P-value $= .02$, but this happens very infrequently. If, instead, we had computed $t = 2.63$, then $2.47 < 2.63 < 2.76$ gives $.01 < P\text{-value} < .02$. These bounds would also have been correct if we had computed $t = -2.63$. ■

Although our t table, like t tables in all other texts, limits us to bounds on the P-value, almost any good statistical computer package is programmed to calculate the P-value resulting from a t test. With this information, the t table is unnecessary—a conclusion can be drawn simply by comparing the P-value to the selected level of significance.

EXAMPLE 23

An automobile manufacturer who wishes to advertise that one of its models achieves 30 mi/gal decides to carry out a fuel efficiency test. Six nonprofessional drivers are selected and each one drives a car from Phoenix to Los Angeles. The resulting miles-per-gallon figures are $x_1 = 27.2$, $x_2 = 29.3$, $x_3 = 31.2$, $x_4 = 28.4$, $x_5 = 30.3$, and $x_6 = 29.6$. Assuming that fuel efficiency (mi/gal) under these circumstances is normally distributed, does the data contradict the claim that true average fuel efficiency is (at least) 30?

With μ denoting true average fuel efficiency, the hypotheses of interest are $H_0 : \mu = 30$ versus $H_a : \mu < 30$ (the alternative statement is the contradiction of prior belief). We used MINITAB to perform a t test; the output is given below. SEMEAN denotes the standard error of the mean, which is $s/\sqrt{n} = .57$. Then $t = (29.33 - 30)/.57 = -1.16$. The P-value for this lower-tailed test is .15, the smallest level at which H_0 can be rejected. Thus even at level of significance .10, H_0 cannot be rejected ($\alpha < P$-value). The data does not contradict the prior belief.

```
TEST OF MU = 30.0 VS MU L.T. 30.0

                N        MEAN      STDEV    SE MEAN       T       P VALUE
                6       29.33       1.41       0.57    -1.16          0.15
```

■

EXERCISES

8.43 Let μ denote the true average surface area covered by 1 gal of a certain paint. $H_0 : \mu = 400$ is to be tested against $H_a : \mu > 400$. Assuming that coverage is normally distributed, give the appropriate test statistic and rejection region for each given sample size and significance level.

a. $n = 10, \alpha = .05$

b. $n = 18, \alpha = .01$

c. $n = 25, \alpha = .001$

d. $n = 50, \alpha = .10$

8.44 Suppose that the amount of air pressure in new tires of a certain type sold by a particular store is normally distributed with mean value μ. For testing $H_0 : \mu = 30$ versus $H_a : \mu \neq 30$, give the test statistic and rejection region for each given sample size and significance level listed in Exercise 8.43.

8.45 If the P-value for a particular hypothesis test was reported to be between .025 and .05, would H_0 be rejected at significance level .01? At level .05? At level .1?

8.46 A researcher collected data in order to test $H_0 : \mu = 17$ versus $H_a : \mu > 17$. Place bounds on the P-value for each of the given t test statistic values and associated degrees of freedom.

a. $t = 1.84$, df $= 14$

b. $t = 3.74$, df $= 25$

c. $t = 2.42$, df $= 13$

d. $t = 1.32$, df $= 8$

e. $t = 2.67$, df $= 45$

8.47 Place bounds on the P-value for a two-tailed t test for each case.

a. $t = 2.3$, df $= 6$

b. $t = -3.0$, df $= 14$

c. $t = 4.2$, df $= 24$

d. $t = -1.3$, df $= 17$

8.48 The National Bureau of Standards had previously reported the value of selenium content in NBS orchard leaves to be .08 ppm. The paper "A Neutron Activation Method for Determining Submicrogram Selenium in Forage Grasses" (*Soil Sci. Soc. Amer. J.* (1978):57–60) reported the following selenium content for five determinations:

.072 .073 .080 .078 .088

a. Construct a normal probability plot for this data.

b. What assumption about the selenium content distribution must you be willing to make in order to use a t test for testing $H_0 : \mu = .08$ versus $H_a : \mu \neq .08$?

c. Based on your plot from (a), do you feel comfortable in making the assumption of normality required by the t test? Explain why or why not.

d. Use a t test at level .01 to test the hypotheses stated in (b).

8.49 The accompanying radiation readings (mR/h) were obtained from television display areas in a sample of 10 department stores ("Many Color TV Set Lounges Show Highest Radiation" *J. Environ. Health* (1969):359–60).

.40 .48 .60 .15 .50 .80 .50 .36 .16 .89

The recommended limit for this type of radiation exposure is .5 mR/h. Assuming that the observations come from a normal distribution with mean μ (the true average amount of radiation in television display areas in all department stores), test $H_0 : \mu = .5$ versus $H_a : \mu > .5$ using a level .1 test.

8.50 The times of first sprinkler activation for a series of tests with fire prevention sprinkler systems using an aqueous film-forming foam were (in s)

27 41 22 27 23 35 30 33 24 27 28 22 24

(See "Use of AEFF in Sprinkler Systems" *Fire Technology* (1976):5). The system has been designed so that the true average activation time is supposed to be at most 25 s. Does the data strongly indicate that the design specifications have not been met?

a. Test the relevant hypotheses using a significance level of .05. What assumptions are you making about the distribution of activation times?

b. Obtain upper and lower bounds for the P-value associated with the test in (a).

8.51 The IQ of adults is thought to be normally distributed with mean 100. Suppose 10 randomly selected prisoners convicted of felony offenses had IQ's of 100, 135, 108, 94, 111, 96, 99, 104, 109, and 120. Using a level .05 test, can you conclude that the mean IQ of those convicted of felony offenses is significantly different from that of the general population?

8.52 Gymnastics is a sport that relies entirely on judges to determine winners. As a result, psychological factors that might affect the way a judge scores a gymnastics event are of interest. The paper "Judging Bias Induced by Viewing Contrived Videotapes: A Function of Selected Psychological Variables" (*J. Sport Psych.* (1983): 427–37) reports on a study of how information on other judges scoring affects assigned score on an event.

a. Twelve gymnastics judges were shown a videotape of a floor exercise in which, through editing, high falsified scores appeared in the background at the end of the tape. Each judge was then asked to score the routine. When originally scored, the routine was given an average score of 7.32. The average score assigned by the sample of 12 judges was 7.40 and the sample standard deviation was .19. Does this data suggest that viewing the high falsified scores results in a higher mean score for the routine? Use a level .05 test.

b. A second group of 12 judges was shown the same routine (which had an unadjusted mean score of 7.32) with low falsified scores. The average score assigned by the sample of 12 judges was 6.92 and the sample standard deviation was .29. Is there evidence that viewing the low false scores results in a lower mean score? Use a level. 05 test.

8.53 Federal officials are currently investigating the problems associated with disposal of hazardous wastes. One disposal site is the abandoned Stringfellow acid pits in Riverside County, California. The EPA had sampled water from 11 wells in nearby Glen Avon. Radiation levels from 38 to 67 pCi/L were observed (*Los Angeles Times,* May 31, 1984). The EPA standard for maximum allowable radiation level for drinking water is 15 pCi. Suppose that the sample of 11 wells had resulted in a sample mean radiation level of 52.5 pCi and a sample standard deviation of 8.

a. Use a level .01 test to determine whether the data strongly suggests that the mean radiation level exceeds the EPA standard.

b. Place a bound on the P-value associated with the test statistic in (a).

8.54 A number of veterinary procedures on pigs require the use of a general anesthetic. To evaluate the effects of a certain anesthetic, it was administered to four pigs and various bodily functions were measured ("Xylazine-Ketamine-Oxymorphone: An Injectable Anesthetic Combination in Swine" *J. Amer. Vet. Med. Assoc.* (1984):182–84). Average normal heart rate for pigs is considered to be 114 beats per minute. The heart rates for the four pigs under anesthesia were 116, 85, 118, and 118. Use a level .10 test to determine whether the anesthetic results in a mean heart rate that differs significantly from the mean normal heart rate.

8.55 A certain type of soil was determined to have a natural mean pH value of 8.75. The authors of the paper "Effects of Brewery Effluent on Agricultural Soil and Crop Plants" (*Environ. Pollution* (1984):341–51) treated soil samples with various dilutions of an acidic effluent. Five soil samples were treated with a solution of 25% water and 75% effluent. The mean and standard deviation of the 5 pH measurements were 8.00 and .05, respectively.

a. Does this data indicate that at this concentration the effluent results in a mean pH that exceeds the natural pH of the soil? Use a level .01 test.

b. Place bounds on the P-value associated with the test statistic in (a). Use the bounds on the P-value to decide whether the conclusion reached in (a) would have been different if a significance level of .10 had been chosen.

8.6 Large-Sample Hypothesis Tests for a Population Proportion

Let π denote the proportion of individuals or objects in a specified population that possess a certain property—those that are successes as opposed to failures. If a random sample of n individuals or objects is selected from the population, then

$$p = \begin{pmatrix}\text{sample proportion} \\ \text{of successes}\end{pmatrix} = \frac{\text{number of successes in the sample}}{n}$$

is the natural statistic for making inferences about π. The computed value of p for a particular sample gives a point estimate of π. The large-sample confidence interval for π in Chapter 7 was also based on p. The following are important properties of p's sampling distribution:

1. $\mu_p = \pi$, so p is an unbiased statistic for estimating π.

2. $\sigma_p = \sqrt{\pi(1 - \pi)/n}$ (which decreases in value as n increases).

3. When n is sufficiently large, the sampling distribution is approximately normal (n is sufficiently large if both $n\pi \geq 5$ and $n(1 - \pi) \geq 5$).

Properties 1–3 imply that when n is large, the sampling distribution of the standardized variable

$$z = \frac{\text{statistic} - \text{mean value}}{\text{standard deviation}} = \frac{p - \pi}{\sqrt{\pi(1 - \pi)/n}}$$

is well approximated by the standard normal curve.

A null hypothesis regarding a population proportion will specify a particular value of π. Consider as an example $H_0 : \pi = .8$. When H_0 is true, $\mu_p = .8$ and $\sigma_p = \sqrt{(.8)(.2)/n}$. Using these values in the aforementioned z yields the test statistic $z = (p - .8)/\sqrt{(.8)(.2)/n}$. That is, the test statistic results from standardizing p under the assumption that H_0 is true. Since the approximate sampling distribution of z when H_0 is true is the z curve, a level α test is determined by using an appropriate z critical value to specify the rejection region. Whether the region is upper-tailed, lower-tailed, or two-tailed depends on which of the three inequalities, $>$, $<$, or $\neq$, appears in H_a. Notice that the value of π specified by H_0 determines the standard deviation $\sqrt{\pi(1 - \pi)/n}$. This standard deviation does not have to be estimated from the sample as it was in the confidence interval formula.

Summary of Large-Sample z Tests for π

Null hypothesis: $H_0 : \pi =$ hypothesized value

Test statistic: $$z = \frac{p - \text{hypothesized value}}{\sqrt{(\text{hypothesized value})(1 - \text{hypothesized value})/n}}$$

Alternative hypothesis	*Rejection region*
$H_a : \pi >$ hypothesized value	Reject H_0 if $z > z$ critical value (upper-tailed test)
$H_a : \pi <$ hypothesized value	Reject H_0 if $z < -z$ critical value (lower-tailed test)

$H_a : \pi \neq$ hypothesized value	Reject H_0 either if $z > z$ critical value or if $z < -z$ critical value (two-tailed test)

z critical values corresponding to the most frequently used levels of significance appear in the bottom row of Table IV. The test is appropriate if both n(hypothesized value) ≥ 5 and $n(1 -$ hypothesized value) ≥ 5.

Before looking at several examples, a word of warning is in order. Small-sample tests about π are *not* based on the t distribution but instead on the binomial distribution discussed in Chapter 5. So in the rare π problem in which n is small, don't simply replace the z critical value above by a t critical value and use the same test statistic.

EXAMPLE 24

The article "Statistical Evidence of Discrimination" (*J. Amer. Stat. Assoc.* (1982): 773–83) discussed the court case *Swain* v. *Alabama* (1965), in which it was alleged that there was discrimination against Blacks in grand jury selection. Census data suggested that 25% of those eligible for grand jury service were Black, yet a random sample of 1050 called to appear for possible duty yielded only 177 Blacks. Using a level .01 test, does this data argue strongly for a conclusion of discrimination?

1. The population characteristic of interest here is $\pi =$ the true proportion of all those called for possible service who are Black.

2. $H_0 : \pi = .25$.

3. $H_a : \pi < .25$ (discrimination exists).

4. Since $n \cdot$ (hypothesized value) $= 1050(.25) \geq 5$ and $n \cdot (1 -$ hypothesized value) $= 1050(.75) \geq 5$, the large-sample test is appropriate. The test statistic is $z = (p - .25)/\sqrt{(.25)(.75)/n}$.

5. The inequality in H_a implies the use of a lower-tailed test, with H_0 rejected if $z < -z$ critical value. From the bottom row of Table IV, z critical value $= 2.33$ for a one-tailed level .01 test. The rejection region is then $z < -2.33$.

6. The denominator of z is $\sqrt{(.25)(.75)/1050} = .0134$ and $p = 177/1050 = .169$, so

$$z = \frac{.169 - .250}{.0134} = \frac{-.081}{.0134} = -6.04$$

7. Since $-6.04 < -2.33$, H_0 is rejected at level .01. Evidence of discrimination seems very clear. Unfortunately, the court looked only at the numerator difference $-.081$ rather than z itself. In the court's view, the difference was not large enough to establish a prima facie (without further examination) case. ■

EXAMPLE 25

Environmental problems associated with leaded gasolines are well known. Many motorists have tampered with emission-control devices in order to save money by purchasing leaded rather than unleaded gas. A *Los Angeles Times* article (March 17, 1984) reported that 15% of all California motorists have engaged in such tampering. Suppose that a random sample of 200 cars from a particular county is obtained, and the emission control devices of 21 are found to have been tampered with. Does this suggest that the proportion of cars in this county with tampered devices differs from the statewide proportion? We use a test with level of significance .05.

1. $\pi =$ the proportion of cars in this county whose emission control devices have been tampered with.

2. $H_0 : \pi = .15$.

3. $H_a : \pi \neq .15$.

4. Since $(200)(.15) \geq 5$ and $(200)(.85) \geq 5$, the z test can be used. The test statistic is

$$z = \frac{(p - .15)}{\sqrt{(.15)(.85)/n}}$$

5. Table IV shows that the critical value for a two-tailed level .05 test is 1.96, so H_0 will be rejected if either $z > 1.96$ or $z < -1.96$.

6. $\sqrt{(.15)(.85)/200} = .0252$ and $p = 21/200 = .105$, so

$$z = \frac{(.105 - .150)}{.0252} = \frac{-.045}{.0252} = -1.79$$

7. Since -1.79 is neither greater than 1.96 nor less than -1.96, H_0 cannot be rejected at level .05. The data does not suggest that the proportion of cars in this county having devices that have been tampered with differs from the statewide proportion. ■

P-Values

Because the test procedures discussed here are based on the z distribution and critical values, exact P-values and bounds are obtained using the same method described in Section 8.4 for large-sample z tests concerning μ.

EXAMPLE 26

An article in the April 6, 1983, *Los Angeles Times* reported on a study carried out on 53 learning-impaired youngsters at Massachusetts General Hospital. The right side of the brain was found to be larger than the left side in 22 of the children. The proportion of the general population with brains having larger right sides is known to be .25 (25%). Does this provide strong evidence for concluding, as the article claims, that the proportion of learning-impaired youngsters with brains having larger right sides exceeds the proportion in the general population? Let's test the appropriate hypotheses by computing the P-value and reaching a decision accordingly.

The hypotheses to be tested are $H_0 : \pi = .25$ versus $H_a : \pi > .25$, where π denotes the proportion of learning-disabled children with brains having larger right than left sides. With $p = 22/53 = .415$ and $\sqrt{(.25)(.75)/53} = .0595$, $z = (.415 - .25)/.0595 = 2.8$. The P-value for this upper-tailed test is then

$$P\text{-value} = \begin{pmatrix} \text{area under the } z \text{ curve} \\ \text{to the right of } 2.8 \end{pmatrix} = .0026$$

Thus at any significance level that exceeds .0026, H_0 is rejected. In particular, H_0 is rejected at level .01 in favor of the conclusion that there is a greater tendency among learning-impaired children to have brains with larger right sides than for the general population.

To obtain bounds rather than the exact P-value, the bottom row of Table IV is used. The critical value for a level .005 one-tailed test is 2.58 and for a level .001 test is 3.09. Since $2.58 < z = 2.8 < 3.09$, $.001 < P\text{-value} < .005$. The exact P-value .0026 of course satisfies these inequalities. Again H_0 is rejected at level .01 (but not at level .001). ■

EXERCISES

8.56 About 10% of the U.S. population is left handed. A random sample of size 200 is to be selected. Describe the approximate sampling distribution of p, the sample proportion who are left handed (where it is centered, how much it spreads out, and what shape it has).

8.57 A large appliance dealer sells both VHS and Beta VCR's. The trend seems to be toward VHS's, so the general manager is thinking of dropping Beta VCR's. Let π denote the proportion of recent VCR sales that are Betas. It is decided to test $H_0: \pi = .25$ versus $H_a: \pi < .25$ and drop Betas if H_0 can be rejected at level .01. If $n = 100$ recent VCR purchases are randomly sampled and 21 are Betas, what decision is appropriate?

8.58 A plan for an executive traveler's club has been developed by an airline on the premise that 5% of its current customers would qualify for membership. Let π denote the proportion of current customers who would, in fact, qualify. A random sample of 500 customers yields $p = .08$ as the sample proportion of qualifiers. Use this data to test $H_0: \pi = .05$ versus $H_a: \pi \neq .05$ at level of significance .10.

8.59 A telephone company is trying to decide whether some new lines in a large community should be installed underground. Because a small surcharge will be added to telephone bills to pay for the extra installation costs, the company has decided to survey customers and proceed only if the survey strongly indicates that more than 60% of all customers favor underground installation. If 118 of 160 customers surveyed favor underground installation in spite of the surcharge, what should the company do? Test using significance level .05.

8.60 The incidence of a certain type of chromosome defect in the U.S. adult male population is believed to be 1 in 80. A random sample of 600 individuals in U.S. penal institutions reveals 12 who have such defects. Can it be concluded that the incidence rate of this defect among prisoners differs from the presumed rate for the entire adult male population?

a. State and test the relevant hypotheses using $\alpha = .05$. What type of error might you have made in reaching a conclusion?

b. What P-value is associated with this test?

8.61 To test the ability of auto mechanics to identify simple engine problems, an automobile with a single such problem was taken in turn to 72 different car repair facilities. Only 42 of the 72 mechanics who examined the car correctly identified the problem. Does this strongly indicate that the true proportion of mechanics who could identify this problem is less than .75? Compute a P-value and use it to draw your conclusion.

8.62 Scientists think that robots will play a crucial role in factories in the next 20 years. Suppose that in an experiment to determine whether the use of robots to weave computer cables is feasible, a robot was used to assemble 500 cables. The cables were examined and there were 14 defectives. If human assemblers have a defect rate of .03 (3%), does this data support the hypothesis that the proportion of defectives is lower for robots than humans? Use a .01 significance level.

8.63 The psychological impact of subliminal advertising has been the subject of much speculation. To decide whether people believe that the use of subliminal advertising is ethical, the authors of the paper "Public Perceptions of Subliminal Advertising" (*J. Adver.* (January 1984):40–44) conducted a survey of 145 residents of Washington, D.C. Of those surveyed, 58 felt that the use of subliminal advertising was acceptable. Does this data provide sufficient evidence to conclude that fewer than half of Washington's residents find subliminal advertising acceptable? Use a level .05 test.

8.64 A U.S. House of Representatives subcommittee has been hearing testimony on a possible link between problem pregnancies and working with video display terminals (VDT's). A survey of employees of United Airlines who work full time on VDT's found that of 48 pregnancies, 15 resulted in miscarriage (*Los Angeles Times*, March 11, 1984). According to the March of Dimes, there is a 10% miscarriage rate for the general population.

a. Does the data strongly indicate that the miscarriage rate of women who work full time on VDT's is higher than that of the general population? Use a level .01 test.

b. On the basis of your work in (a), could it be concluded that full-time work on VDT's tends to cause miscarriages? Explain.

c. If the results of the hypothesis test in (a) are to be reported using a P-value, what value should be given?

8.65 Medical researchers have searched for an effective method of treating virus infections. One such infection, herpes virus encephalitis (an inflammation of the brain), has a mortality rate of 70% (*Newsweek*, August 22, 1977). A study sponsored by the National Institute of Allergy and Infectious Diseases involved 28 victims of herpes virus encephalitis. The subjects received a new medication, ara-A. Five of the 28 subsequently died.

a. Does this data strongly suggest that ara-A is an effective treatment (i.e., lowers the mortality rate) for this particular virus? Use a significance level of .01.

b. What P-value is associated with the test statistic value computed in (a)?

8.66 The use of animation in television advertisements is becoming more widespread. A content analysis of 2454 advertisements appearing at varying times over a 1-week period on the three major networks revealed that 236 used some form of animation (*J. Adver.* (April 1983):20–25). Does this data support the hypothesis that the proportion of advertisements using animation is less than .1? Use a level .1 test.

8.67 A federal agency maintains that drunken driving is involved in 50% of all fatal car accidents. A medical examiner in Fulton County, Georgia, conducted blood alcohol tests on 46 accident victims and found that 37 would have been classified as legally intoxicated (Associated Press, January 9, 1984). Let π denote the true proportion of fatal car accidents involving drunken driving for Fulton County. Use a level .01 test to decide whether this data suggests that π exceeds .5.

8.68 A researcher believes that the percentage of left-handed people among college graduates is higher than the 10% figure for the entire population. Let π denote the proportion of left-handed people among all college graduates, and consider the hypotheses $H_0 : \pi = .10$ versus $H_a : \pi > .10$. A random sample of 10,000 college graduates yields 1080 left-handed people. Is this a statistically significant result? (Use $\alpha = .01$.) Is there any practical significance in what was observed? Comment.

8.7 Type II Error Probabilities for Selected Tests (Optional)

The test procedures presented in this chapter are designed to control the probability of a type I error (rejecting H_0 when H_0 is true) at the desired level α. However, little has been said so far about β, the probability of a type II error (not rejecting H_0 when H_0 is false). Remember that to control α, we needed to know the sampling distribution of the test statistic when H_0 is true. This allowed us to select a critical value to capture tail area α under the sampling distribution curve. Computation of β is more difficult than determination of α because the sampling distribution of a test statistic when H_0 is false is usually substantially more complicated than when H_0 is true. Fortunately, statisticians have managed to surmount these difficulties in the case of some commonly used test procedures. Here we consider the determination of β for two important tests: (1) the t test and (2) the large-sample z test for testing $H_0 : \pi =$ hypothesized value. In the former case, β can be read from a set of graphs specially constructed for this purpose, whereas in the latter case, areas under the standard normal curve are used.

The t Test

While α is a single number (equal to P(rejecting H_0 when H_0 is true)), the same is not true of β. Instead, there is a value of β for each value of the population characteristic for which H_a is true. Consider, for example, testing $H_0 : \mu = 100$ versus $H_a : \mu > 100$ using the t test at level $\alpha = .05$. Then there is a value of β for $\mu = 101$, a value for $\mu = 105$, a value for $\mu = 110$, and a value for any other μ that exceeds 100. Similarly, when the hypotheses are $H_0 : \mu = 100$ versus $H_a : \mu \neq 100$, there is not only a value of β for each μ exceeding 100 but also for each μ less than 100.

The value of β for a specified alternative value of μ depends on the level of significance α. We have already commented and will see from the graphs of β that β increases when α is made smaller. This implies that for any fixed alternative value, β for a level .01 test is larger than β for a level .05 test. In addition, β depends on the number of degrees of freedom, $n - 1$. For any fixed level α, it should be easier for the test to detect a specific departure from H_0 when n is large than when n is small. This is indeed the case; for a fixed alternative value, β decreases as $n - 1$ increases.

There is unfortunately one other quantity on which β depends—the population standard deviation σ. Consider testing $H_0 : \mu = 100$ versus $H_a : \mu > 100$ and focus on the alternative value $\mu = 110$. Figure 6 pictures the population distribution both when $\mu = 110$, $\sigma = 10$ and when $\mu = 110$, $\sigma = 2.5$. In both cases, H_0 is false and we would like β to be small. When

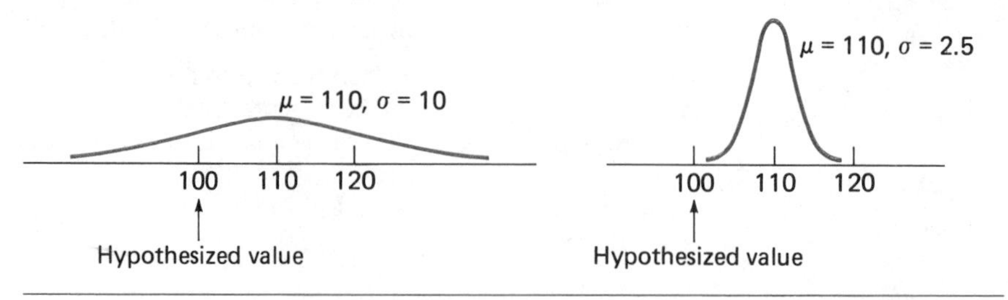

FIGURE 6 TWO NORMAL POPULATION DISTRIBUTIONS FOR WHICH $H_0 : \mu = 100$ IS FALSE BECAUSE $\mu = 110$

$\sigma = 2.5$, virtually all the sampled observations considerably exceed 100, making it rather obvious that H_0 is false. But when $\sigma = 10$, there is a good chance that some of the observations will fall considerably below 110, resulting in a sample whose $\overline{x}$ value does not strongly contradict H_0. That is, it will be easier to detect a departure from H_0 when σ is small (in which case β will be small) than when σ is relatively large.

Once α is specified and n is fixed, the determination of β at a particular alternative value of μ requires that a value of σ be chosen, since each different value of σ yields a different value of β. If the investigator can specify a range of plausible values for σ, then using the largest such value will give a pessimistic β (one on the high side).

Figure 7 pictures three different β curves for a one-tailed t test (appropriate for $H_a : \mu >$ hypothesized value or for $H_a : \mu <$ hypothesized value). A more complete set of curves for both one- and two-tailed tests when $\alpha = .05$ and when $\alpha = .01$ appear in Table V of the appendices. To determine β, first compute the quantity

$$d = \frac{|\text{alternative value} - \text{hypothesized value}|}{\sigma}$$

Then locate d on the horizontal axis, move directly up to the curve for $n - 1$ df, and move over to the vertical axis to read β.

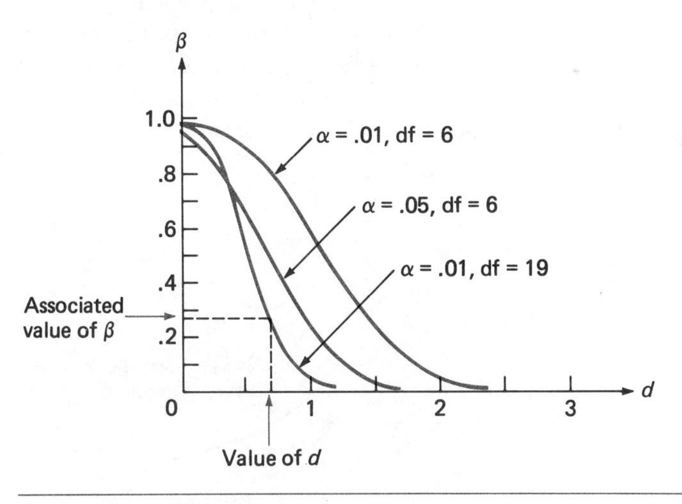

FIGURE 7 β CURVES FOR THE ONE-TAILED t TEST

EXAMPLE 27

Consider testing $H_0 : \mu = 100$ versus $H_a : \mu > 100$ and focus on the alternative value $\mu = 110$. For $\sigma = 10$, $d = |110 - 100|/10 = 10/10 = 1$. If $n = 7$ $(n - 1 = 6)$ and a level .01 test is used, Figure 7 gives $\beta \approx .6$. The interpretation is that if $\sigma = 10$ and a level .01 test based on $n = 7$ is used, when H_0 is false because $\mu = 110$, roughly 60% of all samples will result in erroneously not rejecting H_0! If a level .05 test is used instead, then $\beta \approx .3$, which is still rather large. Using a level .01 test with $n = 20$ (df $= 19$) yields, from Figure 7, $\beta \approx .05$. At the alternative 110, for $\sigma = 10$ the level .01 test based on $n = 20$ has smaller β than the level .05 test with $n = 7$. Substantially increasing n counterbalances using the smaller α.

Now consider the alternative $\mu = 105$, again with $\sigma = 10$, so that $d = |105 - 100|/10 = 5/10 = .5$. Then from Figure 7, $\beta \approx .95$ when $\alpha = .01$, $n = 7$; $\beta \approx .7$ when $\alpha = .05$, $n = 7$; and $\beta \approx .65$ when $\alpha = .01$, $n = 20$. These values of β are all quite large; with $\sigma = 10$, $\mu = 105$ is too close to the hypothesized value of 100 for any of these three tests to have a good chance of detecting such a departure from H_0. A substantial decrease in β necessitates using a much larger sample size. For example, from Table V, $\beta \approx .08$ when $\alpha = .05$ and $n = 40$.

The curves in Figure 7 also give β when testing $H_0 : \mu = 100$ versus $H_a : \mu < 100$. If the alternative value $\mu = 90$ is of interest and $\sigma = 10$, $d = |90 - 100|/10 = 10/10 = 1$ and values of β are the same as those given in the first paragraph of this example. For the alternative $\mu = 95$, $d = |95 - 100|/10 = .5$ and values of β are as given in the second paragraph. ■

Since curves for only selected degrees of freedom appear in Table V, other degrees of freedom require a visual approximation. For example, the 27-df curve (for $n = 28$) would lie between the 19- and 29-df curves, which do appear, and would be closer to the latter. This type of approximation is adequate because it is the general magnitude of β—large, small, or moderate—that is of primary concern.

The curves can also be used to find a value of $n - 1$ (and thus n) for which β has a specified value at a particular alternative. Find the point at which a vertical line through the value of d and a horizontal line through the specified value of β intersect. The nearest curve passing below this point gives the necessary value of $n - 1$. This is pictured in Figure 8, where visual approximation is used to identify the closest curve.

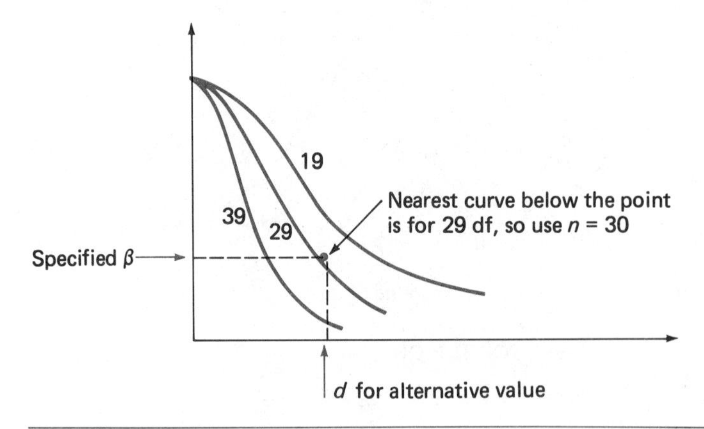

FIGURE 8 DETERMINING $n - 1$ FOR A SPECIFIED d AND β

EXAMPLE 28 A farm supply store that packs its own fertilizer in 50-lb bags is being investigated for underfilling the bags. With μ denoting the true average net weight of all bags, the hypotheses of interest are $H_0 : \mu = 50$ versus $H_a : \mu < 50$. Only if H_0 can be rejected will the store be formally accused of cheating its customers. To avoid an erroneous accusation of cheating (a very serious error), a small level of significance is appropriate. Suppose that $\alpha = .01$ is selected. The investigators feel that the alternative value $\mu = 49.8$ represents a substantial departure from H_0, so they are concerned that β should be small for this alternative. Assuming that net weight is normally distributed and that $\sigma = .25$, what would β be if the test is based on the contents of $n = 15$ bags? What value of n would be required to have $\beta = .05$ for this alternative?

The d value for this alternative and σ is $d = |49.8 - 50|/.25 = .20/.25 = .8$. Using the β curves for a level .01 one-tailed test gives $\beta \approx .33$ from the $n - 1 = 15 - 1 = 14$-df curve. To obtain $\beta = .05$ at this alternative (value of d), a much larger value of n than 15 is clearly required. A horizontal line at height $\beta = .05$ and vertical line through $d = .8$ intersect right on the 29-df curve. Thus $n = 30$ is sufficient to obtain a suitably small β at the alternative of interest, provided that $\sigma = .25$. If σ is actually larger than .25, then $n = 30$ will, of course, not suffice. ■

The β curves in Table V are those for t tests. When the alternative of interest corresponds to a value of d relatively close to zero, β for a t test may be rather large. One might ask whether there is another type of test that has the same level of significance α as does the t test and smaller values of β. For example, in Example 28 we had $\beta \approx .33$ for a hypothesized value of 50, alternative $\mu = 49.8$, $\sigma = .25$ (so $d = .8$), and a level .01 one-tailed t test. Is there some other test procedure, based on a different test statistic and rejection region, that has $\alpha = .01$ and $\beta < .33$ under these circumstances? The following result provides the answer to this question.

> When the population distribution is normal, the level α t test for testing hypotheses about μ has smaller β than does any other test procedure that has the same level of significance α.

Stated another way, among all tests with level of significance α, the t test makes β as small as it can possibly be. In this sense, the t test is a best test. Statisticians have also shown that when the population distribution is not too far from a normal distribution, no test procedure can improve on the t test (have the same α and substantially smaller β) by very much. But when the population distribution is believed to be very nonnormal (heavy-tailed, highly skewed, or multimodal), the t test should not be used. Then it's time to consult your friendly neighborhood statistician, who could provide you with alternative methods of analysis.

The z Test for a Population Proportion

The z tests concerning a population proportion π are among the very few tests in statistics for which β, the probability of a type II error, is easily calculated. To see how this is done, consider testing $H_0: \pi = .5$ versus $H_a: \pi > .5$ based on a random sample of size $n = 100$. The denominator of the test statistic z is $\sqrt{(.5)(.5)/100} = .05$. The test with $\alpha = .01$ rejects H_0 if $(p - .5)/.05 > 2.33$. This inequality is equivalent to $p > (.05)(2.33) + .5 = .617$. H_0 will then not be rejected if $p \leq .617$.

Consider now the alternative value $\pi = .7$. When .7 is the true value of π, the sampling distribution of p is approximately normal with $\mu_p = .7$ and

$\sigma_p = \sqrt{(.7)(.3)/100} = .0458$. In this case, a type II error results if $p \le .617$, so

$$\beta = P(\text{type II error})$$
$$= P\begin{pmatrix} p \le .617 \text{ when } p \text{ is approximately} \\ \text{normal with } \mu = .7, \sigma = .0458 \end{pmatrix}$$

This probability can be computed simply by standardizing (subtracting .7 and dividing by .0458) to obtain the z score and the corresponding z curve area, as shown in Figure 9. Table IV gives β = area to the left of $-1.8 = .0359$. That is, when $\pi = .7$, only 3.6% of all samples with $n = 100$ will result in erroneously not rejecting H_0. This error probability is small because $\pi = .7$ is rather far from .5, and such a departure from H_0 is easy to detect when $n = 100$.

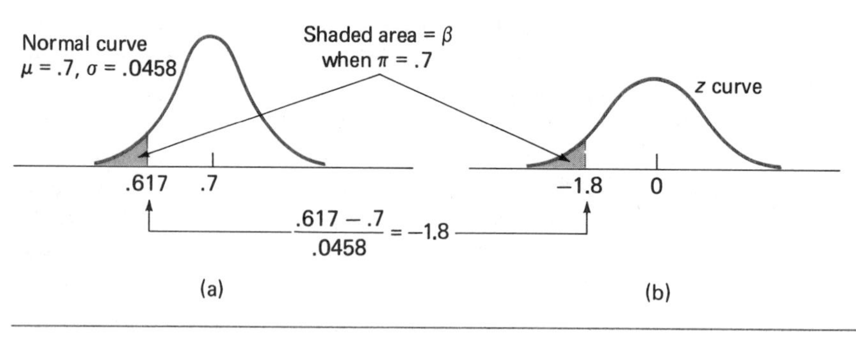

FIGURE 9 $\beta = P(\text{TYPE II ERROR WHEN } \pi = .7)$
(a) Approximate sampling distribution of p when $\pi = .7$
(b) β represented as a z curve area

EXAMPLE 29

A package delivery service advertises that at least 90% of all packages brought to its office by 9 A.M. for delivery in the same city are delivered by noon that day. Let π denote the proportion of all such packages actually delivered by noon. The hypotheses of interest are $H_0: \pi = .9$ versus $H_a: \pi < .9$, where the alternative hypothesis states that the company's claim is untrue. The value $\pi = .8$ represents a substantial departure from the company's claim. If the hypotheses are tested at level .01 using a sample of $n = 225$ packages, what is the probability that the departure from H_0 represented by this alternative value will go undetected?

At significance level .01, H_0 is rejected if $z < -2.33$. This is equivalent to rejecting H_0 if $p < .9 - 2.33\sqrt{(.9)(.1)/225} = .853$. When $\pi = .8$, p has approximately a normal distribution with $\mu_p = .8$ and $\sigma_p = \sqrt{(.8)(.2)/225} = .0267$. Then β is the area under this curve to the right of .853. Standardizing .853 gives $(.853 - .8)/.0267 = 1.98 \approx 2.0$, so

$$\beta = \begin{pmatrix} \text{area under the } z \text{ curve} \\ \text{to the right of 2.0} \end{pmatrix} = .0228$$

When $\pi = .8$ and a level .01 test is used, less than 3% of all samples consisting of 225 packages will result in H_0 being erroneously not rejected. ■

There are formulas available for the sample size n necessary to achieve a specified β at a particular alternative value. Consult a more advanced text for details.

EXERCISES

8.69 The amount of shaft wear after a fixed mileage was determined for each of seven internal combustion engines, resulting in a mean of .0372 in and a standard deviation of .0125 in.

a. Assuming that the distribution of shaft wear is normal, test at level .05 the hypotheses $H_0 : \mu = .035$ versus $H_a : \mu > .035$.

b. Using $\sigma = .0125$, $\alpha = .05$, and Table V, what is the value of β, the probability of a type II error, when $\mu = .04$?

8.70 Optical fibers are used in telecommunications to transmit light. Current technology allows production of fibers that will transmit light about 50 km (*Research at Rensselaer,* 1984). Researchers are trying to develop a new type of glass fiber that will increase this distance. In evaluating a new fiber, it would be of interest to test $H_0 : \mu = 50$ versus $H_a : \mu > 50$, with μ denoting the true average transmission distance for the new optical fiber.

a. If a level .05 test is to be used and 10 measurements are to be made ($n = 10$), what would be the rejection region for the test?

b. Assuming $\sigma = 10$ and $n = 10$, use Table V to find β, the probability of a type II error, for each of the given alternative values of μ when a level .05 test is employed.

i. 52 ii. 55 iii. 60 iv. 70

8.71 The city council in a large city has become concerned about the trend toward exclusion of renters with children in apartments within the city. The housing coordinator has decided to select a random sample of 125 apartments and determine for each whether or not children would be permitted. Let π be the true proportion of apartments that prohibit children. If π exceeds .75, the city council will consider appropriate legislation.

a. If 102 of the 125 sampled exclude renters with children, would a level .05 test lead you to the conclusion that more than 75% of all apartments exclude children?

b. What is the probability of a type II error when $\pi = .8$?

8.72 Refer to the executive travelers' club described in Exercise 8.58.

a. What is the probability of concluding that the premise is correct when, in fact, the true proportion of qualifiers is .10?

b. Repeat (a) when the true proportion of qualifiers is only .03.

c. Repeat (b) when a level .01 test is used (rather than level .10).

d. Repeat (c) when $n = 1000$ (rather than $n = 500$).

8.73 Refer to Exercise 8.65.

a. Suppose that the actual mortality rate for this new medication is 60%. What is β, the probability of a type II error, for the test?

b. Repeat (a) for an actual mortality rate of 60% and a sample size of $n = 100$ and then for an actual mortality rate of 50% when $n = 100$.

8.74 Let π denote the proportion of defective components in a very large shipment sent to a computer manufacturer. The manufacturer decides to test $H_0 : \pi = .10$ versus $H_a : \pi < .10$ and accept the lot only if H_0 can be rejected at level .05 based on a random sample of 100 components. Compute the probability of failing to reject H_0 when the actual percentage of defectives is as given.

a. 7% **b.** 8% **c.** 9%

KEY CONCEPTS

Hypothesis An assertion about the value of a population characteristic. (p. 301)

Null hypothesis H_0 The hypothesis initially assumed to be true. It has the form H_0 : population characteristic = hypothesized value. (p. 302)

Alternative hypothesis H_a Specifies a claim that is contradictory to H_0 and is judged the more plausible claim when H_0 is rejected. It takes one of the three forms

$$H_a : \text{population characteristic} \begin{pmatrix} > \\ < \\ \neq \end{pmatrix} \text{hypothesized value.} \qquad \text{(p. 302)}$$

Test procedure A method for deciding whether or not to reject H_0 in favor of H_a. (p. 306)

Test statistic The quantity computed from sample data and used to make a decision. (p. 306)

Rejection region All values of the test statistic for which H_0 would be rejected. The rejection region is either upper-tailed (consisting only of large values of the test statistic), lower-tailed, or two-tailed, depending on the form of H_a. (p. 306)

Type I error Rejecting H_0 when H_0 is true. The probability of a type I error is denoted by α. (p. 309)

Type II error Not rejecting H_0 when H_0 is false. The probability of a type II error is denoted by β, and there is a different value of β for each different alternative value of the population characteristic. (p. 309)

Level of significance The maximum tolerable type I error probability. A level α test is one for which the type I error probability is controlled at the specified α. (p. 310)

P-value The smallest significance level α at which H_0 can be rejected. (p. 326)

Specific Test Procedures

Population Characteristic	*When Appropriate*	*Test Statistic*	*Rejection Region*
μ	n large	$z = \dfrac{\bar{x} - \text{hypothesized value}}{s/\sqrt{n}}$	Based on a z critical value (Section 8.3)
μ	Normal population	$t = \dfrac{\bar{x} - \text{hypothesized value}}{s/\sqrt{n}}$	Based on a t critical value with $n - 1$ df (Section 8.5)
π	n large	$z = \dfrac{p - \text{hypothesized value}}{\sqrt{\dfrac{(\text{hyp. value})(1 - \text{hyp. value})}{n}}}$	Based on a z critical value (Section 8.6)

SUPPLEMENTARY EXERCISES

8.75 A Norwegian study of 105 males born in 1962 with birth weights of 2500 g or less was described in the article "Males with Low Birthweight Examined at 18 Years of Age" (*J. Amer. Med. Assoc.* (1984):3248). When examined in 1981 by the

Norwegian military draft board, 7 of the 105 were declared unfit for military service. The Norwegian draft board declared 6.2% (a proportion of .062) of all 18 year olds examined in 1981 unfit for military service. Does the data provide sufficient evidence to indicate that the true proportion of males with birthweight of 2500 g or less who are unfit is higher than that of the general population? Use a .01 significance level.

8.76 A standard method for recovering minerals and metals from biological materials results in a mean copper recovery of 63 ppm when used to treat oyster tissue. A new treatment method was described in the paper "Simple Sample Digestion of Sewage and Sludge for Multi-Element Analysis" (*J. Environ. Sci. and Health* (1984): 959–72). Suppose this new treatment is used to treat $n = 40$ bits of oyster tissue, resulting in a sample mean copper recovery and a sample standard deviation of 62.6 ppm and 3.7 ppm, respectively. Is there evidence to suggest that the mean copper recovery is lower for the new method than for the standard? Use a .01 significance level.

8.77 Past experience has indicated that the true response rate is 40% when individuals are approached with a request to fill out and return a particular questionnaire in a stamped and addressed envelope. An investigator believes that if the person distributing the questionnaire is stigmatized in some obvious way, potential respondents would feel sorry for the distributor and thus tend to respond at a rate higher than 40%. To investigate this theory, a distributor is fitted with an eyepatch. Of the 200 questionnaires distributed by this individual, 109 were returned. Does this strongly suggest that the response rate in this situation does exceed the rate in the past?

- **a.** State and test the appropriate hypotheses at significance level .05.
- **b.** Compute the P-value for this data and then use it to carry out a test.

8.78 The drug cibenzoline is currently being investigated for possible use in controlling cardiac arrhythmia. The paper "Quantification of Cibenzoline in Human Plasma by Gas Chromatography-Negative Ion Chemical-Ionization Mass Spectrometry" (*J. Chromatography* (1984):403–09) describes a new method of determining the concentration of cibenzoline in a solution. After 5 ng of cibenzoline was added to a solution, the concentration was measured by the new method. This process was repeated three times, resulting in $n = 3$ concentration readings. The sample mean and standard deviation were reported to be 4.59 ng and .08 ng, respectively. Does this data suggest that the new method produces a mean concentration reading that is too small (less than 5 ng)? Use a .05 significance level and test the appropriate hypotheses.

8.79 The increasing number of senior citizens has made this group an attractive target market for retailers. An understanding of how the elderly feel about various consumer problems is, therefore, important to retailers. The paper "Consumer Problems and Complaint Actions of Older Americans" (*J. of Retailing* (1981):107–23) reported that in a sample of 404 elderly individuals who shop for grocery items, 270 were satisfied with their purchases, whereas 134 were dissatisfied. Suppose that the proportion of all individuals who are satisfied with grocery items is .8 (a value suggested in the paper). Does the sample data suggest that the proportion of elderly people who are satisfied is smaller than the proportion of all individuals who are satisfied? State the relevant hypotheses and carry out the appropriate test using a .01 significance level.

8.80 Police departments across the country have recently voiced concern that too many calls to the 911 emergency telephone number are not true emergencies. Suppose that the police chief in a particular city is contemplating an advertising campaign to warn of the consequences of abusing the 911 number. Because of the cost, the campaign can be justified only if more than 25% of all 911 calls are not emergencies. A random sample of 200 recent calls to the 911 number is selected, and it is determined that 56 were nonemergency calls. Does this sample data support going ahead with the ad campaign? Test the relevant hypotheses using significance level .10.

8.81 To investigate whether sudden infant death syndrome (SIDS) might be related to an imbalance between peptides affecting respiration, the authors of the paper "Post-Mortem Analysis of Neuropeptides in Brains from Sudden Infant Death Victims" (*Brain Research* (1984):279–85) measured cortex met-enkephalin levels (pmol/g wet weight) in brain tissue of 12 SIDS victims. The resulting sample mean and standard deviation were 7.66 and 3.78, respectively. The mean level for children who are not victims of SIDS was reported to be 7.48. Using a .05 significance level, test to determine if the true mean met-enkephalin level of SIDS victims is higher than that of children who are not victims of SIDS.

8.82 The effect of discharging wastewater from a dairy processing plant into groundwater on the growth of kidney beans was examined in the paper "Effect of Industrial Dairy Processing Effluent on Soil and Crop Plants" (*Environ. Pollution* (1984):97–106). The wastewater was rich in bicarbonates and calcium, so it was thought that irrigating with a 50% solution of wastewater would promote growth. Suppose that 40 kidney bean plants are irrigated with this mixture, resulting in a sample mean root length of 5.46 cm and a sample standard deviation of .55. The mean root length for kidney bean plants irrigated with uncontaminated water is known to be 5.20. Does this data support the hypothesis that irrigation with the 50% wastewater solution results in a mean root length that is greater than 5.20? Use a .05 significance level.

8.83 The paper referenced in Exercise 8.82 also gave information on root length for pearl millet. When irrigated with uncontaminated water, the mean root length is 6.40. A sample of 40 plants irrigated with a 50% wastewater solution resulted in a sample mean length of 4.76 cm and a sample standard deviation of .48 cm. Does the data strongly suggest that irrigation with the wastewater mixture results in a mean root length that differs from 6.40? Use a .05 level test.

8.84 A student organization uses the proceeds from a particular soft-drink dispensing machine to finance its activities. The price per can had been \$.40 for a long time, and the average daily revenue during that period had been \$50.00. The price was recently increased to \$.45 per can. A random sample of $n = 20$ days subsequent to the price increase yielded a sample average revenue and sample standard deviation of \$47.30 and \$4.20, respectively. Does this data suggest that the true average daily revenue has decreased from its value prior to the price increase? Test the appropriate hypotheses using $\alpha = .05$.

8.85 A hot-tub manufacturer advertises that with its heating equipment, a temperature of 100°F can be achieved in at most 15 min. A random sample of 32 tubs is selected and the time necessary to achieve a 100°F temperature is determined for each tub. The sample average time and sample standard deviation are 17.5 min and 2.2 min, respectively. Does this data cast doubt on the company's claim?

a. Carry out a test of hypotheses using significance level .05.

b. Compute the P-value, and use it to reach a conclusion at level .05.

8.86 When n is large and the population distribution is normal, the statistic s has approximately a normal distribution with mean value σ and standard deviation $\sigma/\sqrt{2n}$. This suggests that a test of $H_0 : \sigma =$ hypothesized value can be based on the z statistic

$$z = \frac{s - \text{hypothesized value}}{(\text{hypothesized value})/\sqrt{2n}}$$

Suppose that a sample of $n = 50$ ball bearings manufactured for a certain purpose is selected, and the diameter of each one is determined. The resulting sample standard deviation is $s = .03$. If the true standard deviation σ exceeds .025, an adjustment in the production process must be made. Does the sample data suggest that such an adjustment is necessary? State and test the appropriate hypotheses at significance level .01.

8.87 Suppose that the average number of checks written by a bank's noncommercial customers during a certain month is 35. Let μ denote the average number of checks for all customers who have a card for the bank's automatic teller. A random sample of 60 such customers results in a sample average of 31.7 checks written during the month and a sample standard deviation of 6.4. Does this data suggest that μ differs from the average number of checks written by all customers? Test the appropriate hypotheses using $\alpha = .01$.

REFERENCES

The books by Freedman et. al. and by Moore listed in earlier chapter references are excellent sources. Their orientation is primarily conceptual with a minimum of mathematical development, and both sources offer many valuable insights.

CHAPTER NINE

9 Inferences Using Two Independent Samples

INTRODUCTION

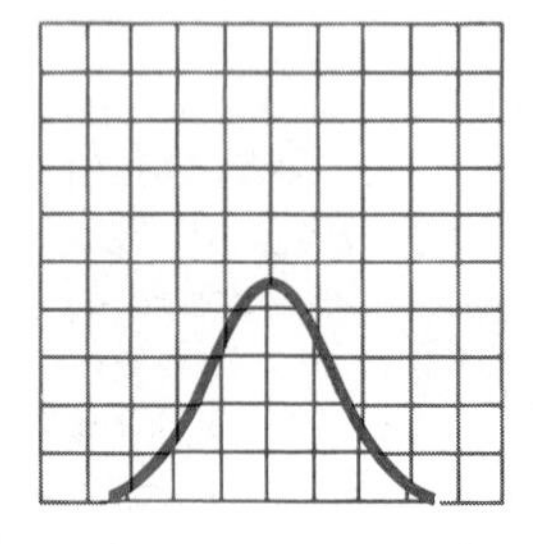

MANY INVESTIGATIONS are carried out in order to compare characteristics of two different populations. A group of health care practitioners might, for example, wish to compare the average hospital stay for those patients having private health insurance with the average stay for those covered by Medicare. Or a consumer organization might wish to know whether the proportion of one manufacturer's washing machines that require no major repairs during the first 5 years of ownership differs from the corresponding proportion for a second manufacturer's machines. The first example involves a comparison of two populations means, μ_1 and μ_2, where subscripts are used to differentiate the populations. The comparison in the second example is between the proportion of successes, π_1, in a first population and the corresponding proportion, π_2, in a second population. This chapter discusses inferences about a difference between two population means or between two population proportions when the samples from the two different populations are selected independently of one another. The methods are based on the difference between sample means or sample proportions. By way of introduction, the first section deals with the sampling distribution of a difference between two random variables.

9.1 The Sampling Distribution of a Difference

An investigator often wishes to compare the way in which values of the same numerical variable are distributed in two different populations. As an example, the variable of interest might be fuel efficiency (in miles per gallon), with the comparison being between all Mazda GLC automobiles and all Toyota Tercels. Or a study might focus on how the distribution of lifetimes of General Electric 60-W light bulbs differs from the distribution of lifetimes for bulbs manufactured by Westinghouse.

Let x denote the value of the variable for an individual or object selected at random from the first population. Similarly, let y denote the value of the variable for an individual randomly chosen from the second population. In the first example, x and y denote fuel efficiency for a randomly selected Mazda GLC and Toyota Tercel, respectively. Both x and y are random variables. The mean value μ_x and standard deviation σ_x of the x distribution, which locate the center and measure the spread about the center, respectively, are important distributional characteristics. The corresponding characteristics for the y distribution are μ_y and σ_y.

Now consider the difference $x - y$, which is often the most natural way to compare x with y. It, too, is a random variable. If, for example, a randomly chosen Mazda GLC has $x = 32.7$ and a randomly selected Toyota Tercel has $y = 30.8$, then $x - y = 1.9$. For $x = 29.9$ and $y = 31.2$, $x - y = -1.3$, and there are many other possible values of $x - y$. Information about the x distribution and about the y distribution can be used to obtain information about the distribution of $x - y$. Prior to stating several general rules, the following summaries of simulation experiments should help you develop an intuitive feel for relationships between the x, y, and $x - y$ distributions.

EXAMPLE 1

Suppose that height at maturity for one type of pine tree is normally distributed with mean value 70 ft and standard deviation 15 ft, whereas height for a second species is normally distributed with mean value 50 ft and standard deviation 10 ft. Let x denote the height of a randomly selected tree of the first type (so $\mu_x = 70$, $\sigma_x = 15$) and let y denote the height of a randomly chosen tree of the second type ($\mu_y = 50$, $\sigma_y = 10$). Then $x - y$ is the difference between two normally distributed variables having different mean values and different standard deviations. These two normal curves are pictured at the top of Figure 1.

We used MINITAB to select 1000 x values from this x distribution, the first few of which were 65.15, 68.41, 53.94, and 79.75. MINITAB was again used to independently select 1000 y values, the first few of which were 48.11, 46.36, 63.47, and 47.63. We then formed 1000 $x - y$ differences. The first observed difference was $65.15 - 48.11 = 17.04$. The second, third, and fourth $x - y$ values were 22.05, -9.53 (from $53.94 - 63.47$), and 32.12, respectively. The negative difference is not surprising: Figure 1 shows a substantial overlap between the x and y distributions, so there is a reasonable chance of an x value being paired with a y value that is larger. In fact, 124 of the 1000 observed $x - y$ values were negative. The smallest observed difference was -31.88 and the largest was 81.47. A sample histogram of all 1000 observed differences appears in Figure 1.

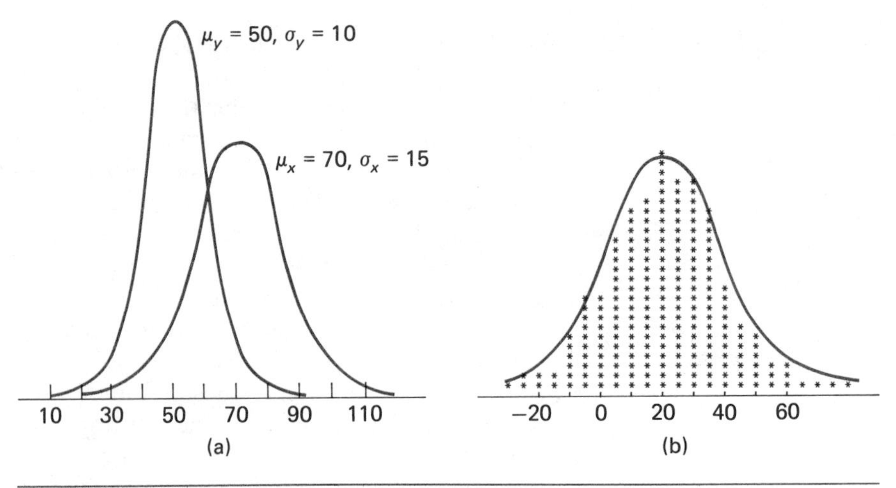

FIGURE 1

THE SAMPLING DISTRIBUTION OF $x - y$ WHEN THE x AND y DISTRIBUTIONS ARE NORMAL
(a) The x and y Distributions
(b) A Sample Histogram Based on 1000 $x - y$ Values from MINITAB

The sample histogram is centered roughly at 20. If the histogram had been based on an arbitrarily large sample of $x - y$ values, the center would be exactly 20. That is, with μ_{x-y} denoting the mean value (center) of the $x - y$ distribution,

$$\mu_{x-y} = 20 = 70 - 50 = \mu_x - \mu_y$$

The difference between the centers of the x and y distributions determines the center of the $x - y$ distribution. As far as spread is concerned, the sample histogram spreads out much more than does the y normal curve and somewhat more than does the x curve. There appears to be more variability in $x - y$ values than in either x or y alone—each variable contributes to variability in the differences. We shall shortly see how σ_{x-y}, the standard deviation of the $x - y$ distribution, is determined by σ_x and σ_y. Lastly, the sample histogram is quite well approximated by a normal curve. To emphasize this we superimposed a normal curve with mean value 20 and standard deviation 18 on the histogram. A much larger sample would yield a histogram that resembles this normal curve even more closely. ■

EXAMPLE 2

Figure 2 presents two population distribution curves that are quite skewed. Each one is derived from a lognormal model. This type of model is very useful in engineering and science applications.

Centers and spreads are not as easily identified and compared in the case of skewed distributions as they are for symmetric distributions. The figure certainly suggests that $\mu_y < \mu_x$, but locating these values visually is not easy. Moreover, it is not obvious by visual inspection that $\sigma_x < \sigma_y$—the reverse looks to be the case! The greater variability in y is entirely due to behavior in the extreme upper tail. Eventually, the x curve crosses the y curve and remains below it. As a result, when we used MINITAB to generate both 1000 x values and 1000 y values from these distributions, the largest x value was only 37.78, whereas the largest y value was 75.24.

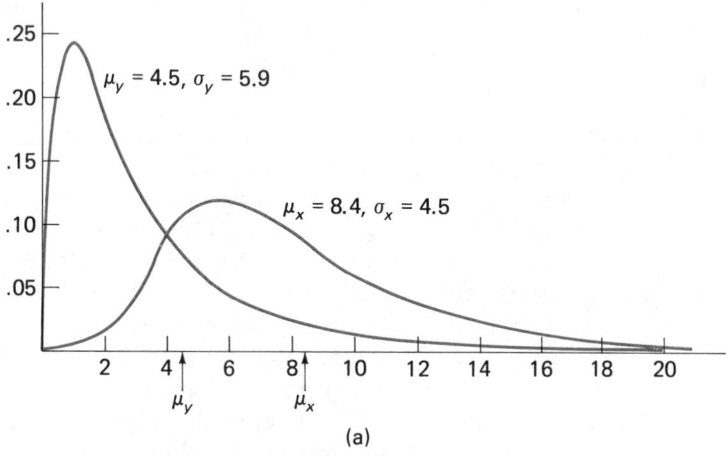

(a)

```
EACH * REPRESENTS 10 OBSERVATIONS

MIDDLE OF     NUMBER OF
INTERVAL      OBSERVATIONS
 -70.00           1    *
 -65.00           1    *
 -60.00           0
 -55.00           0
 -50.00           1    *
 -45.00           0
 -40.00           0
 -35.00           0
 -30.00           1    *
 -25.00           4    *
 -20.00           7    *
 -15.00          15    **
 -10.00          20    **
  -5.00          61    *******
   0.00         232    ************************
   5.00         395    ****************************************
  10.00         188    *******************
  15.00          55    ******
  20.00          11    **
  25.00           6    *
  30.00           1    *
  35.00           1    *
```

(b)

FIGURE 2

THE SAMPLING DISTRIBUTION OF $x - y$ WHEN THE x AND y DISTRIBUTIONS ARE SKEWED
(a) The x and y Distributions
(b) A Sample Histogram of 1000 $x - y$ Values from MINITAB

A sample histogram of the resulting 1000 $x - y$ values appears at the bottom of Figure 2. The sample average of these 1000 differences is approximately 3.9, the value of $\mu_x - \mu_y$. That is, as in Example 1, the sample histogram of the $x - y$ values is centered at $\mu_x - \mu_y$, the difference between the means of the populations sampled. It is difficult to compare spreads visually because the shape of the $x - y$ histogram is rather different from the two population distribution shapes. However, the sample standard deviation of the 1000 $x - y$ values was roughly 7.7, suggesting that there is more variability in the difference than in x or y individually. Finally, it is interesting to note that the $x - y$ histogram is much more symmetric than is either the x or

y distribution. This happens when the x and y distributions are similar in terms of general shape and skew. ■

Some General Rules

The foregoing examples illustrated how characteristics of the x and y distributions determine corresponding characteristics of the $x - y$ distribution. In each example an x observation was obtained in a manner completely independent of the way in which y was observed. Recall from Chapter 6 that two random variables x and y are independent if knowing the value of one variable doesn't affect how values of the other variable are distributed. Suppose, for example, that x and y are the pressures of left front tires on two different randomly selected automobiles. Information about the value of x has no bearing on the value y might have, so x and y are independent. However, now suppose that x and y refer to the pressures of left and right front tires, respectively, on the same randomly selected car. Then knowing that x is relatively small (e.g., 25 lb/in.^{2}) would suggest that y is also, since the front tires on the same car tend to have similar values of pressure. In this case x and y are dependent. The amount of variability in the $x - y$ distribution depends on whether or not x and y are independent.

1. For any two random variables x and y,
$$\mu_{x-y} = \begin{pmatrix}\text{the mean value of} \\ \text{the } x - y \text{ distribution}\end{pmatrix} = \mu_x - \mu_y$$
2. If x and y are independent random variables, then
$$\sigma^2_{x-y} = \begin{pmatrix}\text{the variance of} \\ \text{the } x - y \text{ distribution}\end{pmatrix} = \sigma^2_x + \sigma^2_y$$
$$\sigma_{x-y} = \begin{pmatrix}\text{the standard deviation} \\ \text{of the } x - y \text{ distribution}\end{pmatrix} = \sqrt{\sigma^2_x + \sigma^2_y}$$
3. If both the x and y distributions are (at least approximately) normal, then the $x - y$ distribution is also (at least approximately) normal.

According to (1), the center of the $x - y$ distribution is exactly the difference between the x and y centers. The second rule says that adding the two individual variances yields the $x - y$ variance, so that variability in $x - y$ is greater than either the x or y variability separately. When x and y are dependent, the formula for σ^2_{x-y} involves more than just σ^2_x and σ^2_y. Variability in $x - y$ is usually less in this case than in the case of independence.

EXAMPLE 3

Suppose that x denotes the height of a randomly selected male undergraduate and that $\mu_x = 69$ in. and $\sigma_x = 3$ in. Let y denote the height of a randomly chosen (independently of x) female student, with $\mu_y = 66$ in. and $\sigma_y = 2.5$ in. Then $\mu_{x-y} = 69 - 66 = 3$, $\sigma^2_{x-y} = (3)^2 + (2.5)^2 = 15.25$, and $\sigma_{x-y} = \sqrt{15.25} = 3.91$. Furthermore, if the x and y distributions are normal, then the $x - y$ distribution is also, as pictured in Figure 3. Suppose that x and y were husband's and wife's heights for a randomly selected married couple.

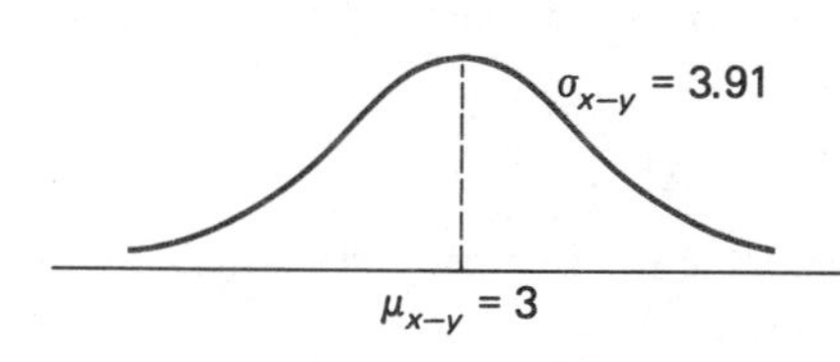

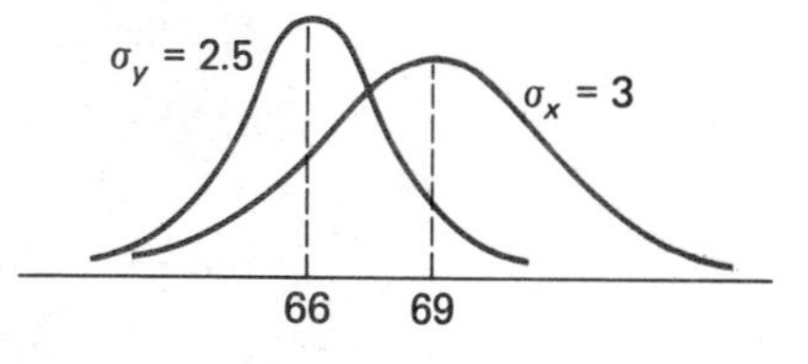

FIGURE 3 DISTRIBUTIONS OF HEIGHTS AND THE DIFFERENCE BETWEEN HEIGHTS FOR EXAMPLE 3

Then x and y would no longer be independent (tall men tend to marry tall women), so the above value of σ_{x-y} would no longer be correct. ■

EXERCISES

9.1 Suppose that x and y are independent random variables with $\mu_x = 100$, $\sigma_x = 10$, $\mu_y = 50$, and $\sigma_y = 20$. What are the values of μ_{x-y} and σ_{x-y}? What can be said about these values if x and y are not independent?

9.2 Let x be a random variable whose distribution has mean 15 and standard deviation 3. Similarly, let y be a random variable with mean 10 and standard deviation 4. Suppose that x and y are independent.

- **a.** What is the mean of the $x - y$ distribution?
- **b.** What is the standard deviation of the $x - y$ distribution?
- **c.** Under what circumstances would the $x - y$ distribution be normal in shape?

9.3 Suppose that the average annual starting salary offered to new computer science graduates across the country is \$25,800 and the standard deviation is \$1500. Suppose that for new business majors, the average salary is \$22,600 and the standard deviation is \$2400. Let x and y denote the salaries of a randomly selected computer science major and business major, respectively.

- **a.** Where is the distribution of $x - y$, the difference between salaries of the two randomly selected individuals, centered?
- **b.** What is the standard deviation of the $x - y$ distribution?
- **c.** Suppose that two different computer science majors are independently selected and that x_1 and x_2 denote their two starting salaries. What are the mean value and standard deviation of $x_1 - x_2$, the difference between the two salaries?

9.4 Suppose that readers of the daily *Boston Globe* spend an average of 30 min reading the paper and that the standard deviation of daily reading time is 20 min, whereas these figures for readers of the Sunday *Boston Globe* are 75 min and 60 min, respectively.

- **a.** Does reading time for either the daily or Sunday paper appear to be normally distributed? Explain.
- **b.** If x denotes the daily reading time of a randomly selected reader and y denotes the Sunday reading time of another reader selected independently of the first one, what are the mean value and the standard deviation of $x - y$?
- **c.** If x and y denote daily reading times for two different independently selected readers, what are the mean value and standard deviation of $x - y$?
- **d.** Suppose that x and y are the daily and Sunday reading times, respectively, for a single randomly selected reader. What is the mean value of $x - y$? How do you think the standard deviation of $x - y$ compares to what you calculated in (b)?

9.5 Suppose that x is a random variable whose probability distribution is given by

x	1	2	3	4	5	6
$p(x)$	.1	.2	.4	.1	.1	.1

and y is a random variable whose probability distribution is

y	1	2	3	4
$p(y)$	.25	.25	.25	.25

a. Use the methods of Chapter 5 to find μ_x, σ_x, μ_y, and σ_y.

b. If x and y are independent, what are the values of the mean and standard deviation of the $x - y$ distribution?

9.6 Let x and y be the random variables whose probability distributions are given in Exercise 9.5. Assume that x and y are independent. One possible value of $x - y$ is -2, which occurs if either $x = 1$ and $y = 3$ (with probability $(.1)(.25) = .025$) or $x = 2$ and $y = 4$ (with probability $(.2)(.25) = .05$). Thus $P(x - y = -2) = .025 + .05 = .075$. In a similar manner, compute the probabilities associated with all other possible $x - y$ values. Then draw the corresponding probability histogram. Does the picture look consistent with μ_{x-y} and σ_{x-y} computed in Exercise 9.5?

9.2 Large-Sample Inferences Concerning a Difference Between Two Population Means

We have used μ and σ to denote the mean value and standard deviation of x in the single population under study. The symbols n, $\overline{x}$, and s were used to represent the sample size, sample mean, and sample standard deviation, respectively. When comparing two different populations based on a random sample from each one, it is necessary to use notation that distinguishes between the characteristics for the first population and sample and those of the second population and sample. The best way to do this is to use subscripts on the symbols previously used.

Notation

	Mean Value	Variance	Standard Deviation
Population 1	μ_1	σ_1^2	σ_1
Population 2	μ_2	σ_2^2	σ_2

	Sample Size	Mean	Variance	Standard Deviation
Sample from Population 1	n_1	$\overline{x}_1$	s_1^2	s_1
Sample from Population 2	n_2	$\overline{x}_2$	s_2^2	s_2

Our focus will be on $\mu_1 - \mu_2$, the difference between the two population means. Because $\overline{x}_1$ provides a point estimate of μ_1 and $\overline{x}_2$ yields an estimate of μ_2, it is natural to use $\overline{x}_1 - \overline{x}_2$ as an estimate of $\mu_1 - \mu_2$.

EXAMPLE 4

Young children often use tapping, counting on fingers, tally marks, or counting aloud to aid in doing elementary arithmetic. Many educators feel that such crutches hinder a child's ability to develop skills for mastering more-complex concepts and computations. The paper "The Relationship of Finger Counting to Certain Pupil Factors" (*J. Educ. Research* (November 1976): 81–83) reported on a comparative study involving counters (those using aids such as the ones listed above) and noncounters at the fourth-grade level. A sample of counters and another of noncounters were selected. An arithmetic achievement test was administered and a grade-placement score was then determined. Let

μ_1 = true average grade placement for all fourth-grade counters

μ_2 = true average grade placement for all fourth-grade noncounters

The relevant sample quantities are as follows:

Noncounters	Sample Size	Sample Mean	Sample SD
Counters	$n_1 = 93$	$\overline{x}_1 = 2.96$	$s_1 = .80$
Noncounters	$n_2 = 55$	$\overline{x}_2 = 3.24$	$s_2 = .78$

The difference $\overline{x}_1 - \overline{x}_2 = 2.96 - 3.24 = -.28$ is a point estimate for $\mu_1 - \mu_2$. The estimate is negative simply because counters were labeled as population 1 and noncounters as population 2 rather than the other way around. The labeling is arbitrary as long as further work is consistent with the choice of labels. ■

Properties of the Sampling Distribution of $\overline{x}_1 - \overline{x}_2$

The statistic $\overline{x}$ was used to make inferences about μ in one-sample situations. A confidence interval and test procedure for μ were based on facts about $\overline{x}$'s sampling distribution—$\mu_{\overline{x}} = \mu$, $\sigma_{\overline{x}} = \sigma/\sqrt{n}$, and approximate normality of $\overline{x}$ for large n. Similarly, the construction of a confidence interval and test procedure for $\mu_1 - \mu_2$ requires information about the sampling distribution of $\overline{x}_1 - \overline{x}_2$. This information is based on the rules from Section 9.1 concerning a difference $x - y$. Here $\overline{x}_1$ plays the role of x and $\overline{x}_2$ plays the role of y.

Consider one random sample of n_1 observations from a population with a mean μ_1 and standard deviation σ_1 and a second random sample of size n_2, chosen independently of the first sample, from another population with mean μ_2 and standard deviation σ_2. Then

1. $$\mu_{\overline{x}_1-\overline{x}_2} = \begin{pmatrix}\text{the mean value} \\ \text{of } \overline{x}_1 - \overline{x}_2\end{pmatrix} = \mu_{\overline{x}_1} - \mu_{\overline{x}_2} = \mu_1 - \mu_2$$

Thus the $\overline{x}_1 - \overline{x}_2$ sampling distribution is always centered at the value of $\mu_1 - \mu_2$, so $\overline{x}_1 - \overline{x}_2$ is an unbiased statistic for estimating $\mu_1 - \mu_2$.

2. $\sigma^2_{\bar{x}_1-\bar{x}_2} = \begin{pmatrix}\text{variance of}\\ \bar{x}_1 - \bar{x}_2\end{pmatrix} = \sigma^2_{\bar{x}_1} + \sigma^2_{\bar{x}_2} = \frac{\sigma_1^2}{n_1} + \frac{\sigma_2^2}{n_2}$

and

$$\sigma_{\bar{x}_1-\bar{x}_2} = \begin{pmatrix}\text{standard deviation}\\ \text{of } \bar{x}_1 - \bar{x}_2\end{pmatrix} = \sqrt{\frac{\sigma_1^2}{n_1} + \frac{\sigma_2^2}{n_2}}$$

3. When n_1 and n_2 are both large, $\bar{x}_1$ and $\bar{x}_2$ each have approximately normal distributions (the central limit theorem), so the sampling distribution of $\bar{x}_1 - \bar{x}_2$ is also approximately normal (even if the two population distributions themselves are not normal).

With large sample sizes, property (3) implies that $\bar{x}_1 - \bar{x}_2$ can be standardized to obtain a variable whose sampling distribution is approximately the standard normal (z) curve. Unfortunately, the values of σ_1^2 and σ_2^2 will rarely be known, but if n_1 and n_2 are both large (typically at least 30), s_1^2 and s_2^2 can be used in their places. This gives the following key result on which large-sample tests and confidence intervals are based.

When n_1 and n_2 are both large, the sampling distribution of

$$z = \frac{\bar{x}_1 - \bar{x}_2 - (\mu_1 - \mu_2)}{\sqrt{\frac{s_1^2}{n_1} + \frac{s_2^2}{n_2}}}$$

is described approximately by the z curve.

Test Procedures

In a test designed to compare two population means, the null hypothesis will claim that $\mu_1 - \mu_2$ has a particular numerical value. The alternative hypothesis will involve the same hypothesized value along with one of the three inequalities, $>$, $<$, or $\neq$. As an example, let μ_1 and μ_2 denote true average fuel efficiencies (mi/gal) for a certain type of car equipped with 4-cylinder and 6-cylinder engines, respectively. The hypotheses under consideration might be $H_0: \mu_1 - \mu_2 = 4$ versus $H_a: \mu_1 - \mu_2 > 4$. This null hypothesis claims that average efficiency for a 4-cylinder engine exceeds average efficiency for a 6-cylinder engine by exactly 4 mi/gal. The alternative hypothesis states that the difference between true average efficiencies is more than 4 mi/gal.

A test statistic is obtained by replacing $\mu_1 - \mu_2$ in z by the hypothesized value, which appears in H_0. Thus the numerator of the z statistic for testing $H_0: \mu_1 - \mu_2 = 4$ is $(\bar{x}_1 - \bar{x}_2) - 4$. When H_0 is true, the sampling distribution of this z statistic is approximately the standard normal (z) curve. The type I error probability can now be controlled by using an appropriate z critical value. Continuing with the fuel efficiency example, H_0 should be rejected in favor of H_a if $\bar{x}_1 - \bar{x}_2$ considerably exceeds 4 (its mean value when H_0 is true). This is equivalent to rejecting H_0 when z is a large positive number (an upper-tailed test). Thus at level of significance .05, H_0 would be rejected in favor of H_a if $z > 1.645$ (since 1.645 captures area .05 in the upper tail of the z curve).

Summary of Large-Sample z Tests for $\mu_1 - \mu_2$

Null hypothesis: $H_0 : \mu_1 - \mu_2 = \text{hypothesized value}$

Test statistic: $$z = \frac{\bar{x}_1 - \bar{x}_2 - \text{hypothesized value}}{\sqrt{\frac{s_1^2}{n_1} + \frac{s_2^2}{n_2}}}$$

Alternative hypothesis	*Rejection region*
$H_a : \mu_1 - \mu_2 > \text{hypothesized value}$	$z > z$ critical value (upper-tailed test)
$H_a : \mu_1 - \mu_2 < \text{hypothesized value}$	$z < -z$ critical value (lower-tailed test)
$H_a : \mu_1 - \mu_2 \neq \text{hypothesized value}$	Either $z > z$ critical value or $z < -z$ critical value (two-tailed test)

The appropriate z critical value is determined by the choice of significance level. Critical values for the most frequently used values of α appear in the bottom row of Table IV.

EXAMPLE 5

Rotating work shifts have become a popular alternative to fixed shifts in certain manufacturing industries and in nursing. Industrial psychologists are interested in the extent to which people who work different types of shifts also spend their nonwork time in different ways. The paper "Work and Nonwork Experience of Employees on Fixed and Rotating Shifts" (*J. Voc. Behavior* (1982):282–93) reported the accompanying data on time spent alone per week for a random sample of fixed-shift nurses and another random sample of rotating-shift nurses. Does this data suggest that the average time spent alone by all nurses working fixed shifts differs from the average time for all rotating-shift nurses? Let's test the relevant hypotheses using level of significance $\alpha = .01$.

Type of Shift	Sample Size	Sample Mean	Sample SD
Fixed	245	11.23	6.77
Rotating	180	13.72	8.91

1. μ_1 = average time per week spent alone by all fixed-shift nurses, μ_2 = average time per week spent alone by all rotating-shift nurses, and $\mu_1 - \mu_2$ = the difference in population average times.

2. $H_0 : \mu_1 - \mu_2 = 0$ (that is, $\mu_1 = \mu_2$, which says that there is no difference in average times).

3. $H_a : \mu_1 - \mu_2 \neq 0$ (there is a difference in average times).

4. Test statistic:

$$z = \frac{\bar{x}_1 - \bar{x}_1 - 0}{\sqrt{\frac{s_1^2}{n_1} + \frac{s_2^2}{n_2}}}$$

5. Because H_a contains the inequality $\neq$, the appropriate test is two-tailed. The last row of Table IV gives z critical value = 2.58 for a level .01 two-tailed test. Thus H_0 will be rejected if either $z > 2.58$ or $z < -2.58$.

6. Computed $z = \dfrac{11.23 - 13.72}{\sqrt{(6.77)^2/245 + (8.91)^2/180}} = \dfrac{-2.49}{.793} = -3.14$

7. Since $-3.14 < -2.58$, z does fall in the lower tail of the two-tailed rejection region, so H_0 is rejected at level .01. The data strongly suggests that the average amount of time per week spent alone differs for nurses on the two types of shifts. ■

EXAMPLE 6

A number of studies have focused on the question of whether children born to women who smoke differ physiologically from children born to mothers who don't smoke. The paper "Placental Transfer of Lead, Mercury, Cadmium, and Carbon Monoxide in Women" (*Environ. Research* (1978): 494–503) reported on results from one such investigation. The accompanying data is on blood-lead concentration (μg/100 mL) for a sample of newborns whose mothers smoked and a second sample of newborn children whose mothers did not smoke. Does the data suggest that average concentration for smokers' newborn children exceeds that for nonsmokers by more than .5 μg/100 mL? Test the appropriate hypotheses at level of significance .10.

Mother	Sample Size	Sample Mean	Sample SD
Smoker	109	8.9	3.3
Nonsmoker	333	8.1	3.5

1. μ_1 = average lead concentration for all newborns born to smoking mothers, μ_2 = average lead concentration for all newborns born to nonsmoking mothers, and $\mu_1 - \mu_2$ = the difference in average concentrations.

2. $H_0 : \mu_1 - \mu_2 = .5$ (this says that average concentration for smokers' children exceeds that for nonsmokers' children by exactly .5).

3. $H_a : \mu_1 - \mu_2 > .5$.

4. Test statistic:

$$z = \frac{\bar{x}_1 - \bar{x}_2 - .5}{\sqrt{\dfrac{s_1^2}{n_1} + \dfrac{s_2^2}{n_2}}}$$

5. Using level of significance .10, the upper-tailed critical value is 1.28. H_0 will be rejected if $z > 1.28$.

6. Computed $z = \dfrac{8.9 - 8.1 - .5}{\sqrt{(3.3)^2/109 + (3.5)^2/333}} = \dfrac{.3}{.37} = .81.$

7. Since .81 is not greater than 1.28, H_0 cannot be rejected at level .10. The data does not give strong support to the claim that μ_1 exceeds μ_2 by more than .5. ■

Rather than specifying a particular significance level α and carrying out a test at that level, the P-value can be computed and reported. Because the large-sample procedure just described is a z test, the P-value is computed exactly as it was for z tests in Chapter 8.

EXAMPLE 7

The test in Example 6, which compared blood-lead level concentration in children born to smokers to that for children of nonsmokers, was upper-tailed and resulted in $z = .81$ (for $H_0 : \mu_1 - \mu_2 = .5$ versus $H_a : \mu_1 - \mu_2 > .5$). Thus

$$P\text{-value} \approx \text{area under the } z \text{ curve to the right of } .8 = .2119$$

The P-value is quite large, so even at level .20, H_0 could not be rejected.

The authors of the paper actually tested $H_0 : \mu_1 - \mu_2 = 0$ versus $H_a : \mu_1 - \mu_2 > 0$ (this alternative states that average concentration in smokers' newborns exceeds that in nonsmokers' newborns). For these hypotheses, it is easily verified that $z \approx 2.2$, so

$$P\text{-value} = \text{area under the } z \text{ curve to the right of } 2.2 = .0139$$

Thus H_0 would be rejected at level .05 (P-value $< \alpha$) but not at level .01. ■

EXAMPLE 8

In Example 5, which involved a comparison of rotating-shift with fixed-shift nurses, the test was two-tailed. The computed value of z was approximately -3.1, so

$$P\text{-value} = 2 \cdot \begin{pmatrix} \text{area under the } z \text{ curve} \\ \text{to the right of } 3.1 \end{pmatrix} = 2(.001) = .002$$

This implies that H_0 can be rejected at level .01 but not at level .001. ■

Comparing Treatments

Often an experiment is carried out in order to compare two different treatments or to compare the effect of a treatment with the effect of no treatment (treatment versus control). For example, an agricultural experimenter might wish to compare weight gains for animals put on two different diets. Let μ_1 denote the expected weight gain (expected response) for an animal on diet (treatment) 1. That is, if the population of all animals were placed on diet 1, μ_1 would be the population average weight gain. This population does not actually exist, but we can conceptualize it—and the observed weight gains constitute a random sample from this conceptual population. Similarily, μ_2 can be viewed either as the expected weight gain for an animal fed diet 2 or as the population average weight gain for the conceptual population consisting of all animals that could receive diet 2. Again, the observed weight gains represent a random sample from this conceptual population. The important point is that our two-sample z test, as well as other two-sample procedures, can be applied to compare conceptual populations.

EXAMPLE 9

The paper "Testing vs. Review: Effects on Retention" (*J. Educ. Psych.* (1982):18–22) reported on an experiment designed to compare several different methods ("treatments") for enhancing retention of material just studied. After high school students studied a brief history text, each one either took a test (method 1) or spent equivalent time reviewing selected passages (method 2). Two weeks afterward, each student took a retention test. Summary data from the experiment appears below. Does retention appear to be better for one method than for the other?

Treatment	Sample Size	Sample Mean	Sample SD
Method 1 (test)	31	12.4	4.5
Method 2 (review)	34	11.0	3.1

Let μ_1 denote the true average retention score for all high school students who might be assigned to method 1 (the mean of a conceptual population), and define μ_2 analogously for method 2. The appropriate hypotheses are $H_0 : \mu_1 - \mu_2 = 0$ versus $H_a : \mu_1 - \mu_2 \neq 0$ (no difference in treatments versus a difference). The computed value of the z statistic is

$$z = \frac{12.4 - 11.0}{\sqrt{(4.5)^2/31 + (3.1)^2/34}} = \frac{1.4}{.967} \approx 1.4$$

Since the test is two-tailed,

$$P\text{-value} = 2 \cdot \left(\begin{array}{c} \text{area under the } z \text{ curve to} \\ \text{the right of } 1.4 \end{array} \right) = 2(.0808) \approx .16$$

Thus, even at level of significance .10, H_0 cannot be rejected, so it certainly cannot be rejected at any smaller level, such as .05 or .01. Retention of material does not seem to depend on which method is used. If many more experiments yield results of this type, maybe testing will go out of fashion! ■

Comparisons and Causation

If the assignment of treatments to the individuals or objects used in a comparison of treatments is not made by the investigators, the study is said to be **observational.** As an example, the article "Lead and Cadmium Absorption Among Children Near a Nonferrous Metal Plant" (*Environ. Research* (1978):290–308) reported data on blood-lead concentrations for two different samples of children. The first sample was drawn from a population residing within 1 km of a lead smelter, while those in the second sample were selected from a rural area much farther from the smelter. It was the parents of children, rather than the investigators, who determined whether the children would be in the close-to-smelter group or the far-from-smelter group. As a second example, a letter in the May 19, 1978, *J. of the Amer. Med. Assoc.* reported on a comparison of remaining lifetimes after medical school graduation for doctors having an academic affiliation and doctors in private practice (the letter writer's stated objective was to see whether "publish or perish" really meant "publish and perish"). Here again, an investigator did not start out with a group of doctors and assign some to academic and others to nonacademic careers. The doctors themselves selected their groups.

The difficulty with drawing conclusions based on an observational study is that a statistically significant difference may be due to some underlying factors that have not been controlled rather than to conditions that define the groups. Does the type of medical practice itself have an effect on longevity, or is the observed difference in lifetimes caused by other factors, which themselves led graduates to choose academic or nonacademic careers? Similarly, is the observed difference in blood-lead concentration levels due to proximity to the smelter? Perhaps there are other physical and socioeconomic factors related both to choice of living area and to concentration.

In general, rejection of $H_0 : \mu_1 - \mu_2 = 0$ in favor of $H_a : \mu_1 - \mu_2 > 0$ suggests that on the average, higher values of the variable are *associated* with individuals in the first population or receiving the first treatment than with those in the second population or receiving the second treatment. But association does not imply causation. Strong statistical evidence for a causal relationship can be built up over time through many different comparative studies that point to the same conclusions (as in the many investigations linking smoking to lung cancer). A **randomized controlled experiment,** in which investigators assign subjects in some prescribed random fashion to the treatments or conditions being compared, is particularly effective in suggesting causality. With such random assignment, the investigator and other interested parties will have more confidence in the conclusion that an observed difference was caused by the difference in treatments or conditions. Such carefully controlled studies are more easily carried out in the hard sciences than in social science contexts, which may explain why the use of statistical methods is less controversial in the former than in the latter disciplines.

A Confidence Interval

A large-sample confidence interval for $\mu_1 - \mu_2$ can be obtained from the same z variable on which the test procedures were based. When n_1 and n_2 are both large, approximately 95% of all samples from the two populations will be such that

$$-1.96 < \frac{\bar{x}_1 - \bar{x}_2 - (\mu_1 - \mu_2)}{\sqrt{\dfrac{s_1^2}{n_1} + \dfrac{s_2^2}{n_2}}} < 1.96$$

The 95% confidence interval results from manipulating to isolate $\mu_1 - \mu_2$ in the middle (just as isolating μ in $-1.96 < (\bar{x} - \mu)/(s/\sqrt{n}) < 1.96$ led to the large-sample interval for μ in Chapter 7):

$$\bar{x}_1 - \bar{x}_2 - 1.96\sqrt{\frac{s_1^2}{n_1} + \frac{s_2^2}{n_2}} < \mu_1 - \mu_2 < \bar{x}_1 - \bar{x}_2 + 1.96\sqrt{\frac{s_1^2}{n_1} + \frac{s_2^2}{n_2}}$$

A confidence level other than 95% is achieved by using an appropriate z critical value in place of 1.96.

The **large-sample confidence interval for $\mu_1 - \mu_2$** is

$$\bar{x}_1 - \bar{x}_2 \pm (z \text{ critical value}) \sqrt{\frac{s_1^2}{n_1} + \frac{s_2^2}{n_2}}$$

The z critical values associated with the most frequently used confidence levels appear in the bottom row of Table IV.

EXAMPLE 10

Much attention has been focused in recent years on merger activity among business firms. Many business analysts are interested in knowing how various characteristics of merged firms compare to those of nonmerged firms. The article "Abnormal Returns from Merger Profiles" (*J. Finan. Quant. Analysis* (1983):149–62) reported the accompanying sample data on price earnings ratios for two samples of firms.

Type of Firm	Sample Size	Sample Mean	Sample SD
Merged	44	7.295	7.374
Nonmerged	44	14.666	16.089

Let μ_1 and μ_2 denote the true average price earning ratios for all merged and unmerged firms, respectively. Then a 99% confidence interval for the difference $\mu_1 - \mu_2$ between true average price earnings ratios is

$$7.295 - 14.666 \pm (2.58)\sqrt{\frac{(7.374)^2}{44} + \frac{(16.089)^2}{44}}$$

$$= -7.371 \pm (2.58)(2.668) \quad = -7.371 \pm 6.883 = (-14.254 - .488)$$

Thus we are highly confident that $\mu_1 - \mu_2$ is between -14.25 and $-.488$ (that is, that the true average price earnings ratio is less for merged than for nonmerged firms by between roughly .5 and 14.3). ■

EXERCISES

9.7 Consider two populations for which $\mu_1 = 30$, $\sigma_1 = 2$, $\mu_2 = 25$, and $\sigma_2 = 3$. Suppose that two independent random samples of sizes $n_1 = 40$ and $n_2 = 50$ are selected. Describe the approximate sampling distribution of $\bar{x}_1 - \bar{x}_2$ (center, spread, and shape).

9.8 An article in the November 1983 *Consumer Reports* compared various types of batteries. The average lifetimes of Duracell Alkaline AA batteries and Eveready Energizer Alkaline AA batteries were given as 4.1 h and 4.5 h, respectively. Suppose that these are the population average lifetimes.

a. Let $\bar{x}_1$ be the sample average lifetime of 100 Duracell batteries and $\bar{x}_2$ be the sample average lifetime of 100 Eveready batteries. What is the mean value of $\bar{x}_1 - \bar{x}_2$ (that is, where is the sampling distribution of $\bar{x}_1 - \bar{x}_2$ centered)? How does your answer depend on the specified sample sizes?

b. Suppose that population standard deviations of lifetime are 1.8 h for Duracell batteries and 2.0 h for Eveready batteries. With the sample sizes as given in (a), what is the variance of the statistic $\bar{x}_1 - \bar{x}_2$, and what is its standard deviation?

c. For the sample sizes as given in (a), draw a picture of the approximate sampling distribution curve of $\bar{x}_1 - \bar{x}_2$ (include a measurement scale on the horizontal axis). Would the shape of the curve necessarily be the same for sample sizes of 10 batteries of each type? Explain.

9.9 How do employees who work for the same company for many years compare to those who leave after a shorter tenure? The paper "The Role of Performance in the Turnover Process" (*Academy of Management J.* (1982):137–47) reported on a study of oil company employees. Let population 1 consist of all long-term employees and population 2 consist of all those who terminated employment within 15 years of being hired. One performance characteristic discussed in the paper was initial performance rating.

a. Suppose that μ_1, the true average initial performance rating for those in population 1, is 3.50 and that $\mu_2 = 3.25$. Let $\bar{x}_1$ and $\bar{x}_2$ denote sample average ratings for independent random samples of sizes $n_1 = 100$ and $n_2 = 150$, respectively. What is the mean value of $\overline{x}_1 - \overline{x}_2$? How does this depend on the given sample sizes?

b. Suppose that $\sigma_1 = .6$ and $\sigma_2 = .5$. What is the standard deviation of $\overline{x}_1 - \overline{x}_2$ when $n_1 = 100$ and $n_2 = 150$? Would doubling the sample sizes (to 200 and 300) result in a standard deviation which is half that for the original sample sizes? By what factor would the sample sizes have to be increased to halve the standard deviation of $\overline{x}_1 - \overline{x}_2$, and why?

c. Using the values in (a) and (b), what is the approximate probability that $\overline{x}_1 - \overline{x}_2 < 0$ (i.e., that the sample average rating for short-term employees exceeds that for long-term employees)? (*Hint:* Because n_1 and n_2 are large, $\overline{x}_1 - \overline{x}_2$ has approximately a normal distribution. Use the results of (a) and (b) to standardize.)

9.10 Let μ_1 and μ_2 denote true average tread lives for two different brands of size FR78-15 radial tires. Test $H_0 : \mu_1 - \mu_2 = 0$ versus $H_a : \mu_1 - \mu_2 \neq 0$ at level .05 using the following data: $n_1 = 40$, $\overline{x}_1 = 36{,}500$, $s_1 = 2200$, $n_2 = 40$, $\overline{x}_2 = 33{,}400$, and $s_2 = 1900$.

9.11 Let μ_1 denote true average tread life for a certain brand of FR78-15 radial tire, and let μ_2 denote the true average tread life for bias-ply tires of the same brand and size. Test $H_0 : \mu_1 - \mu_2 = 10{,}000$ versus $H_a : \mu_1 - \mu_2 > 10{,}000$ at level .01 using the following data: $n_1 = 40$, $\overline{x}_1 = 36{,}500$, $s_1 = 2200$, $n_2 = 40$, $\overline{x}_2 = 23{,}800$, and $s_2 = 1500$.

9.12 Leucocyte (white blood cell) counts in thoroughbred horses have recently been studied as a possible aid to the diagnosis of respiratory viral infections. The accompanying data on neutrophils (the most numerous kind of leucocyte) was reported in a comparative study of counts in horses of different ages ("Leucocyte Counts in the Healthy English Thoroughbred in Training" *Equine Vet. J.* (1984):207–09).

Age	Sample Size	Sample Mean	Sample Standard Deviation
2-year olds	197	51	5.6
4-year olds	77	56	4.3

Does this data suggest that the true average neutrophil count for 4-year-olds exceeds that for 2-year-olds? Let μ_1 and μ_2 denote the true average counts for 2- and 4-year-old horses, respectively. Carry out a test of $H_0 : \mu_1 - \mu_2 = 0$ versus $H_a : \mu_1 - \mu_2 < 0$ using level of significance .001. Be sure to give the test statistic, rejection region, computations, and conclusion (stated in the problem context).

9.13 The accompanying data on water salinity (%) was obtained during a study of seasonal influence of Amazon River water on biological production in the western tropical Atlantic ("Influence of Amazon River Discharge on the Marine Production System off Barbados, West Indies" *J. Marine Research* (1979):669–81). Let μ_1 denote the true average salinity level of water samples collected during the summer, and define μ_2 analogously for samples collected during the winter. Compute a 99% confidence interval for $\mu_1 - \mu_2$. Is it necessary to make any assumptions about the two salinity distributions? Explain.

Period	Sample Size	Sample Mean	Sample Standard Deviation
Summer	51	33.40	.428
Winter	54	35.39	.294

9.14 A comparison of salaries paid in large manufacturing companies (over $350 million in sales) and those paid for comparable jobs by the banking industry was presented in the paper "Bank Compensation: The Next Major Change Forced by Deregulation" (*J. Retail Banking* (Spring 1983):1–5). The accompanying data is compatible with the findings in this paper.

	Programmers	Accounting Clerks
Manufacturing		
n	50	40
$\bar{x}$	19,760	14,600
s	870	360
Banks		
n	45	35
$\bar{x}$	18,350	12,900
s	680	460

a. Use a level .01 test to determine if there is a significant difference between the average salary paid to programmers in manufacturing and in banking.

b. Compute the P-value associated with the z value in (a).

c. Estimate the difference in mean salary paid to accounting clerks in manufacturing and in banking using first a point estimate and then a 99% confidence interval.

9.15 Some astrologists have speculated that people born under certain sun-signs are more extroverted than people born under other signs. The accompanying data was taken from the paper "Self-Attribution Theory and the Sun-Sign" (*J. Social Psych.* (1984):121–26). The Eysenck Personality Inventory (EPI) was used to measure extroversion and neuroticism.

		Extroversion		Neuroticism	
Sun-sign	n	$\bar{x}$	s	$\bar{x}$	s
Water signs	59	11.71	3.69	12.32	4.15
Other signs	186	12.53	4.14	12.23	4.11
Winter signs	73	11.49	4.28	11.96	4.22
Summer signs	49	13.57	3.71	13.27	4.04

a. Is there sufficient evidence to indicate that those born under water signs have a lower mean extroversion score than those born under other (nonwater) signs? Use a level .01 test.

b. Does the data strongly suggest that those born under winter signs have a lower mean extroversion score than those born under summer signs? Use a level .05 test.

c. Does the data indicate that those born under water signs differ significantly from those born under other signs with respect to mean neuroticism score? Use $\alpha = .05$.

d. Do those born under winter signs differ from those born under summer signs with respect to mean neuroticism score? Compute the P-value associated with the test statistic and use it to state a conclusion for significance level .01.

9.16 Recorded speech can be compressed and played back at a faster rate. The paper "Comprehension by College Students of Time-Compressed Lectures" (*J. Exper. Educ.* (Fall 1975):53–56) gave the results of a study designed to test comprehension of time-compressed speech. Fifty students listened to a 60-min lecture and took a comprehension test. Another 50 students heard the same lecture time-compressed to 40 min. The sample mean and standard deviation of comprehension scores for the normal speed group were 9.18 and 4.59, respectively, and those for the time-compressed group were 6.34 and 4.93, respectively.

a. Use a level .01 test to determine if the true mean comprehension score for students hearing a time-compressed lecture is significantly lower than the true mean score for students who hear a lecture at normal speed.

b. Estimate the difference in true mean comprehension score for normal and time-compressed lectures using a 95% confidence interval.

9.17 In a study of attrition among college students 587 students were followed through their college years ("The Prediction of Voluntary Withdrawal from College: An Unsolved Problem" *J. Exper. Educ.* (Fall 1980):29–45). Of the 587 in the sample, 87 withdrew from college for various reasons (although none were dismissed for academic reasons). The accompanying table gives summary statistics on SAT scores for the "persisters" and the "withdrawals."

	SAT Verbal Score	
	Persisters	Withdrawals
n	500	87
$\bar{x}$	491	503
s	80.6	78.8

Is there sufficient evidence to indicate a difference in true mean SAT verbal scores for students who withdraw from college and those who graduate? Perform the relevant hypothesis test using a significance level of .10.

9.18 The earning gap between men and women has been the subject of many investigations. One such study ("Sex, Salary, and Achievement: Reward-Dualism in Academia" *Sociology of Educ.* (1981):71–85) gave the accompanying information for samples of male and female college professors. Suppose that these statistics were based on samples of size 50 chosen from the populations of all male and all female college professors.

	Males		Females	
	Mean	Standard Deviation	Mean	Standard Deviation
Salary	1634.10	715.00	1091.80	418.80
Years at university	7.93	8.04	6.25	7.65

a. Does the data strongly suggest that the mean salary for women is lower than the mean salary for men? Use a level .05 test. Does the resulting conclusion by itself point to discrimination against female professors, or are there some possible nondiscriminating explanations for the observed difference?

b. If the results of the test in (a) were to be summarized using a P-value, what value should be reported?

c. Estimate the difference in mean number of years at the university for men and women using a 90% confidence interval. Based on your interval, do you think that there is a significant difference between the true mean number of years at the university for men and women? Explain.

9.19 Should quizzes be given at the beginning or end of a lecture period? The paper "On Positioning the Quiz: An Empirical Analysis" (*Accounting Review* (1980): 664–70) provides some insight. Two sections of an introductory accounting class were given identical instructions to read and study assigned text materials. Three quizzes and a final exam were given during the term, with one section taking quizzes at the beginning of the lecture and the other section taking quizzes at the end. Final exam scores for the two groups are summarized. Does the data indicate that there is a significant difference in the true mean final exam score for students who take quizzes at the beginning of class and those who take quizzes at the end? Use a .01 significance level.

	Quiz at Beginning	Quiz at End
Sample size	40	40
Mean	143.7	131.7
Standard deviation	21.2	20.9

9.20 Celebrity endorsement of products is a common advertising technique. A group of 98 people was shown an ad containing a celebrity endorsement, and a second group of 98 was shown the same ad but using an unknown actor ("Effectiveness of Celebrity Endorsees" *J. Ad. Research* (1983):57–62). Each participant rated the commercial's believability on a scale of 0 (not believable) to 10 (very believable). Results were as follows.

	Believability of Ad	
	Mean	Standard Deviation
Celebrity ad	3.82	2.63
Noncelebrity ad	3.97	2.51

Is there sufficient evidence to indicate that use of a celebrity endorsement results in a true mean believability rating that differs from the true mean rating for noncelebrity endorsements? Use a level .05 test.

9.21 Referring to Example 6, suppose that the labels 1 and 2 are reversed, so that μ_1 now refers to the average lead concentration for all newborns born to nonsmoking mothers. How do the hypotheses (statements about $\mu_1 - \mu_2$), computations, and conclusions change?

9.3 Small-Sample Inferences Concerning a Difference Between Two Normal Population Means

The z test and confidence interval discussed in Section 9.2 were large-sample procedures. Their use required no specific assumption about the population distributions, as long as n_1 and n_2 were large enough to ensure that $\bar{x}_1$ and $\bar{x}_2$ had approximately normal sampling distributions. When at least one of the sample sizes is small, the z procedures are not appropriate. The approach taken in this section is to make several specific assumptions about the population distributions and then describe procedures that are known to perform very well when these assumptions are met.

Basic Assumptions in this Section

1. Both population distributions are normal.
2. The two population standard deviations are identical ($\sigma_1 = \sigma_2$), with σ denoting the common value.

Figure 4 pictures four different pairs of population distributions. Only Figure 4(a) is consistent with the basic assumptions. If you believe that the relevant picture for your problem is given in Figure 4(b)—normal population distributions but σ_1 substantially different from σ_2—then the methods of this section are inappropriate. There are several possible inferential procedures in this case, but (unlike the procedures of this section) there is still controversy among statisticians about which one should be used. The distributions in Figure 4(c) have exactly the same shape and spread, with one shifted so that it is centered to the right of the other one. Inferential procedures appropriate in this case are presented in Section 9.5. Finally, inference involving population distributions with very different shapes and spreads, as in Figure 4(d), can be very complicated. Good advice from a statistician is particularly important here.

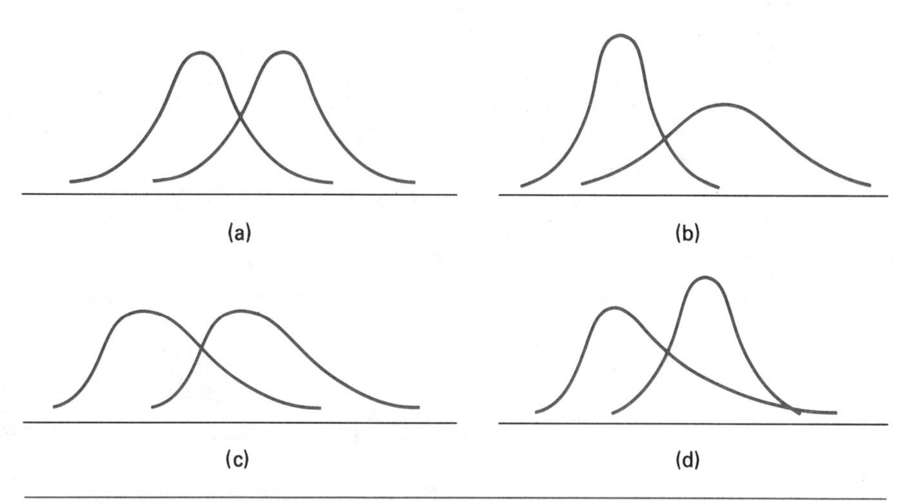

FIGURE 4 FOUR POSSIBLE POPULATION DISTRIBUTION PAIRS
(a) Both normal, $\sigma_1 = \sigma_2$
(b) Both normal, $\sigma_1 \neq \sigma_2$
(c) Same shape and spread, differing only in location
(d) Very different shapes and spreads

In Chapter 6 we pointed out that when the population distribution is normal, the sampling distribution of the sample mean $\overline{x}$ is normal for *any* sample size (small or large). This fact was important in Chapters 7 and 8 in the development of the one-sample t confidence interval and test. Here both population distributions are assumed normal, so the sampling distributions of both $\overline{x}_1$ and $\overline{x}_2$ are normal for any n_1 and n_2. This implies that the sampling distribution of $\overline{x}_1 - \overline{x}_2$ is also normal for any sample sizes.

The facts concerning the mean value and standard deviation of $\bar{x}_1 - \bar{x}_2$ used in Section 9.2 remain valid here. The $\bar{x}_1 - \bar{x}_2$ distribution is centered at $\mu_1 - \mu_2$ (because $\mu_{\bar{x}_1-\bar{x}_2} = \mu_1 - \mu_2$). We are still assuming that the two random samples are obtained independently of one another, but now the common standard deviation σ is used in place of σ_1 and σ_2, so

$$\sigma_{\bar{x}_1-\bar{x}_2} = \sqrt{\frac{\sigma^2}{n_1} + \frac{\sigma^2}{n_2}}$$

If $\bar{x}_1 - \bar{x}_2$ is now standardized by subtracting $\mu_{\bar{x}_1-\bar{x}_2}$ and dividing by $\sigma_{\bar{x}_1-\bar{x}_2}$, the result is a variable whose sampling distribution is the standard normal curve:

$$z = \frac{\bar{x}_1 - \bar{x}_2 - (\mu_1 - \mu_2)}{\sqrt{\frac{\sigma^2}{n_1} + \frac{\sigma^2}{n^2}}}$$

Estimating σ^2

Before this standardized variable can be used as a test statistic or manipulated to yield a confidence interval, it is necessary to estimate σ^2. An obvious way to estimate σ^2 from the first sample alone is to use s_1^2, the sample variance of the n_1 observations in that sample. Similarly, the sample variance for the n_2 observations in the second sample gives an estimate of σ^2. Intuitively, combining s_1^2 and s_2^2 in some fashion should lead to an estimate that is better than either one individually. A first thought might be to use $(s_1^2 + s_2^2)/2$, the ordinary average of the two variances. However, if n_1 is larger than n_2, s_1^2 will tend to be closer to σ^2 than will s_2^2. That is, the sample variance associated with the larger sample size should be more accurate than the one based on the smaller sample size. This suggests that a weighted average should be used, with the variance from a larger sample receiving heavier weight.

DEFINITION

The statistic for estimating the common variance σ^2 is

$$s_p^2 = \left(\frac{n_1 - 1}{n_1 + n_2 - 2}\right) s_1^2 + \left(\frac{n_2 - 1}{n_1 + n_2 - 2}\right) s_2^2$$

The computed value of s_p^2 is called the **pooled estimate of σ^2** (*pooled* is synonymous with *combined*).

The two quantities in parentheses in the formula for s_p^2 always have a sum of 1. If $n_1 = n_2$, $s_p^2 = (\frac{1}{2})s_1^2 + (\frac{1}{2})s_2^2$, the ordinary average of the two sample variances. But whenever $n_1 \neq n_2$, the s^2 based on the larger sample size will receive more weight (be multiplied by a larger number) than will the other s^2. The multipliers of s_1^2 and s_2^2 in s_p^2 might at first seem unnatural. The reason for using them has to do with degrees of freedom. The first sample contributes $n_1 - 1$ df to the estimation of σ^2, and the second contributes $n_2 - 1$ df. The two samples are independent, so the number of degrees of freedom associated with s_p^2 is the sum $(n_1 - 1) + (n_2 - 1) = n_1 + n_2 - 2$.

EXAMPLE 11

The paper "Anthropometric and Physical Performance Characteristics of Male Volleyball Players" (*Canadian J. Appl. Sports Sci.* (1982):182–88) reported on a comparison of Russian and Finnish volleyball players. One of the variables studied was the height increase in body center of gravity (cm) during a vertical jumping test. Summary data appears in the accompanying table.

Sample	Sample Size	Sample Mean	Sample SD
Finns	$n_1 = 14$	$\bar{x}_1 = 46.0$	$s_1 = 3.2$
Russians	$n_2 = 10$	$\bar{x}_2 = 49.4$	$s_2 = 4.3$

The authors of the paper assumed for purposes of analysis that both height-increase population distributions (Finns and Russians) were normal with the same standard deviation σ. The pooled estimate of σ^2 is then

$$s_p^2 = \left(\frac{14 - 1}{14 + 10 - 2}\right)(3.2)^2 + \left(\frac{10 - 1}{14 + 10 - 2}\right)(4.3)^2$$

$$= \left(\frac{13}{22}\right)(10.24) + \left(\frac{9}{22}\right)(18.49) = 6.05 + 7.56 = 13.61$$

This gives $s_p = \sqrt{13.61} = 3.69$ as the pooled estimate of σ. Notice that $s_p^2 = 13.61$ is between $s_1^2 = 10.24$ and $s_2^2 = 18.49$. It is closer to s_1^2 because s_1^2 is based on a larger sample size than is s_2^2. ■

A major result of this section is that when s_p^2 replaces σ^2 in the standardized z variable, the new variable has a t distribution.

> The sampling distribution of the standardized variable
>
> $$t = \frac{\bar{x}_1 - \bar{x}_2 - (\mu_1 - \mu_2)}{\sqrt{\dfrac{s_p^2}{n_1} + \dfrac{s_p^2}{n_2}}}$$
>
> is the t distribution with $n_1 + n_2 - 2$ df (provided that the basic assumptions of this section are met).

Testing Hypotheses about $\mu_1 - \mu_2$

Just as in the previous section, the null hypothesis has the form $H_0 : \mu_1 - \mu_2 =$ hypothesized value. The test statistic results from replacing $\mu_1 - \mu_2$ in t by the hypothesized value (which often is zero). The sampling distribution of the test statistic when H_0 is true is the t distribution with $n_1 + n_2 - 2$ df. The type I error probability is then controlled by using an appropriate t critical value to specify the rejection region.

> **Summary of the Pooled t Test**
>
> *Null hypothesis:* $H_0 : \mu_1 - \mu_2 =$ hypothesized value
>
> *Test statistic:* $$t = \frac{\bar{x}_1 - \bar{x}_2 - \text{hypothesized value}}{\sqrt{\dfrac{s_p^2}{n_1} + \dfrac{s_p^2}{n_2}}}$$

Alternative hypothesis	*Rejection region*
$H_a : \mu_1 - \mu_2 >$ hypothesized value	$t > t$ critical value
$H_a : \mu_1 - \mu_2 <$ hypothesized value	$t < -t$ critical value
$H_a : \mu_1 - \mu_2 \neq$ hypothesized value	Either $t > t$ critical value or $t < -t$ critical value

The test is based on $n_1 + n_2 - 2$ df, so the t critical value is obtained from that row and the appropriate column (depending on choice of α and whether the test is one-tailed or two-tailed) of Table IV.

EXAMPLE 12

British health officials have recently expressed concern about problems associated with vitamin D deficiency among certain immigrants. Doctors have conjectured that such a deficiency could be related to the amount of fiber in a person's diet. The chemical compound H-labeled 25-hydroxyvitamin D_3 (25(OH)D_3) is the major circulating form of vitamin D, and its plasma half-life is intimately related to the body's vitamin D level. An experiment to compare plasma half-lives for two groups of healthy individuals, one placed on a normal diet and the other on a high-fiber diet, resulted in the following data (from "Reduced Plasma Half-Lives of Radio-Labelled 25(OH)D_3 in Subjects Receiving a High-Fibre Diet" *Brit. J. Nutrit.* (1983):213–16).

Normal diet	19.1	24.0	28.6	29.7	30.0	34.8	
High-fiber diet	12.0	13.0	13.6	20.5	22.7	23.7	24.8

Does this data suggest that true average plasma half-life for persons placed on a normal diet exceeds true average half-life for those on a high-fiber diet? Assuming that the two plasma half-life distributions are normal with equal variances, let's test the appropriate hypotheses at level of significance .01 using the pooled t test.

1. $\mu_1 - \mu_2 =$ the difference between true average plasma half-lives for those on normal diets and those on high-fiber diets.
2. $H_0 : \mu_1 - \mu_2 = 0$.
3. $H_a : \mu_1 - \mu_2 > 0$ (equivalently, $\mu_1 > \mu_2$).
4. Test statistic:

$$t = \frac{\bar{x}_1 - \bar{x}_2 - 0}{\sqrt{s_p^2/n_1 + s_p^2/n_2}}$$

5. The form of H_a (>) implies that an upper-tailed test is appropriate. With $\alpha = .01$ and df $= n_1 + n_2 - 2 = 6 + 7 - 2 = 11$, Table IV gives the t critical value as 2.72. H_0 will be rejected in favor of H_a if $t > 2.72$.
6. The first (normal diet) sample yields $\bar{x}_1 = 27.70$ and $s_1^2 = 29.63$. The summary quantities for the second sample are $\bar{x}_2 = 18.61$ and $s_2^2 = 30.80$. Then

$$s_p^2 = \frac{5}{11}(29.63) + \frac{6}{11}(30.80) = 30.27$$

so

$$t = \frac{27.70 - 18.61}{\sqrt{30.27/6 + 30.27/7}} = \frac{9.09}{3.06} = 2.97$$

7. The value $t = 2.97$ is in the rejection region for a level .01 test $(2.97 > 2.72)$, so we reject H_0 in favor of H_a. The sample data rather strongly suggests that true average plasma half-life is higher for individuals on a normal diet than for those on a high-fiber diet. ■

The pooled t test is highly recommended by virtually all statisticians when the two basic assumptions of this section are at least approximately satisfied. Normal probability plots can be used to check the plausibility of the normality assumptions. There is a formal test procedure, called an F test, for testing $H_0 : \sigma_1^2 = \sigma_2^2$ versus $H_a : \sigma_1^2 \neq \sigma_2^2$. Some statisticians suggest that this test be carried out as a preliminary to the pooled t test, with the latter test used only if $H_0 : \sigma_1^2 = \sigma_2^2$ is not rejected. But for technical reasons we, along with many other statisticians, do not recommend this approach.* Instead, suppose that an investigator believes prior to collecting data that variability is roughly the same in the two populations or for the two treatments. Then if calculated values of s_1^2 and s_2^2 are of roughly the same order of magnitude, use of the pooled t test is reasonable (particularly if the two sample sizes are not too different). This was certainly the case in the previous example.

P-Values

Rather than specifying a rejection region corresponding to a fixed significance level, we can compute the value of t and then obtain an upper and/or lower bound on the P-value. If t is between two consecutive critical values in the $n_1 + n_2 - 2$ df row of Table IV, the associated levels of significance (along the bottom margin) give an upper and a lower bound. An upper bound results when t exceeds the largest critical value, and only a lower bound is available when t is smaller than the smallest tabulated value.

EXAMPLE 13

Monosodium glutamate (MSG), a commonly used seasoning in certain types of food, is known to produce brain damage in various mammals. The paper "Monosodium Glutamate Administration to the Newborn Reduces Reproductive Ability in Female and Male Mice" (*Science* (1977):452–54) reported on a study to investigate possible effects of MSG on reproductive ability. The accompanying data refers to weight of ovaries (mg) both for a sample of rats treated with MSG and for a control sample of similar but untreated rats.

	Sample Size	Sample Mean	Sample SD
MSG treatment	10	29.35	4.55
Control	10	21.86	10.09

The second sample variance ($(10.09)^2 = 101.81$) is almost five times as large as the first sample variance. This is somewhat unusual for small samples from normal populations with $\sigma_1 = \sigma_2$ but is still within the realm of possibility. Furthermore, when $n_1 = n_2$, the pooled t test works well even when σ_1

*Both the pooled t test and the F test are based on the assumption of normal population distributions. But the F test is much more sensitive to departures from this assumption than is the t test. A significant value of the F statistic might result not because σ_1^2 and σ_2^2 differ greatly but because the population distributions are slightly nonnormal.

and σ_2 differ somewhat from one another. The investigators used the pooled t test procedure to test for a difference between true average weights for MSG-treated and untreated rats. The relevant hypotheses are $H_0 : \mu_1 - \mu_2 = 0$ versus $H_a : \mu_1 - \mu_2 \neq 0$, so critical values for a two-tailed test are used to obtain bounds on the P-value. Since

$$s_p^2 = \frac{(10 - 1)}{(10 + 10 - 2)}(4.55)^2 + \frac{(10 - 1)}{(10 + 10 - 2)}(10.09)^2 = 61.26$$

then

$$t = \frac{29.35 - 21.86}{\sqrt{61.26/10 + 61.26/10}} = \frac{7.49}{3.50} = 2.14$$

From the 18 df row of the t table, the two-tailed critical values for $\alpha = .05$ and $.02$ are 2.10 and 2.55, respectively. Therefore, $.02 < P\text{-value} < .05$. Since 2.14 is very close to 2.10, it would actually be better to report $P\text{-value} \approx .05$. Using $\alpha = .05$, H_0 would just barely be rejected, but at level of significance .01 the data does not suggest that μ_1 and μ_2 differ. ■

The pooled t test can be carried out using any of the general-purpose statistical packages of computer programs. Standard output includes both the computed value of t and a calculated P-value (not just an upper and/or lower bound). With this output, no further calculation is necessary in order to draw a conclusion at a specified level of significance α. One simply compares α to the given P-value.

EXAMPLE 14

(*Example* 12 *continued*) We used MINITAB to analyze the normal-fiber and high-fiber plasma half-life data presented in Example 12, resulting in the accompanying output. The P-value of .0064 is the smallest level at which H_0 can be rejected. Since $\alpha = .01$ exceeds .0064, $H_0 : \mu_1 - \mu_2 = 0$ is rejected at this level in favor of $H_a : \mu_1 - \mu_2 > 0$. This is exactly the conclusion reached earlier by using the appropriate t critical value. Since .001 is smaller than the P-value, H_0 could not be rejected at level of significance .001.

```
                      N       MEAN      STDEV
          Normal      6      27.70       5.44
          Hifiber     7      18.61       5.55
TTEST MU normal = MU hifiber (VS GT): T=2.97 P=0.0064 DF=11.0
```
■

The number of degrees of freedom on which the pooled t test is based, $n_1 + n_2 - 2$, may well exceed 30 even when n_1 and/or n_2 is small. For example, with $n_1 = 16$ and $n_2 = 21$, $n_1 + n_2 - 2 = 35$, whereas $n_1 + n_2 - 2 = 52$ when $n_1 = 33$ and $n_2 = 21$. In such cases, we do not revert to the large-sample z test discussed in the previous section. That test should generally be used only when both sample sizes exceed 30. Provided that the basic assumptions (normality and $\sigma_1 = \sigma_2$) are met, the pooled t test is still the recommended procedure. However, once past 30 df, Table IV contains only four further rows of critical values—those for 40 df, 60 df, 120 df, and the z critical values. Consider, for example, carrying out an upper-tailed test based on 35 df with $\alpha = .05$. The necessary critical value is between 1.70, the 30 df value, and 1.68, the 40 df value. A reasonable choice is to interpolate and use 1.69. Alternatively, once past 30 df many books recommend using the z critical value, here 1.645. Since the four values 1.70, 1.69, 1.68, and 1.645 are so close, it generally makes little difference which one is used to specify the test procedure.

A Small-Sample Confidence Interval

A small-sample confidence interval for $\mu_1 - \mu_2$ is easily obtained from the basic t variable of this section. Both the derivation of and formula for the interval are very similar to that of the large-sample z interval discussed in the previous section.

> **The pooled t confidence interval for $\mu_1 - \mu_2$,** which is valid when both population distributions are normal and $\sigma_1 = \sigma_2$, is
>
> $$\bar{x}_1 - \bar{x}_2 \pm (t \text{ critical value}) \sqrt{\frac{s_p^2}{n_1} + \frac{s_p^2}{n_2}}$$
>
> The t critical value is based on $n_1 + n_2 - 2$ df. Critical values associated with the most frequently used confidence levels appear in Table IV.

EXAMPLE 15

Much research effort has been expended in studying possible causes of the pharmacological and behavioral effects resulting from smoking marijuana. The article "Intravenous Injection in Man of Δ^9 THC and 11-OH-Δ^9 THC" (*Science* (1982):633) reported on a study of two chemical substances thought to be instrumental in marijuana's effects. Subjects were given one of the two substances in increasing amounts and asked to say when the effect was first perceived. Data values are necessary dose to perception per kilogram of body weight.

Δ^9 THC	19.54	14.47	16.00	24.83	26.39	11.49
			($\bar{x}_1 = 18.79$, $s_1 = 5.91$)			
11-OH-Δ^9 THC	15.95	25.89	20.53	15.52	14.18	16.00
			($\bar{x}_2 = 18.01$, $s_2 = 4.42$)			

From these values, $s_p^2 = 27.23$. Assuming normality of the two distributions and $\sigma_1 = \sigma_2$, a 95% confidence interval for $\mu_1 - \mu_2$ uses the t critical value 2.23 from the 10 df row of the t table. The interval is

$$18.79 - 18.01 \pm 2.23\sqrt{\frac{27.23}{6} + \frac{27.23}{6}} = .78 \pm 6.72$$

$$= (-5.94, 7.50)$$

This interval is rather wide because s_p^2 is large and the two sample sizes are small. Notice that the interval includes 0, so 0 is one of the many plausible values for $\mu_1 - \mu_2$. ■

Transforming Data (Optional)

It is sometimes the case that whereas the two population distributions are quite skewed and have different spreads, suitably transforming values results in distributions that are approximately normal with roughly equal spreads. An appropriate transformation can be suggested either by theoretical arguments or else through trying several of the power transformations discussed in Chapter 2. The most frequently used transformation for this is the logarithmic transformation. In environmental applications in which x denotes pollutant concentration, $\log(x)$ often has approximately a normal distribution. As another example, in reliability studies in which x represents time until component failure, empirical evidence sometimes suggests that $\log(x)$ follows a normal distribution.

Once a transformation that yields approximately normal distributions with equal spreads is identified, the transformation is applied to every observation in each of the two samples. Let μ_1 and μ_2 denote the mean values of the transformed variables in the two populations. The pooled t test can then be used on the transformed data to test $H_0 : \mu_1 - \mu_2 = 0$ ($\mu_1 = \mu_2$). This null hypothesis is equivalent to the claim that the population distributions for the untransformed variables are identical (and in particular have the same mean value). The tricky part of the analysis is in selecting a transformation; applying the t test to transformed data is no more difficult than using it on untransformed data. An example of this sort is given in Exercise 9.37.

β for the Pooled t Test (Optional)

In Chapter 8 we showed how β, the probability of a type II error, could be determined for a one-sample t test using the curves in Table V. These same curves can be used to obtain β for the pooled t test when either $\alpha = .05$ or $.01$. Suppose, for example, that μ_1 is the true average fracture toughness of high purity steel specimens and that μ_2 is the true average fracture toughness of commercial steel specimens. An investigator wishes to test $H_0 : \mu_1 - \mu_2 = 0$ versus $H_a : \mu_1 - \mu_2 > 0$ using a level .05 test based on sample sizes $n_1 = 15$ and $n_2 = 15$ (so the probability of rejecting H_0 when H_0 is true is only .05). The value $\mu_1 - \mu_2 = 5$ might be viewed as a substantial departure from H_0, so the investigator would want the probability of not rejecting H_0 when $\mu_1 - \mu_2 = 5$ to be small. This probability is exactly β for the value $\mu_1 - \mu_2 = 5$. The value of β for the one-sample t test depended on σ, the population standard deviation. Similarly, β for the pooled t test depends on σ, the common value of the two population standard deviations. To determine β, first specify the level of significance α, the two sample sizes, and a plausible value of σ. Then select an alternative value of $\mu_1 - \mu_2$ for which β is desired and compute

$$d = \frac{|\text{alternative value} - \text{hypothesized value}|}{\sigma} \cdot \sqrt{\frac{n_1 n_2}{(n_1 + n_2)(n_1 + n_2 - 1)}}$$

Now locate the appropriate set of curves in Table V ($\alpha = .05$ or $.01$ for a one- or two-tailed test), move over to the value of d on the horizontal axis, move up to the $n_1 + n_2 - 2$ df curve, and finally look over to the vertical axis to read the value of β.

EXAMPLE 16

It is very important that fabric used in children's clothing be fire resistant. Suppose that an investigator wishes to compare burn times of two different fabrics. Fabric specimens of a certain size will be used in the experiment, and the burn time for each one will be recorded. Assuming that burn times are normally distributed with $\sigma_1 = \sigma_2$, the pooled t test is appropriate for testing $H_0 : \mu_1 - \mu_2 = 0$ versus $H_a : \mu_1 - \mu_2 \neq 0$. The chosen sample sizes are $n_1 = 15$ and $n_2 = 15$. If the true standard deviation of each fabric's burn time is $\sigma = 1.50$ sec and μ_1 actually exceeds μ_2 by 2 sec (alternative value = $\mu_1 - \mu_2 = 2$), then

$$d = \frac{2}{1.5}\sqrt{\frac{(15)(15)}{(30)(29)}} = .68$$

The number of degrees of freedom for the test is $n_1 + n_2 - 2 = 28$. The curve for 28 df does not appear in Table V, but it is very close to the 29 df curve, which does appear. Using a level $\alpha = .05$ two-tailed test gives $\beta \approx .08$, whereas using $\alpha = .01$ gives $\beta \approx .2$. At level .05, if the difference between μ_1 and μ_2 is 2 rather than 0, two samples of size 15 will result in H_0 being incorrectly not rejected only 8% of the time. The test has good ability to detect this sort of departure from H_0. Of course, if the actual value of σ is greater than 1.5, the value of d would be smaller and β would be larger. The greater the variability in the populations, the more difficult it is to draw the right conclusion. ■

EXERCISES

9.22 Let μ_1 and μ_2 denote true average stopping distances for two different types of cars traveling at 50 mi/h. Assuming normality and $\sigma_1 = \sigma_2$, test $H_0 : \mu_1 - \mu_2 = 0$ versus $H_a : \mu_1 - \mu_2 \neq 0$ at level .05 using the following data: $n_1 = 6$, $\overline{x}_1 = 122.7$, $s_1 = 5.59$, $n_2 = 5$, $\overline{x}_2 = 129.3$, and $s_2 = 5.25$.

9.23 Suppose that μ_1 and μ_2 are true mean stopping distances at 50 mi/h for cars of a certain type equipped with disk brakes and with pneumatic brakes, respectively. Use the pooled t test at significance level .01 to test $H_0 : \mu_1 - \mu_2 = -10$ versus $H_a : \mu_1 - \mu_2 < -10$ for the following data: $n_1 = 6$, $\overline{x}_1 = 115.7$, $s_1 = 5.03$, $n_2 = 6$, $\overline{x}_2 = 129.3$, and $s_2 = 5.38$.

9.24 In an experiment to study the effects of exposure to ozone, 20 rats were exposed to ozone in the amount of 2 ppm for a period of 30 days. The average lung volume for these rats was determined to be 9.28 mL with a standard deviation of .37, whereas the average lung volume for a control group of 17 rats with similar initial characteristics was 7.97 mL with a standard deviation of .41 ("Effect of Chronic Ozone Exposure on Lung Elasticity in Young Rats" *J. Appl. Physiology* (1974): 92–97). Does this data indicate that there is an increase in true average lung volume associated with ozone? Letting μ_1 and μ_2 denote the true average lung volumes for the exposed and unexposed conditions, test $H_0 : \mu_1 - \mu_2 = 0$ versus $H_a : \mu_1 - \mu_2 > 0$. Use a level .01 test.

9.25 A paper in the *J. of Nervous and Mental Disorders* ((1968):136–46) reported the following data on the amount of dextroamphetamine excreted by a sample of children having organically related disorders and a sample of children with nonorganic disorders (dextroamphetamine is a drug commonly used to treat hyperkinetic children). Use a level .05 test to decide whether the data suggests a difference in mean dextroamphetamine excretion for children with organic and nonorganic disorders.

Organic	17.53	20.60	17.62	28.93	27.10
Nonorganic	15.59	14.76	13.32	12.45	12.79

9.26 Stomach concentrations of DDT in juvenile and adult meadow voles and short-tailed shrews were compared in the paper "Species and Age Differences in Accumulation of C1-DDT by Voles and Shrews in the Field" (*Environ. Pollution* (1984):327-40). Summary statistics are given.

DDT residue (mg/kg)

	Meadow Voles			Short-tailed Shrews		
	n	$\bar{x}$	s	n	$\bar{x}$	s
Adult	28	5.6	.4	24	13.4	1.2
Juvenile	22	7.8	.9	31	13.0	1.6

a. Does the data suggest that the true mean stomach concentration for adult shrews exceeds that of juvenile shrews? Use a level .05 test.

b. Estimate the difference in true mean stomach DDT concentration between adult and juvenile meadow voles using a 95% confidence interval.

9.27 Does language used by children in speaking to dogs resemble the language used in speaking to adults? This question was investigated by the authors of the paper "Doggerel: Motherese in a New Context" (*J. Child Language* (1982):229–37). The number of words in an utterance was noted for a set of 8 utterances to dogs and for 8 utterances to adults. The mean length of utterances to dogs was 3.59 words and the standard deviation was .52, whereas the mean and standard deviation for utterances to adults were 9.36 and .46, respectively. Does this data provide sufficient evidence to conclude that the mean length of utterances to dogs is less than the mean length of utterances to adults? Use a level .01 test.

9.28 Unionization of university faculty is a fairly recent phenomenon. There has been much speculation about the effect of collective bargaining on faculty and student satisfaction. Eighteen unionized and 23 nonunionized campuses participated in a study described in the paper "The Relationship Between Faculty Unionism and Organizational Effectiveness" (*Academy of Management J.* (1982):6–24). The participating schools were scored on a number of dimensions. Summary statistics on faculty satisfaction and student academic development are given.

	Student Academic Development		Faculty Satisfaction	
Unionized (n = 18)	$\bar{x}$ = 3.71	s = .49	$\bar{x}$ = 4.49	s = .56
Nonunionized (n = 23)	$\bar{x}$ = 4.36	s = .88	$\bar{x}$ = 4.85	s = .39

a. Let μ_1 be the mean score on student academic development for all unionized schools and μ_2 be the corresponding mean for nonunionized schools. Use a level .05 test to determine if there is a significant difference in mean student academic development score between unionized and nonunionized colleges.

b. Does the data indicate that unionized and nonunionized schools differ significantly with respect to mean faculty satisfaction score? Use a significance level of .05.

c. What P-value is associated with the value of the test statistic computed in (b)?

9.29 Anthropologists are interested in the extent to which members of any given population exhibit genetic diversity. The paper "Genetic, Acclimatizational, and Anthropometric Factors in Hand Cooling Among North and South Chinese" (*Amer. J. Phys. Anthro.* (1975):31–38) reported on a comparison of responses to cold between northern and southern Chinese. The accompanying summary data refers to temperature of the left forefinger after a 30-min immersion period in 5°C water. Test at level .05 to see whether there is a difference in true average temperature between the two Chinese subpopulations. (*Note:* On the basis of their analyses, the investigators suggested the presence of a genetic component in the cold responses of Continental Asian populations.)

Sample	Sample Size	Mean	Standard Deviation
Northern Chinese	16	7.61	1.01
Southern Chinese	13	6.83	.75

9.30 Measurements on a number of physiological variables for samples of 8 male and 8 female adolescent tennis players were reported in "Physiological and Anthropometric Profiles of Elite Prepubescent Tennis Players" (*Sportsmed.* (1984):111–16). Results are summarized in the accompanying table.

	Boys		Girls	
	$\bar{x}$	s	$\bar{x}$	s
Shoulder flexibility	214.4	12.9	216.3	20.0
Ankle flexibility	71.4	4.6	72.5	8.9
Grip strength	23.9	2.5	22.2	4.1

a. Estimate the difference in true mean grip strength for boys and girls using a 95% confidence interval. Does the confidence interval indicate that precise information about this difference is available?

b. Construct 95% confidence intervals for the difference between boys and girls with respect to both true mean shoulder flexibility and ankle flexibility. Interpret each of the intervals.

9.31 A study concerning the sublethal effects of insecticides ("Effects of Sublethal Doses of DDT and Three Other Insecticides on Tribolium Confusum" *J. Stored Products Research* (1983):43–50) reported the accompanying data on oxygen consumption for flour beetles 10 days after DDT treatment and for a control sample of untreated beetles.

Group	Sample Size	Sample Mean	Sample Standard Deviation
Untreated	25	5.02	.94
Treated	25	4.37	.98

a. Does the use of the pooled t test seem reasonable in determining whether DDT treatment results in a decrease in average oxygen consumption? Explain.

b. Test the hypotheses suggested in (a) using a significance level of .05.

c. Obtain as much information as you can about the P-value for the test in (b), and use the result to reach a conclusion at significance level .01.

9.32 Information concerning the extent to which various agricultural products are affected by drought can have a great bearing on planting decisions. The paper "Varietal Differences in the Response of Potatoes to Repeated Periods of Water Stress in Hot Climates" (*Potato Research* (1983):315–21) reported the accompanying data on tuber yield (g/plant) for one particular potato variety.

Treatment	Sample Size	Mean	Standard Deviation
Unstressed	6	376	78.4
Stressed	12	234	65.8

a. Assuming that the yield distributions for the two treatments have the same variance σ^2, compute the pooled estimate of this variance.

b. Does the data suggest that true average yield under water stress is less than for the unstressed condition? Test at level of significance .01.

c. Based on the computed value of the test statistic in (b), what can you say about the P-value?

9.33 Referring to Exercise 9.32, suppose that the investigator wished to know whether water stress lowered true average yield by more than 50 g/plant. State and test the appropriate hypotheses at level .01.

9.34 Use the data in Exercise 9.32 to estimate the true difference in mean yield for the stressed and unstressed conditions with a 90% confidence interval.

9.35 Do certain behaviors result in a severe drain on energy resources because a great deal of energy is expended in comparison to energy intake? The paper "The Energetic Cost of Courtship and Aggression in a Plethodontid Salamander" (*Ecology* (1983):979–83) reported on one of the few studies carried out in this area. The accompanying data is on oxygen consumption (mL/g/h) for male-female salamander pairs (the determination of consumption values was rather complicated, and it is partly for this reason that so few studies of this type have been carried out). Compute a 95% confidence interval for the difference between true average consumption for noncourting pairs and true average consumption for courting pairs. What assumptions about the two consumption distributions are necessary for the validity of the interval?

Behavior	Sample Size	Sample Mean	Sample Standard Deviation
Noncourting	11	.072	.0066
Courting	15	.099	.0071

9.36 The establishment and maintenance of vegetation is an important component in most metalliferous mine waste-reclamation schemes. The ecological consequences for animal populations exposed to pollutants at contaminated reclamation sites is of concern. The paper "Cadmium in Small Mammals from Grassland Established on Metalliferous Mine Waste" (*Environ. Pollution* (1984):153–62) reported on cadmium concentration in various body organs for several species of small mammals. The accompanying data is for skull concentration in samples of one such species both at a reclamation site and at a control site.

Site	Sample Size	Sample Mean	Sample Standard Deviation
Control	20	.59	.13
Reclamation	21	.72	.18

a. Would you recommend using a z test from the previous section to analyze this data? Why or why not?

b. Use the test procedure discussed in this section with $\alpha = .05$ to decide whether true mean cadmium concentrations in skulls differ at the two sites.

c. Suppose that the difference in true average concentrations is actually .2 rather than 0. What is β, the probability of a type II error (not rejecting H_0 and thus failing to detect such a difference), when a level .05 test is used with the above sample sizes and $\sigma = .20$?

9.37 The paper "Pine Needles as Sensors of Atmospheric Pollution" (*Environ. Monitoring* (1982):273–86) reported on the use of neutron activity analysis to determine pollutant concentrations in pine needles. According to the paper's authors,

"These observations strongly indicated that for those elements which are determined well by the analytical procedures, the distribution of concentrations is lognormal. Accordingly, in tests of significance the logarithms of concentrations will be used." The given data refers to bromine concentration in needles taken from a site near an oil-fired steam plant and from a relatively clean site. The summary values are means and standard deviations of the log-transformed observations. Let μ_1 be the true average log concentration at the first site and define μ_2 analogously for the second site. Test for equality of the two concentration distribution means against the alternative that they are different by using the pooled t test at level .05 with the log-transformed data.

Site	Sample Size	Mean Log Concentration	Standard Deviation of Log Concentration
Steam plant	8	18	4.9
Clean	9	11	4.6

9.4 Large-Sample Inferences Concerning a Difference Between Two Population Proportions

Large-sample methods for drawing conclusions about a single population proportion π (the proportion of successes in the population) were discussed in Chapters 7 and 8. Both the confidence interval and test procedure were based on the sample proportion of successes p resulting from a random sample of size n. Many investigations are carried out to compare the proportion of successes in one population (or resulting from one treatment) with the proportion of successes in another population (or from another treatment). As in the case of means, the subscripts 1 and 2 are used to distinguish between the two population proportions, sample sizes, and sample proportions. The success proportion for the first population is denoted by π_1, the size of the random sample selected from that population by n_1, and the sample proportion of successes—(number of successes)/n_1—by p_1. The symbols π_2, n_2, and p_2 represent the analogous quantities for the second population and random sample from that population. The two samples are assumed to have been selected independently of one another, so that p_1 and p_2 are independent.

The statistic p_1 provides a point estimate for π_1, whereas the statistic p_2 gives a point estimate of π_2. Suppose, for example, that the first sample consisting of $n_1 = 120$ individuals yields 72 successes, whereas the second sample of size $n_2 = 100$ results in 53 successes. Then $p_1 = 72/120 = .600$ and $p_2 = 53/100 = .530$. The obvious estimate of $\pi_1 - \pi_2$, the difference between the two population proportions, is $p_1 - p_2 = .600 - .530 = .070$. If $p_1 = .162$ and $p_2 = .208$, the estimate of $\pi_1 - \pi_2$ would be $.162 - .208 = -.046$. The difference $p_1 - p_2$ will be negative whenever p_1 is less than p_2, just as $\pi_1 - \pi_2$ will have a negative value if π_1 is smaller than π_2.

The statistic $p_1 - p_2$ is the basic tool for making inferences about $\pi_1 - \pi_2$. The test procedure and confidence interval to be presented are based on certain facts about the sampling distribution of this statistic.

Properties of the Sampling Distribution of $p_1 - p_2$

1. $\mu_{p_1-p_2} = \mu_{p_1} - \mu_{p_2} = \pi_1 - \pi_2$, so that $p_1 - p_2$ is an unbiased statistic for estimating $\pi_1 - \pi_2$. That is, whatever the value of $\pi_1 - \pi_2$, the sampling distribution of $p_1 - p_2$ will be centered at that value.
2.

$$\sigma^2_{p_1-p_2} = \sigma^2_{p_1} + \sigma^2_{p_2} = \frac{\pi_1(1 - \pi_1)}{n_1} + \frac{\pi_2(1 - \pi_2)}{n_2}$$

and

$$\sigma_{p_1-p_2} = \sqrt{\frac{\pi_1(1 - \pi_1)}{n_1} + \frac{\pi_2(1 - \pi_2)}{n_2}}$$

3. If both n_1 and n_2 are large [$n_1\pi_1 \geq 5$, $n_1(1 - \pi_1) \geq 5$, $n_2\pi_2 \geq 5$, and $n_2(1 - \pi_2) \geq 5$], then p_1 and p_2 each have approximately normal sampling distributions, so their difference, $p_1 - p_2$, also has approximately a normal sampling distribution. This, along with (1) and (2), implies that the sampling distribution of

$$z = \frac{p_1 - p_2 - (\pi_1 - \pi_2)}{\sqrt{\dfrac{\pi_1(1 - \pi_1)}{n_1} + \dfrac{\pi_2(1 - \pi_2)}{n_2}}}$$

is described approximately by the standard normal (z) curve.

A Large-Sample Test Procedure

Comparisons of π_1 and π_2 are usually based on large samples, so we restrict ourselves to such cases here. The most general null hypothesis of interest has the form $H_0 : \pi_1 - \pi_2 =$ hypothesized value. However, the appropriate test statistic when the hypothesized value is something other than zero differs somewhat from the test statistic used for $H_0 : \pi_1 - \pi_2 = 0$. Since this latter H_0 is almost always the relevant one in applied problems, we'll focus exclusively on it.

Our basic testing principle has been to use a procedure that controls the probability of a type I error at the desired level α. This requires using a test statistic whose sampling distribution is known when H_0 is true. That is, the test statistic should be developed under the assumption that $\pi_1 = \pi_2$ (as specified by the null hypothesis). In this case, π can be used to denote the common value of the two population proportions. The z variable obtained by standardizing $p_1 - p_2$ then simplifies to

$$z = \frac{p_1 - p_2}{\sqrt{\dfrac{\pi(1 - \pi)}{n_1} + \dfrac{\pi(1 - \pi)}{n_2}}}$$

Unfortunately, this cannot serve as a test statistic because the denominator can't be computed. H_0 says that there is a common value π but doesn't specify what that value is. A test statistic can be obtained, though, by first estimating π from the sample data and then using this estimate in the denominator of z.

When $\pi_1 = \pi_2$, either p_1 or p_2 separately gives an estimate of the common proportion π. A better estimate than either of these is a weighted average of the two, in which more weight is given to the sample proportion based on the larger sample.

DEFINITION

The **combined estimate of the common success proportion** is

$$p_c = \left(\frac{n_1}{n_1 + n_2}\right) p_1 + \left(\frac{n_2}{n_1 + n_2}\right) p_2$$

EXAMPLE 17

Many investigators have studied the effect of the wording of questions on survey responses. Consider the following two versions of a question concerning gun control:

1. Would you favor or oppose a law that would require a person to obtain a police permit before purchasing a gun?
2. Would you favor or oppose a law that would require a person to obtain a police permit before purchasing a gun, or do you think that such a law would interfere too much with the right of citizens to own guns?

Let π_1 denote the proportion of all adults who would respond *favor* when asked Question 1, and define π_2 similarly for Question 2. The paper "Attitude Measurement and the Gun Control Paradox" (*Public Opinion Quarterly* (1977–78):427–38) reported the accompanying sample data.

Sample size	$n_1 = 615$	$n_2 = 585$
Number who favor (success)	463	403
Sample proportion	$p_1 = \frac{463}{615} = .753$	$p_2 = \frac{403}{585} = .689$

Suppose that $\pi_1 = \pi_2$ and let π denote the common value. Then the combined estimate of π is (since $n_1 + n_2 = 1200$)

$$p_c = \frac{615}{1200}(.753) + \frac{585}{1200}(.689) = .722$$

■

The test statistic for testing $H_0: \pi_1 - \pi_2 = 0$ results from using p_c in place of π in the standardized variable z given above. This z statistic has approximately a standard normal distribution when H_0 is true, so a test that has desired level of significance α uses the appropriate z critical value to specify the rejection region.

Summary of Large-Sample z Tests for $\pi_1 - \pi_2$

Null hypothesis: $H_0: \pi_1 - \pi_2 = 0$

Test statistic: $$z = \frac{p_1 - p_2}{\sqrt{\frac{p_c(1 - p_c)}{n_1} + \frac{p_c(1 - p_c)}{n_2}}}$$

Alternative hypothesis	*Rejection region*
$H_a : \pi_1 - \pi_2 > 0$	$z > z$ critical value
$H_a : \pi_1 - \pi_2 < 0$	$z < -z$ critical value
$H_a : \pi_1 - \pi_2 \neq 0$	Either $z > z$ critical value or $z < -z$ critical value

The z critical values corresponding to the most frequently used significance levels appear in the bottom row of Table IV. The test should be used only when all the quantities n_1p_1, $n_1(1 - p_1)$, n_2p_2, and $n_2(1 - p_2)$ are at least 5.

EXAMPLE 18

Looking again at the gun control survey questions introduced in Example 17, the extra phrase in Question 2 reminding individuals of the right to bear arms might tend to elicit a smaller proportion of favorable responses than would the first question without the phrase. Does the data suggest that this is indeed the case? Let's state and test the relevant hypotheses using level of significance .01.

1. $\pi_1 - \pi_2$ is the difference between the true proportions of favorable responses to Questions 1 and 2.
2. $H_0 : \pi_1 - \pi_2 = 0$ $(\pi_1 = \pi_2)$.
3. $H_a : \pi_1 - \pi_2 > 0$ $(\pi_1 > \pi_2)$, so that the extra phrase does result in proportionately fewer favorable responses.
4. Test statistic:

$$z = \frac{p_1 - p_2}{\sqrt{p_c(1 - p_c)/n_1 + p_c(1 - p_c)/n_2}}$$

5. The appropriate test is upper-tailed, and for $\alpha = .01$, Table IV gives the z critical value as 2.33. H_0 will be rejected in favor of H_a if $z > 2.33$.
6. $n_1 = 615$, $n_2 = 585$, $p_1 = .753$, $p_2 = .689$, and $p_c = .722$, so

$$z = \frac{.753 - .689}{\sqrt{(.722)(.278)/615 + (.722)(.278)/585}} = \frac{.064}{.0259} = 2.47$$

7. Since $2.47 > 2.33$, H_0 is rejected at level .01. Inclusion of the extra phrase about the right to bear arms does seem to result in fewer favorable responses than would be elicited without the phrase. ■

Because the large-sample procedure here is a z test, we can compute a P-value exactly as we did for other z tests. For an upper-tailed test, the P-value is the area under the z curve to the right of the computed z. In the last example, $z = 2.47 \approx 2.5$, so from Table I, the P-value is approximately .006. Since $.006 < .01$, H_0 would be rejected at level .01, but H_0 could not be rejected at level .005 or level .001.

EXAMPLE 19

Some defendants in criminal proceedings plead guilty and are sentenced without a trial, whereas others plead innocent, are subsequently found guilty, and then are sentenced. In recent years legal scholars have speculated as to whether sentences of those who plead guilty differ in severity from sentences for those who plead innocent and are subsequently judged guilty. Consider the accompanying data on defendants from San Francisco County accused of robbery, all of whom had previous prison records ("Does It Pay to Plead Guilty? Differential Sentencing and the Functioning of Criminal Courts" *Law and Society Rev.* (1981–82):45–69).

	Plea	
	Guilty	Not Guilty
Number judged guilty	$n_1 = 191$	$n_2 = 64$
Number sentenced to prison	101	56
Sample proportion	$p_1 = .529$	$p_2 = .875$

Does this data suggest that the proportion of all defendants in these circumstances who plead guilty and are sent to prison differs from the proportion who are sent to prison after pleading innocent and being found guilty?

Let π_1 and π_2 denote the two population proportions. The hypotheses of interest are $H_0: \pi_1 - \pi_2 = 0$ versus $H_a: \pi_1 - \pi_2 \neq 0$. Rather than fix a level of significance α, state the corresponding rejection region, and carry out the test, let's compute a P-value. The combined estimate of the common success proportion is

$$p_c = \frac{191}{255}(.529) + \frac{64}{255}(.875) = .616$$

The computed value of the test statistic is then

$$z = \frac{.529 - .875}{\sqrt{(.616)(.384)/191 + (.616)(.384)/64}} = \frac{-.346}{.070} = -4.94 \approx -5$$

The P-value for a two-tailed z test is

$$\begin{aligned} P\text{-value} &= 2\cdot(\text{area under } z \text{ curve to left of } -5) \\ &= 2\cdot(.00000029) = .00000058 \end{aligned}$$

This P-value is so miniscule that at any reasonable level α, H_0 should be rejected. The data very strongly suggests that $\pi_1 \neq \pi_2$ and, in particular, that initially pleading guilty may be a good strategy as far as avoiding prison is concerned.

The cited article also reported data on defendants in several other counties. The authors broke down the data by type of crime (burglary or robbery) and by nature of prior record (none, some but no prison, and prison). In every case the conclusion was the same: among defendants judged guilty, those who pleaded that way were less likely to receive prison sentences. ■

A statistically significant result should not be construed as manifest evidence of causality. Statistical analysis can identify association, but reason for the association may not be at all obvious from the analysis. The next example makes this point very nicely.

EXAMPLE 20

The article "Is There Sex Bias in Graduate Admissions?" (*Science* (1975): 398–404) discussed data on admission to graduate programs at the University of California. As the title indicates, interest focused on the possibility of sexual discrimination in admission decisions. For ease of understanding, consider the accompanying contrived data, which exhibits the same properties as did the original data.

	Male	Female
Number of applicants	100	100
Number admitted	60	40
Sample proportion	.60	.40

Let π_1 and π_2 denote the true long-run proportions of male and female admissions, respectively. Then the no-discrimination hypothesis is $H_0: \pi_1 - \pi_2 = 0$ and the discrimination hypothesis is $H_a: \pi_1 - \pi_2 \neq 0$ (after all, before the data is collected, the possibility of admission rates favoring either males or females must be entertained). The value of the combined estimate is $p_c = .50$, from which

$$z = \frac{.60 - .40}{\sqrt{(.50)(.50)/100 + (.50)(.50)/100}} = \frac{.20}{.071} \approx 2.8$$

The P-value for a two-tailed test is

$$P\text{-value} = 2 \cdot \begin{pmatrix}\text{area under } z \text{ curve} \\ \text{to the right of } 2.8\end{pmatrix} = 2 \cdot (.0026) = .0052$$

Since $.0052 < .01$, H_0 should be rejected at level .01 in favor of H_a. The admission rate appears to differ for the two sexes, and the rate for males appears to be significantly higher than that for females.

It is tempting to conclude that sexual discrimination is at the root of the observed difference in admission rates. Suppose, though, that the university maintains graduate programs in two areas, technology and fine arts. Consider the following data on admissions to these two program areas.

	Technology		Fine Arts	
	Male	Female	Male	Female
Number of applicants	80	20	20	80
Number admitted	56	16	4	24
Sample proportion	.70	.80	.20	.30

Which of the two individual programs discriminates against females? The surprising answer is that neither one does! Within each program the females appear to have the edge in admission rate, and yet the overall rate substantially favors the males.

This seemingly paradoxical result can be explained by looking carefully at the data for the two individual programs. First, it is much easier for either sex to gain admission to the technology program. Second, most of the male applicants applied to the technology program, whereas most of the females who applied did so to the fine arts program. So the significant result from the aggregated data is not due to any obvious sexual discrimination but instead to the fact that women applied primarily to a highly competitive program, whereas most males applied to the program with a high acceptance rate. If there had not been this imbalance in applicants to the two programs, the aggregated data would not have suggested discrimination. ■

A Confidence Interval

A large-sample confidence interval for $\pi_1 - \pi_2$ is a special case of the general z interval formula

$$\text{point estimate} \pm (z \text{ critical value})(\text{estimated standard deviation})$$

The statistic $p_1 - p_2$ gives a point estimate of $\pi_1 - \pi_2$, and the standard deviation of this statistic is

$$\sigma_{p_1 - p_2} = \sqrt{\frac{\pi_1(1 - \pi_1)}{n_1} + \frac{\pi_2(1 - \pi_2)}{n_2}}$$

An estimated standard deviation is obtained by using the sample proportions p_1 and p_2 in place of π_1 and π_2, respectively, under the square root symbol. Notice that this estimated standard deviation differs from the one used earlier in the test statistic. Here there isn't a null hypothesis that claims $\pi_1 = \pi_2$, so there is no common value of π to estimate.

> **A large-sample confidence interval for $\pi_1 - \pi_2$** is given by
>
> $$p_1 - p_2 \pm (z \text{ critical value}) \sqrt{\frac{p_1(1 - p_1)}{n_1} + \frac{p_2(1 - p_2)}{n_2}}$$
>
> The interval is valid whenever n_1p_1, $n_1(1 - p_1)$, n_2p_2, and $n_2(1 - p_2)$ are all at least 5.

EXAMPLE 21

A person released from prison before completing the original sentence is placed under the supervision of a parole board. If that person violates specified conditions of good behavior during the parole period, the board can order a return to prison. To what extent does the frequency of parole violation depend on type of crime and various other factors? The paper "Impulsive and Premeditated Homicide: An Analysis of the Subsequent Parole Risk of the Murderer" (*J. Criminal Law and Criminology* (1978): 108–14) reported the accompanying data on parole behavior. One sample of individuals had served time in prison for impulsive murder and the other sample has served time for premeditated murder.

	Impulsive	Premeditated
Sample size	$n_1 = 42$	$n_2 = 40$
Number with no violation	13	22
Sample proportion	$p_1 = .310$	$p_2 = .550$

Let π_1 denote the proportion of all impulsive murderers who successfully complete parole, and define π_2 analogously for premeditated murderers. The sample sizes are large enough for the large-sample interval to be valid ($n_1p_1 = 42(.310) = 13 \geq 5$, $n_1(1 - p_1) = 42(.690) = 29 \geq 5$, etc.). A 98%

confidence interval for $\pi_1 - \pi_2$ uses the z critical value 2.33. The resulting interval is

$$.310 - .550 \pm 2.33\sqrt{\frac{(.310)(.690)}{42} + \frac{(.550)(.450)}{40}}$$

$$= -.240 \pm (2.33)(.106) = -.240 \pm .247 = (-.487, .077)$$

This interval includes zero, so at the 98% level of confidence one of the plausible values of $\pi_1 - \pi_2$ is 0. The interval is quite wide because of the relatively small sample sizes. ■

EXERCISES

9.38 Let π_1 and π_2 denote the proportions of all male and female shoppers, respectively, who buy only name-brand grocery products (as opposed to generic or store brands).

a. Test $H_0 : \pi_1 - \pi_2 = 0$ versus $H_a : \pi_1 - \pi_2 \neq 0$ at level .05 using the following data:

$n_1 = 200$, number of successes (only name-brand purchases) $= 87$, $n_2 = 300$, number of successes $= 96$.

b. Use the data of (a) to compute a 95% confidence interval for $\pi_1 - \pi_2$.

9.39 Is someone who switches brands because of a financial inducement less likely to remain loyal than someone who switches without inducement? Let π_1 and π_2 denote the true proportions of switchers to a certain brand with and without inducement, respectively, who subsequently make a repeat purchase. Test $H_0 : \pi_1 - \pi_2 = 0$ versus $H_a : \pi_1 - \pi_2 < 0$ using $\alpha = .01$ and the following data:

$n_1 = 200$, number of successes $= 30$, $n_2 = 600$, number of successes $= 180$

(Similar data appears in "Impact of Deals and Deal Retraction on Brand Switching," *J. Marketing* (1980):62–70.)

9.40 Using an electronic process called time compression, a 30-s television commercial can be broadcast in its entirety in only 24 s. There is no shift in voice pitch and subjects are not aware that commercials have been altered. The article "Reducing the Costs of TV Commercials by Use of Time Compressions" (*J. Marketing Research* (1980):52–57) reported on a study involving recall ability for subjects watching compressed as compared to noncompressed commercials. For one commercial, 15 of the 57 subjects viewing the normal version could subsequently recall the commercial, while 32 of the 74 subjects viewing the compressed version could subsequently recall it. Does this data suggest any difference between true recall proportions for the two versions?

a. Verify that the sample sizes are large enough to justify using the large-sample procedures.

b. Carry out a test at level .05 to answer the question posed.

c. Compute the P-value. Based on this value, what would you conclude at level .10?

9.41 Ionizing radiation is being given increasing attention as a method for preserving horticultural products. The paper "The Influence of Gamma-Irradiation on the Storage Life of Red Variety Garlic" (*J. Food Processing and Preservation* (1983): 179–83) reported that 153 of 180 irradiated garlic bulbs were marketable (no external sprouting, rotting, or softening) 240 days after treatment, while only 119 of 180 untreated bulbs were marketable after this length of time. Does this data suggest that the true proportion of marketable irradiated bulbs exceeds that for untreated bulbs? Test the relevant hypotheses at level .01.

9.42 How do driving habits relate to involvement in accidents? Two scientists at General Motors' Research Laboratories sampled 8000 drivers and determined for each one whether or not *following headway*—the time interval a driver allows between his or her car and the immediately preceding car—was at most 1 s. Of those who drove with a headway of at most 1 s, 57.9% had been involved in traffic accidents, while only 49.7% of those with a headway of more than 1 s had any accident involvement. Does this data indicate that the proportion of tailgaters involved in accidents is higher than the corresponding proportion of nontailgaters? Assume for purposes of analysis that there were 4000 drivers in each of the two samples, and test the relevant hypotheses at level .05. (Source: *Los Angeles Times,* February 1, 1981.)

9.43 The paper "The Association of Marijuana Use with Outcome of Pregnancy" (*Amer. J. Public Health* (1983):1161–64) reported the accompanying data on incidence of major malfunctions among newborns both for mothers who were marijuana users and mothers who were nonusers. Does this data suggest that the true proportion of newborns with a major malfunction is greater for mothers who use marijuana than for mothers who are not marijuana users?

	User	Nonuser
Sample size	1246	11,178
Number of major malfunctions	42	294

a. Use a test with level of significance .05.

b. Compute the P-value and then use it to reach a conclusion at level .05.

9.44 In a nationwide study directed by Detroit's Henry Ford Hospital, 780 persons with stable heart disease were treated. Half of the subjects were treated with drugs and half underwent bypass surgery. After 6 years, 351 of those treated with drugs and 359 of those who underwent bypass surgery were still alive. Use a level .01 test to determine if there is sufficient evidence to indicate that surgery is a more effective treatment.

9.45 The rising tide of national and regional loyalties around the world has great bearing on the survival of minority (secondary) languages. One of the few comparative studies in this area was reported in the article "Language Maintenance and Shift in a Breton and Welsh Sample" (*Word* (1983):67–88). A random sample of 86 Welsh bilingual adults yielded 76 who spoke Welsh fluently, while another random sample of 77 bilingual adults from Brittany resulted in 57 who spoke Breton fluently (both languages are southern Celtic in origin). Does this data point to the conclusion that the true proportion of fluent speakers among Welsh bilingual adults differs from the corresponding proportion for Breton bilingual adults?

a. Carry out a test of hypotheses using a 5% level of significance.

b. The paper reported that the P-value was less than .01. Do you agree? Answer by computing the P-value.

9.46 The Associated Press (June 13, 1976) reported the results of a comparative study on childbirth methods. A group of 129 women who had been trained in the Lamaze method and a second group of equal size without Lamaze training were studied. Of the women with Lamaze training, 104 had spontaneous, normal deliveries, compared to 61 of those without Lamaze training. Does the data strongly suggest that the true proportion of women with Lamaze training who have spontaneous, normal deliveries exceeds that for women without such training? Use a level

.01 test. Does your analysis constitute proof that Lamaze training tends to cause spontaneous, normal deliveries?

9.47 Are teenagers who smoke less healthy than those who don't smoke? In a study of 54 high school students who had been regular smokers for more than 2 years and another 54 who had never smoked, it was found that 21 of the smokers and 13 of the nonsmokers reported persistent coughing (*San Luis Obispo Telegram-Tribune,* March 25, 1984).

a. Does the data suggest that a significant difference exists between the true proportions for smokers and nonsmokers who experience persistent coughs? Use a level .05 test.

b. Calculate the P-value associated with the test statistic in (a). Use the P-value to decide whether the conclusion of the test in (a) would have been different if a significance level of .01 had been used.

9.48 The positive effect of water fluoridation on dental health is well documented. One study that validates this is described in the paper "Impact of Water Fluoridation on Children's Dental Health: A Controlled Study of Two Pennsylvania Communities" (*Amer. Stat. Assoc. Proc. of the Social Statistics Section* (1981): 262–65). Two communities were compared. One had adopted fluoridation in 1966, while the other had no fluoridation program. Of 143 children from the town without fluoridated water, 106 had decayed teeth, while 67 of 119 children from the town with fluoridated water had decayed teeth. Let π_1 denote the true proportion of children drinking fluoridated water who have decayed teeth and let π_2 denote the analogous proportion for children drinking unfluoridated water. Estimate $\pi_1 - \pi_2$ using a 90% confidence interval. Does the interval contain 0? Interpret the interval.

9.49 Sudden infant death syndrome (SIDS) has baffled researchers for many years. To determine if mothers who smoke cigarettes have an increased risk of losing a child to SIDS, researchers at the National Institute of Child Health did an extensive investigation of 400 SIDS deaths and of 800 healthy babies (*Los Angeles Times,* February 2, 1984). The study showed that 320 of the mothers of healthy babies smoked, whereas 280 of the mothers of babies that had died of SIDS were smokers. Let π_1 and π_2 denote the true proportion of mothers who smoke for healthy babies and SIDS babies, respectively. Estimate $\pi_1 - \pi_2$ using a 99% confidence interval and interpret your results.

9.50 What psychological factors contribute to the success of competitive athletes? Numerous possibilities are examined in the paper "Elite Divers and Wrestlers: A Comparison Between Open- and Closed-Skill Athletes" (*J. Sport Psych.* (1983): 390–409). Competitive divers participating in qualifying trials were asked whether they exercised within 1 h of a competition. Suppose that of 20 qualifying divers, 7 exercise within 1 h of a competition, while 12 of 25 nonqualifying divers exercise within 1 h of a meet. Is there sufficient evidence to indicate that the true proportion of qualifying divers who exercise within 1 h of competition differs from the corresponding proportion for nonqualifying divers? Use a level .10 test.

9.51 As part of a class project, two college students from Florida found that people are willing to help strangers in quite surprising circumstances (Associated Press, May 10, 1984). The two students splashed their faces and hands and rinsed their mouths with gin. They told passersby that they were too drunk to unlock their car doors and asked for help. One student was dressed in a business suit, whereas the other wore a dirty T-shirt. Of 50 people approached by the student in the suit, 21 helped unlock the car door and aided him in getting into the car! The student in the T-shirt also approached 50 people, and was assisted by 23 of them! Does the data suggest that the true proportion of people that would assist a well-dressed drunk into a car differs significantly from the true proportion who would assist a drunk in dirty clothes? Use a level .05 test.

9.52 The *San Francisco Chronicle* (May 27, 1983) reported the results of a poll designed to assess public opinion on legalized gambling. Of 750 Californians interviewed, 578 favored a state lottery. A similar survey in 1971 found that of 750 people contacted, 518 favored a state lottery.

a. Does the data strongly suggest that the true 1983 proportion of Californians who favor a state lottery exceeds the corresponding 1971 proportion? Use a level .01 test.

b. What P-value is associated with the value of the test statistic in (a)?

9.53 The article "New Stance Taken on Blood Cholesterol" (*Los Angeles Times,* February 12, 1984) states: "Ten thousand patients have been treated surgically and 10,000 comparable patients have been treated medically. Five years later, 91% of the surgical patients are alive and 90% of the medical patients are alive. No difference."

a. Use the information above to test $H_0 : \pi_1 - \pi_2 = 0$, where π_1 is the true proportion of patients who survive 5 years when treated surgically and π_2 is defined analogously for those treated medically.

b. Does the result of your hypothesis test in (a) agree with the statement of "no difference" that appeared in the article? Do you think this is a case of statistical rather than practical significance? Explain.

9.5 Distribution-Free Procedures for Inferences Concerning a Difference Between Two Population Means

One approach to making inferences about $\mu_1 - \mu_2$ when n_1 and n_2 are small is to assume that the two population distributions are normal with $\sigma_1 = \sigma_2$ and then use the pooled t test or confidence interval developed in Section 9.3. In some situations, though, the normality assumption may not be reasonable. The procedures to be presented in this section are valid under the following less-restrictive conditions on the population distributions.

Basic Assumptions in this Section

The two population distributions have the same shape and spread. The only possible difference between the distributions is that one may be shifted to one side of the other.

Distributions consistent with these assumptions are pictured in Figure 4(c). A procedure that is valid whenever the basic assumptions are met is one whose use does not depend on any overly specific assumptions about the population distributions. Such a procedure is said to be **distribution-free** (some texts use the phrase *nonparametric* instead of distribution-free). The pooled t test is not distribution-free because its use is predicated on the specific assumption of (at least approximate) normality.

Inferences about $\mu_1 - \mu_2$ will be made from information in two independent random samples, one consisting of n_1 observations from the first population and the other consisting of n_2 observations from the second population. Suppose that we are trying to decide whether or not the two distributions are identical (because of our basic assumptions, this becomes $\mu_1 = \mu_2$ versus $\mu_1 \neq \mu_2$). When they are identical, each of the $n_1 + n_2$ observations is actually drawn from the same population distribution. The distribution-free procedure presented here is based on regarding the $n_1 + n_2$ observations as a single data set and assigning ranks to the ordered values. The assignment is easiest when there are no ties among the $n_1 + n_2$ values (each observation is different from every one of the others), so assume for the moment that this is the case. Then the smallest among the $n_1 + n_2$ values receives rank 1, the second smallest rank 2, and so on, until finally the largest value is assigned rank $n_1 + n_2$.

EXAMPLE 22

An experiment to compare fuel efficiencies for two types of subcompact automobiles was carried out by first randomly selecting $n_1 = 5$ cars of type 1 and $n_2 = 5$ cars of type 2. Each car was then driven from Phoenix to Los Angeles by a nonprofessional driver, after which the fuel efficiency (in mi/gal) was determined. The resulting data, with observations in each sample ordered from smallest to largest, appears below.

Type 1	39.3	41.0	42.4	43.0	44.4
Type 2	37.8	39.0	39.8	40.7	42.1

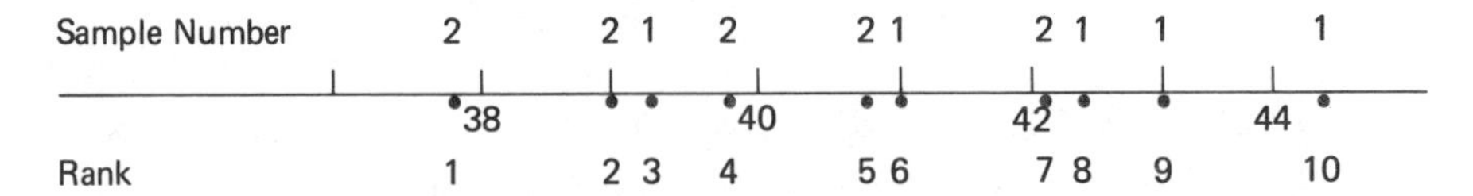

The ranks of the five observations in the first sample are 3, 6, 8, 9, 10. If these five observations had all been larger than every value in the second sample, the corresponding ranks would have been 6, 7, 8, 9, and 10. On the other hand, if all five sample 1 observations had been less than each value in the second sample, the ranks would have been 1, 2, 3, 4, and 5. The ranks of the five observations in the first sample might be any set of five numbers from among 1, 2, 3, . . . , 9, 10—there are actually 252 possibilities. ■

Testing Hypotheses

Let's first consider testing $H_0: \mu_1 - \mu_2 = 0$ $(\mu_1 = \mu_2)$ versus $H_a: \mu_1 - \mu_2 \neq 0$ $(\mu_1 \neq \mu_2)$. When H_0 is true, all $n_1 + n_2$ observations in the two samples are actually drawn from the same population distribution. One would then expect that the observations in the first sample would be intermingled with those of the second sample when plotted on a measurement axis. In this case, the ranks of the observations should also be intermingled. For example, with $n_1 = 5$ and $n_2 = 5$, the set of sample 1 ranks 2, 3, 5, 8, 10 would be consistent with $\mu_1 = \mu_2$, as would the set 1, 4, 7, 8, 9. However, when $\mu_1 = \mu_2$, it would be quite unusual for all five values from sample 1 to be smaller than every value in sample 2, resulting in the set 1, 2, 3, 4, 5 of sample 1 ranks. Similarly, we would not expect to observe 6, 7, 8, 9, 10 as the sample 1 ranks when the population distributions are identical.

A convenient measure of the extent to which the ranks are intermingled is the sum of the sample 1 ranks. These ranks in Example 22 were 3, 6, 8, 9, and 10, so the rank sum is $3 + 6 + 8 + 9 + 10 = 36$. The smallest possible rank sum when $n_1 = n_2 = 5$ is $1 + 2 + 3 + 4 + 5 = 15$, and the largest possible sum is $6 + 7 + 8 + 9 + 10 = 40$. If μ_1 and μ_2 differ greatly, we would expect the rank sum to be near either its smallest possible value (if $\mu_1 < \mu_2$) or its largest possible value (if $\mu_1 > \mu_2$). This suggests selecting both an upper-tail critical value and a lower-tail critical value and then rejecting H_0 in favor of H_a if either *rank sum* $\geq$ *upper-tail critical value* or *rank sum* $\leq$ *lower-tail critical value.* The two critical values should be chosen so that the probability of rejecting H_0 when H_0 is true (the type I error probability) has a specified value α. This requires information about the sampling distribution of the rank sum statistic when H_0 is true.

Consider again the case $n_1 = n_2 = 5$. There are 252 different sets of 5 from among the 10 ranks 1, 2, 3, . . . , 9, 10. The key point is that when H_0 is true, any one of these 252 sets has the same chance of being the sample 1 ranks as does any other set because all 10 observations come from the same population distribution. The chance under H_0 that any particular set occurs is $\frac{1}{252}$ (because the possibilities are equally likely). Table 1 displays the seven sets of sample 1 ranks yielding the smallest rank sum values and the seven sets yielding the largest rank sum values. Each one of the other 238 possible rank sets has a rank sum value between 19 and 36. Suppose that we agree to reject H_0 either if rank sum $\geq$ 37 or if rank sum $\leq$ 18. The probability of a type I error is then

$$\begin{aligned}\alpha &= P(\text{rank sum} \geq 37 \text{ or rank sum} \leq 18 \text{ when } H_0 \text{ is true})\\ &= P(\text{rank set is one of the 14 displayed in Table 1 when } H_0 \text{ is true})\\ &= \frac{14}{252} = .056 \approx .05\end{aligned}$$

That is, when the above test procedure is used, roughly 5% of all possible samples will result in H_0 being incorrectly rejected. The approximate level of significance for the test is .05.

TABLE 1

THE SEVEN RANK SETS THAT HAVE THE SMALLEST RANK SUMS AND THE SEVEN THAT HAVE THE LARGEST RANK SUMS WHEN $n_1 = 5$, $n_2 = 5$

Sample 1 Ranks	Rank Sum	Sample 1 Ranks	Rank Sum
1 2 3 4 5	15	6 7 8 9 10	40
1 2 3 4 6	16	5 7 8 9 10	39
1 2 3 4 7	17	4 7 8 9 10	38
1 2 3 5 6	17	5 6 8 9 10	38
1 2 3 4 8	18	3 7 8 9 10	37
1 2 3 5 7	18	4 6 8 9 10	37
1 2 4 5 6	18	5 6 7 9 10	37

The alternative hypothesis $H_a : \mu_1 - \mu_2 > 0$ states that the first population distribution is shifted to the right of the second one. Evidence supporting this claim is provided by samples for which most sample 1 observations are larger than most observations in the second sample, resulting in large sample 1 ranks and a relatively large rank sum value. Thus for this alternative, an upper-tailed test—reject H_0 in favor of H_a if rank sum $\geq$ upper-tail critical value—is appropriate. In the case $n_1 = n_2 = 5$, Table 1 implies that the upper-tailed test with critical value 39 has $\alpha = P(\text{type I error}) = P(\text{rank sum} \geq 39 \text{ when } H_0 \text{ is true}) = \frac{2}{252} = .008 \approx .01$. There are 12 sets of rankings with rank sum greater than or equal to 36, so using the upper-tail critical value 36 gives $\alpha = P(\text{rank sum} \geq 36 \text{ when } H_0 \text{ is true}) = \frac{12}{252} = .048 \approx .05$. A lower-tailed test is appropriate when the alternative hypothesis is $H_a: \mu_1 - \mu_2 < 0$. The test that rejects H_0 in favor of this H_a when rank sum ≤ 16 has (in the case $n_1 = n_2 = 5$) $\alpha \approx .01$, whereas using critical value 19 gives $\alpha \approx .05$.

It is not possible to obtain exactly level .05 or .01 because the rank sum statistic has a discrete sampling distribution that, at least for n_1 and n_2 small, associates probability with relatively few possible values. This contrasts with z and t tests for which these levels could be achieved exactly. The reason for this is that the t and z sampling distributions are continuous, so probability is the area under a curve, and a critical value that captures area .05 or .01 can always be found.

Summary of the Rank Sum Test*

Null hypothesis: $H_0: \mu_1 - \mu_2 = 0$

Test statistic: rank sum = the sum of ranks assigned to the n_1 observations in the first sample

Alternative hypothesis	*Rejection region*
$H_a: \mu_1 - \mu_2 > 0$	Rank sum $\geq$ upper-tail critical value
$H_a: \mu_1 - \mu_2 < 0$	Rank sum $\leq$ lower-tail critical value
$H_a: \mu_1 - \mu_2 \neq 0$	Either rank sum $\geq$ upper-tail critical value or rank sum $\leq$ lower-tail critical value

The critical values for the upper-, lower-, and two-tailed tests are given in Table VI for the levels of significance closest to $\alpha = .05$ and $\alpha = .01$.

EXAMPLE 23

The extent to which an infant's health is affected by parental smoking is an important public health concern. The paper "Measuring the Exposure of Infants to Tobacco Smoke" (*New Engl. J. of Med.* (1984):1075–78) reported on a study in which various measurements were taken both from a random sample of infants who had been exposed to household smoke and from a sample of unexposed infants. The accompanying data consists of observations on urinary concentration of cotanine, a major metabolite of nicotine (the values constitute a subset of the original data and were read from a plot that appeared in the paper). Does the data suggest that true average cotanine level is higher for exposed than for unexposed infants? The investigators used the rank sum test to analyze the data, so we do also.

*This test procedure is often called the *Wilcoxon rank sum test* or the *Mann-Whitney test,* after the statisticians who developed it. Some sources use a slightly different (but equivalent) test statistic formula and set of critical values.

Unexposed ($n_1 = 7$)	8	11	12	14	20	43	111	
Rank	1	2	3	4	5	7	11	
Exposed ($n_2 = 8$)	35	56	83	92	128	150	176	208
Rank	6	8	9	10	12	13	14	15

1. $\mu_1 - \mu_2$ is the difference between true average cotanine concentration for unexposed and exposed infants.
2. $H_0 : \mu_1 - \mu_2 = 0$.
3. $H_a : \mu_1 - \mu_2 < 0$ (unexposed average is less than exposed average).
4. Test statistic: rank sum = sum of the sample 1 ranks.
5. The form of H_a dictates the use of a lower-tailed test. The critical value for a level .01 test is, from Table VI, 36, so H_0 should be rejected at level .01 if rank sum ≤ 36.
6. Rank sum $= 1 + 2 + 3 + 4 + 5 + 7 + 11 = 33$.
7. Since $33 \leq 36$, the computed value of the test statistic falls in the rejection region. H_0 is therefore rejected at level .01 in favor of the conclusion that $\mu_1 < \mu_2$. Infants exposed to cigarette smoke do seem to have higher contanine levels than do unexposed infants. ■

The test procedure just described is easily modified to handle a hypothesized value other than zero. Consider as an example testing $H_0 : \mu_1 - \mu_2 = 5$. This hypothesis is equivalent to $H_0 : (\mu_1 - 5) - \mu_2 = 0$. That is, if 5 is subtracted from each population 1 value, then according to H_0, the distribution of the resulting values coincides with the population 2 distribution. This suggests that if the hypothesized value of 5 is first subtracted from each sample 1 observation, the test can then be carried out as before.

> To test $H_0 : \mu_1 - \mu_2 =$ hypothesized value, subtract the hypothesized value from each observation in the first sample and then determine the ranks of these when combined with n_2 observations from the second sample.

EXAMPLE 24

Reconsider the contanine concentration data introduced in Example 23. Suppose a researcher wished to know whether average concentration for exposed children exceeds that for unexposed children by more than 25. Recalling that μ_1 is the true average concentration for unexposed children, the exposed average exceeds the unexposed average by exactly 25 when $\mu_1 - \mu_2 = -25$ and by more than 25 when $\mu_1 - \mu_2 < -25$. The hypotheses of interest are, therefore, $H_0 : \mu_1 - \mu_2 = -25$ versus $H_a : \mu_1 - \mu_2 < -25$. These can be tested by first subtracting -25 (equivalently, adding 25) to each sample 1 observation.

Sample 1							
Unexposed	8	11	12	14	20	43	111
Unexposed − (−25)	33	36	37	39	45	68	136
Rank	1	3	4	5	6	8	12

Sample 2								
Exposed	35	56	83	92	128	150	176	208
Rank	2	7	9	10	11	13	14	15

The resulting rank sum is $1 + 3 + 4 + 5 + 6 + 8 + 12 = 39$. The critical value for a level $\alpha = .01$ lower-tailed test is still 36. Since 39 isn't less than or equal to 36, $H_0 : \mu_1 - \mu_2 = -25$ cannot be rejected in favor of $H_a : \mu_1 - \mu_2 < -25$ at level .01. Sample evidence does not suggest that the difference between concentration levels exceeds 25. ■

Frequently the $n_1 + n_2$ observations in the two samples are not all different from one another. When this occurs, the rank assigned to each observation in a tied group is the average of the ranks that would be assigned if the values in the group all differed slightly from one another. Consider, for example, the 10 ordered values 5.6, 6.0, 6.0, 6.3, 6.8, 7.1, 7.1, 7.1, 7.9, and 8.2. If the two 6.0 values differed slightly from each other, they would be assigned ranks 2 and 3. Therefore, each one is assigned rank $(2 + 3)/2 = 2.5$. If the three 7.1 observations were all slightly different, they would receive ranks 6, 7, and 8, so each of the three is assigned rank $(6 + 7 + 8)/3 = 7$. The ranks for the above 10 observations are then 1, 2.5, 2.5, 4, 5, 7, 7, 7, 9, and 10. If the proportion of tied values is quite large, it is recommended that the rank sum statistic be multiplied by a *correction factor*. Several of the chapter references contain details on this.

The P-value for a given set of data and test procedure was defined earlier as the smallest level α at which H_0 could be rejected. Thus it makes sense to speak of the P-value when the rank sum test is used. However, because Table VI gives critical values only for $\alpha = .05$ and .01, an exact P-value or even an accurate bound is not available unless the computed rank sum value coincides with or is very close to a tabulated critical value. There are more detailed tables from which the P-value can be determined.

A Normal Approximation

Table VI contains critical values only for $n_1 \leq 8$ and $n_2 \leq 8$. There are more extensive tables of critical values for other sample-size combinations, but when both sample sizes exceed 8, an alternative approach is based on the following approximation.

If $n_1 > 8$ and $n_2 > 8$, the distribution of the rank sum statistic when H_0 is true is well approximated by a normal distribution having mean value $n_1(n_1 + n_2 + 1)/2$ and standard deviation $\sqrt{n_1 n_2(n_1 + n_2 + 1)/12}$. This implies that the standardized variable

$$z = \frac{\text{rank sum} - n_1(n_1 + n_2 + 1)/2}{\sqrt{n_1 n_2 (n_1 + n_2 + 1)/12}}$$

can serve as a test statistic. The rejection region uses a z critical value that depends on α and is upper-, lower-, or two-tailed according to whether H_a contains the inequality $>$, $<$, or $\neq$.

EXAMPLE 25

To compare interest rates offered by California banking institutions with those offered by Midwestern banks, 10 institutions from each region were randomly selected and the current interest rate on a 6-month certificate of deposit was determined for each one. Does the accompanying data suggest that true average rates differ for the two regions? Let's analyze the data using the rank sum test.

Region	M	M	M	C	M	C	M	M	C	C
Rate	8.7	8.8	9.0	9.0	9.2	9.25	9.4	9.5	9.5	9.5
Rank	1	2	3.5	3.5	5	6	7	9	9	9

Region	M	M	C	M	C	M	C	C	C	C
Rate	9.75	9.75	9.8	10	10	10.2	10.25	10.4	10.5	10.5
Rank	11.5	11.5	13	14.5	14.5	16	17	18	19.5	19.5

1. Let μ_1 = true average rate for California banks, μ_2 = the true average rate for Midwestern banks, and $\mu_1 - \mu_2$ = the difference in average rates.
2. $H_0 : \mu_1 - \mu_2 = 0$.
3. $H_a : \mu_1 - \mu_2 \neq 0$.
4. Test statistic:

$$z = \frac{\text{rank sum} - n_1(n_1 + n_2 + 1)/2}{\sqrt{n_1 n_2 (n_1 + n_2 + 1)/12}}$$

5. The form of H_a necessitates using a two-tailed test. For $\alpha = .05$, the bottom row of Table IV gives the z critical value as 1.96. H_0 should be rejected in favor of H_a either if $z > 1.96$ or if $z < -1.96$.
6. With $n_1 = n_2 = 10$, $n_1(n_1 + n_2 + 1)/2 = 105$ and $\sqrt{n_1 n_2(n_1 + n_2 + 1)/12} = \sqrt{(10)(10)(21)/12} = 13.23$. The rank sum value is $3.5 + 6 + 9 + 9 + 13 + 14.5 + 17 + 18 + 19.5 + 19.5 = 129$, so

$$z = \frac{129 - 105}{13.23} = \frac{24}{13.23} = 1.81$$

7. Since 1.81 is not in the rejection region, H_0 should not be rejected. The sample data does not suggest that average rates for the two regions differ. ■

Ordering the observations can be tedious, especially when n_1 and n_2 are large. Fortunately all the standard statistical computer packages have a rank sum option that does the ordering, ranking, and computations automatically.

Comparing the Rank Sum and Pooled *t* Tests

The basic assumptions of this section (identical shapes and spreads) are satisfied when both population distributions are normal with $\sigma_1 = \sigma_2$—the "home ground" of the pooled t test. Although the rank sum test can be used in this situation, statisticians favor the pooled t test. This is because when both tests are used with the same prescribed level of significance (e.g., $\alpha = .05$), the pooled t test has smaller type II error probabilities than does the rank sum test. However, even on the pooled t test's home ground, the rank sum doesn't fare too badly. Roughly speaking, the rank sum test requires slightly larger sample sizes than what is required by the pooled t test in order to obtain the same β's.

When the population distributions are nonnormal but satisfy the basic assumptions stated earlier in the section, the situation is different. The pooled t

test can now suffer by comparison with the rank sum test in two different ways. First, at least for small samples the actual level α for the t test may be quite different from the α selected by the investigator. This is because the t statistic no longer has a t distribution with $n_1 + n_2 - 2$ df when H_0 is true (it does only for normal population distributions), so the t critical values tabulated in Table IV no longer capture the prescribed tail areas. For example, with $n_1 + n_2 - 2 = 15$, the critical value that captures upper-tail area .05 under the t curve is 1.75. But if H_0 is rejected when $t > 1.75$, the actual α may be .02 rather than .05—which would make β higher than the investigator believed it to be. The rank sum test does not have this difficulty because the tabulated critical values give the specified α whenever the basic assumptions are satisfied—outside as well as inside "normal land" (that mythical country in which t tests rule).

For moderate to large samples, the specification of α is correct for either test. Here, though, when both tests use the same α, the rank sum test may have substantially smaller β than does the pooled t test (remember, we are no longer in normal land). This is particularly true when the population distributions have markedly nonnormal shapes, such as substantial skews or very heavy tails compared to a normal curve. In summary, the rank sum test performs almost as well as the pooled t test for the case of normal distributions and may substantially outperform the t test in other cases—provided that the distributions still have the same shapes and spreads. When this is not the case, comparisons between these two tests and various others are necessary, and statisticians still don't have all the answers.

A Confidence Interval for $\mu_1 - \mu_2$

A confidence interval based on the rank sum statistic is not nearly as familiar to users of statistical methods as is the hypothesis-testing procedure. This is unfortunate because the confidence interval has the same virtues to recommend it as does the rank sum test. It is valid under more general circumstances than is the pooled t interval, and its performance as compared to the pooled t interval parallels the performance of the rank sum test vis-a-vis the pooled t test.

The key to obtaining a confidence interval lies in exploiting a relationship between confidence intervals and two-tailed tests. Consider as an example the one-sample t test for testing $H_0 : \mu =$ *hypothesized value* versus $H_a : \mu \neq$ *hypothesized value* and the one-sample t interval $\bar{x} \pm (t \text{ critical value})(s/\sqrt{n})$. If $n = 12$, $\bar{x} = 52.6$, and $s = 6.3$, the 95% interval is $52.6 \pm (2.20)(6.3/\sqrt{12}) = 52.6 \pm 4.0 = (48.6, 56.6)$. For a two-tailed level .05 test, the computed value of the test statistic $t = (\bar{x} - \text{hypothesized value})/(s\sqrt{n})$ is compared to critical values $+2.20$ and -2.20. Suppose that the hypothesized value is 50.0, a number inside the confidence interval. Then $t = (52.6 - 50)/(6.3/\sqrt{12}) = 1.43$. This value is not in the rejection region, so H_0 is not rejected at level .05. Similarly, if the hypothesized value is any other number in the 95% interval, H_0 cannot be rejected. However, if the hypothesized value is outside the confidence interval (e.g., 46 or 57), then it is easily checked that H_0 would be rejected. So the confidence interval consists of all hypothesized values for which $H_0 : \mu =$ *hypothesized value* is not rejected in favor of $H_a : \mu \neq$ *hypothesized value*.

Suppose that a level .05 test is available for testing

H_0 : population characteristic = hypothesized value

versus

H_a : population characteristic $\neq$ hypothesized value

Then a 95% confidence interval for the population characteristic consists of all hypothesized values for which H_0 cannot be rejected. A 99% interval is associated with a level $\alpha = .01$ test.

The form of the rank sum confidence interval can be most easily understood if an alternative expression for the rank sum statistic is first presented. Recall that the test involving a particular hypothesized value (not necessarily zero) is carried out by first subtracting the hypothesized value from each sample 1 observation and then ranking and summing the n_1 ranks. Suppose instead that each observation in the second sample is subtracted from every observation in the first sample. This gives a set of n_1n_2 differences. Then it can be shown that

$$\text{rank sum} = \frac{n_1(n_1 + 1)}{2} + \begin{pmatrix} \text{number of differences that are greater than} \\ \text{or equal to the hypothesized value} \end{pmatrix}$$

(the smallest possible rank sum value is $1 + 2 + \cdots + n_1 = n_1(n_1 + 1)/2$).

For example, if $n_1 = n_2 = 4$, there are 16 differences and rank sum = 10 + (number of differences $\geq$ hypothesized value). Table VI gives 11 and 25 as the lower- and upper-tail critical values when $\alpha = .05$. Thus H_0 will not be rejected if $12 \leq$ rank sum ≤ 24, i.e., if $2 \leq$ (number of differences $\geq$ hypothesized value) ≤ 14. Figure 5 illustrates the interval of hypothesized values for which this is the case. The 95% confidence interval for $\mu_1 - \mu_2$ then extends from the second smallest difference to the second largest difference.

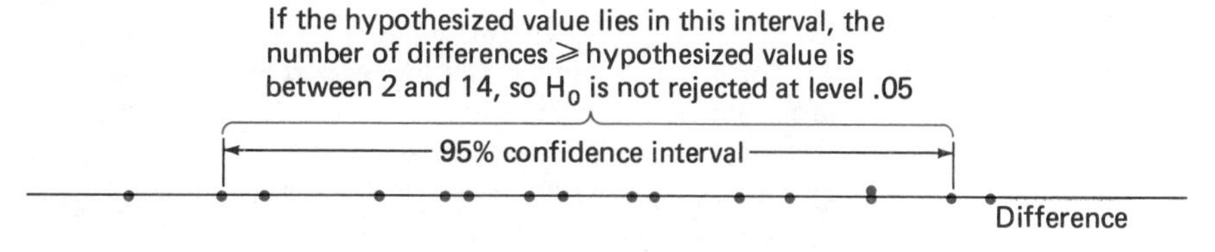

FIGURE 5 16 DIFFERENCES AND THE 95% RANK SUM CONFIDENCE INTERVAL WHEN $n_1 = n_2 = 4$

The **rank sum confidence interval for $\mu_1 - \mu_2$** is based on the n_1n_2 differences that result from subtracting each sample 2 observation from each sample 1 observation. The lower and upper limits of the interval are the dth smallest and the dth largest differences, respectively. Table VII gives values of d corresponding to 90%, 95%, and 99% confidence.

EXAMPLE 26

The paper "Some Mechanical Properties of Impregnated Bark Board" (*Forest Products J.* (1977):31–38) reported the accompanying observations on crushing strength for epoxy-impregnated bark board (sample 1) and bark board impregnated with another polymer (sample 2). The sample values are displayed along the left and top margins in the accompanying table, and the differences appear in the main part of the table. From Table VII, a 95% confidence interval for $\mu_1 - \mu_2$, the difference between true average strengths, when $n_1 = 6$ and $n_2 = 5$ necessitates using $d = 5$. The five smallest and five largest differences are, respectively, 4350, 4470, 4610, 4730, 4830, and 8220, 8480, 8740, 9530, 9790. The confidence interval is then (4830, 8220).

Differences

		Second sample				
		4590	4850	5640	6390	6510
First sample	10,860	6270	6010	5220	4470	4350
	11,120	6530	6270	5480	4730	4610
	11,340	6750	6490	5700	4950	4830
	12,130	7540	7280	6490	5740	5620
	13,070	8480	8220	7430	6680	6560
	14,380	9790	9530	8740	7990	7870

■

In practice it is often not necessary to compute all $n_1 n_2$ differences but only some of the smallest and largest ones. Also, when n_1 and n_2 both exceed values for which d is tabulated, d is given approximately by the following formula (based on the normal approximation for the rank sum statistic):

$$d \approx \frac{n_1 n_2}{2} - \left(\begin{matrix} z \text{ critical value for} \\ \text{desired confidence level} \end{matrix}\right) \sqrt{\frac{n_1 n_2 (n_1 + n_2 + 1)}{12}}$$

EXERCISES

9.54 The urinary fluoride concentration (ppm) was measured both for a sample of livestock that had been grazing in an area previously exposed to fluoride pollution and for a similar sample that had grazed in an unpolluted region. Does the data indicate strongly that the true average fluoride concentration for livestock grazing in the polluted region is larger than for the unpolluted region? Assume that the distributions of urinary fluoride concentration for both grazing areas have the same shape and spread, and use a level .05 rank sum test.

Polluted	21.3	18.7	23.0	17.1	16.8	20.9	19.7
Unpolluted	14.2	18.3	17.2	18.4	20.0		

9.55 A modification has been made to the process for producing a certain type of *time-zero* film (film that begins to develop as soon as a picture is taken). Because the modification involves extra cost, it will be incorporated only if sample data strongly indicates that the modification has decreased true average developing time by more than 1 s. Assuming that both developing-time distributions differ only with respect to location, if at all, use the rank sum test at level .05 on the given data to test the appropriate hypotheses.

Original process	8.6	5.1	4.5	5.4	6.3	6.6	5.7	8.5
Modified process	5.5	4.0	3.8	6.0	5.8	4.9	7.0	5.7

9.56 The study reported in "Gait Patterns During Free Choice Ladder Ascents" (*Human Movement Sci.* (1983):187–95) was motivated by publicity concerning the increased accident rate for individuals climbing ladders. A number of different gait patterns were used by subjects climbing a portable straight ladder according to specified instructions. The ascent times for seven subjects who used a lateral gait and six subjects who used a four-beat diagonal gait are given.

Lateral	.86	1.31	1.64	1.51	1.53	1.39	1.09
Diagonal	1.27	1.82	1.66	.85	1.45	1.24	

a. Use the rank sum test to see if the data suggests any difference in the true average ascent times for the two gaits.

b. Compute a 95% confidence interval for the difference between the true average gait times.

9.57 The paper "Histamine Content in Sputum from Allergic and Non-Allergic Individuals" (*J. of Appl. Physiology* (1969):535–39) reported the accompanying data on sputum histamine level (μg/g) for a sample of 9 individuals classified as allergics and another sample of 13 individuals classified as nonallergics.

Allergics	67.6	39.6	1651.0	100.0	65.9	1112.0	31.0	102.4	64.7				
Nonallergics	34.3	27.3	35.4	48.1	5.2	29.1	4.7	41.7	48.0	6.6	18.9	32.4	45.5

a. Define the two populations to be compared.

b. Does the data indicate that there is a difference in true mean sputum histamine level between allergics and nonallergics? Use a level .01 rank sum test (as did the authors of the paper).

9.58 Many college professors are concerned about plagiarism by students. The paper "The Use of the Cloze Testing Procedure for Detecting Plagiarism" (*J. of Exper. Educ.* (1982):127) reported the results of an experiment designed to investigate one potential method for identifying papers that used plagiarized material. A cloze test involves reproducing a passage of text with certain words omitted. The author of the paper is to fill in the missing words. The test is then scored by counting the number of errors. The authors of this particular paper hypothesized that students should be better able to reproduce their own writing than something they had copied, and so nonplagiarists would tend to make fewer errors on a cloze test. To determine if this were in fact the case, sections of an English composition class were asked to write papers of 6 to 7 pages in length. Two sections were given specific instructions to plagiarize someone else's writing, and two other sections were given no special instructions. When papers were submitted, the purpose of the study was explained to the nonplagiarizing sections, and any student who had used the work of someone else was asked to withdraw his or her paper with no penalty. (Three students confessed!) Cloze tests were prepared for each paper submitted and given 1 week later. The numbers of errors on the cloze test for these students were as follows.

No plagiarism	1	1	2	2	3	3	3	4	4	4	4	4
	4	5	5	5	5	6	6	6	6	6	7	7
	7	8	8	9	9	9	10	10	10	11	13	
Plagiarism	1	2	3	4	4	4	5	6	6	6	7	7
	7	8	8	8	9	9	9	9	9	10	10	10
	10	11	12	13	13	13	13	14	14	15	17	17
	18	19	19									

Use the large-sample rank sum statistic at level .01 to determine whether the mean number of errors for students who plagiarize is significantly higher than the mean number of errors for students who do not plagiarize.

9.59 A blood-lead level of 70 mg/mL has been commonly accepted as safe. However, researchers have noted that some neurophysiological symptoms of lead poisoning appear in people whose blood-lead levels are below 70 mg/mL. The paper "Subclinical Neuropathy at Safe Levels of Lead Exposure" (*Arch. Environ. Health* (1975):180) gives the following nerve-conduction velocities for a group of workers who were exposed to lead in the workplace but whose blood-lead levels were below 70 mg/mL and for a group of controls who had no exposure to lead.

Exposed to lead	46	46	43	41	38	36	31
Control	54	50.5	46	45	44	42	41

Use a level .05 rank sum test to determine if there is a significant difference in mean conduction velocity between workers exposed to lead and those not exposed to lead.

9.60 The effectiveness of antidepressants in treating the eating disorder bulimia was examined in the paper "Bulimia Treated with Imipramine: A Placebo-Controlled Double-Blind Study" (*Amer. J. Psych.* (1983):554–58). A group of patients diagnosed as bulimic were randomly assigned to one of two treatment groups, one receiving imipramine and the other a placebo. One of the variables recorded was frequency of binging. The authors chose to analyze the data using a rank sum test because it makes no assumption of normality. They state: "Because of the wide range of some measures, such as frequency of binges, the rank sum is more appropriate and somewhat more conservative." Data consistent with the findings of this paper is given below.

Number of Binges During One Week								
Placebo	8	3	15	3	4	10	6	4
Imipramine	2	1	2	7	3	12	1	5

Does this data strongly suggest that imipramine is effective in reducing the mean number of binges per week? Use a level .05 rank sum test.

9.61 Researchers have noted that chickens fed a diet that is lacking in sodium and calcium become more active. To determine whether a sodium deficiency causes an increase in pecking activity, the authors of the paper "An Increase in Activity of Domestic Fowls Produced by Nutritional Deficiency" (*Animal Behavior* (1973):10–17) observed 17 chickens who were deprived of sodium and 15 control chickens. They counted the number of pecks for each bird during a fixed period of time. Does the data strongly indicate that the mean number of pecks is higher for chickens whose diet lacks sufficient sodium? Use a large-sample rank sum test with $\alpha = .01$.

Sodium deprived	0	0	0	2	17	58	67	67	68	74	79
	85	92	95	97	150	181					
Control	0	0	0	0	0	8	13	13	20	33	34
	57	60	64	78							

9.62 The accompanying data resulted from an experiment to compare the effects of vitamin C in orange juice and in synthetic ascorbic acid on the length of odontoblasts in guinea pigs over a 6-week period ("The Growth of the Odontoblasts of the Incisor Tooth as a Criterion of the Vitamin C Intake of the Guinea Pig" *J. Nutr.* (1947):491–504). Use the rank sum test at level .01 to decide whether or not true average length differs for the two types of vitamin C intake.

Orange juice	8.2	9.4	9.6	9.7	10.0	14.5	15.2
	16.1	17.6	21.5				
Ascorbic acid	4.2	5.2	5.8	6.4	7.0	10.1	11.2
	11.3	11.5					

9.63 In an experiment to compare the bond strength of two different adhesives, each adhesive was used in five bondings of two surfaces, and the force necessary to separate the surfaces was determined for each bonding. For adhesive 1, the resulting values were 229, 286, 245, 299, and 259, whereas the adhesive 2 observations were 213, 179, 163, 247, and 225. Let μ_1 and μ_2 denote the true average bond strengths of adhesives 1 and 2, respectively. Use a 90% distribution-free confidence interval to estimate $\mu_1 - \mu_2$.

9.64 The article "A Study of Wood Stove Particulate Emissions" (*J. Air Poll. Control Assoc.* (1979):724–28) reported the following data on burn time (h) for samples of oak and pine. Estimate the difference between mean burn time for oak and mean burn time for pine using a 95% distribution-free confidence interval. Interpret the interval.

Oak	1.72	.67	1.55	1.56	1.42	1.23	1.77	.48
Pine	.98	1.40	1.33	1.52	.73	1.20		

KEY CONCEPTS

Sampling distribution of $\bar{x}_1 - \bar{x}_2$ (based on two independently selected random samples) The distribution is centered at $\mu_{\bar{x}_1-\bar{x}_2} = \mu_1 - \mu_2$ and has standard deviation $\sigma_{\bar{x}_1-\bar{x}_2} = \sqrt{\sigma_1^2/n_1 + \sigma_2^2/n_2}$. If both populations are normal, the $\bar{x}_1 - \bar{x}_2$ distribution is normal. Even if one or both population distributions are not normal, the $\bar{x}_1 - \bar{x}_2$ distribution is approximately normal when both n_1 and n_2 are large. (p. 359)

Sampling distribution of $p_1 - p_2$ (based on two independently selected random samples) The distribution is centered at $\mu_{p_1-p_2} = \pi_1 - \pi_2$ and has standard deviation $\sigma_{p_1-p_2} = \sqrt{\pi_1(1 - \pi_1)/n_1 + \pi_2(1 - \pi_2)/n_2}$. If both n_1 and n_2 are large, the sampling distribution is approximately normal. (p. 384)

Distribution-free procedure A procedure whose validity does not depend on any overly specific assumptions about the population distribution. (p. 393)

Hypothesis Testing Procedures

Population Characteristic	*Appropriate When*	*Test Procedure*
$\mu_1 - \mu_2$	Both samples are large	Two-sample z test (Section 9.2)
$\mu_1 - \mu_2$	Both population distributions are normal, $\sigma_1 = \sigma_2$	Pooled t test (Section 9.3)
$\mu_1 - \mu_2$	Both populations have the same shape and spread (but not necessarily normal)	Rank sum test (Section 9.5)
$\mu_1 - \mu_2$	Both populations have the same shape and spread; n_1 and n_2 are both larger than 8	Large-sample rank sum z test (Section 9.5)
$\pi_1 - \pi_2$	Both samples are large	Two-sample z test (Section 9.4)

Confidence Intervals

Population Characteristic	*Appropriate When*	*Confidence Interval*
$\mu_1 - \mu_2$	Both samples are large	Two-sample z interval (Section 9.2)
$\mu_1 - \mu_2$	Both population distributions are normal, $\sigma_1 = \sigma_2$	Pooled t interval (Section 9.3)

$\mu_1 - \mu_2$	Both populations have the same shape and spread	Distribution-free interval (Section 9.5)
$\pi_1 - \pi_2$	Both samples are large	Two-sample z interval (Section 9.4)

SUPPLEMENTARY EXERCISES

9.65 Meteorologists classify storms as either single-peak or multiple-peak. The total number of lightning flashes was recorded for seven single-peak and four multiple-peak storms, resulting in the given data ("Lightning Phenomenology in the Tampa Bay Area" *J. Geophys. Research* (1984): 11,789–805).

Single-peak	117	56	19	40	82	69	80
Multiple-peak	229	197	242	430			

a. Does the data suggest that the true mean number of lightning flashes differs for the two types of storms? Use a .05 significance level.

b. What assumptions about the distribution of number of flashes for each of the two types of storms are necessary in order that your test in (a) be valid?

9.66 The Mekranoti Indians of Central Brazil support "kupry," unmarried women with children who provide sexual services in return for gifts. The paper "Paid Sex Specialists Among the Mekranots" (*J. Anthro. Research* (1984):394–405) compared samples of kupry women and nonkupry women on various characteristics. The paper reported that 6 of the 13 kupry women and 7 of the 56 nonkupry women lost their mothers before reaching puberty. Does this suggest that the true proportion who lost their mothers before reaching puberty is higher for kupry than for nonkupry women? Use a .05 significance level.

9.67 The results of a study on job satisfaction among tenure-track faculty members and librarians employed by the California State University System were described in the paper "Job Satisfaction Among Faculty and Librarians: A Study of Gender, Autonomy, and Decision Making Opportunities" (*J. Library Admin.* (1984):43–56). Random samples of 115 male and 105 female academic employees were selected. Each participant completed the Minnesota Satisfaction Questionnaire (MSQ) and was assigned a satisfaction score. The resulting mean and standard deviation were 75.43 and 10.53 for the males and 72.54 and 13.08 for the females. Does the data strongly suggest that male and female academic employees differ with respect to mean score on the MSQ? Use a .01 significance level.

9.68 The paper "An Evaluation of Football Helmets Under Impact Conditions" (*Amer. J. Sports Med.* (1984):233–37) reported that when 44 padded football helmets and 37 suspension-type helmets were subjected to an impact test (a drop of 1.5 m onto a hard surface), 5 of the padded and 24 of the suspension-type helmets showed damage. Using a .01 significance level, test appropriate hypotheses to determine if there is a difference between the two helmet types with respect to the true proportion of each type that would be damaged by a 1.5-m drop onto a hard surface.

9.69 The paper "Post-Mortem Analysis of Neuropeptides in Brains from Sudden Infant Death Victims" (*Brain Research* (1984):279–85) reported age (in days) at death for infants who died of sudden infant death syndrome (SIDS). Assuming that age at death for SIDS victims is normally distributed, use the given data to construct a 95% confidence interval for the difference in the true mean age at death for female and male SIDS victims. Interpret the resulting interval. How does the interpretation depend on whether zero is included in the interval?

Age at Death (Days)							
Females	55	120	135	154	54		
Males	56	60	60	60	105	140	147

9.70 The paper "Chronic 60 Hz Electric Field Exposure Induced Subtle Bioeffects on Serum Chemistry" (*J. Environ. Sci. and Health* (1984):865–85) described an experiment to assess the effects of exposure to a high-intensity electric field. A group of 45 rats exposed to an electric field from birth to 120 days of age was compared to a control group of 45 rats with no exposure. Summary quantities for various blood characteristics are given. Conduct the hypothesis tests necessary to determine whether the experimental treatment differs from no treatment with respect to true mean glucose, potassium, protein, or cholesterol levels. Use a .05 significance level for each test.

	Control		Experimental	
Treatment	Mean	*s*	Mean	*s*
Glucose (mg/dL)	136.30	12.70	139.20	16.10
Potassium (mg/dL)	7.44	.53	7.62	.54
Total protein (gm/dL)	6.63	.27	6.61	.34
Cholesterol (mg/dL)	69.00	11.40	67.80	9.38

9.71 The paper "Dyslexic and Normal Readers' Eye Movements" (*J. Exper. Psych.* (1983):816–25) reported data on number of eye movements while reading a particular passage for 34 dyslexic and 36 normal readers. The sample mean number of total movements and corresponding sample standard deviation were 8.6 and .30, respectively, for dyslexics and 9.2 and .16 for normal readers.

a. Does the data indicate a significant difference between dyslexic and normal readers with respect to true average number of eye movements? Use a level .10 test.

b. Place bounds on the P-value associated with the test statistic in (a). Would your conclusions have been any different at significance levels .05 or .01?

9.72 The paper "The Effects of Education on Self-Esteem of Male Prison Inmates" (*J. Correctional Educ.* (1982):12-18) described the result of an experiment designed to ascertain whether mathematics education increases the self-esteem of prison inmates. Two random samples of sizes 40 were selected from the population of prison inmates at Angola, Louisiana. One sample was designated as a control group and the other as an experimental group. Inmates in the experimental group received 18 weeks of mathematics tutoring, whereas those in the control group were not tutored. Both groups were given the Self-Esteem Inventory (SEI) at the beginning and end of the 18-week period. The mean and standard deviation of the change in SEI score were 2.9 and 5.4 for the experimental group and -1.3 and 5.6 for the control group.

a. Does the data provide sufficient evidence to conclude that mathematics tutoring results in a higher mean change in SEI score? Test the relevant hypotheses using a .01 significance level.

b. What is the P-value associated with the test in (a)?

9.73 Nine observations of surface-soil pH were made at each of two different locations at the Central Soil Salinity Research Institute experimental farm, and the resulting data appeared in the article "Sodium-Calcium Exchange Equilibria in Soils as Affected by Calcium Carbonate and Organic Matter" (*Soil Sci.* (1984):109). Does the data suggest that the true mean soil pH values differ for the two locations? Test the appropriate hypotheses using a .05 significance level. Be sure to state any assumptions necessary for the validity of your test.

Site	pH								
Location A	8.53	8.52	8.01	7.99	7.93	7.89	7.85	7.82	7.80
Location B	7.85	7.73	7.58	7.40	7.35	7.30	7.27	7.27	7.23

9.74 Reconsider the soil pH data given in Exercise 9.73.

a. Use a distribution-free procedure with significance level .05 to determine if the true mean soil pH is the same for both locations.

b. Construct and interpret a 95% distribution-free confidence interval for the difference in true mean soil pH values for the two locations.

9.75 Two different methods (ampul and hot-plate) for recovering metals from sewage were compared in the paper "Simple Sample Digestion of Sewage Sludge for Multi-Element Analysis" (*J. Environ. Sci. and Health* (1984):959–72). Both methods were used to treat oyster tissue and the metal and mineral recovery were recorded. Answer the following questions assuming that each method was used on 10 tissue specimens. Be sure to state any assumptions that must be true in order for the inferential procedure applied to be valid.

a. For iron, the mean recovery and standard deviation were 16 ppm and 2.5 ppm for the ampul method and 17.7 ppm and 1.2 ppm for the hot-plate method. Does the evidence suggest that the two methods differ with respect to mean iron recovery? Use a .05 significance level.

b. The sample mean and standard deviation for copper recovery were 62.6 ppm and 3.7 ppm for the ampul method and 65.0 and 3.8 for the hot-plate method. Estimate the true difference in mean copper recovery rate for the two methods using a 90% confidence interval. Does the interval include zero? Interpret the interval.

9.76 An electronic implant that stimulates the auditory nerve has been used to restore partial hearing to a number of deaf people. In a study of implant acceptability (*Los Angeles Times,* January 29, 1985), 250 adults born deaf and 250 adults who went deaf after learning to speak were followed for a period of time after receiving an implant. Of those deaf from birth, 75 had removed the implant, while only 25 of those who went deaf after learning to speak had done so. Does this suggest that the true proportion who remove the implants differs for those that were born deaf and those that went deaf after learning to speak? Test the relevant hypotheses using a .01 significance level.

9.77 Oxygen consumption of adult and neonatal (1-day-old) rabbits after exposure to streptococci, a leading cause of bacterial infections in human newborns, was compared in the paper "Oxidative Metabolism of Neonatal and Adult Rabbit Lung Macrophanges Stimulated with Opsonized Group B Streptococci" (*Infection and Immunity* (1985):26–30). Basal oxygen consumption (in nanomoles) was recorded for two independently chosen samples of size 5, resulting in the given summary quantities.

Group	Sample Mean	Sample Standard Deviation
1-day-old	18.5	1.1
Adult	19.6	1.3

a. Does the data suggest that true mean oxygen consumption differs for 1-day-olds and adults? Use a .05 significance level.

b. What assumptions about the oxygen consumption distributions for adults and 1-day-olds must be true in order for your test in (a) to be valid?

9.78 The accompanying data appeared in the article "Effect of Exogenous Oestradiol-17B on Gonadatrophin Secretion in Postpartum Beef Cows" (*J. Reprod. and Fertility* (1984):473–78). Twelve cows between 10 and 17 days postpartum received one or two silicon rubber implants, each containing 45 mg of oestradiol (6 cows received one implant and the other 6 received two). The time (in days) to first rise in milk progesterone concentration was recorded. The experiment was terminated after 50 days, at which time 3 cows had not yet shown a rise in progesterone level.

	Time (Days)					
One implant	38	23	19	33	>50	>50
Two implants	24	34	30	42	35	>50

Use a distribution-free procedure to determine whether the data suggests a difference in true mean number of days to rise in progesteone level between cows receiving one implant and those receiving two implants. Assume that those sample values listed as > 50 are all tied, and assign each one the rank of 11 (the average of ranks 10, 11, and 12). Use a significance level of .05.

9.79 Two different filling operations used in a ground-beef packing plant were described in the paper "Evaluating Variability in Filling Operations" (*Food Tech.* (1984):51–55). Both filling operations were set to fill packages with 1400 g of ground beef. A random sample of size 30 was taken from each filling operation. The resulting means and standard deviations were 1402.24 g and 10.97 g for operation 1 and 1419.63 g and 9.96 g for operation 2.

a. Using a .05 significance level, is there sufficient evidence to indicate that the true mean weight of the packages produced differs for the two operations?

b. Does the data from operation 1 suggest that the true mean weight of packages produced by operation 1 is higher than 1400 g? Use a .05 significance level.

9.80 Is reading computer output on a terminal screen more tiring than reading output on paper? This question is addressed in the paper "Doing the Same Work with Hard Copy and with Cathode-Ray Tube (CRT) Computer Terminals" (*Human Factors* (1984):323–37). One measure of eye fatigue used was number of blinks during a 1-min-interval. Twenty-four clerk-typists were hired from a temporary employment agency to do proof reading. Twelve were asked to proofread material from a paper copy, and the other 12 proofread the same text on a CRT screen. After 1 h, the number of blinks during a 1-min period was recorded. The average number of blinks was 6.70 for the paper group and 9.32 for the CRT group. Suppose that the corresponding sample standard deviations were 1.2 and 1.4 for the paper and CRT groups, respectively.

a. Does the data suggest that the mean number of blinks per minute differs for those working with paper and those working with the CRT? Use a .05 level of significance.

b. The number of blinks was also recorded after 6 h of work. The resulting averages were 9.62 and 12.21 for the paper and CRT groups, respectively. Suppose that the corresponding sample standard deviations are 1.5 and 1.6. Does the data suggest that the mean number of blinks after 6 h of work differs for the two groups? Use a .05 level of significance. Do you reach the same conclusion as in the test of (a)?

9.81 A study of trace-element concentrations in human hair obtained from residents of different parts of SriLanka was described in the paper "Environmental

Significance of Trace Elements in Human Hair—A Case Study from SriLanka" (*Intl. J. Environ. Studies* (1984):41–48). Hair samples were taken from 15 people living in a rural village and from 11 urban university students. Each hair sample was analyzed for zinc (ppb), resulting in the accompanying data. Is there sufficient evidence to indicate that the true mean zinc concentration differs for urban and rural Sri-Lankans? Use a .05 level of significance. Give an upper and/or lower bound on the P-value associated with the observed value of the test statistic.

Zinc Concentration

Rural	3619	1104	243	658	673	598	648	918	133
	289	250	304	555	640	933			
Urban	1120	230	4200	1200	1400	750	2101	430	690
	600	834							

9.82 The discharge of industrial wastewater into rivers affects water quality. To assess the effect on water quality of a particular power plant, 24 water specimens were taken 16 km upstream and 4 km downstream of the plant. Alkalinity (mg/L) was determined for each specimen, resulting in the given summary quantities. Does the data suggest that the true mean alkalinity is higher downstream than upstream? Use a .05 significance level.

Location	n	Mean	Standard Deviation
Upstream	24	75.9	1.83
Downstream	24	183.6	1.70

REFERENCES

Daniel, Wayne. *Applied Nonparametric Statistics.* Boston: Houghton Mifflin, 1978. (An elementary presentation of distribution-free methods, including the rank sum test discussed in the last section of this chapter.)

Devore, Jay. *Probability and Statistics for Engineering and the Sciences.* Monterey, CA: Brooks/Cole, 1982. (Contains a somewhat more comprehensive treatment of the inferential material presented in this and the previous two chapters, though the notation is a bit more mathematical than that of the present text.)

Mosteller, Frederick, and Richard Rourke. *Sturdy Statistics.* Reading, MA: Addison-Wesley, 1973. (A very readable intuitive development of distribution-free methods, including those based on ranks.)

10 Inferences Using Paired Data

INTRODUCTION

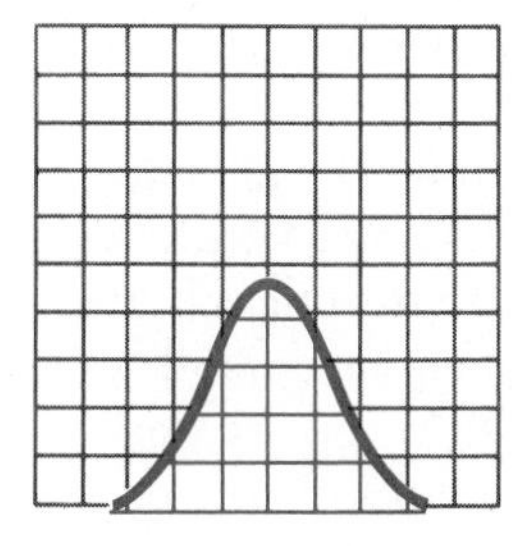

THE METHODS PRESENTED in Chapter 9 are appropriate when the individuals or objects in the sample from the first population are selected independently of those that constitute the sample from the second population. There are many situations, however, in which the two samples under study are related. One way in which this might occur is when the samples consist of pairs of observations on the same person or object. For example, to study the effectiveness of a speed-reading course, the reading speed of subjects would be tested prior to taking the class and again after completion of the course. This gives rise to two samples—one from the population of individuals who have not taken this particular course (the "before" measurements) and one from the population of individuals who have had such a course (the "after" measurements). Data of this type is said to be paired. The two samples are not independently chosen, since the selection of individuals from the first (before) population completely determines which individuals make up the sample from the second (after) population. This dependence invalidates the use of methods from Chapter 9 to draw conclusions about $\mu_1 - \mu_2$. In this chapter, we introduce methods for drawing inferences from paired data and also discuss why in some applications an experiment resulting in paired data might be preferable to an experiment involving independent samples.

10.1 Pairing versus Independent Samples

Two samples are said to be *independent* if the selection of the individuals or objects that make up one of the samples has no bearing on the selection of those in the other sample. However, in some situations an experiment with independent samples is not the best way to obtain information concerning any possible difference between the populations. For example, suppose that an investigator wants to determine if regular aerobic exercise affects blood pressure. A random sample of people who jog regularly and a second random sample of people who do not exercise regularly are selected independently of one another. The researcher then uses the pooled t test to conclude that a significant difference exists between the average blood pressures for joggers and nonjoggers. Is it reasonable to think that jogging influences blood pressure? It is known that blood pressure is related to both diet and body weight. Might it not be the case that joggers in the sample tend to be leaner and adhere to a healthier diet than the nonjoggers and that this might account for the observed difference? On the basis of this study, the researcher wouldn't be able to rule out the possibility that the observed difference in blood pressure is explained by weight differences between the two samples and that aerobic exercise in and of itself has no effect.

One way to avoid this difficulty would be to match subjects by weight. The researcher would find pairs of subjects so that the jogger and nonjogger in each pair were similar in weight (although weights for different pairs might vary widely). The factor *weight* could then be ruled out as a possible explanation for an observed difference in average blood pressure between the two groups. Matching the subjects by weight results in two samples for which each observation in the first sample is coupled in a meaningful way with a particular observation in the second sample. Such samples are said to be **paired.**

Experiments can be designed to yield paired data in a number of different ways. Some studies involve using the same group of individuals with measurements recorded both before and after some intervening treatment. Others use naturally occurring pairs such as twins or husbands and wives, and some construct pairs by matching on factors whose effect might otherwise obscure differences (or the lack of them) between the two populations of interest (as might weight in the jogging example). Paired samples often provide more information than would independent samples because extraneous effects are screened out.

EXAMPLE 1

It has been hypothesized that strenuous physical activity affects hormone levels. The paper "Growth Hormone Increase During Sleep After Daytime Exercise" (*J. of Endocrinology* (1974):473–78) reported the results of an experiment involving six healthy male subjects. For each participant, blood samples were taken during sleep on two different nights using an indwelling venous catheter. The first blood sample was drawn after a day that included no strenuous activities, and the second was drawn after a day during which the subject engaged in strenuous exercise. The resulting data on growth hormone level appears below. The samples are paired rather than independent since both samples are comprised of the same men.

Growth Hormone Level (mg/mL)

Subject	1	2	3	4	5	6
Postexercise	13.6	14.7	42.8	20.0	19.2	17.3
Control	8.5	12.6	21.6	19.4	14.7	13.6

Let μ_1 denote the mean nocturnal growth hormone level for the population of all healthy males who participated in strenuous activity on the previous day. Similarly, let μ_2 denoted the mean nocturnal hormone level for the population consisting of all healthy males whose activities on the previous day did not include any strenuous physical exercise. The hypotheses of interest are then

$$H_0 : \mu_1 - \mu_2 = 0 \quad \text{versus} \quad H_a : \mu_1 - \mu_2 \neq 0$$

Notice that in each of the six data pairs, the postexercise hormone level is higher than the corresponding control level. Intuitively this suggests that there may be a difference between the population means.

However, if the pooled t test for two independent samples is (incorrectly) employed, the resulting t test statistic is 1.28. This value does not allow for rejection of the hypothesis that $\mu_1 - \mu_2 = 0$ even at level of significance .10. This result might surprise you at first, but remember that this test procedure ignores the fact that the samples are paired. Two plots of the data are given in Figure 1. The first one ignores the pairing, and the two samples look quite similar. The plot in which pairs are identified does suggest a difference, since for each pair the exercise observation exceeds the no-exercise observation. Disregarding the paired nature of the samples results in a loss of information. Nocturnal growth hormone levels vary substantially from one individual to another. It is this variability that obscures the difference in hormone level associated with strenuous exercise.

Nocturnal growth hormone level

```
         x x   x x x                    x        Exercise
-------------------------------------------------
   •   •  • •    •  •                            No exercise
```

(a)

```
         3 1   2 6  4                   5        Exercise
-------------------------------------------------
   1   3  2 6    4  5                            No exercise
```

(b)

FIGURE 1

TWO PLOTS OF THE PAIRED DATA FROM EXAMPLE 1
(a) Pairing Ignored
(b) Pairs Identified ■

In Example 1 an independent-samples experiment would not have been very effective in assessing whether exercise affects growth hormone levels. Hormone levels differ quite a bit from one person to another. With independent samples, we wouldn't know if any observed difference in the mean hormone level was a result of one sample by chance containing mostly people whose growth hormone level is naturally high. By using a paired-samples experiment, we are able to rule out this possibility.

EXAMPLE 2

Trace metals in drinking water affect the flavor, and unusually high concentrations can pose a health hazard. The paper "Trace Metals of South Indian River" (*Environ. Studies* (1982):62–66) reported trace-metal concentrations for both surface water and bottom water at six different river locations. Data on zinc concentration is given below.

	Concentration (mg/L)	
Location	Bottom Water	Top Water
1	.430	.415
2	.266	.238
3	.567	.390
4	.531	.410
5	.707	.605
6	.716	.609

Although zinc concentration varies widely from one location to another, in every case the zinc concentration in bottom water is higher than that of top water. This suggests that bottom water differs from top water in mean zinc concentration. Since this data is paired (by location), a method of analysis appropriate for paired data should be employed. ■

These two examples suggest that the methods of inference developed for independent samples aren't adequate for dealing with paired samples. When samples are paired in a meaningful way, it is natural to focus on the differences between the observations making up each pair. In both Example 1 and Example 2, the fact that all such differences are positive suggests that $H_0 : \mu_1 - \mu_2 = 0$ should probably be rejected. In the next section we see that inferences about $\mu_1 - \mu_2$ using paired samples can be based on these differences.

EXERCISES

10.1 Discuss the difference between independent samples and paired data.

10.2 Give an example of an experiment involving a comparison of two different brands of automobile tires that would result in paired data.

10.3 Suppose you were interested in investigating the effect of a drug that is to be used in the treatment of patients who have glaucoma in both eyes. A comparison between the mean reduction in eye pressure for this drug and for a standard treatment is desired. Both treatments are applied directly to an eye.

- **a.** Describe how you would go about collecting data for your investigation.
- **b.** Does your method result in paired data?
- **c.** Can you think of a reasonable method of collecting data that would result in independent samples? Would such an experiment be as informative as a paired experiment? Comment.

10.4 Suppose you were interested in comparing the yield of two different irrigation methods (constant drip and intermittent) for watering avacado seedlings. Describe two possible methods of collecting data—one that would result in paired data and one that would result in independent samples.

10.5 Two different brands of shoes designed for marathon running are to be compared. Suppose that 10 pairs of each brand, in any sizes, can be made available. How might an experiment resulting in paired data be carried out? How might an independent-samples experiment be carried out? Which would you recommend, and why?

10.6 Two different underground pipe coatings for preventing corrosion are to be compared. Effect of a coating (as measured by maximum depth of corrosion penetration on a piece of pipe) may vary with depth, orientation, soil type, pipe composition, etc. Describe how an experiment that filters out the effects of these extraneous factors could be carried out.

10.7 Describe how you would use 10 test specimens to see if two different scales tend to give the same weight. Would an independent-samples experiment make sense here? Comment.

10.2 Inferences Concerning a Difference Between Two Population Means

An investigator who wishes to compare two population means is usually interested in estimating or testing a hypothesis about the difference $\mu_1 - \mu_2$. When sample observations from the first population are paired in some meaningful way with sample observations from the second population, inferences can be based on the differences between the two observations within the sampled pairs.

Suppose that there are n pairs in the sample. The sample mean and sample variance of n observed x values were previously defined as $\overline{x} = \Sigma x/n$ and $s^2 = \Sigma(x - \overline{x})^2/(n - 1)$. If we compute the difference between the first and second observation for each pair, the result is a set of n sample differences. The **sample average difference $\overline{x}_d$** and **sample variance of the differences s_d^2** are then given by

$$\overline{x}_d = \frac{\Sigma(\text{difference})}{n} \qquad s_d^2 = \frac{\Sigma(\text{difference} - \overline{x}_d)^2}{n - 1}$$

The numerator of s_d^2 can be computed as $\Sigma(\text{difference})^2 - n\overline{x}_d^2$.

EXAMPLE 3

The given data on the zinc concentration for top and bottom water at six locations was first presented in Example 2. Since the data is paired (two measurements at each location), we can form differences by subtracting top-water zinc concentration from that of the bottom water for each location. The resulting differences are displayed below the data pairs.

Location	1	2	3	4	5	6
Bottom water	.430	.266	.567	.531	.707	.716
Top water	.415	.238	.390	.410	.605	.609
Difference	.015	.028	.177	.121	.102	.107

The sample average difference and sample variance of the differences are

$$\bar{x}_d = \frac{.015 + .028 + .177 + .121 + .102 + .107}{6} = .0917$$

and

$$s_d^2 = \frac{(.015 - .0917)^2 + (.028 - .0917)^2 + \cdots + (.107 - .0917)^2}{6 - 1}$$

$$= .00368$$

Then $s_d = \sqrt{s_d^2} = \sqrt{.00368} = .061.$ ■

We can regard the n sample differences as having been selected from a large population of differences. In Example 3, this population consists of differences (*bottom-water concentration* $-$ *top-water concentration*) at all locations (from which the six actually sampled were chosen). Let μ_d and σ_d denote the mean value and standard deviation of the difference population. The relationship between μ_d and the two individual population means μ_1 and μ_2 is given by one of the rules concerning the sampling distribution of a difference discussed at the outset of Chapter 9. With $x - y$ denoting a randomly selected difference, the rule states that $\mu_{x-y} = \mu_x - \mu_y$. This says that the mean value of a difference is the difference of the two individual mean values. This rule is valid whether or not x and y are independent. For paired data, the relevant difference is between an observation from the first population (with mean μ_1) and one from the second population (with mean μ_2). Applying the above rule gives

$$\mu_d = \mu_1 - \mu_2$$

That is, the average in the population of differences is the difference between the two individual population means. For example, if the true average bottom-water concentration is $\mu_1 = .60$ and true average top-water concentration is $\mu_2 = .50$, then the population average difference is $\mu_d = .60 - .50 = .10$.

Because of this rule, inferences about $\mu_1 - \mu_2$ when data is paired are equivalent to inferences about μ_d, the mean value of the population of all differences. Since inferences about μ_d can be based on the n observed sample differences, the original two-sample problem becomes a familiar one-sample problem.

The Paired t Test

In a test of hypotheses, the null hypothesis is stated as H_0: μ_d = hypothesized value. This is equivalent to the hypothesis H_0: $\mu_1 - \mu_2$ = hypothesized value. The hypothesized value is most frequently zero, in which case H_0 claims that $\mu_1 = \mu_2$. The alternative hypothesis $\mu_d >$ hypothesized value corresponds to the hypothesis $\mu_1 - \mu_2 >$ hypothesized value. If this alternative hypothesis is correct, we expect sample differences to exceed the hypothesized value since the population mean difference exceeds the hypothesized value. The alternative hypotheses H_a: $\mu_d <$ hypothesized value and H_a: $\mu_d \neq$ hypothesized value are also equivalent to corresponding claims about $\mu_1 - \mu_2$.

When it is reasonable to assume that the population of differences is approximately normal, the one-sample t test based on the differences is the recommended test procedure. The difference population will generally be normal when each of the two individual populations is normal. A normal probability plot of the differences is helpful in assessing the plausibility of this assumption.

Summary of the Paired t Test for Testing Hypotheses about μ_d ($\mu_1 - \mu_2$)

Null hypothesis: $H_0 : \mu_d =$ hypothesized value

$$\text{Test statistic: } t = \frac{\bar{x}_d - \text{hypothesized value}}{s_d/\sqrt{n}}$$

where n is the number of differences or pairs.

Alternative hypothesis	*Rejection region*
$H_a : \mu_d >$ hypothesized value	$t > t$ critical value (upper-tailed test)
$H_a : \mu_d <$ hypothesized value	$t < -t$ critical value (lower-tailed test)
$H_a : \mu_d \neq$ hypothesized value	Either $t > t$ critical value or $t < -t$ critical value (two-tailed test)

Once the level of significance α has been specified, the appropriate t critical value is obtained from the corresponding column and $n - 1$ df row of Table IV.

EXAMPLE 4

The zinc data of Example 3 can be used to test for any difference between mean zinc concentration in top and bottom water. We use the paired t test with level of significance .05.

1. $\mu_d = \mu_1 - \mu_2 =$ mean difference between bottom- and top-water zinc concentrations.

2. $H_0 : \mu_d = 0$.

3. $H_a : \mu_d \neq 0$.

4. Test statistic:

$$t = \frac{\bar{x}_d - 0}{s_d/\sqrt{n}} = \frac{\bar{x}_d}{s_d/\sqrt{n}}$$

5. The nature of H_a implies that a two-tailed rejection region should be used. With level of significance .05 and df $= n - 1 = 6 - 1 = 5$, Table IV gives 2.57 as the appropriate critical value. The null hypothesis will be rejected if either $t > 2.57$ or $t < -2.57$.

6. The values of $\bar{x}_d$ and s_d were previously computed to be $\bar{x}_d = .0917$ and $s_d = .061$.

Substituting these values into the test-statistic formula yields

$$t = \frac{.0917}{.061/\sqrt{6}} = \frac{.0917}{.0249} = 3.68$$

7. The value $t = 3.68$ exceeds 2.57, implying that H_0 should be rejected in favor of H_a. The data strongly suggests that mean zinc concentration is not the same for bottom and top water. ■

Use of the pooled t test on the data in Example 4 is incorrect because the top- and bottom-water samples are not independent. Inappropriate use of the pooled t in this setting would result in a computed t value of

$$t = \frac{(\bar{x}_1 - \bar{x}_2) - 0}{\sqrt{s_p^2/n_1 + s_p^2/n_2}} = \frac{.5362 - .4445}{\sqrt{.025/6 + .025/6}} = \frac{.0917}{.091} = 1.01$$

The hypothesis of equal mean concentrations for top and bottom water would not be rejected. When the pairing is ignored, the difference between top- and bottom-water concentrations is obscured by the variability in zinc concentration from one location to another.

The numerators $\bar{x}_d$ and $\bar{x}_1 - \bar{x}_2$ of the two test statistics are always equal. The difference between paired t and pooled t lies in the denominator. The variability in differences is usually much smaller than variability in each sample separately (because measurements in a pair tend to be similar). As a result, the paired t value is usually larger in magnitude than is the pooled t—3.68 versus 1.01 in the example just considered. Pairing typically reduces variability that might otherwise obscure small but nevertheless significant differences.

EXAMPLE 5

Researchers have long been interested in the effects of alcohol on the human body. The authors of the paper "Effects of Alcohol on Hypoxia" (*J. Amer. Med. Assoc.* (December 13, 1965):135) examined the relationship between alcohol intake and the time of useful consciousness during high-altitude flight. Ten male subjects were taken to a simulated altitude of 25,000 ft and given several tasks to perform. Each was carefully observed for deterioration in performance due to lack of oxygen, and the time at which useful consciousness ended was recorded. Three days later, the experiment was repeated 1 h after the subjects had ingested .5 cc of 100-proof whiskey per pound of body weight. The time (in seconds) of useful consciousness was again recorded. The resulting data appears below.

	Time		
Subject	No Alcohol	Alcohol	Difference
1	261	185	76
2	565	375	190
3	900	310	590
4	630	240	390
5	280	215	65
6	365	420	−55
7	400	405	−5
8	735	205	530
9	430	255	175
10	900	900	0

Since the samples are paired rather than independent, we use the paired t test to determine if the data supports the hypothesis that ingestion of the stated amount of alcohol reduces the mean time of useful consciousness at high altitudes. A normal probability plot based on the sample of differences looks approximately linear, suggesting that the assumption of normality is reasonable. We have selected a .05 significance level for this test.

1. μ_d = difference between the true mean time of useful consciousness when no alcohol is consumed and when .5 cc alcohol per pound of body weight is ingested.
2. $H_0 : \mu_d = 0.$
3. $H_a : \mu_d > 0.$
4. Test statistic:

$$t = \frac{\overline{x}_d}{s_d/\sqrt{n}}$$

5. A one-tailed rejection region is appropriate. With level of significance .05 and df $= n - 1 = 9$, Table IV gives the critical value 1.83. H_0 will be rejected if $t > 1.83$.
6. From the 10 observed differences, $\overline{x}_d = 195.6$ and $s_d = 230.53$. Then

$$t = \frac{195.6}{230.53/\sqrt{10}} = \frac{195.6}{72.9} = 2.68$$

7. The value $t = 2.68$ exceeds the critical value 1.83, so the null hypothesis should be rejected at level .05. There is sufficient evidence to indicate that ingestion of .5 cc of whiskey per pound of body weight reduces the average time of useful consciousness. ■

Methods for computing β, the probability of making a type II error, for the one-sample t test were presented in Chapter 8. The curves given in Table V can be employed in a similar fashion to determine β for the paired t test. Also, when the sample size (the number of differences) is large, the paired t test reduces to a one-sample z test, and the normality assumption is no longer necessary.

An alternate approach to initially fixing α when testing hypotheses is provided by P-values. The value of the paired t test statistic is computed and then an upper and/or lower bound on the associated P-value is obtained using Table IV. The investigator can then see what conclusion would be appropriate at any of the standard significance levels.

EXAMPLE 6

Douglas fir trees are an important source of wood products, so anything that affects tree growth is of interest to the lumber industry. To determine if fluoride contamination stunts growth, the authors of the paper "Patterns of Fluoride Accumulation and Growth Reduction Exhibited by Douglas Fir in the Vicinity of an Aluminum Reduction Plant" (*Environ. Pollution* (1984): 221–35) compared trees within 8 km of an aluminum-processing plant to those farther than 8 km away. Tree growth for a given year can be determined by measuring the width of the trunk ring formed during that year. The accompanying data was collected for the years 1962 to 1975. The aluminum-processing plant went into operation in 1967.

Preoperational Years			Operational Years		
Mean Ring Size			Mean Ring Size		
Year	Within 8 km	Beyond 8 km	Year	Within 8 km	Beyond 8 km
1962	3.3	3.3	1967	2.6	3.2
1963	3.4	3.4	1968	2.4	3.1
1964	3.0	3.1	1969	2.0	2.9
1965	2.9	3.0	1970	2.3	3.2
1966	2.9	2.9	1971	2.2	3.1
			1972	2.1	2.9
			1973	1.9	2.6
			1974	2.2	2.8
			1975	2.0	2.9

To determine whether mean growth rates for trees within 8 km and those beyond 8 km differed significantly during the preoperation period, we test

$$H_0 : \mu_d = 0 \quad \text{versus} \quad H_a : \mu_d \neq 0$$

The *within 8 – beyond 8* differences for the five preoperational years are 0, 0, −.1, −.1, and 0, yielding $\bar{x}_d = -.04$ and $s_d = .0548$.

The paired t test statistic is

$$t = \frac{\bar{x}_d - 0}{s_d/\sqrt{n}} = \frac{-.04 - 0}{.0548/\sqrt{5}} = \frac{-.04}{.0245} = -1.633$$

Since this is a two-tailed test, a bound on the P-value is determined from the two-tailed critical values in the t table. Using the 4 df row of Table IV, $t = 1.63$ falls between the critical values 1.53 ($\alpha = .20$) and 2.13 ($\alpha = .10$). Therefore, $.10 < P\text{-value} < .20$. For level of significance .1 or smaller, the null hypothesis would not be rejected, suggesting no difference between the preoperational mean growth rates.

We also carried out this test using MINITAB. There is no paired t command, so the analysis involves first computing the differences and then requesting a one-sample t test. Resulting computer output is given below. Note that MINITAB has calculated the P-value to be .18, which is consistent with the bounds obtained from the t table.

```
ROW  within8  beyond8  diff

  1      3.3      3.3   0.0
  2      3.4      3.4   0.0
  3      3.0      3.1  -0.1
  4      2.9      3.0  -0.1
  5      2.9      2.9   0.0


TEST OF MU = 0 VS MU N.E. 0

              N     MEAN    STDEV  SE MEAN       T   P VALUE
diff          5  -0.0400   0.0548    0.024   -1.63      0.18
```

A similar analysis of the postoperational years is appropriate for determining whether growth of trees near the site was reduced when the plant became operational. The hypotheses of interest are $H_0 : \mu_d = 0$ versus $H_a : \mu_d < 0$ (mean growth of trees within 8 km is less than mean growth for trees farther away). The observed differences are −.6, −.7, −.9, −.9, −.9, −.8, −.7, −.6, and −.9, resulting in $\bar{x}_d = -.778$ and $s_d = .13$. The corresponding value of the paired t statistic is

$$t = \frac{-.778 - 0}{.13/\sqrt{9}} = \frac{-.778}{.0433} = -17.97$$

For a one-tailed test, the 8 df row of Table IV implies that P-value < .0005.

The accompanying MINITAB output indicates that the P-value is zero when rounded to four decimal places. A P-value this small very strongly indicates a reduction in growth rate near the plant.

ROW	within8	beyond8	diff
1	2.6	3.2	-0.6
2	2.4	3.1	-0.7
3	2.0	2.9	-0.9
4	2.3	3.2	-0.9
5	2.2	3.1	-0.9
6	2.1	2.9	-0.8
7	1.9	2.6	-0.7
8	2.2	2.8	-0.6
9	2.0	2.9	-0.9

TEST OF MU = 0 VS MU L.T. 0

	N	MEAN	STDEV	SE MEAN	T	P VALUE
diff	9	-0.778	0.130	0.043	-17.93	0.0000

■

A Confidence Interval for μ_d

The t confidence interval for μ given in Chapter 7 is easily adapted to obtain an interval estimate for μ_d.

> When it is reasonable to assume that the difference population is (approximately) normal, a **confidence interval for μ_d** is
>
> $$\bar{x}_d \pm (t \text{ critical value}) \cdot \frac{s_d}{\sqrt{n}}$$
>
> For a specified confidence level, the $n - 1$ df row of Table IV gives the appropriate t critical value.

EXAMPLE 7

Cushing's disease is characterized by muscular weakness due to adrenal or pituitary dysfunction. In order to provide effective treatment, it is important to detect childhood Cushing's disease as early as possible. Age at onset of symptoms and age at diagnosis for 15 children suffering from the disease were given in the paper "Treatment of Cushing's Disease in Childhood and Adolescence by Transphenoidal Microadenomectomy" (*N. Engl. J. Med.* (April 5, 1984):889). Since early diagnosis is crucial for successful treatment, the length of time between onset of symptoms and diagnosis is of interest. Let μ_d

be the mean difference between age at onset and age at diagnosis (so that μ_d is a negative number). We use the data below to estimate μ_d with a 90% confidence interval.

Age (months)								
Patient	1	2	3	4	5	6	7	8
Onset	84	90	96	108	126	144	156	63
Diagnosis	108	102	151	123	156	204	170	84
Difference	−24	−12	−55	−15	−30	−60	−14	−21
Patient	9	10	11	12	13	14	15	
Onset	119	120	132	144	144	144	144	
Diagnosis	167	132	157	197	205	213	224	
Difference	−48	−12	−25	−53	−61	−69	−80	

The summary values are $\bar{x}_d = -38.6$ and $s_d = 23.18$. With $n - 1 = 14$ df, the t critical value for a 90% confidence level is 1.76. The interval is

$$-38.6 \pm 1.76(23.18/\sqrt{15}) = -38.6 \pm 10.53 = (-49.13, -28.07)$$

Based on the sample data, we can be 90% confident that the mean elapsed time between onset of symptoms and diagnosis of childhood Cushing's disease is between 28.07 and 49.13 months. It appears that a great deal of time generally passes before Cushing's disease is diagnosed. This is probably due to the fact that childhood Cushing's disease is very rare and so may be overlooked as a possibility until other more common illnesses are ruled out as causes of the symptoms. ■

EXERCISES

10.8 Twelve infants paired according to birth weight were used to compare an enriched formula with a standard formula. Weight gains (g) are given.

Pair	Enriched Formula	Standard Formula
1	3604	3140
2	2950	3100
3	3344	2810
4	4022	3761
5	4316	3774
6	3077	2630

a. Let μ_d denote the true average difference in weight gains between the two formulas (enriched − standard). What alternative hypothesis suggests that the enriched formula is more effective than the standard formula in increasing weight? Test $H_0 : \mu_d = 0$ against this alternative at significance level .05.

b. Why do you think a paired experiment was chosen for this study?

10.9 In a study of memory recall, eight people were given 10 min to memorize a list of 20 nonsense words. Each was asked to list as many of the words as he or she could remember both 1 h and 24 h later.

	Number of Words Recalled	
Subject	1 h Later	24 h Later
1	14	10
2	12	4
3	18	14
4	7	6
5	11	9
6	9	6
7	16	12
8	15	12

Is there evidence to suggest that the mean number of words recalled after 1 h exceeds the mean recall after 24 h by more than 3? Use a level .01 test.

10.10 A large amount of alcohol is known to reduce reaction time. To investigate the effects of small amounts of alcohol, reaction time was recorded for seven individuals before and after 2 oz of 90-proof alcohol was consumed by each. Does the data below suggest that 2 oz of alcohol reduces mean reaction time? Use a significance level of .05.

	Reaction Time (s)						
Subject	1	2	3	4	5	6	7
Before	.6	.8	.4	.7	.8	.9	.7
After	.7	.8	.6	.8	.8	.8	.9

10.11 Dentists make many people nervous (even more so than statisticians!). To see if such nervousness elevates blood pressure, the blood pressure and pulse rates of 60 subjects were measured in a dental setting and in a medical setting ("The Effect of the Dental Setting on Blood Pressure Measurement" *Am. J. Public Health* (1983): 1210–14). For each subject, the difference (dental-setting blood pressure minus medical-setting blood pressure) was formed. The analogous differences were also formed for pulse rates. Summary data is given below.

	Mean Difference	Standard Deviation of Differences
Systolic blood pressure	4.47	8.77
Pulse (beats/min)	−1.33	8.84

a. Does testing hypotheses here require an assumption of normality (about the blood pressure or pulse rate distributions)? Why or why not?

b. Does the data strongly suggest that true mean blood pressure is higher in a dental setting than in a medical setting? Use a level .01 test.

c. Is there sufficient evidence to indicate that true mean pulse rate in a dental setting differs from the true mean pulse rate in a medical setting? Use a significance level of .05.

10.12 The paper "A Supplementary Behavioral Program to Improve Deficient Reading Performance" (*J. Abnormal Child Psych.* (1973):390–99) reported the results of an experiment in which seven pairs of children reading below grade level were obtained by matching so that within each pair the two children were equally deficient in reading ability. Then one child from each pair received experimental training, while the other received standard training. Based on the accompanying improvement scores, does the experimental training appear to be superior to the standard training? Use a .1 significance level.

Pair	1	2	3	4	5	6	7
Experimental	.5	1.0	.6	.1	1.3	.1	1.0
Control	.8	1.1	−.1	.2	.2	1.5	.8

10.13 Samples of both surface soil and subsoil were taken from eight randomly selected agricultural locations in a particular county. The soil samples were analyzed to determine both surface pH and subsoil pH, with the following results.

Location	1	2	3	4	5	6	7	8
Surface pH	6.55	5.98	5.59	6.17	5.92	6.18	6.43	5.68
Subsoil pH	6.78	6.14	5.80	5.91	6.10	6.01	6.18	5.88

a. Compute a 90% confidence interval for the true average difference between surface and subsoil pH for agricultural land in this county.

b. What assumptions have you made about the underlying pH distributions?

10.14 The paper "Selection of a Method to Determine Residual Chlorine in Sewage Effluents" (*Water and Sewage Works* (1971):360–64) reported the results of an experiment in which two different methods, MSI and SIB, for determining chlorine content (mg/L) were used on water specimens.

Specimen	1	2	3	4	5	6	7	8
MSI method	.39	.84	1.76	3.35	4.69	7.70	10.52	10.92
SIB method	.36	1.35	2.56	3.92	5.35	8.33	10.70	10.91

Construct a 99% confidence interval for the difference in true average residual chlorine readings between the two methods.

10.15 Many people who quit smoking complain of weight gain. The results of an investigation into the relationship between smoking cessation and weight gain are given in the paper "Does Smoking Cessation Lead to Weight Gain?" (*Amer. J. Public Health* (1983):1303–05). Three hundred twenty-two subjects who successfully participated in a program to quit smoking were weighed at the beginning of the program and again 1 year later. The mean change in weight was 5.15 lb and the standard deviation of the weight changes was 11.45 lb. Is there sufficient evidence to conclude that the true mean change in weight is positive? Use $\alpha = .05$.

10.16 The paper "Evaluation of the Deuterium Dilution Technique Against the Test-Weighing Procedure for the Determination of Breast Milk Intake" (*Amer. J. Clinical Nutr.* (1983):996–1003) compares an isotopic and a test-weighing method for determining milk intake in breast-fed infants. The volume (mL) of breast milk ingested during a 48-h period was estimated by the two methods for each of 14 infants. Does the accompanying data indicate a difference in the mean intake determination for the two methods? Use a level .05 test.

Infant	1	2	3	4	5	6	7
Isotopic method	1509	1418	1561	1556	2169	1760	1098
Test-weighing	1498	1254	1336	1565	2000	1318	1410

Infant	8	9	10	11	12	13	14
Isotopic method	1198	1479	1281	1414	1954	2174	2058
Test-weighing	1129	1342	1124	1468	1604	1722	1518

10.17 Using the data in Exercise 10.16, estimate the true mean difference in intake determination for the two methods with a 90% confidence interval. Interpret the interval.

10.18 A famous paper on the effects of marijuana smoking ("Clinical and Psychological Effects of Marijuana in Man" *Science* (1968):1234–41) described the results of an experiment in which the change in heartbeat rate was measured for nine subjects who had never used marijuana before. Measurements were taken both 15 min after smoking at a low-dose level and 15 min after smoking a placebo (untreated) cigarette.

Subject	1	2	3	4	5	6	7	8	9
Placebo	16	12	8	20	8	10	4	−8	8
Low dose	20	24	8	8	4	20	28	20	20

Does the data suggest that marijuana smoking leads to a greater increase in heartbeat rate than does smoking a placebo cigarette? Test using $\alpha = .01$.

10.19 Referring to Example 5, suppose that the positions of the two data columns had been reversed, so that the (alcohol–no alcohol) differences had been computed (e.g., $185 - 261 = -76$). How would this relabeling change the hypotheses, analysis, and conclusion?

10.3 Distribution-Free Procedures for Inferences Concerning a Difference Between Two Population Means

In the previous section, the paired t test and paired t confidence interval were used to make inferences about μ_d, the population mean difference. These methods are appropriate when it is reasonable to assume that the difference population (from which the sample differences were randomly selected) is normal in shape. Since this may not always be the case, in this section we present an alternate test procedure, called the *signed-rank test,* and an associated confidence interval. These procedures are also based on the sample differences, but their validity requires only that the difference distribution be symmetric in shape. Symmetry is a weaker condition than normality (any normal distribution is symmetric, but there are many symmetric distributions which are not normal), so the signed-rank procedures are more widely valid than are the paired t procedures. Since the signed-rank procedures do not depend on specific distributional assumptions such as normality, they are distribution-free. A sufficient condition for the difference distribution to be symmetric is that the two population distributions (from which the first and second observations in each pair are drawn) are identical with respect to shape and spread.

As with the paired t test, we begin by forming differences. Next the absolute values of the differences are assigned ranks (this amounts to ignoring any negative signs when ranking). We then associate a + or a − sign with each rank, depending on whether the corresponding difference is positive or negative. For example, with $n = 5$, the differences might be −17, 12, 3, 10, and −6. The ordered absolute differences are then 3, 6, 10, 12, and 17, and the corresponding signed ranks are 1, −2 (because the difference is −6 rather than 6), 3, 4, and −5. If there are ties in the differences, the average of appropriate ranks is assigned as with the rank-sum test in Chapter 9.

The signed-rank test statistic for testing $H_0 : \mu_d = 0$ is the sum of the signed ranks. A large positive sum suggests that $\mu_d > 0$, since if this were the case, most differences would be positive and larger in magnitude than the few negative differences; most of the ranks, and especially the larger ones, would then be positively signed. Similarly, a large negative sum would suggest $\mu_d < 0$. A signed-rank sum near zero would be compatible with $H_0 : \mu_d = 0$.

EXAMPLE 8

Treatment of terminal renal failure involves surgical removal of a kidney (a nephrectomy). The paper "Hypertension in Terminal Renal Failure, Observations Pre and Post Bilateral Nephrectomy" (*J. Chronic Diseases* (1973): 471–501) gave the accompanying blood pressure readings for five terminal renal patients before and 2 months after surgery.

Diastolic Blood Pressure

Patient	1	2	3	4	5
Before surgery	107	102	95	106	112
After surgery	87	97	101	113	80
Difference	20	5	−6	−7	32

We can determine whether the mean blood pressure before surgery exceeds the mean blood pressure two months after surgery by testing $H_0 : \mu_1 - \mu_2 = 0$ versus $H_a : \mu_1 - \mu_2 > 0$, where μ_1 denotes the true mean diastolic blood pressure for patients in renal failure and μ_2 denotes the true mean blood pressure for patients 2 months after surgery (equivalent hypotheses are $H_0 : \mu_d = 0$ and $H_a : \mu_d > 0$, where μ_d is the mean difference in blood pressure).

A normal probability plot for this set of differences appears below. Since the plot appears to be more S-shaped than linear, the assumption of a normal difference population is questionable. If it is reasonable to assume that the difference distribution is symmetric, a test based on the signed ranks can be used.

The absolute values of the differences and the corresponding ranks are as follows.

Absolute difference	5	6	7	20	32
Rank	1	2	3	4	5

Associating the appropriate sign with each rank then yields signed ranks 1, −2, −3, 4, and 5, and a signed-rank sum of 5.

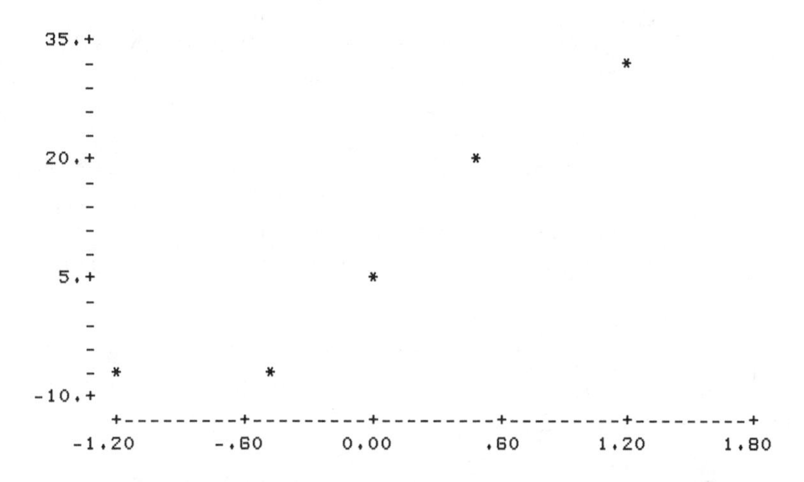

The largest possible value for this sum would be 15, occurring only when all differences are positive. There are 32 possible ways to associate signs with ranks 1, 2, 3, 4, and 5, and 10 of them have rank sums of at least 5. When the null hypothesis $H_0 : \mu_d = 0$ is true, each of the 32 possible assignments is equally likely to occur, and so $P(\text{signed-rank sum} \geq 5) = \frac{10}{32} = .3125$. Therefore, the observed sum of 5 is compatible with H_0—it does not provide evidence that H_0 should be rejected. ■

Testing Hypotheses Using Signed Ranks

Suppose we are interested in testing $H_0 : \mu_d = 0$. Given a set of n pairs of observations, ranking the absolute differences requires using ranks 1 to n. Since each rank could then be designated as either a plus or a minus, there are 2^n different possible sets of signed ranks. When the null hypothesis is true, each of the 2^n signed rankings has the same chance of occurring. Examining these different signed rankings and the associated sums gives information about how the signed-rank sum behaves when the null hypothesis is true. In particular, by looking at the distribution of the sum when H_0 is true, we can determine which values are unusual enough to suggest rejection of H_0.

For example, when $n = 5$ there are 2^5 different signed-rank sets. A few of these and the associated sums are

$$1, 2, 3, 4, 5 \qquad \text{sum} = 15$$
$$-1, 2, -3, 4, 5 \qquad \text{sum} = 7$$
$$-1, -2, 3, 4, -5 \qquad \text{sum} = -1$$

By systematically listing all 32 possible signed rankings, the following information is obtained:

Signed-rank sum	15	13	11	9	7	5	3	1
Number of rankings yielding sum	1	1	1	2	2	3	3	3

Signed-rank sum	−1	−3	−5	−7	−9	−11	−13	−15
Number of rankings yielding sum	3	3	3	2	2	1	1	1

If we were to reject $H_0 : \mu_1 - \mu_2 = 0$ in favor of $H_a : \mu_1 - \mu_2 \neq 0$ whenever we observed a signed-rank sum greater than or equal to 13 or less than or equal to -13, the probability of an incorrect rejection would be $4/32 = .125$ (since 4 of the possible signed rankings result in sums in the rejection region). Therefore, when $n = 5$, using the indicated rejection region gives a test with significance level .125.

For values of n larger than 5, finding the exact distribution of the signed-rank sum when H_0 is true is tedious and time consuming, so tables have been developed. For selected sample sizes, Table VIII in the appendices gives critical values for the signed-rank test for levels of significance closest to the usual choices of .01, .05, and .10.

Summary of the Signed-Rank Test*

Null hypothesis: $H_0 : \mu_d = 0$

Test statistic: signed-rank sum

Alternative hypothesis	*Rejection region*
$H_a : \mu_d > 0$	signed-rank sum $\geq$ critical value
$H_a : \mu_d < 0$	signed-rank sum $\leq$ $-$critical value
$H_a : \mu_d \neq 0$	either signed-rank sum $\geq$ critical value or signed-rank sum $\leq$ $-$critical value

Selected critical values are given in Table VIII.

EXAMPLE 9

Some swimming races are won by less than .001 s. As a result, a technique that might give a competitive swimmer even a slight edge is given careful consideration. To determine which of two racing starts, the hole entry or the flat entry, is faster, the authors of the paper "Analysis of the Flat vs. the Hole Entry" (*Swimming Technique* (Winter 1980):112–17) studied 10 college swimmers. A number of variables were measured for each type of start. The data for time to water entry appears below.

Swimmer	1	2	3	4	5	6	7	8	9	10
Flat entry	1.13	1.11	1.18	1.26	1.16	1.41	1.43	1.25	1.33	1.36
Hole entry	1.07	1.03	1.21	1.24	1.33	1.42	1.35	1.32	1.31	1.33
Difference	.06	.08	−.03	.02	−.17	−.01	.08	−.07	.02	.03

The authors of the paper used a level .05 signed-rank test to determine if there is a significant difference between the mean time to water entry for the two entry methods. Ordering the absolute differences results in the following assignment of signed ranks:

Difference	−.01	.02	.02	.03	−.03	.06	−.07	.08	.08	−.17
Signed rank	−1	2.5	2.5	4.5	−4.5	6	−7	8.5	8.5	−10

*Equivalent forms of the test statistic sometimes used are the sum of positive ranks, the sum of negative ranks, or the smaller of the sum of positive ranks and the sum of negative ranks. However, Table VIII should not be used to obtain critical values for these statistics.

1. Let μ_d denote the mean difference in time to water entry for flat and hole entry.

2. $H_0 : \mu_d = 0$.

3. $H_a : \mu_d \neq 0$.

4. Test statistic: signed-rank sum.

5. With $n = 10$ and $\alpha = .05$, Table VIII gives 39 as the critical value for a two-tailed test. Therefore, H_0 will be rejected if either signed-rank sum ≥ 39 or signed-rank sum ≤ -39.

6. Signed-rank sum $= -1 + 2.5 + \cdots + (-10) = 10$.

7. Since 10 does not fall in the rejection region, we do not reject H_0. There is not sufficient evidence to indicate that the mean time to water entry differs for the two methods. ■

The next example illustrates how zero differences are handled when performing a signed-rank test. Since zero is considered to be neither positive nor negative, zero values are generally excluded from a signed-rank analysis, and the sample size is reduced accordingly.

EXAMPLE 10

Two assay methods for measuring the level of vitamin B_{12} in red blood cells were compared in the paper "Noncobalimin Vitamin B_{12} Analogues in Human Red Cells, Liver and Brain" (*Am. J. of Clinical Nutr.* (1983):774–77). Blood samples were taken from 15 healthy adults, and, for each blood sample, the B_{12} level was determined using both methods. The resulting data is given below.

Subject	1	2	3	4	5	6	7	8
Method 1	204	238	209	277	197	227	207	205
Method 2	204	238	198	253	180	209	217	204
Difference	0	0	11	24	17	18	−10	1

Subject	9	10	11	12	13	14	15
Method 1	131	282	76	194	120	92	114
Method 2	137	250	82	165	79	100	107
Difference	−6	32	−6	29	41	−8	7

We assume that the difference distribution is symmetric and proceed with a signed-rank test to determine whether there is a significant difference between the two methods for measuring B_{12} content. A significance level of .05 will be used.

Two of the observed differences are zero. Eliminating the two zeros reduces the sample size from 15 to 13. Ordering the nonzero absolute differences results in the following assignment of signed ranks:

Difference	1	−6	−6	7	−8	−10	11	17	18	24	29	32	41
Signed rank	1	−2.5	−2.5	4	−5	−6	7	8	9	10	11	12	13

1. μ_d is the mean difference in B_{12} determination for the two methods.

2. $H_0 : \mu_d = 0$.

3. $H_a : \mu_d \neq 0$.

4. Test statistic: signed-rank sum.

5. The form of H_a indicates that a two-tailed test should be used. With $n = 13$ and $\alpha = .05$, Table VIII gives a critical value of 57 (corresponding to an actual significance level of .048). Therefore, H_0 will be rejected if either signed-rank sum ≥ 57 or signed-rank sum ≤ -57.

6. Signed-rank sum $= 1 + (-2.5) + (-2.5) + \cdots + 13 = 59$.

7. Since 59 falls in the rejection region, H_0 is rejected in favor of H_a. We conclude that there is a significant difference in measurement of B_{12} level in red blood cells for the two assay methods. ■

The procedure described above for testing $H_0 : \mu_d = 0$ can be easily adapted to test $H_0 : \mu_d =$ hypothesized value, where the hypothesized value is something other than zero.

> To test $H_0 : \mu_d =$ hypothesized value, subtract the hypothesized value from each difference prior to assigning signed ranks.

EXAMPLE 11

Tardive dyskinesia is a syndrome that sometimes follows long-term use of antipsychotic drugs. Symptoms include abnormal involuntary movements. In an experiment to evaluate the effectiveness of the drug Deanol in reducing symptoms, Deanol and a placebo treatment were each administered for four weeks to 14 patients. A Total Severity Index (TSI) score was used to measure improvement (larger TSI scores indicate greater improvement). The accompanying data comes from "Double Blind Evaluation of Deanol in Tardive Dyskinesia" (*J. Amer. Med. Assoc.* (1978):1997–98). Let's use this data and a significance level of .01 to determine if the mean TSI score for people treated with Deanol exceeds the mean placebo TSI score by more than 1.

TSI SCORES

Patient	1	2	3	4	5	6	7
Deanol	12.4	6.8	12.6	13.2	12.4	7.6	12.1
Placebo	9.2	10.2	12.2	12.7	12.1	9.0	12.4
Difference	3.2	−3.4	.4	.5	.3	−1.4	−.3

Patient	8	9	10	11	12	13	14
Deanol	5.9	12.0	1.1	11.5	13.0	5.1	9.6
Placebo	5.9	8.5	4.8	7.8	9.1	3.5	6.4
Difference	0	3.5	−3.7	3.7	3.9	1.6	3.2

1. Let μ_d denote the mean difference in TSI score between Deanol and the placebo treatment.

2. $H_0 : \mu_d = 1$.

3. $H_a : \mu_d > 1$.

4. Test statistic: signed-rank sum.

5. An upper-tailed test is indicated by the form of H_a. For $n = 14$ and $\alpha = .01$, Table VIII gives a critical value of 73. Therefore, H_0 will be rejected in favor of H_a at level .01 if the signed-rank sum equals or exceeds 73.

6. Subtracting 1 from each difference results in the following set of values.

2.2	−4.4	−.6	−.5	−.7	−2.4	−1.3
−1	2.5	−4.7	2.7	2.9	.6	2.2

Ordering these values and associating signed ranks yields:

Sign	−	−	+	−	−	−	+
Absolute difference	.5	.6	.6	.7	1	.3	2.2
Signed rank	−1	−2.5	2.5	−4	−5	−6	7.5

Sign	+	−	+	+	+	−	−
Absolute difference	2.2	2.4	2.5	2.7	2.9	4.4	4.7
Signed rank	7.5	−9	10	11	12	−13	−14

Then signed-rank sum $= -1 + (-2.5) + 2.5 + \cdots + (-14) = -4$.

7. Since $-4 < 73$, we fail to reject H_0. There is not sufficient evidence to indicate that the mean TSI score for the drug Deanol exceeds the mean TSI score for a placebo treatment by more than 1. ■

A Normal Approximation

Signed-rank critical values for sample sizes up to 20 are given in Table VIII. For larger sample sizes, the distribution of the signed-rank statistic when H_0 is true can be approximated by a normal distribution.

If $n > 20$, the distribution of the signed-rank sum when H_0 is true is well approximated by the normal distribution with mean 0 and standard deviation $\sqrt{n(n+1)(2n+1)/6}$. This implies that the standardized statistic

$$z = \frac{\text{signed-rank sum}}{\sqrt{n(n+1)(2n+1)/6}}$$

has approximately a standard normal distribution. This z statistic can be used as a test statistic, with the rejection region based on an appropriate z critical value.

EXAMPLE 12

The exercise capability of people suffering chronic airflow obstruction (CAO) is severely limited. In order to determine maximum exercise ventilation under two different experimental conditions, 21 patients suffering from CAO exercised to exhaustion under each condition. Ventilation was then measured. The data is from "Exercise Performance with Added Dead Space in Chronic Airflow Obstruction" (*J. Applied Physiology* (1984):1020–23).

Ventilation

Patient	1	2	3	4	5	6	7	8	9	10	11
Condition 1	62	57	56	55	50.5	50	47.2	43.5	40	40	41
Condition 2	52	46	51	52.4	55	51	43	40	34.2	34	33
Difference	10	11	5	2.6	−4.5	−1	4.2	3.5	5.8	6	8

Patient	12	13	14	15	16	17	18	19	20	21
Condition 1	33	31	28	27.1	27.5	27	25	19.2	17.5	12
Condition 2	32	38	26	28	28	18	21	18	16	15
Difference	1	−7	2	−.9	−.5	9	4	1.2	1.5	−3

Does this data suggest that the mean ventilation is different for the two experimental conditions? Let's analyze the data using a level .05 signed-rank test.

1. Let μ_d denote the true mean difference in ventilation between experimental conditions 1 and 2.
2. $H_0 : \mu_d = 0$.
3. $H_a : \mu_d \neq 0$.
4. Test statistic:

$$z = \frac{\text{signed-rank sum}}{\sqrt{n(n + 1)(2n + 1)/6}}$$

5. With $\alpha = .05$, the bottom row of Table IV gives the z critical value for a two-tailed test as 1.96. H_0 will be rejected in favor of H_a if $z > 1.96$ or $z < -1.96$.
6. Ordering the absolute differences and assigning signed ranks yields −1, −2, −3.5, 3.5, 5, 6, 7, 8, −9 10, 11, 12, −13, 14, 15, 16, −17, 18, 19, 20, 21. The signed-rank sum is $-1 + (-2) + \cdots + 21 = 140$, and the denominator of z is $\sqrt{n(n + 1)(2n + 1)/6} = \sqrt{(21)(22)(43)/6} = 57.54$, so

$$z = \frac{140}{57.54} = 2.43$$

7. Since 2.43 falls in the rejection region, we reject H_0 in favor of H_a. The sample data does suggest that the true mean ventilation rate differs for the two experimental conditions. ■

Comparing the Paired *t* and Signed-Rank Tests

In order for the paired t test to be an appropriate method of analysis, it must be assumed that the underlying difference distribution is normal. Proper use of the signed-rank test requires only that the difference distribution be symmetric. Since a normal distribution is symmetric, when the distribution of differences is normal, either the paired t or signed-rank test could be used. In this case, however, for a fixed significance level and sample size, the paired t

test gives a slightly smaller type II error probability. Therefore, when the assumption of a normal difference distribution is met, the paired t test would be the preferred method for testing hypotheses about $\mu_1 - \mu_2$ using paired data. However, when the difference distribution is symmetric but not necessarily normal, the signed-rank test may prove to be a better choice.

A Distribution-Free Confidence Interval for μ_d

The distribution-free confidence interval for $\mu_1 - \mu_2$ discussed in Chapter 9 consisted of all hypothesized values for which $H_0 : \mu_1 - \mu_2 =$ hypothesized value could not be rejected by the rank-sum test. Similarly, the signed-rank sum confidence interval consists of those values for which $H_0 : \mu_d =$ hypothesized value cannot be rejected by the signed-rank test. Unfortunately, in order to see the relation between the test procedure and confidence interval formula clearly, the test statistic must first be expressed in a different form, one that involves taking averages of all pairs of sample differences. We ask you to take it on faith that the procedure described below is correct (or consult one of the chapter references).

A **signed-rank confidence interval for μ_d** is based on all possible pairwise averages of sample differences (including the average of each difference with itself). The limits of the interval are the dth smallest and dth largest average, where the value of d is obtained from Table IX in the appendices and depends on the specified confidence level and the sample size.

EXAMPLE 13

Elevated levels of growth hormone are characteristic of diabetic control. The paper "Importance of Raised Growth Hormone Levels in Medicating the Metabolic Derangements of Diabetes" (*N. Engl. J. of Med.* (March 29, 1981):810–15) reported the results of a comparison of growth hormone levels for a conventional treatment and an insulin pump treatment for diabetes. Five diabetic patients participated in the study, with each patient receiving both treatments over a period of time. The resulting data is given. It would be useful to estimate the difference between mean growth hormone level for the two treatments.

Growth Hormone Level (mg/mL)

Patient	1	2	3	4	5
Conventional	10	16	17	20	10
Pump	9	7	8	8	6
Difference	1	9	9	12	4

To compute the required pairwise averages, it is convenient to arrange the differences along the top and left of a rectangular table. Then the averages of the corresponding pairs of differences can be calculated and entered at the intersection of each row and column on or above the diagonal of the table.

Pairwise Averages

		Difference				
		1	4	9	9	12
Difference	1	1	2.5	5	5	6.5
	4	—	4	6.5	6.5	8
	9	—	—	9	9	10.5
	9	—	—	—	9	10.5
	12	—	—	—	—	12

Arranging the pairwise averages in order yields

1, 2.5, 4, 5, 5, 6.5, 6.5, 6.5, 8, 9, 9, 9, 10.5, 10.5, 12

With a sample size of 5 and a 90% confidence level, Table IX gives $d = 2$ (corresponding to an actual confidence level of 87.5%). The confidence interval for $\mu_d = \mu_1 - \mu_2$ is then (2.5, 10.5).

As you can see, the calculations required in obtaining the pairwise averages can be tedious, especially for larger sample sizes. Fortunately, many of the standard computer packages calculate both the signed-rank sum and the signed-rank confidence interval. An approximate 90% signed-rank confidence interval from MINITAB appears below.

```
      ESTIMATED
N        CENTER   CONFIDENCE INTERVAL
5         6.500  (     2.500,   10.500)
```

■

The Signed-Rank Test for Single-Sample Problems

Although we have introduced the signed-rank test in a two-sample context, it can also be used to test $H_0 : \mu =$ hypothesized value, where μ is the mean value of a single population. In this setting, rather than forming differences and then associating signed ranks, a single sample is used and the hypothesized value from H_0 is subtracted from each observed sample value. Signed ranks are then associated with the resulting values. The rest of the test procedure (test statistic and rejection region) remains the same.

EXERCISES

10.20 In an experiment to study the way in which different anesthetics affected plasma epinephrine concentration, 10 dogs were selected and concentration was measured while they were under the influence of the anesthetics isoflurane and halothane ("Sympathoadrenal and Hemodynamic Effects of Isoflurane, Halothane, and Cyclopropane in Dogs" *Anesthesiology* (1974):465–70). The resulting data is as follows.

Dog	1	2	3	4	5	6	7	8	9	10
Isoflurane	.28	.51	1.00	.39	.29	.36	.32	.69	.17	.33
Halothane	.30	.39	.63	.38	.21	.88	.39	.51	.32	.42

Use a level .05 signed-rank test to see whether the true mean epinephrine concentration differs for the two anesthetics. What assumptions must be made about the epinephrine concentration distributions?

10.21 The accompanying data refers to the concentration of the radioactive isotope strontium-90 in samples of nonfat and 2% fat milk from five dairies. Does the data strongly support the hypothesis that the true mean strontium-90 concentration is higher for 2% fat milk than for nonfat? Use a level .05 signed-rank test.

	Concentration				
Dairy	1	2	3	4	5
Nonfat	6.4	5.8	6.5	7.7	6.1
2%	7.1	9.9	11.2	10.5	8.8

10.22 Both a gravimetric and a spectrophotometric method are under consideration for determining phosphate content of a particular material. Six samples of the material are obtained, each is split in half, and a determination is made on each half using one of the two methods, resulting in the following data. Use an approximate 95% distribution-free confidence interval to estimate the true mean difference for the two techniques. Interpret the interval.

Sample	1	2	3	4	5	6
Gravimetric	54.7	58.5	66.8	46.1	52.3	74.3
Spectrophotometric	55.0	55.7	62.9	45.5	51.1	75.4

10.23 The paper "Growth Hormone Treatment for Short Stature" (*N. Engl. J. Med.* (October 27, 1983):1016–22) gives the accompanying data on height velocity before growth hormone therapy and during growth hormone therapy for 14 children with hypopituitarism.

Child	1	2	3	4	5	6	7
Before	5.3	3.8	5.6	2.0	3.5	1.7	2.6
During	8.0	11.4	7.6	6.9	7.0	9.4	7.9

Child	8	9	10	11	12	13	14
Before	2.1	3.0	5.5	5.4	2.1	3.0	2.4
During	7.4	7.4	7.5	11.8	6.4	8.8	5.0

a. Use a level .05 signed-rank test to decide if growth hormone therapy is successful in increasing the mean height velocity.

b. What assumptions about the height velocity distributions must be made in order that the analysis in (a) be valid?

10.24 The paper "Analysis of the Flat vs. the Hole Entry" cited in Example 9 of this section also gave the accompanying data on time from water entry to first stroke and initial velocity. The authors of the paper used signed-rank tests to analyze the data.

a. Use a level .01 test to ascertain whether there is a significant difference in true mean time from entry to first stroke for the two entry methods.

b. Does the data suggest a difference in true mean initial velocity for the two entry methods? Use a level .05 signed-rank test.

	Time from Entry to First Stroke		Initial Velocity	
Swimmer	Hole	Flat	Hole	Flat
1	1.18	1.06	24.0	25.1
2	1.10	1.23	22.5	22.4
3	1.31	1.20	21.6	24.0
4	1.12	1.19	21.4	22.4
5	1.12	1.29	20.9	23.9
6	1.23	1.09	20.8	21.7
7	1.27	1.09	22.4	23.8
8	1.08	1.33	22.9	22.9
9	1.26	1.27	23.3	25.0
10	1.27	1.38	20.7	19.5

10.25 The paper "Effects of a Rice-rich versus Potato-rich Diet on Glucose, Lipoprotein, and Cholesterol Metabolism in Noninsulin-Dependent Diabetics" (*Amer. J. Clinical Nutr.* (1984):598–606) gave the accompanying data on cholesterol synthesis rate for eight diabetic subjects. Subjects were fed a standardized diet with potato or rice as the major carbohydrate source. Participants received both diets for specified periods of time, with cholesterol synthesis rate (mmol/day) measured at the end of each dietary period. The analysis presented in this paper used the signed-rank test. Use such a test with significance level .05 to determine whether the true mean cholesterol synthesis rate differs significantly for the two sources of carbohydrates.

Cholesterol Synthesis Rate

Subject	1	2	3	4	5	6	7	8
Potato	1.88	2.60	1.38	4.41	1.87	2.89	3.96	2.31
Rice	1.70	3.84	1.13	4.97	.86	1.93	3.36	2.15

10.26 The following pre- and postoperative lung capacities for 22 patients who underwent surgery as treatment for tuberculosis kyphosis of the spine appeared in the paper "Tuberculosis Kyphosis, Correction with Spinal Osteotomy, Halo-Pelvic Distractor, and Anterior and Posterior Fusion" (*J. Bone Joint Surgery* (1974): 1419–34). Does the data suggest that surgery increases the mean lung capacity? Use a level .05 large-sample signed-rank test.

Patient	1	2	3	4	5	6	7	8
Preoperative	1540	1160	1870	1980	1520	3155	1485	1150
Postoperative	1620	1500	2220	2080	2160	3040	2030	1370

Patient	9	10	11	12	13	14	15	16
Preoperative	1740	3260	4950	1440	1770	2850	2860	1530
Postoperative	2370	4060	5070	1680	1750	3730	3430	1570

Patient	17	18	19	20	21	22
Preoperative	3770	2260	3370	2570	2810	2990
Postoperative	3750	2840	3500	2640	3260	3100

10.27 Using the data of Exercise 10.23, estimate the true mean difference in height velocity before and during growth hormone therapy with a 90% distribution-free confidence interval.

10.28 The signed-rank test can be adapted for use in testing $H_0: \mu =$ hypothesized value, where μ is the mean of a single population (see the last part of this section). Suppose that the time required to process a request at a bank's automated teller machine is recorded for each of 10 randomly selected transactions, resulting in the following times (in min): 1.4, 2.1, 1.9, 1.7, 2.4, 2.9, 1.8, 1.9, 2.6, 2.2. Use the one-sample version of the signed-rank test and a .05 significance level to decide if the data indicates that the true mean processing time exceeds 2 min.

KEY CONCEPTS

Paired samples Two samples for which each observation in the first sample is paired in a meaningful way with a particular observation in the second sample. (p. 412)

Hypothesis Testing Procedures

Population Characteristic	*Appropriate When*	*Test Statistic*
μ_d	The difference distribution is normal (or $n > 30$)	Paired t test (Section 10.2)
μ_d	The difference distribution is symmetric, $n \leq 20$	Signed-rank test (Section 10.3)
μ_d	The difference distribution is symmetric, $n > 20$	Large-sample signed-rank test (Section 10.3)

Confidence Intervals

Population Characteristic	*Appropriate When*	*Confidence Interval*
μ_d	The difference distribution is normal (or $n > 30$)	Paired t interval (Section 10.2)
μ_d	The difference distribution is symmetric	Signed-rank interval (Section 10.3)

SUPPLEMENTARY EXERCISES

10.29 Recent evidence suggests that nonsmokers who live with smokers and are therefore exposed to what researchers call *sidestream smoke* (as opposed to *mainstream smoke,* which is inhaled by the smoker directly from the cigarette) may have an increased risk of lung disease. Mainstream (M) and sidestream (S) yields (in mg) of tar, nicotine, and carbon monoxide for eight brands of nonfilter cigarettes appeared in the paper "Yields of Tar, Nicotine, and Carbon Monoxide in the Sidestream Smoke from Fifteen Brands of Canadian Cigarettes" (*Amer. J. Public Health* (1984): 228–31).

Use this data and a .05 significance level to answer the following questions.

a. Does the data strongly suggest that the mean tar yield of sidestream smoke is higher than that of mainstream smoke?

b. Is there sufficient evidence to suggest that the mean nicotine yield is higher for sidestream smoke than for mainstream smoke?

c. Use a distribution-free procedure to estimate the mean difference in carbon monoxide yield between sidestream and mainstream smoke. Based on the resulting interval, do you think that the mean carbon monoxide yield is higher for sidestream than mainstream smoke? Explain.

	Tar		Nicotine		Carbon Monoxide	
Brand	S	M	S	M	S	M
A	15.8	18.5	2.8	1.2	40.5	18.6
B	16.9	17.0	2.7	1.1	59.8	20.5
C	21.6	17.2	3.7	1.2	42.9	16.8
D	18.8	19.4	2.8	1.0	42.0	17.8
E	29.3	15.6	4.3	1.1	60.8	19.8
F	20.7	16.4	3.9	1.2	45.1	16.4
G	18.9	13.3	3.3	1.0	43.9	13.1
H	25.0	10.2	4.6	1.0	67.3	12.4

10.30 Many researchers have investigated the relationship between stress and reproductive efficiency. One such study is described in the paper "Stress or Acute Adrenocorticotrophin Treatment Suppresses LHRH-Induced LH Release in the Ram" (*J. Repro. and Fertility* (1984):385–93). Seven rams were used in the study, and LH (luteinizing hormone) release (ng/min) was recorded before and after treatment with ACTH (adrenocorticotrophin, a drug that results in stimulation of the adrenal gland). Use the accompanying data and $\alpha = .01$ to determine if there is a significant reduction in mean LH release following treatment with ACTH.

LH Release (ng/min)

Ram	1	2	3	4	5	6	7
Before	2400	1400	1375	1325	1200	1150	850
After	2250	1425	1100	800	850	925	700

10.31 The accompanying 1982 and 1983 net earnings (in millions of dollars) of 10 food and beverage firms appeared in the article "Capital Expenditures Report" (*Food Engr.* (1984):93–101). Is there sufficient evidence to indicate that mean net earnings increased from 1982 to 1983? Assume that the 10 companies represent a random sample of all food and beverage firms. Perform the appropriate hypothesis test using a .01 significance level.

Firm	1983	1982	Firm	1983	1982
Coors	89.0	33.0	Nestle	113.4	93.7
ConAgra	28.7	20.6	Beatrice	292.1	320.8
ADM	110.2	155.0	Carnation	155.0	137.6
Heinz	237.5	214.3	Hershey	100.2	94.2
General MIlls	130.7	134.1	Procter & Gamble	58.5	66.5

10.32 Fifteen healthy male Indian students who lived at sea level and attended a mountaineering course in the Himalayan Mountains participated in a study on the effect of altitude on the level of thyroid stimulating hormone (TSH) in blood plasma ("Comparative Studies of Moderate and High Altitude Stress on Humans: Studies on Plasma T_3, T_4, TSH and Cortisol Levels" *Inter. J. Environ. Studies* (1984):277–82). The researchers hypothesized that the TSH level would increase at high altitude. TSH level (μu/mL) was recorded for each subject at sea level and again in the mountains. Suppose that the mean and standard deviation of the 15 difference values were .19 and .01, respectively.

a. Does the data support the researchers' hypothesis? Test the appropriate hypotheses using a .01 significance level.

b. Compute and interpret a 95% confidence interval for the true mean difference in TSH level.

10.33 The article "Action of Drugs on Movements of the Rat During Swimming" (*J. Human Movement Studies* (1984):225–230) described the effects of the drug ephedrine. Rats were placed in a swimming apparatus where swimming movement triggered rotation of an exercise wheel. The number of revolutions during a fixed time interval was recorded both before and after administration of a dose of 5 mg per kilogram of body weight of ephedrine. The resulting data appears below.

Before	15	30	3	16	11
After	6	5	3	6	2

a. Does the data suggest that ephedrine reduces the true mean number of revolutions? Test using a .05 significance level.

b. Use a distribution-free procedure to test the same hypotheses as in (a). Does the distribution-free test lead to the same conclusion?

10.34 The paper "Cardiac Output in Preadolescent Competitive Swimmers and in Untrained Normal Children" (*J. Sports Med.* (1983):291–99) reported the results of an experiment designed to assess the effect of athletic training on cardiac output. Sixteen children participated in the study. Eight of the subjects were trained competitive swimmers. The other eight children were normal healthy untrained children selected from a large group of volunteers. An untrained subject was chosen to match each trained subject with respect to age, height, weight, and body surface area, giving eight matched pairs of subjects. Resting heart rate (beats/min) and cardiac output (L/min) were measured for each child, resulting in the accompanying data.

	Heart Rate		Cardiac Output	
Pair	Trained	Untrained	Trained	Untrained
1	90	95	3.2	2.9
2	85	75	5.9	5.4
3	75	80	4.2	3.4
4	120	65	7.4	2.8
5	95	82	5.5	4.3
6	105	80	4.5	4.8
7	85	100	4.3	4.3
8	75	85	5.3	4.9

a. Is there sufficient evidence to indicate a difference between trained and untrained children with respect to mean resting heart rate? Use a .05 significance level.

b. Does the data suggest that mean resting cardiac output differs for trained and untrained children? Test the appropriate hypotheses using a .01 significance level.

c. Explain why the researchers used paired samples rather than independent samples.

REFERENCES

See references at the end of Chapter 9.

11 Simple Linear Regression and Correlation*

INTRODUCTION

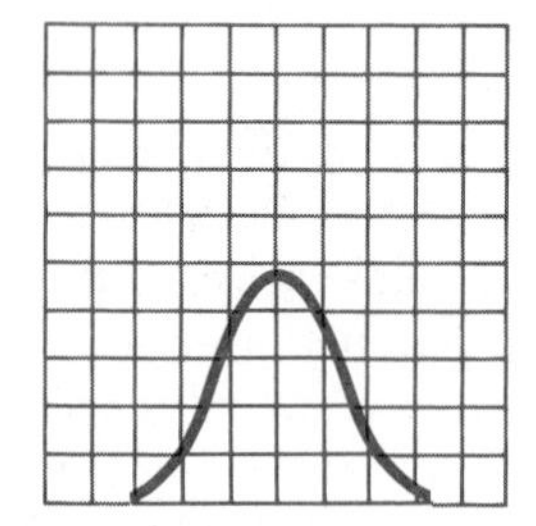

A RELATIONSHIP between two numerical variables x and y is said to be **deterministic** if the value of y is completely and uniquely determined by the value of x. As an example, a recreational facility might assess a yearly membership fee of $100 and then charge $5 per visit. Then with $x =$ number of visits in a year and $y =$ a member's annual cost, the value of y is completely determined by x through the equation $y = 100 + 5x$. There is no uncertainty in the value of y once x is known, and two individuals with the same x value also have the same y value.

Often x and y are related but not in a deterministic fashion. For example, let $x =$ child's age and $y =$ vocabulary size. Then it is clear that an increase in the value of x tends to be associated with an increase in y. However, there are many factors other than age that affect vocabulary size—parents' educational level, number of siblings, home environment, amount of schooling, etc.—so that two children with identical x values often have very different y values. As another example, x and y might represent the amount of a catalyst used in a chemical reaction and the reaction time, respectively. There will typically be a strong relationship between x and y, but the reaction time also depends on various other experimental conditions. Thus repeating the experiment with the same amount of catalyst could yield somewhat different reaction times.

When there is uncertainty in the value of the dependent variable y for any given value of the independent variable x, the relationship between x and y

*Chapter 4 contains prerequisite material.

can be described by a **probabilistic model.** Such a model specifies for any particular x value how the associated values of y are distributed. That is, the entire population of all (x, y) pairs can be regarded as consisting of many distinct populations, a different y population for each x value (for example, all vocabulary sizes when $x = 2$, all vocabulary sizes when $x = 3$, and so on). A probabilistic model simultaneously describes all these y population distributions. The most important aspects of such a model include the way in which the mean, or average, y value depends on x (how does average vocabulary size increase with age, or how is mean reaction time affected by a change in the amount of catalyst used?) and the amount of variability in y for each x.

The simplest and most widely used probabilistic model is the simple linear regression model. This model extends the idea, first introduced in Chapter 4, of fitting a straight line to sample data $(x_1, y_1), \ldots, (x_n, y_n)$. There the best-fit line was viewed as a descriptive summary of an approximate relationship between x and y values in the data set. In this chapter we show how the sample data can be used to draw conclusions about characteristics of the simple linear regression model by constructing confidence intervals, testing hypotheses, and making predictions. We also consider how the sample correlation coefficient r, which measures the extent to which x and y are linearly related in the sample, can be used to draw conclusions about a corresponding relationship in the population consisting of all (x, y) pairs.

11.1 The Simple Linear Regression Model

The *simple linear regression model* formalizes the notion of an approximate straight-line relationship between x and y. A key element of the model is a line, called the *true*, or *population, regression line,* which summarizes how the variables are related. This line represents the line of population averages; its height above a specific x value is the average value of y for all (x, y) pairs in the population with that particular x value. When x is fixed and a single y observation is made, the resulting point typically deviates from the population regression line because an observation on any random variable usually differs from the variable's mean value. The possibility of such a deviation is what distinguishes this probabilistic model from a deterministic linear model, where there is no deviation from the line. The points $(x_1, y_1), \ldots, (x_n, y_n)$ resulting from a sample of n observations are scattered about the true regression line in a random manner, some falling quite close to the line and others farther away.

Consider, for example, the variables x = first year mileage (in thousands of miles) for a certain car model used by a rental company and y = resulting maintenance cost. Suppose that for x values between 25 and 50 (usual first year mileages), x and y are related according to the simple linear regression model with population regression line $y = 35 + 5x$. The height of this line when $x = 40$ is 235, so the average, or expected, maintenance cost for cars driven 40,000 mi is $235. If a car having $x = 40$ is selected, the resulting maintenance cost might be $y = 265$, corresponding to the point (40, 265), which lies above the true regression line. Another car with the same x value might yield $y = 230$, 5 below the average cost for that x. The line $35 + 5x$ gives the *average* y value for any particular x value, but observed y's will deviate from the population averages by random amounts. The line has slope 5, which represents the average change in y associated with a 1-unit increase in x. If two cars have x values that differ by 1 (1000 extra miles), the average extra cost for the higher-mileage car is $5.

In addition to using a model based on a straight line rather than a curve, simple linear regression also involves other assumptions concerning the distribution of y values at any given x. Here is the precise model description.

DEFINITION

The **simple linear regression model** assumes that there is a line with slope β and y intercept α, called the **true, or population, regression line.** When x is fixed and an observation on y is made,

$$y = \alpha + \beta x + e$$

where e is a normally distributed random variable (random deviation from the line) with mean value zero and variance σ^2. It follows that for any fixed x value, y itself has a normal distribution with variance σ^2 (standard deviation σ) not depending on x and

$$\begin{pmatrix}\text{mean } y \text{ value} \\ \text{for fixed } x\end{pmatrix} = \begin{pmatrix}\text{height of the population} \\ \text{regression line above } x\end{pmatrix} = \alpha + \beta x$$

The slope β is the average change in y associated with a one unit increase in x, and the y intercept α is the height of the population regression line when $x = 0$.*

The salient features of the simple linear regression model are illustrated in Figure 1. For any particular x, the mean value of y is the height of the true regression line $y = \alpha + \beta x$ above that x value. The distribution of y at that x is specified by a normal curve centered at the corresponding mean value and having standard deviation σ. The standard deviation is the same for every x, so when x changes from x_1 to x_2 to x_3, the only aspect of the y distribution that changes is the mean value.

The model involves three population characteristics: α and β, the coefficients in the population regression line, and σ, which describes the extent to which a y observation might deviate from its mean value. If the value of σ is quite small, an observed point (x, y) will almost surely fall quite close to the population line. With a large value of σ, the point (x, y) might fall quite far above or below the line. The use of α and β is consistent with the choice of Greek letters to denote population characteristics.

*Although the symbols are the same, α and β have nothing to do with type I and II error probabilities in hypothesis testing.

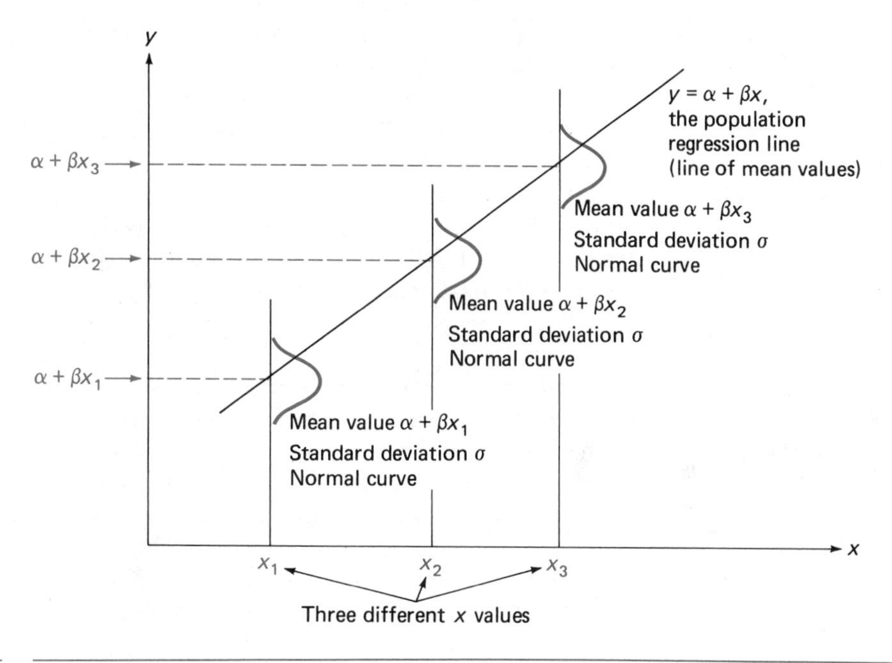

FIGURE 1 ILLUSTRATION OF THE SIMPLE LINEAR REGRESSION MODEL

EXAMPLE 1

It is often important to have a preliminary assessment of material strength. The article "Some Field Experience in the Use of an Accelerated Method in Estimating 28-Day Strength of Concrete" (*J. Amer. Concrete Inst.* (1969): 895) suggests the simple linear regression model as a reasonable way to relate y = 28-day standard cured strength of concrete (lb/in.^{2}) to x = accelerated strength (lb/in.^{2}). Suppose that the true regression line has slope $\beta = 1.25$ and y intercept $\alpha = 1800$ and that $\sigma = 350$ lb/in.^{2} Then for any fixed x value, y has a normal distribution with mean value $1800 + 1.25x$ and standard deviation 350. For example, when $x = 2000$, 28-day strength is normally distributed with mean value $1800 + (1.25)(2000) = 4300$ and standard deviation 350. What is the chance that a batch with accelerated strength 2000 will result in 28-day strength that exceeds 5000? Standardizing 5000 gives $z = (5000 - 4300)/350 = 2.0$, so

$$P(y > 5000 \text{ when } x = 2000) = \begin{pmatrix} \text{area under the } z \text{ curve} \\ \text{to the right of 2.0} \end{pmatrix} = .0228$$

That is, only 2.28% of all concrete batches with accelerated strength 2000 will yield a 28-day strength in excess of 5000.

If $x = 2500$, then y has a normal distribution with mean value $1800 + (1.25)(2500) = 4925$ and standard deviation 350. In this case, standardization and the use of Table I yield .42 as the probability that 28-day strength exceeds 5000.

The slope $\beta = 1.25$ is the average increase in 28-day strength associated with a $1 =$ lb/in.^{2} increase in accelerated strength. If two batches of concrete have accelerated strengths that differ by 1 lb/in.^{2}, their 28-day strengths can be expected to differ by 1.25 lb/in.^{2} Notice that for $x = 0$, the height of the true regression line is $\alpha + \beta(0) = \alpha = 1800$. Yet it doesn't make sense to predict 1800 as the 28-day strength when accelerated strength is 0. The use of the simple linear regression model for *all* x values, and in particular for $x = 0$, is unrealistic here and in many other situations. The model is usually assumed to be valid only for x values within some interval (e.g., between 1000 and 4000 lb/in.^{2}). If x is too small or large, use of the model can result in very inaccurate or even nonsensical predictions. In this example, it is best to think of α simply as the height of the true regression line rather than as the predicted y value when $x = 0$. ■

Estimating the Population Regression Line

In some applications, particularly in scientific contexts, the simple linear regression model can be justified by theoretical reasoning. The random deviation e might then represent measurement error or the effects of varying experimental conditions. Much more frequently, however, the appropriateness of the model is suggested by sample data consisting of n independently obtained pairs $(x_1, y_1), (x_2, y_2), \ldots, (x_n, y_n)$. The simple linear regression model should be used only when a scatter plot of the data shows a linear pattern. Furthermore, to be consistent with the assumption that the random deviation variance σ^2 is the same for every x value, the vertical spread of points should not be greater at one end of the plot than at the other end. Figure 2 exhibits three different scatter plots. The simple linear regression model is reasonable only for the variables x and y that give rise to the first plot.

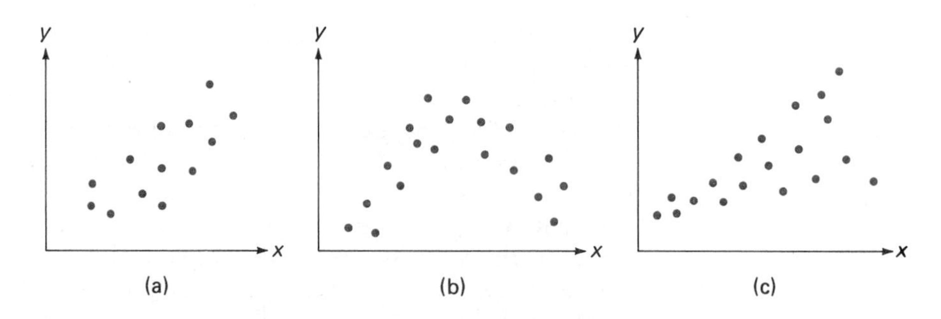

FIGURE 2 SOME COMMONLY ENCOUNTERED PATTERNS IN SCATTER PLOTS
(a) A Scatter Plot Consistent with the Simple Linear Regression Model
(b) A Scatter Plot that Suggests a Nonlinear Probabilistic Model
(c) A Scatter Plot that Suggests that Variability in y Is Not the Same for All x Values

The values of the slope β and y intercept α of the population regression line are almost never known to an investigator. Instead, these values have to be estimated from sample data. The method for doing this utilizes the principle of least squares discussed in Chapter 4. Recall that the least squares line is the line about which the sum of squared vertical deviations is smaller than would be the case for any other line. The slope b and y intercept a of the least squares line will be our point estimates of β and α, respectively. The population regression line $y = \alpha + \beta x$ is then estimated by the least squares line

$\hat{y} = a + bx$ (we use y in the equation for the population line and $\hat{y}$ in the equation for the least squares line).

DEFINITION

The point estimates of β and α are the slope and y intercept, respectively, of the least squares line, given by

$$b = \frac{\Sigma(x - \bar{x})(y - \bar{y})}{\Sigma(x - \bar{x})^2}$$

$$a = \bar{y} - b\bar{x}$$

The **estimated regression line** is then $\hat{y} = a + bx$ (which is just the least squares line). As in Chapter 4, the numerator and denominator of b can be calculated from the computing formulas $\Sigma xy - n\bar{x}\bar{y}$ and $\Sigma x^2 - n\bar{x}^2$, respectively.*

EXAMPLE 2

Landslides are common events in tree-growing regions of the Pacific Northwest, so their effect on timber growth is of special concern to foresters. The paper "Effects of Landslide Erosion on Subsequent Douglas-fir Growth and Stocking Levels in the Western Cascades, Oregon" (*Soil Sci. Society of Amer. J.* (1984):667–71) reported on the results of a study in which growth in a landslide area was compared with growth in a previously clear-cut area. Here we present data on clear-cut growth, with x = tree age (years) and y = 5-year height growth (cm).

Observation	1	2	3	4	5	6	7	8
x	5	9	9	10	10	11	11	12
y	70	150	260	230	255	165	225	340

Observation	9	10	11	12	13	14	15	16
x	13	13	14	14	15	15	18	18
y	305	335	290	340	225	300	380	400

The scatter plot displayed in Figure 3 shows a pronounced linear pattern with reasonably uniform variability in all regions of the plot. Because of this, the authors of the paper proceeded with a simple linear regression analysis. The calculations necessary to obtain $\Sigma(x - \bar{x})(y - \bar{y})$ and $\Sigma(x - \bar{x})^2$ are given in the accompanying tabular format ($\bar{x} = 12.3125$ and $\bar{y} = 266.875$); $\Sigma(y - \bar{y})^2$ is not needed at present but will be used shortly to compute an es-

* $\Sigma(x - \bar{x})^2$, $\Sigma(y - \bar{y})^2$, and $\Sigma(x - \bar{x})(y - \bar{y})$ appear quite frequently in this chapter. The corresponding computational formulas can always be used to calculate these quantities.

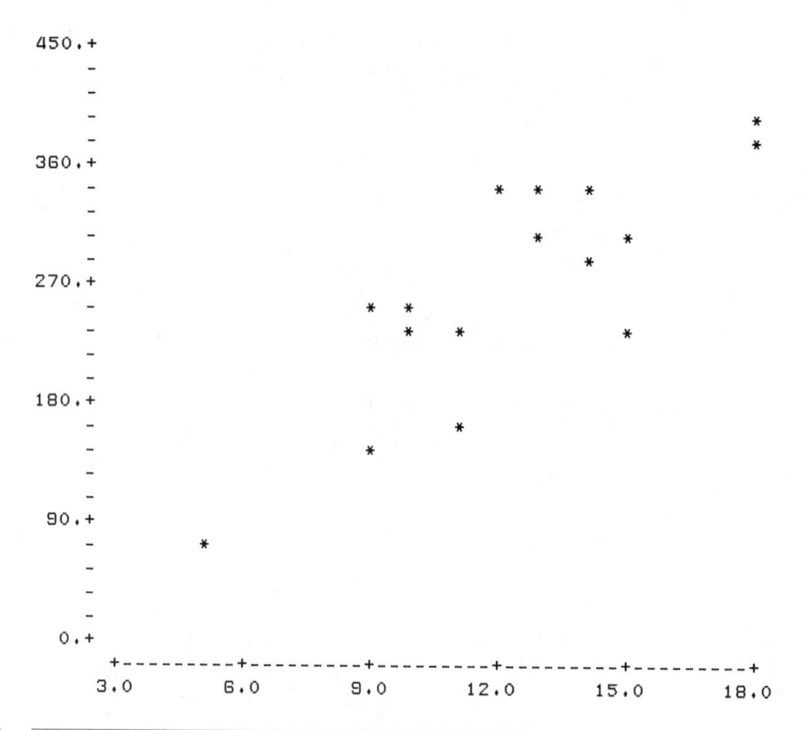

FIGURE 3 MINITAB SCATTER PLOT FOR DATA OF EXAMPLE 2

	$x - \bar{x}$	$y - \bar{y}$	$(x - \bar{x})^2$	$(x - \bar{x})(y - \bar{y})$	$(y - \bar{y})^2$
1	−7.3125	−196.875	53.4727	1439.65	38759.8
2	−3.3125	−116.875	10.9727	387.15	13659.8
3	−3.3125	−6.875	10.9727	22.77	47.3
4	−2.3125	−36.875	5.3477	85.27	1359.8
5	−2.3125	−11.875	5.3477	27.46	141.0
6	−1.3125	−101.875	1.7227	133.71	10378.5
7	−1.3125	−41.875	1.7227	54.96	1753.5
8	−0.3125	73.125	0.0977	−22.85	5347.3
9	0.6875	38.125	0.4727	26.21	1453.5
10	0.6875	68.125	0.4727	46.84	4641.0
11	1.6875	23.125	2.8477	39.02	534.8
12	1.6875	73.125	2.8477	123.40	5347.3
13	2.6875	−41.875	7.2227	−112.54	1753.5
14	2.6875	33.125	7.2227	89.02	1097.3
15	5.6875	113.125	32.3477	643.40	12797.3
16	5.6875	133.125	32.3477	757.15	17722.3
			175.4382	3740.60	116794.0
			↑ $\Sigma(x - \bar{x})^2$	↑ $\Sigma(x - \bar{x})(y - \bar{y})$	↑ $\Sigma(y - \bar{y})^2$

timate of σ. The estimated slope and y intercept of the true regression line are

$$b = \text{estimate of } \beta = \frac{\Sigma(x - \bar{x})(y - \bar{y})}{\Sigma(x - \bar{x})^2} = \frac{3740.60}{175.4382} = 21.321468$$

$$a = \text{estimate of } \alpha = \bar{y} - b\bar{x} = 266.875 - (21.321468)(12.3125)$$
$$= 4.354$$

The estimated regression line (least squares line) is $\hat{y} = 4.354 + 21.321x$.

If the computational formulas are used to calculate the numerator and denominator of b, a tabular format with columns containing x, y, x^2, xy, and y^2 values (the y^2 values for later use) should be employed. It is easy to check that $\Sigma x^2 = 2601$ and $\Sigma xy = 56{,}315$, so

$$\Sigma(x - \bar{x})^2 = \Sigma x^2 - n\bar{x}^2 = 2601 - 16(12.3125)^2 = 175.4375$$

$$\Sigma(x - \bar{x})(y - \bar{y}) = \Sigma xy - n\bar{x}\bar{y} = 56{,}315 - 16(12.3125)(266.875) = 3740.6250$$

These values differ slightly from the corresponding values given earlier (due to rounding), but they lead to almost exactly the same values of b and a (21.322 and 4.352, respectively). ■

The estimated slope b and y intercept a are both statistics. Their values depend on which sample of pairs $(x_1, y_1), \ldots, (x_n, y_n)$ is selected. The $n = 16$ observations of Example 2 resulted in $b = 21.321$, but another sample might have given $b = 19.250$ and yet another sample, $b = 24.576$. Like any statistic, b exhibits sampling variability. In the next section some facts about the sampling distribution of the statistic b are presented. These facts, in turn, provide information about the precision with which b estimates β. Similar comments apply to a as the statistic for estimating α.

Using the Estimated Regression Line

Once the equation of the estimated regression line has been obtained, it can be used for two distinct purposes:

1. To estimate the average y value associated with a particular value of x
2. To predict the value of y that will result from a single observation made with x fixed at a particular value

For example, data might suggest that the relationship between x = child's age and y = vocabulary size is well described by the simple linear regression model. The investigator might then wish to estimate the average vocabulary size for all 6-year-old children (average y value when $x = 6$). Alternatively, someone might approach the investigator with a particular 6-year-old child and ask for a prediction of that child's vocabulary size. As another example, a chemist might wish either to estimate the average yield (y) for all possible experimental runs in which process temperature (x) is set at 100°C or predict yield for a single experimental run to be carried out at 100°C.

Although the two uses just described are different, the (point) estimate of the average y and prediction of a single y value are computed in exactly the same way: by simply substituting the specified x value in the estimated regression equation and doing the required arithmetic. That is, both the estimate and the prediction are given by the height of the estimated line above the specified x value.

EXAMPLE 3

When anthropologists analyze human skeletal remains, an important piece of information is living stature. Since skeletons are usually quite incomplete, inferences about stature are commonly based on statistical methods that utilize measurements on small bones. The paper "The Estimation of Adult Stature from Metacarpal Bone Length" (*Amer. J. Phys. Anthropology* (1978): 113–20) presented data to validate one such method. Consider the accompanying representative data, where x = metacarpal bone I length (cm) and y = stature (cm).

Observation	1	2	3	4	5	6	7	8	9	10
x	45	51	39	41	52	48	49	46	43	47
y	171	178	157	163	183	172	183	172	175	173

A scatter plot strongly suggests the appropriateness of a linear relationship. It is easily verified that $\bar{x} = 46.10$, $\bar{y} = 172.70$, $\Sigma(x - \bar{x})(y - \bar{y}) = 271.30$, and $\Sigma(x - \bar{x})^2 = 158.90$, from which

$$b = \frac{\Sigma(x - \bar{x})(y - \bar{y})}{\Sigma(x - \bar{x})^2} = \frac{271.30}{158.90} = 1.707363$$

$$a = \bar{y} - b\bar{x} = 172.70 - (1.707363)(46.10) = 93.99$$

This gives $\hat{y} = 93.99 + 1.707x$ as the estimated regression line. An estimate of the average stature for all population members having metacarpal bone length 45 cm results from substituting $x = 45$ into the estimated equation:

$$\begin{pmatrix}\text{estimate of average } y \\ \text{when } x = 45\end{pmatrix} = a + b(45) = 93.99 + (1.707)(45)$$
$$= 170.81$$

If a particular population member with metacarpal bone length 45 is selected, that individual's predicted stature would be

$$\begin{pmatrix}\text{predicted } y \text{ value} \\ \text{when } x = 45\end{pmatrix} = a + b(45) = 93.99 + (1.707)(45) = 170.81$$

The sample contains the observation (45, 171), which falls almost exactly on the estimated line (because $\hat{y} = 170.81$ when $x = 45$). The predicted stature when $x = 48$ (also in the sample) is $93.99 + (1.707)(48) = 175.93$, so the difference between the observed and predicted values (a residual) is $172 - 175.93 = -3.93$. This difference is negative because (48, 172) lies below the estimated line. The estimate of average stature when $x = 44$, an x value not contained in the sample, is $93.99 + (1.707)(44) = 169.10$. It would be dangerous to compute an estimate or prediction for an x value much smaller than 39 or much larger than 52. Without sample data for such values, there is no hard evidence that the estimated linear relationship can be extrapolated very far. ■

Both a and b are subject to sampling variability, and for a fixed x value, so is the statistic $a + bx$. If the estimated coefficients are $b = 4.125$ and $a = 103.75$, then the estimate of average y when $x = 10$ is $103.75 + (4.125)(10) = 145.00$. Another sample might yield $b = 3.980$ and $a = 106.90$. Then the value of the statistic $a + b(10)$ would be $106.90 + (3.980)(10) = 146.70$. In Section 11.3 we discuss the sampling distribution of $a + bx$ for a fixed x value. Its properties are then used to obtain a confidence interval for average y and a prediction interval for a single y. These intervals are very informative in assessing the precision of estimates or predictions obtained from the estimated regression line.

Estimating σ^2

The value of σ^2 reflects the extent to which observed points tend to fall close to or far away from the true regression line $y = \alpha + \beta x$. Although the equation of the true line is not available, information about σ^2 can be obtained by assessing the extent to which points in the scatter plot fall close to or far away from the estimated line $\hat{y} = a + bx$. Figure 4 shows two different scatter plots, each with the estimated line superimposed. The fact that the points in Figure 4(a) all lie very close to $\hat{y} = a + bx$ suggests a small value of σ^2. The large spread of points about the estimated line in the second plot indicates that σ^2 is relatively large.

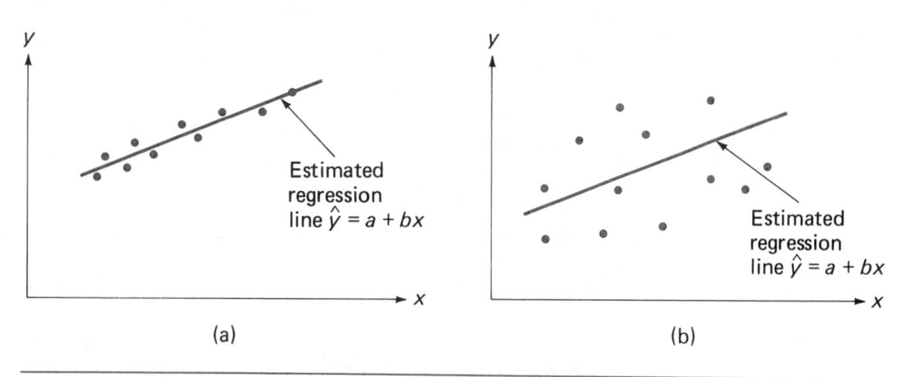

FIGURE 4

VARIATION ABOUT THE ESTIMATED REGRESSION LINE
(a) A Plot that Suggests a Small Value of σ^2
(b) A Plot that Suggests a Large Value of σ^2

Recall from Chapter 4 that the *predicted values* corresponding to observations at $x_1, x_2, \ldots, x_n$ are $\hat{y}_1 = a + bx_1$, $\hat{y}_2 = a + bx_2, \ldots, \hat{y}_n = a + bx_n$, and the *residuals* are $y_1 - \hat{y}_1$, $y_2 - \hat{y}_2, \ldots, y_n - \hat{y}_n$. The residuals are just the vertical deviations from the estimated regression line. A positive residual corresponds to a point above the estimated line ($y > \hat{y}$ implies $y - \hat{y} > 0$) and a negative residual results from a point below the estimated line. The estimate of σ^2 is based on the sum of squared residuals SSResid $= \Sigma(y - \hat{y})^2$ introduced in Chapter 4. A computational formula for this sum of squares is

$$\text{SSResid} = \Sigma(y - \bar{y})^2 - b \cdot \Sigma(x - \bar{x})(y - \bar{y})$$

The statistic for estimating the variance σ^2 is

$$s_e^2 = \frac{\text{SSResid}}{n - 2}$$

The estimate of σ is the **estimated standard deviation s_e** $= \sqrt{s_e^2}$. It is customary to call $n - 2$ the number of degrees of freedom associated with estimating σ^2 in simple linear regression.

To understand the rationale for dividing SSResid by $n - 2$, recall that in one-sample problems discussed in Chapters 7 and 8, σ^2 was estimated by $s^2 = \Sigma(x - \bar{x})^2/(n - 1)$. There the number of degrees of freedom was $n - 1$ because before σ^2 could be estimated, the single population characteristic μ

first had to be estimated, entailing the loss of 1 df. In simple linear regression, before σ^2 can be estimated, the two population characteristics α and β must be estimated (by a and b, respectively), entailing a loss of 2 df. This implies that t tests and t confidence intervals in simple linear regression are based on $n - 2$ df.

EXAMPLE 4

For the tree-growth data introduced in Example 2, we calculated $\Sigma(x - \bar{x})(y - \bar{y}) = 3740.60$, $\Sigma(y - \bar{y})^2 = 116{,}794$, and $b = 21.321468$ (recall from Chapter 4 that to ensure accuracy in computing SSResid from the computational formula, as many digits as possible should be used in b). The sample size was $n = 16$, so s_e^2 and s_e will be based on $n - 2 = 14$ df. We now have

$$\begin{aligned}\text{SSResid} &= \Sigma(y - \bar{y})^2 - b\Sigma(x - \bar{x})(y - \bar{y}) \\ &= 116{,}794 - (21.321468)(3740.6) \\ &= 37{,}038.92\end{aligned}$$

from which

$$s_e^2 = \frac{37{,}038.92}{14} = 2645.64 \quad \text{and} \quad s_e = \sqrt{2645.64} = 51.44 \; \blacksquare$$

The estimated variance s_e^2 is used in confidence-interval and test-statistic formulas presented in the next several sections.

The estimated standard deviation s_e is a measure of absolute variation in the data expressed in the same unit of measurement as y (e.g., with y = 5-year growth in centimeters, $s_e = 51.44$ cm in Example 4). A small value of s_e suggests that observed points fall quite close to the true line and that the line $\hat{y} = a + bx$ should provide accurate estimates and predictions. The coefficient of determination, first introduced in Chapter 4, is another informative measure for assessing how effectively the simple linear regression model has explained the observed variation in y.

> Let SSTo $= \Sigma(y - \bar{y})^2$, the total sum of squares. Then the **coefficient of determination, r^2,** is given by
>
> $$r^2 = 1 - \frac{\text{SSResid}}{\text{SSTo}}$$
>
> It is the proportion of variation in y that has been explained by the simple linear regression model, and $100r^2$ is the percentage of total variation explained by the model.

Recall that a large value of r^2 does not necessarily imply that s_e is small. If total variation is quite large, most of it could be explained by the model even while there is a substantial amount of variation in absolute terms (large σ as indicated by large s_e). The use of the linear regression model will be most successful when r^2 is quite large and s_e is quite small.

EXAMPLE 5

(*Example 4 continued*) For the tree-growth data, we found that SSResid = 37,038.92, SSTo = 116,794, and s_e = 51.44. The coefficient of determination is

$$r^2 = 1 - \frac{37{,}038.92}{116{,}794} = .683$$

Thus the model explains 68.3% of observed variation, a reasonably large percentage. Yet the value $s_e = 51.44$ suggests a substantial amount of variation in y for any particular value of x. ■

EXAMPLE 6

The Arabian Sea suffers from oxygen depletion to a greater extent than almost any other part of an open ocean. Dentrification, the process by which nitrate and nitrite are reduced to other forms of nitrogen, is one step in the oxidation of organic matter to carbon dioxide and consequent reduction of oxygen. The paper "Evidence for and Rate of Dentrification in the Arabian Sea" (*Deep Sea Research* (1978):431–35) reported on a study in which water samples were selected, and x = salinity (%) and y = nitrate level (μM/L) were determined. A simple linear regression of y on x was then carried out. The accompanying data is a subset of that contained in the article.

x	35.43	36.10	35.74	35.30	35.40	35.91	35.48	36.28
y	30.0	24.2	25.4	29.8	30.7	24.0	28.5	22.7

Based on the summary quantities $n = 8$, $\bar{x} = 35.7050$, $\bar{y} = 26.9125$, $\Sigma(x - \bar{x})^2 = .91320$, $\Sigma(x - \bar{x})(y - \bar{y}) = -7.6745$, and $\Sigma(y - \bar{y})^2 = 70.60875$, regression calculations are

$$\begin{aligned}
b &= \frac{\Sigma(x - \bar{x})(y - \bar{y})}{\Sigma(x - \bar{x})^2} = \frac{-7.6745}{.9132} = -8.403964 \\
a &= \bar{y} - b\bar{x} = 26.9125 - (-8.403964)(35.7050) \\
&= 26.9125 + 300.0635 = 326.976 \\
\hat{y} &= a + bx = 326.976 - 8.404x \\
\text{SSResid} &= \Sigma(y - \bar{y})^2 - b\Sigma(x - \bar{x})(y - \bar{y}) \\
&= 70.60875 - (-8.403964)(-7.6745) \\
&= 70.60875 - 64.49622 \approx 6.113 \\
s_e^2 &= \frac{\text{SSResid}}{n - 2} = \frac{6.113}{6} = 1.019 \qquad s_e = \sqrt{1.019} = 1.009 \\
r^2 &= 1 - \frac{\text{SSResid}}{\text{SSTo}} = 1 - \frac{6.113}{70.60875} = .913
\end{aligned}$$

Thus the simple linear regression model explains 91.3% of total variation in nitrate level by relating it to salinity, and in absolute terms $s_e = 1.009$ is rather small. We conclude that the simple linear regression model provides an effective method for estimating average nitrate level and predicting nitrate level at any specified value of salinity. ■

EXERCISES

11.1 Data presented in the paper "Manganese Intake and Serum Manganese Concentration of Human Milk-Fed and Formula-Fed Infants" (*Amer. J. Clinical Nutr.* (1984):872–78) suggests that a simple linear regression model is reasonable for describing the relationship between y = serum manganese (Mn) and x = Mn intake (μg/kg/day). Suppose that the true regression line is $y = -2 + 1.4x$ and that $\sigma = 1.2$. Then for a fixed x value, y has a normal distribution with mean $-2 + 1.4x$ and standard deviation 1.2.

a. What is the mean value of serum Mn when Mn intake is 4.0? When Mn intake is 4.5?

b. What is the probability that an infant whose Mn intake is 4.0 will have serum Mn greater than 5?

c. Approximately what proportion of infants whose Mn intake is 5 will have a serum Mn greater than 5? Less than 3.8?

11.2 Suppose that a simple linear regression model is appropriate for describing the relationship between y = house price and x = house size (ft^2) for houses in a large city. The true regression line is $y = 23{,}000 + 47x$ and $\sigma = 5000$.

a. What is the average (i.e., expected) change in price associated with one extra square foot of space? With 100 extra square feet of space?

b. What proportion of 1800-ft^2 homes would be priced over \$110,000? Under \$100,000?

11.3 The relationship between x = development age (days) and y = egg diameter (mm) was examined in the paper "Brooding Behavior by the Mountain Dusky Salamander" (*Herpetologica* (1984):105–09). The accompanying data is representative of that given in a scatter plot in the paper. Construct a scatter plot of y versus x. Is the resulting scatter plot consistent with the assumptions of the simple linear regression model? What characteristics of the plot support your answer?

x	5	5	6	8	8	8	9	10	11	15	20
y	2.8	3.0	3.0	3.0	3.1	3.3	3.2	3.4	3.2	3.1	2.9

x	21	21	23	25	25	30	36	40	40	43	43
y	3.5	3.0	3.0	3.4	3.8	3.5	3.8	3.4	3.5	3.7	3.8

x	43	45	45	46	47	48	49	50	50
y	3.8	4.0	4.0	4.0	4.2	4.4	4.4	4.4	4.4

11.4 The authors of the paper "Age, Spacing and Growth Rate of Tamarix as an Indication of Lake Boundary Fluctuations at Sebkhet Kelbia, Tunisia" (*J. Arid Environ.* (1982):43–51) used a simple linear regression model to describe the relationship between y = vigor (average width in centimeters of last two annual rings) and x = stem density (stems/m^2). Data on which the estimated model was based is as follows.

x	4	5	6	9	14	15	15	19	21	22
y	.75	1.20	.55	.60	.65	.55	0	.35	.45	.40

a. Construct a scatter plot for the data.

b. Summary quantities are $\Sigma x = 130$, $\Sigma x^2 = 2090$, $\Sigma y = 5.5$, $\Sigma y^2 = 3.875$, and $\Sigma xy = 59.95$. Find the estimated regression line and draw it on your scatter plot.

c. What is your estimate of the average change in vigor associated with a 1-unit increase in stem density?

d. What would you predict vigor to be for a plant whose density was 17 stems/m^2?

e. Would you use the estimated regression line from (b) to predict vigor when density was 30 stems/m^2? Why or why not?

11.5 Use the salamander egg data of Exercise 11.3 to answer the following questions.

a. What is the equation of the estimated regression line?

b. What would you estimate the mean egg diameter of 10-day-old eggs to be?

c. Predict the diameter of a particular 10-day-old egg.

d. Would you use the estimated regression line of (a) to predict y when $x = 0$? Explain your answer.

11.6 The data below on x = advertising share and y = market share for a particular brand of cigarettes during 10 randomly selected years appeared in the paper "Testing Alternative Econometric Models on the Existence of Advertising Threshold Effect" (*J. Marketing Research* (1984):298–308).

x	.103	.072	.071	.077	.086	.047	.060	.050	.070	.052
y	.135	.125	.120	.086	.079	.076	.065	.059	.051	.039

a. Construct a scatter plot for this data. Do you think the simple linear regression model would be appropriate for describing the relationship between x and y?

b. Calculate the equation of the estimated regression line and use it to obtain the predicted market share when advertising share is .09.

c. Compute r^2. How would you interpret this value?

11.7 Periodic measurements of salinity and water flow were taken in North Carolina's Pamlico Sound, resulting in the given data (*J. Amer. Stat. Assoc.* (1980): 828–38).

Water flow (x)	23	24	26	25	30	24	23	22
Salinity (y)	7.6	7.7	4.3	5.9	5.0	6.5	8.3	8.2

Water flow (x)	22	24	25	22	22	22	24
Salinity (y)	13.2	12.6	10.4	10.8	13.1	12.3	10.4

a. Find the equation of the estimated regression line.

b. What would you predict salinity to be when water flow is 25?

c. Estimate the mean salinity for times when the water flow is 29.

11.8 Using the water data and the estimated regression line from Exercise 11.7, compute SSResid, s_e (the estimate of σ), and r^2. How would you interpret these values? Do you think predictions based on the estimated regression line would be very accurate? Explain.

11.9 The article "Sex Ratio Variation in Odocoileus: A Critical Review" (*J. Wildlife Mgmt.* (1983):573–82) reported the accompanying data on y = percent male fawns and x = fawns/doe for 29 groups of white-tailed, mule, and black-tailed deer. The paper reported an estimated regression line of $\hat{y} = 75.58 - 15.55x$.

x	1.03	1.43	1.79	1.74	0.95	1.47	0.54	1.76	1.76	1.09
y	57.6	60.1	45.7	35.6	68.4	41.5	75.0	51.3	50.0	60.5

x	1.97	1.65	1.69	1.41	1.68	1.26	1.70	1.92	1.25	1.71
y	57.0	44.0	52.3	40.7	58.4	43.8	48.8	52.9	48.6	53.1

x	1.22	1.24	1.44	1.62	1.47	1.64	1.85	1.19	1.32
y	51.9	56.0	51.9	57.6	50.0	44.9	44.1	65.7	59.1

a. Use the given estimated regression line to obtain the residuals. Are there any unusually large residuals?

b. Compute SSResid and use it to obtain an estimate of σ^2.

c. Calculate r^2 for this data set.

d. Interpret the values obtained for s_e^2 and r^2.

11.10 Data on density and vigor of Tamarix was presented in Exercise 11.4. In addition to the summary quantities given there, $\Sigma(y - \bar{y})^2 = .85$.

a. Compute the residuals. Are there any unusually large deviations from the estimated regression line?

b. Estimate σ^2 using s_e^2.

c. Calculate and interpret the value of r^2 for this data set.

11.2 Inferences Concerning the Slope of the Population Regression Line

The slope β in the simple linear regression model is the average or expected change in the dependent variable y associated with a 1-unit increase in the value of the independent variable x. Examples include the average change in vocabulary size associated with an age increase of 1 year, the expected change in yield associated with the use of an additional gram of catalyst, and the average change in annual maintenance expense associated with using a word processing system for one additional hour per week (all presuming that the simple linear regression model is appropriate).

For any specified population, the value of β will be a fixed number, but this value will almost never be known to an investigator. Instead, a sample of n independently selected observations $(x_1, y_1), \ldots, (x_n, y_n)$ will be available, and inferences concerning β are based on this data. In particular, substitution into the formula for b given in Section 11.1 yields a point estimate of β (its most plausible value based on the given sample). As with any point estimate, though, it is desirable to have some indication of how accurately b estimates β. In some situations, the value of the statistic b may vary greatly from sample to sample, so b computed from a single sample is quite likely to be far from β. In other situations, almost all possible samples may yield b values quite close to β, so the error of estimation is almost sure to be small. To proceed further, we need some facts about the sampling distribution of the statistic b—information concerning the shape of the sampling distribution curve, where the curve is centered relative to β, and how much the curve spreads out about its center.

Properties of the Sampling Distribution of b

1. The mean value of b is β. That is, $\mu_b = \beta$, so the sampling distribution of b is always centered at the value of β. Thus b is an unbiased statistic for estimating β.

2. The variance of the statistic b, denoted by σ_b^2, is given by

$$\sigma_b^2 = \frac{\sigma^2}{\Sigma(x - \overline{x})^2}$$

The standard deviation of the statistic b, σ_b, is the square root of σ_b^2.

3. The assumption that the random deviation e in the model has a normal distribution implies that b itself is normally distributed.

The sampling distribution of b is thus described by a normal curve centered at the value of β. The variance σ_b^2—or, equivalently, the standard deviation σ_b—determines the extent to which the curve spreads out about β.

Even though b is an unbiased statistic, if σ_b^2 is large, an estimate (value of b) rather far from β may result. For σ_b^2 to be small, it is necessary that the numerator, σ^2, be small and that the denominator, $\Sigma(x - \overline{x})^2$, be large. The first condition says simply that β will be more accurately estimated when there is little variability about the population regression line (small σ^2) than when there is a great deal of variability in y for fixed x (large σ^2). The second condition has an important implication when the investigator can select x values at which to make observations on y (as would be the case with amounts of catalyst at which yield will be observed). The quantity $\Sigma(x - \overline{x})^2$ is a measure of how much the x values spread out. Since $\Sigma(x - \overline{x})^2$ appears in the denominator of σ_b^2, a greater spread in x values results in smaller σ_b^2 and greater accuracy in estimation. Of course, if the x values are spread out too much, the simple linear regression model may no longer be valid at the extremes (e.g., vocabulary size does not continue to increase linearly with age among teenagers—and many parents think it begins to decrease). Figure 5 provides intuition into why spreading out the x values yields precision in estimating β. When x values are close, small changes in observed y's can drastically change b. Such changes in y change b only slightly when x values are spread out.

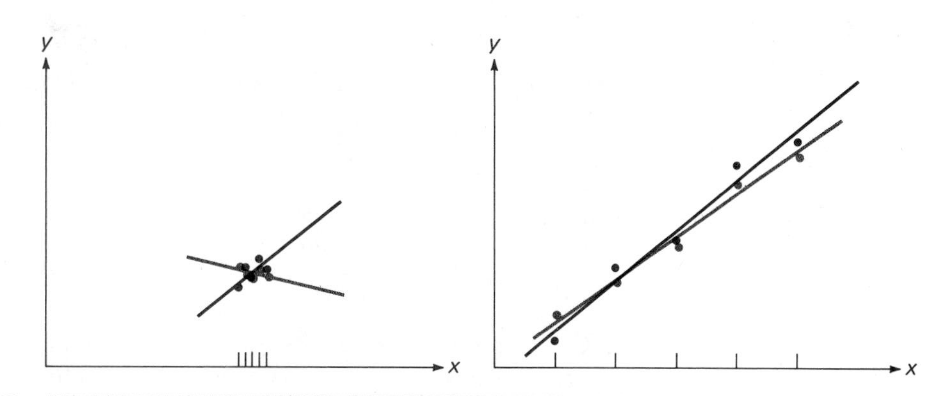

FIGURE 5

THE EFFECT ON b OF CHANGES IN THE y VALUES
(a) When x Values are Close Together, Small Changes in y's Substantially Change b
(b) When x Values are Spread Out, Small Changes in y's Have Little Effect on b

EXAMPLE 7

Reconsider the situation described in Example 1, in which x = accelerated strength of concrete and y = 28-day cured strength. Suppose that the simple linear regression model is valid for x between 1000 and 4000 and that $\beta = 1.25$ and $\sigma = 350$. Consider first an experiment in which $n = 7$ and the x values at which observations are made are $x_1 = 1000$, $x_2 = 1500$, $x_3 = 2000$, $x_4 = 2500$, $x_5 = 3000$, $x_6 = 3500$, and $x_7 = 4000$. Then $\bar{x} = 2500$ and $\Sigma(x - \bar{x})^2 = 7{,}000{,}000$, so $\sigma_b^2 = \sigma^2/\Sigma(x - \bar{x})^2 = (350)^2/7{,}000{,}000 = .0175$, and $\sigma_b = \sqrt{.0175} = .132$. Thus the statistic b has a normal probability distribution with mean value $\mu_b = \beta = 1.25$ and standard deviation $\sigma_b = .132$. The chance that such an experiment results in an estimated slope b that is between 1.00 and 1.50 is obtained by standardizing these two limits and using the z table. Since $(1.00 - 1.25)/.132 = -1.9$ and $(1.50 - 1.25)/.132 = 1.9$,

$$P(1.00 < b < 1.50) = \begin{pmatrix}\text{area under the } z \text{ curve} \\ \text{between } -1.9 \text{ and } 1.9\end{pmatrix}$$
$$= .9713 - .0287 = .9426$$

That is, roughly 94% of all samples resulting from this experiment yield an estimated slope that is within .25 of the population value $\beta = 1.25$.

Now consider an experiment in which $n = 11$ and a single y observation is made at x values 2000, 2100, 2200, 2300, 2400, 2500, 2600, 2700, 2800, 2900, and 3000. For this experiment, $\bar{x} = 2500$ and $\Sigma(x - \bar{x})^2 = 1{,}100{,}000$, so $\sigma_b^2 = (350)^2/1{,}100{,}000 = .1114$ and $\sigma_b = .334$. Even though the sample size is larger here than in the earlier experiment, σ_b is more than twice the previous value of .132. This is because the 7 values in the first experiment are much more spread out than the 11 values in the second one. It is easily checked that $P(1.00 < b < 1.50) = .5160$, whereas $P(b < .50$ or $b > 2.00) = 1 - P(.50 \le b \le 2.00) = .0278$. Almost 3% of all samples taken as described result in an estimate of β that is more than .75 from the true value 1.25. ■

Calculating σ_b^2 requires that the value of σ^2 be known. As with α and β, this is virtually never the case. However, replacing σ^2 in the expression for σ_b^2 by its estimate s_e^2 gives the estimated variance of the statistic b.

The **estimated variance of the statistic *b***, denoted by s_b^2, is given by

$$s_b^2 = \frac{s_e^2}{\Sigma(x - \bar{x})^2}$$

where $s_e^2 = \text{SSResid}/(n - 2)$. The **estimated standard deviation of *b***, denoted by s_b, is the square root of s_b^2.

In Chapters 7 and 8 we used $s_{\bar{x}} = s/\sqrt{n}$, the estimated standard deviation of the statistic $\bar{x}$, to construct a confidence interval for μ and test hypotheses concerning μ. The key result there was that the sampling distribution of the standardized variable $t = (\bar{x} - \mu)/(s/\sqrt{n})$ was a t distribution with $n - 1$ df. In the same way, s_b will be used to construct a confidence interval for β

and test hypotheses concerning β. Here is the result on which these procedures are based.

> The probability distribution of the standardized variable
>
> $$t = \frac{b - \beta}{s_b}$$
>
> is the t distribution with $n - 2$ df.

A Confidence Interval for β

The derivation of a t confidence interval for β starting from the above result parallels the argument that led to the t interval for μ, so we omit the details.

> A **confidence interval for β,** the slope of the population regression line, has the form
>
> $$b \pm (t \text{ critical value}) \cdot s_b$$
>
> where $s_b = \sqrt{s_e^2/\Sigma(x - \overline{x})^2}$ and the t critical value is based on $n - 2$ df. Table IV gives critical values corresponding to the most frequently used confidence levels.

Notice how similar this interval is to the interval $\overline{x} \pm (t \text{ critical value})(s/\sqrt{n})$ for μ. This interval is centered at the point estimate $\overline{x}$ for μ, and the interval for β is also centered at the corresponding point estimate b. Each interval extends out from the estimate an amount that depends on the variability of the statistic used to estimate the population characteristic of interest. When s_b is small, the interval will be narrow and the investigator will have rather precise knowledge of β.

EXAMPLE 8

Durable-press cotton fabric is produced by a chemical reaction involving formaldehyde. For economic reasons, finished fabric usually receives its first wash at home rather than at the manufacturing plant. Because the pH of in-home wash water varies greatly from location to location, textile researchers are interested in how pH affects different fabric properties. The paper "Influence of pH in Washing on the Formaldehyde-Release Properties of Durable-Press Cotton (*Textile Research J.* (1981): 263–70) reported the accompanying data, read from a scatter plot, on x = wash-water pH and y = formaldehyde release (in ppm). The scatter plot suggested the appropriateness of the simple linear regression model. The slope β in this context is the average change in formaldehyde release associated with a 1-unit pH increase.

Observation	1	2	3	4	5	6	7	8	9
x	5.3	6.8	7.1	7.1	7.2	7.6	7.6	7.7	7.7
y	545	770	780	790	680	760	790	795	935

Observation	10	11	12	13	14	15	16	17	18
x	7.8	7.9	8.1	8.6	9.1	9.2	9.4	9.4	9.5
y	780	935	830	1015	1190	1030	1045	1250	1075

Summary quantities are $\bar{x} = 7.9500$, $\bar{y} = 888.6111$, $\Sigma(x - \bar{x})^2 = 20.685$, $\Sigma(x - \bar{x})(y - \bar{y}) = 3121.25$, and $\Sigma(y - \bar{y})^2 = 568{,}340.2801$. Then $b = 3121.25/20.685 = 150.894368$ and $a = 888.6111 - (150.894368)(7.95) = -310.9991$, so the estimated regression line is $\hat{y} = -311.00 + 150.894x$. Notice that use of this line for the pH value $x = 2$ yields a predicted formaldehyde release of -9.21, which is impossible. The smallest x value in the sample is 5.3, so it would be dangerous to extrapolate the approximate linear relationship much below this value.

The computational formula for residual sum of squares yields SSResid $= \Sigma(y - \bar{y})^2 - b\Sigma(x - \bar{x})(y - \bar{y}) = 568{,}340.2801 - (150.894368)(3121.25) = 97{,}361.234$. Then $s_e^2 = (97{,}361.234)/(18 - 2) = 6085.077$, $s_b^2 = s_e^2/\Sigma(x - \bar{x})^2 = 6085.077/20.685 = 294.178$, and $s_b = \sqrt{294.178} = 17.152$. Relative to the magnitude of b itself, this estimated standard deviation is not particularly large. The 95% confidence interval based on 16 df requires t critical value $= 2.12$. The resulting interval is

$$b \pm (t \text{ critical value}) \cdot s_b = 150.89 \pm (2.12)(17.15)$$
$$= 150.89 \pm 36.36 = (114.53, 187.25)$$

The investigator can be quite confident that formaldehyde release will increase on average by between 114.53 ppm and 187.25 ppm when pH is increased by 1 unit. ■

Output from any of the standard statistical computer packages routinely includes the computed values of a, b, SSResid, s_e, SSTo, r^2, and s_b. Figure 6 displays MINITAB output for the data of Example 8. The format from other

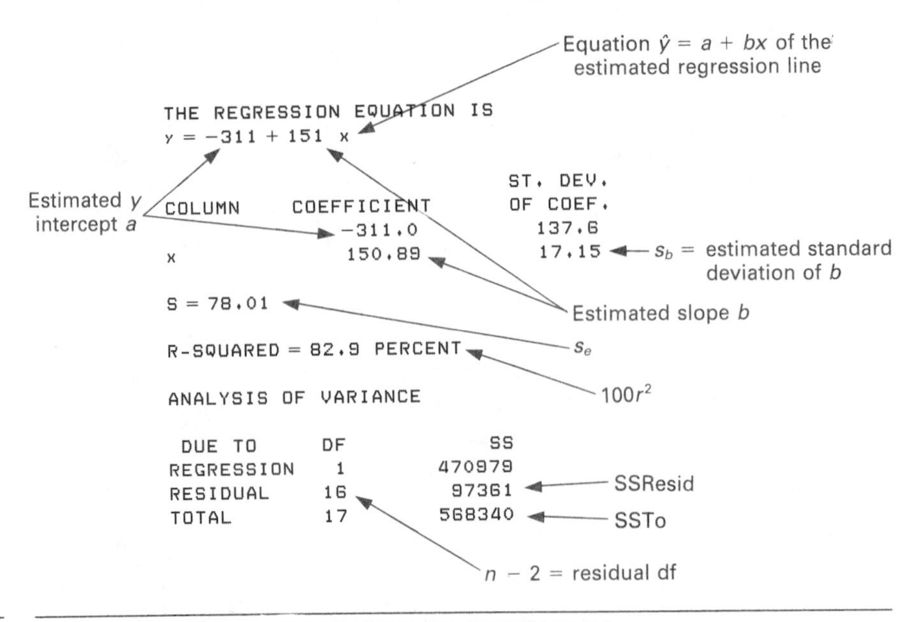

FIGURE 6 PARTIAL MINITAB OUTPUT FOR THE DATA OF EXAMPLE 8

packages is very similar. Rounding will occasionally lead to small discrepancies between hand-calculated and computer-calculated values, but there are no such discrepancies for this example.

Hypothesis Tests Concerning β

Hypotheses about β can be tested using a t test very similar to the t tests discussed in earlier chapters. The null hypothesis states that β has a specific hypothesized value. The t statistic results from standardizing b, the point estimate of β, under the assumption that H_0 is true. The sampling distribution of this statistic when H_0 is true is the t distribution with $n - 2$ df. The level of significance (type I error probability) is then controlled through the use of an appropriate t critical value from Table IV.

Summary of Hypothesis Tests Concerning β

Null hypothesis: $H_0 : \beta = \text{hypothesized value}$

Test statistic: $t = \dfrac{b - \text{hypothesized value}}{s_b}$

Alternative hypothesis	*Rejection region*
$H_a : \beta >$ hypothesized value	$t > t$ critical value
$H_a : \beta <$ hypothesized value	$t < -t$ critical value
$H_a : \beta \neq$ hypothesized value	Either $t > t$ critical value or $t < -t$ critical value

EXAMPLE 9

When machine speed can be varied in a manufacturing setting, interest frequently centers on how speed is related to various other process characteristics. The article "Gas Turbines for Process Improvement of Industrial Thermal Power Plants" (*Combustion* (April 1976):35–41) presented the accompanying data for a high-performance tissue machine used by paper mills to process paper. The independent variable x is machine speed and the dependent variable y is temperature in the drying hood.

x	1000	1100	1200	1250	1300	1400	1450
y	220	280	350	375	450	470	500

It is easily verified that $b = .640141$, $a = -417.75$, $s_e = 17.59$, and $r^2 = .976$. A scatter plot and the very high r^2 value strongly support using the simple linear regression model to describe the relationship between x and y.

Suppose that the machine had been designed so that the average temperature change associated with a 1-unit increase in speed was at most .5 (i.e., $\beta \leq .5$). Does the data suggest that this design specification has not been met? Let's carry out a test of the implied hypotheses using level of significance .01. In the statement of the problem, the alternative hypothesis will be the claim that the specification is violated, since prior belief is in the validity of the specification.

1. Population characteristic: β = the average change in hood temperature associated with a 1-unit increase in machine speed.
2. Null hypothesis: $H_0 : \beta = .5$.
3. Alternative hypothesis: $H_a : \beta > .5$.
4. Test statistic:

$$t = \frac{b - .5}{s_b}$$

5. Rejection region: $t > t$ critical value; since $n = 7$, the test is based on $n - 2 = 5$ df, and Table IV gives the one-tailed level .01 critical value 3.37.
6. Computations: the estimated standard deviation of the statistic b is

$$s_b = \sqrt{s_e^2 \,/\, \Sigma(x - \overline{x})^2} = \sqrt{(17.59)^2/152{,}142.860} = .0451, \text{ so}$$

$$t = \frac{.640 - .5}{.0451} = \frac{.140}{.0451} = 3.10$$

7. The computed test statistic value 3.10 does not exceed the critical value 3.37, so H_0 is not rejected. At level of significance .01, the data does not strongly suggest a specification violation. Further examination of the 5 df row of the t table yields the bounds $.01 < P\text{-value} < .025$. ■

A Test for Model Utility

The population regression line $y = \alpha + \beta x$ gives the average value of y for any particular x value. When $\beta = 0$, this average does not depend on x, so knowledge of x provides no information about y. The simple linear regression model specifies a useful relationship between x and y only if β is something other than zero. This suggests that before the model is employed to draw conclusions about y, the hypothesis $H_0 : \beta = 0$ should be tested. If this null hypothesis cannot be rejected at a reasonably small level of significance, the investigator must continue to search for a useful alternative model. Such a model may involve either a nonlinear relationship or the introduction of additional independent variables (multiple regression).

DEFINITION

The **model utility test** in simple linear regression is a test of the hypothesis $H_0 : \beta = 0$. The usual alternative hypothesis is $H_a : \beta \neq 0$, which says that there is a useful linear relationship between x and y. The test statistic is then the ***t* ratio** $t = b/s_b$ based on $n - 2$ df.

EXAMPLE 10

Dairy scientists have recently carried out several studies on protein biosynthesis in milk and the accompanying decomposition of nucleic acids into various constituents. The paper "Metabolites of Nucleic Acids in Bovine Milk" (*J. Dairy Sci.* (1984):723–28) reported the accompanying data on x = milk production (kg/day) and y = milk protein (kg/day) for Holstein-Friesan cows.

Observation	1	2	3	4	5	6	7
x	42.7	40.2	38.2	37.6	32.2	32.2	28.0
y	1.20	1.16	1.07	1.13	.96	1.07	.85

Observation	8	9	10	11	12	13	14
x	27.2	26.6	23.0	22.7	21.8	21.3	20.2
y	.87	.77	.74	.76	.69	.72	.64

To see whether the simple linear regression model provides useful information for the prediction of protein secretion from knowledge of milk production, let's carry out the model utility test at level of significance .01.

1. β = the average change in protein secretion associated with a 1-kg increase in milk production.
2. $H_0 : \beta = 0$.
3. $H_a : \beta \neq 0$.
4. Test statistic:

$$t = \frac{b - \text{hypothesized value}}{s_b} = \frac{b}{s_b}$$

5. From Table IV, the two-tailed t critical value for a level .01 test based on 12 df is 3.06, so H_0 will be rejected in favor of H_a if either $t > 3.06$ or $t < -3.06$.
6. The estimated regression coefficients are $b = .024576$ and $a = .175571$, and SSResid = .021200, $s_e = .0420$, SSTo = .481436, and $r^2 = .956$ (this very large r^2 value certainly suggests a useful model). With $\Sigma(x - \overline{x})^2 = 762.012140$, $s_b = .00152$, and the computed t ratio is $t = .0246/.00152 = 16.18$.
7. Since 16.18 greatly exceeds the upper-tailed critical value 3.06, H_0 is easily rejected in favor of H_a. The utility of the simple linear regression model is clearly confirmed. ■

Several commonly used statistical computer packages automatically compute both the value of the t ratio b/s_b and the P-value for the alternative hypothesis H_a: $\beta \neq 0$. The utility of the model can then be judged simply by comparing the P-value to the chosen level of significance, as discussed in earlier chapters.

EXERCISES

11.11 Explain the difference between b and β.

11.12 The accompanying summary quantities were computed using representative values from data collected in a study of y = pavement deflection and x = temperature (°F) (*Transpor. Engr. J.* (1977):75–85).

$$n = 15 \qquad \Sigma x = 1425 \qquad \Sigma y = 10.68$$
$$\Sigma x^2 = 139{,}037.25 \qquad \Sigma y^2 = 7.8518 \qquad \Sigma xy = 987.645$$

a. What is the equation of the estimated regression line?
b. Construct a 95% confidence interval for β and interpret the resulting interval.
c. Would a 99% confidence interval for β have been narrower or wider than the interval in (b)? Why?

11.13 The paper "Bumblebee Response to Variation in Nectar Availability" (*Ecology* (1981):1648–61) reported a positive linear relationship between y, a measure of bumblebee abundance, and x, a measure of nectar availability. Representative data is given.

x	3	8	11	10	23	23	30	35
y	4	6	12	18	11	24	22	37

a. Assuming that the simple linear regression model is valid, estimate β, the true average change in abundance associated with a 1-unit increase in availability, using a 90% confidence interval.

b. Does the interval in (a) support the paper's claim of the existence of a positive linear relationship between x and y? Explain.

11.14 Data on x = development age (days) and y = egg diameter (mm) for salamander eggs was given in Exercise 11.3. Summary quantities computed from this data are:

$$n = 31 \quad \Sigma(x - \bar{x})^2 = 8191.42 \quad \Sigma(y - \bar{y})^2 = 7.48$$
$$\Sigma(x - \bar{x})(y - \bar{y}) = 217.26$$

a. Calculate b, the point estimate of β.

b. Construct a 95% confidence interval for the expected change in egg diameter associated with a 1-day increase in age.

c. Based on your interval in (b), would you conclude that there is a linear relationship between age and diameter? Explain.

11.15 The paper "Effects of Enhanced UV-B Radiation on Ribulose-1, 5-Biphosphate, Carboxylase in Pea and Soybean" (*Environ. and Exper. Botany* (1984): 131–43) included the accompanying pea plant data, with y = sunburn index and x = distance (cm) from an ultraviolet light source.

x	18	21	25	26	30	32	36	40	40	50	51	54	61	62	63
y	4.0	3.7	3.0	2.9	2.6	2.5	2.2	2.0	2.1	1.5	1.5	1.5	1.3	1.2	1.1

Estimate the mean change in the sunburn index associated with an increase of 1 cm in distance using a 99% confidence interval.

11.16 Data on per capita availability of crude and dietary fiber is given in the paper "Estimation of Per Capita Crude and Dietary Fiber Supply in 38 Countries" (*Am. J. Clin. Nutr.* (1984):821–29). Let x and y denote crude and dietary fiber availability, respectively (both measured in g/day). The following summary quantities were computed using the 38 observations given in the paper: $\bar{x} = 142.5$, $\Sigma x^2 = 1{,}024{,}625.83$, $\bar{y} = 473.0$, $\Sigma y^2 = 11{,}492{,}957.01$, $\Sigma xy = 2{,}961{,}190.58$. Using a significance level of .05, test to determine if the simple linear regression model provides useful information for predicting dietary fiber availability from knowledge of crude fiber availability.

11.17 The accompanying data on x = soil pH and y = Cl^- ion retention (mL/100 g) is from the paper "Single Equilibration Method for Determination of Cation and Anion Retention by Variable Charge Soils" (*Soil Sci. Plant Nutr.* (1984): 71–76).

x	6.15	6.11	5.88	6.45	5.80	6.06	5.83	6.33	7.35
y	0.14	0.37	1.47	1.12	2.08	1.79	3.18	2.15	0.51

x	8.18	7.69	7.29	6.53	5.01	5.34	6.19	5.81
y	0.32	0.76	2.13	2.75	6.69	5.59	2.87	4.22

a. Compute the slope and intercept of the estimated regression line.

b. Does the data indicate that the simple linear regression model is useful for predicting ion retention from pH? Use a level .01 test.

c. What can you say about the P-value associated with the computed value of the test statistic in (b)?

11.18 The paper "Technology, Productivity and Industry Structure" (*Techn. Forecasting and Social Change* (1983):1–13) included the accompanying data on x = research and development expenditure and y = growth rate for eight different industries.

x	1.90	3.96	2.44	.88	.37	−.90	.49	1.01
y	2024	5038	905	3572	1157	327	378	191

a. Would a simple linear regression model provide useful information for predicting growth rate from research and development expenditure? Use a .05 level of significance.

b. What can be said about the P-value associated with the test statistic value in (a)?

c. Use a 90% confidence interval to estimate the average change in growth rate associated with a 1-unit increase in expenditure. Interpret the resulting interval.

11.19 The accompanying data on fish survival and ammonia concentration is taken from the paper "Effects of Ammonia on Growth and Survival of Rainbow Trout in Intensive Static-Water Culture" (*Trans. Amer. Fisheries Soc.* (1983):448–54). Let x = ammonia exposure (mg/L) and y = percent survival.

x	10	10	20	20	25	27	27	31	50
y	85	92	85	96	87	80	90	59	62

a. Estimate the slope and intercept of the true regression line.

b. Is the simple linear regression model useful for predicting survival from knowledge of ammonia exposure?

c. Predict percent survival when ammonia exposure is 30.

d. Estimate β using a 90% confidence interval.

11.20 It is possible to construct a confidence interval for α, the intercept of the true regression line. The y intercept of the least squares line, a, is a point estimate of α. It can be shown that $t = (a - \alpha)/s_a$ has a t distribution with $n - 2$ df. The estimated standard deviation of a is the square root of s_a^2, where $s_a^2 = s_e^2(\Sigma x^2)/n[\Sigma(x - \bar{x})^2]$. A confidence interval for α is then

$$a \pm (t \text{ critical value})s_a$$

Use the data from Exercise 11.19 to construct a 95% confidence interval for α.

11.21 A paper in the 1943 issue of the *Journal of Experimental Psychology* entitled "A Study of the Relationship Between Hypnotic Susceptibility and Intelligence" reported on a study in which for each of 32 subjects, both an intelligence score x and a hypnotic susceptibility score y were obtained. The computed values for summary quantities were:

$$\Sigma(x - \bar{x})^2 = 4929.22 \qquad \Sigma(y - \bar{y})^2 = 1531.88$$
$$\Sigma(x - \bar{x})(y - \bar{y}) = 1192.69$$

Assuming that the simple linear regression model is appropriate, test $H_0 : \beta = 0$ versus $H_a : \beta > 0$ to determine if there is a *positive* relationship between intelligence and hypnotic susceptibility. Use level of significance .01.

11.22 In anthropological studies, an important characteristic of fossils is cranial capacity. Frequently skulls are at least partially decomposed, so it is necessary to use other characteristics to obtain information about capacity. One such measure that has been used is the length of the lambda-opisthion chord. A paper that appeared in the 1971 *American Journal of Physical Anthropology*, entitled "Vertesszollos and the Presapiens Theory," reported the following data for $n = 7$ *Homo erectus* fossils.

x (chord length in mm)	78	75	78	81	84	86	87
y (capacity in cm^3)	850	775	750	975	915	1015	1030

Suppose that from previous evidence, anthropologists had believed that for each 1-mm increase in chord length, cranial capacity would be expected to increase by 20 cm^3. Does this new experimental data strongly contradict prior belief? That is, should $H_0 : \beta = 20$ be rejected in favor of $H_a : \beta \neq 20$? Use a .05 level of significance.

11.23 The article "Hydrogen, Oxygen, and Nitrogen in Cobalt Metal" (*Metallurgia* (1969):121–27) contains a plot of the following data pairs, where x = pressure of extracted gas (in microns) and y = extraction time (min).

x	40	130	155	160	260	275	325	370	420	480
y	2.5	3.0	3.1	3.3	3.7	4.1	4.3	4.8	5.0	5.4

a. Suppose that the investigators had believed prior to the experiment that $\beta = .006$. Does the data contradict this prior belief? Use a significance level of .10.

b. Give an upper and/or lower bound on the P-value associated with the test statistic value in (a).

c. Compute and interpret a 95% confidence interval for the slope of the true regression line.

11.24 Suppose that the unit of measurement for the independent variable x is changed through multiplication by a *scale factor c*, so that each x value becomes cx. For example, in Exercise 11.19, ammonia content can be changed from mg/L to g/L by multiplying each x value by .001.

a. How does this scale change affect $\Sigma(x - \bar{x})(y - \bar{y})$, $\Sigma(x - \bar{x})^2$, and b? (*Hint:* When each x is multiplied by c, $\bar{x}$ is also.)

b. Such a scale change does not affect y at all, so SSResid and s_e are unchanged. How does the scale change affect s_b? (*Hint:* $\Sigma(x - \bar{x})^2$ appears in s_b.)

c. Use the results of (a) and (b) to show that the t ratio b/s_b for testing model utility is not affected by a change in x units. Is this a reasonable property for the test statistic to have?

11.25 **a.** Suppose that a single y observation is made at each of the x values 5, 10, 15, 20, and 25. If $\sigma = 4$, what is σ_b, the standard deviation of the statistic b?

b. Now suppose that a second y observation is made at every x value listed in (a) (a total of 10 observations). Is the resulting value of σ_b half of what it was in (a)?

c. How many observations at each x value in (a) are required to yield a σ_b value that is half the value calculated in (a)? Verify your conjecture.

11.26 In some studies an investigator has n (x, y) pairs sampled from one population and m (x, y) pairs from a second population. Let β and β' denote the slopes of the first and second population lines, respectively, and let b and b' denote the estimated slopes calculated from the first and second samples, respectively. The investigator may then wish to test the null hypothesis $H_0 : \beta - \beta' = 0$ (i.e., $\beta = \beta'$) against an appropriate alternative hypothesis. Suppose that σ^2, the variance about the population line, is the same for both populations. Then this common variance can be estimated by

$$s^2 = \frac{\text{SSResid} + \text{SSResid}'}{n + m - 4}$$

where SSResid and SSResid′ are the residual sums of squares for the first and second samples, respectively. With SS_x and SS'_x denoting $\Sigma(x - \bar{x})^2$ for the first and second samples, respectively, the test statistic is

$$t = \frac{b - b'}{\sqrt{\dfrac{s^2}{SS_x} + \dfrac{s^2}{SS'_x}}}$$

When H_0 is true, this statistic has a t distribution based on $n + m - 4$ df.

The given data is a subset of the data in the paper "Diet and Foraging Mode of *Bufo marinus* and *Leptodactylus ocellatus*" (*J. Herpetology* (1984):138–46). The independent variable x is body length (cm) and the dependent variable y is mouth width (cm), with $n = 9$ observations for one type of nocturnal frog and $m = 8$ observations for a second type. Test at level .05 to see whether or not the slopes of the true regression lines for the two different frog populations are identical.

Leptodactylus ocellatus	*x*	3.8	4.0	4.9	7.1	8.1	8.5	8.9	9.1	9.8
	y	1.0	1.2	1.7	2.0	2.7	2.5	2.4	2.9	3.2
Bufa marinus	*x*	3.8	4.3	6.2	6.3	7.8	8.5	9.0	10.0	
	y	1.6	1.7	2.3	2.5	3.2	3.0	3.5	3.8	

11.27 Referring to Exercise 11.19, suppose the researcher's objective had been to estimate β as accurately as possible. Would an experiment with $n = 8$ and $x_1 = 10$, $x_2 = 10$, $x_3 = 10$, $x_4 = 30$, $x_5 = 30$, $x_6 = 50$, $x_7 = 50$, and $x_8 = 50$ have been preferable to the one carried out? Explain.

11.3 Inferences Based on the Estimated Regression Line

We have seen how the estimated regression line $\hat{y} = a + bx$ gives, for any specified x value, either an estimate of the corresponding average y value or a prediction of a single y value. How precise is the resulting estimate or prediction—that is, how close might $a + bx$ be to the actual mean value $\alpha + \beta x$ or to a particular y observation? Because both a and b vary in value from sample to sample (each one is a statistic), for fixed x the statistic $a + bx$ also has different values for different samples. The way in which this statistic varies in value with different samples is summarized by its sampling distribution. Properties of the sampling distribution are used to obtain both a confidence interval formula for $\alpha + \beta x$ and a prediction interval formula for a particular y observation. The narrowness of the corresponding interval conveys information about the precision of the estimate or prediction.

Properties of the Sampling Distribution of $a + bx$ for a Fixed x Value

Let x^* denote a particular value of the independent variable x. Then the sampling distribution of the statistic $a + bx^*$ has the following properties:

1. The mean value of $a + bx^*$ is $\alpha + \beta x^*$, so $a + bx^*$ is an unbiased statistic for estimating the average y value when $x = x^*$. That is, for any x^*, the sampling distribution of $a + bx^*$ is centered at $\alpha + \beta x^*$.
2. The variance of the statistic $a + bx^*$, denoted by $\sigma^2_{a+bx^*}$, is given by

$$\sigma^2_{a+bx^*} = \sigma^2\left[\frac{1}{n} + \frac{(x^* - \bar{x})^2}{\Sigma(x - \bar{x})^2}\right]$$

The standard deviation, σ_{a+bx^*}, is the square root of this expression.

3. The assumption that the random deviation e in the model has a normal distribution implies that $a + bx^*$ is normally distributed.

The expression for $\sigma^2_{a+bx^*}$ depends on σ^2 (which reflects variation about the population line), the sample size n, and the difference between x^* and $\bar{x}$. As σ^2 increases, so does $\sigma^2_{a+bx^*}$—greater variability about the true regression line implies greater variability in the estimate, or prediction, $a + bx^*$. In general, increasing n gives greater precision in $a + bx^*$ (though a larger sample size might actually give less precision than a smaller one if the x values in the larger sample are close together and those in the smaller sample are quite spread out). Finally, the average $\bar{x}$ locates the center of the x values at which observations were made. The closer x^* is to $\bar{x}$—i.e., the nearer the specified x value is to the center of the data—the smaller will be $(x^* - \bar{x})^2$ and thus $\sigma^2_{a+bx^*}$. That is, there is less variability in $a + bx^*$ when x^* is close to $\bar{x}$ than when x^* is far from $\bar{x}$.

EXAMPLE 11

In Examples 1 and 7 we considered the regression of y = 28-day strength of concrete on x = accelerated strength. There we assumed that $\alpha + \beta x = 1800 + 1.25x$ and $\sigma = 350$ ($\sigma^2 = 122{,}500$). Suppose that $n = 7$ observations are made at the equally spaced x values $x_1 = 1000$, $x_2 = 1500, \ldots, x_7 = 4000$, so that $\bar{x} = 2500$ and $\Sigma(x - \bar{x})^2 = 7{,}000{,}000$. For $x^* = 2000$, the statistic $a + b(2000)$ has a normal distribution with mean value $\alpha + \beta(2000) = 1800 + (1.25)(2000) = 4300$ and standard deviation

$$\sigma_{a+b(2000)} = \sqrt{(122{,}500)\left[\frac{1}{7} + \frac{(2000 - 2500)^2}{7{,}000{,}000}\right]} = \sqrt{21{,}875.00} = 147.90$$

Since it is highly likely that a normally distributed variable falls within 2 standard deviations of its mean value and $2(147.90) \approx 300$, most samples with $n = 7$ and the x values just described yield values of a and b for which the estimate $a + b(2000)$ is within 300 of the true mean value 4300.

The x value 3600 is further from $\bar{x}$ than is 2000, so $\sigma_{a+b(3600)}$ is larger than $\sigma_{a+b(2000)}$. The statistic $a + b(3600)$ has a normal distribution with mean value $\alpha + \beta(3600) = 6300$ and standard deviation

$$\sigma_{a+b(3600)} = \sqrt{(122{,}500)\left[\frac{1}{7} + \frac{(3600 - 2500)^2}{7{,}000{,}000}\right]} = 196.66$$

There is a substantial amount of variability in the statistic $a + b(3600)$ because σ is large, n is small, and 3600 is rather far from $\bar{x}$. ■

Replacing σ^2 by its estimate s_e^2 in the formula for $\sigma^2_{a+bx^*}$ yields the estimated variance of $a + bx^*$.

The **estimated variance of the statistic $a + bx^*$**, denoted by $s^2_{a+bx^*}$, is given by

$$s^2_{a+bx^*} = s_e^2 \left[\frac{1}{n} + \frac{(x^* - \bar{x})^2}{\Sigma(x - \bar{x})^2}\right]$$

The **estimated standard deviation of $a + bx^*$**, denoted by s_{a+bx^*}, is the square root of $s^2_{a+bx^*}$.

The key result on which a confidence interval for $\alpha + \beta x^*$ is based (as well as procedures for testing hypotheses about $\alpha + \beta x^*$) involves the standardization of $a + bx^*$ using this estimated standard deviation.

The probability distribution of the standardized variable

$$t = \frac{a + bx^* - (\alpha + \beta x^*)}{s_{a+bx^*}}$$

is the t distribution with $n - 2$ df.

Inferences About the Mean Value $\alpha + \beta x^*$

Previous z and t standardized variables were manipulated to give confidence intervals of the form (point estimate) ± (critical value)(estimated standard deviation). An analogous argument leads immediately to the following interval.

A **confidence interval for $\alpha + \beta x^*$**, the average y value when x has value x^*, is

$$a + bx^* \pm (t \text{ critical value}) \cdot s_{a+bx^*}$$

where the t critical value is based on $n - 2$ df. Table IV gives critical values corresponding to the most frequently used confidence levels.

EXAMPLE 12

Example 2 presented a sample of $n = 16$ observations on x = tree age (years) and y = subsequent 5-year growth (cm). A scatter plot shows a pronounced linear pattern, $r^2 = .683$, and the model utility test of Section 11.2 rejects $H_0 : \beta = 0$ in favor of $H_a : \beta \neq 0$ at any reasonable significance level. The simple linear regression model appears useful for extracting information about 5-year growth from knowledge of age. The sampled x values range from 5 to 18, so it is reasonable to compute a confidence interval for average growth when $x = 10$. A 95% confidence interval requires the following quantities: $\bar{x} = 12.3125$, $\Sigma(x - \bar{x})^2 = 175.4382$, $a = 4.35$, $b = 21.32$, $s_e^2 = 2645.64$ (from Example 4), and t critical value = 2.15 (based on $16 - 2 = 14$ df). Then $a + bx^* = 4.35 + (21.32)(10) = 217.55$ cm, and the estimated standard deviation of $a + bx^*$ is

$$s_{a+b(10)} = \sqrt{(2645.64)\left(\frac{1}{16} + \frac{(10 - 12.3125)^2}{175.4382}\right)} = 15.68$$

Substitution into the confidence interval formula yields

$$217.55 \pm (2.15)(15.68) = 217.55 \pm 33.71 = (183.84, 251.26)$$

We can be quite confident that average 5-year growth for all 10-year-old trees of the type included in the study is between 183.84 and 251.26 cm. The width of this interval emphasizes that the point estimate 217.55 is not very precise. The reason for this is clearly the relatively large value of s_e. With this much inherent variability about the regression line, precise estimation of $\alpha + \beta x^*$ requires many more than 16 observations. ■

EXAMPLE 13

(*Example* 10 *continued*) The $n = 14$ observations on x = milk production and y = milk protein yielded x values between 20.2 and 42.7 with $\bar{x} = 29.564$. Let's use the data to compute a 99% confidence interval for average milk protein when milk production is 30 kg/day—i.e., for $\alpha + \beta(30)$. Since $a = .176$ and $b = .0246$, the point estimate of average milk protein when $x = 30$ is $a + b(30) = .914$. The estimated standard deviation of $a + b(30)$ is

$$s_{a+b(30)} = \sqrt{s_e^2\left(\frac{1}{n} + \frac{(30 - \bar{x})^2}{\Sigma(x - \bar{x})^2}\right)} = \sqrt{(.001767)\left(\frac{1}{14} + \frac{(.436)^2}{762.012}\right)}$$

$$= \sqrt{(.001767)(.071429 + .000249)} = .0113$$

Because 30 is so close to $\bar{x}$, the contribution of the term involving $(30 - \bar{x})^2$ is negligible compared to $1/n$. This would not be the case for an x value such as 40.

The t critical value for 99% confidence based on 12 df is 3.06. The confidence interval is

$$a + b(30) \pm (t \text{ critical value}) \cdot s_{a+b(30)} = .914 \pm (3.06)(.0113)$$
$$= .914 \pm .035 = (.879, .949)$$

Even with the very high confidence level, this interval is relatively narrow, partly because $x = 30$ is very close to $\bar{x}$ and partly because the model fits the data so well ($s_e = .0420$ and $r^2 = .956$). ■

Some statistical computer packages, such as SAS, automatically compute the confidence interval limits as soon as the user inputs x^*. Other packages, such as MINITAB, compute both $a + bx^*$ and s_{a+bx^*} upon request. Once the t critical value is obtained from a table, the interval is easily computed.

The standardized t variable just given leads immediately to a test statistic for testing hypotheses about $\alpha + \beta x^*$. If the null hypothesis is H_0 : $\alpha + \beta x^* =$ hypothesized value, replacing $\alpha + \beta x^*$ in the numerator of t by the hypothesized value yields a t statistic based on $n - 2$ df. For example, consider the null hypothesis that average temperature (y) equals 400 when machine speed (x) is 1200—that is, $H_0 : \alpha + \beta(1200) = 400$. The appropriate test statistic is $t = [a + b(1200) - 400]/s_{a+b(1200)}$. The test is upper-, lower-, or two-tailed depending on which alternative hypothesis is relevant. Several exercises illustrate the use of this test procedure.

A Prediction Interval for a Single y

Suppose that an investigator is contemplating making a single observation on y when x has the value x^* at some future time. Let y^* denote the resulting future observation. The predicted value of this future observation based on available data is $a + bx^*$, so the prediction error is $a + bx^* - y^*$. Contrast this with the estimation error $a + bx^* - (\alpha + \beta x^*)$ when $a + bx^*$ is used to estimate the mean value $\alpha + \beta x^*$. In the estimation error, only $a + bx^*$ is subject to sampling variability, since $\alpha + \beta x^*$ is a fixed (albeit unknown) number. However, both $a + bx^*$ and the observation y^* in the prediction error are subject to sampling variability. This implies that there is more uncertainty associated with predicting a single value y^* than with estimating a mean value $\alpha + \beta x^*$.

We can obtain an assessment of how precise the prediction $a + bx^*$ is by computing a prediction interval for y^*. If the resulting interval is narrow, there is little uncertainty in y^*, and the prediction $a + bx^*$ is quite precise. The interpretation of a prediction interval is very similar to the interpretation of a confidence interval. A 95% prediction interval for y^* is one for which 95% of all possible samples would yield interval limits capturing y^*; only 5% of all samples would give an interval that did not include y^*.

The justification for the prediction interval is based on a standardized t variable similar to the one that led to the confidence interval. Consider the difference $a + bx^* - y^*$. Because the observation (x^*, y^*) has not yet been made, it is independent of the data used to compute a and b. Thus the two quantities in the difference are independent of one another. Both $a + bx^*$ and y^* have mean value $\alpha + \beta x^*$ (the first because $a + bx^*$ is an unbiased statistic and the second because y^* is a single observation from the model when $x = x^*$). The mean value of their difference is then zero, so nothing need be subtracted when $a + bx^* - y^*$ is standardized. Furthermore, because of independence, the variance of the difference is the sum of $\sigma^2_{a+bx^*}$ and $\sigma^2_{y^*}$. The variance $\sigma^2_{a+bx^*}$ was given earlier, and its estimate is $s^2_{a+bx^*}$. The variance $\sigma^2_{y^*}$ is just σ^2 (since y^* is a single observation from the model), estimated by s^2_e. These facts, along with properties of the normal distribution, yield the following result.

Let y^* denote a single observation made when $x = x^*$. Then the standardized variable

$$t = \frac{a + bx^* - y^*}{\sqrt{s_e^2 + s_{a+bx^*}^2}}$$

has a t distribution based on $n - 2$ df.

Comparing this t with the earlier t used to obtain the confidence interval, y^* replaces $\alpha + \beta x^*$ as expected, and the standard deviation in the denominator is larger because of the extra uncertainty in y^*. Manipulation of this t variable (exactly as earlier standardized variables were manipulated to obtain confidence intervals) yields the desired prediction interval.

A **prediction interval for y^***, a single y observation made when $x = x^*$, has the form

$$a + bx^* \pm (t \text{ critical value}) \sqrt{s_e^2 + s_{a+bx^*}^2}$$

The prediction interval and confidence interval are centered at exactly the same place, $a + bx^*$. The inclusion of s_e^2 under the square root symbol makes the prediction interval wider—often substantially so—than the confidence interval.

EXAMPLE 14

In Example 13 we computed a 99% confidence interval for average milk protein when milk production is 30 kg/day. Suppose that a single cow is randomly selected and its milk production on that day is found to be 30. Let's compute a 99% prediction interval for y^*, the amount of protein in this milk. As before, $a + bx^* = .176 + (.0246)(30) = .914$. In addition, $s_e^2 = .001767$, $s_{a+bx(30)}^2 = (.0113)^2 = .000128$, and the appropriate t critical value based on 12 df is 3.06 (the same value used for a 99% confidence interval). Substitution gives the interval

$$.914 \pm (3.06)\sqrt{.001767 + .000128} = .914 \pm (3.06)(.0435)$$
$$= .914 \pm .133 = (.781, 1.047)$$

We can be quite confident that an individual cow whose milk production is 30 will have a milk protein yield of between .781 and 1.047.

The confidence interval for the mean value $\alpha + \beta(30)$ was $.914 \pm .035$. The prediction interval is almost four times as wide as the confidence interval. Even with a rather precise estimate of $\alpha + \beta(30)$, there is a relatively wide range of plausible values for y^*. ■

EXERCISES

11.28 Explain the difference between a confidence interval and a prediction interval. How can a prediction level of 95% be interpreted?

11.29 The sugar content of certain types of fruit is a critical factor in determining when harvesting should begin. One method for assessing sugar content involves taking a measurement using a refractometer. The paper "Use of Refractometer to Determine Soluble Solids of Astringent Fruits of Japanese Persimmons" (*J. Horticultural Sci.* (1983):241–46) examined the relationship between y = total sugar content (%) and x = refractometer reading for persimmons. The estimated regression equation for predicting total sugar content from refractometer reading was given in the paper as $\hat{y} = -7.52 + 1.15x$. Suppose that $n = 50$, $\bar{x} = 17$, $s_e^2 = 1.1$, and $\Sigma(x - \bar{x})^2 = 112.5$.

a. Use a 95% confidence interval to estimate the mean percent of sugar for all persimmons with a refractometer reading of 18.

b. Construct a 90% prediction interval for the percent of sugar of an individual persimmon with refractometer reading 20.

c. Would a 90% prediction interval for percent of sugar when the refractometer reading is 15 be narrower or wider than the interval of (b)? Answer without computing the interval.

11.30 High blood-lead levels are associated with a number of different health problems. The paper "A Study of the Relationship Between Blood Lead Levels and Occupational Lead Levels" (*Am. Stat.* (1983):471) gave data on x = air-lead level ($\mu g/m^3$) and y = blood-lead level ($\mu g/dL$). Summary quantities (based on a subset of the data given in a plot appearing in the paper) are:

$$n = 15 \qquad \Sigma x = 1350 \qquad \Sigma y = 600$$
$$\Sigma x^2 = 155{,}400 \qquad \Sigma y^2 = 24{,}869.33 \qquad \Sigma xy = 57{,}760$$

a. Find the equation of the estimated regression line.

b. Estimate the mean blood-lead level for people who work where the air-lead level is 100 $\mu g/m^3$ using a 90% interval.

c. Construct a 90% prediction interval for the blood-lead level of a particular person who works where the air-lead level is 100 $\mu g/m^3$.

d. Explain the difference in interpretation of the intervals computed in (b) and (c).

11.31 The paper "Digestive Capabilities in Elk Compared to White-Tailed Deer" (*J. Wildlife Mgmt.* (1982):22–29) examined the relationship between y = digestible amount of detergent-solubles (g) and x = amount of detergent-solubles in feed (%). Data for white-tailed deer is given.

x	30	40	40	48	56	60
y	15	28	27	29	33	38

a. Assuming that the simple linear regression model is appropriate, find a 95% confidence interval for the mean digestible amount of detergent-solubles when feed is composed of 36% detergent-solubles.

b. Would a 95% confidence interval for the mean y value when x is 46 be wider or narrower than the interval in (a)? Explain.

11.32 The shelf life of packaged food depends on many factors. Dry cereal is considered to be a moisture-sensitive product (no one likes soggy cereal!) with the shelf life determined primarily by moisture content. In a study of shelf life of one particular brand of cereal, x = time on shelf (stored at 73°F and 50% relative humidity) and y = moisture content were recorded. The resulting data is from "Computer Simulation Speeds Shelf Life Assessments" (*Package Engr.* (1983):72–73).

x	0	3	6	8	10	13	16
y	2.8	3.0	3.1	3.2	3.4	3.4	3.5

x	20	24	27	30	34	37	41
y	3.1	3.8	4.0	4.1	4.3	4.4	4.9

a. Summary quantities are:

$\Sigma x = 269 \quad \Sigma x^2 = 7445 \quad \Sigma y = 51 \quad \Sigma y^2 = 190.78$

$\Sigma xy = 1081.5$

Find the equation of the estimated regression line for predicting moisture content from time on the shelf.

b. Does the simple linear regression model provide useful information for predicting moisture content from knowledge of time?

c. Find a 95% interval for the moisture content of an individual box of cereal that has been on the shelf 30 days.

d. According to the paper, taste tests indicate that this brand of cereal is unacceptably soggy when the moisture content exceeds 4.1. Based on your interval in (c), do you think that a box of cereal that has been on the shelf 30 days will be acceptable? Explain.

11.33 For the cereal data of Exercise 11.32, the average x value is 19.21. Would a 95% confidence interval with $x^* = 20$ or $x^* = 17$ be wider? Explain. Answer the same question for a prediction interval.

11.34 The number of viable rhizobia (a small soil bacteria that forms nodules on the roots of legumes and aids in fixing nitrogen) per clover seed at the time of sowing is of interest to crop scientists. The paper "Survival of Rhizobia on Commercially Lime-Pelleted White Clover and Lucerne Seed" (*N. Zeal. J. Exp. Ag.* (1983): 275–78) gave the data below on x = time stored (in weeks) and y = number of viable rhizobia per seed for clover.

x	1	8	12	16	20	24	32	44
y	41	40	35	32	28	28	25	24

a. Find the equation of the estimated regression line.

b. Use a 95% confidence interval to estimate the mean number of rhizobia per seed for seeds stored 18 weeks.

c. Estimate the mean number of rhizobia for seed stored 22 weeks using a 95% confidence interval.

11.35 An experiment to measure y = magnetic relaxation time in crystals (μs) as a function of x = strength of the external magnetic field (KG) resulted in the following data ("An Optical Faraday Rotation Technique for the Determination of Magnetic Relaxation Times" *IEEE Trans. Magnetics* (1968):175–78).

x	11.0	12.5	15.2	17.2	19.0	20.8
y	187	225	305	318	367	365

x	22.0	24.2	25.3	27.0	29.0
y	400	435	450	506	558

Summary quantities are

$\Sigma x = 223.2 \qquad \Sigma(x - \bar{x})^2 = 348.569$

$\Sigma(x - \bar{x})(y - \bar{y}) = 6578.718 \qquad \Sigma(y - \bar{y})^2 = 126{,}649.636$

a. Use the given information to compute the equation of the estimated regression line.

b. Is the simple linear regression model useful for predicting relaxation time from knowledge of the strength of the magnetic field?

c. Estimate the mean relaxation time when the field strength is 18 using a 95% confidence interval.

11.36 An article "Performance Test Conducted for a Gas Air-Conditioning System" (*Am. Soc. Heating, Refrigerating, and Air Cond. Engr.* (1969):54) reported the following data on maximum outdoor temperature (x) and hours of chiller operation per day (y) for a 3-ton residential gas air-conditioning system.

x	72	78	80	86	88	92
y	4.8	7.2	9.5	14.5	15.7	17.9

Suppose that the system is actually a prototype model, and the manufacturer does not wish to produce this model unless the data strongly indicates that when maximum outdoor temperature is 82, the true average number of hours of chiller operation is less than 12. The appropriate hypotheses are then $H_0 : \alpha + \beta(82) = 12$ versus $H_a : \alpha + \beta(82) < 12$. Use the statistic $t = [a + b(82) - 12]/s_{a+b(82)}$, which has a t distribution based on $n - 2$ df when H_0 is true, to test the above hypotheses at significance level .01.

11.37 The paper "The Incorporation of Uranium and Silver by Hydrothermally Synthesized Galena" (*Econ. Geol.* (1964):1003–24) reported on the determination of silver content of galena crystals grown in a closed hydrothermal system over a range of temperatures. With x = crystallization temperature (°C) and y = silver content (%), the data is:

x	398	292	352	575	568	450	550
y	.15	.05	.23	.43	.23	.40	.44

x	408	484	350	503	600	600
y	.44	.45	.09	.59	.63	.60

Summary quantities are:

$$\Sigma x = 6130 \qquad \Sigma x^2 = 3{,}022{,}050$$
$$\Sigma y = 4.73 \qquad \Sigma y^2 = 2.1785 \qquad \Sigma xy = 2418.74$$

Does the true average silver content when temperature equals 400°C appear to differ significantly from .25? Test the appropriate hypotheses at the .01 level of significance.

11.38 Occasionally an investigator may wish to compute a confidence interval for α, the y intercept of the true regression line, or test hypotheses about α. The estimated y intercept is just the height of the estimated line when $x = 0$: $a + \text{b}(0) = a$. This implies that the estimated standard deviation of the statistic a, s_a, results from substituting $x^* = 0$ in the formula for s_{a+bx^*}. The desired confidence interval is then $a \pm (t \text{ critical value})s_a$, whereas a test statistic is given by the formula $t = (a - \text{hypothesized value})/s_a$.

a. The paper "Comparison of Winter-Nocturnal Geostationary Satellite Infrared-Surface Temperature with Shelter-Height Temperature in Florida" (*Remote Sensing of the Environ.* (1983):313–27) used the simple linear regression model to relate surface temperature as measured by a satellite (y) to actual air temperature (x) as determined from a thermocouple placed on a traversing vehicle. Selected data is given (read from a scatter plot in the paper).

x	−2	−1	0	1	2	3	4	5	6	7
y	−3.9	−2.1	−2.0	−1.2	0	1.9	.6	2.1	1.2	3.0

Estimate the true regression line.

b. Compute the estimated standard deviation s_a. Carry out a test at level of significance .05 to see whether the y intercept of the true regression line differs from zero.

c. Compute a 95% confidence interval for α. Does the result indicate that $\alpha = 0$ is plausible? Explain.

11.4 Inferences Concerning the Population Correlation Coefficient

Methods from correlation analysis are used when the objective in studying two variables is to assess the strength of any relationship between them. Whereas in regression analysis, the distinction between the independent variable x and dependent variable y is important, the assignment of labels x and y to the two variables in a correlation analysis is arbitrary. Let (x_1, y_1), (x_2, y_2), . . . , (x_n, y_n) denote a sample consisting of n (x, y) pairs. In Chapter 4 the **sample correlation coefficient r** was introduced as a descriptive measure of how strongly the sample x and y values are linearly related. The defining formula was

$$r = \frac{\Sigma(x - \bar{x})(y - \bar{y})}{(n - 1)s_x s_y}$$

We now suppose that these n pairs were randomly selected from a population of (x, y) pairs. Then there is a **population correlation coefficient ρ** analogous to the sample measure r that measures the strength of association between x and y values in the entire population. The relationship between r and ρ is similar to the relationship between $\bar{x}$ and μ, between p and π, and between b and β. The first-listed quantity in each pair is a statistic whose value varies from sample to sample. Just as one sample might give $\bar{x} = 57.3$ and another (from the same population) might give $\bar{x} = 61.2$, so a first sample of pairs might yield $r = .57$, a second might give $r = .65$, a third $r = .48$, and so on. On the other hand, the second quantity in each case has a fixed value characteristic of the population being studied—e.g., $\mu = 60.0$ or $\beta = 12.5$ or $\rho = .55$—but its value is typically unknown to the investigator.

The sample and population correlation coefficients satisfy a number of similar properties. Here are the most important ones:

1. Interchanging the variable labels x and y does not change values of r and ρ.

2. A change in the unit of measurement for either variable (e.g., height from feet to inches or centimeters) has no effect on the values of r or ρ.

3. The only possible values for both r and ρ are numbers between -1 and 1. Thus the strongest possible positive or direct relationship (large x values asso-

ciated with large y values and small x values with small y values) is indicated by $r = 1$ (for the sample) or $\rho = 1$ (for the population). The strongest possible negative or inverse relationship (large x values associated with small y values and vice-versa) is indicated by $r = -1$ or $\rho = -1$.

4. Both r and ρ measure only the strength of any linear relationship—the extent to which pairs of values in the sample or population fall close to a straight line. The most extreme values 1 and -1 occur only when all pairs fall exactly on a straight line (which slopes upward or downward, respectively).

5. The values $r = 0$ and $\rho = 0$ indicate the absence only of any linear relationship between x and y. The two variables may be strongly related but in a nonlinear manner, which would not be detected by r or ρ.

Properties 4 and 5 limit the utility of ρ and r. Investigators are often interested not just in detecting linear association but in detecting association of *any* kind. When there is no association of any type between x and y values, statisticians say that the two variables are *independent*. In general, $\rho = 0$ is not equivalent to the independence of x and y. However, there is one special, yet frequently occurring, situation in which the two conditions ($\rho = 0$ and independence) are identical. This is when the pairs in the population have what is called a **bivariate normal distribution.** The essential feature of such a distribution is that for *any* fixed x value, the distribution of associated y values is normal, *and* for any fixed y value, the distribution of x values is normal. As an example, suppose that height x and weight y have a bivariate normal distribution in the American adult male population (there is good empirical evidence for this). Then when $x = 68$ in., weight y has a normal distribution, when $x = 72$ in., weight is normally distributed, when $y = 160$ lb, height x has a normal distribution, when $y = 175$ lb, height has a normal distribution, and so on. In this example, of course, x and y are not independent, since a large height value tends to be associated with large weight values and a small height value with small weight values.

There is no really good way to check the assumption of bivariate normality, especially when the sample size n is small. A partial check can be based on the following property: If (x, y) has a bivariate normal distribution, then x alone has a normal distribution and so does y. This suggests doing a normal probability plot of $x_1, x_2, \ldots, x_n$ and a separate normal probability plot of $y_1, \ldots, y_n$. If either plot shows a substantial departure from a straight line, bivariate normality is a questionable assumption. If both plots are reasonably straight, bivariate normality is plausible, although no guarantee can be given.

Just as the sampling distribution of the statistic $\bar{x}$ depends on μ and the sampling distribution of b depends on β, the sampling distribution of r depends on the value of ρ. If n is large, most samples give a value of r quite close to that of ρ, but when n is small there is considerable sampling variability in r. Unfortunately, even when the population distribution is bivariate normal, the sampling distribution of r is difficult to describe (although statisticians have managed a general description). This makes the development of a confidence interval formula and general hypothesis testing procedures for ρ somewhat complicated. Here we present only one inferential procedure, that for testing $\rho = 0$ (independence) in a bivariate normal population.

A Test for Independence in a Bivariate Normal Population

Null hypothesis: $H_0 : \rho = 0$ (x and y are independent)

Test statistic: $t = \dfrac{r}{\sqrt{(1 - r^2)/(n - 2)}}$

Alternative hypothesis	*Rejection region*
$H_a : \rho > 0$ (positive dependence)	$t > t$ critical value
$H_a : \rho < 0$ (negative dependence)	$t < -t$ critical value
$H_a : \rho \neq 0$ (dependence)	Either $t > t$ critical value or $t < -t$ critical value

The t critical value is based on $n - 2$ df.

EXAMPLE 15

In some locations there is a strong association between concentrations of two different pollutants. The paper "The Carbon Component of the Los Angeles Aerosol: Source Apportionment and Contributions to the Visibility Budget" (*J. Air Pollution Control Fed.* (1984):643–50) reported the accompanying data on O_3 (ozone) concentration x (ppm) and secondary carbon concentration y (μg/m^3).

Observation	1	2	3	4	5	6	7	8
x	.066	.088	.120	.050	.162	.186	.057	.100
y	4.6	11.6	9.5	6.3	13.8	15.4	2.5	11.8

Observation	9	10	11	12	13	14	15	16
x	.112	.055	.154	.074	.111	.140	.071	.110
y	8.0	7.0	20.6	16.6	9.2	17.9	2.8	13.0

Separate normal probability plots of the 16 x's and 16 y's appear quite straight. Assuming that the population distribution is bivariate normal, let's test at level .05 to see whether or not O_3 concentration and secondary carbon concentration are independent.

1. $\rho =$ the correlation between O_3 and secondary carbon concentrations in the population from which the given 16 observations were selected.
2. $H_0 : \rho = 0$.
3. $H_a : \rho \neq 0$.
4. Test statistic:

$$t = \frac{r}{\sqrt{(1 - r^2)/(n - 2)}}$$

5. The t critical value for a two-tailed test based on $n - 2 = 14$ df is 2.15, so H_0 will be rejected in favor of H_a if either $t > 2.15$ or $t < -2.15$.
6. Summary quantities are $\bar{x} = .1035$, $\Sigma(x - \bar{x})^2 = .025516$, $\bar{y} = 10.6625$, $\Sigma(y - \bar{y})^2 = 434.5375$, and $\Sigma(x - \bar{x})(y - \bar{y}) = 2.3826$. Thus $s_x = .0412$, $s_y = 5.3823$, and

$$r = \frac{2.3826}{15(.0412)(5.3823)} = .716$$

is the point estimate for ρ. The computed value of t is then

$$t = \frac{.716}{\sqrt{(1 - .513)/14}} = 3.84$$

7. Since $3.84 > 2.15$, the null hypothesis of independence is rejected at level .05. The data strongly suggests a linear relationship between concentrations of the two pollutants. ■

In the context of regression analysis, the hypothesis of no linear relationship ($H_0 : \beta = 0$) was tested using the t ratio b/s_b. Some algebraic manipulation shows that $r/\sqrt{(1 - r^2)/(n - 2)} = b/s_b$, so the two test procedures are completely equivalent. The reason for using the formula for t that involves r is that when interest lies only in correlation, the extra effort involved in computing regression quantities b, a, SSResid, s_e, and s_b need not be expended.

It is possible to construct a confidence interval for ρ either from a formula or from charts especially constructed for that purpose. Also, hypotheses about ρ in which the hypothesized value is something other than zero (e.g., $H_0 : \rho = .5$ versus $H_a : \rho > .5$) can be tested. Finally, there are other tests for independence, such as one based on Spearman's correlation coefficient. Appropriate references can be consulted for details.

EXERCISES

11.39 Discuss the difference between r and ρ.

11.40 **a.** If the sample correlation coefficient is equal to 1, is it necessarily true that $\rho = 1$?

b. If $\rho = 1$, is it necessarily true that $r = 1$?

11.41 The paper "Alcohol Consumption and Diabetes Mellitus Mortality in Different Countries" (*Am. J. Public Health* (1983):1316) gave data on y = per capita alcohol consumption and x = diabetes mellitus mortality rates for men in 19 countries. Summary quantities are

$$\Sigma(x - \bar{x})^2 = 287.54 \qquad \Sigma(y - \bar{y})^2 = 232.21$$
$$\Sigma(x - \bar{x})(y - \bar{y}) = 132.86$$

Does this data provide sufficient evidence to conclude that there is a positive linear relationship between alcohol consumption and diabetes mortality rate? Test the appropriate hypotheses using a .01 significance level.

11.42 In a study of bacterial concentration in surface and subsurface water ("Pb and Bacteria in a Surface Microlayer" *J. Marine Research* (1982):1200–06), the following data was obtained.

	Concentration ($\times 10^6$/ml)								
Surface	48.6	24.3	15.9	8.29	5.75	10.8	4.71	8.26	9.41
Subsurface	5.46	6.89	3.38	3.72	3.12	3.39	4.17	4.06	5.16

Summary quantities are:

$$\Sigma x = 136.02 \quad \Sigma x^2 = 3602.65 \quad \Sigma y = 39.35 \quad \Sigma y^2 = 184.27$$
$$\Sigma xy = 673.65$$

a. Using a significance level of .05, determine whether the data supports the hypothesis of a linear relationship between surface and subsurface concentration.

b. Give an upper and/or lower bound on the P-value associated with the computed test statistic value in (a).

11.43 Physical properties of six flame-retardant fabric samples were investigated in the paper "Sensory and Physical Properties of Inherently Flame-Retardant Fabrics" (*Textile Research* (1984):61–68). Use the accompanying data and a .05 significance level to determine if a linear relationship exists between stiffness and thickness.

Stiffness (mg-cm)	7.98	24.52	12.47	6.92	24.11	35.71
Thickness (mm)	.28	.65	.32	.27	.81	.57

11.44 The April 11, 1983, issue of *Advertising Age* gave the accompanying data on x = memory size (K) and y = retail price (in dollars) for 13 of the many home computer systems on the market. Does this data suggest a linear relationship between x and y? Use a .10 significance level.

x	2	1	4	5	16	16	16
y	80	80	100	200	300	300	400

x	16	64	128	32	48	64
y	450	595	795	995	679	899

11.45 The paper "Chronological Trend in Blood Lead Levels" (*N. Engl. J. Med.* (1983):1373–77) gave the data below on y = average blood-lead level of white children age 6 months to 5 years and x = amount of lead used in gasoline production (in 1000 tons) for ten 6-month periods.

x	48	59	79	80	95	95	97	102	102	107
y	9.3	11.0	12.8	14.1	13.6	13.8	14.6	14.6	16.0	18.2

a. Construct separate normal probability plots for x and y. Do you think that it is reasonable to assume that the (x, y) pairs are from a bivariate normal population?

b. Does the data provide sufficient evidence to conclude that there is a linear relationship between blood-lead level and the amount of lead used in gasoline production? Use $\alpha = .01$.

11.46 Two methods of measuring the amount of protein in sugars used in the manufacture of soft drinks were compared in the paper "Detection of Floc-Producing Sugars by a Protein Dye-Binding Method" (*J. Agric. Food Chem.* (1982):340–41). Five sugar specimens were analyzed using both a dye method and another method known as the Kjeldahl method.

Specimen	1	2	3	4	5
Dye method	.366	.370	.004	.303	.004
Kjeldahl method	.205	.509	.002	.253	.004

a. Construct separate normal probability plots for the data from each method. Does bivariate normality of the (x, y) pairs seem like a reasonable assumption?

b. Test appropriate hypotheses to determine if there is a significant correlation between protein determinations by the two methods. Use $\alpha = .05$.

c. Give an upper and/or lower bound for the P-value associated with the test statistic value in (b).

11.47 The paper "The Mechanics of Swimming Muskrats" (*J. Experimental Biology* (1984):183–201) contained a scatter plot of y, the arc (in degrees) through which the hind feet were swept during the power phase, versus x, the swimming velocity (m/s). Selected data is given.

x	.25	.30	.35	.40	.45	.50	.50
y	98	92	87	97	101	116	96

x	.55	.55	.60	.65	.70	.75
y	115	114	110	115	123	133

a. Compute the value of the sample correlation coefficient.

b. Does the data suggest that muskrats increase swimming speed in a linear fashion by increasing the sweep arc of their hind feet? State and test the appropriate hypotheses at level of significance .05.

c. What can you say about the P-value corresponding to the test statistic value computed in (b)?

d. How would your conclusion change if x were expressed in feet per second? Explain.

11.48 A sample of $n = 500$ (x, y) pairs was collected and a test of $H_0 : \rho = 0$ versus $H_a : \rho \neq 0$ was carried out. The resulting P-value was computed to be .00032.

a. What conclusion would be appropriate at level of significance .001?

b. Does this small P-value indicate that there is a very strong linear relationship between x and y (a value of ρ that differs considerably from zero)? Explain.

11.49 A sample of $n = 10{,}000$ (x, y) pairs resulted in $r = .022$. Test $H_0 : \rho = 0$ versus $H_a : \rho \neq 0$ at level .05. Is the result statistically significant? Comment on the practical significance of your analysis.

11.5 Checking Model Adequacy

The inferential methods discussed in previous sections assumed that the observations in the sample came from the simple linear regression model $y = \alpha + \beta x + e$. With $x_1, x_2, \ldots, x_n$ denoting the x values at which n observations are made, the relationship between the x_i's and resulting y_i's is $y_1 = \alpha + \beta x_1 + e_1$, $y_2 = \alpha + \beta x_2 + e_2, \ldots, y_n = \alpha + \beta x_n + e_n$. Here e_1 is the random deviation from the population regression line $y = \alpha + \beta x$ corresponding to the first observation, e_2 is the random deviation corresponding to the second observation, and so on. The key model assumptions are that each e_i is normally distributed with mean value zero and the same variance σ^2, and that the e_i's are independent of one another. These properties imply that the y_i's themselves are independent of one another and are normally distributed with mean values $\alpha + \beta x_1, \alpha + \beta x_2, \ldots, \alpha + \beta x_n$ and the same variance σ^2.

Inferences based on the simple linear regression model continue to be reliable when model assumptions are slightly violated (e.g., mild nonnormality of the random deviation distribution). However, use of an estimated model in the face of grossly violated assumptions can result in very misleading conclusions being drawn. Therefore, it is desirable to have available easily applied methods for identifying such gross violations and suggesting how a satisfactory model can be obtained.

Residual Analysis

If the deviations $e_1, e_2, \ldots, e_n$ from the population line were available, they could be examined for any inconsistencies. For example, a normal probability plot would suggest whether the normality assumption was tenable. But since $e_1 = y_1 - (\alpha + \beta x_1)$, $e_2 = y_2 - (\alpha + \beta x_2)$, etc., these deviations can be calculated only if the equation of the population line is known. In practice, this will never be the case. Instead, diagnostic checks must be based on the residuals $y_1 - \hat{y}_1 = y_1 - [a + bx_1]$, $y_2 - \hat{y}_2, \ldots, y_n - \hat{y}_n$, which are the deviations from the estimated line.

Before a sample has been selected, any particular residual $y_i - \hat{y}_i$ is a random variable because its value varies from sample to sample. When all model assumptions are met, the mean, or expected, value of any residual is zero. Any observation that gives a very large positive or negative value should be examined carefully for any anomalous circumstances, such as a recording error or exceptional experimental conditions. Identifying residuals with unusually large magnitudes is made easier by calculating **standardized residuals.** Recall that standardization of a quantity means subtracting its expected value (zero here) and dividing by its standard deviation. The value of a standardized residual tells how many standard deviations the corresponding residual lies from its expected value zero.

It might seem that because each e_i has standard deviation σ, which is estimated by s_e, the appropriate way to standardize the residuals is to divide each one by s_e. This is indeed done by some analysts and statistical computer packages, but it is not the correct standardization. The standard deviation of the ith residual $y_i - \hat{y}_i$ in simple linear regression is the square root of σ^2 multiplied by an expression depending on the values $x_1, x_2, \ldots, x_n$ at which observations were made. The exact formula for this standard deviation needn't concern us—statisticians know it, and it has been programmed into MINITAB and some other statistical packages (with s_e^2 used to estimate σ^2). What is important is that residuals corresponding to different x_i's have different standard deviations, and the further x_i is from $\overline{x}$, the *smaller* will be the standard deviation of $y_i - \hat{y}_i$. This is because when the principle of least squares is used to obtain the estimated (best-fit) line, observations far from the center of the data are more influential than those close to the center.

EXAMPLE 16

Suppose that y = commuting time is related to x = commuting distance according to the simple linear regression model with $\sigma^2 = 100$. If $n = 5$ observations are to be made at the x values $x_1 = 5$, $x_2 = 10$, $x_3 = 15$, $x_4 = 20$, and $x_5 = 25$, it can be shown that the standard deviations of $y_1 - \hat{y}_1, \ldots, y_5 - \hat{y}_5$ are 6.32, 8.37, 8.94, 8.37, and 6.32, respectively. If $x_1 = 5$, $x_2 = 10$, $x_3 = 15$, $x_4 = 20$, and $x_5 = 50$, these standard deviations become 7.87, 8.49, 8.83, 8.94, and 2.83. For this latter choice of x values, there is much less variability in $y_5 - \hat{y}_5$ than in the other four residuals. ■

When each residual is divided by its correct standard deviation (using s_e^2 in place of σ^2), the result is the set of standardized residuals. In Example 16, when the first set of x values is used, if $y_1 - \hat{y}_1 = 8$ and $s_e^2 = 100$, then the first standardized residual is $8/6.32 = 1.27$. If n is reasonably large and no x_i is too far from any others, the standard deviation of each residual is approximately σ, so standardizing using just s_e is then approximately correct.

In Chapter 5 we discussed constructing a normal probability plot to see if the n observations in a random sample could plausibly have come from a normal population distribution. To check the assumption that $e_1, e_2, \ldots, e_n$ all come from the same normal distribution, we need a normal probability plot based on the residuals. Each residual does have a normal distribution when the model assumptions are satisfied, but the variances are different. Fortunately, standardizing solves this problem, so *to check the assumption of normality we recommend a normal probability plot of the standardized residuals.*

EXAMPLE 17

Reconsider the $n = 16$ observations on x = age of tree and y = 5-year growth introduced in Example 2. The residuals, their standard deviations, and the standardized residuals (residual/standard deviation of residual) are given in the accompanying table. Notice that except for $x = 5$ and $x = 18$, the two most extreme values, the residuals have roughly equal standard deviations. The residual with the largest magnitude, -99.2, initially seems quite extreme, but the corresponding standardized residual is only -2.04. That is, the residual is approximately 2 standard deviations below its expected value zero, which is not terribly unusual in a sample this size. On the standardized scale, no residual here is surprisingly large. Before standardization, there are some large residuals simply because there appears to be a substantial amount of variability about the true regression line ($s_e = 51.43$, $r^2 = .683$).

Figure 7 displays a normal probability plot of the standardized residuals. Few plots are straighter than this one! The plot casts no doubt on the normality assumption.

Observation	x	y	$\hat{y}$	Residual	Standard Deviation of Residual	Standardized Residual
1	5	70	111.0	−41.0	40.9	−1.00
2	9	150	196.2	−46.2	48.1	−.96
3	9	260	196.2	63.8	48.1	1.33
4	10	230	217.6	12.4	49.0	.25
5	10	255	217.6	37.4	49.0	.76
6	11	165	238.9	−73.9	49.5	−1.49
7	11	225	238.9	−13.9	49.5	−.28
8	12	340	260.2	79.8	49.8	1.60
9	13	305	281.5	23.5	49.7	.47
10	13	335	281.5	53.5	49.7	1.08
11	14	290	302.9	−12.9	49.4	−.26
12	14	340	302.9	37.1	49.4	.75
13	15	225	324.2	−99.2	48.7	−2.04
14	15	300	324.2	−24.2	48.7	−.50
15	18	380	388.1	−8.1	44.6	−.18
16	18	400	388.1	11.9	44.6	.27

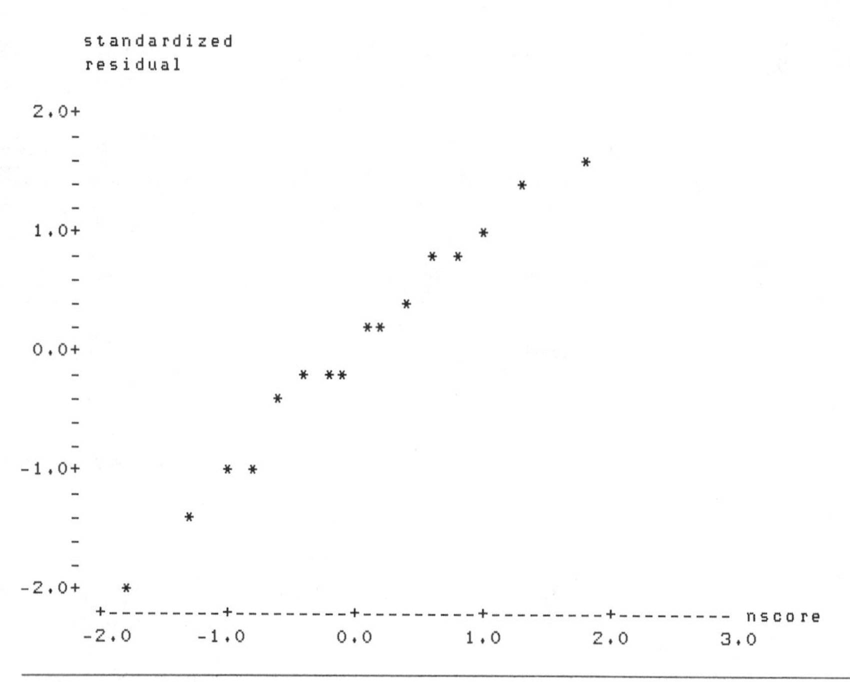

FIGURE 7 NORMAL PROBABILITY PLOT OF THE STANDARDIZED RESIDUALS FROM EXAMPLE 17 (FROM MINITAB) ■

In Chapter 4 we recommended examining a plot of the (x, residual) pairs for any unusual behavior or patterns. Because it is easier to identify outliers after standardization, many analysts prefer a plot of (x, standardized residual) pairs, called a **standardized residual plot.** A plot such as the one pictured in Figure 8(a) is desirable, since no point lies much outside the horizontal band between -2 and 2 (no unusually large residual corresponding to an outlying observation), there is no point far to the left or right of the others (no observation that might greatly influence the fit), and there is no pattern to indicate that the model should somehow be modified. When the plot has the appearance of Figure 8(b), the fitted model should be changed to incorporate curvature (a nonlinear model). Many such models can be fit by first using the methods of Chapter 4 to transform x and/or y and then performing a linear regression on the transformed values. The details are beyond the scope of this text.

The increasing spread in Figure 8(c) as one moves from left to right suggests that the variance of y is not the same at each x value but rather increases with x. A straight-line model may still be appropriate, but the best-fit line should be selected by using weighted least squares rather than ordinary least squares as described earlier. This involves giving more weight to observations for which y appears to have little variability and less weight to observations in the region exhibiting high variability. A specialized regression analysis text or a knowledgable statistician should be consulted for details.

The standardized residual plots of Figures 8(d) and 8(e) show an extreme outlier and potentially influential observation, respectively. Consider deleting the observation corresponding to such a point from the data set and refitting the same model. Substantial changes in estimates and various other quantities

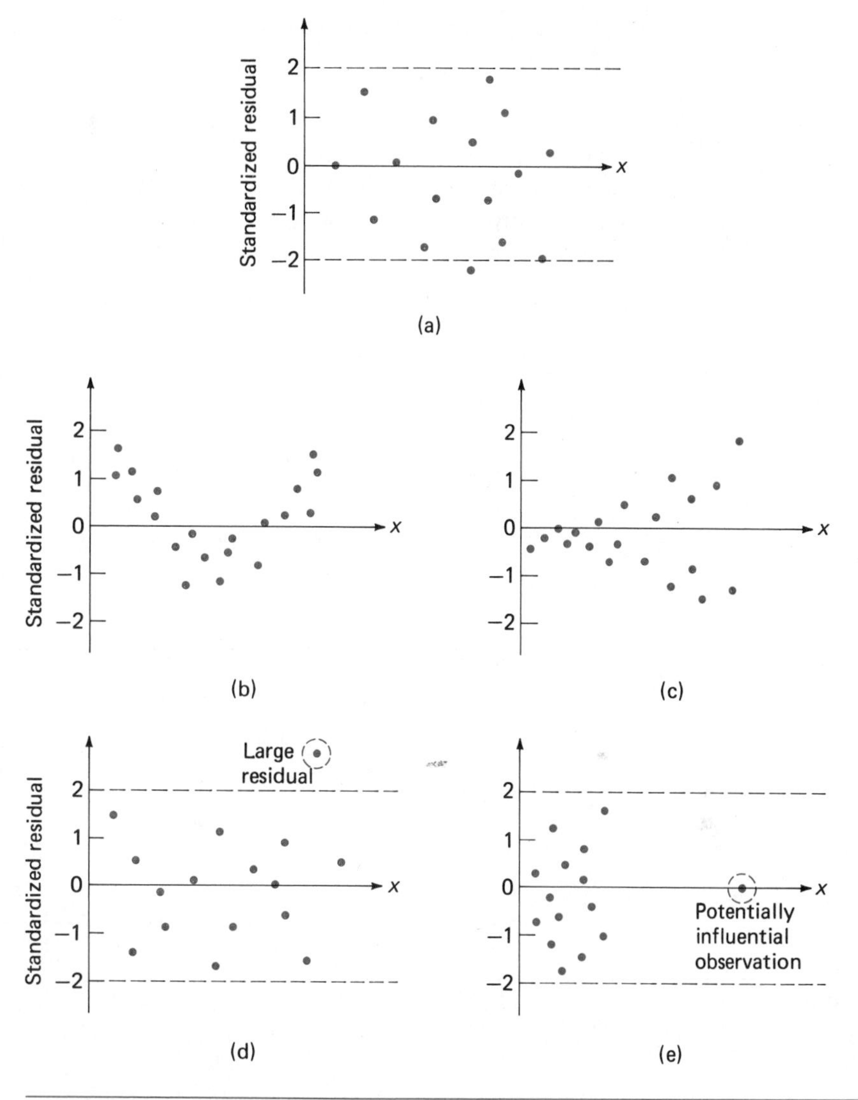

FIGURE 8 EXAMPLES OF RESIDUALS PLOTS
(a) A Satisfactory Plot
(b) A Plot Suggesting that a Curvilinear Regression Model Is Needed
(c) A Plot Indicating Nonconstant Variance
(d) A Plot Showing a Large Residual
(e) A Plot Showing a Potentially Influential Observation

warn of instability in the data. The investigator should certainly carry out a more careful analysis and perhaps collect more data before drawing any firm conclusions. Changes in the estimated coefficients a and b are easily examined. Suppose that for the full data set, $b = 150$ and $s_b = 5$, whereas a deletion results in $b = 135$. The resulting change in b is $150 - 135 = 15$, or $15/5 = 3$ standard deviations. Because the estimated standard deviation s_a is almost always available from computer output, the standardized change in the intercept a can also be computed. In general, a change of more than 1 standard deviation in a coefficient is cause for concern. Improved computing power has allowed statisticians to develop and implement various other diagnostics of this type in recent years.

EXAMPLE 18

Figure 9 displays a standardized residual plot for the tree age–5-year growth data of Example 17. The first observation was at $x_1 = 5$ and the corresponding standardized residual was -1.00, so the first plotted point is $(5, -1.00)$. Other points are similarly obtained and plotted. The plot shows no unusual behavior that might call for model modification or further analysis.

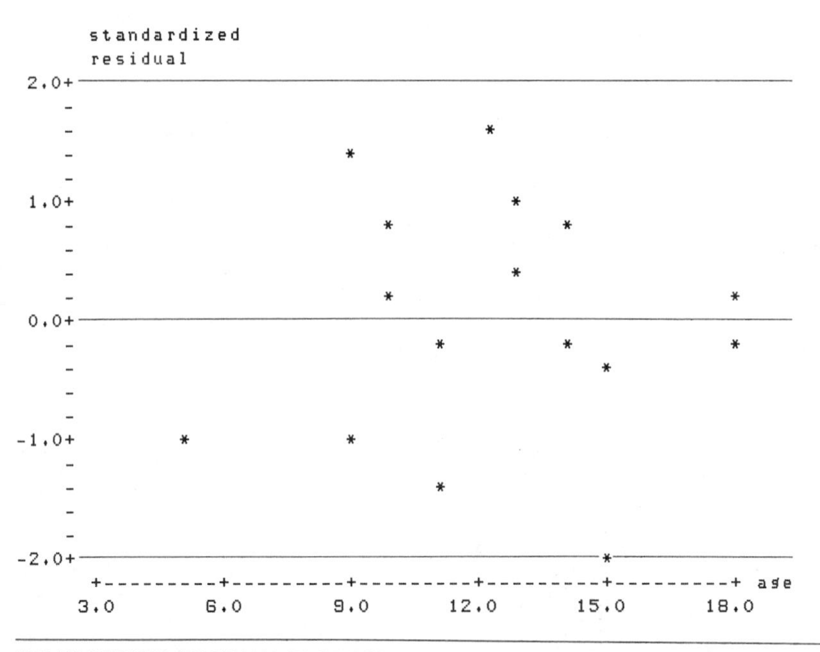

FIGURE 9 STANDARDIZED RESIDUAL PLOT FOR THE DATA OF EXAMPLES 17 AND 18 (FROM MINITAB) ■

EXAMPLE 19

The accompanying data on x = exposure time of lettuce seeds to an ethylene solution and y = ethylene content of the seeds appeared in the paper "Ethylene Synthesis in Lettuce Seeds: Its Physiological Significance" (*Plant Physiology* (1972):719–22). Figure 10 displays a scatter plot and a plot of the standardized residuals from a simple linear regression fit. Without the plots, one might think the simple linear regression model reasonably satisfactory since $r^2 = .793$ (though $\hat{y}$ is negative for $x = 90$ or 100). However, curvature in the scatter plot is obvious, and this curvature is magnified in the standardized residual plot. A curvilinear model should be fit to this data.

x	2	10	20	30	40	50	60	70	80	90	100
y	408	274	196	137	90	78	51	40	30	22	15

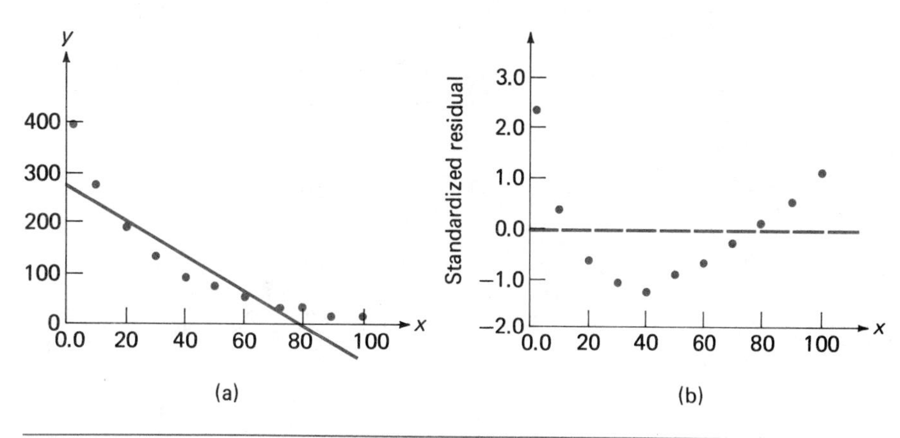

FIGURE 10 PLOTS FOR THE DATA OF EXAMPLE 19
(a) Scatter Plot
(b) Standardized Residual Plot

■

EXAMPLE 20

The paper "Snow Cover and Temperature Relationships in North America and Eurasia" (*J. of Climate and Appl. Meteorology* (1983):460–69) explored the relationship between October-November continental snow cover (x, in millions of km^2) and December-February temperature (y, in °C). The given data refers to Eurasia during the $n = 13$ time periods 1969–70, 1970–71, . . . , 1981–82. A simple linear regression done by the authors yielded $r^2 = .52$ ($r = -.72$), suggesting a substantial linear relationship. This is confirmed by a model utility test.

x	13.00	12.75	16.70	18.85	16.60	15.35	13.90
y	−13.5	−15.7	−15.5	−14.7	−16.1	−14.6	−13.4
Standardized residual	−.11	−2.19	−.36	1.23	−.91	−.12	.34

x	22.40	16.20	16.70	13.65	13.90	14.75
y	−18.9	−14.8	−13.6	−14.0	−12.0	−13.5
Standardized residual	−1.54	.04	1.25	−.28	1.54	.58

The scatter plot and standardized residual plot are displayed in Figure 11. There are no unusual patterns, though one standardized residual, −2.19, is a bit on the large side. However, the most interesting feature is the observation (22.40, −18.9) corresponding to a point far to the right of the others in these plots. This observation may have had a substantial influence on all aspects of the fit. The estimated slope when all 13 observations are included is $b = -.459$, and $s_b = .133$. When the potentially influential observation is deleted, the estimate of β based on the remaining 12 observations is $b = -.228$. Thus deletion has caused the estimated slope to change by $[-.459 - (-.228)]/.133 = -1.74$ standard deviations, which is quite large. Additionally, r^2 based just on the 12 observations is only .13, and the t ratio for β is not significant. Evidence for a linear relationship is much less conclusive in light of this analysis. The investigators should seek a climatological ex-

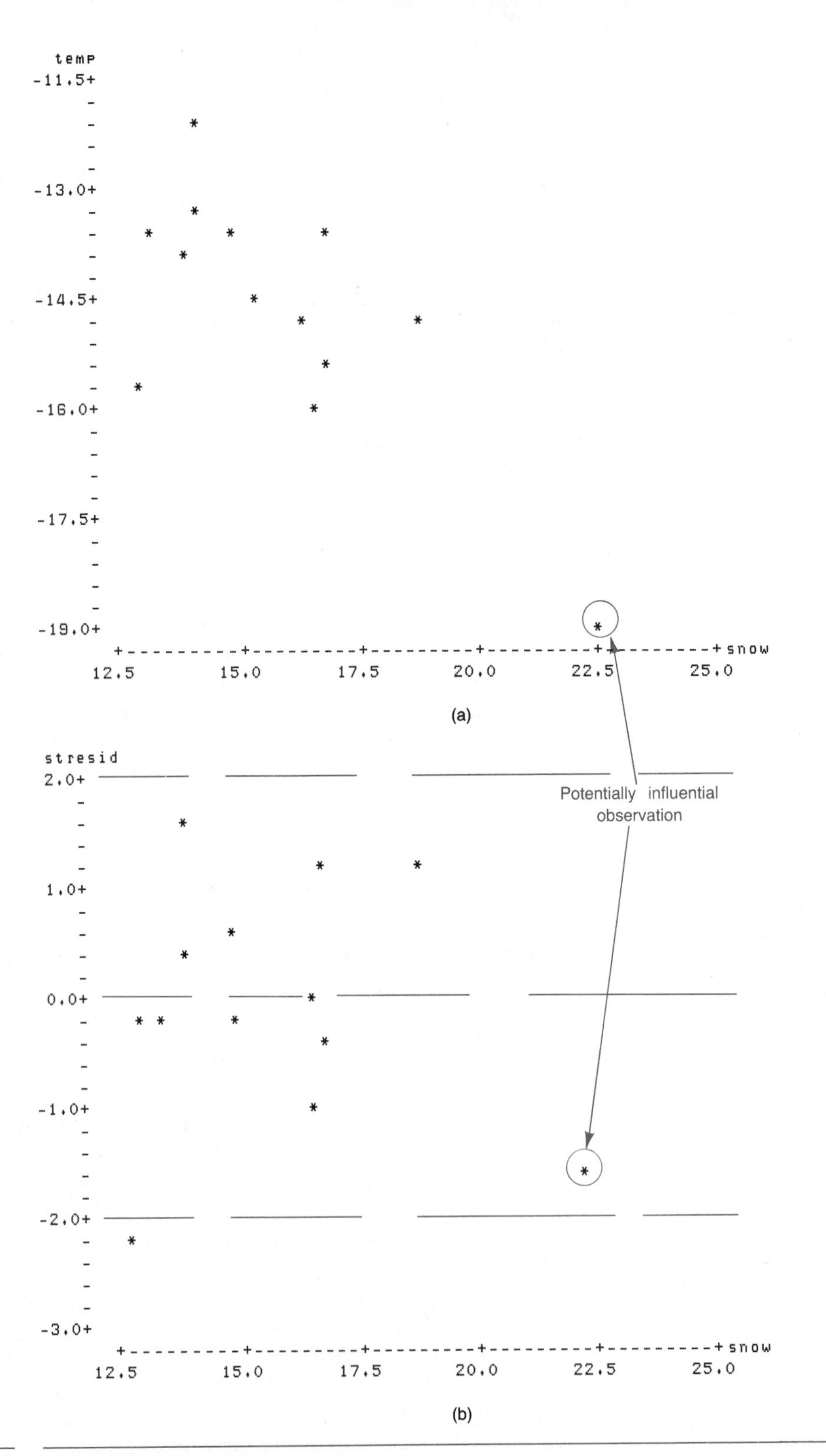

FIGURE 11

PLOTS FOR THE DATA OF EXAMPLE 20 (FROM MINITAB)
(a) Scatter Plot
(b) Standardized Residual Plot

planation for the influential observation and collect more data which can be used in seeking an effective relationship. ■

When the distribution of the random deviation e has heavier tails than does the normal distribution, observations with large standardized residuals are not that unusual. Such observations can have great effects on the estimated regression line when the least squares approach is used. In recent years, statisticians have proposed a number of alternative methods—called **robust,** or **resistant,** methods—for fitting a line. Such methods give less weight to outlying observations than does least squares without deleting them from the data set. The most widely used robust procedures require a substantial amount of computation, so a good computer program is necessary. Associated confidence-interval and hypothesis-testing formulas are still in the developmental stage.

EXERCISES

11.50 What are the assumptions of the simple linear regression model? For each of the assumptions, what technique would you use to detect violations?

11.51 The accompanying data (first described in Exercise 11.4) on y = vigor (average width in centimeters of last two annual rings) and x = stem density (stems/m^2) was used to obtain an estimated regression line of $\hat{y} = .925 - .0289x$. Also given are the standardized residuals.

x	4	5	6	9	14
y	.75	1.20	.55	.60	.65
Standardized residual	−.28	1.92	−.90	−.28	.54

x	15	15	19	21	22
y	.55	0	.35	.45	.40
Standardized residual	.24	−2.05	−.12	.60	.52

a. What assumptions are required in order that the simple linear regression model be appropriate?

b. Construct a normal probability plot of the standardized residuals. Does the assumption that the random deviation distribution is normal appear to be reasonable? Explain.

c. Construct a standardized residual plot. Are there any unusually large residuals?

d. Is there anything about the standardized residual plot that would cause you to question the use of a simple linear regression model to describe the relationship between x and y?

11.52 The article "Effects of Gamma Radiation on Juvenile and Mature Cuttings of Quaking Aspen" (*Forest Sci.* (1967):240–45) reported the following data on x = exposure time to radiation (kR/16 h) and y = dry weight of roots (mg $\times$ 10^{-1}).

x	0	2	4	6	8
y	110	123	119	86	62

a. Construct a scatter plot for this data. Does the plot suggest that a simple linear regression model might be appropriate?

b. The estimated regression line for this data is $\hat{y} = 127 - 6.65x$ and the standardized residuals are as given.

x	0	2	4	6	8
Standardized residual	−1.55	.68	1.25	−.05	−1.06

Construct a standardized residual plot. What does the plot suggest about the adequacy of the simple linear regression model?

11.53 Carbon aerosols have been identified as a contributing factor in a number of air quality problems. In a chemical analysis of diesel engine exhaust, x = mass ($\mu g/cm^2$) and y = elemental carbon ($\mu g/cm^2$) were recorded ("Comparison of Solvent Extraction and Thermal Optical Carbon Analysis Methods: Application to Diesel Vehicle Exhaust Aerosol" *Environ. Sci. Tech.* (1984):231–34). The estimated regression line for this data set is $\hat{y} = 31 + .737x$. Given are the observed x and y values and the corresponding standardized residuals.

x	164.2	156.9	109.8	111.4	87.0	82.9	78.9
y	181	156	115	132	96	90	86
Standardized residual	2.52	0.82	0.27	1.64	0.08	−0.18	−0.27

x	161.8	230.9	106.5	97.6	79.7	100.8	387.8
y	170	193	110	94	77	88	310
Standardized residual	1.72	−0.73	0.05	−0.77	−1.11	−1.49	−0.89

x	118.7	248.8	102.4	64.2	89.4	117.9	135.0
y	106	204	98	76	89	130	141
Standardized residual	−1.07	−0.95	−0.73	−0.20	−0.68	1.05	0.91

x	108.1	89.4	76.4	131.7
y	102	91	97	128
Standardized residual	−0.75	−0.51	0.85	0.00

a. Construct a normal probability plot of the standardized residuals. Does the assumption of normality of the random deviation distribution seem reasonable?

b. Construct a standardized residual plot. Are there any unusually large residuals? Do you think that there are any influential observations?

c. Is there any pattern in the standardized residual plot that would indicate that the simple linear regression model is not appropriate?

d. Based on your plot in (b), do you think that it is reasonable to assume that the variance of y is the same at each x value? Explain.

11.54 An investigation of the relationship between traffic flow x (thousands of cars per 24 h) and lead content y of bark on trees near the highway ($\mu g/g$ dry weight) yielded the accompanying data. A simple linear regression model was fit, and the resulting estimated regression line was $\hat{y} = 28.7 + 33.3x$. Both residuals and standardized residuals are also given.

x	8.3	8.3	12.1	12.1	17.0
y	227	312	362	521	640
Residual	−78.1	6.9	−69.6	89.4	45.3
Standardized residual	−0.99	0.09	−0.81	1.04	0.51

x	17.0	17.0	24.3	24.3	24.3
y	539	728	945	738	759
Residual	−55.7	133.3	107.2	−99.8	−78.8
Standardized residual	−0.63	1.51	1.35	−1.25	−0.99

a. Plot the residuals against x. Does the resulting plot suggest that a simple linear regression model is an appropriate choice? Explain your reasoning.

b. Plot the standardized residuals against x. Does the plot differ significantly in general appearance from the plot in (a)?

11.55 The paper "Sex Ratio Variation in Odocoileus: A Critical Review" (*J. Wildlife Mgmt.* (1983):573–82) gave the accompanying data on x = number of fawns per doe and y = percent of male fawns. Standardized residuals based on a simple linear regression model are given. Use appropriate techniques to assess the adequacy of the simple linear regression model.

x	1.03	1.43	1.79	1.74	0.95	0.54	1.76	1.09
y	57.6	60.1	45.7	35.6	68.4	75.0	51.3	60.5
Standardized residual	−0.42	0.89	−0.11	−1.58	0.86	0.87	0.56	0.12

x	1.65	1.69	1.41	1.68	1.26	1.70	1.92
y	44.0	52.3	40.7	58.4	43.8	48.8	52.9
Standardized residual	−0.68	0.52	−1.68	1.30	−1.65	0.08	1.22

11.56 The sum of the residuals from the least squares line is always zero (except for the effects of rounding), even when the data did not come from the simple linear regression model. Is this true of the standardized residuals? (*Hint:* See Exercise 11.52.)

11.57 Is it possible for all the standardized residuals to be positive? Negative? Explain.

11.58 Consider the following four (x, y) data sets; the first three have the same x values, so these values are listed only once (from Frank Anscombe, "Graphs in Statistical Analysis" *Amer. Statistician* (1973):17–21).

Data set	1–3	1	2	3	4	4
Variable	x	y	y	y	x	y
	10.0	8.04	9.14	7.46	8.0	6.58
	8.0	6.95	8.14	6.77	8.0	5.76
	13.0	7.58	8.74	12.74	8.0	7.71
	9.0	8.81	8.77	7.11	8.0	8.84
	11.0	8.33	9.26	7.81	8.0	8.47
	14.0	9.96	8.10	8.84	8.0	7.04
	6.0	7.24	6.13	6.08	8.0	5.25
	4.0	4.26	3.10	5.39	19.0	12.50
	12.0	10.84	9.13	8.15	8.0	5.56
	7.0	4.82	7.26	6.42	8.0	7.91
	5.0	5.68	4.74	5.73	8.0	6.89

For each of these data sets, the values of the summary quantities $\bar{x}$, $\bar{y}$, $\Sigma(x - \bar{x})^2$, $\Sigma(y - \bar{y})^2$, and $\Sigma(x - \bar{x})(y - \bar{y})$ are identical, so all quantities computed from these will be identical for the four sets—the estimated regression line, SSResid, s_e^2, r^2, and so on. The summary quantities provide no way of distinguishing among the four data sets. Based on a scatter plot and a residual plot for each set, comment on the appropriateness or inappropriateness of fitting a simple linear regression model.

KEY CONCEPTS

Simple linear regression model $y = \alpha + \beta x + e$, where e is a normally distributed random deviation with mean value 0 and variance σ^2. (p. 442)

Population regression line $y = \alpha + \beta x$, where β is the slope (average change in y associated with a 1-unit increase in x) and α is the y intercept. The height of the line above any specified x value is the average y value for all pairs (x, y) having that x value. (p. 442)

Estimated regression line $\hat{y} = a + bx$, where the estimated slope b and y intercept a result from applying the principle of least squares. (p. 445)

Estimated standard deviation, s_e $s_e = \sqrt{\text{SSResid}/(n - 2)}$, where SSResid = $\Sigma(y - y)^2$; s_e is the point estimate of σ. (p. 449)

Coefficient of determination, r^2 Given by $1 - \text{SSResid}/\text{SSTo}$, where $\text{SSTo} = \Sigma(y - \bar{y})^2$, it is the proportion of variation in observed y values that can be explained by the simple linear regression model. (p. 450)

Sampling distribution of b A normal distribution with mean value β (so b is an unbiased statistic for estimating β) and standard deviation $\sqrt{\sigma^2/\Sigma(x - \bar{x})^2}$. The estimated standard deviation of b is $s_b = \sqrt{s_e^2/\Sigma(x - \bar{x})^2}$. (p. 454)

Model utility test A test of the hypotheses $H_0: \beta = 0$ vs. $H_a: \beta \neq 0$. The test statistic is the t ratio b/s_b. (p. 460)

Sampling distribution of $a + bx^*$ (x^* is a specified value of x) A normal distribution with mean value $\alpha + \beta x^*$ (so $a + bx^*$ is an unbiased statistic for estimating $\alpha + \beta x^*$) and standard deviation $\sqrt{\sigma^2[1/n + (x^* - \bar{x})^2/\Sigma(x - \bar{x})^2]}$. The estimated standard deviation s_{a+bx^*} results from replacing σ^2 by s_e^2 in this expression. (p. 466)

Correlation coefficient The sample correlation coefficient r gives a point estimate of the population correlation coefficient ρ. Both r and ρ measure the extent to which there is a linear association between x and y. (p. 474)

Standardized residual A residual divided by its estimated standard deviation. (p. 480)

Standardized residual plot A plot of (x, standardized residual) pairs that is useful for checking the appropriateness of the fitted model and identifying any unusual observations. (p. 482)

Hypothesis Testing Procedures

Population Characteristic	*Appropriate When*	*Test Statistic*
β	Basic assumptions of the simple linear regression model are met	$t = \dfrac{b - \text{hypothesized value}}{s_b}$ (Section 11.2)
ρ	Population is bivariate normal	$t = \dfrac{r}{\sqrt{(1 - r^2)/(n - 2)}}$ (Section 11.4)

Confidence and Prediction Intervals

Interval for	*Appropriate When*	*Interval*
β	Basic assumptions of the simple linear regression model are met	$b \pm (t \text{ critical value})s_b$ (Section 11.2)
$\alpha + \beta x^*$ (average y value at a fixed value x^*)	Same as above	Confidence interval for $\alpha + \beta x^*$ (Section 11.3)
y^* (individual y value at a fixed value x^*)	Same as above	Prediction interval for y^* (Section 11.3)

11.59 Silane coupling agents have been used in the rubber industry to improve the performance of fillers in rubber compounds. The accompanying data on y = tensile modulus (in MPa, a measure of silane coupling effectiveness) and x = bound rubber content (%) appeared in the paper "The Effect of the Structure of Sulfur-Containing Silane Coupling Agents on Their Activity in Silica-Filled SBR" (*Rubber Chem. and Tech.* (1984):675–85).

x	16.1	31.5	21.5	22.4	20.5	28.4	30.3	25.6	32.7	29.2	34.7
y	4.41	6.81	5.26	5.99	5.92	6.14	6.84	5.87	7.03	6.89	7.87

a. Construct a scatter plot for this data. Does the plot look linear?

b. Find the equation of the estimated regression line.

c. Using a .01 significance level, does the data suggest the existence of a linear relationship between y and x?

d. Compute and interpret the values of r^2 and s_e.

e. Use a 95% confidence interval to estimate the true mean tensile modulus when bound rubber content is 20%.

11.60 The relationship between depth of flooding and the amount of flood damage was examined in the paper "Significance of Location in Computing Flood Damage" (*J. Water Resources Planning and Management* (1985):65–81). The data on x = depth of flooding (feet above first-floor level) and y = flood damage (as a percent of structure value) was obtained using a sample of flood insurance claims.

x	1	2	3	4	5	6	7	8	9	10	11	12	13
y	10	14	26	28	29	41	43	44	45	46	47	48	49

a. Does the data suggest the existence of a positive correlation between flood depth and damage? Test using a .05 significance level.

b. Compute the equation of the estimated regression line for predicting flood damage from flood depth.

c. Construct a scatter plot and draw the estimated regression line on the plot. Does it look as though a straight line provides an adequate description of the relationship between y and x? Explain.

11.61 Eye weight (g) and cornea thickness (μm) were recorded for nine randomly selected calves, and the resulting data from the paper "The Collagens of the Developing Bovine Cornea" (*Exper. Eye Research* (1984):639–52) is given. Use this data and a .05 significance level to test the null hypothesis of no correlation between eye weight and cornea thickness against the alternative hypothesis of a positive correlation.

Eye weight	.2	1.4	2.2	2.7	4.9	5.3	8.0	8.8	9.6
Thickness	416	673	733	801	957	1035	883	736	567

11.62 Eight surface soil samples were analyzed to determine physiochemical properties. The data below on x = calcium-sodium exchange rate and y = percent of sodium in the soil was read from a scatter plot that appeared in the paper "Sodium-Calcium Exchange Equilibria in Soils as Affected by Calcium Carbonate and Organic Matter" (*Soil Sci.* (1984):109).

x	.641	.611	.463	.375	.260	.184	.182	.089
y	3.4	3.0	3.0	2.2	2.2	2.0	1.9	1.6

a. Find the equation of the estimated regression line.

b. Compute the predicted values and the corresponding residuals. Use the residuals to calculate SSResid and s_e.

c. Using a significance level of .05, test to determine if the data suggests the existence of a linear relationship between the variables x and y.

d. Compute and interpret the values of r and r^2.

11.63 The paper "Statistical Comparison of Heavy Metal Concentrations in Various Louisiana Sediments" (*Environ. Monitoring and Assessment* (1984):163–70) gave the accompanying data on depth (m), zinc concentration (ppm), and iron concentration (%) for 17 core samples.

Core	Depth	Zinc	Iron
1	.2	86	3.4
2	2.0	77	2.9
3	5.8	91	3.1
4	6.5	86	3.4
5	7.6	81	3.2
6	12.2	87	2.9
7	16.4	94	3.2
8	20.8	92	3.4
9	22.5	90	3.1
10	29.0	108	4.0
11	31.7	112	3.4
12	38.0	101	3.6
13	41.5	88	3.7
14	60.0	99	3.5
15	61.5	90	3.4
16	72.0	98	3.5
17	104.0	70	4.8

a. Using a .05 significance level, test appropriate hypotheses to determine if a correlation exists between depth and zinc concentration.

b. Using a .05 significance level, does the data strongly suggest a correlation between depth and iron concentration?

c. Calculate the slope and intercept of the estimated regression line relating y = iron concentration and x = depth.

d. Use the estimated regression equation to construct a 95% prediction interval for the iron concentration of a single core sample taken at a depth of 50 m.

e. Compute and interpret a 95% interval estimate for the true average iron concentration of core samples taken at 70 m.

11.64 The given observations on x = body weight (kg) and y = water intake (L/day) for n = 7 subjects appeared in the paper "Validation of a Metabolic Model for Tritium" (*Radiation Research* (1984):503–09). Calculate the correlation coefficient r and use it to test the null hypothesis of no correlation between body weight and water intake. Use a .01 significance level.

Subject	1	2	3	4	5	6	7
x	95	52	73	50	82	68	60
y	3.94	1.03	1.71	1.75	1.76	2.01	.97

11.65 The concentration of various gases in ancient atmospheres can be deduced by analyzing air bubbles embedded in polar ice cores. The accompanying data on y = concentration of methane (ppb) and x = age of ice (years) is part of a data set that appeared in the paper "Gas- and Aqueous-Phase Chemistry of HO_2 in Liquid

Water Clouds" (*J. Geophys. Research* (1984):11,589–98). This data was used to compute $r = -.485$ and an estimated regression line of $\hat{y} = 705.376 - .039x$. Calculate the 10 residuals and construct a plot of the residuals versus the corresponding x values. Discuss the features of this plot as they relate to the assumptions of the linear regression model.

x	510	1380	3140	660	840	1100	1180	1280	1300	1550
y	600	620	590	800	670	690	650	650	630	640

11.66 The given scatter diagram, based on 34 sediment samples with x = sediment depth (cm) and y = oil and grease content (mg/kg), appeared in the paper "Mined Land Reclamation Using Polluted Urban Navigable Waterway Sediments" (*J. Environ. Quality* (1984):415–22). Discuss the effect that the observation (30, 33,000) will have on the estimated regression line. If this point were omitted, what can you say about the slope of the estimated regression line? What do you think will happen to the slope if this observation is included in the computations?

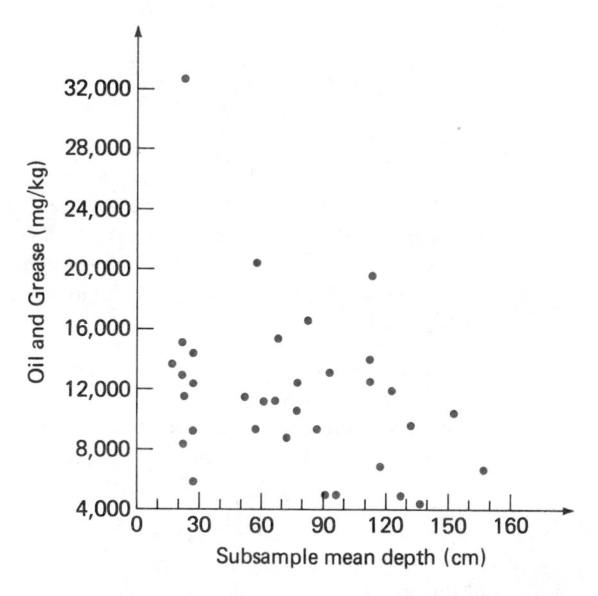

11.67 The paper "Improving Fermentation Productivity with Reverse Osmosis" (*Food Tech.* (1984):92–96) gave the following data (read from a scatter plot) on y = glucose concentration (g/L) and x = fermentation time (days) for a blend of malt liquor.

x	1	2	3	4	5	6	7	8
y	74	54	52	51	52	53	58	71

a. Use the data to calculate the estimated regression line.

b. Does the data indicate a linear relationship between y and x? Test using a .10 significance level.

c. Using the estimated regression line of (a), compute the residuals and construct a plot of the residuals versus x.

d. Based on the plot in (c), do you think that a linear model is appropriate for describing the relationship between y and x? Explain.

11.68 The employee relations manager of a large company was concerned that raises given to employees during a recent period might not have been based strictly on objective performance criteria. A sample of $n = 20$ employees was selected, and the values of x, a quantitative measure of productivity, and y, the percentage salary increase, were determined for each one. A computer package was used to fit the simple linear regression model, and the resulting output gave P-value $= .0076$ for the model utility test. Does the percentage raise appear to be linearly related to productivity? Explain.

11.69 Give a brief answer, comment, or explanation for each of the following:

a. What is the difference between $e_1, e_2, \ldots, e_n$ and the n residuals?

b. The simple linear regression model states that $y = \alpha + \beta x$.

c. Does it make sense to test hypotheses about b?

d. SSResid is always positive.

e. A student reported that a data set consisting of $n = 6$ observations yielded residuals 2, 0, 5, 3, 0, and 1 from the least squares line.

f. A research report included the following summary quantities obtained from a simple linear regression analysis:
$\Sigma(y - \bar{y})^2 = 615, \Sigma(y - \hat{y})^2 = 731$

REFERENCES

Neter, John, William Wasserman, and Michael Kutner. *Applied Linear Statistical Models.* Homewood, IL: Richard D. Irwin, 1985. (The first half of this book gives a comprehensive up-to-date treatment of regression analysis—simple linear, multiple, and nonlinear—without overindulging in mathematical development; a highly recommended reference.)

Younger, Mary Sue. *A Handbook for Linear Regression.* Boston: Duxbury, 1985. (A good, thorough introduction to many aspects of regression; particularly recommended for its discussion of various statistical computer packages.)

12 Multiple Regression Analysis

INTRODUCTION

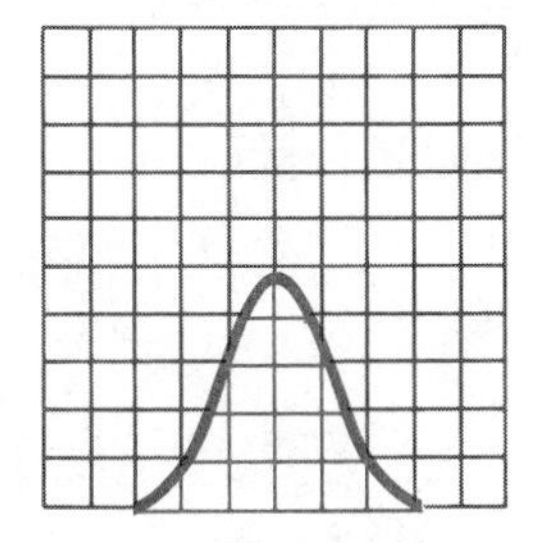

THE GENERAL OBJECTIVE of regression analysis is to establish a useful relationship between a dependent variable y and one or more independent (predictor) variables. The simple linear regression model $y = \alpha + \beta x + e$ discussed in Chapter 11 has been used successfully by many investigators in a wide variety of disciplines to relate y to a single predictor variable x. However, in many situations there will not be a strong relationship between y and any single predictor variable, but knowing the values of several independent variables may considerably reduce uncertainty concerning the associated y value. For example, some variation in house prices in a large city can certainly be attributed to house size, but knowledge of size by itself would not usually enable a bank appraiser to accurately assess (predict) a home's value. Price is also determined to some extent by other variables such as age, lot size, number of bedrooms and bathrooms, distance from schools, and so on. As another example, let y denote a quantitative measure of wrinkle resistance in cotton cellulose fabric. Then the value of y may depend upon such variable quantities as curing temperature, curing time, concentration of formaldehyde, and concentration of the catalyst sulfur dioxide used in the production process.

In this chapter we extend the regression methodology developed in the previous chapter to *multiple regression models*, those models that include at least two predictor variables. Fortunately, many of the concepts developed in the context of simple linear regression carry over to multiple regression with little or no modification. The calculations required to fit such a model and make further inferences are *much* more tedious than those for simple linear regression, so a computer is an indispensable tool for doing multiple regression analysis. The computer's ability to perform a huge number of computations in a very short time has spurred the development of new methods for analyzing large data sets with many predictor variables. These include techniques for fitting numerous alternative models and choosing between them, tools for identifying influential observations, and both algebraic and graphical diagnostics designed to reveal potential violations of model assumptions. A single chapter can do little more than scratch the surface of this important and beautiful subject area. For more extensive expositions, refer to the sources listed in the Chapter 11 references.

12.1 Multiple Regression Models

The distinction between deterministic and probabilistic models when relating y to two or more independent variables is as important as it was in the case of a single predictor variable x. Generally speaking, the relationship is deterministic if the value of y is completely determined with no uncertainty once values of the independent variables have been specified. Consider, for example, a school district in which teachers with no prior teaching experience and no college credits beyond a bachelor's degree start at an annual salary of \$16,000. Suppose that for each year of teaching experience up to 20 years, a teacher receives an additional \$800 per year and that each unit of postcollege coursework up to a limit of 75 results in an extra \$60 per year. Define three variables by

y = salary of a teacher who has at most 20 years teaching experience and at most 75 postcollege units
x_1 = number of years teaching experience
x_2 = number of postcollege units

Previously, x_1 and x_2 denoted the first two observations on the single variable x, but they now represent two different variables.

The value of y is entirely determined by values of x_1 and x_2 through the equation

$$y = 16{,}000 + 800x_1 + 60x_2$$

Thus if $x_1 = 10$ and $x_2 = 30$, $y = 16{,}000 + (800)(10) + (60)(30) = 16{,}000 + 8000 + 1800 = 25{,}800$. If two different teachers both have the same x_1 values and the same x_2 values, they will also have identical y values.

We previously discussed representing the equation $y = \alpha + \beta x$ geometrically as a straight line plotted on a two-dimensional coordinate system with a horizontal x axis and a vertical y axis. The equation $y = 16{,}000 + 800x_1 + 60x_2$ can be visualized geometrically as a plane (flat surface) plotted on a three-dimensional coordinate system (see Figure 1). The point $(x_1, x_2, y) = (10, 30, 25{,}800)$ lies on the plane because $y = 25{,}800$ when $x_1 = 10$ and $x_2 = 30$. To locate this point, go out a distance 10 along the x_1 axis and then

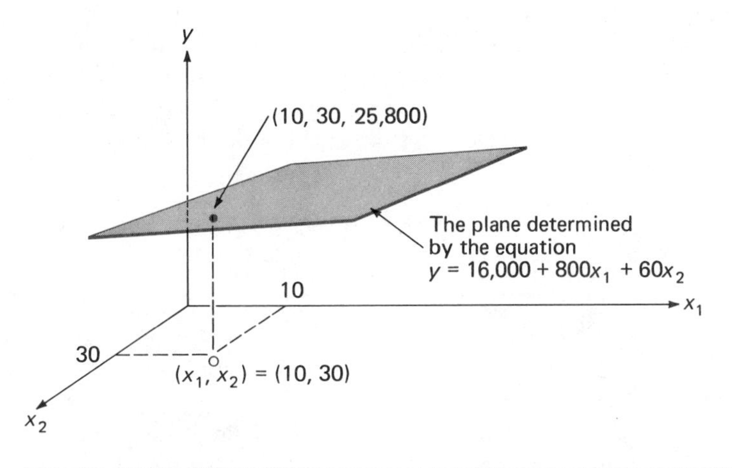

FIGURE 1 GEOMETRIC REPRESENTATION OF THE EQUATION $y = 16{,}000 + 800x_1 + 60x_2$

a distance 30 parallel to the x_2 axis to the point marked with an **o**. Finally move up (off the ground, so to speak) from the point **o** a distance 25,800 parallel to the vertical axis to the point marked ●. Any other point on the pictured plane is located by first substituting values of x_1 and x_2 into the equation to obtain the height of the plane above the (x_1, x_2) point "on the ground." Thus (5, 50, 23,000) is the point on the plane at height 23,000 above the ground point (5, 50). Because $y = 16{,}000$ when $x_1 = 0$ and $x_2 = 0$, the constant 16,000 in the equation is the y intercept.

The equation relating the variables y, x_1, and x_2 just defined is a special case of the general deterministic relationship

$$y = \alpha + \beta_1 x_1 + \beta_2 x_2$$

The geometric representation of this equation is a plane in which the constant term α is the y intercept (the height of the plane when $x_1 = 0$ and $x_2 = 0$). The coefficient β_1 is the change in y when x_1 increases by 1 unit while x_2 is held fixed. Thus in the example involving teacher salaries, $\beta_1 = 800$, so salary increases by 800 when experience increases by 1 year while the number of postcollege units remains fixed. Similarly, β_2 gives the change in y when x_2 increases by one unit while x_1 is held fixed.

It is rarely the case, except in some scientific contexts, that y can be deterministically related to independent variables x_1 and x_2. Even in science, a theoretical model might postulate a deterministic relationship, but such things as measurement error and variation in experimental conditions introduce some uncertainty about y observations. In the same way that the simple linear regression model generalizes the deterministic linear relationship $y = \alpha + \beta x$, the following probabilistic model generalizes the deterministic model $y = \alpha + \beta_1 x_1 + \beta_2 x_2$.

DEFINITION

The simplest **multiple regression model** with two independent variables x_1 and x_2 is given by the model equation

$$y = \alpha + \beta_1 x_1 + \beta_2 x_2 + e$$

The random deviation e is assumed to be normally distributed with mean value 0 and variance σ^2 regardless of the values of x_1 and x_2. This implies that for fixed x_1 and x_2 values, y itself has a normal distribution with variance σ^2 (standard deviation σ) and

$$\begin{pmatrix}\text{mean value of } y \text{ for} \\ \text{fixed } x_1 \text{ and } x_2 \text{ values}\end{pmatrix} = \alpha + \beta_1 x_1 + \beta_2 x_2$$

This model is similar in spirit to the simple linear regression model, and there is an analogous geometric interpretation. Consider the plane whose height above the point (x_1, x_2) is $\alpha + \beta_1 x_1 + \beta_2 x_2$. This plane is called the *population* (or *true*) *regression plane*. Without the random deviation e, every observed point (x_1, x_2, y) would have to lie exactly on the plane. The presence of the random deviation in the model equation allows observed points to fall above or below the plane. The height of the plane above (x_1, x_2) is now the average, or expected, value of y for these values of the independent variables, but the observed value will almost always deviate from the expected value by a random amount. This is illustrated in Figure 2. The coefficients β_1 and β_2 are often called *population regression coefficients*. The coefficient β_1 gives the average change in y associated with a 1-unit increase in x_1 when the value of x_2 is held constant, and β_2 has an analogous interpretation. Just as the value of σ described the amount of variability about the population regression line in simple linear regression, σ here reflects the amount of variability about the population regression plane. If σ is small, almost all observed points will lie on or close to the plane. When σ is large, it is quite likely that some observations will fall rather far above or below the plane. The value of σ indicates the amount by which a typical observation deviates from the plane.

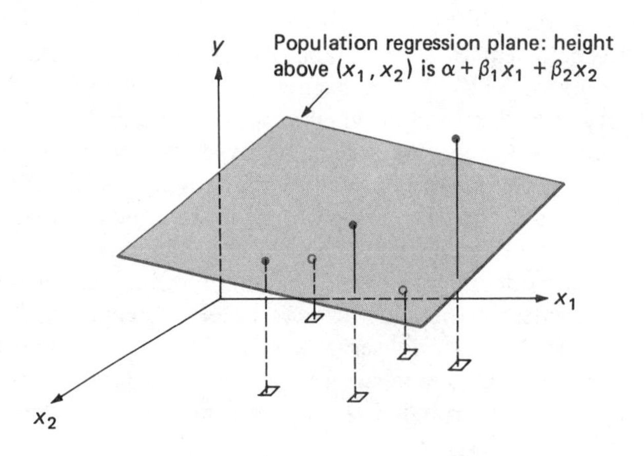

FIGURE 2 GEOMETRIC REPRESENTATION OF OBSERVATIONS FROM THE MODEL
$y = \alpha + \beta_1 x_1 + \beta_2 x_2 + e$
[□ = (x_1, x_2), a point on the ground
● = observation for which (x_1, x_2, y) lies above the regression plane
o = observation for which (x_1, x_2, y) lies below the regression plane]

EXAMPLE 1

Soil and sediment adsorption, the extent to which chemicals collect in a condensed form on the surface, is an important characteristic because it influences the effectiveness of pesticides and various agricultural chemicals. The paper "Adsorption of Phosphate, Arsenate, Methanearsonate, and Cacodylate by Lake and Stream Sediments: Comparisons with Soils" (*J. of Environ. Qual.* (1984):499–504) suggested the appropriateness of the given regression model for relating y = phosphate adsorption index to x_1 = amount of extractable iron and x_2 = amount of extractable aluminum. Suppose that $\alpha = -7.00$, $\beta_1 = .10$, $\beta_2 = .35$, and $\sigma = 4.5$, so that the model equation is

$$y = -7.00 + .10x_1 + .35x_2 + e$$

Then for any specified values of x_1 and x_2, the mean value of adsorption is $-7.00 + .10x_1 + .35x_2$. When $x_1 = 150$ and $x_2 = 40$, true average adsorption is $-7.00 + (.10)(150) + (.35)(40) = 22.00$. An observation on adsorption made for these values of x_1 and x_2 may result in a y value less than 22 (an observed point below the regression plane, resulting from a negative deviation e) or one that exceeds 22.00 (a point above the plane, corresponding to $e > 0$). With $\sigma = 4.5$, it would not be very surprising to observe a y value as small as 13.00 (2 standard deviations below the mean value) or as large as 31.00 when $x_1 = 150$ and $x_2 = 40$.

Because $\beta_1 = .10$, an increase of 1 μM/g in extractable iron while extractable aluminum remains fixed (at any value) is associated with an average increase of .10 in adsorption. Similarly, if extractable aluminum is increased by 1 μM/g while extractable iron is held fixed, the associated average increase in adsorption is .35. ■

This model involving two predictor variables, x_1 and x_2, can be extended to the case in which there are more than two predictor variables.

DEFINITION

A **general additive multiple regression model,** which relates a dependent variable y to k predictor variables $x_1, x_2, \ldots, x_k$, is given by the model equation

$$y = \alpha + \beta_1 x_1 + \beta_2 x_2 + \cdots + \beta_k x_k + e$$

The random deviation e is assumed to be normally distributed with mean value 0 and variance σ^2 for any values of $x_1, \ldots, x_k$. This implies that for fixed $x_1, x_2, \ldots, x_k$ values, y has a normal distribution with variance σ^2 and

$$\begin{pmatrix} \text{mean } y \text{ value for fixed} \\ x_1, \ldots, x_k \text{ values} \end{pmatrix} = \alpha + \beta_1 x_1 + \beta_2 x_2 + \cdots + \beta_k x_k$$

The β_i's are called **population regression coefficients,** and $\alpha + \beta_1 x_1 + \cdots + \beta_k x_k$ is often referred to as the **population regression function.**

A geometric interpretation of this k-predictor model is no longer possible except in certain special cases, but algebraic properties are analogous to those for the two-predictor model. In particular, the value of β_i gives the average change in y when x_i increases by 1 unit while the values of all other predictors are held fixed.

EXAMPLE 2

The paper "The Value of Information for Selected Appliances" (*J. of Marketing Research* (1980):14–25) suggests the plausibility of the general multiple regression model for relating the dependent variable y = the price of an air conditioner to $k = 3$ independent variables: x_1 = Btu-per-hour rating, x_2 = energy efficiency ratio, and x_3 = number of settings. Suppose that the model equation is

$$y = -70 + .025x_1 + 20x_2 + 7.5x_3 + e$$

and that $\sigma = 20$. Then the regression function is

$$\begin{pmatrix}\text{mean value of } y \text{ for} \\ \text{fixed values of } x_1, x_2, x_3\end{pmatrix} = -70 + .025x_1 + 20x_2 + 7.5x_3$$

When $x_1 = 6000$, $x_2 = 8.0$, and $x_3 = 5$, the mean value of price is $-70 + (.025)(6000) + (20)(8.0) + (7.5)(5) = \$277.50$. With $2\sigma = 40$, a y observation for these values of x_1, x_2, and x_3 is quite likely to be between \$237.50 and \$317.50. Because $\beta_2 = 20$, the average price increase associated with a 1-unit increase in energy efficiency ratio x_2 while both Btu rating x_1 and number of settings x_3 remain fixed is \$20. Similar interpretations apply to the regression coefficients $\beta_1 = .025$ and $\beta_3 = 7.5$. ■

A Special Case: Polynomial Regression

Consider again the case of a single independent variable x, and suppose that a scatter plot of the n sample (x, y) pairs has the appearance of Figure 3. The simple linear regression model is clearly not appropriate, but it does look as though a parabola (quadratic function) with equation $y = \alpha + \beta_1 x + \beta_2 x^2$ would provide a very good, though not perfect, fit to the data for appropriately chosen values of α, β_1, and β_2. Just as the inclusion of the random deviation e in simple linear regression allowed an observation to deviate from the population regression line by a random amount, adding e to this quadratic function yields a probabilistic model in which an observation is allowed to fall above or below the parabola. The model equation is $y = \alpha + \beta_1 x + \beta_2 x^2 + e$.

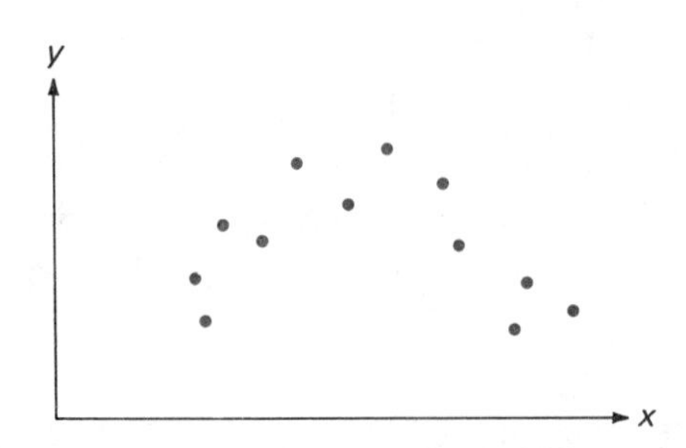

FIGURE 3 A SCATTER PLOT THAT SUGGESTS THE APPROPRIATENESS OF A QUADRATIC PROBABILISTIC MODEL

Let's rewrite the model equation by using x_1 to denote x and x_2 to denote x^2. The model equation then becomes $y = \alpha + \beta_1 x_1 + \beta_2 x_2 + e$, which is a special case of the general multiple regression model with $k = 2$. You may wonder about the legitimacy of allowing one predictor variable to be a mathematical function of another predictor—here $x_2 = (x_1)^2$. However, there is absolutely nothing in the general multiple regression model that prevents this. *In the model $y = \alpha + \beta_1 x_1 + \cdots + \beta_k x_k + e$, it is permissible to have several predictors that are mathematical functions of other predictors.* For example, starting with the two independent variables x_1 and x_2, we could create a model with $k = 4$ predictors in which x_1 and x_2 themselves are the first two predictor variables and $x_3 = (x_1)^2$, $x_4 = x_1 x_2$ (we shortly discuss the consequences of using a predictor such as x_4). In particular, the general polynomial regression model begins with a single independent variable x and creates predictors $x_1 = x$, $x_2 = x^2$, $x_3 = x^3, \ldots, x_k = x^k$ for some specified value of k.

DEFINITION

The **kth degree polynomial regression model**

$$y = \alpha + \beta_1 x + \beta_2 x^2 + \cdots + \beta_k x^k + e$$

is a special case of the general multiple regression model with $x_1 = x$, $x_2 = x^2, \ldots, x_k = x^k$. The regression function (mean value of y for fixed values of the predictors) is $\alpha + \beta_1 x + \cdots + \beta_k x^k$. The most important special case other than simple linear regression ($k = 1$) is the **quadratic regression model**

$$y = \alpha + \beta_1 x + \beta_2 x^2 + e$$

This model replaces the line of mean values $\alpha + \beta x$ in simple linear regression with a parabolic curve of mean values $\alpha + \beta_1 x + \beta_2 x^2$. If $\beta_2 > 0$, the curve opens upward, whereas if $\beta_2 < 0$, the curve opens downward (see Figure 4). A less frequently encountered special case is that of cubic regression, in which $k = 3$.

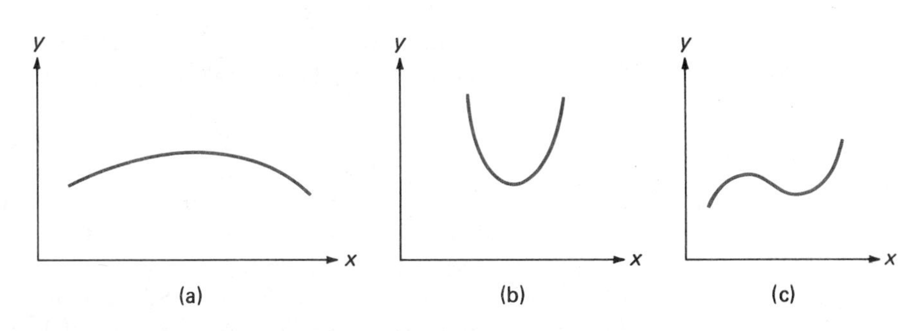

FIGURE 4 GRAPHS OF POLYNOMIAL REGRESSION FUNCTIONS
(a) Quadratic Regression Model with $\beta_2 < 0$
(b) Quadratic Regression Model with $\beta_2 > 0$
(c) Cubic Regression Model with $\beta_3 > 0$

EXAMPLE 3

The technique of manual defoliation has been used to simulate pest infestation in order to gain an understanding of how crop yield is affected by such infestation. The paper "Effect of Manual Defoliation on Pole Bean Yield" (*J. Econ. Entom.* (1984):1019–23) suggested the appropriateness of a quadratic regression model for relating yield y to defoliation level x. Suppose that the model is $y = 20 + 10x - 25x^2 + e$. Then the mean value of yield for a fixed defoliation level x (the regression function) is $20 + 10x - 25x^2$, whose graph is similar to that in Figure 4(a). The mean yield when $x = .50$ is $20 + 10(.5) - 25(.5)^2 = 18.75$. If $\sigma = 1.5$, it is quite likely that a yield observation made when $x = .50$ would be between 15.75 and 21.75. ■

Notice that the interpretation of β_i previously given for the general multiple regression model cannot be applied in polynomial regression. This is because all predictors are functions of the single variable x, so $x_i = (x)^i$ cannot be increased by 1 unit without changing the values of all the other predictor variables as well. In general, the interpretation of regression coefficients requires extra care when some predictor variables are mathematical functions of other variables.

Interaction Between Variables

Suppose that an industrial chemist is interested in the relationship between product yield (y) from a certain chemical reaction and two independent variables, x_1 = reaction temperature and x_2 = pressure at which the reaction is carried out. The chemist initially suggests that for temperature values between 80 and 110 in combination with pressure values ranging from 50 to 70, the relationship can be well described by the probabilistic model

$$y = 1200 + 15x_1 - 35x_2 + e$$

The regression function, which gives the mean y value for any specified values of x_1 and x_2, is then $1200 + 15x_1 - 35x_2$. Consider this mean y value for three different particular temperature values:

$$\text{when } x_1 = 90, \text{ mean } y \text{ value} = 1200 + (15)(90) - 35x_2 = 2550 - 35x_2$$
$$\text{when } x_1 = 95, \text{ mean } y \text{ value} = 2625 - 35x_2$$
$$\text{when } x_1 = 100, \text{ mean } y \text{ value} = 2700 - 35x_2$$

Graphs of these three mean value functions (each a function only of pressure x_2, since the temperature value has been specified) are shown in Figure 5(a). Each graph is a straight line, and the three lines are parallel, each one having slope -35. Because of this, the average change in yield when pressure x_2 is increased by 1 unit is -35 irrespective of the fixed temperature value.

Since chemical theory suggests that the decline in average yield when pressure x_2 increases should be more rapid for a high temperature than for a low temperature, the chemist now has reason to doubt the appropriateness of the proposed model. Rather than the lines being parallel, the line for temperature 100 should be steeper than the line for temperature 95, and that line in turn should be steeper than the one for $x_1 = 90$. A model that has this property includes, in addition to predictors x_1 and x_2 separately, a third predictor variable $x_3 = x_1x_2$. One such model is

$$y = -4500 + 75x_1 + 60x_2 - x_1x_2 + e$$

which has regression function $-4500 + 75x_1 + 60x_2 - x_1x_2$. When

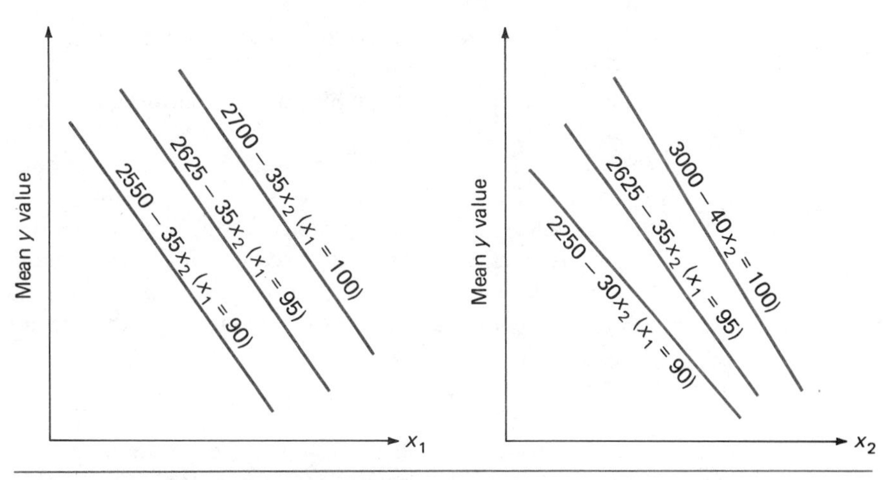

FIGURE 5 GRAPHS OF MEAN VALUE FUNCTIONS
(a) Mean y Values for the Model $y = 1200 + 15x_1 - 35x_2 + e$
(b) Mean y Values for the Model $y = -4500 + 75x_1 + 60x_2 - x_1x_2 + e$

$x_1 = 100$, the mean y value is $-4500 + (75)(100) + 60x_2 - 100x_2 = 3000 - 40x_2$, but the mean y value is $2625 - 35x_2$ when $x_1 = 95$ and $2250 - 30x_2$ when $x_1 = 90$. These are graphed in Figure 5(b), where it is clear that the three slopes are different. In fact, each different value of x_1 yields a different slope, so the average change in yield associated with a 1-unit increase in x_2 depends on the value of x_1. When this is the case, the two variables are said to *interact*.

DEFINITION	If the change in the mean y value associated with a 1-unit increase in one independent variable depends on the value of a second independent variable, there is **interaction** between these two variables. When the variables are denoted by x_1 and x_2, such interaction can be modeled by including x_1x_2, the product of the variables that interact, as a predictor variable.

The general equation for a multiple regression model based on two independent variables x_1 and x_2 that also includes an interaction predictor is

$$y = \alpha + \beta_1x_1 + \beta_2x_2 + \beta_3x_1x_2 + e$$

This model has a geometric interpretation similar to the regression plane for the simpler model involving x_1 and x_2 alone ($\alpha + \beta_1x_1 + \beta_2x_2 + \beta_3x_1x_2$ specifies a surface, more complicated than a plane, such that observations fall above or below the surface by random amounts), but the details are beyond the scope of our discussion. The important point is that when x_1 and x_2 do interact, this model will usually give a much better fit to resulting sample data—and thus explain more variation in y—than would the no-interaction model. Failure to consider a model with interaction too often leads an investigator to conclude incorrectly that there is no strong relationship between y and a set of independent variables.

When there are more than two independent variables, more than a single interaction predictor can be included in the model. If, for example, there are three independent variables under consideration, one possible model would have regression function $\alpha + \beta_1x_1 + \beta_2x_2 + \beta_3x_3 + \beta_4x_4 + \beta_5x_5 + \beta_6x_6$, where $x_4 = x_1x_2$, $x_5 = x_1x_3$, and $x_6 = x_2x_3$. One could even include a three-way interaction predictor $x_7 = x_1x_2x_3$ (the product of all three independent variables), although in practice this is rarely done. In applied work, quadratic terms, such as x_1^2, x_2^2, etc., are often included to model a curved relationship between y and several independent variables. For example, a frequently used model involving just two independent variables x_1 and x_2 but $k = 5$ predictors is the *full quadratic model*

$$y = \alpha + \beta_1x_1 + \beta_2x_2 + \beta_3x_1x_2 + \beta_4x_1^2 + \beta_5x_2^2 + e$$

This model replaces the straight lines of Figure 5 with parabolas (each one is the graph of the regression function for different values of x_2 when x_1 has a fixed value). With four independent variables, one could examine a model containing four quadratic predictors and six two-way interaction predictor variables. Clearly, with just a few independent variables, one could examine a great many different multiple regression models. In the last section of this chapter we briefly discuss methods for selecting one model from a number of competing models.

Qualitative Predictor Variables

Up to this point we have explicitly considered the inclusion of only quantitative (numerical) predictor variables in a multiple regression model. Using a simple numerical coding, qualitative (categorical) variables can also be incorporated into a model. Let's focus first on a dichotomous variable, one with just two possible categories—male or female, U.S. or foreign manufacture, a house with or without a view, etc. With any such variable we associate a numerical variable x whose possible values are 0 and 1, where 0 is identified with one category (e.g., married) and $x = 1$ with the other possible category (not married). This 0-1 variable is often called a **dummy,** or **indicator, variable.**

EXAMPLE 4

The paper "Estimating Urban Travel Times: A Comparative Study" (*Transportation Research* (1980):173–75) considered relating the dependent variable y = travel time between locations in a certain city and the independent variable *distance between locations* for two types of vehicles, passenger cars and trucks. Let

$$x_1 = \begin{cases} 1 \text{ if the vehicle is a truck} \\ 0 \text{ if the vehicle is a passenger car} \end{cases}$$

$$x_2 = \text{distance between locations}$$

One possible multiple regression model is

$$y = \alpha + \beta_1x_1 + \beta_2x_2 + e$$

while a second possibility is a model with an interaction term,

$$y = \alpha + \beta_1x_1 + \beta_2x_2 + \beta_3x_1x_2 + e$$

The regression function for the no-interaction model is $\alpha + \beta_1x_1 + \beta_2x_2$, and for the interaction model it is $\alpha + \beta_1x_1 + \beta_2x_2 + \beta_3x_3$, where $x_3 = x_1x_2$. Considering separately the cases $x_1 = 0$ and $x_1 = 1$ yields

$$\text{No interaction: } \begin{cases} \text{average time} = \alpha + \beta_2 x_2 & \text{when } x_1 = 0 \text{ (cars)} \\ \text{average time} = \alpha + \beta_1 + \beta_2 x_2 & \text{when } x_1 = 1 \text{ (trucks)} \end{cases}$$

$$\text{Interaction: } \begin{cases} \text{average time} = \alpha + \beta_2 x_2 & \text{when } x_1 = 0 \\ \text{average time} = \alpha + \beta_1 + (\beta_2 + \beta_3) x_2 & \text{when } x_1 = 1 \end{cases}$$

For each model, the graph of the average time, when regarded as a function of distance, is a line for either type of vehicle (Figure 6). In the no-interaction model, the coefficient on x_2 is β_2 both when $x_1 = 0$ and when $x_1 = 1$, so the two lines are parallel, although their intercepts are different (unless $\beta_1 = 0$). With interaction, the lines not only have different intercepts but also different slopes (unless $\beta_3 = 0$). For this latter model, the change in average travel time when distance increases by 1 unit depends on which type of vehicle is involved—the two variables *vehicle type* and *travel time* interact. Indeed, data collected by the authors of the paper suggested the presence of interaction.

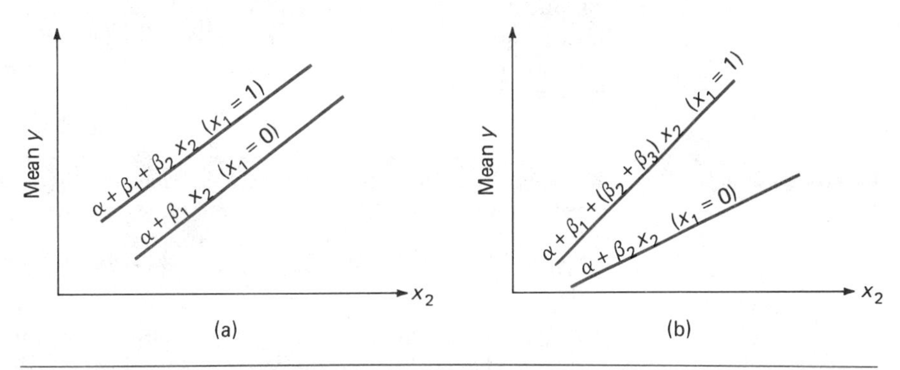

FIGURE 6 REGRESSION FUNCTIONS FOR MODELS WITH ONE QUALITATIVE VARIABLE (x_1) AND ONE QUANTITATIVE VARIABLE (x_2)
(a) No Interaction
(b) Interaction ■

You might think that the way to handle a three-category situation is to define a single numerical variable with coded values such as 0, 1, and 2 corresponding to the three categories. This is incorrect because it imposes an ordering on the categories that is not necessarily implied by the problem context. The correct approach to modeling a categorical variable with three categories is to define *two* different 0-1 variables.

EXAMPLE 5

The paper "The Effect of Ownership on the Organization Structure in Small Firms" (*Admin. Sci. Quarterly* (1984):232–37) considers three different management status categories—owner-successors, professional managers, and owner-founders—in relating degree of horizontal differentiation (y) to firm size. Let

$$x_1 = \begin{cases} 1 & \text{if the firm is managed by owner-successors} \\ 0 & \text{otherwise} \end{cases}$$

$$x_2 = \begin{cases} 1 & \text{if the firm is managed by professional managers} \\ 0 & \text{otherwise} \end{cases}$$

$$x_3 = \text{firm size}$$

Thus $x_1 = 1, x_2 = 0$ indicates owner-successors, $x_1 = 0, x_2 = 1$, professional managers, and $x_1 = x_2 = 0$, owner-founders ($x_1 = x_2 = 1$ is not allowed). The author then suggests an interaction model of the form

$$y = \alpha + \beta_1 x_1 + \beta_2 x_2 + \beta_3 x_3 + \beta_4 x_1 x_3 + \beta_5 x_2 x_3 + e$$

This model allows the change in mean differentiation when size increases by 1 to be different for all three management status categories. ■

In general, incorporating a categorical variable with c possible categories into a regression model requires the use of $c - 1$ indicator variables. Thus even one such categorical variable can add many predictors to a model.

Nonlinear Multiple Regression Models

Many nonlinear relationships can be put in the form $y = \alpha + \beta_1 x_1 + \cdots + \beta_k x_k + e$ by transforming one or more of the variables. An appropriate transformation could be suggested by theory or by various plots of the data (e.g., residual plots after fitting a particular model). There are also relationships that cannot be linearized by transformations, in which case more complicated methods of analysis must be used. A general discussion of nonlinear regression is beyond the scope of this text, but you can learn more by consulting one of the sources listed in the chapter references.

EXERCISES

12.1 Explain the difference between a deterministic and a probabilistic model. Give an example of a dependent variable y and two or more independent variables that might be related to y deterministically. Give an example of a dependent variable y and two or more independent variables that might be related to y in a probabilistic fashion.

12.2 The paper "The Influence of Temperature and Sunshine on the Alpha-Acid Contents of Hops" (*Ag. Meteorology* (1974):375–82) used a multiple regression model to relate y = yield of hops to x_1 = mean temperature between date of coming into hop and date of picking and x_2 = mean percentage of sunshine during the same period. The model equation proposed is

$$y = 415.11 - 6.60x_1 - 4.50x_2 + e$$

a. Suppose that this equation does indeed describe the true relationship. What mean yield corresponds to a temperature of 20 and a sunshine percentage of 40?

b. What is the mean yield when the mean temperature and percentage of sunshine are 18.9 and 43, respectively?

12.3 The multiple regression model $y = \alpha + \beta_1 x_1 + \beta_2 x_2 + e$ can be used to describe the relationship between y = profit margin of a savings and loan company and x_1 = net revenues and x_2 = number of branch offices. Based on data given in the paper "Entry and Profitability in a Rate-Free Savings and Loan Market" (*Quarterly Review of Econ. and Business* (1978):87–95) a reasonable model equation is

$$y = 1.565 + .237x_1 - .0002x_2 + e$$

a. How would you interpret the values of β_1 and β_2?

b. If the number of branch offices remains fixed and net revenue increases by 1, what is the average change in profit margin?

c. What would the mean profit margin be for a savings and loan with a net revenue of 4.0 and 6500 branch offices?

12.4 The paper "Readability of Liquid Crystal Displays: A Response Surface" (*Human Factors* (1983):185–90) used a multiple regression model with four independent variables, where

y = error percentage for subjects reading a four-digit liquid crystal display
x_1 = level of backlight (ranging from 0 to 122 cd/m^2)
x_2 = character subtense (ranging from .025° to 1.34°)
x_3 = viewing angle (ranging from 0° to 60°)
x_4 = level of ambient light (ranging from 20 to 1500 lx)

The model equation suggested in the paper is

$$y = 1.52 + .02x_1 - 1.40x_2 + .02x_3 - .0006x_4 + e$$

a. Assume that this is the correct equation. What is the mean value of y when $x_1 = 10$, $x_2 = .5$, $x_3 = 50$, and $x_4 = 100$?

b. What mean error percentage is associated with a backlight level of 20, character subtense of .5, viewing angle of 10, and ambient light level of 30?

12.5 The multiple regression model

$$y = .69 + 4.70x_1 + .00041x_2 - .72x_3 + .023x_4 + e$$

where

y = stock purchase tender premium (as a percentage of closing market price one week prior to offer date)
x_1 = fee per share (as a percentage of price)
x_2 = percentage of shares sought
x_3 = relative change in Dow Jones industrial average
x_4 = volume of shares traded (as a percentage of those outstanding)

is based on data appearing in the paper "Factors Influencing the Pricing of Stock Repurchase Tenders" (*Quarterly Rev. of Econ. and Business* (1978): 31–39).

a. What is the mean stock tender premium when $x_1 = 2$, $x_2 = .1$, $x_3 = 1.2$, and $x_4 = 6$?

b. If the variables x_1, x_3, and x_4 are held fixed, would the mean tender premium increase or decrease as the percentage of shares sought increases? Explain.

12.6 The article "Pulp Brightness Reversion: Influence of Residual Lignin on the Brightness Reversion of Bleached Sulfite and Kraft Pulps" (*TAPPI* (1964):653–62) proposed a quadratic regression model to describe the relationship between x = degree of delignification during the processing of wood pulp for paper and y = total chlorine content. Suppose that the actual model is

$$y = 220 + 75x - 4x^2 + e$$

a. Graph the regression function $220 + 75x - 4x^2$ over x values ranging between 2 and 12 (substitute $x = 2, 4, 6, 8, 10$, and 12 to find points on the graph, and connect them with a smooth curve).

b. Would mean chlorine content be higher for a degree of delignification value of 8 or 10?

c. What is the change in mean chlorine content when the degree of delignification increases from 8 to 9? From 9 to 10?

12.7 The relationship between yield of maize, date of planting, and planting density was investigated in the article "Development of a Model for Use in Maize Replant Decisions" (*Agronomy J.* (1980):459–64). Letting

y = percent maize yield
x_1 = planting date (days after April 20)
x_2 = planting density (plants/ha),

the regression model with both quadratic terms ($y = \alpha + \beta_1 x_1 + \beta_2 x_2 + \beta_3 x_3 + \beta_4 x_4 + e$, where $x_3 = x_1^2$ and $x_4 = x_2^2$) provides a good description of the relationship between y and the independent variables.

a. If $\alpha = 21.09$, $\beta_1 = .653$, $\beta_2 = .0022$, $\beta_3 = -.0206$, and $\beta_4 = .00004$, write out the regression function.

b. Use the regression function in (a) to determine the true mean yield for a plot planted on May 6 with a density of 41,180 plants/ha.

c. Would the mean yield be higher for a planting date of May 6 or May 22 (for the same density)?

12.8 Consider a regression analysis with two independent variables.

a. Explain what is meant by the statement "there is no interaction between the independent variables."

b. Explain what is accomplished by including an interaction term in a regression model.

12.9 Suppose that the variables y, x_1, and x_2 are related by the regression model

$$y = 1.8 + .1x_1 + .8x_2 + e$$

Then

$$\text{mean } y \text{ value} = 1.8 + .1x_1 + .8x_2$$

a. Construct a graph (similar to that of Figure 5) showing the relationship between y and x_2 for fixed values 10, 20, and 30 of x_1.

b. Construct a graph depicting the relationship between y and x_1 for fixed values 50, 55, and 60 of x_2.

c. What aspect of the graphs in (a) and (b) can be attributed to the lack of an interaction between x_1 and x_2?

12.10 Suppose the interaction term $.03x_3$, where $x_3 = x_1x_2$, is added to the regression equation in Exercise 12.9. Using this new model, construct the graphs described in 12.9(a) and 12.9(b). How do they differ from those obtained in Exercise 12.9?

12.11 Consider a regression analysis with three independent variables x_1, x_2, and x_3. Write out the following regression models.

a. The model that includes all independent variables but no quadratic or interaction terms

b. The model that includes all independent variables and all quadratic terms

c. All models that include all independent variables, no quadratic terms, and exactly one interaction term

d. The model that includes all independent variables, all quadratic terms, and all interaction terms

12.12 The paper "The Value and the Limitations of High-Speed Turbo-Exhausters for the Removal of Tar-Fog from Carburetted Water-Gas" (*Soc. Chem. Industry J.* (1946):166–68) presented data on y = tar content (grains/100 ft^3) of a gas stream as a function of x_1 = rotor speed (rev/min) and x_2 = gas inlet temperature (°F). A regression model using x_1, x_2, $x_3 = x_2^2$, and $x_4 = x_1x_2$ was suggested:

$$\text{mean } y \text{ value} = 86.8 - .123x_1 + 5.09x_2 - .0709x_3 + .001x_4$$

a. According to this model, what is the mean y value if $x_1 = 3200$ and $x_2 = 57$?

b. For this particular model, does it make sense to interpret the values of any individual β (β_1, β_2, β_3, or β_4) in the way we have previously suggested? Explain.

12.13 How many dummy variables would be needed to incorporate a nonnumerical variable with four categories? Suppose that you wanted to incorporate size class of car with four categories (subcompact, compact, midsize, and luxury) into a regression equation where y = gas mileage, x_1 = age of car, and x_2 = engine size. Define the necessary dummy variables and write out the complete model equation.

12.2 Fitting a Model and Assessing Its Utility

In Section 12.1 we discussed and contrasted multiple regression models containing several different types of predictors. Let's now suppose that a particular set of k predictor variables $x_1, x_2, \ldots, x_k$ has been selected for inclusion in the model

$$y = \alpha + \beta_1 x_1 + \beta_2 x_2 + \cdots + \beta_k x_k + e$$

It is then necessary to estimate the model coefficients $\alpha, \beta_1, \ldots, \beta_k$ and the regression function $\alpha + \beta_1 x_1 + \cdots + \beta_k x_k$ (mean y value for specified values of the predictors), assess the model's utility, and perhaps use the estimated model to make further inferences. All this, of course, requires sample data. As before, n denotes the number of observations in the sample. With just one predictor variable, the sample consisted of n (x, y) pairs. Now each observation consists of $k + 1$ numbers: a value of x_1, a value of $x_2, \ldots,$ a value of x_k, and the associated value of y. The n observations are assumed to have been selected independently of one another.

EXAMPLE 6

Example 1 cited an article that reported on a regression analysis with dependent variable y = phosphate adsorption index of a sediment and $k = 2$ predictor variables, x_1 = amount of extractable iron, and x_2 = amount of extractable aluminum. The analysis was based on the following sample of $n = 13$ observations, each one consisting of a triple (x_1, x_2, y):

Observation	x_1 = Extractable Iron	x_2 = Extractable Aluminum	y = Adsorption Index
1	61	13	4
2	175	21	18
3	111	24	14
4	124	23	18
5	130	64	26
6	173	38	26
7	169	33	21
8	169	61	30
9	160	39	28
10	244	71	36
11	257	112	65
12	333	88	62
13	199	54	40

■

The principle of least squares, which was used to estimate the model coefficients α and β in simple linear regression, is also used to estimate coefficients in multiple regression. Let $a, b_1, \ldots, b_k$ denote the estimates of $\alpha, \beta_1, \ldots, \beta_k$, so that the estimated regression function is $a + b_1x_1 + b_2x_2 + \cdots + b_kx_k$. If the values of $x_1, x_2, \ldots x_k$ corresponding to a single sample observation are substituted into the estimated regression function, a prediction of the associated y value results. For example, the first observation in the data set of Example 6 has $x_1 = 61$ and $x_2 = 13$, so the corresponding predicted y value is $a + b_1(61) + b_2(13)$. The deviation between the observed and predicted y value is then $4 - [a + b_1(61) + b_2(13)]$. In a similar manner, we can write an expression for the deviation between any observed y and the corresponding predicted y value. The principle of least squares then says to use as estimates of α, β_1, and β_2 the values of a, b_1, and b_2 that minimize the sum of these squared deviations.

DEFINITION

According to the principle of least squares, the fit of a particular estimated regression function $a + b_1x_1 + \cdots + b_kx_k$ to the observed data is measured by the sum of squared deviations between the observed y values and the y values predicted by the estimated function:

$$\Sigma[y - (a + b_1x_1 + \cdots + b_kx_k)]^2$$

The **least squares estimates** of $\alpha, \beta_1, \ldots, \beta_k$ are those values of $a, b_1, \ldots, b_k$ that make this sum of squared deviations as small as possible.

The least squares estimates for a given data set are obtained by solving a system of $k + 1$ equations in the $k + 1$ unknowns $a, b_1, \ldots, b_k$ (called the *normal equations*). In the case $k = 1$, simple linear regression, there are only two equations, and we gave their general solution—the expressions for b and a—in Chapter 11. For $k \geq 2$, it is not as easy to write general expressions for the estimates without using advanced mathematical methods (which we are trying hard to avoid). Fortunately, the computer saves us! Formulas for the estimates have been programmed into all the commonly used statistical computer packages. After the data has been properly entered, a regression command instructs the computer to perform all the required calculations quickly and accurately and then print out the estimates, as well as much additional information.

EXAMPLE 7

Figure 7 displays MINITAB output from a regression command requesting that the model $y = \alpha + \beta_1x_1 + \beta_2x_2 + e$ be fit to the phosphate adsorption data of Example 6 (we named the dependent variable HPO, an abbreviation for the chlorate H_2PO_4 (dihydrogen phosphate), and the two predictor variables x_1 and x_2 were named Fe and Al, the standard abbreviations for iron and aluminum, respectively). In the remainder of this section and in the next one, we discuss many aspects of this output. For now, focus on the column labeled COEFFICIENT in the table near the top of the figure. The three numbers in this column are the estimated model coefficients: $a = -7.351$ (the estimate of the constant term α), $b_1 = .11273$ (the estimate of the coefficient β_1), and $b_2 = .34900$ (the estimate of the coefficient β_2). Thus the estimated regression

```
THE REGRESSION EQUATION IS
HPO = - 7.35 + 0.113 FE + 0.349 AL

                                    ST. DEV.     T-RATIO =
COLUMN       COEFFICIENT           OF COEF.      COEF/S.D.
                -7.351               3.485         -2.11
FE              0.11273             0.02969         3.80
AL              0.34900             0.07131         4.89

S = 4.379

R-SQUARED = 94.8 PERCENT
R-SQUARED = 93.8 PERCENT, ADJUSTED FOR D.F.

ANALYSIS OF VARIANCE

 DUE TO       DF           SS          MS=SS/DF
REGRESSION     2         3529.9          1765.0
RESIDUAL      10          191.8            19.2
TOTAL         12         3721.7

FURTHER ANALYSIS OF VARIANCE
SS EXPLAINED BY EACH VARIABLE WHEN ENTERED IN THE ORDER GIVEN
 DUE TO       DF           SS
REGRESSION     2         3529.9
FE             1         3070.5
AL             1          459.4

                        Y      PRED. Y    ST.DEV.
 ROW       FE         HPO       VALUE    PRED. Y   RESIDUAL     ST.RES.
   1       61        4.00        4.06       2.43      -0.06       -0.02
   2      175       18.00       19.71       2.31      -1.71       -0.46
   3      111       14.00       13.54       1.72       0.46        0.11
   4      124       18.00       14.66       1.67       3.34        0.83
   5      130       26.00       29.64       2.62      -3.64       -1.04
   6      173       26.00       25.41       1.41       0.59        0.14
   7      169       21.00       23.22       1.56      -2.22       -0.54
   8      169       30.00       32.99       1.60      -2.99       -0.73
   9      160       28.00       24.30       1.30       3.70        0.89
  10      244       36.00       44.94       1.71      -8.94       -2.22R
  11      257       65.00       60.71       3.20       4.29        1.44
  12      333       62.00       60.90       3.19       1.10        0.37
  13      199       40.00       33.93       1.29       6.07        1.45

R DENOTES AN OBS. WITH A LARGE ST. RES.
```

FIGURE 7 MINITAB OUTPUT FOR THE REGRESSION ANALYSIS OF EXAMPLE 7

function is

$$\left(\begin{array}{c}\text{estimated mean value of } y\\ \text{for specified } x_1 \text{ and } x_2 \text{ values}\end{array}\right) = -7.351 + .11273x_1 + .34900x_2$$

This equation is given (with coefficients rounded slightly) using the named variables at the very top of the MINITAB output. ■

EXAMPLE 8

In Section 12.1 we pointed out that the quadratic regression model $y = \alpha + \beta_1 x + \beta_2 x^2 + e$ based on a single independent variable x can be regarded as a special case of the general additive multiple regression model with $k = 2$ and predictor variables $x_1 = x$ and $x_2 = x^2$. The accompanying data appeared in the paper "Determination of Biological Maturity and Effect of Harvesting and Drying Conditions on Milling Quality of Paddy" (*J. of Ag.*

Engr. Research (1975):353–61). The dependent variable y is yield (kg/ha) of paddy, a grain farmed in India, and x is the number of days after flowering at which harvesting took place.

	$x_1 = x$	$x_2 = x^2$	y		$x_1 = x$	$x_2 = x^2$	y
1	16	256	2508	9	32	1024	3823
2	18	324	2518	10	34	1156	3646
3	20	400	3304	11	36	1296	3708
4	22	484	3423	12	38	1444	3333
5	24	576	3057	13	40	1600	3517
6	26	676	3190	14	42	1764	3241
7	28	784	3500	15	44	1936	3103
8	30	900	3883	16	46	2116	2776

A scatter plot of the 16 (x, y) observations given in Figure 8 shows curvature resembling the downward-opening parabola of Figure 4(a) (Section 12.1). On the basis of this plot, we agreed with the paper's authors that a quadratic regression model was appropriate for relating y to x. Figure 8 gives MINITAB output resulting from first entering x and y values, then requesting that the x^2 values be computed, and finally using a regression command with $x_1 = x$ and $x_2 = x^2$ as the two predictor variables. The resulting estimated model coefficients are $a = -1070.4$, $b_1 = 293.48$ (the estimated coefficient on the linear term), and $b_2 = -4.5358$ (the estimated coefficient on the quadratic term). This gives as the equation of the estimated regression function

$$\begin{pmatrix}\text{estimated mean } y \text{ value} \\ \text{for a specified value of } x\end{pmatrix} = -1070.4 + 293.48x - 4.5358x^2$$

The negative value of b_2 implies that a plot of the estimated regression function opens downward.

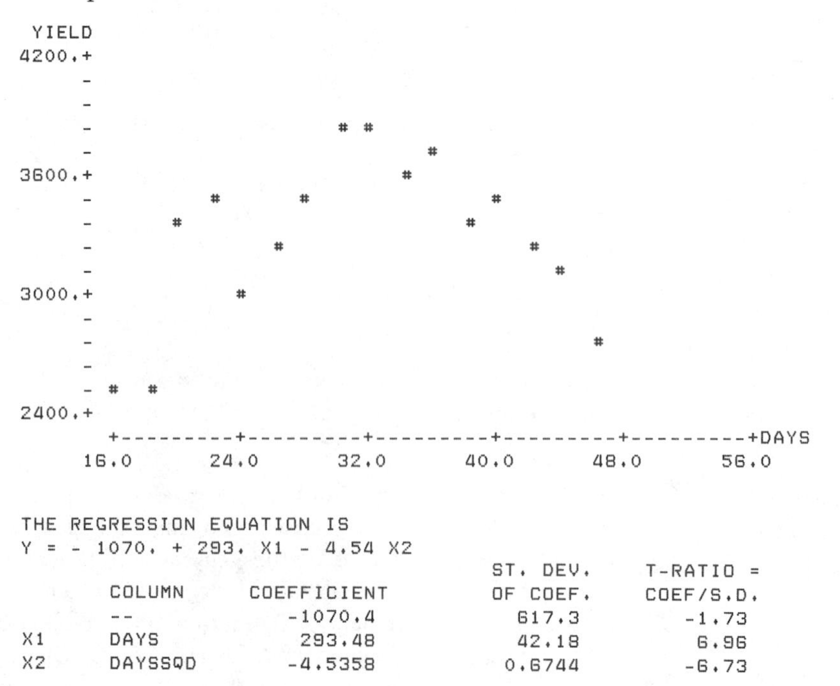


```
THE REGRESSION EQUATION IS
Y = - 1070. + 293. X1 - 4.54 X2
                                          ST. DEV.      T-RATIO =
        COLUMN        COEFFICIENT         OF COEF.      COEF/S.D.
        --              -1070.4            617.3          -1.73
X1      DAYS             293.48            42.18           6.96
X2      DAYSSQD         -4.5358            0.6744         -6.73
```

```
THE ST. DEV. OF Y ABOUT REGRESSION LINE IS
S = 203.9
WITH (  16- 3) =   13 DEGREES OF FREEDOM

R-SQUARED = 79.4 PERCENT
R-SQUARED = 76.2 PERCENT, ADJUSTED FOR D.F.

ANALYSIS OF VARIANCE

 DUE TO        DF           SS        MS=SS/DF
REGRESSION      2      2084779         1042390
RESIDUAL       13       540388           41568
TOTAL          15      2625168

FURTHER ANALYSIS OF VARIANCE
SS EXPLAINED BY EACH VARIABLE WHEN ENTERED IN THE ORDER GIVEN

 DUE TO        DF           SS
REGRESSION      2      2084779
DAYS            1       204526
DAYSSQD         1      1880253

            X1          Y      PRED. Y    ST.DEV.
ROW       DAYS      YIELD       VALUE     PRED. Y    RESIDUAL    ST.RES.
  1       16.0     2508.0      2464.2      135.6        43.8      0.29 X
  2       18.0     2518.0      2742.7      104.8      -224.7     -1.28
  3       20.0     3304.0      2984.9       83.0       319.1      1.71
  4       22.0     3423.0      3190.9       71.3       232.1      1.22
  5       24.0     3057.0      3360.6       68.4      -303.6     -1.58
  6       26.0     3190.0      3494.0       70.7      -304.0     -1.59
  7       28.0     3500.0      3591.1       74.2       -91.1     -0.48
  8       30.0     3883.0      3651.9       76.4       231.1      1.22
  9       32.0     3823.0      3676.4       76.4       146.6      0.78
 10       34.0     3646.0      3664.6       74.2       -18.6     -0.10
 11       36.0     3708.0      3616.6       70.7        91.4      0.48
 12       38.0     3333.0      3532.3       68.4      -199.3     -1.04
 13       40.0     3517.0      3411.6       71.3       105.4      0.55
 14       42.0     3241.0      3254.7       83.0       -13.7     -0.07
 15       44.0     3103.0      3061.5      104.8        41.5      0.24
 16       46.0     2776.0      2832.1      135.6       -56.1     -0.37 X

X DENOTES AN OBS. WHOSE X VALUE GIVES IT LARGE INFLUENCE.
```

FIGURE 8 MINITAB OUTPUT FOR THE DATA AND MODEL OF EXAMPLE 8

Is the Model Useful?

The utility of an estimated model can be assessed by examining the extent to which predicted y values based on the estimated regression function are close to the y values actually observed.

DEFINITION

The ith predicted value $\hat{y}_i$ is obtained by taking the values of the predictor variables $x_1, x_2, \ldots, x_k$ for the ith sample observation and substituting these values into the estimated regression function. Doing this successively for $i = 1, 2, 3, \ldots, n$ yields the **predicted values** $\hat{y}_1, \hat{y}_2, \ldots, \hat{y}_n$. The **residuals** are then the differences $y_1 - \hat{y}_1, y_2 - \hat{y}_2, \ldots, y_n - \hat{y}_n$ between the observed and predicted y values.

The predicted values and residuals are defined here exactly as they were in simple linear regression, but computation of the $\hat{y}_i$'s is more tedious because there is more than one predictor. Fortunately, the $\hat{y}_i$'s and $y_i - \hat{y}_i$'s are automatically computed and printed out by all good statistical computer packages. Consider Figure 7, which contains MINITAB output for the phosphate adsorption data. Since the first y observation, $y_1 = 4$, was made with $x_1 = 61$ and $x_2 = 13$, the first predicted value is

$$\hat{y}_1 = -7.351 + .11273(61) + .34900(13) \approx 4.06$$

as given in the output. The first residual is then $y_1 - \hat{y}_1 = 4 - 4.06 = -.06$. The other predicted values and residuals are computed in a similar fashion. The sum of residuals from a least squares fit should, except for rounding effects, be zero. The sum in Example 7 is $-.01$.

As in simple linear regression, the sum of squared residuals is the basis for several important summary quantities that are indicative of a model's utility.

DEFINITION

The **residual sum of squares, SSResid,** and **total sum of squares, SSTo,** are given by

$$\text{SSResid} = \Sigma(y - \hat{y})^2 \qquad \text{SSTo} = \Sigma(y - \bar{y})^2$$

where $\bar{y}$ is the mean of the y observations in the sample. The number of degrees of freedom associated with SSResid is $n - (k + 1)$, because $k + 1$ df are lost in estimating the $k + 1$ coefficients $\alpha, \beta_1, \ldots, \beta_k$. An estimate of the random deviation variance σ^2 is given by

$$s_e^2 = \frac{\text{SSResid}}{n - (k + 1)}$$

and s_e is the estimate of σ. The **coefficient of multiple determination,** R^2**,** interpreted as the proportion of variation in observed y values that is explained by the fitted model, is

$$R^2 = 1 - \frac{\text{SSResid}}{\text{SSTo}}$$

Generally speaking, a desirable model is one that results in both a large R^2 value and a small s_e value. However, there is a catch: These two conditions can be achieved by fitting a model that contains a large number of predictors. Such a model may be successful in explaining y variation, but it almost always specifies a relationship that is unrealistic and difficult to interpret. What we really want is a simple model, one with relatively few predictors whose roles are easily interpreted, which also does a good job of explaining variation in y.

All statistical computer packages include R^2 and s_e in their output, and most give SSResid also. In addition, some packages compute a quantity called *adjusted* R^2, which involves a downward adjustment of R^2. If a large R^2 has been achieved through using just a few model predictors, adjusted R^2 will differ little from R^2. However, the adjustment can be substantial when either a great many predictors (relative to the number of observations) have been used or when R^2 itself is small to moderate (which could happen even when there is no relationship between y and the predictors).

EXAMPLE 9

(*Example 7 continued*) Looking again at Figure 7, which contains MINITAB output for the adsorption data fit by a two-predictor model, residual sum of squares is

$$\text{SSResid} = (-.06)^2 + (-1.71)^2 + (.46)^2 + \cdots + (6.07)^2 = 191.8337$$

This value also appears in rounded form in the RESIDUAL row and SS column of the table headed ANALYSIS OF VARIANCE. The associated number of degrees of freedom is $n - (k + 1) = 13 - (2 + 1) = 10$, which appears in the DF column just to the left of SSResid. The sample average y value is $\bar{y} = 29.85$, so

$$\text{SSTo} = \Sigma(y - \bar{y})^2 = (4 - 29.85)^2 + (18 - 29.85)^2 + \cdots + (40 - 29.85)^2 = 3721.6923$$

This value, rounded to one decimal place, appears in the TOTAL row and SS column of the ANALYSIS OF VARIANCE table just under the value of SSResid. The values of s_e^2, s_e, and R^2 are then

$$s_e^2 = \frac{\text{SSResid}}{n - (k + 1)} = \frac{191.83}{10} = 19.18 \approx 19.2$$

(in the MS column of the MINITAB output)

$$s_e = \sqrt{s_e^2} = \sqrt{19.18} = 4.379$$

(S = 4.379 appears near the top of the output)

$$R^2 = 1 - \frac{\text{SSResid}}{\text{SSTo}} = 1 - \frac{191.83}{3721.69} = 1 - .052 = .948$$

Thus the percentage of variation explained is $100R^2 = 94.8\%$, which appears on the ouput as R-SQUARED = 94.8 PERCENT. Adjusted R^2 is .938, not much less than R^2, since a high R^2 value was achieved with only two predictor variables. The values of R^2, adjusted R^2, and s_e all suggest that the chosen model has been very successful in relating y to the predictors. ■

The F Test for Model Utility

In the simple linear model with regression function $\alpha + \beta x$, if $\beta = 0$, there is no useful linear relationship between y and the single predictor variable x. Similarly, if all k coefficients $\beta_1, \beta_2, \ldots, \beta_k$ are zero in the general k-predictor multiple regression model, there is no useful linear relationship between y and *any* of the predictor variables $x_1, x_2, \ldots, x_k$ included in the model. Before using an estimated model to make further inferences (e.g., predictions and estimates of mean values), it is desirable to confirm the model's utility through a formal test procedure.

Recall that SSTo is a measure of total variation in the sample data and that SSResid measures the amount of total variation that has not been explained by the fitted model. The difference SSTo − SSResid, which we shall denote by SSRegr and call **regression sum of squares,** is interpreted as the amount of total variation that has been explained by the model. Intuitively, the model should be judged useful if SSRegr is large relative to SSResid, and this is achieved by using a small number of predictors relative to the sample size. A statistic based on the ratio of SSRegr to SSResid is used to test model utility. The procedure description involves a new type of probability distribution called an F distribution, so we first digress briefly to describe some general properties of F distributions.

An F distribution always arises in connection with a ratio in which the numerator involves one sum of squares and the denominator involves a second sum of squares (SSRegr and SSResid, respectively, for the model utility test). Each sum of squares has associated with it a specified number of degrees of freedom, so a particular F distribution is determined by fixing values of numerator df and denominator df. There is a different F distribution for each different numerator df–denominator df combination (each df must be a whole number). For example, there is an F distribution based on 4 numerator df and 12 denominator df, another F distribution based on 3 numerator df and 20 denominator df, and so on. A typical F curve for fixed numerator and denominator df appears in Figure 9.

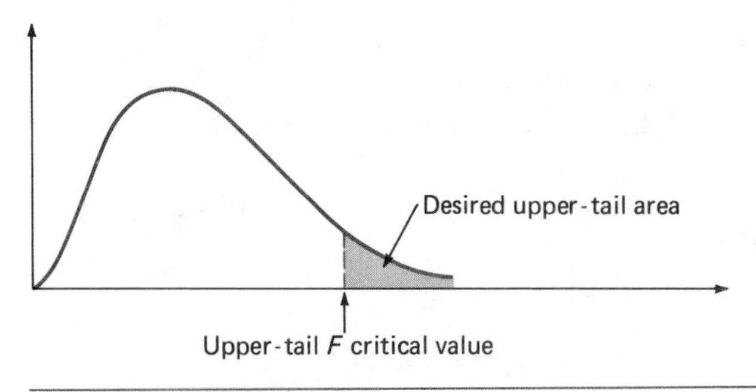

FIGURE 9 A TYPICAL F PROBABILITY DISTRIBUTION CURVE

All F tests presented in this book are upper-tailed, so associated rejection regions require only upper-tailed F critical values, those values that capture specified areas in the upper tail of the appropriate F curves. A particular F critical value depends not only on the desired tail area but also on both numerator and denominator df. A tabulation of F critical values therefore consumes much more space than was the case for t critical values, which depended on only a single number of degrees of freedom. Table XI(a) in the appendices gives the critical values for tail area .05 and Table XI(b) gives those for tail area .01. Different columns of the tables are identified with different numerator df and different rows are associated with different denominator df. For example, the .05 critical value for the F distribution with 4 numerator df and 6 denominator df is found in the column of Table XI labeled 4 and the row labeled 6 to be 4.53. The .05 critical value for 6 numerator and 4 denominator df is 6.16 (so don't accidentally interchange numerator and denominator df!).

The number of degrees of freedom associated with SSRegr is k, the number of predictors in the model, and df for SSResid is $n - (k + 1)$. The model utility F test is based on the following distributional result.

When all k β_i's have value zero in the model $y = \alpha + \beta_1 x_1 + \cdots + \beta_k x_k + e$, the statistic

$$F = \frac{\text{SSRegr}/k}{\text{SSResid}/[n - (k + 1)]}$$

has an F probability distribution based on k numerator df and $n - (k + 1)$ denominator df.

The value of F tends to be larger when at least one β_i is not zero than when all the β_i's are zero, since more variation is typically explained by the model in the former case than in the latter. An F statistic value far out in the upper tail of the associated F distribution can be more plausibly attributed to at least one nonzero β_i than to something very unusual having occurred when all β_i's are zero. This is why the model utility F test is upper-tailed.

The F Test for Utility of the Model $y = \alpha + \beta_1 x_1 + \cdots + \beta_k x_k + e$

Null hypothesis: $H_0 : \beta_1 = \beta_2 = \cdots = \beta_k = 0$ (There is no useful linear relationship between y and any of the predictors.)

Alternative hypothesis: H_a : At least one among $\beta_1, \ldots, \beta_k$ is not zero. (There is a useful linear relationship between y and at least one of the predictors.)

Test statistic:

$$F = \frac{\text{SSRegr}/k}{\text{SSResid}/[n - (k + 1)]}$$

where SSRegr = SSTo − SSResid. Dividing both SSRegr and SSResid by SSTo yields the equivalent formula

$$F = \frac{R^2/k}{(1 - R^2)/[n - (k + 1)]}$$

Rejection region: $F > F$ critical value. The F critical value is based on k numerator df and $n - (k + 1)$ denominator df, respectively. Table XI(a) gives critical values corresponding to level of significance .05 and Table XI(b) gives level .01 critical values.

The null hypothesis is the claim that the model is not useful. Unless H_0 can be rejected at a small level of significance, the model has not demonstrated its utility, in which case the investigator must search further for a model that can be judged useful. The alternative formula for F allows the test to be carried out when only R^2, k, and n are available, as is frequently the case in published articles.

EXAMPLE 10

The model fit to the phosphate adsorption data introduced in Example 7 involved $k = 2$ predictors. Let's carry out the model utility test at level .05 using information from the MINITAB output displayed in Figure 7.

1. The model is $y = \alpha + \beta_1 x_1 + \beta_2 x_2 + e$, where y = phosphate adsorption, x_1 = extractable Fe, and x_2 = extractable Al.
2. $H_0 : \beta_1 = \beta_2 = 0$.
3. H_a : At least one of the two β_i's is not zero.

4. Test statistic:

$$F = \frac{\text{SSRegr}/k}{\text{SSResid}/[n - (k + 1)]}$$

5. Rejection region: $F > F$ critical value. With numerator df $= k = 2$ and denominator df $= n - (k + 1) = 13 - 3 = 10$, Table XI(a) gives the level .05 critical value as 4.10. Thus H_0 will be rejected if $F > 4.10$.

6. Directly from the table below ANALYSIS OF VARIANCE in Figure 7, the SS column gives SSRegr = 3529.9 and SSResid = 191.8. Thus

$$F = \frac{3529.9/2}{191.8/10} = \frac{1764.95}{19.18} = 92.02$$

7. Since $92.02 > 4.10$, H_0 can be decisively rejected. The utility of the fitted model has been confirmed. ■

EXAMPLE 11

Example 2 cited an article that reported on a regression analysis with $y =$ the price of an air conditioner. Data was collected on three independent variables—$x_1 =$ Btu rating, $x_2 =$ energy efficiency ratio, and $x_3 =$ number of settings—and the model with just these three predictors was fit. The resulting estimated regression function, based on $n = 19$ observations, was

$$\begin{pmatrix}\text{estimated mean } y \text{ value} \\ \text{for specified } x_1, x_2, x_3\end{pmatrix} = -68.326 + .023x_1 + 19.729x_2 + 7.653x_3$$

and R^2 for this model was .84. The high R^2 value certainly suggests a useful model, but let's carry out a formal test at level of significance .01.

1. The fitted model was $y = \alpha + \beta_1 x_1 + \beta_2 x_2 + \beta_3 x_3 + e$.
2. $H_0: \beta_1 = \beta_2 = \beta_3 = 0$.
3. H_a: At least one of the three β_i's is not zero.
4. Test statistic:

$$F = \frac{R^2/k}{(1 - R^2)/[n - (k + 1)]}$$

5. Rejection region: $F > F$ critical value. For $k = 3$ and $n = 19$, there are 3 numerator df and $19 - 4 = 15$ denominator df. Table XI(b) then gives the level .01 critical value as 5.42, so H_0 will be rejected if $F > 5.42$.

6. $R^2 = .84$ and $1 - R^2 = .16$, so

$$F = \frac{.84/3}{.16/15} = \frac{.280}{.0107} = 26.2$$

7. The computed F value does fall in the rejection region ($26.2 > 5.42$), so H_0 is rejected in favor of the conclusion that the postulated model is indeed useful. ■

In the next section we presume that a model has been judged useful by the F test and then show how the estimated coefficients and regression function can be used to draw further conclusions. However, you should realize that in many applications, there will be more than one model whose utility could be confirmed by the F test. Suppose, for example, that data has been collected on six independent variables $x_1, x_2, \ldots, x_6$. Then a model with all six of these as predictors might be judged useful, another useful model might have only x_1, x_3, x_5, and x_6 as predictors, and yet another useful model might incorporate the six predictors x_1, x_4, x_6, $x_7 = x_1x_4$, $x_8 = x_1^2$, and $x_9 = x_6^2$. In the last section of this chapter we briefly consider strategies for selecting a model.

EXERCISES

12.14 The paper "The Influence of Mount St. Helens Ash on Wheat Growth and Phosphorous, Sulfur, Calcium, and Magnesium Uptake" (*J. Envir. Quality* (1984): 91–96) used a quadratic regression model ($y = \alpha + \beta_1x_1 + \beta_2x_2 + e$, where $x_1 = x$ and $x_2 = x^2$) to describe the relationship between y = biomass production of wheat (g/pot) and x = percent volcanic ash in the soil. Data from a greenhouse experiment in which x ranged between 0 and 75 was used to estimate α, β_1, and β_2. The resulting least squares estimates a, b_1, and b_2 were .067, .054, and $-.00052$, respectively.

a. Write out the estimated regression equation.

b. What would you predict biomass to be when the percent of volcanic ash is 20? When it is 40?

c. Graph the parabola $\hat{y} = .067 + .054x_1 - .00052x_2$. Use values of x ranging between 0 and 75. Based on this graph, at approximately what point does an increasing percent of volcanic ash begin to have a detrimental effect on average biomass?

12.15 In order to predict the demand for imports in Jamaica, the coefficients α, β_1, and β_2 in a multiple regression model with

y = import volume
x_1 = expenditures on personal consumption
x_2 = price of import/domestic price

were estimated using 19 years of data. The resulting estimated regression equation, $\hat{y} = -58.9 + .20x_1 - .10x_2$, appeared in the article "Devaluation and the Balance of Payments Adjustment in a Developing Economy: An Analysis Relating to Jamaica" (*Appl. Econ.* (1981):151–65).

a. How would you interpret the value $b_1 = .20$?

b. If personal consumption and import prices remain stable, would the predicted import volume increase or decrease with a reduction in domestic price?

c. The paper reported that the value of R^2 associated with this regression equation was .96. How would you interpret this value?

12.16 When coastal power stations take in large quantities of cooling water, it is inevitable that a number of fish are drawn in with the water. Various methods have been designed to screen out the fish. The paper "Multiple Regression Analysis For Forecasting Critical Fish Influxes at Power Station Intakes" (*J. Appl. Ecol.* (1983): 33–42) examined intake fish catch at an English power plant and several other variables thought to affect fish intake:

y = fish intake (number of fish)
x_1 = water temperature (°C)
x_2 = number of pumps running
x_3 = sea state (taking values 0, 1, 2, or 3)
x_4 = speed (knots)

Part of the data given in the paper was used to obtain the estimated regression equation

$$\hat{y} = 92 - 2.18x_1 - 19.20x_2 - 9.38x_3 + 2.32x_4 \quad \text{(based on } n = 26\text{)}$$

a. SSRegr = 1486.9 and SSResid = 2230.2 were also calculated. Use this information to carry out the model utility test with a significance level of .05.

b. What information (in addition to the result of the model utility test in (a)) would you need in order to decide whether the estimated regression equation should be used to predict fish catch? Explain how you would use the requested information.

12.17 The paper "Microbial Ecology of the Soils of Indian Desert" (*Ag. Ecosystems and Environ.* (1983):361–69) reported data on characteristics of Indian desert soils. The following variables were included: y = soil pH, x_1 = electrical conductivity (mho/cm), x_2 = organic carbon (%), x_3 = nitrogen (%), x_4 = phosphorus (ppm), and x_5 = annual rainfall (mm). The n = 10 observations were used to compute SSRegr = .170368 and SSResid = .213632. Use the model utility test and a significance level of .05 to decide if the independent variables as a group provide any information that is useful for predicting soil pH.

12.18 The article "The Undrained Strength of Some Thawed Permafrost Soils" (*Canad. Geotech. J.* (1979):420–27) contained the accompanying data on y = shear strength of sandy soil (kPa), x_1 = depth (m) and x_2 = water content (%). The predicted values and residuals were computed using the estimated regression equation $\hat{y} = -151.36 - 16.22x_1 + 13.48x_2 + .094x_3 - .253x_4 + .492x_5$, where $x_3 = x_1^2$, $x_4 = x_2^2$, and $x_5 = x_1x_2$.

y	x_1	x_2	Predicted y	Residual
14.7	8.9	31.5	23.35	−8.65
48.0	36.6	27.0	46.38	1.62
25.6	36.8	25.9	27.13	−1.53
10.0	6.1	39.1	10.99	−.99
16.0	6.9	39.2	14.10	1.90
16.8	6.9	38.3	16.54	.26
20.7	7.3	33.9	23.34	−2.64
38.8	8.4	33.8	25.43	13.37
16.9	6.5	27.9	15.63	1.27
27.0	8.0	33.1	24.29	2.71
16.0	4.5	26.3	15.36	.64
24.9	9.9	37.8	29.61	−4.71
7.3	2.9	34.6	15.38	−8.08
12.8	2.0	36.4	7.96	4.84

a. Use the given information to compute SSResid, SSTo, and SSRegr.

b. Calculate R^2 for this regression model. How would you interpret this value?

c. Use the value of R^2 from (b) and a .05 level of significance to conduct the appropriate model utility test.

12.19 Partial MINITAB output for the data and variables described in Exercise 12.12 is given below.

COLUMN	COEFFICIENT	ST. DEV. OF COEF.	T-RATIO = COEF/S.D.
	86.85	85.39	1.02
X1	-0.12297	0.03276	-3.75
X2	5.090	1.969	2.58
X3	-0.07092	0.01799	-3.94
X4	0.0015380	0.0005560	2.77

```
S = 4.784

R-SQUARED = 90.8 PERCENT
R-SQUARED = 89.4 PERCENT, ADJUSTED FOR D.F.

ANALYSIS OF VARIANCE

 DUE TO       DF          SS        MS=SS/DF
REGRESSION     4      5896.6          1474.2
RESIDUAL      26       595.1            22.9
TOTAL         30      6491.7
```

a. What is the estimated regression equation?

b. Using a .01 significance level, perform the model utility test (for explanation of the MINITAB output, see Example 9).

c. Interpret the values of R^2 and s_e given in the output.

12.20 The paper "The Caseload Controversy and the Study of Criminal Courts" (*J. Criminal Law and Criminology* (1979):89–101) used a multiple regression analysis to help assess the impact of judicial caseload on the processing of criminal court cases. Data was collected in the Chicago criminal courts on the following variables:

y = number of indictments

x_1 = number of cases on the docket

x_2 = number of cases pending in criminal court trial system

The estimated regression equation (based on $n = 367$ observations) was $\hat{y} = 28 - .05x_1 - .003x_2 + .00002x_3$, where $x_3 = x_1x_2$.

a. The reported value of R^2 was .16. Conduct the model utility test. Use a .05 significance level.

b. Given the results of the test in (a), does it surprise you that the R^2 value is so low? Can you think of a possible explanation for this?

12.21 The paper "Readability of Liquid Crystal Displays: A Response Surface" (*Human Factors* (1983):185–90) used the estimated regression equation

$$\hat{y} = 1.52 + .02x_1 - 1.40x_2 + .02x_3 - .0006x_4$$

to describe the relationship between y = error percentage for subjects reading a four-digit liquid crystal display and the independent variables x_1 = level of backlight, x_2 = character subtense, x_3 = viewing angle, and x_4 = level of ambient light. From a table given in the paper, SSRegr = 19.2, SSResid = 20, and $n = 30$.

a. Does the estimated regression equation specify a useful relationship between y and the independent variables? Use the model utility test with a .05 significance level.

b. Calculate R^2 and s_e for this model. Interpret these values.

c. Do you think that the estimated regression equation would provide reasonably accurate predictions of error rate? Explain.

12.22 Factors affecting breeding success of puffins were examined in the paper "Breeding Success of the Common Puffin on Different Habitats at Great Island, Newfoundland" (*Ecol. Monographs* (1972):239–66). Data given in the paper was used to estimate the regression model

$$y = \alpha + \beta_1x_1 + \beta_2x_2 + \beta_3x_3 + \beta_4x_4 + e$$

where y = puffin nest density (number per 9 m^2), x_1 = grass cover (%), x_2 = mean soil depth (cm), x_3 = angle of slope (degrees), and x_4 = distance from cliff edge (m).

The estimated regression equation (using least squares) was

$$\hat{y} = 12.3 - .0186x_1 - .0430x_2 + .224x_3 - .182x_4$$

and SSRegr = 1650.02, SSResid = 264.19, and $n = 38$.

a. Do the independent variables provide information that is useful for predicting y? Use the model utility test with a significance level of .01.

b. What quantities would you want to calculate in order to assess the accuracy of predictions based on the estimated regression equation? Explain your choices.

c. Calculate R^2 and s_e and interpret these values.

d. For this same data set, would the equation $\hat{y} = 12 - .02x_1 - .05x_2 + .2x_3 - .2x_4$ result in a larger or smaller sum of squared residuals than the estimated regression equation given? Explain.

12.23 The paper "Effect of Manual Defoliation on Pole Bean Yield" (*J. Econ. Ent.* (1984):1019–23) used a quadratic regression model to describe the relationship between y = yield (kg/plot) and x = defoliation level (a proportion between zero and one). The estimated regression equation based on $n = 24$ was $\hat{y} = 12.39 + 6.67x_1 - 15.25x_2$, where $x_1 = x$ and $x_2 = x^2$. The paper also reported that R^2 for this model was .902. Does the quadratic model specify a useful relationship between y and x? Carry out the appropriate test using a .01 level of significance.

12.24 Suppose that a multiple regression data set consists of $n = 20$ observations. For what values of k, the number of model predictors, would the corresponding model with $R^2 = .90$ be judged useful at significance level .05? Does such a large R^2 value necessarily imply a useful model? Explain.

12.25 *This problem requires the use of a computer package.* Use the data given in Exercise 12.18 to verify that the true regression equation

$$\text{mean } y \text{ value} = \alpha + \beta_1x_1 + \beta_2x_2 + \beta_3x_3 + \beta_4x_4 + \beta_5x_5$$

is estimated by

$$\hat{y} = -151.36 - 16.22x_1 + 13.48x_2 + .094x_3 - .253x_4 + .492x_5$$

12.26 *This problem requires the use of a computer package.* The accompanying data resulted from a study of the relationship between y = brightness of finished paper and the independent variables x_1 = hydrogen peroxide (% by weight), x_2 = sodium hydroxide (% by weight), x_3 = silicate (% by weight), and x_4 = process temperature ("Advantages of CE-HDP Bleaching for High Brightness Kraft Pulp Production" *TAPPI* (1964):170A–73A).

x_1	x_2	x_3	x_4	y	x_1	x_2	x_3	x_4	y
.2	.2	1.5	145	83.9	.1	.3	2.5	160	82.9
.4	.2	1.5	145	84.9	.5	.3	2.5	160	85.5
.2	.4	1.5	145	83.4	.3	.1	2.5	160	85.2
.4	.4	1.5	145	84.2	.3	.5	2.5	160	84.5
.2	.2	3.5	145	83.8	.3	.3	.5	160	84.7
.4	.2	3.5	145	84.7	.3	.3	4.5	160	85.0
.2	.4	3.5	145	84.0	.3	.3	2.5	130	84.9
.4	.4	3.5	145	84.8	.3	.3	2.5	190	84.0
.2	.2	1.5	175	84.5	.3	.3	2.5	160	84.5
.4	.2	1.5	175	86.0	.3	.3	2.5	160	84.7
.2	.4	1.5	175	82.6	.3	.3	2.5	160	84.6
.4	.4	1.5	175	85.1	.3	.3	2.5	160	84.9
.2	.2	3.5	175	84.5	.3	.3	2.5	160	84.9
.4	.2	3.5	175	86.0	.3	.3	2.5	160	84.5
.2	.4	3.5	175	84.0	.3	.3	2.5	160	84.6
.4	.4	3.5	175	85.4					

a. Find the estimated regression equation for the model that includes all independent variables, all quadratic terms, and all interaction terms.

b. Using a .05 significance level, perform the model utility test.

c. Interpret the values of the following quantities: SSResid, R^2, s_e.

12.27 *This problem requires the use of a computer package.* The cotton aphid poses a threat to cotton crops in Iraq. The accompanying data on

y = infestation rate (aphids/100 leaves)

x_1 = mean temperature (°C)

x_2 = mean relative humidity

appeared in the paper "Estimation of the Economic Threshold of Infestation for Cotton Aphid" (*Mesopotamia J. Ag.* (1982):71–75). Use the data to find the estimated regression equation and assess the utility of the multiple regression model

$$y = \alpha + \beta_1 x_1 + \beta_2 x_2 + e$$

y	x_1	x_2	y	x_1	x_2
61	21.0	57.0	77	24.8	48.0
87	28.3	41.5	93	26.0	56.0
98	27.5	58.0	100	27.1	31.0
104	26.8	36.5	118	29.0	41.0
102	28.3	40.0	74	34.0	25.0
63	30.5	34.0	43	28.3	13.0
27	30.8	37.0	19	31.0	19.0
14	33.6	20.0	23	31.8	17.0
30	31.3	21.0	25	33.5	18.5
67	33.0	24.5	40	34.5	16.0
6	34.3	6.0	21	34.3	26.0
18	33.0	21.0	23	26.5	26.0
42	32.0	28.0	56	27.3	24.5
60	27.8	39.0	59	25.8	29.0
82	25.0	41.0	89	18.5	53.5
77	26.0	51.0	102	19.0	48.0
108	18.0	70.0	97	16.3	79.5

12.3 Inferences Based on an Estimated Model

In the previous section we discussed estimating the coefficients $\alpha, \beta_1, \ldots, \beta_k$ in the model $y = \alpha + \beta_1 x_1 + \cdots + \beta_k x_k + e$ (using the principle of least squares) and then showed how the utility of the model could be confirmed by application of the model utility F test. If $H_0: \beta_1 = \cdots = \beta_k = 0$ cannot be rejected at a reasonably small level of significance, it must be concluded that the model does not specify a useful relationship between y and any of the predictor variables $x_1, \ldots, x_k$. The investigator must search further for a model that does describe a useful relationship, perhaps by introducing different predictors or making variable transformations. Only if H_0 can be rejected is it appropriate to proceed further with the chosen model and make inferences based on the estimated coefficients $a, b_1, \ldots, b_k$ and the estimated regression function $a + b_1 x_1 + \cdots + b_k x_k$. Here we shall consider two different types of inferential problems. One type involves drawing a conclusion about an individual regression coefficient β_i—either computing a confidence interval for

β_i or testing a hypothesis concerning β_i. The second type of problem involves first fixing values of $x_1, \ldots, x_k$ and then computing either a point estimate or confidence interval for the corresponding mean y value, testing a hypothesis about this mean value, or predicting a future y value (using a single number or a prediction interval).

Inferences About Regression Coefficients

A confidence interval for and hypothesis test concerning the slope coefficient β in simple linear regression were based on facts about the sampling distribution of the statistic b used to obtain a point estimate of β. The model assumptions implied that b had a normal distribution with mean value β (so b was an unbiased statistic for estimating β) and standard deviation σ_b. Because σ_b involved the unknown σ^2 as well as the sampled x values, we replaced σ^2 by s_e^2 to obtain the estimated standard deviation s_b. The key result on which inferential procedures were based was that the standardized variable $t = (b - \beta)/s_b$ had a t distribution with $n - 2$ df.

Inferences concerning a particular β_i are based on results analogous to those just described. The difficulty is that formulas for b_i and its standard deviation σ_{b_i} are quite complicated and cannot be stated concisely except by using some advanced mathematical notation. Fortunately, these formulas have been programmed into all the most popular statistical computer packages, so the calculations are performed automatically once the sample data is entered. We have only to use the output intelligently. (That's not always easy!)

Properties of the Sampling Distribution of b_i

Let b_i denote the statistic for estimating (via the principle of least squares) the coefficient β_i in the model $y = \alpha + \beta_1 x_1 + \cdots + \beta_k x_k + e$. Assumptions about this model given in Section 12.1 imply that

1. b_i has a normal distribution.
2. b_i is an unbiased statistic for estimating β_i ($\mu_{b_i} = \beta_i$).
3. The standard deviation of b_i, σ_{b_i}, involves σ^2 and a complicated function of all the values of $x_1, x_2, \ldots, x_k$ in the sample. The estimated standard deviation s_{b_i} results from replacing σ^2 by s_e^2 in the formula for σ_{b_i}.

The standardized variable

$$t = \frac{b_i - \beta_i}{s_{b_i}}$$

then has a t distribution with $n - (k + 1)$ df.

Because the formula for s_{b_i} is quite complicated (as is the formula for b_i itself), we won't give it here. However, a good statistical computer package will provide both the estimated coefficients $a, b_1, \ldots, b_k$ and their estimated standard deviations $s_a, s_{b_1}, \ldots, s_{b_k}$. These values can then be used to compute a confidence interval for β_i or to test a hypothesis about β_i.

Inferences Concerning a Single β_i

A confidence interval for β_i is

$$b_i \pm (t \text{ critical value}) \cdot s_{b_i}$$

A procedure for testing hypotheses about β_i is given by

Null hypothesis: $H_0 : \beta_i =$ hypothesized value

Test statistic: $t = \dfrac{b_i - \text{hypothesized value}}{s_{b_i}}$

Alternative hypothesis	*Rejection region*
$H_a : \beta_i >$ hypothesized value	$t > t$ critical value
$H_a : \beta_i <$ hypothesized value	$t < -t$ critical value
$H_a : \beta_i \neq$ hypothesized value	Either $t > t$ critical value or $t < -t$ critical value

The t critical value is based on $n - (k + 1)$ df and is selected from Table IV to yield the desired confidence level or level of significance. When the hypothesized value is 0, the test statistic $t = b_i/s_{b_i}$ is often called a ***t* ratio.**

MINITAB displays the b_i's in the COEFFICIENT column near the top of the output, and the s_{b_i}'s appear right next to them in the column headed ST. DEV. OF COEF. (See Figure 7 or Figure 8). In addition, MINITAB includes the value of each ratio b_i/s_{b_i}, appropriate for testing $H_0 : \beta_i = 0$, in the T-RATIO column.

EXAMPLE 12

Reconsider the regression of phosphate adsorption (y) against extractable iron (x_1) and extractable aluminum (x_2), with the sample data given in Example 6 and MINITAB output displayed in Figure 7. Suppose that the investigator had requested a 95% confidence interval for β_1, the average change in phosphate adsorption when extractable iron increases by one unit and extractable aluminum is held fixed. The necessary quantities are $b_1 = .11273$, $s_{b_1} = .02969$, and $n - (k + 1) = 13 - 3 = 10$. From Table IV, the appropriate t critical value for 95% confidence is 2.23, so the confidence interval is

$$.11273 \pm (2.23)(.02969) = .11273 \pm .06621 \approx (.047, .179)$$ ■

EXAMPLE 13

After fitting a quadratic model (one involving predictors x and x^2) to sample data, it is often of interest to see whether the inclusion of the quadratic predictor is important. That is, one often wishes to test $H_0 : \beta_2 = 0$ versus $H_a : \beta_2 \neq 0$. If H_0 cannot be rejected, the fit of the simple linear regression model (the model when $\beta_2 = 0$) is not much improved upon by including the quadratic term. Let's carry out this test at level of significance .01 for the paddy yield data introduced in Example 8. A MINITAB printout appears in Figure 8, from which we see that $b_2 = -4.5358$, $s_{b_2} = .6744$, and the t ratio is -6.73.

1. The model is $y = \beta_0 + \beta_1 x + \beta_2 x^2 + e$.
2. $H_0 : \beta_2 = 0$.

3. $H_a : \beta_2 \neq 0$.

4. Test statistic: $t = b_2/s_{b_2}$, the t ratio for β_2.

5. Rejection region: $n = 16$ and $k = 2$, so df $= 16 - 3 = 13$. From Table IV, the t critical value for a two-tailed level .01 test based on 13 df is 3.01. H_0 will be rejected at this level of significance either if $t > 3.01$ or if $t < -3.01$.

6. The computed value of t is the t ratio -6.73.

7. Since $-6.73 < -3.01$, H_0 is rejected. The data strongly suggests that $\beta_2 \neq 0$, so it is important to retain the quadratic predictor. ■

EXAMPLE 14

Our analysis of the phosphate adsorption data introduced in Example 6 has so far focused on the model $y = \alpha + \beta_1 x_1 + \beta_2 x_2 + e$ in which x_1 (extractable iron) and x_2 (extractable aluminum) affect the response separately. Suppose that the researcher wishes to investigate the possibility of interaction between x_1 and x_2 through fitting the model $y = \alpha + \beta_1 x_1 + \beta_2 x_2 + \beta_3 x_3 + e$, where $x_3 = x_1 x_2$. We list a few of the sample values of y, x_1, and x_2, along with the corresponding values of x_3. In practice, a statistical computer package would automatically compute x_3 values upon request once x_1 and x_2 values had been entered, so hand computations would not be necessary. Figure 10 displays partial MINITAB output resulting from a request to fit this model. Let's use the output to see whether inclusion of the interaction predictor is justified.

Observation	y	x_1	x_2	$x_3 = x_1 x_2$
1	4	61	13	793
2	18	175	21	3,675
3	14	111	24	2,664
⋮	⋮	⋮	⋮	⋮
13	40	199	54	10,746

```
THE REGRESSION EQUATION IS
HPO = - 2.37 + 0.0828 FE + 0.246 AL +0.000528 FEAL

                                ST. DEV.       T-RATIO =
COLUMN     COEFFICIENT          OF COEF.       COEF/S.D.
              -2.368             7.179           -0.33
FE             0.08279           0.04818          1.72
AL             0.2460            0.1481           1.66
FEAL           0.0005278         0.0006610        0.80

S = 4.461

R-SQUARED = 95.2 PERCENT
R-SQUARED = 93.6 PERCENT, ADJUSTED FOR D.F.
```

FIGURE 10

MINITAB OUTPUT FOR MODEL WITH INTERACTION FIT TO THE PHOSPHATE ADSORPTION DATA (x_1 = FE, x_2 = AL, $x_3 = x_1 x_2$ = FEAL)

1. The model is $y = \alpha + \beta_1 x_1 + \beta_2 x_2 + \beta_3 x_3 + e$, where $x_3 = x_1 x_2$.

2. $H_0 : \beta_3 = 0$ (the interaction predictor does not belong in the model)

3. $H_a : \beta_3 \neq 0$.

4. Test statistic: $t = b_3/s_{b_3}$, the t ratio for β_3.

5. Rejection region: $n = 13$ and $k = 3$ for this model, so $n - (k + 1) = 9$. Using level of significance .05, the t critical value for a two-tailed test is 2.26. H_0 will be rejected, justifying inclusion of the interaction predictor variable, only if either $t > 2.26$ or $t < -2.26$.

6. Directly from the T-RATIO column in Figure 10, $t = .80$.

7. Since .80 is not in the rejection region, H_0 cannot be rejected at level of significance .05. In fact, P-value > .20 (since $t = .80$ does not exceed the smallest t critical value in the 9 df row of Table IV). So sample evidence suggests that the interaction predictor should be eliminated from the model. ■

An interesting aspect of the computer output for the interaction model in the above example is that the t ratios for β_1, for β_2, and for β_3 (1.72, 1.66, and .80) are all relatively small, yet $R^2 = .952$ is quite large. The high R^2 value suggests a useful model (this can be confirmed by the model utility test), yet the smallness of each t ratio might tempt us to conclude that all three β_i's are zero. This sounds like a contradiction, but it involves a misinterpretation of the t ratios. For example, the t ratio for β_3—i.e., b_3/s_{b_3}—tests $H_0 : \beta_3 = 0$ *when x_1 and x_2 are included in the model.* The smallness of a given t ratio suggests that the associated predictor can be dropped from the model *as long as the other predictors are retained.* The fact that all t ratios are small in this example does not, therefore, allow us simultaneously to delete all predictors. The data does suggest deleting x_3 because the model that includes x_1 and x_2 has already been found to give a very good fit to the data and is simple to interpret. In Section 12.4 we comment further on why it might happen that all t ratios are small even when the model seems very useful.

The model utility test amounts to testing a simultaneous claim about the values of all β_i's—that they are all zero. There is also an F test for testing a hypothesis involving a specified subset consisting of at least two β_i's. For example, we might fit the model $y = \alpha + \beta_1 x_1 + \beta_2 x_2 + \beta_3 x_1 x_2 + \beta_4 x_1^2 + \beta_5 x_2^2 + e$ and then wish to test $H_0 : \beta_3 = \beta_4 = \beta_5 = 0$ (which says that the second-order predictors contribute nothing to the model). Please see one of the chapter references for further details.

Inferences Based on the Estimated Regression Function

The estimated regression line $\hat{y} = a + bx$ in simple linear regression was used both to estimate the mean y value when x had a specified value and to predict the associated y value for a single observation made at a particular x value. The estimated regression function for the model $y = \alpha + \beta_1 x_1 + \cdots + \beta_k x_k + e$ can be used in the same two ways. When fixed values of the predictor variables $x_1, x_2, \ldots, x_k$ are substituted into the estimated regression function

$$a + b_1 x_1 + b_2 x_2 + \cdots + b_k x_k$$

the result can be used either as a point estimate of the corresponding mean y value or as a prediction of the y value that will result from a single observation when the x_i's have the specified values.

EXAMPLE 15

Precise information concerning bus transit times is important when making transportation planning decisions. The paper "Factors Affecting Running Time on Transit Routes" (*Transportation Research* (1983):107–13) reported on an empirical study based on data gathered in Cincinnati, Ohio. The variables of interest were

y = running time per mile during the morning peak period (s)

x_1 = number of passenger boardings per mile

x_2 = number of passenger alightings per mile

x_3 = number of signalized intersections per mile

x_4 = proportion of a route on which parking is allowed

The values of x_1 and x_2 were not necessarily equal because observations were made over partial segments of routes (so not all passengers entered or exited on the segments). The estimated regression function was

$$169.50 + 5.07x_1 + 4.53x_2 + 6.61x_3 + 67.70x_4$$

Consider the predictor variable values $x_1 = 4.5$, $x_2 = 5.5$, $x_3 = 5$, and $x_4 = .1$. Then a point estimate for true average running time per mile when the x_i's have these values is

$$169.50 + (5.07)(4.5) + (4.53)(5.5) + (6.61)(5) + (67.70)(.1) = 257.05$$

This value 257.05 is also the predicted running time per mile for a single trip when x_1, x_2, x_3, and x_4 have the given values. ■

Remember that before the sample observations $y_1, y_2, \ldots, y_n$ are obtained, a and all b_i's are statistics (because they all involve the y_i's). This implies that for fixed values of $x_1, x_2, \ldots, x_k$, the estimated regression function $a + b_1x_1 + \cdots + b_kx_k$ is a statistic (its value varies from sample to sample). To obtain a confidence interval for the mean y value or test a hypothesis about this mean y value for specified $x_1, \ldots, x_k$ values, we need some facts about the sampling distribution of this statistic.

Properties of the Sampling Distribution of $a + b_1x_1 + \cdots + b_kx_k$

Assumptions about the model $y = \alpha + \beta_1x_1 + \cdots + \beta_kx_k + e$ given in Section 12.1 imply that for fixed values of $x_1, x_2, \ldots, x_k$, the statistic $a + b_1x_1 + \cdots + b_kx_k$ satisfies the following properties:

1. It has a normal distribution.
2. Its mean value is $\alpha + \beta_1x_1 + \cdots + \beta_kx_k$—that is, the statistic is unbiased for estimating the mean y value when $x_1, \ldots, x_k$ are fixed as stated above.
3. The standard deviation of this statistic involves σ^2 and a very complicated function of all the sample predictor variable values. The estimated standard deviation of the statistic results from replacing σ^2 by s_e^2 in this function.

The standardized variable

$$t = \frac{a + b_1x_1 + \cdots + b_kx_k - (\alpha + \beta_1x_1 + \cdots + \beta_kx_k)}{\text{estimated standard deviation of } a + b_1x_1 + \cdots + b_kx_k}$$

then has a t distribution with $n - (k + 1)$ df.

While the formula for the estimated standard deviation of $a + b_1x_1 + \cdots + b_kx_k$ is complicated, it has been programmed into the most widely used statistical computer packages. Once the values of $x_1, \ldots, x_k$ have been specified by the user, an appropriate request results in the estimated standard deviation being computed and printed out. A hypothesis concerning $\alpha + \beta_1x_1 + \cdots + \beta_kx_k$ can then be tested by replacing $\alpha + \beta_1x_1 + \cdots + \beta_kx_k$ in t by the hypothesized value to obtain a t test statistic based on $n - (k + 1)$ df. We omit the details and instead give just a confidence interval.

For fixed values of $x_1, x_2, \ldots, x_k$, a **confidence interval for the mean y value**—i.e., for $\alpha + \beta_1x_1 + \cdots + \beta_kx_k$—is

$$a + b_1x_1 + \cdots + b_kx_k \pm \begin{pmatrix} t \text{ critical} \\ \text{value} \end{pmatrix} \begin{pmatrix} \text{estimated standard deviation} \\ \text{of } a + b_1x_1 + \cdots + b_kx_k \end{pmatrix}$$

where the t critical value is based on $n - (k + 1)$ df.

EXAMPLE 16

When MINITAB is used to perform a multiple regression analysis, the confidence interval just given is especially easy to obtain for values of $x_1, x_2, \ldots, x_k$ for which an observation on y is part of the sample. In this case, the value of $a + b_1x_1 + \cdots + b_kx_k$ appears in the column labeled PRED. Y VALUE, and the estimated standard deviation of $a + b_1x_1 + \cdots + b_kx_k$ appears in the next column over, which is labeled ST. DEV. PRED. Y. For example, looking again at Example 6, in which the phosphate adsorption data was displayed, in the ninth observation $x_1 = 160$ and $x_2 = 39$. By using the MINITAB output of Figure 7, a 95% confidence interval for $\alpha + \beta_1(160) + \beta_2(39)$, the mean value of phosphate adsorption when extractable iron = 160 and extractable aluminum = 39 is easily calculated:

$a + b_1(160) + b_2(39) = 24.30$ (ROW 9, PRED. Y VALUE column)

estimated standard deviation of $a + b_1(160) + b_2(39) = 1.30$

(ROW 9, ST. DEV. PRED. Y column)

The t critical value for 10 df and 95% confidence is 2.23, and the interval is

$$24.30 \pm (2.23)(1.30) = 24.30 \pm 2.90 = (21.40, 27.20)$$

When the (x_1, x_2) pair of interest is not one for which the sample contains an observation, there is a way to obtain both $a + b_1x_1 + b_2x_2$ and the estimated standard deviation from MINITAB. Some other packages will automatically compute the confidence interval limits once x_i values have been specified by the user. ■

Recall that the prediction interval formula for a single y observation in simple linear regression was similar to the confidence interval formula for a mean y value. However, while both intervals were centered at the same place—$a + bx$—the prediction interval was typically much wider because of the extra uncertainty associated with prediction. These properties carry over to the prediction interval and confidence interval in multiple regression.

A **prediction interval for a single y observation** from the model $y = \alpha + \beta_1 x_1 + \cdots + \beta_k x_k + e$ is

$$a + b_1x_1 + \cdots + b_kx_k \pm (t \text{ critical value}) \cdot \sqrt{s_e^2 + \begin{bmatrix}\text{estimated standard deviation} \\ \text{of } a + b_1x_1 + \cdots + b_kx_k\end{bmatrix}^2}$$

EXAMPLE 17

Reconsider the data on paddy yield (y) and time between flowering and harvesting (x) presented in Example 8. MINITAB output from fitting the quadratic regression model $y = \alpha + \beta_1 x + \beta_2 x^2 + e$ (so $x_1 = x$, $x_2 = x^2$) appeared in Figure 8. The value $x = 30$ is one for which the sample contains an observation (corresponding to $x_1 = 30$, $x_2 = 900$), so the PRED. Y. VALUE and ST. DEV. PRED. Y columns of the output give

$$a + b_1x + b_2x^2 = -1070.4 + (293.48)(30) - 4.5358(900) = 3651.9$$
$$\text{estimated standard deviation of } a + b_1x + b_2x^2 = 76.4$$

The t critical value for 95% confidence based on $16 - (2 + 1) = 13$ df is 2.16. A 95% confidence interval for true average yield when $x = 30$ is then

$$3651.9 \pm (2.16)(76.4) = 3651.9 \pm 165.0 = (3486.9, 3816.9)$$

A 95% prediction interval for a single yield observation when $x = 30$ is (because $s_e = 203.9$)

$$\begin{aligned} 3651.9 \pm (2.16)\sqrt{(203.9)^2 + (76.4)^2} &= 3651.9 \pm (2.16)(217.7) \\ &= 3651.9 \pm 470.2 \\ &= (3181.7, 4122.10) \end{aligned}$$

Clearly the prediction interval is much wider than the confidence interval. ■

The danger of extrapolation in simple linear regression is that if the x value of interest is much outside the interval of x values in the sample, the postulated model might no longer be valid, and even if it were, $\alpha + \beta x$ could still be quite poorly estimated (s_{a+bx} large). There is a similar danger in multiple regression, but it is not always obvious when the x_i values of interest involve an extrapolation from sample data. As an example, suppose that a single y observation is made for each of the following 13 (x_1, x_2) pairs:

Observation	1	2	3	4	5	6	7	8	9	10	11	12	13
x_1	0	0	0	0	0	0	5	5	5	10	−5	−5	−10
x_2	10	5	0	−5	−10	−5	0	5	0	−5	0	5	0

The x_1 values range between -10 and 10, as do the x_2 values. After fitting the model $y = \alpha + \beta_1 x_1 + \beta_2 x_2 + e$, we might then want a confidence interval for the mean y value when $x_1 = 10$ and $x_2 = 10$. However, whereas each of these values separately is within the range of sample x_i values, Figure 11 shows that the point $(x_1, x_2) = (10, 10)$ is actually far from (x_1, x_2) pairs in the sample. Thus a conclusion about y when $x_1 = 10$ and $x_2 = 10$ involves a substantial extrapolation. In particular, the estimated standard deviation of

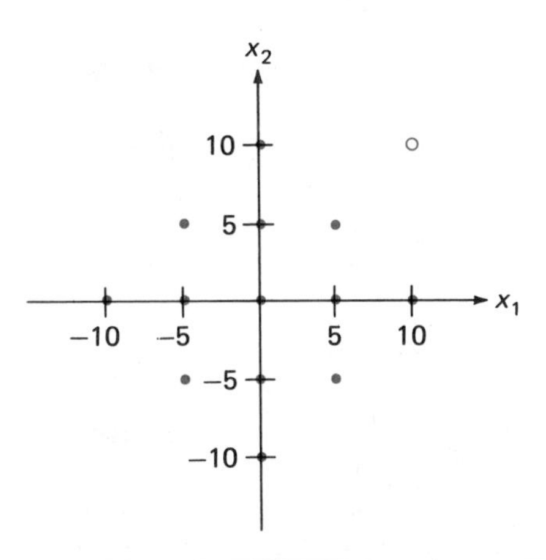

FIGURE 11

THE DANGER OF EXTRAPOLATION
● Denotes an (x_1, x_2) pair for which the sample contains a y observation
○ Denotes an (x_1, x_2) pair well outside the sample region, although the individual values $x_1 = 10$ and $x_2 = 10$ are within x_1 and x_2 sample ranges separately

$a + b_1(10) + b_2(10)$ would probably be quite large even if the model were valid near this point.

When more than two predictor variables are included in the model, we cannot see from a plot like that of Figure 11 whether the $x_1, x_2, \ldots, x_k$ values of interest involve an extrapolation. It is then best to compare the estimated standard deviation of $a + b_1x_1 + \cdots + b_kx_k$ for the values of interest with the estimated standard deviations corresponding to $x_1, x_2, \ldots, x_k$ values in the sample (those in the ST. DEV. PRED. Y column of the MINITAB output). Extrapolation is indicated by a value of the former standard deviation (at the x_i values of interest) that is much larger than the standard deviations for sampled values.

EXERCISES

12.28 Explain why it is preferable to perform a model utility test before using an estimated regression model to make predictions or to estimate the mean y value for specified values of the independent variables.

12.29 Twenty-six observations appearing in the paper "Multiple Regression Analysis for Forecasting Critical Fish Influxes at Power Station Intakes" (see Exercise 12.16) were used to fit a multiple regression model relating y = number of fish at intake to the independent variables x_1 = water temperature (°C), x_2 = number of pumps running, x_3 = sea state (taking values 0, 1, 2, or 3), and x_4 = speed (knots). Partial MINITAB output is given.

```
THE REGRESSION EQUATION IS
Y = 92.0 - 2.18 X1 - 19.2 X2 - 9.38 X3 + 2.32 X4

                                 ST. DEV.     T-RATIO =
COLUMN       COEFFICIENT         OF COEF.     COEF/S.D.
                   91.98           42.07          2.19
  X1              -2.179           1.087         -2.00
  X2             -19.189           9.215         -2.08
  X3              -9.378           4.356         -2.15
  X4              2.3205          0.7686          3.02
```

```
S = 10.53

R-SQUARED = 39.0 PERCENT
R-SQUARED = 27.3 PERCENT, ADJUSTED FOR D.F.

ANALYSIS OF VARIANCE

 DUE TO       DF          SS          MS=SS/DF
REGRESSION     4       1486.9           371.7
RESIDUAL      21       2330.2           111.0
TOTAL         25       3817.1
```

a. Construct a 95% confidence interval for β_3, the coefficient of x_3 = number of pumps running. Interpret the resulting interval.

b. Construct a 90% confidence interval for the mean change in y associated with a 1° increase in temperature when number of pumps, sea state, and speed remain fixed.

12.30 The estimated regression equation

$$\hat{y} = 28 - .05x_1 - .003x_2 + .00002x_3$$

where

y = number of indictments disposed of in a given month

x_1 = number of cases on judge's docket

x_2 = number of cases pending in the criminal trial courts

$x_3 = x_1x_2$

appeared in the paper "The Caseload Controversy and the Study of Criminal Courts" (*J. Criminal Law and Criminology* (1979):89–101). This equation was based on n = 367 observations. The b_i's and their associated standard deviations are given in the accompanying table.

i	b_i	*Estimated Standard Deviation of* b_i
1	−.05	.03
2	−.003	.0024
3	.00002	.000009

Is inclusion of the interaction term important? Test $H_0 : \beta_3 = 0$ using a .05 level of significance.

12.31 Does exposure to air pollution result in decreased life expectancy? This question was examined in the paper "Does Air Pollution Shorten Lives?" (*Statistics and Public Policy,* Reading, MA: Addison-Wesley, 1977). Data on

y = total mortality rate (deaths per 10,000)

x_1 = mean suspended particle reading (μg/m^3)

x_2 = smallest sulfate reading ([μg/m^3] × 10)

x_3 = population density (people/mi^2)

x_4 = (percent nonwhite) × 10

x_5 = (percent over 65) × 10

for the year 1960 was recorded for n = 117 randomly selected standard metropolitan statistical areas. The estimated regression equation was

$$\hat{y} = 19.607 + .041x_1 + .071x_2 + .001x_3 + .041x_4 + .687x_5$$

a. For this model, $R^2 = .827$. Using a .05 significance level, perform a model utility test.

b. The estimated standard deviation of b_1 was .016. Calculate and interpret a 90% confidence interval for β_1.

c. Given that the estimated standard deviation of b_4 is .007, determine if percent nonwhite is an important variable in the model. Use a .01 significance level.

d. In 1960, the values of x_1, x_2, x_3, x_4, and x_5 for Pittsburgh were 166, 60, 788, 68, and 95, respectively. Use the given regression equation to predict Pittsburgh's mortality rate. How does your prediction compare with the actual 1960 value of 103 deaths per 10,000?

12.32 The accompanying data was obtained from a study of a certain method for preparing pure alcohol from refinery streams ("Direct Hydration of Olefins" *Indus. and Engr. Chem.* (1961):209–11). The independent variable x is volume hourly space velocity and the dependent variable y is amount of isobutylene converted.

x	1	1	2	4	4	4	6
y	23.0	24.5	28.0	30.9	32.0	33.6	20.0

MINITAB output—the result of fitting a quadratic regression model where $x_1 = x$ and $x_2 = x^2$—is given. Would a linear regression have sufficed? That is, is the quadratic term important? Use a level .05 test.

```
THE REGRESSION EQUATION IS
Y =      13.6 +    11.4 X1  -    1.72 X2

                                    ST. DEV.      T-RATIO =
              COEFFICIENT           OF COEF.      COEF/S.D.
                   13.636              1.896           7.19
X1                 11.406              1.356           8.41
X2                -1.7155             0.2036          -8.42

THE ST. DEV. OF Y ABOUT REGRESSION LINE IS
S = 1.428
WITH (    7- 3) =     4 DEGREES OF FREEDOM

R-SQUARED = 94.7 PERCENT
R-SQUARED = 92.1 PERCENT, ADJUSTED FOR D.F.
```

12.33 The paper "Bank Full Discharge of Rivers" (*Water Resources J.* (1978): 1141–54) reported data on y = discharge amount (m^2/s), x_1 = flow area (m^2), and x_2 = slope of the water surface (m/m) obtained at $n = 10$ floodplain stations. A multiple regression model using x_1, x_2, and $x_3 = x_1x_2$ was fit to this data, and the resulting MINITAB output appears below.

```
THE REGRESSION EQUATION IS
Y = -    3.14 +    1.70 X1 +    96.1 X2
    +    8.39 X3

                                    ST. DEV.      T-RATIO =
              COEFFICIENT           OF COEF.      COEF/S.D.
                    -3.14              14.54          -0.22
X1                  1.697              1.431           1.19
X2                   96.1              702.7           0.14
X3                    8.4              199.0           0.04

THE ST. DEV. OF Y ABOUT REGRESSION LINE IS
S = 17.58
WITH (   10- 4) =      6 DEGREES OF FREEDOM
```

```
R-SQUARED = 73.2 PERCENT
R-SQUARED = 59.9 PERCENT, ADJUSTED FOR D.F.

ANALYSIS OF VARIANCE

 DUE TO        DF          SS          MS=SS/DF
REGRESSION      3      5073.4          1691.1
RESIDUAL        6      1854.1           309.0
TOTAL           9      6927.5
```

a. Perform the model utility test using a significance level of .05.

b. Is the interaction term important? Test using a .05 significance level.

c. Does it bother you that the model utility test indicates a useful model but all values in the t-ratio column of the output are small? Explain.

12.34 If an estimated regression equation results in a large R^2 value, must all independent variables (the x_i's) have large t ratios (those appropriate for testing $H_0 : \beta_i = 0$) associated with them? Explain.

12.35 In the paper "An Ultracentrifuge Flour Absorption Method" (*Cereal Chem.* (1978):96–101), the authors discussed the relationship between water absorption for wheat flour and various characteristics of the flour. A multiple regression model was used to relate y = absorption (%) to x_1 = flour protein (%) and x_2 = starch damage (Farrand units). MINITAB output based on $n = 28$ observations is given. Use a significance level of .05 for all tests requested.

```
THE REGRESSION EQUATION IS
Y =      19.4 +   1.44 X1 +   .336 X2

                                 ST. DEV.        T-RATIO =
         COEFFICIENT             OF COEF.        COEF/S.D.
              19.440               2.188              8.88
X1            1.4423               0.2076             6.95
X2           0.33563               0.01814           18.51

THE ST. DEV. OF Y ABOUT REGRESSION LINE IS
S = 1.094
WITH (  28-  3) =   25 DEGREES OF FREEDOM

R-SQUARED = 96.4 PERCENT
R-SQUARED = 96.2 PERCENT, ADJUSTED FOR D.F.

ANALYSIS OF VARIANCE

 DUE TO        DF          SS          MS=SS/DF
REGRESSION      2     812.380          406.190
RESIDUAL       25      29.928            1.197
TOTAL          27     842.307
```

a. Using the model utility test, determine if the regression equation specifies a useful relationship between y and the independent variables.

b. Conduct tests for each of the following two sets of hypotheses.
   i. $H_0 : \beta_1 = 0$ versus $H_a : \beta_1 \neq 0$
   ii. $H_0 : \beta_2 = 0$ versus $H_a : \beta_2 \neq 0$

c. Based on the results of (b), would you conclude that both independent variables are important? Explain.

d. An estimate of the mean water absorption for wheat with 10.2% protein and a starch damage of 20 is desired. The estimated standard deviation of $a + b_1(10.2) + b_2(20)$ is .318. Use this to compute a 95% confidence interval for $\alpha + \beta_1(10.2) + \beta_2(20)$.

e. Compute a 90% confidence interval for $\alpha + \beta_1(11.7) + \beta_2(57)$ if the estimated standard deviation of $a + b_1(11.7) + b_2(57)$ is .522. Interpret the resulting interval.

f. A single shipment of wheat is received. For this particular shipment, $x_1 = 11.7$ and $x_2 = 57$. Predict the water absorption for this shipment (a single y value) using a 90% interval.

12.36 The paper "Predicting Marathon Time from Anaerobic Theshold Measurements" (*The Physician and Sportsmed.* (1984): 95–98) gave data on y = maximum heart rate (beats/min), x_1 = age, and x_2 = weight (kg) for $n = 18$ marathon runners. The estimated regression equation for the model $y = \alpha + \beta_1 x_1 + \beta_2 x_2 + e$ was $\hat{y} = 179 - .8x_1 + .5x_2$ and SSRegr = 649.75, SSResid = 538.03.

a. Is the model useful for predicting maximum heart rate? Use a significance level of .10.

b. Predict the maximum heart rate of a particular runner who is 43 years old and weighs 65 kg using a 99% interval. The estimated standard deviation of $a + b_1(43) + b_2(65)$ is 3.52.

c. Use a 90% interval to estimate the average maximum heart rate for all marathon runners who are 30 years old and weigh 77.2 kg. The estimated standard deviation of $a + b_1(30) + b_2(77.2)$ is 2.97.

d. Would a 90% prediction interval for a single 30-year-old runner weighing 77.2 kg be wider or narrower than the interval computed in (c)? Explain. (You need not compute the interval.)

12.37 The effect of manganese (Mn) on wheat growth is examined in the article "Manganese Deficiency and Toxicity Effects on Growth, Development and Nutrient Composition in Wheat" (*Agronomy J.* (1984):213–17). A quadratic regression model was used to relate y = plant height (cm) to x = log (added Mn), with μM as the units for added Mn. The accompanying data was read from a scatter diagram appearing in the paper. Also given is MINITAB output, where $x_1 = x$ and $x_2 = x^2$. Use a .05 significance level for any hypothesis tests needed to answer the questions.

x	−1	−.4	0	.2	1	2	2.8	3.2	3.4	4
y	32	37	44	45	46	42	42	40	37	30

```
THE REGRESSION EQUATION IS
Y =      41.7 +   6.58 X1  -   2.36 X2

                                ST. DEV.     T-RATIO =
              COEFFICIENT       OF COEF.     COEF/S.D.
                  41.7422        0.8522         48.98
X1                  6.581         1.002          6.57
X2                -2.3621        0.3073         -7.69

THE ST. DEV. OF Y ABOUT REGRESSION LINE IS
S = 1.963
WITH (  10- 3) =    7 DEGREES OF FREEDOM

R-SQUARED =  89.8 PERCENT
R-SQUARED =  86.9 PERCENT, ADJUSTED FOR D.F.

ANALYSIS OF VARIANCE

 DUE TO        DF          SS          MS=SS/DF
REGRESSION      2      237.520         118.760
RESIDUAL        7       26.980           3.854
TOTAL           9      264.500
```

a. Is the quadratic model useful for describing the relationship between y and x?

b. Are both the linear and quadratic terms important? Could either one be eliminated from the model? Explain.

c. Give a 95% confidence interval for the mean y value when $x = 2$. The estimated standard deviation of $a + b_1(2) + b_2(4)$ is 1.037. Interpret the resulting interval.

d. Estimate the mean height for wheat treated with 10 μM of Mn using a 90% interval. (*Note:* The estimated standard deviation of $a + b_1(1) + b_2(1)$ is 1.031 and log (10) = 1.)

12.38 *This problem requires the use of a computer package.* A study on the effect of applying fertilizer in bands is described in the paper "Fertilizer Placement Effects on Growth, Yield, and Chemical Composition of Burley Tobacco" (*Agronomy J.* (1984):183–88). The accompanying data was taken from a scatter diagram appearing in the paper, with y = plant Mn (μg/g dry weight) and x = distance from the fertilizer band (cm). The authors suggest a quadratic regression model.

x	0	10	20	30	40
y	110	90	76	72	70

a. Use a suitable computer package to find the estimated quadratic regression equation.

b. Perform the model utility test.

c. Interpret the values of R^2 and s_e.

d. Are both the linear and quadratic terms important? Carry out the necessary hypothesis tests and interpret the results.

e. Find a 90% confidence interval for the mean plant Mn for plants that are 30 cm from the fertilizer band.

12.4 Other Issues in Multiple Regression

Primary objectives in multiple regression include the estimation of a mean y value, prediction of an as-yet unobserved y value, and gaining insight into how changes in predictor variable values impact y. Often an investigator has data on a number of predictor variables that might be incorporated into a model to be used for such purposes. Some of these may actually be unrelated or only very weakly related to y or may contain information that duplicates information provided by other predictors. If all these predictor variables are included in the model, many model coefficients have to be estimated. This reduces the number of degrees of freedom associated with SSResid, leading to a deterioration in the degree of precision associated with other inferences (e.g., wide confidence and prediction intervals). A model with many predictors can also be cumbersome to use and difficult to interpret.

In this section we first introduce some guidelines and procedures for selecting a set of useful predictors. Before a model is decided upon, the analyst should examine the data carefully for evidence of unusual observations or potentially troublesome patterns. It is important to identify unusually deviant or influential observations and to look for possible inconsistencies with model assumptions. Our discussion of multiple regression closes with a brief mention of some diagnostic methods that facilitate such an examination.

Variable Selection

Suppose that an investigator has data on p predictor variables $x_1, x_2, \ldots, x_p$, which are candidates for use in building a model. Some of these variables might be specified functions of others—e.g., $x_3 = x_1x_2$, $x_4 = x_1^2$, etc. The objective is then to select a set of these predictors which in some sense specifies a best model (of the general additive form considered in the two previous sections). Fitting a model that consists of a specified k predictors requires that $k + 1$ model coefficients (α and the k corresponding β's) be estimated. Generally speaking, the number of observations, n, should be at least twice the number of predictors in the largest model under consideration in order to ensure reasonably accurate coefficient estimates and a sufficient number of degrees of freedom associated with SSResid.

If p is not too large, a good statistical computer package can rather quickly fit a model based on each different subset of the p predictors. Consider as an example the case $p = 4$. There are two possibilities for each predictor—it could be included or not included in a model—so the number of possible models in this case is $2 \cdot 2 \cdot 2 \cdot 2 = 2^4 = 16$ (including the model with all four predictors and the model with only the constant term and none of the four predictors). These 16 possibilities are displayed in the accompanying table.

Predictors included	None	x_1	x_2	x_3	x_4	x_1, x_2	x_1, x_3	x_1, x_4	x_2, x_3	x_2, x_4
Number of predictors in model	0	1	1	1	1	2	2	2	2	2

Predictors included	x_3, x_4	x_1, x_2, x_3	x_1, x_2, x_4	x_1, x_3, x_4	x_2, x_3, x_4	x_1, x_2, x_3, x_4
Number of predictors in model	2	3	3	3	3	4

More generally, when there are p candidate predictor variables, the number of possible models is 2^p. The number of possible models is, therefore, substantial if p is even moderately large—e.g., 1024 possible models when $p = 10$ and 32,768 possibilities when $p = 15$.

Model selection methods can be divided into two types. There are those based on fitting every possible model, computing one or more summary quantities from each fit, and comparing these quantities to identify the most satisfactory models. Several of the most powerful statistical computer packages have an "all subsets" option, which will give limited output from fitting each possible model. Methods of the second type are appropriate when p is so large that it is not feasible to examine all subsets. These methods are often referred to as *automatic selection,* or *stepwise, procedures.* The general idea is to either begin with the p predictor model and delete variables one by one until all remaining predictors are judged important, or begin with no predictors and add predictors until no predictor not in the model seems important. With modern-day computing power, the value of p for which examination of all subsets is feasible is surprisingly large, so automatic selection procedures are not as important as they once were.

Suppose then that p is small enough so that all subsets can be fit. What characteristic(s) of the estimated models should be examined in the search for a best model? An obvious and appealing candidate is the coefficient of multiple determination, R^2, which measures the proportion of observed y variation explained by the model. Certainly a model with a large R^2 value is preferable to another model containing the same number of predictors but which has a much smaller R^2 value. Thus if the model with predictors x_1 and x_2 has $R^2 = .765$ and the model with predictors x_1 and x_3 has $R^2 = .626$, the latter model would almost surely be eliminated from further consideration.

However, using R^2 to choose between models containing different numbers of predictors is not so straightforward because adding a predictor to a model can never decrease the value of R^2. In particular, the model containing all p candidate predictors is guaranteed to have an R^2 value at least as large as R^2 for any model that includes some but not all of these predictors. More generally, let $R^2_{(1)}$ denote the largest R^2 value for any model containing just one predictor variable, $R^2_{(2)}$ denote the largest R^2 value for any two-predictor model, and so on. Then $R^2_{(1)} \leq R^2_{(2)} \leq \cdots \leq R^2_{(p-1)} \leq R^2_{(p)}$. When statisticians base model selection on R^2, the objective is not simply to find the model with the largest R^2 value—the model with p predictors does that. Instead we should look for a model containing relatively few predictors but which has a large R^2 value and is such that no other model containing more predictors gives much of an improvement in R^2. Suppose, for example, that $p = 5$ and that $R^2_{(1)} = .427$, $R^2_{(2)} = .733$, $R^2_{(3)} = .885$, $R^2_{(4)} = .898$, and $R^2_{(5)} = .901$. Then the best three-predictor model appears to be a good choice, since it substantially improves on the best one- and two-predictor models, whereas very little is gained by using the best-four predictor model or all five predictors.

A small increase in R^2 resulting from the addition of a predictor to a model may be offset by the increased complexity of the new model and the reduction in degrees of freedom associated with SSResid. This has led statisticians to introduce a quantity called **adjusted R^2**, which can either decrease or increase when a predictor is added to a model.* It follows that adjusted R^2 for the best k-predictor model (the one with coefficient of multiple determination $R^2_{(k)}$) may be larger than adjusted R^2 for the best model based on $k + 1$ predictors. Adjusted R^2 formalizes the notion of diminishing returns as more predictors are added—small increases in R^2 are outweighed by corresponding decreases in degrees of freedom associated with SSResid. A reasonable strategy in model selection is to identify the model with the largest value of adjusted R^2 (the corresponding number of predictors k is often much smaller than p) and then consider only that model and any others whose adjusted R^2 values are nearly as large.

EXAMPLE 18

The paper "Anatomical Factors Influencing Wood Specific Gravity of Slash Pines and the Implications for the Development of a High-Quality Pulpwood" (*TAPPI* (1964):401–04) reported the results of an experiment in which 20 samples of slash pine wood were analyzed. A primary objective was to relate wood specific gravity (y) to various other wood characteristics. The independent variables on which observations were made were x_1 = number of fibers/mm^2 in springwood, x_2 = number of fibers/mm^2 in summerwood,

*When n observations are used to fit a model that contains k predictors,

$$\text{adjusted } R^2 = 1 - \left[\frac{n-1}{n-(k+1)}\right] \cdot \frac{\text{SSResid}}{\text{SSTo}}$$

Since the quantity in square brackets exceeds 1, the number subtracted from 1 is larger than SSResid/SSTo, so adjusted R^2 is smaller than R^2. R^2 itself must be between 0 and 1, but adjusted R^2 can be negative.

x_3 = springwood %, x_4 = % springwood light absorption, and x_5 = % summerwood light absorption. The data is displayed in the accompanying table. Consider $x_1, x_2, \ldots, x_5$ as the set of potential predictors ($p = 5$, with no derived predictors such as squares or interaction terms as candidates for inclusion). Then there are $2^5 = 32$ possible models, among which 5 consist of a single predictor, 10 involve two predictors, 10 others involve three predictors, and 5 include four predictor variables. We used a statistical computer package to fit each possible model and extracted both R^2 and adjusted R^2 from the output. To save space, results appear for all one- and four-predictor models but only for the five best two- and three-predictor models.

It is immediately clear that the best three-predictor models offer considerable improvement with respect to both R^2 and adjusted R^2 over any model containing one or two predictors and that the model with all five predictors is inferior to the best four-predictor model. This latter model has the largest value of adjusted R^2 (.709), but the best three-predictor model is not far behind. The five models that seem most appealing to us are color-tinted in the display of results. The selection of a single model would have to be based on a more detailed comparison of these five models, with special attention given to residuals and other diagnostics (to be mentioned shortly) that bear on model adequacy. In particular, it is dangerous to embrace the model with the largest adjusted R^2 value automatically and reject all other models out of hand.

	x_1	x_2	x_3	x_4	x_5	y
1	573	1059	46.5	53.8	84.1	0.534
2	651	1356	52.7	54.5	88.7	0.535
3	606	1273	49.4	52.1	92.0	0.570
4	630	1151	48.9	50.3	87.9	0.528
5	547	1135	53.1	51.9	91.5	0.548
6	557	1236	54.9	55.2	91.4	0.555
7	489	1231	56.2	45.5	82.4	0.481
8	685	1564	56.6	44.3	91.3	0.516
9	536	1182	59.2	46.4	85.4	0.475
10	685	1564	63.1	56.4	91.4	0.486
11	664	1588	50.6	48.1	86.7	0.554
12	703	1335	51.9	48.4	81.2	0.519
13	653	1395	62.5	51.9	89.2	0.492
14	586	1114	50.5	56.5	88.9	0.517
15	534	1143	52.1	57.0	88.9	0.502
16	523	1320	50.5	61.2	91.9	0.508
17	580	1249	54.6	60.8	95.4	0.520
18	448	1028	52.2	53.4	91.8	0.506
19	476	1057	42.9	53.2	92.9	0.595
20	528	1057	42.4	56.6	90.0	0.568

Models with One Predictor

Predictor	x_3	x_5	x_2	x_4	x_1
R^2	$R^2_{(1)}$ = .564	.106	.053	.020	.008
Adjusted R^2	.539	.057	.001	−.034	−.047

Models with Two Predictors

Predictors	x_3, x_5	x_2, x_3	x_1, x_3	x_3, x_4	x_2, x_5
R^2	$R^2_{(2)} = .655$	.621	.603	.564	.158
Adjusted R^2	.614	.576	.556	.513	.059

Models with Three Predictors

Predictors	x_1, x_3, x_5	x_3, x_4, x_5	x_2, x_3, x_5	x_1, x_2, x_3	x_2, x_3, x_4
R^2	$R^2_{(3)} = .723$	.712	.711	.622	.611
Adjusted R^2	.671	.659	.657	.551	.550

Models with Four Predictors

Predictors	x_1, x_3, x_4, x_5	x_2, x_3, x_4, x_5	x_1, x_2, x_3, x_5	x_1, x_2, x_3, x_4	x_1, x_2, x_4, x_5
R^2	$R^2_{(4)} = .770$	.748	.727	.622	.239
Adjusted R^2	.709	.681	.654	.522	.036

The model with all five predictors included has $R^2_{(5)} = .770$, adjusted $R^2 = .689$. ■

There are various other criteria that have been proposed and used for model selection after fitting all subsets. A chapter reference can be consulted for more details.

When using particular criteria as a basis for model selection, many of the 2^p possible subset models are not serious candidates because of poor criteria values. For example, if $p = 15$ there are 5005 different models consisting of six predictor variables, many of which typically have small R^2 and adjusted R^2 values. An investigator usually wishes to consider only a few of the best models of each different size (a model whose criteria value is close to the best one may be easier to interpret than the best model or may include a predictor that the investigator thinks should be in the selected model). In recent years statisticians have developed computer programs to achieve this without actually fitting all possible models. One version of such a program has been implemented in the BMDP statistical computer package and can be used as long as $p \leq 26$ (there are roughly 67 million possible models when $p = 26$, so fitting them all would be out of the question). The user specifies a number between 1 and 10 as the number of models of each given size for which output will be provided. Output includes R^2, adjusted R^2, estimated model coefficients, and a few other quantities. After the choice of models is narrowed down, the analyst can request more detail on each finalist.

EXAMPLE 19

The accompanying data was taken from the paper "Applying Stepwise Multiple Regression Analysis to the Reaction of Formaldehyde with Cotton Cellulose" (*Textile Research J.* (1984):157–65). The dependent variable y is durable press rating, a quantitative measure of wrinkle resistance. The four independent variables used in the model building process are x_1 = HCHO (formaldehyde) concentration, x_2 = catalyst ratio, x_3 = curing temperature, and x_4 = curing time. In addition to these, the investigators considered as potential predictors x_1^2, x_2^2, x_3^2, x_4^2, and all six cross products $x_1x_2, \ldots, x_3x_4$, a total of $p = 14$ candidates. We display BMDP output for the best three subset models of each size from $k = 4$ predictor variables up to $k = 9$ predictor variables. In addition to R^2 and adjusted R^2, values of another criterion, called *Mallows' CP,* are included. A good model according to this criterion is one that has small CP (for accurate predictions) and CP $\approx k + 1$ (for unbiasedness in estimating model coefficients).

The choice of a best model here is, as often happens, not clear-cut. We certainly don't see the benefit of including more than $k = 8$ predictor variables (after that, adjusted R^2 begins to decrease), nor would we suggest a model with fewer than five predictors (adjusted R^2 is still increasing and CP is large). Based just on this output, the best six-predictor model is a reasonable choice. The corresponding estimated regression function is

$$-1.218 + .9599x_2 - .0373x_1^2 - .0389x_2^2 + .0037x_1x_3 + .019x_1x_4 - .0013x_2x_3$$

Another good candidate is the best seven-predictor model. Although it includes one more predictor than the model just suggested, only one of the seven predictors is a cross-product term (x_1x_3), so model interpretation is somewhat easier (notice, though, that none of the best three models with seven predictors results simply from adding a single predictor to the best six-predictor model). Since every good model includes x_1, x_2, x_3, and x_4 in some predictor, it appears that HCHO concentration, catalyst ratio, curing time, and curing temperature are all important determinants of durable press rating.

Observation	x_1	x_2	x_3	x_4	y	Observation	x_1	x_2	x_3	x_4	y
1	8	4	100	1	1.4	16	4	10	160	5	4.6
2	2	4	180	7	2.2	17	4	13	100	7	4.3
3	7	4	180	1	4.6	18	10	10	120	7	4.9
4	10	7	120	5	4.9	19	5	4	100	1	1.7
5	7	4	180	5	4.6	20	8	13	140	1	4.6
6	7	7	180	1	4.7	21	10	1	180	1	2.6
7	7	13	140	1	4.6	22	2	13	140	1	3.1
8	5	4	160	7	4.5	23	6	13	180	7	4.7
9	4	7	140	3	4.8	24	7	1	120	7	2.5
10	5	1	100	7	1.4	25	5	13	140	1	4.5
11	8	10	140	3	4.7	26	8	1	160	7	2.1
12	2	4	100	3	1.6	27	4	1	180	7	1.8
13	4	10	180	3	4.5	28	6	1	160	1	1.5
14	6	7	120	7	4.7	29	4	1	100	1	1.3
15	10	13	180	3	4.8	30	7	10	100	7	4.6

```
                                               SUBSETS WITH  4 VARIABLES
                                               -------------------------

              ADJUSTED
R-SQUARED     R-SQUARED        CP

  .822152       .793697      13.88   VARIABLE     COEFFICIENT     T-STATISTIC
                                       X2            .719972           6.34
                                       X4            .100399           2.23
                                       X2SQ        -.0353810          -4.50
                                       X1X3         .00136851          4.67
                                      INTERCEPT     -.567677
```

			VARIABLE	COEFFICIENT	T-STATISTIC
.820829	.792161	14.13	X2	.739574	6.48
			X2SQ	-.0366689	-4.66
			X4SQ	.0120970	2.18
			X1X3	.00137952	4.68
			INTERCEPT	-.500064	
.817380	.788160	14.79	VARIABLE	COEFFICIENT	T-STATISTIC
			X2	.725317	6.30
			X2SQ	-.0357036	-4.49
			X1X3	.00129562	4.39
			X3X4	.000620500	2.05
			INTERCEPT	-.469898	

SUBSETS WITH 5 VARIABLES

R-SQUARED	ADJUSTED R-SQUARED	CP	VARIABLE	COEFFICIENT	T-STATISTIC
.866735	.838971	7.39	X2	.755757	7.42
			X1SQ	-.0249950	-3.24
			X2SQ	-.0377513	-5.37
			X1X3	.00261758	4.97
			X1X4	.0194928	3.11
			INTERCEPT	-.751386	
.855228	.825068	9.58	VARIABLE	COEFFICIENT	T-STATISTIC
			X2	.760646	7.17
			X4	.109915	2.64
			X1SQ	-.0173769	-2.34
			X2SQ	-.0379611	-5.19
			X1X3	.00246978	4.56
			INTERCEPT	-.936857	
.855015	.824810	9.62	VARIABLE	COEFFICIENT	T-STATISTIC
			X2	.783083	7.36
			X1SQ	-.0176972	-2.38
			X2SQ	-.0394252	-5.39
			X4SQ	.0134874	2.63
			X1X3	.00250377	4.60
			INTERCEPT	-.876926	

SUBSETS WITH 6 VARIABLES

R-SQUARED	ADJUSTED R-SQUARED	CP	VARIABLE	COEFFICIENT	T-STATISTIC
.884952	.854939	5.92	X2	.959914	6.66
			X1SQ	-.0372760	-3.82
			X2SQ	-.0389469	-5.82
			X1X3	.00368402	4.91
			X1X4	.0192505	3.23
			X2X3	-.00128271	-1.91
			INTERCEPT	-1.21835	
.881662	.850791	6.54	VARIABLE	COEFFICIENT	T-STATISTIC
			X2	.779668	7.93
			X4	.408872	2.98
			X1SQ	-.0313804	-3.40
			X2SQ	-.0394407	-5.81
			X1X3	.00356717	5.12
			X3X4	-.00208494	-2.27
			INTERCEPT	-1.34754	

```
 .880382      .849177     6.79   VARIABLE    COEFFICIENT    T-STATISTIC
                                   X2           .760472         7.71
                                   X1SQ       -.0380752        -3.46
                                   X2SQ       -.0383770        -5.64
                                   X1X3        .00319522        5.13
                                   X1X4        .0384039         2.92
                                   X3X4       -.000908222      -1.62
                                  INTERCEPT   -.632344

                                         SUBSETS WITH  7 VARIABLES
               ADJUSTED
R-SQUARED     R-SQUARED      CP

 .899208      .867138     5.20   VARIABLE    COEFFICIENT    T-STATISTIC
                                   X2           .800294         8.40
                                   X3           .0966644        2.41
                                   X4           .129477         3.49
                                   X1SQ       -.0392306        -3.13
                                   X2SQ       -.0416758        -6.35
                                   X3SQ       -.000388272      -2.70
                                   X1X3        .00435437        4.08
                                  INTERCEPT   -7.27515

 .898832      .866643     5.27   VARIABLE    COEFFICIENT    T-STATISTIC
                                   X2           .825845         8.60
                                   X3           .0974167        2.42
                                   X1SQ       -.0392516        -3.12
                                   X2SQ       -.0433512        -6.57
                                   X3SQ       -.000390110      -2.71
                                   X4SQ        .0158501         3.47
                                   X1X3        .00436212        4.08
                                  INTERCEPT   -7.25606

 .896670      .863792     5.68   VARIABLE    COEFFICIENT    T-STATISTIC
                                   X2           .786050         8.15
                                   X3           .0820370        2.02
                                   X1SQ       -.0413732        -3.21
                                   X2SQ       -.0407510        -6.13
                                   X3SQ       -.000325208      -2.24
                                   X1X3        .00401885        3.79
                                   X1X4        .0195025         3.36
                                  INTERCEPT   -6.06342

                                         SUBSETS WITH  8 VARIABLES
               ADJUSTED
R-SQUARED     R-SQUARED      CP

 .915161      .882841     4.16   VARIABLE    COEFFICIENT    T-STATISTIC
                                   X2           .775486         8.58
                                   X3           .109369         2.86
                                   X4           .455906         2.71
                                   X1SQ       -.0340476        -2.82
                                   X2SQ       -.0404306        -6.52
                                   X3SQ       -.000385008      -2.85
                                   X1X3        .00371235        3.53
                                   X3X4       -.00237915       -1.99
                                  INTERCEPT   -8.68827

 .905701      .869777     5.96   VARIABLE    COEFFICIENT    T-STATISTIC
                                   X2           .845585         8.79
                                   X3           .104772         2.61
                                   X1SQ       -.0364813        -2.89
                                   X2SQ       -.0448477        -6.76
                                   X3SQ       -.000390164      -2.74
                                   X4SQ        .0360597         2.13
                                   X1X3        .00403009        3.69
                                   X3X4       -.00121700       -1.24
                                  INTERCEPT   -7.93925
```

			VARIABLE	COEFFICIENT	T-STATISTIC
.904737	.868446	6.15	X2	.923957	6.59
			X3	.0788275	1.97
			X1SQ	-.0432398	-3.40
			X2SQ	-.0405590	-6.20
			X3SQ	-.000297392	-2.06
			X1X3	.00417478	3.98
			X1X4	.0190484	3.34
			X2X3	-.000964844	-1.33
			INTERCEPT	-6.25679	

SUBSETS WITH 9 VARIABLES

R-SQUARED	ADJUSTED R-SQUARED	CP	VARIABLE	COEFFICIENT	T-STATISTIC
.916768	.879313	5.86	X1	.169779	.62
			X2	.761409	8.06
			X3	.106348	2.72
			X4	.464198	2.72
			X1SQ	-.0395106	-2.62
			X2SQ	-.0394067	-6.06
			X3SQ	-.000358450	-2.50
			X1X3	.00301598	1.95
			X3X4	-.00248516	-2.03
			INTERCEPT	-8.96140	
.916069	.878301	5.99	X2	.826844	5.75
			X3	.106946	2.72
			X4	.425886	2.33
			X1SQ	-.0350603	-2.81
			X2SQ	-.0404567	-6.40
			X3SQ	-.000374296	-2.68
			X1X3	.00380925	3.48
			X2X3	-.000346329	-.47
			X3X4	-.00217879	-1.68
			INTERCEPT	-8.61679	
.916046	.878267	5.99	X2	.769047	8.25
			X3	.103502	2.52
			X4	.410687	2.08
			X1SQ	-.0351573	-2.81
			X2SQ	-.0400069	-6.27
			X3SQ	-.000358782	-2.41
			X1X3	.00357947	3.22
			X1X4	.00847361	.46
			X3X4	-.00243189	-1.98
			INTERCEPT	-8.22400	

■

The most easily understood and implemented automatic selection procedure is referred to as **backward elimination.** It involves starting with the model that contains all p potential predictors and then deleting them one by one until all remaining predictors seem important. The first step is to specify the value of a positive constant t_{out}, which is used to decide whether deletion should be continued. After fitting the p predictor model, the t ratios b_1/s_{b_1}, $b_2/s_{b_2}, \ldots, b_p/s_{b_p}$ are examined. The predictor variable whose t ratio is closest to zero, whether positive or negative, is the obvious candidate for deletion. If this t ratio satisfies $-t_{out} \le t$ ratio $\le t_{out}$, the corresponding predictor is eliminated from the model. Suppose that this is the case. The model with the remaining $p - 1$ predictors is then fit, and again the predictor with the

t ratio closest to zero is eliminated, provided that it satisfies $-t_{out} \le t$ ratio $\le t_{out}$. The procedure continues until, at some stage, no t ratio satisfies $-t_{out} \le t$ ratio $\le t_{out}$ (all are either greater than t_{out} or less than $-t_{out}$). The chosen model is then the last one fit (though some analysts recommend examining other models of the same size). It is customary to use $t_{out} = 2$, since many t critical values for a two-tailed test with level of significance .05 are close to this value.*

EXAMPLE 20

We display MINITAB output resulting from the application of the backward elimination procedure with $t_{out} = 2$ to the specific gravity data introduced in Example 18. The t ratio closest to zero when the model with all five predictors was fit was .12, so the corresponding predictor x_2 was deleted from the model. When the model with the four remaining predictors was fit, the t ratio closest to zero was -1.76, which satisfied $-2 \le -1.76 \le 2$. Thus the corresponding predictor, x_4, was eliminated, leaving x_1, x_3, and x_5 still in the model. The next predictor to be dropped was x_1 because its t ratio, 1.98, was the one closest to zero and (barely) satisfied $-2 \le 1.98 \le 2$. When the model with predictors x_3 and x_5 was fit, neither t ratio satisfied $-2 \le t$ ratio ≤ 2, so the procedure terminated. Looking back to Example 18, this is actually the best model containing just two predictors. However, x_1 just barely met the elimination criterion, so the model with predictors x_1, x_3, and x_5 should also be given serious consideration.

STEP	1	2	3	4
CONSTANT	0.4421	0.4384	0.4381	0.5179
X1	0.00011	0.00011	0.00012	
T-RATIO	1.17	1.95	1.98	
X2	0.00001			
T-RATIO	0.12			
X3	-0.00531	-0.00526	-0.00498	-0.00438
T-RATIO	-5.70	-6.56	-5.96	-5.20
X4	-0.0018	-0.0019		
T-RATIO	-1.63	-1.76		
X5	0.0044	0.0044	0.0031	0.0027
T-RATIO	3.01	3.31	2.63	2.12
S	0.0180	0.0174	0.0185	0.0200
R-SQ	77.05	77.03	72.27	65.50

Unfortunately, the backward elimination method does not always terminate with a model that is the best of its size, and this is also true of other automatic selection procedures. For example, the authors of the paper mentioned in Example 19 used an automatic procedure to obtain a six-predictor model with $R^2 = .77$, whereas all of the 10 best six-predictor models have R^2 values of at least .87. Because of this, we recommend using a statistical computer package which will identify best subsets of different sizes whenever possible.

*Some computer packages base the procedure on the squares of the t ratios, which are F ratios, and continue to delete as long as F ratio $\le F_{out}$ for at least one predictor. The predictor with the smallest F ratio is eliminated. $F_{out} = 4$ corresponds to $t_{out} = 2$.

Checks on Model Adequacy

In Chapter 11 we discussed some informal techniques for checking the adequacy of the simple linear regression model. Most of these were based on plots involving the standardized residuals. The formula for standardizing residuals in multiple regression is quite complicated, but again it has been programmed into many statistical computer packages. Once the standardized residuals resulting from the fit of a particular model have been computed, plots similar to those discussed earlier are useful in diagnosing model defects. A normal probability plot of the standardized residuals that departs too much from a straight line casts doubt on the assumption that the random deviation e has a normal distribution. Plots of the standardized residuals against each predictor variable in the model—i.e., a plot of (x_1, standardized residual) pairs, another of (x_2, standardized residual) pairs, etc.—are analogous to the standardized residual versus x plot discussed and illustrated in the previous chapter. The appearance of any discernible pattern in these plots (e.g., curvature or increasing spread from left to right) points to the need for model modification. If observations have been made over time, a periodic pattern when the standardized residuals are plotted in time order suggests that successive observations were not independent. Models that incorporate dependence of successive observations are substantially more complicated than those we have presented here. They are especially important in econometrics, which involves using statistical methodology to model economic data. To conserve space, we forgo presenting examples of standardized residual plots. (Please consult a chapter reference for examples of those mentioned here as well as other types of plots.)

One other aspect of model adequacy that has received much attention from statisticians in recent years is the identification of any observations in the data set that may have been highly influential in estimating model coefficients. Recall that in simple linear regression, an observation with potentially high influence was one whose x value placed it far to the right or left of the other points in the scatter plot or standardized residual plot. If a multiple regression model involves only two predictor variables x_1 and x_2, an observation with potentially large influence can be revealed by examining a plot of (x_1, x_2) pairs. Any point in this plot that is far away from the others corresponds to an observation which, if deleted from the sample, may cause coefficient estimates and other quantities to change considerably. The detection of influential observations when the model contains at least three predictors is more difficult. Recent research has yielded several helpful diagnostic quantities. One of these has been implemented in MINITAB, and a large value of this quantity automatically results in the corresponding observation being identified as one that might have great influence. Deleting the observation and refitting the model will reveal the extent of actual influence (which depends on how consistent the corresponding y observation is with the rest of the data).

Multicollinearity

When the simple linear regression model is fit using sample data in which the x values are all close to one another, small changes in observed y's can cause the values of the estimated coefficients a and b to change considerably. This may well result in standard deviations σ_b and σ_a that are quite large, so that estimates of β and α from any given sample are likely to differ greatly from the true values. There is an analogous condition that leads to this same type of behavior in multiple regression—a configuration of predictor variable values that is likely to result in poorly estimated model coefficients.

Consider first the case in which a model involving just two predictor variables x_1 and x_2 is to be fit. Suppose that the sample (x_1, x_2) pairs are those displayed in the scatter plot of Figure 12(a) and on the ground in the three-dimensional representation of Figure 12(b). An obvious characteristic of the (x_1, x_2) plot is the strong linear relationship between x_1 and x_2. Each sample observation (x_1, x_2, y) is a point in three dimensions, as pictured in Figure 12(b). Recall that using the principle of least squares to estimate the coefficients α, β_1, and β_2 amounts to finding the plane (flat surface) that has the smallest sum of squared deviations between it and the given points—the best-fit plane. Now suppose that the y observations made at the three points falling above the straight line in Figure 12(a)—the colored points in Figure 12(b)—are increased a bit and the y values at the other three points are decreased somewhat. Such small changes in y considerably changes the tilt of the least squares plane and thus the values of a, b_1 and b_2. That is, even a small amount of variability in y (small σ^2 in the model) translates into a considerable amount of variability in a, b_1, and b_2. Trying to fit a plane here is like trying to rest a cardboard sheet on a picket fence—the result is very unstable.

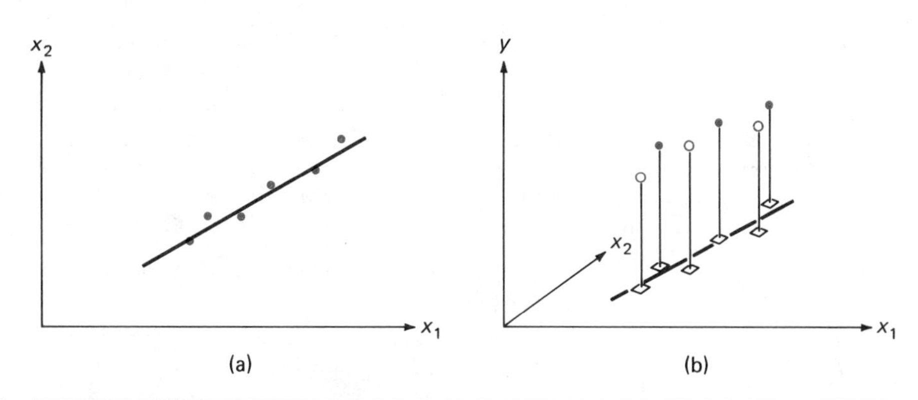

FIGURE 12 AN ILLUSTRATION OF MULTICOLLINEARITY
(a) A Scatter Plot of Predictor Variable Values that Shows a Strong Linear Relationship (Multicollinearity)
(b) y Observations Made at (x_1, x_2) Pairs Shown in (a)

This instability would not have occurred had the sample (x_1, x_2) points been more spread out and not fallen so close to a straight line (try to visualize in this case how small changes in y would not have much effect on the best-fit plane). More generally, when the model to be fit includes k predictors x_1, $x_2, \ldots, x_k$, there is said to be **multicollinearity** if there is a strong linear relationship between values of the predictors. Severe multicollinearity leads to instability of estimated coefficients as well as various other problems. Such a relationship is difficult to visualize when $k > 2$, so statisticians have developed various quantitative indicators to measure the extent of multicollinearity in a data set. The most straightforward approach involves computing R^2 values for regressions in which the dependent variable is taken to be one of the k x's and the predictors are the remaining $k - 1$ x's. For example, when $k = 3$

there are three relevant regressions:

1. dependent variable = x_1, predictor variables = x_2 and x_3
2. dependent variable = x_2, predictor variables = x_1 and x_3
3. dependent variable = x_3, predictor variables = x_1 and x_2

Each regression yields an R^2 value. In general, there are k such regressions and, therefore, k resulting R^2 values. If one or more of these R^2 values is large (close to one), multicollinearity is present. MINITAB will print a message saying that the predictors are highly correlated when at least one of the k R^2 values exceeds .99 and will refuse to include a predictor in the model if the corresponding R^2 value is larger than .9999. Other analysts are more conservative and would judge multicollinearity to be a potential problem if any of these R^2 values exceeded .9.

When predictor variable values are under the control of the investigator, which often happens in scientific experimentation, a careful choice of values will preclude multicollinearity from arising. Multicollinearity does frequently occur in social science and business applications of regression analysis, where data result simply from observation rather than from intervention by an experimenter. Statisticians have proposed various remedies for the problems associated with multicollinearity in such situations, but a discussion would take us beyond the scope of this book. (After all, we want to leave something for your next statistics course!)

EXERCISES

12.39 The paper "The Caseload Controversy and the Study of Criminal Courts" (*J. Criminal Law and Criminology* (1979):89–101) used multiple regression to analyze a data set consisting of observations on the variables

y = length of sentence in trial case (months)
x_1 = seriousness of first offense
x_2 = dummy variable indicating type of trial (bench or jury)
x_3 = number of legal motions
x_4 = measure of delay for confined defendants (0 for those not confined)
x_5 = measure of judge's caseload

The estimated regression equation proposed by the authors is

$$\hat{y} = 12.6 + .59x_1 - 70.8x_2 - 33.6x_3 - 15.5x_5 + .0007x_6 + 3x_7 - 41.5x_8$$

where $x_6 = x_4^2$, $x_7 = x_1x_2$, and $x_8 = x_3x_5$. How do you think the authors might have arrived at this particular model?

12.40 The article "The Analysis and Selection of Variables in Linear Regression" (*Biometrics* (1976):1–49) reports on an analysis of data taken from the 1974 issues of *Motor Trend* magazine. The dependent variable y was gas mileage, there were $n = 32$ observations, and the independent variables were

x_1 = engine type (1 = straight, 0 = V)
x_2 = number of cylinders
x_3 = transmission type (1 = manual, 0 = automatic)
x_4 = number of transmission speeds

x_5 = engine size
x_6 = horsepower
x_7 = number of carburetor barrels
x_8 = final drive ratio
x_9 = weight
x_{10} = quarter-mile time

The R^2 and adjusted R^2 values are given for the best model using k predictors for $k = 1, \ldots, 10$.

k	Variables Included	R^2	Adjusted R^2
1	x_9	.756	.748
2	x_2, x_9	.833	.821
3	x_3, x_9, x_{10}	.852	.836
4	x_3, x_6, x_9, x_{10}	.860	.839
5	$x_3, x_5, x_6, x_9, x_{10}$	.866	.840
6	$x_3, x_5, x_6, x_8, x_9, x_{10}$	.869	.837
7	$x_3, x_4, x_5, x_6, x_8, x_9, x_{10}$	.870	.832
8	$x_3, x_4, x_5, x_6, x_7, x_8, x_9, x_{10}$	.871	.826
9	$x_1, x_3, x_4, x_5, x_6, x_7, x_8, x_9, x_{10}$	.871	.818
10	all independent variables	.871	.809

What model would you select? Explain your choice and the criteria used to reach your decision.

12.41 The paper "Estimation of the Economic Threshold of Infestation for Cotton Aphid" (*Mesopotamia J. Ag.* (1982):71–75) gave $n = 34$ observations on

y = infestation rate (number of aphids per 100 leaves),
x_1 = mean temperature (°C)
x_2 = mean relative humidity

Partial SAS computer output resulting from fitting the model $y = \alpha + \beta_1 x_1 + \beta_2 x_2 + e$ to the data given in the paper is shown. If the method of backward elimination is to be employed, which variable would be the first candidate for elimination? Using the criteria of eliminating a variable if its t ratio satisfies $-2 \leq t$ ratio ≤ 2, can it be eliminated from the model?

```
                                        R-SQUARE          0.5008
                                        ADJ R-SQ          0.4533

                       PARAMETER        STANDARD        T FOR H0:
VARIABLE      DF        ESTIMATE           ERROR      PARAMETER=0

INTERCEP       1       15.667014       84.157420            0.186
TEMP           1       -0.360928        2.330835           -0.155
RH             1        1.659715        0.576946            2.877
```

12.42 The accompanying data appeared in the paper "Breeding Success of the Common Puffin on Different Habitats at Great Island, Newfoundland" (*Ecol. Monographs* (1972):246–52). The variables considered are y = nesting frequency (burrows per 9 m^2), x_1 = grass cover (%), x_2 = mean soil depth (cm), x_3 = angle of slope (degrees), and x_4 = distance from cliff edge (m).

y	x_1	x_2	x_3	x_4	y	x_1	x_2	x_3	x_4
16	45	39.2	38	3	16	60	37.1	35	6
15	65	47.0	36	12	25	60	47.1	35	12
10	40	24.3	14	18	13	85	34.0	23	18
7	20	30.0	16	21	13	90	43.6	12	21
11	40	47.6	6	27	11	20	30.8	9	27
7	80	47.6	9	36	3	85	34.6	6	33
4	80	45.6	7	39	0	30	37.7	8	42
0	15	27.8	8	45	0	75	45.5	5	48
0	0	41.9	8	54	0	15	51.4	8	54
0	20	36.8	5	60	18	40	32.1	36	6
15	40	34.9	31	3	19	40	35.4	37	9
21	60	45.2	37	12	8	90	30.2	11	18
12	95	32.9	24	18	12	80	33.9	9	24
8	50	26.6	11	24	10	80	40.2	11	30
9	80	32.7	10	30	3	75	33.5	7	36
6	80	38.1	5	36	0	65	40.3	10	42
0	60	31.4	5	39	0	80	40.3	12	45
0	70	32.7	2	48	0	50	43.1	13	51
0	35	38.1	8	51	0	50	42.0	3	57

MINITAB output resulting from application of the backward elimination procedure is given. Explain what action was taken at each step and why.

```
    STEP            1          2          3
CONSTANT        12.29      11.45      13.96

X1             -0.019
T-RATIO         -0.70

X2             -0.043     -0.043
T-RATIO         -1.42      -1.42

X3              0.224      0.225      0.176
T-RATIO          2.91       2.94       2.54

X4             -0.182     -0.203     -0.293
T-RATIO         -2.12      -2.55      -6.01

S                2.83       2.81       2.85
R-SQ            86.20      85.99      85.16
```

12.43 *This problem requires use of a computer package.* Using a statistical computer package, compare the best one-, two-, three-, and four-predictor models for the data given in Exercise 12.42. Does this variable-selection procedure lead you to the same choice of model as the backward elimination method employed in Exercise 12.42?

12.44 The formulas for R^2 and adjusted R^2 are

$$R^2 = 1 - \frac{\text{SSResid}}{\text{SSTo}}$$

$$\text{Adjusted } R^2 = 1 - \frac{(n - 1)\text{SSResid}}{[n - (k + 1)]\text{SSTo}}$$

When would adjusted R^2 be substantially smaller than R^2?

12.45 Referring to the formulas given in Exercise 12.44, if $n = 21$ and $k = 10$, for what values of R^2 would adjusted R^2 be negative?

12.46 Suppose you were considering a multiple regression analysis with y = house price, x_1 = number of bedrooms, x_2 = number of bathrooms, and x_3 = total number of rooms. Do you think that multicollinearity might be a problem? Explain.

12.47 *This problem requires use of a computer package.* The accompanying n = 25 observations on y = catch at intake (number of fish), x_1 = water temperature (°C), x_2 = minimum tide height (m), x_3 = number of pumps running, x_4 = speed (knots), x_5 = wind—range of direction (degrees) constitute a subset of the data that appeared in the paper "Multiple Regression Analysis for Forecasting Critical Fish Influxes at Power Station Intakes" (*J. Applied Ecol.* (1983): 33–42). Use the variable-selection procedures discussed in this section to formulate a model.

y	x_1	x_2	x_3	x_4	x_5	y	x_1	x_2	x_3	x_4	x_5
17	6.7	.5	4	10	50	3	15.8	1.6	3	7	120
42	7.8	1.0	4	24	30	7	16.2	.4	3	10	50
1	9.9	1.2	4	17	120	9	15.8	1.2	3	9	60
11	10.1	.5	4	23	30	10	16.0	.8	3	12	90
8	10.0	.9	4	18	20	7	16.2	1.2	3	5	160
30	8.7	.8	4	9	160	12	17.1	.7	3	10	90
2	10.3	1.5	4	13	40	12	17.5	.8	3	12	110
6	10.5	.3	4	10	150	26	17.5	1.2	3	18	130
11	11.0	1.2	3	9	50	14	17.4	.8	3	9	60
14	11.2	.6	3	7	100	18	17.4	1.1	3	13	30
53	12.9	1.8	3	10	90	14	17.8	.5	3	8	160
9	13.2	.2	3	12	50	5	18.0	1.6	3	10	40
4	16.2	.7	3	6	80						

12.48 Given that $R^2 = .723$ for the model containing predictors x_1, x_4, x_5, and x_8 and $R^2 = .689$ for the model with predictors x_1, x_3, x_5, and x_6,

a. What can you say about R^2 for the model containing predictors x_1, x_3, x_4, x_5, x_6, and x_8? Explain.

b. What can you say about R^2 for the model containing predictors x_1 and x_4? Explain.

KEY CONCEPTS

General additive multiple regression model based on k predictors $x_1, x_2, \ldots, x_k$ $y = \alpha + \beta_1 x_1 + \beta_2 x_2 + \cdots + \beta_k x_k + e$, where the random deviation e is normally distributed with mean value zero and variance σ^2 for any choice of $x_1, \ldots, x_k$ values. $\alpha + \beta_1 x_1 + \cdots + \beta_k x_k$ is called the population regression function, and α and the β_i's are the population regression coefficients. (p. 499)

Estimated regression function $a + b_1 x_1 + \cdots + b_k x_k$ Obtained by applying the principle of least squares. The values of the estimated coefficients $a, b_1, b_2, \ldots, b_k$ are chosen to minimize the sum of squared deviations $\Sigma[y - (a + b_1 x_1 + \cdots + b_k x_k)]^2$. (p. 510)

Coefficient of multiple determination, R^2 $R^2 = 1 - (\text{SSResid}/\text{SSTo})$, where $\text{SSTo} = \Sigma(y - \bar{y})^2$ and $\text{SSResid} = \Sigma(y - \hat{y}^2)$ (the predicted values $\hat{y}_1, \ldots, \hat{y}_n$ result from substituting the values of $x_1, \ldots, x_k$ for each sample observation into the estimated regression function). It is interpreted as the proportion of observed variation in y attributable to the relationship between y and the predictors given by the model equation. (p. 514)

F test for model utility A test of $H_0 : \beta_1 = \beta_2 = \cdots = \beta_k = 0$ (which says that there is no useful linear relationship between y and any of the k model predictors) versus H_a : at least one β_i is not zero. The test statistic is the F ratio

$$F = \frac{\text{SSRegr}/k}{\text{SSResid}/[n - (k + 1)]}$$

where SSRegr = SSTo − SSResid, and H_0 is rejected if F exceeds the appropriate F critical value based on k numerator df and $n - (k + 1)$ denominator df. Only if H_0 can be rejected should the model be used to make other inferences. (p. 517)

Inferences about individual β_i's, a mean y value, and a future y value A confidence interval for β_i has the form $b_i \pm (t \text{ critical value})s_{b_i}$ and a t statistic is used to test hypotheses concerning β_i. A confidence interval for mean y has the form $a + b_1x_1 + \cdots + b_kx_k \pm (t$ critical value)(estimated standard deviation), and there is a similar formula for a prediction interval for a single y value. (p. 525–30)

Model selection An investigator may have data on many predictors that could be included in a model. There are two different approaches to choosing a model: (1) Use an *all-subsets procedure* to identify the best models of each different size (one-predictor models, two-predictor models, etc.) and then compare these according to criteria such as R^2 and adjusted R^2; (2) employ an *automatic selection procedure*, which either successively eliminates predictors until a stopping point is reached (backward elimination) or starts with no predictors and adds them successively until the inclusion of additional predictors cannot be justified. (p. 537–45)

SUPPLEMENTARY EXERCISES

12.49 The accompanying data on y = glucose concentration (g/L) and x = fermentation time (days) for a particular blend of malt liquor was read from a scatter plot appearing in the paper "Improving Fermentation Productivity with Reverse Osmosis" (*Food Tech.* (1984):92–96).

x	1	2	3	4	5	6	7	8
y	74	54	52	51	52	53	58	71

a. Construct a scatter plot for this data. Based on the scatter plot, what type of model would you suggest?

b. MINITAB output resulting from fitting a multiple regression model with $x_1 = x$ and $x_2 = x^2$ is given. Does this quadratic model specify a useful relationship between y and x?

```
THE REGRESSION EQUATION IS
Y =       84.5 -    15.9 X1 +    1.77 X2
                                ST. DEV.        T-RATIO =
           COEFFICIENT          OF COEF.        COEF/S.D.
               84.482             4.904           17.23
X1            -15.875             2.500           -6.35
X2              1.7679            0.2712           6.52

THE ST. DEV. OF Y ABOUT REGRESSION LINE IS
S = 3.515
WITH (    8- 3) =     5 DEGREES OF FREEDOM

R-SQUARED = 89.5 PERCENT
R-SQUARED = 85.3 PERCENT, ADJUSTED FOR D.F.
```

c. Could the quadratic term have been eliminated? That is, would a simple linear model have sufficed? Test using a .05 significance level.

12.50 Much interest in management circles has recently focused on how employee compensation is related to various company characteristics. The paper "Determinants of R and D Compensation Strategies" (*Personnel Psych.* (1984):635-50) proposed a quantitative scale for y = base salary for employees of high-tech companies. The following estimated multiple regression equation was then presented:

$$\hat{y} = 2.60 + .125x_1 + .893x_2 + .057x_3 - .014x_4$$

where x_1 = sales volume (in millions of dollars), x_2 = stage in product life cycles (1 = growth, 0 = mature), x_3 = profitability (%), and x_4 = attrition rate (%).

a. There were $n = 33$ firms in the sample and $R^2 = .69$. Is the fitted model useful?

b. Predict base compensation for a growth stage firm with sales volume \$50 million, profitability 8%, and attrition rate 12%.

c. The estimated standard deviations for the coefficient estimates were .064, .141, .014, and .005 for b_1, b_2, b_3, and b_4, respectively. Should any of the predictors be deleted from the model? Explain.

d. β_3 is the difference between average base compensation for growth stage and mature stage firms when all other predictors are held fixed. Use the information in (c) to calculate a 95% confidence interval for β_3.

12.51 A study of total body electrical conductivity was described in the article "Measurement of Total Body Electrical Conductivity: A New Method for Estimation of Body Composition" (*Amer. J. Clinical Nutr.* (1983):735–39). Nineteen observations were given for the variables y = total body electrical conductivity, x_1 = age (years), x_2 = sex (0 = male, 1 = female), x_3 = body mass (kg/m^2), x_4 = body fat (kg), x_5 = lean body mass (kg).

a. The backward elimination method of variable selection was employed, and MINITAB output is given. Explain what occurred at each step in the process.

```
STEPWISE REGRESSION OF    Y       ON  5 PREDICTORS, WITH N =  19

     STEP           1          2          3
CONSTANT       -6.193    -13.285    -15.175

X1               0.31       0.36       0.38
T-RATIO          1.90       2.36       2.96

X2               -7.9       -7.5       -7.0
T-RATIO         -1.93      -1.89      -2.04

X3              -0.43
T-RATIO         -0.72

X4               0.22       0.03
T-RATIO          0.78       0.29

X5              0.365      0.339      0.378
T-RATIO          2.16       2.09       4.18

S                5.16       5.07       4.91
R-SQ            79.53      78.70      78.57
```

b. MINITAB output for the multiple regression model relating y to x_1, x_2, and x_5 is given. Interpret the values of R^2 and s_e.

```
THE REGRESSION EQUATION IS
Y = - 15.2 + 0.377 X1 - 6.99 X2 + 0.378 X5

                                         ST. DEV.       T-RATIO =
COLUMN        COEFFICIENT               OF COEF.       COEF/S.D.
                  -15.175                 9.620           -1.58
X1                 0.3771                0.1273            2.96
X2                 -6.988                 3.425           -2.04
X5                0.37779               0.09047            4.18

S = 4.914

R-SQUARED = 78.6 PERCENT
R-SQUARED = 74.3 PERCENT, ADJUSTED FOR D.F.
```

c. Interpret the value of b_2 in the estimated regression equation.

d. If the estimated standard deviation of $a + b_1(31) + b_2(1) + b_3(52.7)$ is 1.42, give a 95% confidence interval for the true mean total body electrical conductivity of all 31-year-old females whose lean body mass is 52.7 kg.

12.52 The paper "Creep and Fatigue Characteristics of Ferrocement Slabs" (*J. Ferrocement* (1984):309–22) reported data on y = tensile strength (MPa), x_1 = slab thickness (cm), x_2 = load (kg), x_3 = age at loading (days), and x_4 = time under test (days) resulting from stress tests of $n = 9$ reinforced concrete slabs. The backward elimination method of variable selection was applied. Given below is partial MINITAB output. Explain what action was taken at each step in the process. MINITAB output for the selected model is also given. Use the estimated regression equation to predict tensile strength for a slab that is 25 cm thick, 150 days old, and is subjected to a load of 200 kg for 50 days.

```
    STEP          1          2          3
CONSTANT      8.496     12.670     12.989

X1            -0.29      -0.42      -0.49
T-RATIO       -1.33      -2.89      -3.14

X2           0.0104     0.0110     0.0116
T-RATIO        6.30       7.40       7.33

X3           0.0059
T-RATIO        0.83

X4           -0.023     -0.023
T-RATIO       -1.48      -1.53

S             0.533      0.516      0.570
R-SQ          95.81      95.10      92.82

THE REGRESSION EQUATION IS
Y =    13.0 -   .487 X1 + .0116 X2

                               ST. DEV.      T-RATIO =
         COEFFICIENT           OF COEF.      COEF/S.D.
              12.989              3.640           3.57
X1           -0.4867             0.1549          -3.14
X2          0.011569           0.001579           7.33

THE ST. DEV. OF Y ABOUT REGRESSION LINE IS
S = 0.5698
WITH (   9- 3) =    6 DEGREES OF FREEDOM

R-SQUARED = 92.8 PERCENT
R-SQUARED = 90.4 PERCENT, ADJUSTED FOR D.F.
```

12.53 A study of pregnant grey seals involved $n = 25$ observations on the variables y = fetus progesterone level (μg), x_1 = fetus sex (0 = male, 1 = female), x_2 = fetus length (cm), and x_3 = fetus weight (g). MINITAB output for the model using all three independent variables is given ("Gonadotrophin and Progesterone Concentration in Placental of Grey Seals" *J. Repro. and Fertility* (1984):521–28).

```
THE REGRESSION EQUATION IS
Y = -   1.98 -   1.87 X1 +    .234 X2
    + .0001 X3
                               ST. DEV.        T-RATIO =
          COEFFICIENT          OF COEF.        COEF/S.D
               -1.982            4.290            -0.46
X1             -1.871            1.709            -1.09
X2             0.2340           0.1906             1.23
X3           0.000060         0.002020             0.03

THE ST. DEV. OF Y ABOUT REGRESSION LINE IS
S = 4.189
WITH (  25- 4) = 21 DEGREES OF FREEDOM

R-SQUARED = 55.2 PERCENT
R-SQUARED = 48.8 PERCENT, ADJUSTED FOR D.F.

ANALYSIS OF VARIANCE

 DUE TO        DF            SS         MS=SS/DF
REGRESSION      3        454.63           151.54
RESIDUAL       21        368.51            17.55
TOTAL          24        823.15
```

a. Use information from the MINITAB output to test the null hypothesis H_0: $\beta_1 = \beta_2 = \beta_3 = 0$.

b. Using an elimination criteria of $-2 \leq t$ ratio ≤ 2, should any variable be eliminated? If so, which one?

c. MINITAB output for the regression using only x_1 = sex and x_2 = length is given. Would you recommend keeping both x_1 and x_2 in the model? Explain.

```
THE REGRESSION EQUATION IS
Y = -   2.09 -   1.87 X1 +   .240 X2

                               ST. DEV.        T-RATIO =
          COEFFICIENT          OF COEF.        COEF/S.D.
               -2.090            2.212            -0.94
X1             -1.865            1.661            -1.12
X2            0.23952          0.04604             5.20

THE ST. DEV. OF Y ABOUT REGRESSION LINE IS
S = 4.093
WITH (  25- 3) =   22 DEGREES OF FREEDOM

R-SQUARED = 55.2 PERCENT
R-SQUARED = 51.2 PERCENT, ADJUSTED FOR D.F.
```

d. After elimination of both x_3 and x_1, the estimated regression equation is $\hat{y} = -2.61 + .231x_2$. The corresponding values of R^2 and s_e are .527 and 4.116, respectively. Interpret these values.

e. Referring to (d), how would you interpret the value of $b_2 = .231$? Does it make sense to interpret the value of a as the estimate of average progesterone level when length is zero? Explain.

12.54 The authors of the paper "Influence of Temperature and Salinity on Development of White Perch Eggs" (*Trans. Amer. Fisheries Soc.* (1982):396–98) used a quadratic regression model to describe the relationship between y = percent hatch of white perch eggs and x = water temperature (°C). The estimated regression equation was given as $\hat{y} = -41.0 + 14.6x - .55x^2$.

a. Graph the curve corresponding to the estimated regression equation by plotting the (x, y) pairs using x values of 10, 12, 14, 16, 18, 20, 22, and 24 (these are the temperatures used in the paper). Draw a smooth curve through the points in your plot.

b. The authors make the claim that the optimal temperature for hatch is 14°C. Based on your graph in (a), does this statement seem reasonable? Explain.

12.55 A sample of $n = 20$ companies was selected, and the values of y = stock price and $k = 15$ predictor variables (such as quarterly dividend, previous year's earnings, debt ratio, etc.) were determined. When the multiple regression model with these 15 predictors was fit to the data, $R^2 = .90$ resulted.

a. Does the model appear to specify a useful relationship between y and the predictor variables? Carry out a test using significance level .05. (The F critical value for 15 numerator and 4 denominator df is 5.86.)

b. Based on the result of (a), does a high R^2 value by itself imply that a model is useful? Under what circumstances might you be suspicious of a model with a high R^2 value?

c. With n and k as given above, how large would R^2 have to be for the model to be judged useful at the .05 level of significance?

12.56 *This problem requires the use of a computer package.* The paper "Entry and Profitability in a Rate-Free Savings and Loan Market" (*Quarterly Review of Econ. and Business* (1978):87–95) gave the accompanying data on y = profit margin, x_1 = net revenues, and x_2 = number of savings and loan branch offices for a sample of 25 savings and loan associations.

x_1	3.92	3.61	3.32	3.07	3.06	3.11	3.21	3.26	3.42
x_2	7298	6855	6636	6506	6450	6402	6368	6340	6349
y	.75	.71	.66	.61	.70	.72	.77	.74	.90
x_1	3.42	3.45	3.58	3.66	3.78	3.82	3.97	4.07	4.25
x_2	6352	6361	6369	6546	6672	6890	7115	7327	7546
y	.82	.75	.77	.78	.84	.79	.70	.68	.72
x_1	4.41	4.49	4.70	4.58	4.69	4.71	4.78		
x_2	7931	8097	8468	8717	8991	9179	9318		
y	.55	.63	.56	.41	.51	.47	.32		

a. Fit a multiple regression model using both independent variables.

b. Use the F test to determine if the model provides useful information for predicting profit margin.

c. Interpret the values of R^2 and s_e.

d. Would a regression model using a single independent variable (x_1 alone or x_2 alone) have sufficed? Explain.

e. Plot the (x_1, x_2) pairs. Does the plot indicate any sample observation that may have been highly influential in estimating the model coefficients? Explain. Do you see any evidence of multicollinearity? Explain.

REFERENCES

See the references at the end of Chapter 11.

13 The Analysis of Variance

INTRODUCTION

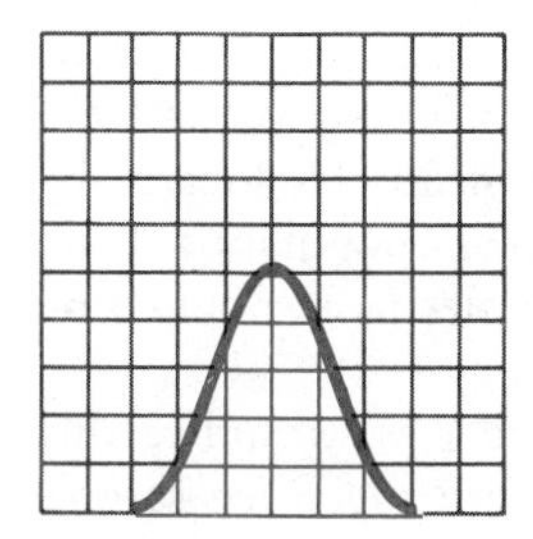

METHODS FOR TESTING $H_0 : \mu_1 - \mu_2 = 0$ (i.e., $\mu_1 = \mu_2$), where μ_1 and μ_2 are the means of two different populations or the true average responses when two different treatments are applied, were discussed in Chapter 9. Many investigations involve a comparison of more than two population or treatment means. For example, let μ_1, μ_2, μ_3, and μ_4 denote true average burn times under specified conditions for four different fabrics used in children's sleepwear. Data from an appropriate experiment could be used to test the null hypothesis that $\mu_1 = \mu_2 = \mu_3 = \mu_4$ (no difference in true average burn times) against the alternative hypothesis that there are differences among the values of the μ's.

The characteristic that distinguishes the populations or treatments from one another is called the **factor** under investigation. In this example, the factor is fabric type. An experiment might be carried out to compare three different methods for teaching reading (three different treatments), in which case the factor of interest is teaching method. Both fabric type and teaching method are qualitative factors. If growth of fish raised in waters having different salinity levels—0%, 10%, 20%, and 30%—is of interest, the factor *salinity level* is quantitative.

A **single-factor analysis of variance** (ANOVA) problem involves a comparison of k population or treatment means $\mu_1, \mu_2, \ldots, \mu_k$. The objective is to test $H_0 : \mu_1 = \mu_2 = \cdots = \mu_k$ against H_a : at least two of the means are different. The analysis is based on k independently selected random samples, one from each population or for each treatment. That is, in the case of populations, the sample from any particular population is selected independently of that from any other population. When comparing treatments, the experimental units (subjects or objects) that receive any particular treatment are chosen independently of the units that receive any other treatment. A comparison of treatments based on independently selected experimental units is often referred to as a **completely randomized design.** Sections 13.1, 13.2, 13.3, and 13.6 discuss various aspects of single-factor ANOVA. In Section 13.4 we present an alternative to a completely randomized design, called a **randomized block design,** which controls for extraneous variation among experimental units in order to obtain a more precise assessment of treatment effects.

Researchers frequently carry out experiments to study the effects of two or more factors on some response variable. For example, a testing organization might wish to investigate whether average coverage (square feet) by a single gallon of paint depends either on which of four brands (factor A) is used or on which of three colors (factor B) is applied. Some concepts and methods for the analysis of data from two-factor experiments are introduced in Section 13.5.

13.1 Single-Factor ANOVA and the F Test

The decision about whether or not the null hypothesis of single-factor ANOVA should be rejected depends on how substantially the samples from the different populations, or treatments, differ from one another. To develop some intuition, it is helpful first to consider a simple example. Figure 1 displays two possible data sets that might arise when observations are selected from each of three populations under study. Each data set consists of five observations from the first population, four from the second, and six observations from the third. For both data sets the three resulting sample means are located by vertical line segments. The means of the two samples from population 1 are identical, and a similar statement holds for the two samples from population 2 and those from population 3.

Almost anyone, after looking at the data set in Figure 1(a), would readily agree that the claim $\mu_1 = \mu_2 = \mu_3$ is false. Not only are the three sample means different, but there is very clear separation between the samples. Put another way, differences between the three sample means are quite large relative to variability within each sample (if all data sets gave such clear-cut messages, there would be many more unemployed statisticians than is currently the case). The situation pictured in Figure 1(b) is much less clear-cut. While the sample means are as different as they were in the first data set, there is now considerable overlap in the three samples. The separation between sample means here can plausibly be attributed to substantial variability in the populations being reflected in the samples rather than to differences between μ_1, μ_2, and μ_3. The phrase *analysis of variance* springs from the idea of analyzing variability in the entire data set (consisting of all k samples) to see how much can be attributed to differences between μ's and how much is due to variability in the individual populations. In Figure 1(a) there is little within-sample variability relative to the amount of between-samples variability,

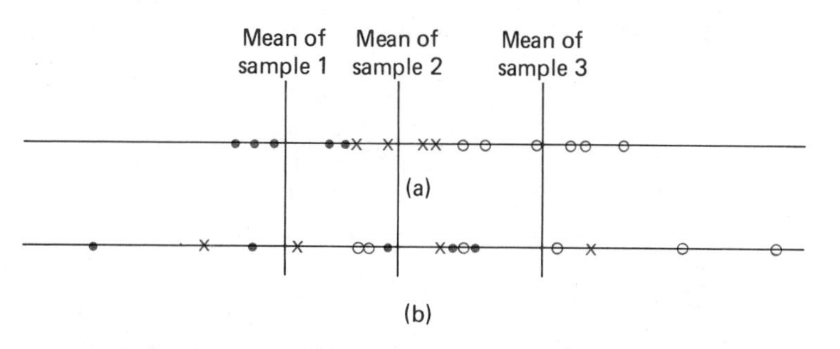

FIGURE 1 TWO POSSIBLE DATA SETS WHEN $k = 3$
● = Observation from Population 1
x = Observation from Population 2
○ = Observation from Population 3

whereas a great deal more of the total variability in Figure 1(b) is due to variation within each sample. If differences between the sample means can be explained by within-sample variability, there is no compelling reason for rejecting H_0.

Notation in single-factor ANOVA is a natural extension of the notation used in Chapter 9 for comparing two population or treatment means.

ANOVA Notation

k = the number of populations or treatments being compared

Population or treatment	1	2	3	. . .	k
Population or treatment mean	μ_1	μ_2	μ_3	. . .	μ_k
Population or treatment variance	σ_1^2	σ_2^2	σ_3^2	. . .	σ_k^2
Sample size	n_1	n_2	n_3	. . .	n_k
Sample mean	$\bar{x}_1$	$\bar{x}_2$	$\bar{x}_3$	. . .	$\bar{x}_k$
Sample variance	s_1^2	s_2^2	s_3^2	. . .	s_k^2

In addition, let $N = n_1 + n_2 + \cdots + n_k$, the total number of observations in the data set, and let $\bar{\bar{x}}$ denote the average of all N observations ($\bar{\bar{x}}$ is the **grand mean**). When $n_1 = n_2 = \cdots = n_k$ (all sample sizes are the same), denote their common value by n; in this case, $N = kn$.

EXAMPLE 1

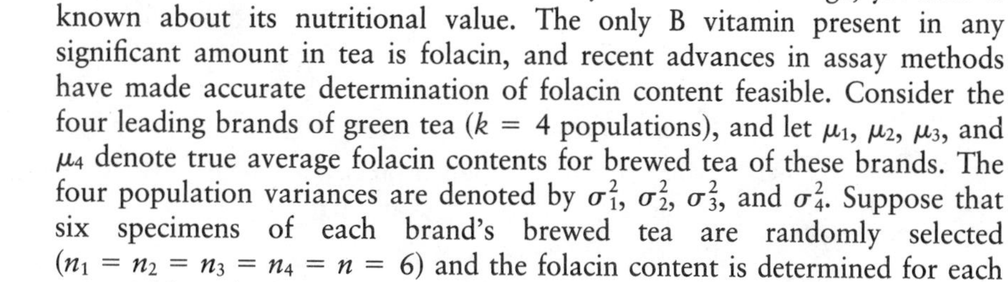

After water, tea is the world's most widely consumed beverage, yet little is known about its nutritional value. The only B vitamin present in any significant amount in tea is folacin, and recent advances in assay methods have made accurate determination of folacin content feasible. Consider the four leading brands of green tea ($k = 4$ populations), and let μ_1, μ_2, μ_3, and μ_4 denote true average folacin contents for brewed tea of these brands. The four population variances are denoted by σ_1^2, σ_2^2, σ_3^2, and σ_4^2. Suppose that six specimens of each brand's brewed tea are randomly selected ($n_1 = n_2 = n_3 = n_4 = n = 6$) and the folacin content is determined for each one, yielding the accompanying data and summary quantities.

Brand	Observations	Sample Mean	Sample Variance
1	7.9, 6.2, 6.6, 8.6, 8.9, 10.1	$\bar{x}_1 = 8.05$	$s_1^2 = 2.16$
2	5.7, 7.5, 9.8, 6.1, 8.4, 9.6	$\bar{x}_2 = 7.85$	$s_2^2 = 3.00$
3	6.8, 7.5, 5.0, 7.4, 5.3, 6.1	$\bar{x}_3 = 6.35$	$s_3^2 = 1.12$
4	6.4, 7.1, 7.9, 4.5, 5.0, 4.0	$\bar{x}_4 = 5.82$	$s_4^2 = 2.41$

Then $N = (4)(6) = 24$ and the grand mean is $\bar{\bar{x}} =$ (total of all 24 observations)/24 = 168.40/24 = 7.02. (This example is based on the paper "Folacin Content of Tea," *J. Amer. Dietetic Assoc.* (1983):627–32. The authors do not give raw data, but their summary values are quite close to those given above.)

Does this data suggest that true average folacin content is the same for all brands? The relevant hypotheses are $H_0 : \mu_1 = \mu_2 = \mu_3 = \mu_4$ versus H_a : at least two of the four μ's are different. To decide whether H_0 should be rejected, it is natural to examine the four sample means $\bar{x}_1$, $\bar{x}_2$, $\bar{x}_3$, and $\bar{x}_4$. Even if all four population means were identical, the four sample means would typically differ somewhat from one another (and thus from the grand mean $\bar{\bar{x}}$) just because of variation in each population and resulting sampling variability. So we must ask whether sampling variability by itself would account for observed discrepancies in the $\bar{x}$'s. If not, the most plausible explanation is that differences in $\bar{x}$'s are attributable to differences among the μ's, and rejection of H_0 is appropriate. ■

In general, the hypotheses to be tested in single-factor ANOVA are

$$H_0 : \mu_1 = \mu_2 = \cdots = \mu_k$$

versus

$$H_a : \text{at least two of the } \mu\text{'s are different}$$

A decision between H_0 and H_a will be based on examining the $\bar{x}$'s to see whether observed discrepancies are small enough to be attributable simply to sampling variability or whether an alternative explanation for the differences is necessary.

The most frequently used inferential procedure in single-factor ANOVA, the F test, is based on two assumptions which are analogous to those on which the two-sample t test of Chapter 9 were based.

Basic ANOVA Assumptions

Each of the k population or treatment response distributions is normal and $\sigma_1^2 = \sigma_2^2 = \cdots = \sigma_k^2$, with σ^2 used to denote the common population variance.

In practice, the test based on these assumptions will work well as long as the assumptions are not too badly violated. Typically, sample sizes are so small that a separate normal probability plot for each sample is of little value in checking normality. A single combined plot results from first subtracting $\overline{x}_1$ from each observation in the first sample, $\overline{x}_2$ from each value in the second sample, etc., and then constructing a normal probability plot of all N deviations. The plot should be reasonably straight. There are formal procedures for testing the equality of population variances, but for reasons discussed in Chapter 9, we do not favor them. Provided that the largest of the k sample variances is not too many times larger than the smallest, the F test can safely be used. In Example 1, the largest variance (3.00) is less than three times the smallest (1.12). For small sample sizes, this is not at all surprising when all four σ^2's are equal.

The Case of Equal Sample Sizes

Notation, computations, and the logic of the F test are most straightforward when all k sample sizes are equal (to n), so we first consider this case. As indicated earlier, if $\overline{x}_1, \overline{x}_2, \ldots, \overline{x}_k$ are all reasonably close to the grand mean $\overline{\overline{x}}$, there is little reason to think H_0 is false. Only samples yielding $\overline{x}_1, \overline{x}_2, \ldots, \overline{x}_k$, some of which differ substantially from $\overline{\overline{x}}$, cast doubt on H_0. We therefore need a measure of how much the $\overline{x}$'s deviate from $\overline{\overline{x}}$.

DEFINITION

A measure of discrepancy between the sample means $\overline{x}_1, \overline{x}_2, \ldots, \overline{x}_k$ is the **mean square for treatments,** denoted by MSTr and defined by

$$\text{MSTr} = \frac{n}{k-1}[(\overline{x}_1 - \overline{\overline{x}})^2 + (\overline{x}_2 - \overline{\overline{x}})^2 + \cdots + (\overline{x}_k - \overline{\overline{x}})^2]$$

Except for the factor $n/(k-1)$, MSTr is the sum of squared deviations of the individual sample averages from the grand mean. The reason for including this factor will be explained shortly.

EXAMPLE 2

The folacin content data of Example 1 yielded $\overline{x}_1 = 8.05$, $\overline{x}_2 = 7.85$, $\overline{x}_3 = 6.35$, $\overline{x}_4 = 5.82$, and $\overline{\overline{x}} = 7.02$. With $k = 4$ and $n = 6$,

$$\begin{aligned}\text{MSTr} &= \frac{6}{4-1}[(8.05 - 7.02)^2 + (7.85 - 7.02)^2 \\ &\qquad + (6.35 - 7.02)^2 + (5.82 - 7.02)^2] \\ &= 2\,(1.06 + .69 + .45 + 1.44) = 7.28\end{aligned}$$

■

The smallest possible value of MSTr is 0, which occurs when $\overline{x}_1, \overline{x}_2, \ldots, \overline{x}_k$ are all equal (and therefore equal to $\overline{\overline{x}}$). The more the $\overline{x}$'s differ from $\overline{\overline{x}}$, the greater the value of MSTr. Furthermore, the $\overline{x}$'s tend to spread out more when H_0 is false (unequal μ's) than when H_0 is true, so MSTr tends to be larger when H_0 is false than when it is true. This suggests that H_0 should be rejected whenever the k samples result in a value of the statistic MSTr that is suitably large. To obtain a more specific decision rule, we need information about how MSTr behaves in repeated sampling both when H_0 is true and when H_0 is false.

The mean value of the statistic MSTr when H_0 is true is σ^2. That is, in this case MSTr is an unbiased statistic* for estimating σ^2. When H_0 is false, MSTr is a biased statistic because it tends to overestimate (be larger than) σ^2.

MSTr is sometimes called the *between-samples estimate* of σ^2, since it is based on the $\bar{x}$'s from different samples. If the value of σ^2 were known, H_0 could be rejected when the calculated value of MSTr is considerably larger than the known value. But, of course, the value of σ^2 is not known. The way out of this dilemma is to compute another estimate of σ^2, one that is reliable whether or not H_0 is true. Then if the between-samples estimate considerably exceeds this second estimate, H_0 can justifiably be rejected.

The second estimate of σ^2 is based on the sample variances $s_1^2, s_2^2, \ldots, s_k^2$. Each s^2 is an unbiased statistic for estimating the population variance σ^2, since it is computed from a single sample drawn from a population with that variance. A more reliable estimate than that given by any individual s^2 is obtained by averaging the k sample variances (an unweighted average is used because each sample contains the same number of observations).

DEFINITION

A statistic for estimating σ^2 that is unbiased whether or not H_0 is true is the **mean square for error,** denoted by MSE and given by

$$\text{MSE} = \frac{s_1^2 + s_2^2 + \cdots + s_k^2}{k}$$

The resulting estimate is often referred to as the *within-samples estimate* of σ^2, since it is based on estimating σ^2 within each sample and combining the results.

EXAMPLE 3

(*Examples* 1 *and* 2 *continued*) The four sample variances for the folacin content data were $s_1^2 = 2.16$, $s_2^2 = 3.00$, $s_3^2 = 1.12$, and $s_4^2 = 2.41$. Averaging these gives

$$\text{MSE} = \frac{2.16 + 3.00 + 1.12 + 2.41}{4} = 2.17$$

We previously found the between-samples estimate to be MSTr = 7.28. The between-samples estimate is more than three times as large as the within-samples estimate. There are two possibilities. Either H_0 is true and MSTr is much larger than MSE just because of sampling variability, or H_0 is false, resulting in a value of MSTr that considerably overestimates σ^2. Which of these is more tenable depends on how frequently the between-samples estimate is at least this much larger than the within-samples estimate when H_0 is true. ■

*The factor $n/(k - 1)$ is included in MSTr precisely to obtain an unbiased statistic when H_0 is true.

A comparison of the two estimates MSTr and MSE is made via the ratio MSTr/MSE. If H_0 is true, the two estimates should agree reasonably well, so the value of the ratio should not be too far from 1. On the other hand, when H_0 is false and the μ's are quite different, the numerator, MSTr, will usually greatly exceed the denominator, and the ratio will be much larger than 1. Therefore, H_0 should be rejected if the ratio considerably exceeds 1. A formal test procedure requires that a cutoff value be chosen and H_0 rejected only if the ratio exceeds the cutoff. The value should be such that MSTr/MSE is unlikely to exceed it when H_0 is true (so that the type I error probability is small). A suitable value is easily obtained once we know something about the sampling distribution of the ratio MSTr/MSE when H_0 is true.

A Necessary Digression: *F* Distributions

Many ANOVA test procedures are based on sampling distributions called *F* distributions.* An *F* distribution always arises in connection with a ratio of two quantities: *F* ratio = numerator/denominator. An *F* distribution is characterized by a number of df associated with the numerator and a number of df associated with the denominator. Thus a particular *F* distribution may have 3 numerator df and 20 denominator df or 5 numerator df and 18 denominator df. As was the case with *t* distributions, each df is a positive whole number.

Figure 2 pictures a typical *F* curve obtained by specifying both numerator and denominator df. The numerator and denominator of an *F* ratio are *never* negative, so an *F* curve has positive height only for positive values of an *F* variable. We need critical values for different *F* distributions, i.e., values that capture specified tail areas underneath *F* curves. Notice that an *F* curve is not symmetric, so knowledge of an upper-tail critical value does not immediately yield the corresponding lower-tailed value. However, our *F* tests will always be upper-tailed (reject H_0 if the *F* ratio *exceeds* a specified critical value), so only upper-tailed critical values need be tabulated. Because the critical value for a given tail area depends on both numerator and denominator degrees of freedom—changing either one changes the critical value—an entire rectangular table is needed for each desired tail area (rather than just a single column, as in the *t* table).

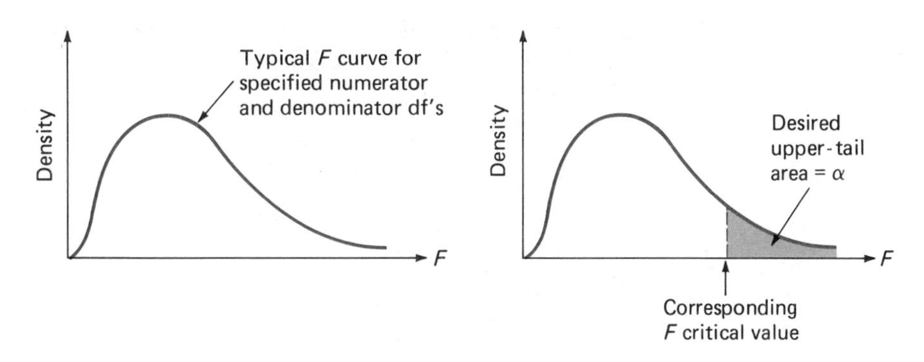

FIGURE 2 AN *F* CURVE AND *F* CRITICAL VALUE

**F* distributions were introduced in Chapter 12 as a basis for the model utility test in multiple regression. However, our discussion here presumes that this is your first confrontation with these distributions. The use of an uppercase *F* is traditional.

Table XI(a) in the appendices gives F critical values that capture upper-tail area .05 and Table XI(b) gives values for area .01. To obtain a critical value for a particular F distribution, go to Table XI(a) or XI(b), depending on the desired tail area. Then locate the column corresponding to numerator df, the row corresponding to denominator df, and finally the critical value at the intersection of this column and row. For example, the critical value for tail area .05, numerator df = 3, and denominator df = 20 is 3.10, the number at the intersection of the column marked 3 and the row marked 20. Similarly the 1% critical value for numerator df = 3 and denominator df = 8 is 7.59, whereas that for numerator df = 8 and denominator df = 3 is 27.49. Do not accidentally reverse numerator and denominator df!

The F Test

The single-factor ANOVA F test uses the F ratio MSTr/MSE as the test statistic. The rejection region for a level α test is specified by an F critical value, with numerator and denominator df determined by the number of populations or treatments, k, and number of observations per sample, n.

> The **single-factor ANOVA F test** utilizes the test statistic
>
> $$F = \frac{\text{MSTr}}{\text{MSE}}$$
>
> where formulas for MSTr and MSE in the equal-sample-sizes case were given earlier. When H_0 is true and previously stated assumptions are satisfied, this statistic has an F distribution with $k - 1$ numerator df and $N - k$ denominator df. A level α test then consists of obtaining the corresponding critical value from Table XI and rejecting H_0 if $F > F$ critical value.

EXAMPLE 4

Let's reconsider the folacin content data introduced in Example 1 and carry out a level .05 test to see whether there are any differences in true average folacin content between the four brands of brewed tea.

1. Population characteristics of interest: μ_1 = true average folacin content for brand 1, and μ_2, μ_3, μ_4 are defined analogously for the other brands.
2. $H_0 : \mu_1 = \mu_2 = \mu_3 = \mu_4$.
3. H_a : at least two among μ_1, μ_2, μ_3, μ_4 are different.
4. Test statistic: F = MSTr/MSE.
5. Rejection region: $k - 1 = 4 - 1 = 3$ and $N - k = 24 - 4 = 20$, so a level .05 test requires the F critical value based on 3 numerator and 20 denominator df from Table XI(a). This critical value is 3.10, so H_0 will be rejected if $F > 3.10$.
6. Computations: From Example 2, MSTr = 7.28, and from Example 3, MSE = 2.17. Thus $F = 7.28/2.17 = 3.35$.
7. The computed F ratio falls in the level .05 rejection region ($3.35 > 3.10$). H_0 is therefore rejected in favor of the conclusion that at least two brands differ with respect to true average folacin content. ■

EXAMPLE 5

An experiment designed to investigate whether rapid eye movement (REM) sleep time depends on the amount of ethanol given in an injection proceeded as follows. Four injection concentrations (treatments, in grams per kilogram body weight) were selected: 0, 1, 2, 4. Twenty rats were then chosen, randomly divided into four groups, and each group was given a different one of the four treatments. The REM sleep time during a 24-hour period was recorded for each rat, resulting in the accompanying data. (This example is based on an experiment described in the article "Relationship of Ethanol Blood Level to REM and Non-REM Sleep Time and Distribution in the Rat" *Life Sci.* (1978):838–46).

	Treatment	Data	$\bar{x}$	s^2
1.	0 g/kg	88.6, 73.2, 91.4, 68.0, 75.2	79.28	103.65
2.	1 g/kg	63.0, 53.9, 69.2, 50.1, 71.5	61.54	87.31
3.	2 g/kg	44.9, 59.5, 40.2, 56.3, 38.7	47.92	89.51
4.	4 g/kg	31.0, 39.6, 45.3, 25.2, 22.7	32.76	91.37
		$\bar{\bar{x}}$ =	55.38	

The sample means differ substantially from one another, but there is also much variability in REM sleep time within each of the four samples. Is sampling variability alone a plausible explanation for observed differences in the $\bar{x}$'s, or does the data suggest that true average time does depend on ethanol concentration? Let's test the appropriate hypotheses using level of significance $\alpha = .01$.

1. Population characteristics: μ_i = true average REM sleep time for treatment i ($i = 1,2,3,4$).
2. $H_0 : \mu_1 = \mu_2 = \mu_3 = \mu_4$.
3. H_a : at least two of the four μ's are different.
4. Test statistic: $F = \text{MSTr}/\text{MSE}$.
5. Rejection region: The critical value for $\alpha = .01$, numerator df $k - 1 = 3$, and denominator df $N - k = 16$ is 5.29. H_0 should be rejected if $F > 5.29$.
6. Computations:

$$\text{MSE} = \frac{103.65 + 87.31 + 89.51 + 91.37}{4} = 92.96$$

and

$$\text{MSTr} = \frac{5}{4 - 1}[(79.28 - 55.38)^2 + (61.54 - 55.38)^2 + (47.92 - 55.38)^2 + (32.76 - 55.38)^2] = 1960.79$$

The computed value of the F ratio is $F = 1960.79/92.96 = 21.09$.

7. The computed F ratio, 21.09, considerably exceeds the critical value 5.29, so H_0 is rejected at level .01 in favor of the conclusion that true average REM sleep time does depend on ethanol concentration. ■

Unequal Sample Sizes

When at least two among the k sample sizes $n_1, n_2, \ldots, n_k$ are different, the F statistic is still MSTr/MSE, but the definitions of both mean squares must be revised.

> In the case of unequal sample sizes,
>
> $$\text{MSTr} = \frac{n_1(\bar{x}_1 - \bar{\bar{x}})^2 + n_2(\bar{x}_2 - \bar{\bar{x}})^2 + \cdots + n_k(\bar{x}_k - \bar{\bar{x}})^2}{k - 1}$$
>
> and
>
> $$\text{MSE} = \frac{(n_1 - 1)s_1^2 + (n_2 - 1)s_2^2 + \cdots + (n_k - 1)s_k^2}{N - k}$$
>
> where $N = n_1 + n_2 + \cdots + n_k$. Numerator and denominator df for the F test are still $k - 1$ and $N - k$, respectively.

MSE is still an unbiased statistic for estimating σ^2 whether or not H_0 is true, but it is now a weighted average of the k sample variances. As before, MSTr is unbiased when H_0 is true but tends to overestimate σ^2 when H_0 is false. So the test statistic F is still the ratio of a between-samples estimate to a within-samples estimate.

EXAMPLE 6

An individual's critical flicker frequency (cff) is the highest frequency (in cycles/s) at which the flicker in a flickering light source can be detected. At frequencies above the cff, the light source appears to be continuous even though it is actually flickering. An investigation carried out to see if true average cff depends on iris color yielded the following data (based on the article "The Effect of Iris Color on Critical Flicker Frequency" *J. General Psych.* (1973): 91–95).

	Iris Color	Data	n	$\bar{x}$	s^2
1.	Brown	26.8, 27.9, 23.7, 25.0, 26.3, 24.8, 25.7, 24.5	8	25.59	1.86
2.	Green	26.4, 24.2, 28.0, 26.9, 29.1	5	26.92	3.40
3.	Blue	25.7, 27.2, 29.9, 28.5, 29.4, 28.3	6	28.17	2.33

Also, $N = 19$, grand total $= 508.30$, and $\bar{\bar{x}} = 508.30/19 = 26.75$.

Let's perform a level .05 test to see if true average cff does indeed depend on iris color.

1. Let μ_1, μ_2, and μ_3 denote true average cff for individuals having brown, green, and blue iris colors, respectively.
2. $H_0 : \mu_1 = \mu_2 = \mu_3$.
3. H_a : at least two among μ_1, μ_2, μ_3 are different.
4. Test statistic: $F =$ MSTr/MSE
5. Rejection region: Numerator df $= k - 1 = 2$ and denominator df $= N - k = 19 - 3 = 16$. For $\alpha = .05$, Table XI(a) gives 3.63 as the desired critical value. H_0 will now be rejected in favor of H_a if $F > 3.63$.

6. Computations:

$$\text{MSTr} = \frac{8(25.59 - 26.75)^2 + 5(26.92 - 26.75)^2 + 6(28.17 - 26.75)^2}{3 - 1} = 11.50$$

$$\text{MSE} = \frac{(8 - 1)(1.86) + (5 - 1)(3.40) + (6 - 1)(2.33)}{19 - 3} = 2.39$$

and $F = 11.50/2.39 = 4.81$.

7. Since $4.81 > 3.63$, we reject H_0 in favor of H_a at level .05. The data does suggest that true average cff depends on iris color, a conclusion agreeing with that given in the paper cited. ■

Other Issues

In the next several sections we discuss ANOVA computations in more detail, present methods for further data analysis when H_0 is rejected (multiple comparisons), develop a distribution-free procedure whose validity does not require the normality assumption, and introduce an alternative experimental design (randomized blocks) that controls for extraneous variation. Here we mention two other aspects of single-factor ANOVA. The first concerns type II error probabilities—the probability of not rejecting H_0 when it is false—for the F test. In Chapters 8–10 we used the curves in Table V to compute β for t tests. There are similar sets of curves for the F test. They are called power curves (power $= 1 - \beta$). Such curves are given and their uses described in sources listed in the chapter references.

Secondly, the F test was introduced to test the equality of $\mu_1, \mu_2, \ldots, \mu_k$ for $k > 2$. However, the test can be used also when $k = 2$, i.e., to test $H_0 : \mu_1 = \mu_2$. The formulas for MSTr and MSE remain valid in this case, and Table XI gives F critical values for $k - 1 = 1$ numerator df and various denominator df. The pooled t test can also be used to test $H_0 : \mu_1 = \mu_2$. It doesn't matter which test is used when the alternative hypothesis is $H_a : \mu_1 \neq \mu_2$. Irrespective of what samples happen to result, the two-test procedures yield exactly the same conclusion—reject H_0 or don't reject H_0—when used at the same level of significance. However, the F test can be used only for $H_a : \mu_1 \neq \mu_2$, while a one-tailed t test can be used for $H_a : \mu_1 > \mu_2$ or $H_a : \mu_1 < \mu_2$. For this reason, the t test is more flexible when $k = 2$.

EXERCISES

13.1 What assumptions about the k population or treatment response distributions must you be willing to make in order for the ANOVA F test to be an appropriate method of analysis?

13.2 State the rejection region for the ANOVA F test in each case.

a. Numerator df = 5 and denominator df = 18.

b. There are three populations and six observations from each.

c. There are six treatments and each is applied to three subjects.

13.3 The paper "Utilizing Feedback and Goal Setting to Increase Performance Skills of Managers" (*Academy of Mgmt. J.* (1979):516–26) reported the results of an experiment to compare three different interviewing techniques for employee evaluations. One method allowed the employee being evaluated to discuss previous evaluations, the second involved setting goals for the employee, and the third did not allow either feedback or goal setting. After the interviews were concluded, the employee evaluated was asked to indicate how satisfied he or she was with the interview (a numerical scale was used to quantify level of satisfaction). The authors used ANOVA to compare the three interview techniques. An F statistic value of 4.12 was reported.

a. Suppose that a total of 33 subjects had been used, with each technique applied to 11 of them. Use this information to test at significance level .05 the null hypothesis of no difference in mean satisfaction level for the three interview techniques.

b. The actual number of subjects on which each technique was used was 45. After studying the F table, explain why the conclusion in (a) still holds.

13.4 Mercury is a very hazardous pollutant, and many studies have been carried out in an attempt to assess its toxic effects. One such study is described in the paper "Comparative Responses of the Action of Different Mercury Compounds on Barley" (*Intl. J. Environ. Studies* (1983):323–27). Ten different concentrations of mercury (0, 1, 5, 10, 50, 100, 200, 300, 400, and 500 mg/L) were compared with respect to their effects on average dry weight (per 100 7-day-old seedlings). The basic experiment was replicated four times for a total of 40 dry-weight observations (four for each treatment level). The paper reported an ANOVA F statistic value of 1.895. Using a significance level of .05 and the usual seven-step procedure, test the null hypothesis that the true mean dry weight is the same for all 10 concentration levels.

13.5 In an experiment to compare tensile strength of five different types of wire, four specimens of each type were tested. MSTr and MSE were computed to be 2573.3 and 1394.2, respectively. Use the ANOVA F test with significance level .01 to test $H_0 : \mu_1 = \mu_2 = \mu_3 = \mu_4 = \mu_5$ versus H_a : at least two of these μ's are different.

13.6 Measurements of athletes' body characteristics have been used to establish optimal playing weights for use in training programs. The paper "Prediction of Body Composition in Female Athletes" (*J. Sports Med.* (1983):333–41) compared average lean body mass (kg) for female basketball players, cross-country runners, and swimmers. From summary quantities given in the paper, MSTr and MSE were calculated to be 803.9 and 27.5, respectively. These figures were based on three random samples, each of size 10. Does this data suggest that the true average lean body mass differs by sport for female athletes? Use a .01 significance level.

13.7 Three different laboratory methods for determining the concentration of a contaminant in water are to be compared. A gallon of well water is divided into nine equal parts, and each is placed in a container. The containers are then randomly assigned to the three methods. Summary quantities are:

	n	$\bar{x}$	s
Method 1	3	1.914	.212
Method 2	3	1.949	.095
Method 3	3	2.327	.198

Use this data and a significance level of .05 to test the null hypothesis of no difference in mean concentration determination for the three methods.

13.8 The paper "Computer-Assisted Instruction Augmented with Planned Teacher/Student Contacts" (*J. Exp. Educ.* (Winter 1980/81):120–26) compared five different methods for teaching descriptive statistics. The five methods were traditional lecture and discussion (L/D), programmed textbook instruction (R), programmed text with lectures (R/L), computer instruction (C), and computer instruction with lectures (C/L). Forty-five students were randomly assigned, nine to each method. After completion of the course, a 1-hour exam was given. In addition, a 10-minute retention test was administered 6 weeks later. Summary quantities are given.

		Exam		Retention Test	
Method	n	$\bar{x}$	s	$\bar{x}$	s
L/D	9	29.3	4.99	30.20	3.82
R	9	28.0	5.33	28.80	5.26
R/L	9	30.2	3.33	26.20	4.66
C	9	32.4	2.94	31.10	4.91
C/L	9	34.2	2.74	30.20	3.53

The grand mean for the exam was 30.82 and the grand mean for the retention test was 29.3.

a. Does the data suggest that there is a difference between the five teaching methods with respect to true mean exam score? Use $\alpha = .05$.

b. Using a .05 significance level, test the null hypothesis of no difference between the true mean retention test scores for the five different teaching methods.

13.9 Growing interest in trout farming has prompted a number of experiments designed to compare various growing conditions. One factor of interest is the salinity of the water. The effect of salinity on the growth of rainbow trout (measured by increase in weight) was examined in the paper "Growth, Training and Swimming Ability of Young Trout Maintained Under Different Salinity Conditions" (*J. Marine Biological Assoc. of U.K.* (1982):699–708). Full-strength seawater (32% salinity), brackish water (18% salinity), and freshwater (.5% salinity) were used, and the following summary quantities were obtained.

Salinity	Number of Fish	Mean Weight Gain	s
Fresh	12	8.078	1.786
18%	12	7.863	1.756
32%	8	6.468	1.339

Does the data provide sufficient evidence to conclude that the mean weight gain is not the same for the three salinity levels? Use a significance level of .01.

13.10 The *fog index* is a measure of reading difficulty based on the average number of words per sentence and the percentage of words with three or more syllables. High values of the fog index are associated with difficult reading levels. Independent random samples of six advertisements were taken from three different magazines and fog indices were computed to obtain the given data ("Readability Levels of Magazine Advertisements" *J. Ad. Research* (1981):45–50).

Scientific American	15.75	11.55	11.16	9.92	9.23	8.20
Fortune	12.63	11.46	10.77	9.93	9.87	9.42
New Yorker	9.27	8.28	8.15	6.37	6.37	5.66

Use a significance level of .01 to test the null hypothesis of no difference between the mean fog index levels for advertisements appearing in the three magazines.

13.11 The paper of Exercise 13.10 also reported the following data for advertisements in the magazines *True Confessions, People Weekly,* and *Newsweek.*

True Confessions	12.89	12.69	11.15	9.52	9.12	7.08
People Weekly	9.50	8.60	8.59	6.50	4.79	4.29
Newsweek	10.21	9.66	7.67	5.12	4.88	3.12

Is there sufficient evidence to conclude that the mean fog indices for advertisements differ for the three magazines? Use $\alpha = .01$.

13.12 Combine the data from Exercises 13.10 and 13.11 and perform an ANOVA F test using all six samples. Use a significance level of .01.

13.13 The paper "Chemical Factors Affecting Soiling and Soil Release from Cotton DP Fabric" (*Amer. Dyestuff Reporter* (1983):25–30) gave the accompanying data on the degree of soiling of fabric copolymerized with three different mixtures of methacrylic acid (MAA).

	Degree of Soiling				
Mixture 1	0.52	1.12	0.90	1.07	1.04
Mixture 2	0.76	0.82	0.80	0.78	0.81
Mixture 3	0.52	1.08	1.07	1.09	0.93

Is there sufficient evidence to indicate that the true mean degree of soiling is not the same for all three MAA mixtures? Use $\alpha = .05$.

13.14 The given observations are tomato yields (kg/plot) for four different levels of electrical conductivity (EC) of the soil. Chosen EC levels were 1.6, 3.8, 6.0, and 10.2 nmhos/cm.

EC Level	Yield				
1.6	59.5	53.3	56.8	63.1	58.7
3.8	55.2	59.1	52.8	54.5	
6.0	51.7	48.8	53.9	49.0	
10.2	44.6	48.5	41.0	47.3	46.1

Use the ANOVA F test with $\alpha = .05$ to test for any differences in true average yield due to the different EC levels.

13.2 ANOVA Computations

The F test developed in the previous section for testing $H_0 : \mu_1 = \mu_2 = \cdots = \mu_k$ is based on the assumption of equal population or treatment variances. The F statistic is MSTr/MSE, a ratio of two different estimates for the common variance σ^2. Although the denominator, MSE, is unbiased whether or not H_0 is true, the numerator, MSTr, is unbiased only when H_0 is true and otherwise tends to overestimate σ^2. When the F value is large, it appears that MSTr is overestimating σ^2, which in turn suggests that H_0 is false. The formulas for MSTr and MSE given in Section 13.1 are somewhat cumbersome. An alternative method for efficiently computing F involves introducing quantities called *sums of squares* and exploiting a simple relationship between them.

DEFINITION	

Total sum of squares (SSTo), treatment sum of squares (SSTr), and **error sum of squares (SSE)** are defined by

$$\text{SSTo} = \sum_{\substack{\text{all } N \\ \text{observations}}} (x - \bar{\bar{x}})^2$$

$$\text{SSTr} = n_1(\bar{x}_1 - \bar{\bar{x}})^2 + n_2(\bar{x}_2 - \bar{\bar{x}})^2 + \cdots + n_k(\bar{x}_k - \bar{\bar{x}})^2$$

$$\text{SSE} = \sum_{\substack{\text{1st} \\ \text{sample}}} (x - \bar{x}_1)^2 + \sum_{\substack{\text{2nd} \\ \text{sample}}} (x - \bar{x}_2)^2 + \cdots + \sum_{\substack{k\text{th} \\ \text{sample}}} (x - \bar{x}_k)^2$$

Each sum of squares has a number of df associated with it. These are $N - 1$ for SSTo, $k - 1$ for SSTr, and $N - k$ for SSE.

The computation of SSTo involves first subtracting the *same* number $\bar{\bar{x}}$ from each observation in the data set to obtain deviations from the grand mean, then squaring all N deviations, and finally adding them together. SSE results from separately computing a sum of squared deviations within each sample—by first subtracting the mean *for that sample* from each observation, squaring, and summing—and then adding these k quantities together. Intuitively, the observations in a particular sample tend to be closer to the mean of that sample than to the grand mean, so it should be the case that SSE $\leq$ SSTo.

EXAMPLE 7

Reconsider the critical flicker frequency data introduced in Example 6 and reproduced here.

Sample	Data	n	$\bar{x}$
1	26.8, 27.9, 23.7, 25.0, 26.3, 24.8, 25.7, 24.5	8	25.59
2	26.4, 24.2, 28.0, 26.9, 29.1	5	26.92
3	25.7, 27.2, 29.9, 28.5, 29.4, 28.3	6	28.17
		N = 19	$\bar{\bar{x}}$ = 26.75

The three sums of squares are

$$\begin{aligned}
\text{SSTo} &= (26.8 - 26.75)^2 + \cdots + (26.4 - 26.75)^2 \\
&\quad + \cdots + (25.7 - 26.75)^2 + \cdots + (28.3 - 26.75)^2 \\
&= 61.31 \\
\text{SSE} &= (26.8 - 25.59)^2 + \cdots + (26.4 - 26.92)^2 \\
&\quad + \cdots + (25.7 - 28.17)^2 + \cdots + (28.3 - 28.17)^2 \\
&= 38.31 \\
\text{SSTr} &= 8(25.59 - 26.75)^2 + 5(26.92 - 26.75)^2 \\
&\quad + 6(28.17 - 26.75)^2 = 23.01
\end{aligned}$$

Associated df are $N - 1 = 18$ for SSTo, $N - k = 19 - 3 = 16$ for SSE, and $k - 1 = 3 - 1 = 2$ for SSTr. Not only is SSE $\leq$ SSTo, but SSTr + SSE = 23.01 + 38.31 = 61.32 $\approx$ SSTo (the slight discrepancy is due to rounding). Similarly, treatment df + error df = 2 + 16 = 18 = total df. ■

The relationships in Example 7 between sums of squares and degrees of freedom holds in general for single-factor ANOVA.

The fundamental identity for single-factor ANOVA is

$$\text{SSTo} = \text{SSTr} + \text{SSE}$$

and the associated df satisfy the same relationship.

This result has both an interesting interpretation and an implication for efficient ANOVA computation. Consider first the interpretation. SSTo, the sum of squared deviations about the grand mean, is a measure of total variability in the data set consisting of all k samples. SSE results from measuring variability separately within each sample and then combining. Such within-sample variability is present regardless of whether or not H_0 is true. The magnitude of SSTr, on the other hand, has much to do with the status of H_0 (true or false). The more the μ's differ from one another, the larger SSTr will tend to be. Thus SSTr represents variation that can (at least to some extent) be explained by any differences between means. An informal paraphrase of the fundamental identity is

total variation = explained variation + unexplained variation

As we shall see momentarily, the F ratio involves SSTr and SSE. However, the two most easily calculated sums of squares are SSTo and SSTr. Once these have been obtained, the fundamental identity implies that SSE can be obtained by subtracting SSTr from SSTo.

Computational Formulas for ANOVA Sums of Squares

$$\text{SSTo} = \sum_{\substack{\text{all } N \\ \text{observations}}} x^2 - N\bar{\bar{x}}^2$$

$$\text{SSTr} = n_1\bar{x}_1^2 + n_2\bar{x}_2^2 + \cdots + n_k\bar{x}_k^2 - N\bar{\bar{x}}^2$$

$$\text{SSE} = \text{SSTo} - \text{SSTr}$$

As in regression, computations using either defining or computational formulas in ANOVA are sensitive to the effects of rounding. To guard against such effects, use as much decimal accuracy as possible. Having calculated SSTr and SSE, the numerator and denominator of F can immediately be obtained.

DEFINITION

A **mean square** is a sum of squares divided by its associated df. For single-factor ANOVA,

$$\text{MSTr} = \frac{\text{SSTr}}{k - 1} \qquad \text{MSE} = \frac{\text{SSE}}{N - k}$$

The ANOVA computations are frequently summarized in a tabular format called an **ANOVA table.** The general form of such a table is displayed in Table 1.

TABLE 1

GENERAL FORMAT FOR A SINGLE-FACTOR ANOVA TABLE

Source of Variation	df	Sum of Squares	Mean Square	F
Treatments	$k-1$	SSTr	$\text{MSTr} = \frac{\text{SSTr}}{k-1}$	$F = \frac{\text{MSTr}}{\text{MSE}}$
Error	$N-k$	SSE	$\text{MSE} = \frac{\text{SSE}}{N-k}$	
Total	$N-1$	SSTo		

EXAMPLE 8

Parents are frequently concerned when their child seems slow to begin walking (although when the child finally walks, the resulting havoc sometimes has the parents wishing they could turn back the clock!). The paper "Walking in the Newborn" (*Science* 176 (1972):314–15) reported on an experiment in which the effects of several different treatments on the age at which a child first walks were compared. Children in the first group were given special walking exercises for 12 minutes per day beginning at age 1 week and lasting 7 weeks. The second group of children received daily exercises but not the walking exercises administered to the first group. The third and fourth groups were control groups—they received no special treatment and differed only in that the third group's progress was checked weekly, whereas the fourth group's progress was checked just once at the end of the study. Observations on age (in months) when the children first walked are given.

Treatment	Age						n	$\bar{x}$
1	9.00	9.50	9.75	10.00	13.00	9.50	6	10.1250
2	11.00	10.00	10.00	11.75	10.50	15.00	6	11.3750
3	11.50	12.00	9.00	11.50	13.25	13.00	6	11.7083
4	13.25	11.50	12.00	13.50	11.50		5	12.3500
							$N = 23$	$\bar{\bar{x}} = \frac{261.00}{23} = 11.3478$

Let's carry out a level .05 test to see whether true average age at which a child first walks depends on which treatment is given.

1. $\mu_1, \mu_2, \mu_3, \mu_4$ are the true average ages of first walking for the four treatments.
2. $H_0: \mu_1 = \mu_2 = \mu_3 = \mu_4$.
3. H_a: at least two among the four μ's are different.
4. Test statistic: $F = \text{MSTr}/\text{MSE}$.
5. Rejection region: $F > F$ critical value; with $\alpha = .05$, treatment df $= k - 1 = 3$ and error df $= N - k = 23 - 4 = 19$, Table XI gives F critical value $= 3.13$.

6. Computations: The computational formulas give

$$\begin{aligned}
\text{SSTo} &= \sum_{\substack{\text{all 23}\\ \text{observations}}} x^2 - N\bar{\bar{x}}^2 \\
&= (9.00)^2 + (9.50)^2 + \cdots + (11.50)^2 - 23(11.3478)^2 \\
&= 3020.25 - 2961.77 = 58.48 \\
\text{SSTr} &= n_1\bar{x}_1^2 + \cdots + n_k\bar{x}_k^2 - N\bar{\bar{x}}^2 \\
&= 6(10.1250)^2 + 6(11.3750)^2 + 6(11.7083)^2 + 5(12.3500)^2 \\
&\quad - 2961.77 \\
&= 2976.56 - 2961.77 = 14.79 \\
\text{SSE} &= 58.48 - 14.79 = 43.69
\end{aligned}$$

The remaining computations are summarized in the accompanying ANOVA table.

Source of Variation	df	Sum of Squares	Mean Square	F
Treatments	3	14.79	4.93	$\frac{4.93}{2.30} = 2.14$
Error	19	43.69	2.30	
Total	22	58.48		

7. The computed F ratio, 2.14, does not exceed the critical value 3.13, so H_0 is not rejected at level of significance .05. The data does not suggest that there are differences in true average responses between the treatments. ■

All the commonly used statistical computer packages will perform a single-factor ANOVA upon request and summarize the results in an ANOVA table. As an example, Table 2 resulted from the use of MINITAB to analyze the data of Example 8. (There are slight differences due to rounding in the hand calculations.)

TABLE 2 AN ANOVA TABLE FROM MINITAB

```
ANALYSIS OF VARIANCE

DUE TO    DF       SS   MS=SS/DF    F-RATIO
FACTOR     3    14.78       4.93       2.14
ERROR     19    43.69       2.30
TOTAL     22    58.47
```

P-Values

As with other test procedures, the P-value for an F test is the smallest level of significance at which H_0 can be rejected. Because our F table contains only 5% and 1% F critical values, information from these tables about the P-value is limited to one of the following three statements:

1. P-value $> .05$ (if $F <$ critical value for $\alpha = .05$)
2. $.01 < P$-value $< .05$ (if F falls between the two critical values)
3. P-value $< .01$ (if $F >$ critical value for $\alpha = .01$).

However, several of the most widely available statistical computer packages will provide an exact P-value for the test. Table 3 displays output from the package SPSS for the data of Example 8. The P-value appears in the far right column under F PROB as .129. This is consistent with our earlier decision not to reject H_0 at level .05. Even at level of significance .10, H_0 would not be rejected because P-value = .129 > .10.

TABLE 3

ANOVA TABLE WITH *P*-VALUE FROM SPSS

ANALYSIS OF VARIANCE

SOURCE	D.F.	SUM OF SQ.	MEAN SQ.	F RATIO	F PROB
BETWEEN GROUPS	3	14.778	4.926	2.142	.129
WITHIN GROUPS	19	43.690	2.299		
TOTAL	22	58.467			

EXERCISES

13.15 In an experiment to investigate the performance of four different brands of sparkplugs intended for use on a 125-cc motorcycle, five plugs of each brand were tested and the number of miles (at a constant speed) until failure was observed. A partially completed ANOVA table is given. Fill in the missing entries and test the relevant hypotheses using a .05 level of significance.

Source of Variation	df	Sum of Squares	Mean Square	F
Treatments				
Error		235,419.04		
Total		310,500.76		

13.16 The partially completed ANOVA table given is taken from the article "Perception of Spatial Incongruity" (*J. Nervous and Mental Disease* (1961):222) in which the abilities of three different groups to identify a perceptual incongruity were assessed and compared. All individuals in the experiment had been hospitalized to undergo psychiatric treatment. There were 21 individuals in the depressive group, 32 individuals in the functional "other" group, and 21 individuals in the brain-damaged group. Complete the ANOVA table. Carry out the appropriate test of hypotheses (use $\alpha = .01$) and interpret your results.

Source of Variation	df	Sum of Squares	Mean Square	F
Treatments		152.18		
Error				
Total		1123.14		

13.17 The paper "Effect of Transcendental Meditation on Breathing and Respiratory Control" (*J. Appl. Physiology* (1984):607–11) reported on an experiment to compare four different groups—alert nonmeditators, relaxed nonmeditators, alert meditators, and meditators while meditating—with respect to breathing characteristics. Sixteen observations (breaths/min) were made for each condition. Data compatible with summary values given in the paper was used to compute SSTr = 136.14 and SSE = 532.26. Construct an ANOVA table. State and test the relevant hypotheses using a .01 level of significance.

13.18 The accompanying summary quantities are representative of data on professional productivity given in the article "Research Productivity in Academia: A Comparative Study of the Sciences, Social Sciences and Humanities" (*Soc. of Educ.* (1981):238–53). Randomly selected faculty members in each of the three disciplines were asked to indicate the number of years that they had been teaching. Suppose that 30 faculty members from each subject area were included in the study. Construct an ANOVA table. Is there sufficient evidence to indicate that the true mean number of years of teaching experience is not the same for the three subject areas? Use a .05 significance level.

Grand mean: $\bar{\bar{x}} = 11.33$

Natural Sciences	Social Sciences	Humanities
$\Sigma(x - \bar{x}_1)^2 = 4468.92$	$\Sigma(x - \bar{x}_2)^2 = 4138.34$	$\Sigma(x - \bar{x}_3)^2 = 4629.44$
$\bar{x}_1 = 12.65$	$\bar{x}_2 = 10.41$	$\bar{x}_3 = 10.93$

13.19 An article in the British scientific journal *Nature* ("Sucrose Induction of Hepatic Hyperplasis in the Rat" (August 25, 1972):461) reported on an experiment in which five groups, each consisting of six rats, were put on diets with different carbohydrates. At the conclusion of the experiment, liver DNA content (mg/g) was determined, with the following results:

$$\Sigma(x - \bar{\bar{x}})^2 = 3.61 \qquad \bar{\bar{x}} = 2.448$$

$$\bar{x}_1 = 2.58 \quad \bar{x}_2 = 2.63 \quad \bar{x}_3 = 2.13 \quad \bar{x}_4 = 2.41 \quad \bar{x}_5 = 2.49$$

Does the data indicate that the true average DNA content is affected by the type of carbohydrate in the diet? Construct an ANOVA table and use a .05 level of significance.

13.20 College students were assigned to various study methods in an experiment to determine the effect of study technique on learning. The given data was generated to be consistent with summary quantities found in the paper "The Effect of Study Techniques, Study Preferences and Familiarity on Later Recall" (*J. Exper. Educ.* (1979):92–95). The study methods compared were reading only, reading and underlining, and reading and taking notes. One week after studying the paper "Love in Infant Monkeys" by Harlow, students were given an exam on the article. Test scores are given in the accompanying table.

Technique	Test Score					
Read only	15	14	16	13	11	14
Read and underline	15	14	25	10	12	14
Read and take notes	18	18	18	16	18	20

a. Compute SSTo, SSTr, and SSE.

b. Construct an ANOVA table.

c. Use a .05 level of significance to test the null hypothesis of no difference between the true mean exam scores for the three study methods.

d. Which of the following statements can be made about the P-value associated with the computed value of the ANOVA F statistic?
  i. $P\text{-value} > .05$
  ii. $.01 < P\text{-value} < .05$
  iii. $P\text{-value} < .01$

e. Based on your answer in (d), would the conclusion of the hypothesis test of (c) have been any different if a .01 significance level had been used? Explain.

13.21 Some investigators think that the concentration (μg/mL) of a particular antigen in supernatant fluids could be related to onset of meningitis in infants. The accompanying data is typical of that given in plots appearing in the paper "Type-Specific Capsular Antigen Is Associated with Virulence in Late-Onset Group B Streptococcal Type III Disease" (*Infection and Immunity* (1984):124–29).

Asymptomatic infants	1.56	1.06	.87	1.39	.71	.87	.95	1.51
Infants with late-onset sepsis	1.51	1.78	1.45	1.13	1.87	1.89	1.07	1.72
Infants with late-onset meningitis	1.21	1.34	1.95	2.00	2.27	.88	1.67	2.57

Construct an ANOVA table and use it to test the null hypothesis of no difference in mean antigen concentrations for the three groups.

13.3 Multiple Comparisons

When $H_0 : \mu_1 = \mu_2$ is rejected in favor of $H_a : \mu_1 \neq \mu_2$ using a two-sample t test, the conclusion that μ_1 and μ_2 are different requires no further amplification. Consider the case of $k = 3$ populations or treatments and the null hypothesis $H_0 : \mu_1 = \mu_2 = \mu_3$. If H_0 is not true, there are four possible groupings of the μ's : (1) $\mu_1 = \mu_2$ and μ_3 differs from these two, (2) $\mu_1 = \mu_3$ and μ_2 differs from these two, (3) $\mu_2 = \mu_3$ and μ_1 differs from these two, and (4) all three μ's are different from one another. After H_0 is rejected by the F test, an investigator would typically want to know which of these four groupings is most plausible.

In the case $k = 4$, there are even more possible groupings of the μ's when H_0 is false. Three possibilities are (1) all four μ's are different from one another, (2) $\mu_1 = \mu_2 = \mu_3$ and μ_4 differs from these three, and (3) $(\mu_1 = \mu_2) \neq (\mu_3 = \mu_4)$, so there are two distinct groups of μ's. Following rejection of H_0, further analysis is appropriate to identify differences among the μ's.

Any procedure for further analyzing the data to identify significant differences among the μ's is called a **multiple comparisons procedure.** A number of multiple comparisons procedures have been developed by statisticians, and there is not always agreement as to which one should be used in any given situation. We present one procedure that is easy to understand and apply. The general idea behind this procedure, as well as a number of others, is first to compute a confidence interval for the difference between each possible pair of μ's. For example, in the case $k = 3$, an interval would be computed for $\mu_1 - \mu_2$, another for $\mu_1 - \mu_3$, and a third for $\mu_2 - \mu_3$. When $k = 4$, there are six differences : $\mu_1 - \mu_2$, $\mu_1 - \mu_3$, $\mu_1 - \mu_4$, $\mu_2 - \mu_3$, $\mu_2 - \mu_4$, and $\mu_3 - \mu_4$. In general, $k(k - 1)/2$ confidence intervals will be computed when there are k populations or treatments to be compared. The number of intervals goes up dramatically as k increases. After all such confidence intervals have been obtained, each one is examined to see whether or not it includes zero. If an interval *does not* include zero, the two corresponding μ's are said to differ significantly from one another. If zero is included in an interval, the two μ's are judged not significantly different from one another. Suppose, for example, that the interval for $\mu_1 - \mu_2$ is $(-.9, 3.5)$, the interval for $\mu_1 - \mu_3$ is $(2.6, 7.0)$, and the interval for $\mu_2 - \mu_3$ is $(1.2, 5.7)$. Then the interval for $\mu_1 - \mu_2$ includes 0, so μ_1 and μ_2 are not significantly different. However, the other two intervals don't include 0, so μ_3 is judged to differ significantly from both μ_1 and μ_2. That is, the true average responses for treatments 1 and 2 appear not to differ, but both differ from the true average response for treatment 3.

Examination of the interval for $\mu_1 - \mu_2$ to see whether or not it includes 0 is equivalent to testing $H_0 : \mu_1 = \mu_2$ versus $H_a : \mu_1 \neq \mu_2$. If the interval does not include 0, H_0 is rejected and the two means are judged to differ. A similar comment applies to any other interval—seeing whether or not zero is included amounts to testing the null hypothesis of equality against the alternative hypothesis that the two means are different. In constructing and examining $k(k - 1)/2$ intervals, we are really performing $k(k - 1)/2$ different hypothesis tests (but these individual tests are carried out only if the F test rejects the overall hypothesis of equality).

The procedure presented here is based on two modifications of the two-sample pooled t interval for $\mu_1 - \mu_2$ introduced in Chapter 9. That interval had the form

$$\bar{x}_1 - \bar{x}_2 \pm (t \text{ critical value})\sqrt{\frac{s_p^2}{n_1} + \frac{s_p^2}{n_2}}$$

where s_p^2 was the pooled estimate of σ^2 obtained by suitably combining the two individual sample variances s_1^2 and s_2^2. The first modification consists of using MSE, the unbiased statistic for estimating σ^2 based on all k samples, in place of s_p^2 in each interval to be computed. In fact, s_p^2 and MSE are identical when $k = 2$.

The rationale for the second modification is a bit more subtle. Suppose that two different 95% confidence intervals, based on independent data sets, are computed. Before the data is obtained, the probability that both intervals capture the corresponding true values is (by independence) $(.95)(.95) = .9025$. The joint, or simultaneous, confidence level is then $100(.9025) = 90.25\%$. When these intervals are used to identify significant differences, the probability that at least one difference between means is incorrectly judged significantly different from zero is $1 - .9025 = .0975$. For six independent 95% intervals, the simultaneous confidence level is $100(.95)^6 \approx 73.5\%$, and the chance of identifying at least one difference when none exists is .265. This is a very high error rate. If an error rate of only .05 is desired, corresponding to simultaneous confidence 95%, then the confidence level for each interval must be substantially larger than 95%.

To achieve a 95% *simultaneous* confidence level, the t critical value in the above formula must be replaced by a larger value (so each interval is wider). Such values are called *Bonferroni t critical values* and appear in Table X in the appendices. The appropriate critical value is selected from the row of that table corresponding to error df and the column corresponding to the number of intervals to be computed. Use of these critical values gives an error rate of at most 5% (even though the intervals are not independent because they are based on the same data set).* The 5% error rate refers to all comparisons rather than to any single comparison, so it is called an *experimentwise error rate.*

* There are Bonferroni t critical values for simultaneous confidence levels other than 95%, but this is the one most frequently used.

The Bonferroni Multiple Comparisons Procedure

When there are k treatments or populations to be compared, first compute the following $k(k-1)/2$ confidence intervals using the appropriate critical value from Table X (based on error df):

$$\text{For } \mu_1 - \mu_2: \quad \bar{x}_1 - \bar{x}_2 \pm \begin{pmatrix}\text{Bonferroni } t \\ \text{critical value}\end{pmatrix} \sqrt{\frac{\text{MSE}}{n_1} + \frac{\text{MSE}}{n_2}}$$

$$\vdots \qquad\qquad\qquad \vdots$$

$$\text{For } \mu_{k-1} - \mu_k: \quad \bar{x}_{k-1} - \bar{x}_k \pm \begin{pmatrix}\text{Bonferroni } t \\ \text{critical value}\end{pmatrix} \sqrt{\frac{\text{MSE}}{n_{k-1}} + \frac{\text{MSE}}{n_k}}$$

Then two μ's are judged to differ significantly if the corresponding interval does not include 0. This procedure guarantees that for (at least) 95% of all data sets, no means will be incorrectly judged significantly different (an experimentwise error rate of at most 5%).

An effective summarizing display is obtained by first listing the $\bar{x}$'s in increasing order, with each population or treatment number appearing just above the corresponding $\bar{x}$. Then for any two population or treatment means judged not significantly different, the corresponding pair of $\bar{x}$'s is underscored with a horizontal line segment. This is done for every such pair. The resulting system of underscoring gives a clear picture of where the significant differences are.

EXAMPLE 9

The paper "Managers' Occupational Histories, Organizational Environments, and Climates for Management Development" (*J. Mgmt. Studies* (1977):58–79) considered four different types of managers: (1) those with high levels of stimulation and support and average levels of public spirit, (2) those having low stimulation, average support, and high public spirit, (3) those with average stimulation, low support, and low public spirit, and (4) those who were low on all three criteria. Let μ_1, μ_2, μ_3, and μ_4 denote true average salary levels for these four types of managers. Summary sample quantities appear below.

Type	1	2	3	4		
n_i	62	52	7	13	$N = 134$	
$\bar{x}_i$	7.87	7.47	5.14	3.69	$\bar{\bar{x}} = 7.17$,	MSE = 2.88

The defining formula gives MSTr = 73.78, so $F = 73.78/2.88 = 25.62$. The F critical value for $\alpha = .05$, numerator df = 3, and denominator df = 130 is (approximately) 2.66. Since $25.62 > 2.66$, we conclude at level of significance .05 that at least two of the μ's differ from one another.

For $k = 4$, the Bonferroni procedure requires that $k(k-1)/2 = 4(3)/2 = 6$ intervals be computed. From Table X, the Bonferroni t critical value for six intervals and 130 df is (approximately) 2.65. The intervals are as follows:

$$\text{For } \mu_1 - \mu_2: \quad 7.87 - 7.47 \pm 2.65\sqrt{\frac{2.88}{62} + \frac{2.88}{52}} = .40 \pm .85$$
$$= (-.45, 1.25)$$

$$\text{For } \mu_1 - \mu_3: \quad 7.87 - 5.14 \pm 2.65\sqrt{\frac{2.88}{62} + \frac{2.88}{7}} = 2.73 \pm 1.79$$
$$= (.94, 4.52)$$

For $\mu_1 - \mu_4$: $4.18 \pm 1.37 = (2.81, 5.55)$

For $\mu_2 - \mu_3$: $2.33 \pm 1.81 = (.52, 4.14)$

For $\mu_2 - \mu_4$: $3.78 \pm 1.39 = (2.39, 5.17)$

For $\mu_3 - \mu_4$: $1.45 \pm 2.11 = (-.66, 3.56)$

Only the intervals for $\mu_1 - \mu_2$ and for $\mu_3 - \mu_4$ include 0, so μ_1 and μ_2 are judged not significantly different, μ_3 and μ_4 are judged not significantly different, but all other pairs of μ's are judged as being significantly different. The four μ's divide naturally into two groups, with μ_1 and μ_2 in the first group and μ_3 and μ_4 in the second group. This is summarized by underscoring:

Type of manager	4	3	2	1
Sample mean	3.69	5.14	7.47	7.87

■

Frequently when a multiple comparison procedure is applied, the populations or treatments do not group into nonoverlapping subsets. This makes interpretation of the results more difficult.

EXAMPLE 10

Summary quantities for the critical flicker frequency (cff) data analyzed in Examples 6 and 7 are $n_1 = 8$, $\bar{x}_1 = 25.59$, $n_2 = 5$, $\bar{x}_2 = 26.92$, $n_3 = 6$, $\bar{x}_3 = 28.17$, and MSE $= 2.39$. The Bonferroni t critical value for $k(k - 1)/2 = 3$ intervals and $N - k = 16$ error df is 2.67. The resulting intervals are $(-3.68, 1.02)$ for $\mu_1 - \mu_2$, $(-4.81, -.35)$ for $\mu_1 - \mu_3$, and $(-3.75, 1.25)$ for $\mu_2 - \mu_3$. Only the interval for $\mu_1 - \mu_3$ does not include 0, so only μ_1 and μ_3 are judged significantly different. The corresponding underscoring is shown below. Although true average cff's for those with brown and blue irises do appear to differ significantly from one another, there is not enough evidence to differentiate either between brown and green or between green and blue. Thus no color stands out by itself either on the high end or on the low end as far as true average cff is concerned.

Color	brown	green	blue
$\bar{x}_i$	25.59	26.92	28.17

■

Often there are no significant differences between any pairs in a group of three or more means. It is customary in this case to underscore the entire group with a single line. The accompanying underscoring illustrates this. Again, there is a nice grouping that makes interpretation easy: μ_1 and μ_4 don't differ significantly, nor do μ_2 and μ_3, μ_2 and μ_5, or μ_3 and μ_5, but any μ from one group differs significantly from any μ in the other group.

2	3	5	1	4
14.2	14.4	14.7	15.9	16.4

Unfortunately, a more typical and more difficult to interpret underscoring pattern is

3	1	2	4	5
25.6	26.9	28.3	29.1	30.7

In this situation μ_3 differs from both μ_4 and μ_5, and μ_1 differs from μ_5, but no other differences are significant.

Believe it or not, something even worse can happen. H_0 can be rejected by the F test, yet the Bonferroni procedure won't identify *any* significant differences (this can happen with other procedures as well)! At this point, it's time to call your friendly neighborhood statistician.

EXERCISES

13.22 Leaf surface area is an important variable in plant gas-exchange rates. The paper "Fluidized Bed Coating of Conifer Needles with Glass Beads for Determination of Leaf Surface Area" (*Forest Sci.* (1980):29–32) included an analysis of dry matter per unit surface area (mg/cm) for trees raised under three different growing conditions. Let μ_1, μ_2, and μ_3 represent the true mean dry matter per unit surface area for the growing conditions 1, 2, and 3, respectively. The given 95% simultaneous confidence intervals are based on summary quantities that appear in the paper.

Difference	Confidence Interval
$\mu_1 - \mu_2$	$(-3.11, -1.11)$
$\mu_1 - \mu_3$	$(-4.06, -2.06)$
$\mu_2 - \mu_3$	$(-1.95, 0.05)$

Which of the four statements below do you think describes the relationship between μ_1, μ_2, and μ_3? Explain your choice.

1. $\mu_1 = \mu_2$ and μ_3 differs from μ_1 and μ_2.
2. $\mu_1 = \mu_3$ and μ_2 differs from μ_1 and μ_3.
3. $\mu_2 = \mu_3$ and μ_1 differs from μ_2 and μ_3.
4. All three μ's are different from one another.

13.23 The accompanying table appeared in the paper "Effect of SO_2 on Transpiration, Chlorophyll Content, Growth and Injury in Young Seedlings of Woody Angiosperms" (*Canadian J. Forest Research* (1980):78–81). Water loss of plants (species: *Acer saccharinum*) exposed to 0, 2, 4, 8, and 16 hours of fumigation was recorded, and a multiple comparison procedure was used to detect differences among the mean water losses for the different fumigation durations. How would you interpret this table?

Duration of fumigation	16	0	8	2	4
Sample mean water loss	27.57	28.23	30.21	31.16	36.21

13.24 The paper of Exercise 13.23 also included a similar table (see page 582) for plants of species *Robinia pseudoacacia*. How would you interpret this table?

Duration of fumigation	0	8	16	4	2
Sample mean water loss	23.52	28.39	32.49	36.54	39.26

13.25 The paper "Growth Response in Radish to Sequential and Simultaneous Exposures of NO_2 and SO_2" (*Environ. Pollution* (1984):303–25) compared a control group (no exposure), a sequential exposure group (plants exposed to one pollutant followed by exposure to the second 4 weeks later) and a simultaneous-exposure group (plants exposed to both pollutants at the same time). The paper states: "Sequential exposure to the two pollutants had no effect on growth compared to the control. Simultaneous exposure to the gases significantly reduced plant growth." Let $\overline{x}_1$, $\overline{x}_2$, and $\overline{x}_3$ represent the sample means for the control, sequential, and simultaneous groups, respectively. Suppose that $\overline{x}_1 > \overline{x}_2 > \overline{x}_3$. Use the given information to construct a table where the sample means are listed in increasing order with those that are not judged to be significantly different underscored.

13.26 The nutritional quality of shrubs commonly used for feed by rabbits was the focus of a study summarized in the paper "Estimation of Browse by Size Classes for Snowshoe Hare" (*J. Wildlife Mgmt.* (1980):34–40). The energy content (cal/g) of three sizes (4 mm or less, 5–7 mm, and 8–10 mm) of serviceberries was studied. Let μ_1, μ_2, and μ_3 denote the true mean energy content for the three size classes. Suppose that 95% simultaneous confidence intervals for $\mu_1 - \mu_2$, $\mu_1 - \mu_3$, and $\mu_2 - \mu_3$ are $(-10, 290)$, $(150, 450)$, and $(10, 310)$, respectively. How would you interpret these intervals?

13.27 Most large companies have established grievance procedures for their employees. One question of interest to employers is why certain groups within a company have higher grievance rates than others. The study described in the paper "Grievance Rates and Technology" (*Academy of Mgmt.* (1979):810–15) distinguished four types of jobs based on level of technology. They were labeled apathetic, erratic, strategic, and conservative. Suppose that a total of 52 work groups were selected (13 of each type) and a measure of grievance rate was determined for each one. An analysis of variance resulted in the given ANOVA table and the sample means.*

Source of Variation	df	Sum of Squares	Mean Square	F
Treatments	3	175.9034	58.6344	5.56
Error	48	506.1936	10.5457	
Total	51	682.0970		

Group	Apathetic	Erratic	Strategic	Conservative
Sample mean	2.96	5.05	8.74	4.91

a. Test the null hypothesis of no difference in mean grievance rate for the four groups. Use a .05 significance level.

b. Construct 95% simultaneous confidence intervals for $\mu_1 - \mu_2$, $\mu_1 - \mu_3$, $\mu_1 - \mu_4$, $\mu_2 - \mu_3$, $\mu_2 - \mu_4$, and $\mu_3 - \mu_4$.

c. Summarize the information given in the intervals in (b) by listing the sample means in increasing order and underscoring those that are not judged to be significantly different. Interpret the resulting diagram.

*All observations were transformed by taking square roots prior to analysis because the authors believed that the distribution of transformed values conformed more closely to the basic ANOVA assumptions than did the orginal distribution.

13.28 The paper "Effect of Transcendental Meditation on Breathing and Respiratory Control," first discussed in Exercise 13.17, investigated breaths per minute under four different experimental conditions. Data compatible with summary quantities given in the paper resulted in the given ANOVA table and sample means.

Source of Variation	df	Sum of Squares	Mean Square	F
Treatments	3	136.14	45.38	5.12
Error	60	532.26	8.87	
Total	63	668.40		

$\bar{x}_1 = 16.72 \quad \bar{x}_2 = 14.73 \quad \bar{x}_3 = 14.89 \quad \bar{x}_4 = 12.60$

a. Verify that $H_0 : \mu_1 = \mu_2 = \mu_3 = \mu_4$ is rejected at level .05.

b. Use the Bonferroni procedure to identify significant differences among the condition means.

13.29 Samples of six different brands of diet or imitation margarine were analyzed to determine the level of physiologically active polyunsaturated fatty acids (PAPFUA, in percent), resulting in the following data:

Imperial	14.1	13.6	14.4	14.3	
Parkay	12.8	12.5	13.4	13.0	12.3
Blue Bonnet	13.5	13.4	14.1	14.3	
Chiffon	13.2	12.7	12.6	13.9	
Mazola	16.8	17.2	16.4	17.3	18.0
Fleischmann's	18.1	17.2	18.7	18.4	

(The above data is fictitious, but the sample means agree with data reported in the January 1975 issue of *Consumer Reports*.)

a. Test for differences among the true average PAPFUA percentages for the different brands. Use $\alpha = .05$.

b. Use the Bonferroni procedure to compute 95% simultaneous confidence intervals for all differences between means.

c. Summarize the confidence intervals of (b) by listing the sample means in increasing order and then underscoring those that are not significantly different. Interpret the resulting display.

13.30 Scores of 24 hard-of-hearing children on a test of basic concepts are given in the accompanying table ("Performance of Young Hearing-Impaired Children on a Test of Basic Concepts" *J. Speech and Hearing Research* (1974):342–51).

Age 6	17	20	24	34	34	38						
Age 7	23	25	27	34	38	47						
Age 8	22	23	26	32	34	34	36	38	38	42	48	50

a. Use an ANOVA F test with significance level .05 to determine if true mean score depends on age.

b. Let μ_1, μ_2, and μ_3 represent the true mean score for 6-, 7-, and 8-year-olds, respectively. A reasonable theory is that the true mean score increases with age ($\mu_1 < \mu_2 < \mu_3$). Use the Bonferroni procedure to compute simultaneous confidence intervals for $\mu_1 - \mu_2$, $\mu_1 - \mu_3$, and $\mu_2 - \mu_3$. Based on these intervals, would you conclude that $\mu_1 < \mu_2 < \mu_3$? Explain.

13.4 The F Test for a Randomized Block Experiment

In Chapter 10 we introduced the concept of a paired experiment for comparing two treatments and showed how the paired t test could be used to analyze data from such an experiment. A paired experiment is preferable to an independent-samples experiment when there is extraneous variation among experimental subjects or objects that may affect the response being studied. For example, time to recover from an illness after a drug is administered could easily be affected by characteristics like age, sex, blood pressure, and general health status. One possible danger in basing a comparison of two drugs on two independent and completely unrelated samples of patients is that variation in such characteristics might obscure a real difference between drugs. On the other hand, an observed difference between the two samples may be due not to any drug effects but instead to a "favorable" allocation of patients to one sample (e.g., all in the first sample might be relatively healthy and all those assigned to the second drug might be in poor health). By pairing patients in such a way that those within any pair are as alike as possible except with respect to the drug administered, the effects of extraneous factors are filtered out and the magnitude of any drug effect can be more easily assessed.

An experiment designed to control for extraneous variation is also desirable in studies involving a comparison of more than two treatments. A comparison of three different drugs could be carried out by dividing patients into groups of three so that within each group the patients are as similar as possible with respect to characteristics that might affect recovery time. Then each drug would be administered to a different patient in each group. As another example, suppose that a publisher has developed four different cover designs for a cookbook that will be sold in supermarkets. To see if sales depend on which cover is used, 24 supermarkets are selected to participate in a study. Books with one cover design will be sold in 6 supermarkets for a 1-month period, books with a second cover design will be sold in a different 6 supermarkets during the same period, and so on. The way to protect against variation in stores and clientele that might affect sales is to separate the 24 supermarkets into six groups of 4 markets each, so that within each group, stores are as much alike as possible. Then each different cover design would be randomly allocated to one of the four stores in every group.

DEFINITION

Suppose that experimental units are first separated into groups consisting of k units in such a way that the units within each group are as similar as possible, and then each unit in a group receives a different treatment. The groups are often called **blocks,** and the experimental design is referred to as a **randomized block design.**

EXAMPLE 11

High energy costs have made consumers and home builders increasingly conscious of whether or not household appliances are energy-efficient. A large developer carried out a study to compare electricity usage for four different residential air-conditioning systems being considered for tract homes. Each system was installed in five homes, and the resulting electricity usage (in

KWh) was monitored for a 1-month period. Because the developer realized that many characteristics of a home could affect usage (e.g., floor space, type of insulation, directional orientation, and type of roof and exterior), care was taken to ensure that extraneous variation in such characteristics did not influence the conclusions. Homes selected for the experiment were grouped into five blocks consisting of four homes each so that the four homes within any given block were as similar as possible. Resulting data is displayed in the accompanying rectangular table, in which rows correspond to the different treatments (air-conditioning systems) and columns correspond to the different blocks.

		Block					
		1	2	3	4	5	Treatment Average
Treatment	1	116	118	97	101	115	109.40
	2	171	131	105	107	129	128.60
	3	138	131	115	93	110	117.40
	4	144	141	115	93	99	118.40
Block Average		142.25	130.25	108.00	98.50	113.25	Grand Mean 118.45

We analyze this data shortly to see whether electricity usage depends on which system is used. ■

The hypotheses of interest and assumptions underlying the analysis are similar to those for a completely randomized design.

Assumptions and Hypotheses

The single observation made on any particular treatment in a given block is assumed to be selected from a normal distribution. The variance of the distribution is σ^2, the same for each block-treatment combination, but the mean value may depend separately both on the treatment applied and on the block.* The hypotheses of interest** are

H_0 : the mean value does not depend on which treatment is given

versus

H_a : the mean value does depend on which treatment is given

The basis for the **randomized block *F* test** is a fundamental identity that shows how the total sum of squares SSTo, again a measure of total variation in the data, breaks down into three parts: (1) error sum of squares, SSE,

*More specifically, it is assumed that there is no interaction between treatment effects and block effects. The concept of interaction is explored further in Section 13.5.

** An observation's mean value now depends both on the treatment and block utilized, so our previous notation $\mu_1, \mu_2, \ldots, \mu_k$ is no longer appropriate for stating hypotheses. In more advanced sources, hypotheses are stated using more complex notation.

which reflects random variation alone, (2) treatment sum of squares, SSTr, whose value reflects both random variation and any differences in treatments, and (3) block sum of squares, SSBl, whose value reflects both random variation and any block differences. Separating out variability due to block differences makes the randomized block test procedure more sensitive to the existence of treatment differences than would be the case for a completely randomized design followed by a single-factor F test. Treatment df is still $k - 1$, but blocking reduces error df as compared to a single-factor ANOVA with the same number of observations. The test statistic is again $F = \text{MSTr}/\text{MSE}$, a ratio of two different estimates of σ^2, whose properties are analogous to properties of the two estimates in single-factor ANOVA.

Summary of the Randomized Block F Test

Notation:

Let k = number of treatments

l = number of blocks

$\bar{x}_1$ = the average of all observations on the first treatment

⋮

$\bar{x}_k$ = the average of all observations on the kth treatment

$\bar{b}_1$ = the average of all observations in the first block

⋮

$\bar{b}_l$ = the average of all observations in the lth block

$\bar{\bar{x}}$ = the average of all bl observations in the experiment (the grand mean)

Sums of squares and associated df are as follows.*

Sum of Squares	Symbol	df	Formula
Total	SSTo	$kl - 1$	$\sum_{\text{all } kl \text{ observations}}(x - \bar{\bar{x}})^2 = \sum_{\text{all } kl \text{ observations}} x^2 - kl\bar{\bar{x}}^2$
Treatments	SSTr	$k - 1$	$l[(\bar{x}_1 - \bar{\bar{x}})^2 + (\bar{x}_2 - \bar{\bar{x}})^2 + \cdots + (\bar{x}_k - \bar{\bar{x}})^2]$
Blocks	SSBl	$l - 1$	$k[(\bar{b}_1 - \bar{\bar{x}})^2 + (\bar{b}_2 - \bar{\bar{x}})^2 + \cdots + (\bar{b}_l - \bar{\bar{x}})^2]$
Error	SSE	$(k - 1)(l - 1)$	By subtraction

SSE is most easily obtained through the use of the fundamental identity

$$\text{SSTo} = \text{SSTr} + \text{SSB}l + \text{SSE}$$

Thus SSE = SSTo − SSTr − SSBl (and error df = total df − treatment df − block df).

*Calculations can be expedited a bit by using the computational formulas

$$\text{SSTr} = l[\bar{x}_1^2 + \bar{x}_2^2 + \cdots + \bar{x}_k^2] - kl\bar{\bar{x}}^2$$
$$\text{SSB}l = k[\bar{b}_1^2 + \bar{b}_2^2 + \cdots + \bar{b}_l^2] - kl\bar{\bar{x}}^2$$

> *Test statistic:* $F = \dfrac{\text{MSTr}}{\text{MSE}}$
>
> where $\text{MSTr} = \dfrac{\text{SSTr}}{k-1}$ and $\text{MSE} = \dfrac{\text{SSE}}{(k-1)(l-1)}$
>
> *Rejection region:* $F > F$ critical value, where the F critical value is based on $k - 1$ numerator df and $(k-1)(l-1)$ denominator df.

Calculations for this F test are usually summarized in an ANOVA table. The table is similar to the one for single-factor ANOVA except that blocks are an extra source of variation, so four rows are included rather than just three.

TABLE 4 THE ANOVA TABLE FOR A RANDOMIZED BLOCK EXPERIMENT

Source of Variation	df	Sum of Squares	Mean Square	F
Treatments	$k-1$	SSTr	$\text{MSTr} = \dfrac{\text{SSTr}}{k-1}$	$F = \dfrac{\text{MSTr}}{\text{MSE}}$
Blocks	$l-1$	SSBl	$\text{MSB}l = \dfrac{\text{SSB}l}{l-1}$	
Error	$(k-1)(l-1)$	SSE	$\text{MSE} = \dfrac{\text{SSE}}{(k-1)(l-1)}$	
Total	$kl-1$			

Table 4 shows a mean square for blocks as well as for treatments and error. Sometimes the F ratio MSBl/MSE is also computed. A large value of this ratio suggests that blocking was effective in filtering out extraneous variation.

EXAMPLE 12

Let's reconsider the electricity usage data given in Example 11 and test at level .05 for the presence of any treatment effects.

H_0 : the mean electricity usage does not depend on which air conditioning system is used.

H_a : the mean electricity usage does depend on which system is used.

Test statistic: $F = \dfrac{\text{MSTr}}{\text{MSE}}$

Rejection region: $F > F$ critical value, where the F critical value is based on $k - 1$ numerator and $(k-1)(l-1)$ denominator df. Since $k = 4$ and $l = 5$, $k - 1 = 3$, and $(k-1)(l-1) = 12$. Table XI then gives the F critical value for $\alpha = .05$ as 3.49.

Computations: From Example 11, $\overline{\overline{x}} = 118.45$, $\overline{x}_1 = 109.40$, $\overline{x}_2 = 128.60$, $\overline{x}_3 = 117.40$, $\overline{x}_4 = 118.40$ (these are the four row averages), $\overline{b}_1 = 142.25$, $\overline{b}_2 = 130.25$, $\overline{b}_3 = 108.00$, $\overline{b}_4 = 98.50$, and $\overline{b}_5 = 113.25$ (the five column averages). Using the individual observations given earlier,

$$\text{SSTo} = \sum_{\substack{\text{all 20}\\ \text{observations}}} x^2 - kl\overline{\overline{x}}^2 = (116)^2 + (118)^2 + \cdots + (99)^2 - (4)(5)(118.45)^2$$
$$= 288{,}203 - 280{,}608.05 = 7594.95$$

The other sums of squares are

$$\begin{aligned}
\text{SSTr} &= l[(\bar{x}_1 - \bar{\bar{x}})^2 + \cdots + (\bar{x}_4 - \bar{\bar{x}})^2] \\
&= 5[(109.4 - 118.45)^2 + (128.6 - 118.45)^2 \\
&\quad + (117.4 - 118.45)^2 + (118.4 - 118.45)^2] \\
&= 930.15, \\
\text{SSB}l &= k[(\bar{b}_1 - \bar{\bar{x}})^2 + \cdots + (\bar{b}_5 - \bar{\bar{x}})^2] \\
&= 4[(142.25 - 118.45)^2 + (130.25 - 118.45)^2 \\
&\quad + \cdots + (113.25 - 118.45)^2] \\
&= 4959.70 \\
\text{SSE} &= \text{SSTo} - \text{SSTr} - \text{SSB}l = 7594.95 - 930.15 - 4959.70 \\
&= 1705.10.
\end{aligned}$$

The remaining calculations are displayed in the accompanying ANOVA table.

Source of Variation	df	Sum of Squares	Mean Square	F
Treatments	3	930.15	310.05	$\frac{310.05}{142.09} = 2.18$
Blocks	4	4959.70	1239.93	
Error	12	1705.10	142.09	
Total	19	7594.95		

Conclusion: $F = 2.18$ does not exceed the critical value 3.49, so H_0 should not be rejected at level .05. Mean electricity usage does not seem to depend on which of the four air-conditioning systems is used. ■

In many studies, all k of the treatments can be applied to the same experimental unit, so there is no need to group different experimental units to form blocks. For example, an experiment to compare effects of four different gasoline additives on automobile engine efficiency could be carried out by selecting just 5 engines and using all four treatments on each one rather than using 20 engines and blocking them. Each engine by itself then constitutes a block. As another example, a manufacturing company might wish to compare outputs for three different packaging machines. Because output could be affected by which machine operator is used, a design that controls for the effects of operator variation is desirable. One possibility is to use 15 operators grouped into homogeneous blocks of 5 operators each, but such homogeneity within each block may be difficult to achieve. An alternative approach is to use only 5 operators and have each one operate all three machines. There are then three observations in each block, all three with the same operator.

EXAMPLE 13

The accompanying data resulted from a study of assembly and implementation times for four different packaged injection systems used to give shots ("An Assessment of Unit Dose Injection Systems" *Amer. J. of Hospital Pharmacy* (1972):61). Ten subjects, all pharmacists and nurses, were selected, and each was timed while using all four systems.

Treatment	Block (Subject) 1	2	3	4	5	6	7	8	9	10	Treatment Average
1. Standard	35.6	31.3	36.2	31.1	39.4	34.7	34.1	36.5	32.2	40.7	35.18
2. Vari-Ject	17.3	16.4	18.1	17.8	18.8	17.0	14.5	17.9	14.6	16.4	16.88
3. Unimatic	24.4	22.4	22.8	21.5	23.3	21.8	23.0	24.1	23.5	31.3	23.81
4. Tubex	25.0	26.0	25.3	24.0	24.2	26.2	24.0	20.9	23.5	36.9	25.60
Block Average	25.575	24.025	25.600	23.600	26.425	24.925	23.900	24.850	23.450	31.325	$\bar{\bar{x}} = 25.3675$

H_0 : mean implementation time does not depend on which injection system is used.

H_a : mean implementation time does depend on which system is used.

Test statistic: $F = \text{MSTr/MSE}$.

Rejection region: For a level .01 test based on $k - 1 = 3$ numerator df and $(k - 1)(l - 1) = (3)(9) = 27$ denominator df, the required F critical value is 4.60, so H_0 will be rejected if $F > 4.60$.

Computations: The required sums of squares are

$$\text{SSTo} = (35.6)^2 + (31.3)^2 + \cdots + (36.9)^2 - (4)(10)(25.3675)^2 = 2051.99$$

$$\text{SSTr} = 10\,[(35.18 - 25.3675)^2 + \cdots + (25.60 - 25.3675)^2] = 1708.03$$

$$\text{SSB}l = 4[(25.575 - 25.3675)^2 + \cdots + (31.325 - 25.3675)^2] = 191.71$$

The remaining calculations appear in the accompanying ANOVA table.

Source of Variation	df	Sum of Squares	Mean Square	F
Treatments	3	1708.03	569.34	100.95
Blocks	9	191.71	21.30	
Error	27	152.25	5.64	
Total	39	2051.99		

Conclusion: Because $F = 100.95 > 4.60 = F$ critical value, H_0 should be rejected at level of significance .01. The data very strongly suggests that average implementation time depends on which injection system is used. ■

Experiments such as the one described in Example 13, in which repeated observations are made on the same experimental unit, are sometimes called *repeated measures designs*. Such designs should not be used when application of the first several treatments somehow affects responses to later treatments. This would be the case if treatments were different methods for learning the same skill, so that if all treatments were given to the same subject, the response to the treatment given last would presumably be much better than the response to the treatment initially applied.

A randomized block experiment is a special case of a two-factor experiment in which one factor, blocks, is created solely to control for extraneous variation. Many statistical computer packages will perform a two-factor ANOVA on request, so data from a randomized block experiment can be analyzed by any such package. If H_0 is rejected by the F test, a multiple comparison procedure can be used to identify significant differences between treatments. In particular, the only modification to the Bonferroni procedure described earlier is that the Bonferroni t critical value now comes from the $(k - 1)(l - 1)$ df row of Table X (because this is now error df). Finally, when $k = 2$, the randomized block F test and two-tailed paired t test are equivalent test procedures.

EXERCISES

13.31 A particular county employs three assessors who are responsible for determining the value of residential property in the county. To see whether or not these assessors differ systematically in their appraisals, 5 houses are selected and each assessor is asked to determine the market value of each house. Explain why a randomized block experiment (with blocks corresponding to the 5 houses) was used rather than a completely randomized experiment involving a total of 15 houses with each assessor asked to appraise 5 different houses (a different group of 5 for each assessor).

13.32 A partially completed ANOVA table for the experiment described in Exercise 13.31 (with houses representing blocks and assessors representing treatments) is given.

Source of Variation	df	Sum of Squares	Mean Square	F
Treatments		11.7		
Blocks		113.5		
Error				
Total		250.8		

a. Fill in the missing entries in the ANOVA table.

b. Use the ANOVA F statistic and a .05 level of significance to test the null hypothesis of no difference between assessors.

13.33 Land-treatment wastewater-processing systems work by removing nutrients and thereby discharging water of better quality. The land used is often planted with a crop such as corn because plant uptake removes nitrogen from the water and sale of the crop helps reduce the costs of wastewater treatment. The concentration of nitrogen in the treated water was observed from 1975 to 1979 under wastewater application rates of none, .05 m/week, and .1 m/week. A randomized block ANOVA was performed with the 5 years serving as blocks. A partially completed ANOVA table is given ("Quality of Percolate Water After Treatment of a Municipal Wastewater Effluent by a Crop Irrigation System" *J. Environ. Quality* (1984):256–64).

Source of Variation	df	Sum of Squares	Mean Square	F
Treatments		1835.2		
Blocks				
Error		206.1		
Total	14	2134.1		

a. Complete the ANOVA table.

b. Is there sufficient evidence to reject the null hypothesis of no difference between the true mean nitrogen concentrations for the three application rates? Use $\alpha = .05$.

13.34 In a comparison of the energy efficiency of three types of ovens—conventional, biradiant, and convection—the energy used in cooking was measured for eight different foods (one-layer caker, two-layer cake, biscuits, bread, frozen pie, baked potatoes, lasagna, and meat loaf). Since a comparison between the three types of ovens is desired, a randomized block (with the eight foods serving as blocks) ANOVA will be used. Suppose calculations result in the quantities SSTo = 4.57, SSTr = 3.97, and SS*Bl* = .2503 (a similar study is described in the paper "Optimizing Oven Radiant Energy Use" *Home Ec. Res. J.* (1980):242–51). Construct an ANOVA table and test the null hypothesis of no difference in mean energy use for the three types of ovens. Use a .01 significance level.

13.35 The article "Rate of Stuttering Adaptation Under Two Electro-Shock Conditions" (*Behavior Res. Therapy* (1967):49–54) gave adaptation scores for three different treatments: no shock (treatment 1), shock following each stuttered word (treatment 2), and shock during each moment of stuttering (treatment 3). These treatments were used on each of 18 stutterers. The 18 subjects were viewed as blocks and the data analyzed using a randomized block ANOVA. Summary quantities are SSTr = 28.78, SS*Bl* = 2977.67, and SSE = 469.55. Construct the ANOVA table and test at significance level .05 to see whether true average adaptation score depends on the treatment given.

13.36 The given table shows average height of cotton plants during 1978–1980 under three different effluent application rates (350, 440, and 515 mm). ("Drip Irrigation of Cotton with Treated Municipal Effluents: Yield Response" *J. Environ. Quality* (1984):231–38)

	Application Rate		
Year	350	440	515
1978	166	176	177
1979	109	126	136
1980	140	155	156

a. This data was analyzed using a randomized block ANOVA with years serving as blocks. Explain why this would be better for comparing treatments than a completely randomized ANOVA.

b. With treatments 1, 2, and 3 denoting the application rates 350, 440, and 515, respectively, summary quantities are:

$\Sigma(x - \bar{\bar{x}})^2 = 4266.0$ $\quad \bar{\bar{x}} = 149.00$

$\bar{x}_1 = 138.33 \quad \bar{x}_2 = 152.33 \quad \bar{x}_3 = 156.33$

$\bar{b}_1 = 173.00 \quad \bar{b}_2 = 123.67 \quad \bar{b}_3 = 150.33$

Construct an ANOVA table. Use a .01 significance level to determine if the true mean height differs for the three effluent rates.

13.37 The paper "Cardiac Output in Preadolescent Competitive Swimmers and in Untrained Normal Children" (*J. Sports Med.* (1983):291–99) gave the accompanying data on heart rate at rest for eight trained and eight untrained swimmers. The trained and untrained swimmers were paired on the basis of physical characteristics so that the members of each pair were as similar as possible.

Subject pair	1	2	3	4	5	6	7	8
Trained	90	85	75	120	95	105	85	75
Untrained	95	75	80	65	82	80	100	85

To determine whether there is a difference in true mean resting heart rate, the data can be analyzed using a randomized block ANOVA with subjects serving as blocks. Summary quantities include $\bar{\bar{x}} = 87$ and $\Sigma(x - \bar{\bar{x}})^2 = 2770$.

a. Construct an ANOVA table and test the null hypothesis of no difference between true mean heart rate of trained and untrained swimmers. Use $\alpha = .05$.

b. A paired t test can be used to test $H_0 : \mu_1 - \mu_2 = 0$, where μ_1 and μ_2 denote the true mean heart rate for trained and untrained swimmers, respectively. Verify that a paired t test with significance level .05 leads you to the same conclusion as the test in (a).

13.38 The paper "Responsiveness of Food Sales to Shelf Space Changes in Supermarkets" (*J. Mktg. Research* (1964):63–67) described an experiment to assess the effect of allotted shelf space on product sales. Two of the products studied were baking soda (a staple product) and Tang (considered to be an impulse product). Six stores (blocks) were used in the experiment and six different shelf-space allotments were tried for one week each. Space allotments of 2, 4, 6, 8, 10, and 12 ft were used for baking soda and 6, 9, 12, 15, 18, and 21 ft were used for Tang. The author speculated that sales of staple goods would not be sensitive to changes in shelf space, whereas sales of impulse products would be affected by changes in shelf space.

a. Data on number of boxes of baking soda sold during a 1-week period is given. Construct an ANOVA table and test the appropriate hypotheses. Use a significance level of .05.

	Shelf Space					
Store	2	4	6	8	10	12
1	36	42	36	40	30	22
2	74	61	65	67	83	84
3	40	58	42	73	69	63
4	43	65	65	41	43	47
5	27	33	35	17	40	26
6	23	31	36	38	42	37

b. Data for Tang sales is also given. Construct an ANOVA table and test the appropriate hypotheses using a .05 significance level.

	Shelf Space					
Store	6	9	12	15	18	21
1	30	35	25	25	38	31
2	47	59	43	62	65	48
3	47	55	48	54	36	54
4	29	19	41	27	33	39
5	17	11	25	23	24	26
6	22	9	19	18	25	22

c. Was the author correct in his speculation that sales of the staple product would not be affected by shelf space allocation, whereas sales of the impulse product would be affected? Explain.

13.39 Four plots were available for an experiment to compare clover accumulation (kg DM/ha) for four different sowing rates ("Performance of Overdrilled Red Clover with Different Sowing Rates and Initial Grazing Managements" *N. Zeal. J. Exp. Ag.* (1984): 71–81). Since the four plots had been grazed differently prior to the experiment and it was thought that this might affect clover accumulation, a randomized block experiment was used with all four sowing rates tried on a section of each plot. Use the given data to test the null hypothesis of no difference in true mean clover accumulation for the different sowing rates.

	Sowing Rate (kg/ha)			
Plot	3.6	6.6	10.2	13.5
1	1155	2255	3505	4632
2	123	406	564	416
3	68	416	662	379
4	62	75	362	564

13.40 The paper "Measuring Treatment Effects Through Comparisons Along Plot Boundaries" (*Forest Sci.* (1980):704–09) reported the results of a randomized block experiment. Five different sources of pine seed were used in each of four blocks. Data is given for plant height (m) and plant diameter (cm).

	Height				Diameter			
Block	I	II	III	IV	I	II	III	IV
Source								
1	7.1	5.8	7.2	6.9	9.4	7.9	9.4	9.4
2	6.2	5.3	7.7	4.7	8.9	7.9	10.7	7.6
3	7.9	5.4	8.6	6.2	10.4	7.9	12.2	9.1
4	9.0	5.9	5.7	7.3	12.2	8.4	7.6	9.4
5	7.0	6.3	4.4	6.1	9.9	8.6	6.6	8.6

a. Does the data provide sufficient evidence to conclude that the true mean height is not the same for all five seed sources? Use a .05 level of significance.

b. Does the data strongly suggest that the true mean diameter is not the same for all seed sources? Use a .05 level of significance.

3.41 The paper of Exercise 13.40 also gave the accompanying data on survival rate for five different seed sources.

Block	I	II	III	IV
Source				
1	62.5	87.5	50.0	70.3
2	50.0	50.0	54.7	59.4
3	93.8	92.2	87.5	87.5
4	96.9	76.6	70.3	65.6
5	56.3	50.0	45.3	56.3

a. Construct an ANOVA table and test the null hypothesis of no difference in true mean survival rates for the five seed sources. Use $\alpha = .05$.

b. Use the Bonferroni procedure (described in Section 13.3) to identify significant differences among the seed sources.

13.5 Two-Factor ANOVA

An investigator will often be interested in assessing the effects of two different factors on a response variable. Consider the following examples.

1. A physical education researcher wishes to know how body density of football players varies with position played (a categorical factor with categories defensive back, offensive back, defensive lineman, and offensive lineman) and level of play (a second categorical factor with categories professional, college division I, college division II, and college division III).
2. An agricultural scientist is interested in seeing how yield of tomatoes is affected by choice of variety planted (a categorical factor with each category corresponding to a different variety) and planting density (a quantitative factor with a level corresponding to each planting density being considered).
3. An applied chemist might wish to investigate how shear strength of a particular adhesive varies with application temperature (a quantitative factor with levels 250°F, 260°F, and 270°F) and application pressure (a quantitative factor with levels 110 lb/in.^{2}, 120 lb/in.^{2}, 130 lb/in.^{2}, and 140 lb/in.^{2}).

Let's call the two factors under study factor A and factor B. Even when a factor is categorical, it simplifies terminology to refer to the categories as levels. Thus in (1) the categorical factor *position played* has four levels. The number of levels of factor A is denoted by k, and l denotes the number of levels of factor B. The rectangular table displayed contains a row corresponding to each level of factor A and a column corresponding to each level of factor B. Each cell in this table corresponds to a particular level of factor A in combination with a particular level of factor B. Because there are l cells in each row and k rows, there are kl cells in the table. The kl different combinations of factor A and factor B levels are often referred to as **treatments.** In (2), if there are three tomato varieties and four different planting densities under consideration, the number of treatments is 12.

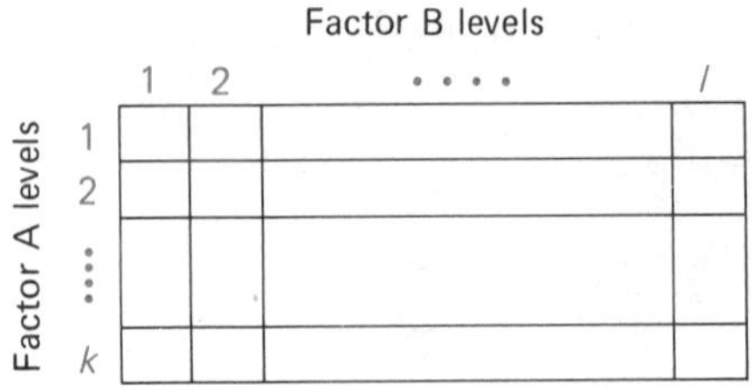

Factor B levels

Factor A levels	1	2	....	l
1				
2				
⋮				
k				

Suppose that an experiment is carried out, resulting in a data set that contains some number of observations for each of the kl treatments. In general, there could be more observations on some treatments than on others, and there may even be a few treatments for which no observations are available. An experimenter may set out to make the same number of observations on each treatment, but on occasion forces beyond the experimenter's control—the death of an experimental subject, malfunctioning equipment, etc.—result in different sample sizes for some treatments. However, such imbalance in sample sizes makes analysis of the data rather difficult. We will restrict consideration to data sets containing the same number of observations on each treatment and let m denote this number.

> **Notation**
>
> k = number of levels of factor A
>
> l = number of levels of factor B
>
> kl = number of treatments (each one a combination of a factor A level and a factor B level)
>
> m = number of observations on each treatment

EXAMPLE 14

An experiment was carried out to assess the effects of tomato variety (factor A, with $k = 3$ levels) and planting density (factor B, with $l = 4$ levels—10, 20, 30, and 40 thousand plants per hectare) on yield. Each of the $kl = 12$ treatments was used on $m = 3$ plots, resulting in the accompanying data set, which consists of $klm = 36$ observations (adapted from "Effects of Plant Density on Tomato Yields in Western Nigeria" *Exper. Ag.* (1976):43–47).

		Density (B)			
		1	2	3	4
Variety (A)	1	7.9, 9.2, 10.5	11.2, 12.8, 13.3	12.1, 12.6, 14.0	9.1, 10.8, 12.5
	2	8.1, 8.6, 10.1	11.5, 12.7, 13.7	13.7, 14.4, 15.4	11.3, 12.5, 14.5
	3	15.3, 16.1, 17.5	16.6, 18.5, 19.2	18.0, 20.8, 21.0	17.2, 18.4, 18.9

Sample average yields for each treatment, each level of factor A, and each level of factor B are important summary quantities. These are displayed in the given rectangular table.

Sample Average Yield for Each Treatment		Factor B 1	2	3	4	Sample Average Yield for Each Level of Factor A
Factor A	1	9.20	12.43	12.90	10.80	11.33
	2	8.93	12.63	14.50	12.77	12.21
	3	16.30	18.10	19.93	18.17	18.13
Sample Average Yield for Each Level of Factor B		11.48	14.39	15.78	13.91	Grand Mean 13.89

A plot of these sample averages is also quite informative. First construct horizontal and vertical axes, and scale the vertical axis in units of the re-

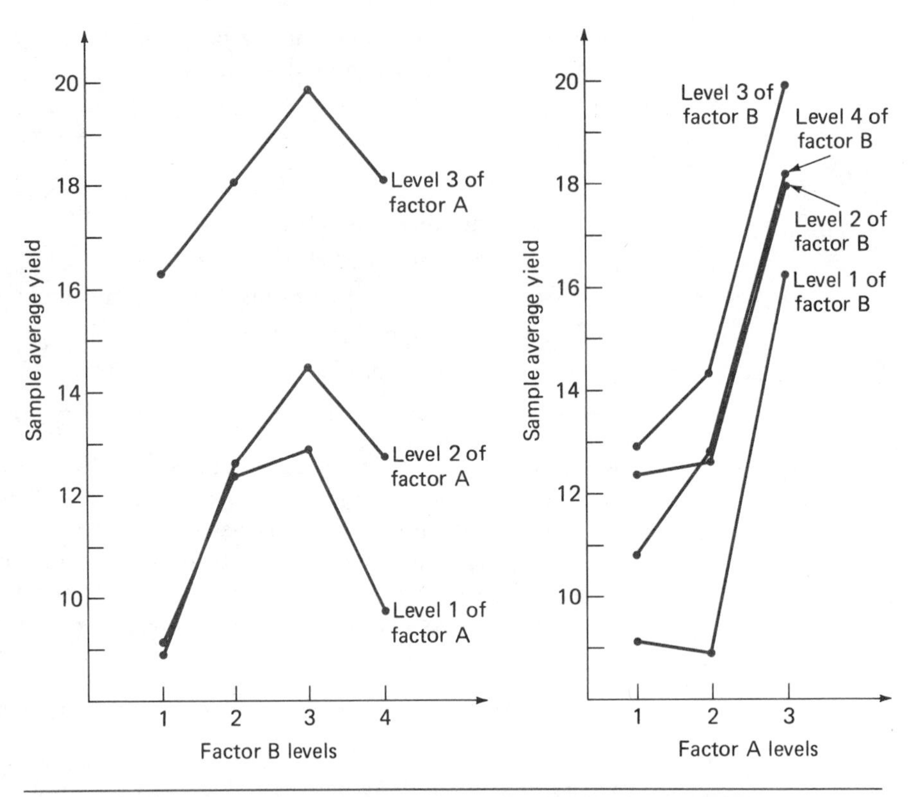

FIGURE 3 GRAPHS OF TREATMENT SAMPLE AVERAGE RESPONSES FOR THE DATA OF EXAMPLE 14.

sponse variable (yield). Then mark a point on the horizontal axis for each level of one of the factors (either A or B can be chosen). Now above each such mark, plot a point for the sample average response for each level of the other factor. Finally, connect all points corresponding to the same level of the other factor by straight line segments. Figure 3 displays one such plot for which factor A levels mark the horizontal axis and another plot for which factor B levels mark this axis, although usually only one of the two plots is constructed. ■

Interaction

An important aspect of two-factor studies involves assessing how simultaneous changes in the levels of both factors affect the response. As a simple example, suppose that an automobile manufacturer is studying engine efficiency (miles per gallon) for two different engine sizes (factor A, with $k = 2$ levels) in combination with two different carburetor designs (factor B, with $l = 2$ levels). Consider the two possible sets of true average responses displayed in Table 5. In Table 5(a), when factor A changes from level 1 to level 2 and factor B remains at level 1 (the change within the first column), the true average response increases by 2. Similarly, when factor B changes from level 1 to level 2 and factor A is fixed at level 1 (the change within the first row), the true average response increases by 3. And when the levels of both factors are changed from 1 to 2, the true average response increases by 5, which is the sum of the two "one-at-a-time" increases. This is because the change in true average response when the level of either factor changes from 1 to 2 is the same for each level of the other factor—the change within either row is 3 and the change within either column is 2. In this case changes in levels of the two factors affect the true average response separately, or in an additive manner.

TABLE 5

TWO POSSIBLE SETS OF TRUE AVERAGE RESPONSES WHEN $k = 2$ AND $l = 2$

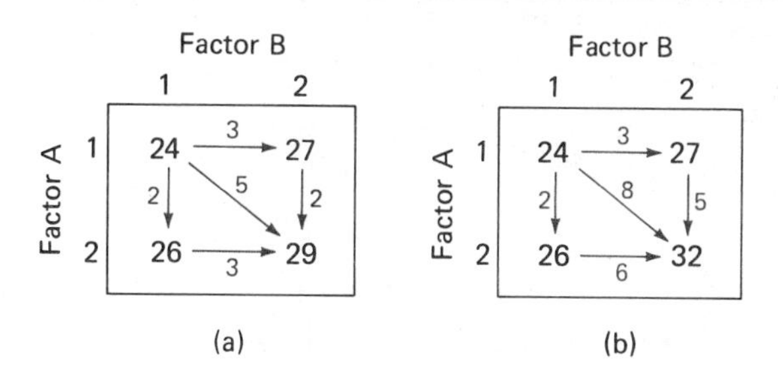

(a) (b)

The changes in true average responses in the first row and in the first column of Table 5(b) are 3 and 2, respectively, exactly as in Table 5(a). However, the change in true average response when the levels of both factors change simultaneously from 1 to 2 is 8, which is much larger than the separate changes suggest. In this case there is interaction between the two factors, so that the effect of simultaneous changes cannot be determined from the individual effects of separate changes. This is because in Table 5(b), the change in going from the first to the second column is different for the two rows, and the change in going from the first to the second row is different for the two columns. That is, *the change in true average response when the level of one factor changes depends on the level of the other factor.* This is not true in Table 5(a).

When there are more than two levels of either factor, a graph of true average responses, similar to that for sample average responses in Figure 3, provides insight into how changes in the level of one factor depend on the level of the other factor. Figure 4 shows several possible such graphs when $k = 4$ and $l = 3$. The most general situation is pictured in Figure 4(a). There the change in true average response when the level of B is changed (a vertical distance) depends on the level of A. An analogous property would hold if the picture were redrawn so that levels of B were marked on the horizontal axis. This is a prototypical picture suggesting **interaction** between the factors—the change in true average response when the level of one factor changes depends on the level of the other factor.

There is no interaction between the factors when the connected line segments are parallel, as in Figure 4(b). Then the change in true average response when the level of one factor changes is the same for each level of the other factor (the vertical distances are the same for each level of factor A). Figure 4(c) illustrates an even more restrictive situation—there is no interaction between factors, and, in addition, the true average response does not depend on the level of factor A. Only when the graph looks like this can it be

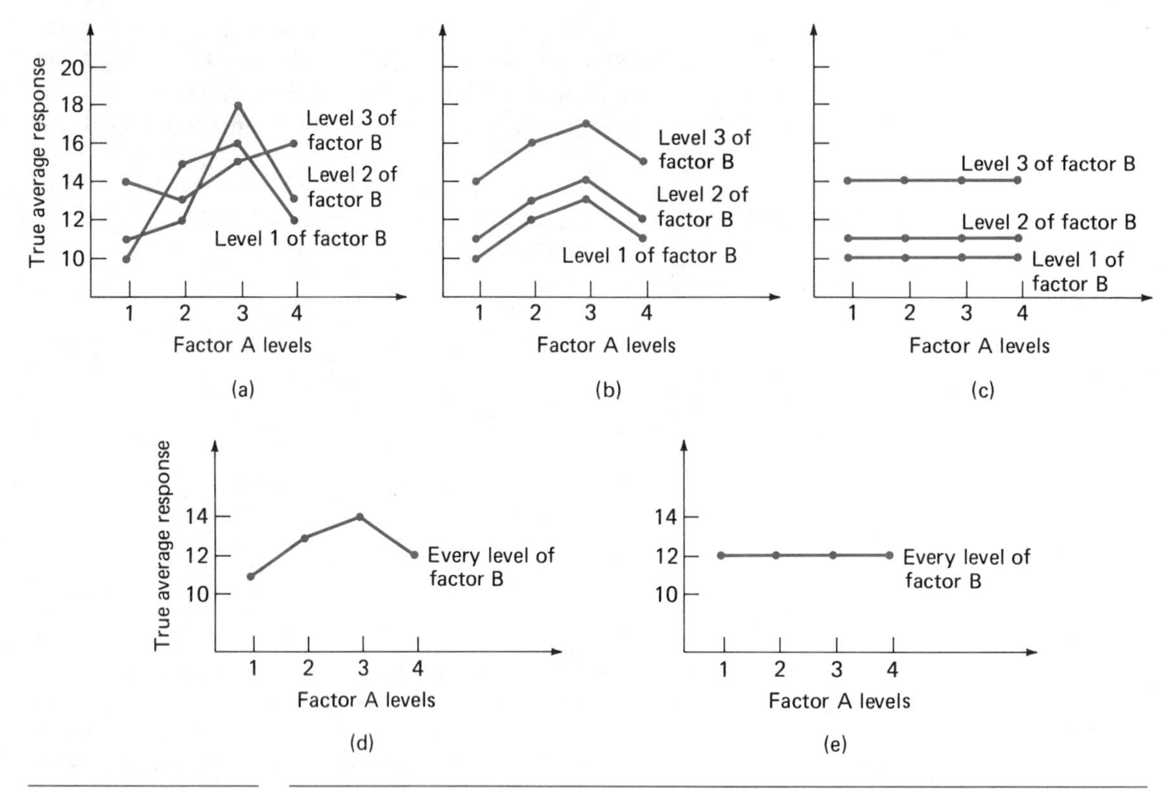

FIGURE 4 SOME GRAPHS OF TRUE AVERAGE RESPONSES

said that factor A has no effect on the responses. Similarly, the graph in Figure 4(d) indicates no interaction and no dependence on the level of factor B. A final case, pictured in Figure 4(e), shows a single set of four points connected by horizontal line segments, which indicates that the true average response is identical for every level of either factor.

> If the graphs of true average responses are connected line segments that are parallel, there is no interaction between the factors. That is, in this case the change in true average response when the level of one factor is changed is the same for each level of the other factor. Special cases of no interaction are (1) the true average response is the same for each level of factor A (no factor A main effects), and (2) the true average response is the same for each level of factor B (no factor B main effects).

The graphs of Figure 4 pictured true average responses, quantities whose values are fixed but unknown to an investigator. Figure 3 contained graphs of the sample average responses based on data resulting from an experiment. These sample averages are, of course, subject to variability because there is sampling variation in the individual observations. If the experiment discussed in Example 14 were repeated, the resulting graphs of sample averages would probably look somewhat different from the graph in Figure 3 and perhaps a great deal different if there is substantial underlying variability in responses. Even when there is no interaction among factors, the connected line segments in the sample mean picture will not typically be exactly parallel and may deviate quite a bit from parallelism in the presence of substantial underlying variability. Similarly, there might actually be no factor A effects (Figure 4(c)), yet the sample graphs would not usually be exactly horizontal. The sample pictures give us insight, but formal inferential procedures are needed in order to draw sound conclusions about the nature of the true average responses for different factor levels.

Hypotheses and *F* Tests

> **Basic Assumption:** The observations on any particular treatment are independently selected from a normal distribution with variance σ^2 (the same variance for each treatment), and samples for different treatments are independent of one another.

Because of the normality assumption, tests based on F statistics and F critical values are appropriate. The necessary sums of squares result from breaking up SSTo $= \Sigma(x - \overline{\overline{x}})^2$ into four parts, which reflect random variation and variation attributable to various factor effects:

$$\text{SSTo} = \text{SSA} + \text{SSB} + \text{SSAB} + \text{SSE}$$

where

1. SSTo is total sum of squares, with associated df $klm - 1$.
2. SSA is the factor A main effect sum of squares, with associated df $k - 1$.
3. SSB is the factor B main effect sum of squares, with associated df $l - 1$.
4. SSAB is the interaction sum of squares, with associated df $(k - 1)(l - 1)$.
5. SSE is error sum of squares, with associated df $kl(m - 1)$.

The formulas for these sums of squares are similar to those given in previous sections, but for our purposes it is not necessary to give them. The standard statistical computer packages will calculate all sums of squares and other necessary quantities. As before, the magnitude of SSE is related entirely to the amount of underlying variability (as specified by σ^2) in the distributions being sampled. It has nothing to do with values of the various true average responses. SSAB reflects in part underlying variability, but its value is also affected by whether or not there is interaction between the factors. Generally speaking, the more extensive the amount of interaction (i.e., the further the graphs of true average responses are from being parallel), the larger the value of SSAB tends to be. The test statistic for testing the null hypothesis that there is no interaction between factors is MSAB/MSE. The numerator and denominator both give reliable (unbiased) estimates of σ^2 when this null hypothesis is true, but MSAB tends to overestimate σ^2 when there is interaction. The appropriate F critical value specifies a rejection region so that the type I error probability is controlled at the desired level.

The absence of factor A effects and of factor B effects are both special cases of no-interaction situations. If the data suggests that interaction is present, it doesn't make sense to investigate effects of one factor without reference to the other factor. Our recommendation is that hypotheses concerning the presence or absence of separate factor effects be tested only if the hypothesis of no interaction is not rejected. Then the factor A main effect sum of squares SSA will reflect random variation and, in addition, any differences between true average responses for different levels of factor A. A similar comment applies to SSB. The F ratio MSA/MSE is the appropriate statistic for testing the null hypothesis that there are no factor A main effects, and MSB/MSE plays a similar role for factor B main effects.

Two-Factor ANOVA Hypotheses and Tests

1. H_0 : there is no interaction between factors.
   H_a : there is interaction between factors.
   Test statistic: F_{AB} = MSAB/MSE.
   Rejection region: $F_{AB} > F$ critical value, where the test statistic and critical value are based on $(k - 1)(l - 1)$ numerator df and $kl(m - 1)$ denominator df.

The following two hypotheses should be tested only if H_0 in (1) is not rejected.

2. H_0 : there are no factor A main effects (true average response is the same for each level of factor A).
   H_a : H_0 is not true.
   Test statistic: F_A = MSA/MSE.
   Rejection region: $F_A > F$ critical value, where the test statistic and critical value are based on $k - 1$ numerator and $kl(m - 1)$ denominator df, respectively.
3. H_0 : there are no factor B main effects.
   H_a : H_0 is not true.
   Test statistic: F_B = MSB/MSE.
   Rejection region: $F_B > F$ critical value, where the test statistic and critical value are based on $l - 1$ numerator and $kl(m - 1)$ denominator df, respectively.

Computations are typically summarized in an ANOVA table, as pictured.

Source of Variation	df	Sum of Squares	Mean Square	F
A Main Effects	$k - 1$	SSA	$\text{MSA} = \dfrac{\text{SSA}}{k - 1}$	$F_A = \dfrac{\text{MSA}}{\text{MSE}}$
B Main Effects	$l - 1$	SSB	$\text{MSB} = \dfrac{\text{SSB}}{l - 1}$	$F_B = \dfrac{\text{MSB}}{\text{MSE}}$
AB Interaction	$(k - 1)(l - 1)$	SSAB	$\text{MSAB} = \dfrac{\text{SSAB}}{(k - 1)(l - 1)}$	$F_{AB} = \dfrac{\text{MSAB}}{\text{MSE}}$
Error	$kl(m - 1)$	SSE	$\text{MSE} = \dfrac{\text{SSE}}{kl(m - 1)}$	
Total	$klm - 1$	SSTo		

EXAMPLE 15

The accompanying ANOVA table resulted from using MINITAB's two-factor ANOVA command on the tomato-yield data of Example 14.

ANALYSIS OF VARIANCE ON YIELD

SOURCE	DF	SS	MS
VARIETY	2	327.60	163.80
DENSITY	3	86.69	28.90
INTERACTION	6	8.03	1.34
ERROR	24	38.04	1.58
TOTAL	35	460.36	

1. To test the no-interaction null hypotheses at level .05, the F critical value based on $(k - 1)(l - 1) = 6$ numerator df and $kl(m - 1) = 24$ denominator df is needed. This value is 2.51, so H_0 will be rejected if $F_{AB} = \text{MSAB}/\text{MSE} > 2.51$. The ANOVA table gives MSAB $= 1.34$ and MSE $= 1.58$, from which $F_{AB} = 1.34/1.58 = .85$. Because .85 does not fall in the rejection region, H_0 should not be rejected. There is no evidence that variety and planting density interact in their effects on yield. It is thus appropriate to carry out further tests concerning the presence of main effects.

2. The null hypothesis of no factor A (variety) main effects will be rejected at level .05 if $F_A = \text{MSA}/\text{MSE} > 3.40$ (the F critical value based on $k - 1 = 2$ numerator and $kl(m - 1) = 24$ denominator df). Since $F_A = 163.80/1.58 = 103.7 > 3.40$, this null hypothesis is rejected. True average yield seems to depend strongly on which variety (level of factor A) is used.

3. The null hypothesis of no factor B (planting density) main effects will be rejected at level .05 if $F_B = \text{MSB}/\text{MSE} > 3.01$, the F critical value for 3 numerator and 24 denominator df. The computed value of F_B is $28.90/1.58 = 18.3$, which falls into the rejection region. True average yield appears to depend separately on both variety and on planting density. ■

After the null hypothesis of no factor A main effects has been rejected, significant differences in factor A levels can be identified by using a multiple comparisons procedure. In particular, the Bonferroni method described earlier can be applied. The quantities $\bar{x}_1, \bar{x}_2, \ldots, \bar{x}_k$ are now the sample average responses for levels $1, \ldots, k$ of factor A, and error df is $kl(m - 1)$. A similar comment applies to factor B main effects and significant differences in factor B levels.

The Case $m = 1$

There is a problem with the foregoing analysis when $m = 1$ (one observation on each treatment). Although we did not give the formula, MSE is an estimate of σ^2 obtained by computing a separate sample variance s^2 for the m observations on each treatment and then averaging these kl sample variances. With only one observation on each treatment, there is no way to estimate σ^2 separately from each of the treatments.

One way out of this dilemma is to assume a priori that there is no interaction between factors. This should, of course, be done only when the investigator has sound reasons, based on a thorough understanding of the problem, for believing that the factors contribute separately to the response. Having made this assumption, what would otherwise be interaction sum of squares can then be used for SSE. The fundamental identity is

$$\text{SSTo} = \text{SSA} + \text{SSB} + \text{SSE}$$

with the four associated df $kl - 1$, $k - 1$, $l - 1$, and $(k - 1)(l - 1)$.

The analysis of data from a randomized block experiment in fact assumed no interaction between treatments and blocks. If SSTr is relabeled SSA and SSBl is relabeled SSB, the formulas for all sums of squares given in Section 13.4 are valid here ($\bar{x}_1, \ldots, \bar{x}_k$ and $\bar{b}_1, \ldots, \bar{b}_l$ are now the sample average responses for factor A levels and factor B levels, respectively). The corresponding ANOVA table is given. F_A is the test statistic for testing the null hypothesis that true average responses are identical for all factor A levels. F_B plays a similar role for factor B main effects.

Source of Variation	df	Sum of Squares	Mean Square	F
Factor A	$k - 1$	SSA	$\text{MSA} = \frac{\text{SSA}}{k-1}$	$F_A = \text{MSA}/\text{MSE}$
Factor B	$l - 1$	SSB	$\text{MSB} = \frac{\text{SSB}}{l-1}$	$F_B = \text{MSB}/\text{MSE}$
Error	$(k - 1)(l - 1)$	SSE	$\text{MSE} = \frac{\text{SSE}}{(k-1)(l-1)}$	
Total	$kl - 1$	SSTo		

EXAMPLE 16

When metal pipe is buried in soil, it is desirable to apply a coating to retard corrosion. Four different coatings are under consideration for use with pipe that will ultimately be buried in three types of soil. An experiment to investigate the effects of these coatings and soils was carried out by first selecting 12 pipe segments and applying each coating to three segments. The segments were then buried in soil for a specified period in such a way that each soil type received one piece with each coating. The resulting data (depth of corrosion) and ANOVA table are given. Assuming that there is no interaction between coating type and soil type, let's test at level .05 for the presence of separate factor A (coating) and factor B (soil) effects.

		Factor B (Soil) 1	2	3	Sample Average
Factor A (Coating)	1	64	49	50	54.33
	2	53	51	48	50.67
	3	47	45	50	47.33
	4	51	43	52	48.67
Sample Average		53.75	47.00	50.00	$\bar{\bar{x}} = 50.25$

Source of Variation	df	Sum of Squares	Mean Square	F
Factor A	3	83.5	27.8	$F_A = \frac{27.8}{20.6} = 1.3$
Factor B	2	91.5	45.8	$F_B = \frac{45.8}{20.6} = 2.2$
Error	6	123.3	20.6	
Total	11	298.3		

The number of denominator df for each F test is 6, and the level .05 F critical values for 3 and 2 numerator df are 4.76 and 5.14, respectively. Neither computed F ratio exceeds the corresponding critical value, so it appears that the true average response (amount of corrosion) depends neither on the coating used nor on the type of soil in which pipe is buried. ■

EXERCISES

13.42 Many students report that test anxiety affects their performance on exams. A study of the effect of anxiety and instructional mode (lecture versus independent study) on test performance was described in the paper "Interactive Effects of Achievement Anxiety, Academic Achievement, and Instructional Mode on Performance and Course Attitudes" (*Home Ec. Research J.* (1980):216–27). Students classified as either belonging to high- or low-achievement anxiety groups (factor A) were assigned to one of the two instructional modes (factor B). Mean test scores for the four treatments (factor level combinations) are given. Use these means to construct a graph (similar to those of Figure 3) of the treatment sample averages. Does the picture suggest the existence of an interaction between factors? Explain.

	Instructional Mode	
Anxiety Group	Lecture	Independent Study
High	145.8	144.3
Low	142.9	144.8

13.43 The accompanying plot appeared in the paper "Group Process—Work Outcome Relationships: A Note on the Moderating Impact of Self-Esteem" (*Academy of Mgmt. J.* (1982):575–85). The response variable was tension, a measure of job strain. The two factors of interest were peer-group interactions (with two levels—high and low) and self-esteem (also with two levels—high and low).

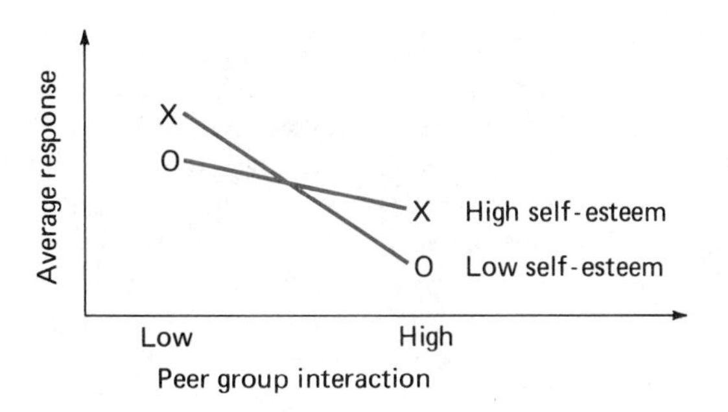

a. Does this plot suggest an interaction between peer-group interaction and self-esteem? Explain.

b. The authors of the paper state: "Peer group interaction had a stronger effect on individuals with lower self-esteem than on those with higher self-esteem." Do you agree with this statement? Explain.

13.44 Explain why the individual effects of factor A or factor B can't be interpreted when an AB interaction is present.

13.45 The accompanying partially completed ANOVA table was taken from the paper "Influence of Temperature and Salinity on Development of White Perch Eggs" (*Trans. Am. Fisheries Soc.* (1982):396–98). The response variable is hatch rate of perch eggs, factor A is water temperature (with eight levels), and factor B is water salinity (with five levels). Three observations were made at each factor level combination, for a total of 120 observations.

Source of Variation	df	Sum of Squares	Mean Square	F
A main effects		102,526		
B main effects		95		
AB interaction		315		
Error				
Total		105,987		

a. Complete the ANOVA table by providing the missing entries.

b. Using a .05 significance level, test the null hypothesis of no AB interaction.

c. If appropriate, conduct hypothesis tests to determine if there are factor A or factor B main effects.

13.46 Three ultrasonic devices (factor A, with levels 20, 30, and 40 kHz) were tested for effectiveness under two test conditions (factor B, with levels *plentiful food supply* and *restricted food supply*). Daily food consumption was recorded for three rats under each factor level combination for a total of 18 observations. Data compatible with summary values given in the paper "Variables Affecting Ultrasound Repellency in Philippine Rats" (*J. Wildlife Mgmt.* (1982):148–55) was used to obtain the sums of squares given in the ANOVA table below. Complete the table and use it to test the relevant hypotheses.

Source of Variation	df	Sum of Squares	Mean Square	F
A main effects		4206		
B main effects		1782		
AB interaction				
Error		2911		
Total		10,846		

13.47 The paper "Learning, Opportunity to Cheat, and Amount of Reward" (*J. Exper. Educ.* (1977):30–40) described a study to determine the effects of expectations concerning cheating during a test and perceived payoff on test performance. Subjects, students at UCLA, were randomly assigned to a particular factor-level combination. Factor A was expectation of opportunity to cheat, with levels high, medium, and low. Those in the high group were asked to study and then recall a list of words. For the first four lists, they were left alone in a room with the door closed and so could look at the original list of words if desired. The medium group was asked to study and recall the list while left alone but with the door open. For the low group, the experimenter remained in the room. For study and recall of a fifth list, the experimenter stayed in the room for all three groups, thus precluding any cheating on the fifth list. Score on the fifth test was the response variable. The second factor (B) under study was the perceived payoff, with a high and a low level. The high payoff group was told that if they scored above average on the test, they would receive 2 hours of credit rather than just 1 hour (subjects were fulfilling a course requirement by participating in experiments). The low group was not given any extra incentive for scoring above the average. The paper gave the following statistics: $F_A = 4.99$, $F_B = 4.81$, $F_{AB} < 1$, error df ≈ 120. Test the null hypothesis of no interaction between the factors. If appropriate, test the null hypotheses of no factor A and no factor B effects. Use $\alpha = .05$.

13.48 Identification of sex in human skeletons is an important part of many anthropological studies. An experiment conducted to determine if measurements of the sacrum could be used to determine sex was described in the paper "Univariate and Multivariate Methods for Sexing the Sacrum" (*Am. J. Phys. Anthro.* (1978):103–10). Sacra from skeletons of individuals whose race (factor A, with two levels—Caucasian and Black) and sex (factor B, with two levels—male and female) were known were measured and the lengths recorded. Data compatible with summary quantities given in the paper was used to compute the following: SSA = 857, SSB = 291, SSAB = 32, SSE = 5541, and error df = 36.

a. Use a significance level of .01 to test the null hypothesis of no interaction between race and sex.

b. Using a .01 significance level, test to determine if the true average length differs for the two races.

c. Using a .01 significance level, test to determine if the true average length differs for males and females.

13.49 The paper "Food Consumption and Energy Requirements of Captive Bald Eagles" (*J. Wildlife Mgmt.* (1982):646–54) investigated mean gross daily energy intake (the response variable) for different diet types (factor A, with three levels—fish, rabbit, and mallard) and temperatures (factor B, with three levels). Summary quantities given in the paper were used to generate data, resulting in SSA = 18,138, SSB = 5182, SSAB = 1737, SSE = 11,291, and error df = 36. Construct an ANOVA table and test the relevant hypotheses.

13.50 The effect of three different soil types and three phosphate application rates on total phosphorus uptake (mg) of white clover was examined in the paper "A Glasshouse Comparison of Six Phosphate Fertilisers" (*N. Zeal. J. Exp. Ag.* (1984): 131–40). Only one observation was obtained for each factor-level combination. Assuming there is no interaction between soil type and application rate, use the data below to construct an ANOVA table and to test the null hypotheses of no main effects due to soil type and of no main effects due to application rate.

	Soil Type		
Application Rate (kgP/ha)	Ramiha	Konini	Wainui
0	1.29	10.42	17.10
75	11.73	21.08	23.69
150	17.63	31.37	32.88

13.6 Distribution-Free ANOVA

The validity of the F tests presented in earlier sections is based on the assumption that observations are selected from normal distributions, all of which have the same variance, σ^2. When this is the case, the type I error probability is controlled at the desired level of significance α by using an appropriate F critical value. Additionally, the test has good ability to detect departures from the null hypothesis—its type II error probabilities are smaller than those for any other test.

Suppose that an F test is used when the basic assumptions are violated. Then there are two potential difficulties. One is that the actual level of significance may be different from what the investigator desires. This is because the test statistic will no longer have an F distribution, so an F critical value won't capture the desired tail area (e.g., .05 or .01) under the actual sampling distribution curve. A second problem is that the test may have rather large type II error probabilities, so that substantial departures from H_0 are likely to go undetected.

Studies have shown that when population or treatment distributions are only mildly nonnormal, neither problem alluded to above is serious enough to warrant abandoning the F test. Statisticians say that the test is *robust* to small departures from normality. However, distributions that are either very skewed or have much heavier tails than the normal distribution do adversely affect the performance of the F test. Here we present test procedures that are valid (have a guaranteed type I error probability) when underlying population or treatment distributions are nonnormal, as long as they have the same shape and spread. These procedures are distribution-free because their validity is guaranteed for a very wide class of distributions rather than just for a particular type of distribution, such as the normal. As was the case with the distribution-free rank-sum test discussed in Chapter 9, the distribution-free ANOVA procedures are based on ranks of the observations.

The Kruskal-Wallis Test for a Completely Randomized Design

As before, k denotes the number of populations or treatments being compared, and $\mu_1, \mu_2, \ldots, \mu_k$ represent the population means or average responses when treatments are administered. The hypotheses to be tested are still $H_0: \mu_1 = \mu_2 = \cdots = \mu_k$ versus H_a : at least two among the k means are different.

> **Basic Assumption:** The k population or treatment distributions all have the same shape and spread.

The distribution-free test to be described here is called the Kruskal-Wallis (K-W) test after the two statisticians who developed it. Suppose that k independent random samples are available, one from each population or treatment. Again let $n_1, n_2, \ldots, n_k$ denote the sample sizes, with $N = n_1 + n_2 + \cdots + n_k$. When H_0 is true, observations in all samples are selected from the same population or treatment-response distribution. Observations in the different samples should then be quite comparable in magnitude. However, when some μ's are different, some samples will consist mostly of relatively small values, whereas others will contain a preponderance of large values.

Let the smallest observation among all N in the k samples be assigned rank 1, the next smallest rank 2, and so on (for the moment let's assume that there are no tied observations). The average of all ranks assigned is $(1 + 2 + 3 + \cdots + N)/N = (N + 1)/2$. If all μ's are equal, the average of the ranks for each of the k samples should be reasonably close to $(N + 1)/2$ (since observations will typically be intermingled, their ranks will be also). On the other hand, large differences between some of the μ's will usually result in some samples having average ranks much below $(N + 1)/2$ (those samples that contain mostly small observations), whereas others will have average ranks considerably exceeding $(N + 1)/2$. The K-W statistic measures the discrepancy between the average rank in each of the k samples and the overall average $(N + 1)/2$.

DEFINITION

Let $\bar{r}_1$ denote the average of the ranks for observations in the first sample, $\bar{r}_2$ denote the average rank for observations in the second sample, and let $\bar{r}_3, \ldots, \bar{r}_k$ denote the analogous rank averages for samples $3, \ldots, k$. Then the K-W statistic is

$$\text{KW} = \frac{12}{N(N+1)}\left[n_1\left(\bar{r}_1 - \frac{N+1}{2}\right)^2 + n_2\left(\bar{r}_2 - \frac{N+1}{2}\right)^2 + \cdots + n_k\left(\bar{r}_k - \frac{N+1}{2}\right)^2\right]$$

EXAMPLE 17

A recent *Newsweek* article (September 24, 1984) reported that average starting salaries for graduating seniors in four business disciplines were ranked as follows: (1) finance, (2) accounting, (3) marketing, and (4) business administration. To gain information on salaries for its own graduates, suppose that a business school selected a random sample of students from each discipline and obtained the accompanying starting salary data. Within each sample, values are listed in increasing order, and the corresponding rank among all $N = 22$ reported salaries appears in parentheses beside each observation.

1. Finance ($n_1 = 5$): 19.4(10), 19.8(12), 20.3(14), 22.3(19), 23.9(22)
2. Accounting ($n_2 = 6$): 18.7(6), 18.9(8), 19.5(11), 20.1(13), 21.8(18), 22.9(20)
3. Marketing ($n_3 = 6$): 16.7(1), 17.6(4), 18.2(5), 20.4(15), 21.4(17), 23.4(21)
4. Business administration ($n_4 = 5$): 16.9(2), 17.3(3), 18.8(7), 19.1(9), 21.0(16)

The average rank in the first sample is $\bar{r}_1 = (10 + 12 + 14 + 19 + 22)/5 = 15.4$, and the other rank averages are $\bar{r}_2 = 12.7$, $\bar{r}_3 = 10.5$, and $\bar{r}_4 = 7.4$. The average of all ranks assigned is $(N + 1)/2 = 23/2 = 11.5$, so

$$\text{KW} = \frac{12}{(22)(23)}[5(15.4 - 11.5)^2 + 6(12.7 - 11.5)^2 + 6(10.5 - 11.5)^2 + 5(7.4 - 11.5)^2] = \frac{12}{(22)(23)}(174.74) = 4.14 \blacksquare$$

H_0 will be rejected when the value of KW is sufficiently large. To specify a critical value that controls the type I error probability, it is necessary to know how KW behaves when H_0 is true. That is, we need information about the sampling distribution of KW when H_0 is true.

There are only a finite number of ways to assign the N ranks, and these all have the same chance of occurring when H_0 is true. Suppose all possibilities are enumerated, KW is computed for each one, and the 5% with the largest KW values are separated out. Rejecting H_0 when the observed allocation of ranks to samples falls within this 5% set then results in a level .05 test. The difficulty with this procedure is that unless N is small, the number of possibilities is quite large, and so enumeration is really out of the question. Fortunately, as long as no n_i is too small, there is an approximate result that saves the day. The approximation is based on a new type of probability distribution called a *chi-squared distribution*. As with t distributions, there is a different chi-squared distribution for each different number of df. Unlike a t curve, a chi-squared curve is not symmetric but instead looks rather like an F curve. Table XII gives upper-tail critical values for various chi-squared distributions.

The Kruskal-Wallis Test

When H_0 is true and either (1) $k = 3$ and each n_i is at least 6, or (2) $k \geq 4$ and each n_i is at least 5, the statistic KW has approximately a chi-squared distribution based on $k - 1$ df. A test with (approximate) level of significance α results from using KW as the test statistic and rejecting H_0 if KW > chi-squared critical value. The chi-squared critical value is obtained from the $k - 1$ df row of Table XII in the column headed by the desired α.

EXAMPLE 18

Let's use the K-W test at level .05 to analyze the salary data introduced in Example 17.

1. Let μ_1, μ_2, μ_3, and μ_4 denote the true average starting salaries for all graduates in the four disciplines, respectively.
2. $H_0 : \mu_1 = \mu_2 = \mu_3 = \mu_4$.
3. H_a : at least two of the four μ's are different.
4. Test statistic:

$$\text{KW} = \frac{12}{N(N+1)}\left[n_1\left(\bar{r}_1 - \frac{N+1}{2}\right)^2 + \cdots + n_4\left(\bar{r}_4 - \frac{N+1}{2}\right)^2\right]$$

5. Rejection region: The number of df for the chi-squared approximation is $k - 1 = 3$. For $\alpha = .05$, Table XII gives 7.82 as the critical value. H_0 will be rejected if KW > 7.82.
6. We previously computed KW as 4.14.
7. The computed KW value 4.14 does not exceed the critical value 7.82, so H_0 should not be rejected. The data does not provide enough evidence to conclude that the true average starting salaries for the four disciplines are different. ■

When there are tied values in the data set, ranks are determined as they were for the rank-sum test—by assigning each tied observation in a group the average of the ranks they would receive if they all differed slightly from one another. Rejection of H_0 by the K-W test can be followed by the use of an appropriate multiple comparison procedure (although not the Bonferroni method described earlier). Also, the most widely used statistical computer packages will perform a K-W test on request.

The K-W test does not require normality, but it does require equal population or treatment-response distribution variances (all distributions must have the same spread). If you encounter a data set in which variances appear to be quite different, you should consult a statistician for advice.

Friedman's Test for a Randomized Block Design

The validity of the randomized block F test rested on the assumption that every observation in the experiment was drawn from a normal distribution with the same variance. The test described here, called Friedman's test, does not require normality.

> **Basic Assumption:** Every observation in the experiment is assumed to have been selected from a distribution having exactly the same shape and spread, but the mean value may depend separately both on the treatment applied and on the block.

The hypotheses are again

H_0 : the mean value does not depend on which treatment is applied

versus

H_a : the mean value does depend on which treatment is applied

The rationale for Friedman's test is quite straightforward. The observations in each block are first ranked separately from 1 to k (since every treatment appears once, there are k observations in any block). Then the rank averages $\bar{r}_1, \bar{r}_2, \ldots, \bar{r}_k$ for treatments $1, 2, \ldots, k$, respectively, are computed. When H_0 is false, some treatments will tend to receive small ranks in most blocks, whereas other treatments will tend to receive mostly large ranks. In this case the $\bar{r}$'s will tend to be rather different. On the other hand, when H_0 is true, all the $\bar{r}$'s will tend to be close to the same value $(k + 1)/2$, the average of ranks $1, 2, \ldots, k$. The test statistic measures the discrepancy between the $\bar{r}$'s and $(k + 1)/2$, with a large discrepancy suggesting that H_0 is false.

> **Friedman's Test**
>
> After ranking observations separately from 1 to k within each of the l blocks, let $\bar{r}_1, \bar{r}_2, \ldots, \bar{r}_k$ denote the resulting rank averages for the k treatments. The test statistic is
>
> $$F_r = \frac{12l}{k(k+1)}\left[\left(\bar{r}_1 - \frac{k+1}{2}\right)^2 + \left(\bar{r}_2 - \frac{k+1}{2}\right)^2 + \cdots + \left(\bar{r}_k - \frac{k+1}{2}\right)^2\right]$$
>
> As long as l is not too small, when H_0 is true F_r has approximately a chi-squared distribution based on $k - 1$ df. The rejection region for a test that has approximate level of significance α is then $F_r >$ chi-squared critical value.

EXAMPLE 19

High-pressure sales tactics of door-to-door salespeople can be quite offensive. Many people succumb to such tactics, sign a purchase agreement, and later regret their actions. In the mid-1970s the Federal Trade Commission implemented regulations clarifying and extending rights of purchasers to cancel such agreements. The accompanying data is a subset of that given in the paper "Evaluating the FTC Cooling-Off Rule" (*J. Consumer Affairs* (1977): 101–06). Individual observations are cancellation rates for each of nine salespeople (the blocks) during each of 4 years. Let's use Friedman's test at level .05 to see if true average cancellation rate depends on the year.

	Salespeople								
Cancellation Rates	1	2	3	4	5	6	7	8	9
1973	2.8	5.9	3.3	4.4	1.7	3.8	6.6	3.1	0.0
1974	3.6	1.7	5.1	2.2	2.1	4.1	4.7	2.7	1.3
1975	1.4	.9	1.1	3.2	.8	1.5	2.8	1.4	.5
1976	2.0	2.2	.9	1.1	.5	1.2	1.4	3.5	1.2

	Salespeople									
Ranks	1	2	3	4	5	6	7	8	9	$\bar{r}_i$
1973	3	4	3	4	3	3	4	3	1	3.11
1974	4	2	4	2	4	4	3	2	4	3.22
1975	1	1	2	3	2	2	2	1	2	1.78
1976	2	3	1	1	1	1	1	4	3	1.89

H_0 : average cancellation rate does not depend on the year.

H_a : average cancellation rates differ for at least two of the years.

Test statistic: F_r

Rejection region: With $\alpha = .05$ and $k - 1 = 3$, chi-squared critical value = 7.82. H_0 will be rejected at level of significance .05 if $F_r > 7.82$.

Computations: Using $(k + 1)/2 = 2.5$,

$$F_r = \frac{(12)(9)}{(4)(5)}[(3.11 - 2.5)^2 + (3.22 - 2.5)^2 + (1.78 - 2.5)^2 + (1.89 - 2.5)^2] = 9.62$$

Conclusion: Since $9.62 > 7.82$, H_0 is rejected in favor of H_a. Cancellation rate does seem to vary with year. ■

EXERCISES

13.51 The paper "The Effect of Social Class on Brand and Price Consciousness for Supermarket Products" (*J. Retailing* (1978):33–42) used the Kruskal-Wallis test to determine if social class (lower, middle, and upper) influenced the importance (scored on a scale of 1 to 7) attached to a brand name when purchasing paper towels. The reported value of the K-W statistic was .17. Use the seven-step hypothesis-testing procedure and a .05 significance level to test the null hypothesis of no difference in the true mean importance score for the three social classes.

13.52 Protoporphyrin levels were determined for three groups of people–a control group of normal workers, a group of alcoholics with sideroblasts in their bone marrow, and a group of alcoholics without sideroblasts. The given data appeared in the paper "Erythrocyte Coproporphyrin and Protoporphyrin in Ethanol-Induced Sideroblastic Erythroporiesis" (*Blood* (1974):291–95). Does the data suggest that normal workers and alcoholics with and without sideroblasts differ with respect to true mean protoporphyrin level? Use the K-W test with a .05 level of significance.

Group	Protoporphyrin Level (mg)										
Normal	22	27	47	30	38	78	28	58	72	56	30
	39	53	50	36							
Alcoholics with sideroblasts	78	172	286	82	453	513	174	915	84	153	780
Alcoholics without sideroblasts	37	28	38	45	47	29	34	20	68	12	
	37	8	76	148	11						

13.53 The given data on phosphorus concentration in topsoil for four different soil treatments appeared in the article "Fertilisers for Lotus and Clover Establishment on a Sequence of Acid Soils on the East Otago Uplands" (*N. Zeal. J. Exper. Ag.* (1984):119–29). Use the K-W test and a .01 significance level to test the null hypothesis of no difference in true mean phosphorus concentration for the four soil treatments.

Treatment	Concentration (mg/g)				
I	8.1	5.9	7.0	8.0	9.0
II	11.5	10.9	12.1	10.3	11.9
III	15.3	17.4	16.4	15.8	16.0
IV	23.0	33.0	28.4	24.6	27.7

13.54 The paper "Physiological Effects During Hypnotically Requested Emotions" (*Psychosomatic Med.* (1963):334–43) reported the following data on skin potential (mV) when the emotions of fear, happiness, depression, and calmness were requested from each of eight subjects.

	Subject (Block)							
Emotion	1	2	3	4	5	6	7	8
Fear	23.1	57.6	10.5	23.6	11.9	54.6	21.0	20.3
Happiness	22.7	53.2	9.7	19.6	13.8	47.1	13.6	23.6
Depression	22.5	53.7	10.8	21.1	13.7	39.2	13.7	16.3
Calmness	22.6	53.1	8.3	21.6	13.3	37.0	14.8	14.8

Does the data suggest that the true mean skin potential differs for the emotions tested? Use a significance level of .05.

13.55 In a test to determine if soil pretreated with small amounts of Basic-H makes the soil more permeable to water, soil samples were divided into blocks and each block received all four treatments under study. The treatments were (1) water with

.001% Basic-H on untreated soil, (2) water without Basic-H on untreated soil, (3) water with Basic-H on soil pretreated with Basic-H, and (4) water without Basic-H on soil pretreated with Basic-H. Using a significance level of .01, determine if mean permeability differs for the four treatments.

	Treatment					Treatment			
Block	1	2	3	4	Block	1	2	3	4
1	37.1	33.2	58.9	56.7	6	25.3	19.3	48.8	37.1
2	31.8	25.3	54.2	49.6	7	23.7	17.3	47.8	37.5
3	28.0	20.2	49.2	46.4	8	24.4	17.0	40.2	39.6
4	25.9	20.3	47.9	40.9	9	21.7	16.7	44.0	35.1
5	25.5	18.3	38.2	39.4	10	26.2	18.3	46.4	36.5

13.56 The data below on amount of food consumed (g) by eight rats after 0, 24, and 72 h of food deprivation appeared in the paper "The Relation Between Differences in Level of Food Deprivation and Dominance in Food Getting in the Rat" (*Psych. Sci.* (1972):297–98). Does the data indicate a difference in the true mean food consumption for the three experimental conditions? Use $\alpha = .01$.

	Rat							
Hours	1	2	3	4	5	6	7	8
0	3.5	3.7	1.6	2.5	2.8	2.0	5.9	2.5
24	5.9	8.1	8.1	8.6	8.1	5.9	9.5	7.9
72	13.9	12.6	8.1	6.8	14.3	4.2	14.5	7.9

13.57 The article "Effect of Storage Temperature on the Viability and Fertility of Bovine Sperm Diluted and Stored in Caprogen" (*N. Zeal. J. Ag. Res.* (1984): 173–77) examined the effect of temperature on sperm survival. Survival data for various storage times is given below. Use Friedman's test with a .05 significance level to determine if storage temperature affects survival (regard time as the blocking factor).

	Storage Time (h)				
Storage Temperature (C°)	6	24	48	120	168
15.6	61.9	59.6	57.0	58.8	53.7
21.1	62.5	60.0	57.4	59.3	54.9
26.7	60.7	55.5	54.5	53.3	45.3
32.2	59.9	48.6	42.6	36.6	24.8

KEY CONCEPTS

Hypotheses in single-factor ANOVA $H_0 : \mu_1 = \mu_2 = \cdots = \mu_k$ versus H_a : at least two μ's are different. (p. 560)

Mean squares in single-factor ANOVA **Mean square for treatments,** MSTr, is a measure of how different the k sample means $\bar{x}_1, \bar{x}_2, \ldots, \bar{x}_k$ are from one another (how much the $\bar{x}$'s deviate from the grand mean, $\bar{\bar{x}}$). MSTr is an unbiased statistic for estimating the common population variance σ^2 when H_0 is true, but tends to overestimate σ^2 when H_0 is false. **Mean square for error,** MSE, measures the amount of variability within the individual samples. It is an unbiased statistic for estimating σ^2 whether or not H_0 is true. (p. 561–62)

F test in single-factor ANOVA   The test statistic is $F = \text{MSTr}/\text{MSE}$. H_0 is rejected if $F > F$ critical value, where the critical value is based on $k - 1$ numerator df and $N - k$ denominator df (N is the total number of observations in the k samples). (p. 564)

Sums of squares in single-factor ANOVA   The three sums of squares SSTo, SSTr, and SSE (with associated df $N - 1$, $k - 1$, and $N - k$, respectively) are related by the fundamental identity SSTo = SSTr + SSE. Computational formulas for SSTo and SSTr can be used to obtain these sums of squares, and then SSE = SSTo − SSTr. Finally, $\text{MSTr} = \text{SSTr}/(k - 1)$ and $\text{MSE} = \text{SSE}/(N - k)$. (p. 571)

Multiple comparisons procedure   A procedure for identifying significant differences between the μ's once H_0 has been rejected by the F test. In the Bonferroni method, a confidence interval is computed for the difference between each pair of μ's ($\mu_1 - \mu_2$, $\mu_1 - \mu_3$, etc.). If an interval does not include zero, the corresponding μ's are judged to be significantly different. (p. 577)

Randomized block design   An experimental design that controls for extraneous variation when comparing treatments. The experimental units are grouped into homogeneous *blocks* so that within each block, the units are as similar as possible. Then each treatment is used on exactly one experimental unit in every block (each treatment appears once in every block). (p. 584)

Randomized block *F* test   The four sums of squares for a randomized block design, SSTo, SSTr, SSBl, and SSE (with df $kl - 1$, $k - 1$, $l - 1$, and $(k - 1)(l - 1)$, respectively), are related by SSTo = SSTr + SSBl + SSE. SSE is usually obtained by subtraction once the other three have been calculated using computing formulas. The null hypothesis is that the true average response does not depend on which treatment is applied. With $\text{MSTr} = \text{SSTr}/(k - 1)$ and $\text{MSE} = \text{SSE}/(k - 1)(l - 1)$, the test statistic is $F = \text{MSTr}/\text{MSE}$. H_0 is rejected if $F > F$ critical value (based on $k - 1$ numerator df and $(k - 1)(l - 1)$ denominator df) (p. 586).

Interaction between factors   Two factors are said to interact if the average change in response associated with changing the level of one factor depends on the level of the other factor. (p. 597)

Two-factor ANOVA   When there are k levels of factor A and l levels of factor B, and m (> 1) observations made for each combination of A–B levels, total sum of squares SSTo can be decomposed into SSA (sum of squares for A main effects), SSB, SSAB (interaction sum of squares) and SSE. Associated df are $klm - 1$, $k - 1$, $l - 1$, $(k - 1)(l - 1)$, and $kl(m - 1)$, respectively. The null hypothesis of no interaction between factors is tested using $F_{AB} = \text{MSAB}/\text{MSE}$, where $\text{MSAB} = \text{SSAB}/(k - 1)(l - 1)$ and $\text{MSE} = \text{SSE}/kl(m - 1)$. If this null hypothesis cannot be rejected, tests for A and B main effects are based on $F_A = \text{MSA}/\text{MSE}$ and $F_B = \text{MSB}/\text{MSE}$, respectively. (p. 600)

Distribution-free ANOVA   The Kruskal-Wallis test and Friedman's test are appropriate for completely a randomized design and a randomized block deşign, respectively. Both tests are based on ranks of the observations. Whereas the F tests require normal population or treatment distributions, these tests require only that the distributions have the same shape and spread. (p. 608–9)

13.58 Are some methods of cooking more economical than others? The paper "Cookery Methods for Vegetables: Influence on Sensory Quality, Nutrient Retention, and Energy Consumption" (*Home Ec. Res. J.* (1984):61–79) gave the energy usage (w-h) required to cook a potato for five different cooking methods. Data compatible with summary quantities given in the paper is given below.

Cooking Method	Energy Usage		
Microwave	201	199	219
Pressure cooker	394	375	381
Electric pressure cooker	359	405	419
Boiling	583	584	553
Baking	1281	1242	1248

a. Construct an ANOVA table. Using a .05 significance level, test to determine if the mean energy use is the same for all cooking methods.

b. Construct appropriate confidence intervals in order to determine which methods differ with respect to mean energy usage.

13.59 The results of a study on the effectiveness of line drying on the smoothness and stiffness of fabric was summarized in the paper "Line-Dried vs. Machine-Dried Fabrics: Comparison of Appearance, Hand, and Consumer Acceptance" (*Home Ec. Res. J.* (1984):27–35). Smoothness scores were given for nine different types of fabric and five different drying methods ((1) machine dry, (2) line dry, (3) line dry followed by 15-min tumble, (4) line dry with softener, and (5) line dry with air movement). Regarding the different types of fabric as blocks, construct an ANOVA table. Using a .05 significance level, test to see if there is a difference in the true mean smoothness scores for the drying methods.

	Drying Method				
Fabric	1	2	3	4	5
Crepe	3.3	2.5	2.8	2.5	1.9
Double knit	3.6	2.0	3.6	2.4	2.3
Twill	4.2	3.4	3.8	3.1	3.1
Twill mix	3.4	2.4	2.9	1.6	1.7
Terry	3.8	1.3	2.8	2.0	1.6
Broadcloth	2.2	1.5	2.7	1.5	1.9
Sheeting	3.5	2.1	2.8	2.1	2.2
Corduroy	3.6	1.3	2.8	1.7	1.8
Denim	2.6	1.4	2.4	1.3	1.6

13.60 A study to determine whether use of concentrated milk proteins affected cheese production was described in the paper "Membrane Processing of Milk on the Dairy Farm" (*Food Technology* (1984):88–90). Samples of size 5 were analyzed to determine the bacteria count for each of four concentration levels (1× concentrate, 1.16× concentrate, 1.48× concentrate, and 1.79× concentrate). The resulting data is given. Construct the appropriate ANOVA table. Does the data suggest that the mean bacteria count is not the same for the four concentrations? Test the relevant hypotheses using a .05 significance level.

Concentration	Bacteria Count				
1X	39	21	48	58	63
1.16X	24	19	34	28	50
1.48X	22	12	54	16	59
1.79X	29	13	36	30	63

13.61 Controlling a filling operation with multiple fillers requires adjustment of the individual units. Data resulting from a sample of size 5 from each pocket of a 12-pocket filler was given in the paper "Evaluating Variability of Filling Operations" (*Food Technology* (1984):51–55). Data for the first 5 pockets is given.

Pocket	Fill (oz)				
1	10.2	10.0	9.8	10.4	10.0
2	9.9	10.0	9.9	10.1	10.0
3	10.1	9.9	9.8	9.9	9.7
4	10.0	9.7	9.9	9.7	9.6
5	10.2	9.8	9.9	9.7	9.8

Use the ANOVA F test to determine if the null hypothesis of no difference in the mean fill weight of the 5 pockets can be rejected. If so, use an appropriate technique to determine where the differences lie.

13.62 *This problem requires the use of a computer package.* The effect of oxygen concentration on fermentation end products was examined in the article "Effects of Oxygen on Pyruvate FormateLyase in Situ and Sugar Metabolism of Streptococcus-mutans and *Streptococcus samguis*" (*Infection and Immunity* (1985):129–34). Four oxygen concentrations (0, 46, 92, and 138 μM) and two types of sugar (galactose and glucose) were used. Below are two observations on amount of ethanol (μmol/mg) for each sugar-oxygen concentration combination. Construct an ANOVA table and test the relevant hypotheses.

Oxygen Concentration	Galactose	Glucose
0	.59, .30	.25, .03
46	.44, .18	.13, .02
92	.22, .23	.07, .00
138	.12, .13	.00, .01

13.63 Eye inflammation can be induced by the endotoxin lipopolysaccharide (LPS). A random sample of 35 rats was randomly divided into five groups of seven rats each (a completely randomized design). Rats within each group received the same dose of LPS. The accompanying data on vascular permeability was read from plots that appeared in the paper "Endotoxin-Induced Uveitis in the Rat: Observations on Altered Vascular Permeability, Clinical Findings, and Histology" (*Exper. Eye Res.* (1984):665–76). Use analysis of variance with $\alpha = .05$ to test the null hypothesis of no difference between the five treatment means.

Treatment Dose (μg)	Vascular Permeability (ocular to serum fluorescence ratio)						
0	8	3	2	1	0	0	0
1	4.5	4	4	3.5	3	2	0
10	5	5	4	3.5	1	0	0
100	13	12	12	9	8	4	2
500	13	12	9	7.5	7	5	4

13.64 Four types of mortars—ordinary cement mortar (OCM), polymer impregnated mortar (PIM), resin mortar (RM), and polymer cement mortar (PCM)—were subjected to a compression test to measure strength. Three strength observations for each mortar type appeared in the paper "Polymer Mortar Composite Matrices for Maintenance-Free Highly Durable Ferrocement" (*J. Ferrocement* (1984):337–45) and are reproduced below. Construct an ANOVA table. Using a .05 significance level, determine whether the data suggests that the true mean strength is not the same for all four mortar types. If you determine that the true mean strengths are not all equal, use the Bonferroni method to identify the significant differences.

Mortar Type	Strength (MPa)		
OCM	32.15	35.53	34.20
PIM	126.32	126.80	134.79
RM	117.91	115.02	114.58
PCM	29.09	30.87	29.80

13.65 Referring to Exercise 13.64, use a distribution-free procedure with a .05 significance level to test the null hypothesis of no difference between the true mean strengths of the four mortars.

13.66 The paper "Clothing Symbolism and the Changing Role of Nurses" (*Home Ec. Res. J.* (1980):294–301) reported on a comparison of five groups of nurses (those who wore their caps (1) all the time, (2) 75% of the time, (3) 50% of the time, (4) 25% of the time, or (5) none of the time) with respect to attitudes toward role symbols. Attitude was measured using the Role Symbol Scale (RSS). A partially completed ANOVA table is given.

Source	df	Sum of Squares	Mean Square	F
Treatments	4	62,222		
Error				
Total	299	4,772,243		

a. Complete the ANOVA table and use it to test the null hypothesis of no difference in true mean RSS scores for the five groups.

b. The paper states "Further analysis . . . indicated that nurses who wore their caps all of the time or 75% of the time had a significantly more favorable attitude (lower mean RSS scores) than nurses who never wore their caps or wore their caps only 25% of the time." Groups means and standard deviations are given. Assuming five equal sample sizes of 60 each, does the application of the Bonferroni method yield conclusions which agree with the statement in the paper?

Group	Sample Mean	Sample Standard Deviation
1	82.17	20.00
2	85.19	19.20
3	98.09	11.71
4	102.26	20.10
5	117.05	20.96

13.67 The effect of nitrogen application on pasture yield was examined in the paper "Response of Pastures to Nitrogen Fertiliser Applied in Autumn and Spring on the West Coast of South Iceland" (*N. Zeal. J. Exper. Ag.* (1983):247–50). Four levels of nitrogen application were used (0, 25, 50, and 100 kg/ha) and the dry-matter production (kg/ha) was recorded on three different cutting dates. Viewing the three cutting dates as blocks, analyze the data below to determine if the mean dry matter production is the same for the four nitrogen levels. Construct an ANOVA table and test the appropriate hypotheses using a .05 level of significance.

	Cut Date		
Nitrogen	I	II	III
0	210	1030	1680
25	390	1130	1570
50	540	1410	1560
100	770	1840	1530

13.68 The accompanying ANOVA table appeared in the paper "Bacteriological and Chemical Variations and Their Inter-Relationships in a Slightly Polluted Water-Body" (*Inter. J. Environ. Studies* (1984):121–29). A water specimen was taken every month for a year at each of 15 designated locations on the Lago di Piediluco in Italy. The ammonia-nitrogen concentration was determined for each specimen and the resulting data analyzed using a two-way ANOVA. The researchers were willing to assume that there was no interaction between the two factors *location* and *month*. Complete the given ANOVA table, and use it to perform the tests required to determine if the true mean concentration differs by location or by month of year. Use a .05 significance level for both tests.

Source	df	Sum of Squares	Mean Square	F
Location	14	.6		
Month	11	2.3		
Error	154			
Total	179	6.4		

13.69 The paper "The Effect of Sewage with Special Reference to Aquatic Insects in the River Kshipra (India)" (*Inter. J. Environ. Studies* (1984):191–208) describes a randomized block experiment designed to evaluate chemical conditions at three sampling stations along the river Kshipra. Water specimens at each station were taken each month for 12 months and chemical oxygen demand was determined. Data for four of the months appears below. Viewing the months as blocks, use a distribution-free procedure to test the null hypothesis of no difference in true mean oxygen demand for the three stations.

	Station		
Month	A	B	C
January	78.9	56.0	113.6
February	51.2	86.4	208.0
March	62.4	73.6	108.8
April	41.6	126.4	133.6

13.70 The concentration of cadmium (Cd) in sediments from lakes around a coal-powered electric plant in Texas appeared in the article "The Analysis of Aqueous Sediments for Heavy Metals" (*J. Environ. Sci. and Health* (1984):911–24). At each lake, one measurement was taken at each of five different depths. Does the mean concentration of Cd vary with depth? Viewing the four lakes as blocks, construct an ANOVA table and test the appropriate hypotheses using a .01 significance level.

	Depth				
Lake	1	2	3	4	5
Oyster Creek (O)	1.0	0.5	1.0	1.0	2.0
Herman (He)	1.0	3.0	0.5	1.0	1.0
Gannoway (G)	3.0	3.0	2.0	1.0	2.0
Horseshoe (Ho)	2.0	1.0	2.0	1.0	2.0

13.71 The water absorption of two types of mortar used to repair damaged cement was discussed in the paper "Polymer Mortar Composite Matrices for Maintenance-Free, Highly Durable Ferrocement" (*J. Ferrocement* (1984):337–45). Specimens of ordinary cement mortar (OCM) and polymer cement mortar (PCM) were submerged for varying lengths of time (5, 9, 24, or 48 h) and water absorption (% by weight) was recorded. With mortar type as factor A (with two levels) and submersion period as factor B (with four levels), three observations were made for each factor-level combination. Data included in the paper was used to compute the given sums of squares. Use this information to construct an ANOVA table, and then use a .05 significance level to test the appropriate hypotheses.

Sum of squares for factor A: 322.667
Sum of squares for factor B: 35.623
Sum of squares for AB interaction: 8.557
Total sum of squares: 372.113

13.72 Suppose that each observation in a single-factor ANOVA data set is multiplied by a constant c (a change in units; e.g., $c = 2.54$ changes observations from inches to centimeters). How does this affect MSTr, MSE, and the test statistic F? Is this reasonable? Explain.

13.73 Is it true that the grand mean $\bar{\bar{x}}$ the ordinary average of $\bar{x}_1, \bar{x}_2, \ldots, \bar{x}_k$, i.e., that

$$\bar{\bar{x}} = \frac{\bar{x}_1 + \cdots + \bar{x}_k}{k}$$

Under what conditions on $n_1, n_2, \ldots, n_k$ will the above relationship be true?

REFERENCES

Hicks, Charles. *Fundamental Concepts in the Design of Experiments*. New York: Holt, Rinehart and Winston, 1982. (Discusses the analysis of data arising from many different types of designed experiments; the focus is more on methods than concepts.)

Neter, John, William Wasserman, and Michael Kutner. *Applied Linear Statistical Models*. Homewood, Ill.: Richard D. Irwin, 1985. (The latter half of the book contains a readable survey of ANOVA and experimental design.)

Ott, Lyman. *An Introduction to Statistical Methods and Data Analysis*. Boston: Duxbury Press, 1984. (A good source for learning more about the methods most frequently employed to analyze experimental data, including various multiple comparison techniques.)

14 The Analysis of Categorical Data and Goodness-of-Fit Methods

INTRODUCTION

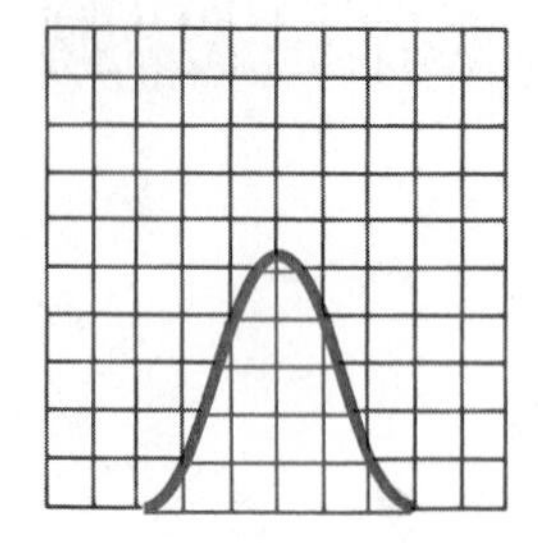

MOST OF THE TECHNIQUES presented in earlier chapters are designed for numerical data. Different methods must be employed when information has been collected on categorical variables (variables whose possible values are not numerical in nature). As with numerical data, categorical data sets can be univariate (consisting of observations on a single categorical variable), bivariate (observations on two categorical variables), or even multivariate. In Section 14.1, we examine some descriptive techniques, and then turn in the remaining sections to inferential methods for univariate and bivariate categorical data sets.

14.1 Summarizing Categorical Data Using One- and Two-Way Frequency Tables

Surveys and research studies often involve making observations on one or more categorical variables. In the case of a single categorical variable, sample data is most frequently summarized in a **one-way frequency table.** Suppose, for example, that a grass seed mixture is composed of rye, blue grass, and bermuda grass seed. A random sample of 1000 seeds is selected and the value of the variable *seed type* is recorded for each one. The first few observed values might be *r, r, bl, be, r, bl,* . . . . Counting the number of seeds of each type (i.e., the number of times each value of the variable appears) might then result in the accompanying one-way table.

	Seed Type		
	Rye	Blue Grass	Bermuda Grass
Frequency	341	298	361

For a categorical variable with K possible values (K different categories), sample data is summarized in a one-way frequency table consisting of K cells displayed horizontally or vertically (this is really nothing more than a frequency distribution as discussed in Chapter 2).

When observations are made on two different categorical variables, the data can also be summarized using a tabular format. As an example, suppose that residents of a particular city can watch national news on ABC, CBS, NBC, or PBS (the public television network) affiliate stations. A researcher wishes to know whether there is any relationship between political philosophy (liberal, moderate, or conservative) and preferred news program among those residents who regularly watch the national news. Let x denote the variable *political philosophy* and y denote the variable *preferred network*. A random sample of 300 regular watchers is to be selected, and each one will be asked for his or her x and y values. The data set is bivariate and might initially be displayed as follows:

Observation	x Value	y Value
1	Liberal	CBS
2	Conservative	ABC
3	Conservative	PBS
⋮	⋮	⋮
299	Moderate	NBC
300	Liberal	PBS

Bivariate categorical data of this sort can most easily be summarized by constructing a **two-way frequency table,** or **contingency table.** This is a rectangular table that consists of a row for each possible value of x (each category specified by this variable) and a column for each possible value of y. There is then a cell in the table for each possible (x, y) combination. Once such a table has been constructed, the number of times each particular (x, y) combination occurs in the data set is determined and these numbers are entered in the corresponding cells of the table. The resulting numbers are called the **observed joint frequencies,** or **observed cell counts.** The table for the *political philosophy–preferred news program* example discussed earlier contains 3 rows and 4 columns (because x and y have 3 and 4 possible values, respectively). Table 1 is one possible table, with the rows and columns labeled with the possible x and y values. These are often referred to as the *row* and *column categories.*

TABLE 1 AN EXAMPLE OF A 3 × 4 FREQUENCY TABLE

	ABC	CBS	NBC	PBS	Row Marginal Total
Liberal	20	20	25	15	80
Moderate	45	35	50	20	150
Conservative	15	40	10	5	70
Column Marginal Total	80	95	85	40	300

Marginal totals are obtained by adding the observed cell counts in each row and also in each column of the table. The row and column marginal totals, along with the total of all observed cell counts in the table—the grand total—have been included in Table 1. The marginal totals provide information on the distribution of observed values for each variable separately. In this example, the row marginal totals reveal that the sample consisted of 80 liberals, 150 moderates, and 70 conservatives. Similarly, column marginal totals indicated how often each of the preferred program categories occurred—80 preferred ABC news, 95 preferred CBS, and so on. The **grand total,** 300, is the number of observations in the bivariate data set, in this case the sample size (though occasionally such a table results from a census of the entire population).

Two-way frequency tables are often characterized by the number of rows and columns in the table (specified in that order—rows first, then columns). Table 1 is called a 3 × 4 table. The smallest two-way frequency table is a 2 × 2 table, which has only two rows and two columns and thus four cells.

Bivariate categorical data arises naturally in two different types of investigations. A researcher may be interested in comparing two or more groups on the basis of a categorical variable and so may obtain a sample separately from each group. For example, data could be collected at a university in order to compare students, faculty, and staff on the basis of primary mode of transportation to campus (car, bicycle, motorcycle, bus, or by foot). One random sample of 200 students, another of 100 faculty, and a third of 150 staff might be chosen, and the selected individuals could be interviewed in order to obtain the needed transportation information. Data from such a study could easily be summarized in a 3×5 two-way frequency table with row categories of student, faculty, and staff and column categories corresponding to the five possible modes of transportation. The observed cell counts could then be used to gain insight into differences and similarities between the three groups with respect to the means of transportation used.

Bivariate data also arises when the values of two different variables are observed for all individuals or items in a single sample. For example, a sample of 500 registered voters might be selected. Each voter could then be asked both if he or she favored a particular property tax initiative and if he or she was a registered Democrat, Republican, or Independent. This would result in a bivariate data set with x representing *political affiliation* (with categories Democrat, Republican, and Independent) and y representing *response* (favor initiative or oppose initiative). The corresponding 3×2 frequency table could then be used to investigate any relationship between position on the tax initiative and political affiliation.

Relative Frequencies

Insight into both types of investigations described above—comparing two or more groups on the basis of a categorical variable or describing the relationship between two categorical variables in a single group—can be gained by working with certain types of relative frequencies. Let's begin by considering an example where a comparison of two groups is desired.

EXAMPLE 1

The Associated Press (March 25, 1984) reported on a British study of the effects of smoking among teenagers. Of 54 smokers (students who had been regular smokers for 2 years), 27 reported persistent coughing. Of 54 nonsmokers, only 13 reported coughing. The data given is a concise description of a bivariate data set consisting of 108 pairs of observations where x = *group* (smoker or nonsmoker) and y = *coughing status* (cough or no cough). The accompanying 2×2 frequency table summarizes this data.

	Cough	No Cough	Row Marginal Total
Smoker	27	27	54
Nonsmoker	13	41	54
Column Marginal Total	40	68	108

Inspection of this table reveals that the number of smokers who reported coughing is much higher than the number of nonsmokers who experience persistent coughing, while the number with no cough is substantially lower for the nonsmoker group. Suppose that the 54 smokers and 54 nonsmokers in the study represent random samples from the population of all British teenage smokers and the population of all British teenage nonsmokers, respectively. Due to sampling variation, we shouldn't necessarily expect the two observed cell counts in the column labeled *cough* to be exactly equal even when there is no difference between the population of smokers and the population of nonsmokers with respect to coughing status. However, if there is no difference between the populations, we would expect the cell counts to be similar. The large differences seen here suggest that the two groups do differ with respect to coughing status.

In this example, the group comparison is simple because the two sample sizes (numbers of smokers and nonsmokers) are identical. Comparing observed cell counts can be misleading when the sample sizes are unequal. Suppose, for example, that 50 smokers and 200 nonsmokers had participated in the study, resulting in the accompanying table.

	Cough	No Cough	Row Marginal Total
Smoker	25	25	50
Nonsmoker	48	152	200
Column Marginal Total	73	177	250

The fact that the observed count for the nonsmoker-cough cell is higher than the observed count for smokers doesn't necessarily indicate that nonsmokers are more likely than smokers to report coughing—it is simply a reflection of the larger sample size for nonsmokers. A reasonable comparison between smokers and nonsmokers necessitates taking the different sample sizes into account. An easy way of doing this is to calculate the proportion of smokers in each of the column categories. These proportions could then be compared to the corresponding proportions for nonsmokers. That is, for each row we can compute relative frequencies by dividing each observed cell count in the row by the appropriate row marginal total. Relative frequencies computed in this way are usually called **row proportions.** The row proportions can be displayed in a two-way table similar to the one used for observed cell counts, as is done below.

Row Proportions

	Cough	No Cough	Row Marginal Total
Smoker	.50	.50	1.00
Nonsmoker	.24	.76	1.00

Note that since the entries in the above table are simply relative frequencies obtained via division by row totals, the row proportions in any row will add to 1. Examination of this table reveals the same differences between smokers and nonsmokers as were observed in the original table based on equal sample sizes. ■

Using row proportions to facilitate making comparisons between groups is appropriate when the groups are used as the row categories in the two-way frequency table. If the groups had instead been used to identify columns, column proportions would be used. **Column proportions** result from dividing each observed cell count in the table by the corresponding column marginal total. The relative frequencies in each column of the table then add to one.

Several types of relative frequencies are also commonly used to summarize bivariate categorical data that results from the measurement of two categorical variables in a single sample. Their use is illustrated in the following example.

EXAMPLE 2

Each individual in a random sample of 325 people participating in a drug program was categorized both with respect to the presence or absence of hypoglycemia and with respect to mean daily dosage of insulin. The given two-way frequency table appeared in the paper "Relation of Body Weight and Insulin Dose to the Frequency of Hypoglycemia" (*J. Amer. Med. Assoc.* (1974):192–94). The variable *dose* is numerical but has been transformed to a categorical variable by grouping doses into five different categories.

		Daily Insulin Dose					Row Marginal Total
		$<.25$	.25–.49	.50–.74	.75–.99	≥ 1.0	
Hypoglycemia	Present	4	21	28	15	12	80
	Absent	40	74	59	26	46	245
Column Marginal Total		44	95	87	41	58	325

An alternate way of presenting this data would be to report the proportion of the 325 observations that fall into each cell. This entails dividing each observed cell count by the total sample size (the grand total). The result is the accompanying set of **joint relative frequencies.**

	Joint Relative Frequencies					Row Marginal Total
	$<.25$	.25–.49	.50–.74	.75–.99	≥ 1.0	
Present	.0123	.0646	.0862	.0462	.0369	.2462
Absent	.1231	.2277	.1815	.0800	.1415	.7538
Column Marginal Total	.1354	.2923	.2677	.1262	.1784	1.0000

From this table, we see that 24.62% of the sample had hypoglycemia, whereas 75.38% did not. The column marginals can be interpreted in a similar fashion: 13.54% of the sample were receiving doses of less than .25, 29.23% were taking insulin doses of between .25 and .49, and so on. The individual cell relative frequencies can also be easily interpreted: The sample was comprised of 1.23% hypoglycemics who were receiving an insulin dose of less than .25, etc. The joint relative frequencies have an advantage over the observed cell counts in that their interpretation is not dependent on sample size.

In addition to joint relative frequencies, it is sometimes insightful to compute row proportions or column proportions. Here the row proportions (displayed in the following table) in the first and second rows give the proportion of hypoglycemics and nonhypoglycemics, respectively, falling into each dosage category.

	Row Proportions					Row Marginal Total
	<.25	.25–.49	.50–.74	.75–.99	>1.0	
Present	.0500	.2625	.3500	.1875	.1500	1.0000
Absent	.1633	.3020	.2408	.1061	.1878	1.0000

If computed for each dosage category, column proportions would provide the proportion with hypoglycemia present and the proportion with hypoglycemia absent. The goal of any particular study will often suggest which one of the three types of proportions discussed—joint relative frequencies, row proportions, or column proportions—is most informative.

The CROSSTABS (short for cross-tabulation, another name for a two-way frequency table) program in the SPSS computer package computes this type of information for a two-way table. SPSS output for this data set is given. SPSS computes row and column percents rather than proportions, and each cell in the displayed table contains four values—an observed cell count, a row percent, a column percent, and a total percent (corresponding to what we have called the joint relative frequency).

```
    H          HYPOGLYCEMIA
BY  D          INSULIN DOSE

             D
   COUNT
   ROW PCT   LT.25      .25-.49  .50-.74  .75-.99  GT1.00     ROW
   COL PCT                                                    TOTAL
   TOT PCT        1.        2.       3.       4.       5.
H
        1.        4        21       28       15       12       80
PRESENT         5.0      26.2     35.0     18.8     15.0     24.6
                9.1      22.1     32.2     36.6     20.7
                1.2       6.5      8.6      4.6      3.7

        2.       40        74       59       26       46      245
ABSENT         16.3      30.2     24.1     10.6     18.8     75.4
               90.9      77.9     67.8     63.4     79.3
               12.3      22.8     18.2      8.0     14.2

     COLUMN      44        95       87       41       58      325
      TOTAL    13.5      29.2     26.8     12.6     17.8    100.0
```

■

Comparing Two or More Groups Using Expected Cell Counts

We have seen that relative frequencies can be used to compare two or more groups on the basis of a categorical variable. When data is summarized in the form of a two-way frequency table with groups as row categories, this is accomplished by computing and comparing row proportions. An alternate method for assessing whether differences between groups exist is often used. This second approach poses this question: What cell counts would have been expected if in fact there were no difference between the groups? If we can answer this question, a comparison can be made between the actual observed cell counts and what would have been expected when there is no difference between groups. Large discrepancies between observed cell counts and what was expected would indicate a difference between groups, whereas small differences would support the theory of no difference between groups. The cell counts that would be expected when no difference between group exists are called the **expected cell counts.**

EXAMPLE 3

Until recently, a number of professions were prohibited from advertising. In 1977, the U.S. Supreme Court ruled that prohibiting doctors and lawyers from advertising violated their right to free speech. The paper "Should Dentists Advertise?" (*J. Adver. Res.* (June 1983):33–38) compared the attitudes of consumers and dentists toward advertising of dental services. Separate samples of 101 consumers and 124 dentists were asked to respond to the following statement: I favor the use of advertising by dentists to attract new patients. Possible responses were strongly agree, agree, neutral, disagree, and strongly disagree. The data presented in the paper appears in the accompanying 2×5 frequency table. The authors were interested in determining whether the two groups—consumers and dentists—differed in their attitudes toward advertising.

	Response					
Group	Strongly Agree	Agree	Neutral	Disagree	Strongly Disagree	Row Marginal Total
Consumers	34	49	9	4	5	101
Dentists	9	18	23	28	46	124
Column Marginal Total	43	67	32	32	51	225
Column % of Total	19.11%	29.78%	14.22%	14.22%	22.67%	

Expected cell counts can be reasoned out in the following manner: There were a total of 225 responses, of which 43 were strongly agree. The proportion of the total responding strongly agree is then $43/225 = .1911$, or 19.11%. If there were no difference in response for consumers and dentists, we would then expect about 19.11% of the consumers and 19.11% of the dentists to have strongly agreed. Therefore, the expected cell count for the consumer–strongly agree cell is $.1911(101) = 19.30$. The expected cell count for the dentist–strongly agree cell is $.1911(124) = 23.70$. Note that the expected cell counts need not be whole numbers. The expected cell counts for the remaining cells can be computed in a similar manner. For example, $67/225 = .2978$, or 29.78%, of all responses were in the agree category, so the expected cell counts for the consumer–agree and the dentist–agree cells are $.2978(101) = 30.08$ and $.2978(124) = 36.93$, respectively.

It is common practice to display the observed cell counts and the corresponding expected cell counts in the same table, with the expected cell counts enclosed in parentheses. Expected cell counts for the remaining six cells have been computed and entered into the table on the next page. Except for small differences due to rounding, each marginal total for expected cell counts is identical to that for the corresponding observed counts.

Observed and Expected Cell Counts

Group	Strongly Agree	Agree	Strongly Neutral	Disagree	Disagree	Row Marginal Total
Consumers	34 (19.30)	49 (30.08)	9 (14.36)	4 (14.36)	5 (22.89)	101
Dentists	9 (23.70)	18 (36.92)	23 (17.64)	28 (17.64)	46 (28.11)	124
Column Marginal Total	43	67	32	32	51	225

A quick comparison of the observed and expected cell counts reveals large discrepancies. What was observed differs greatly from what would be expected for the case of no difference between groups. Consumers and dentists seem to have very different attitudes about dental advertising. It appears that most consumers are more favorable toward advertising since the observed counts in the consumer–strongly agree and consumer–agree cells are substantially higher than expected, whereas the observed cell counts for consumers in the disagree and strongly disagree cells are lower than expected (when no difference exists). The opposite relationship between observed and expected counts is exhibited by the dentists. ■

In Example 3, the expected count for a cell corresponding to a particular group-response combination was computed in two steps. First the response marginal proportion was computed (e.g., $\frac{43}{225}$ for the strongly agree response). This was then multiplied by a marginal group total, e.g., $101(\frac{43}{225})$, for the consumer group. This is equivalent to first multiplying the row and column marginal totals and then dividing by the grand total: $(101)(43)/225$.

To compare two or more groups on the basis of a categorical variable, calculate an **expected cell count** for each cell by selecting the corresponding row and column marginal totals and then computing

$$\text{expected cell count} = \frac{\text{(column marginal total) (row marginal total)}}{\text{grand total}}$$

These quantities represent what would be expected when there is no difference between the groups under study.

EXAMPLE 4

As part of the study reported in the paper "Should Dentists Advertise?" (see Example 3), consumers and dentists were asked to respond to the statement "Advertising by dentists will permit patients to make intelligent choices." Possible responses were the same as in Example 3; the resulting data is given in the accompanying two-way frequency table.

	Strongly Agree	Agree	Neutral	Disagree	Strongly Disagree	Row Marginal Total
Consumers	30	39	13	14	5	101
Dentists	0	6	10	23	85	124
Column Marginal Total	30	45	23	37	90	225

All the most widely used statistical computer packages will compute expected cell counts after the observed counts have been entered. MINITAB output for this data set is given. Comparison of observed and expected cell counts again reveals substantial differences between the responses of consumers and dentists. Consumers tend to feel that advertising provides patients with information that will enable them to make better decisions, whereas dentists tend to disagree with this statement.

```
EXPECTED FREQUENCIES ARE PRINTED BELOW OBSERVED FREQUENCIES
       I  stagrI  agreeI  neut I  disagI  stdisITOTALS
-------I-------I-------I-------I-------I-------I-------
    1  I   30  I   39  I   13  I   14  I    5  I   101
       I   13.5I   20.2I   10.3I   16.6I   40.4I
-------I-------I-------I-------I-------I-------I-------
    2  I    0  I    6  I   10  I   23  I   85  I   124
       I   16.5I   24.8I   12.7I   20.4I   49.6I
-------I-------I-------I-------I-------I-------I-------
TOTALS I     30 I     45 I     23 I     37 I     90 I   225
```

■

Even when groups are identical with respect to the responses under study, small discrepancies between observed and expected counts can easily arise because of sampling variability (just as throwing a fair die 60 times might yield observations differing somewhat from the expected 10 ones, 10 twos, etc.). Only substantial discrepancies point to the conclusion that the groups differ. A formal procedure for deciding whether the conclusion of group differences is justified is based on the hypothesis-testing methods developed in Section 14.3.

Association in a Two-Way Frequency Table

When a two-way frequency table results from classifying individuals or objects according to two categorical variables, interest often centers on describing the way in which the two variables are related. For example, the personnel office of a large company might wish to determine the extent of any relationship between *sex* (male, female) and *promotion status* (promoted within the last 3 years or not promoted within that period) among the company's employees. If a random sample of employees is selected, the data can be summarized in a 2×2 table with rows corresponding to sex and columns corresponding to promotion status. Since a 2×2 table is the simplest form of a two-way frequency table, we first focus attention on such tables.

Most analyses of data in a 2×2 frequency table attempt to do two things: determine if a relationship or association exists between the two variables used to define the table and, if so, describe numerically the extent of the association. In this respect, an analysis of a two-way frequency table parallels a correlation analysis for two numerical variables. If there is no relationship between two categorical variables, they are said to be **independent.** Independence implies that knowing the value (category) of one variable gives no information about the value of the other variable. Independence of sex and promotion status (as described earlier) would mean that knowledge of an individual's sex provides no information on promotion status. Suppose that half the employees of this company have been promoted within the last 3 years. If sex and promotion status are independent, we would expect about half of the male employees to have been promoted and about half of the female employees to have been promoted. On the other hand, suppose that half of all employees have been promoted but that sex and promotion status are related, with men more likely to be promoted. Then more than half the men and fewer than half the women would have received promotions. In this case, knowledge of an individual's sex provides information on promotion status. Variables are said to be *associated*, or *dependent*, when knowledge of one variable's value (category) provides relevant information about the value of the second variable.

A common way of assessing whether an association exists between two variables uses the previously introduced idea of expected cell counts. In determining association, expected cell counts specify what the cell counts are expected to be when there is no association between the two variables. Large discrepancies between the observed and expected cell counts then indicate the presence of association.

To see how expected cell counts are computed when no association is assumed, consider the accompanying 2 × 2 table for the *sex–promotion status* example with observed marginal totals.

	Promoted	Not Promoted	Row Marginal Total
Male			60
Female			40
Column Marginal Total	20	80	100

Then $\frac{20}{100}$, or 20%, of the sampled employees were promoted. With no association between sex and promotion status, we would expect that 20% of the sampled males, or $(.20)(60) = 12$, would have been promoted. That is, the expected count for the male-promoted cell is $(20)(60)/100$. This is the product of the two marginal totals divided by the grand total, exactly as in the group-comparison situation considered earlier.

To assess association between two categorical variables, calculate an **expected cell count** for each cell by using the corresponding row and column marginal totals in the formula

$$\text{expected cell count} = \frac{(\text{column marginal total})(\text{row marginal total})}{\text{grand total}}$$

These expected cell counts represent what we would expect to see when there is *no association* between the variables under study.

EXAMPLE 5

The paper "Impulsive and Premeditated Homicide: An Analysis of the Subsequent Parole Risk of the Murderer" (*J. Criminal Law and Criminology* (1978):108–14) investigated the relationship between the circumstances surrounding a murder and the subsequent parole success of the murderer. A sample of 82 murderers was selected, and each murderer was categorized according to the type of murder (impulsive, premeditated) and parole outcome (success, failure). The authors used the given data to conclude that there was an association between type of murder and parole outcome. Does a comparison of observed and expected cell counts support the authors' conclusions? The expected cell counts for the four cells are:

Cell	Expected Cell Count
Impulsive-success	$\frac{(35)(42)}{82} = 17.93$
Impulsive-failure	$\frac{(47)(42)}{82} = 24.07$
Premeditated-success	$\frac{(35)(40)}{82} = 17.07$
Premeditated-failure	$\frac{(47)(40)}{82} = 22.93$

The expected cell counts appear in parentheses in the accompanying table.

	Success	Failure	Row Marginal Total
Impulsive	13 (17.93)	29 (24.07)	42
Premeditated	22 (17.07)	18 (22.93)	40
Column Marginal Total	35	47	82

The discrepancies between the observed cell counts and those expected when there is no association tend to support the conclusion that there is an association between type of murder and parole outcome. The analysis presented in the paper actually used techniques that are discussed in Section 14.3 to decide that the discrepancies between observed and expected cell counts were too large to be attributed solely to sampling variation. ■

It is sometimes useful to describe quantitatively the degree, or extent, of association between two categorical variables. The correlation coefficient r provided such a measure (of linear association) for numerical variables. A number of different measures of association have been proposed for categorical variables. The most popular measure for 2×2 tables is easily obtained from the four observed cell counts. Let a, b, c, and d represent the observed counts as displayed in the accompanying table:

			Row Marginal Total
	a	b	$a + b$
	c	d	$c + d$
Column Marginal Total	$a + c$	$b + d$	$a + b + c + d$

DEFINITION

The **index of association i***, which measures the extent of association in a 2×2 table, is defined by

$$i = \frac{ad - bc}{\sqrt{(a + b)(c + d)(a + c)(b + d)}}$$

Motivation for i comes from examining the expected cell counts when the variables are independent. In this case, recall that expected cell count = (row total)(column total)/(grand total). The strongest evidence for independence is an observed table in which every observed count is exactly equal to the corresponding expected count. Furthermore, if this condition holds for any one cell, it will also hold automatically for the other three cells (because the row and column totals for expected counts must equal the corresponding totals for observed counts). Focusing on the a cell, independence is most plausible when

$$a = \frac{(a + b)(a + c)}{(a + b + c + d)}$$

Straightforward manipulation gives the equivalent condition $ad = bc$. Thus the data argues most strongly for independence when $ad = bc$, i.e., when $ad - bc = 0$ and $i = 0$. The greater the difference between observed and expected counts, the more $ad - bc$, the numerator of i, will differ from zero. The denominator of i is introduced to yield a measure whose value always lies between -1 and $+1$.

Interpretation of i is similar to that of the correlation coefficient, with $i = 0$ corresponding to no association. The value $i = +1$ is obtained only when $b = c = 0$. This happens when the first category of one variable is always paired with the first category of the other variable and the second category of one is always paired with the second category of the other. A value near $+1$ indicates a strong positive association between the variables, where positive association is defined as a tendency for data to fall in the a and d

*This index is usually called the phi coefficient and is denoted by the Greek letter ϕ. We have used the letter i here in order to reserve Greek letters for population characteristics.

cells. A negative association, as suggested by a negative value of i, indicates that the first category of one variable tends to be paired with the second category of the other (i.e., most observations fall in the b and c cells).

EXAMPLE 6

For the *type of murder–parole outcome* data given in Example 5, $a = 13$, $b = 29$, $c = 22$, and $d = 18$, yielding

$$i = \frac{(13)(18) - (29)(22)}{\sqrt{(13 + 29)(22 + 18)(13 + 22)(29 + 18)}}$$

$$= \frac{234 - 638}{\sqrt{(42)(40)(35)(47)}} = -.243$$

The fact that i is somewhat negative indicates that there is a mild tendency for impulsive murderers to be parole failures and premeditated murderers to be parole successes. ■

In Example 6, there is no natural ordering of either the row or column categories. The data could just as easily have been summarized in a 2×2 table with the order of either the row or the column categories reversed. The resulting table would yield $i = .243$. Changing the order of both row and column categories would give $i = -.243$. In general, rearranging the row or column categories of a 2×2 table may change the sign but not the magnitude of i. The sign of i is important when there is a natural ordering of categories (e.g., when each variable has a low and a high category), since then $i = .243$ and $i = -.243$ have rather different interpretations.

When studying association of two categorical variables, expected cell counts can also be used as an aid to interpreting two-way frequency tables with more than two row or column categories. Expected cell counts are obtained using exactly the same formula as for 2×2 tables, i.e., by multiplying row and column totals and then dividing by the grand total.

EXAMPLE 7

Many research studies have demonstrated an association between smoking habits and coronary heart disease (CHD). One such study is described in the paper "Women Smokers and Sudden Death: The Relationship of Cigarette Smoking to Coronary Disease" (*J. Amer. Med. Assoc.* (1973):1005–10). Autopsy records were examined for 182 cases of sudden and unexpected death. Each victim was classified according to smoking habits (nonsmoker, less than 20 cigarettes per day, 20 or more cigarettes per day) and cause of death (not CHD, CHD). MINITAB was used to compute expected cell counts. Both observed and expected counts appear in the accompanying table.

```
EXPECTED FREQUENCIES ARE PRINTED BELOW OBSERVED FREQUENCIES
           I  NOTCHD I  CHD I  TOTALS
----------I---------I------I--------
NONSMOKER I      81  I   3  I     84
          I    70.6I  13.4I
----------I---------I------I--------
LESS      I      29  I   8  I     37
THAN 20   I    31.1I   5.9I
----------I---------I------I--------
MORE      I      43  I  18  I     61
THAN 20   I    51.3I   9.7I
----------I---------I------I--------
TOTALS    I     153 I  29 I    182
```

The investigator felt that the discrepancies between the observed and expected cell counts were indicative of an association between smoking and cause of death. ■

It is rather difficult to extend the index of association to obtain a quantitative measure of association for tables larger than 2 × 2. We do not attempt to do so in this book.

EXERCISES

14.1 Career, family, and leisure are three important factors affecting quality of life. To determine whether changes have occurred in the importance attached to family life, a study performed in 1970 was replicated in 1980. In each study, samples of male college students were asked to rank career, family, and leisure in order of importance. The numbers ranking family first, second, and third were noted to obtain the given table (this data is compatible with that given in "University Students: A Change in Expectations and Aspirations Over the Decade" *Sociology of Educ.* (1982):223).

	Importance of Family		
Year	Ranked First	Ranked Second	Ranked Third
1970	22	12	16
1980	42	28	30

Since the 1970 and 1980 sample sizes are different, compute row proportions. Based on these proportions, does it look as though there has been a change in attitudes from 1970 to 1980? Explain.

14.2 The paper "The Information Content of Comparative Magazine Advertisements" (*J. Adver.* (1983):10–16) examined the types of ads appearing in different magazines. Ads from four magazines were classified by type to obtain the accompanying two-way table.

	Magazine			
Ad Type	*Ladies Home Journal*	*Newsweek*	*Esquire*	*Reader's Digest*
Strict comparison	38	42	15	38
Implied comparison	197	169	123	153
Noncomparative	520	330	216	554

a. If you wanted to compare the four magazines on the basis of the proportion of each type of ad, would you use row or column proportions? Why?

b. Do you think that the magazines differ with respect to the proportion of each type of ad? Explain.

14.3 Do people tend to marry those of the same religious faith? The accompanying two-way table is the result of classifying married couples according to the husband's religion and wife's religion. This table appeared in the paper "Religious Homogamy and Marital Satisfaction Reconsidered" (*J. Marriage and the Family* (1984):

729–33). The following codes are used: (C) Catholic, (B) Baptist, (M) Methodist, (L) Lutheran, (O) other, and (N) none.

		Wife					
		C	B	M	L	O	N
Husband	C	722	29	24	35	51	16
	B	36	577	32	8	53	2
	M	27	40	345	10	38	3
	L	36	6	11	233	18	5
	O	79	44	45	20	772	17
	N	53	31	17	21	55	67

Compute the joint relative frequencies for this table. Based on these frequencies, do you think that people are more likely to marry someone of the same religion? Explain.

14.4 In order to investigate whether men and women differ in their attitudes toward capital punishment, a sample of men and a sample of women were asked if they favored the death penalty for persons convicted of murder. The responses are summarized in the given table. (This data is compatible with that given in the paper "The Polls: Gender and Attitudes Toward Violence" *Public Opinion Quarterly* (1984):384–96).

	Response		
	Yes	No	No Opinion
Male	71	22	7
Female	32	14	5

a. Use expected cell counts to compare the responses of males and females.

b. The next table is also similar to one that appeared in the same paper. It summarizes response to the question "Do you favor the death penalty for someone convicted of rape?" Use expected cell counts to compare the two groups (males and females).

	Response		
	Yes	No	No Opinion
Male	40	50	10
Female	17	28	5

14.5 The relationship between vocational preferences of adolescents and their birth order was examined in the paper "Birth Order and Vocational Preference" (*J. Exper. Educ.* (1980):15–18). The Self-Directed Search, a measure of occupational preference, was used to identify vocational preference. Samples of 122 ninth-grade firstborn children and 122 laterborn children were used. Compute the expected cell counts (assuming no difference in preference for the two groups) and use them to compare the vocational preferences of first- and laterborn children (data follows).

Vocation Preference	Firstborn	Laterborn
Conventional	38	9
Realistic	26	19
Enterprising	24	15
Social	12	15
Artistic	12	21
Investigative	10	43

14.6 Research has shown that if people find an advertisement distasteful, it affects their attitude toward the product being advertised. The paper "Attitude Toward the Ad: Links to Humor and to Advertising Effectiveness" (*J. Adver.* (1983):34–42) reported the results of a study that involved soliciting opinions on an antismoking advertisement. Each subject was asked to indicate both his or her reaction to the ad and smoking status. The resulting data is given in the accompanying two-way table. If there is no association between smoking status and reaction to the ad, what would the expected cell counts be? Does the data suggest that there is an association between smoking status and reaction? Explain.

	Strongly Dislike	Dislike	Neutral	Like	Strongly Like
Smoker	8	14	35	21	19
Nonsmoker	31	42	78	61	69

14.7 NBC interviewers asked voters if they were more or less likely to vote for Walter Mondale as a result of his choosing Geraldine Ferraro for a running mate in the 1984 presidential campaign. Data compatible with that broadcast during election coverage appears in the table below. Does the data suggest an association between sex and response? Explain.

	Response		
Sex	More Likely	Less Likely	Makes No Difference
Female	19	24	57
Male	12	30	58

14.8 The accompanying table appeared in the paper "Early Employment Situations and Work Role Satisfaction Among Recent College Graduates" (*J. Vocational Behavior* (1984):305–18). Describe the degree of association between income and work satisfaction by computing the index of association.

	Work Satisfaction	
Income	High	Low
High (10,000+)	39	29
Low (<10,000)	21	29

14.9 An advertisement claiming that Listerine antiseptic was effective in treating colds and sore throats was recently judged deceptive by a federal court. The manufacturer was required to run advertisements to correct this misconception. A study to determine the effectiveness of these ads was described in the paper "Marketing's Scarlet Letter: The Theory and Practice of Corrective Advertising" (*J. Mktg.* (1984):11–31). Data compatible with summary values given in the paper appears below. Use the index of association to describe the association between brand preference and whether or not cold and sore throat effectiveness was listed as a key attribute in selecting a brand.

	Sore Throat Effectiveness	
Brand Preference	Factor	Not a Factor
Listerine Users	57	43
Scope Users	15	85

14.2 Hypothesis Testing Using a One-Way Frequency Table

Univariate categorical data sets arise in a variety of different settings. If each printed circuit board in a sample of 50 is classified as either defective or nondefective, a categorical variable with two categories results. If specifications require that the thickness of the selected boards be .06 ± .005 in., boards might be classified into three categories—undersized, meeting specification, and oversized. Each registered voter in a sample of 100 selected from those registered in a particular city might be asked which of five city council members he or she favored for mayor. This would result in a categorical variable with five categories. In this section, we consider testing hypotheses about the proportion of the population falling into each of the possible categories. For example, a marketing researcher might be interested in determining whether four brands of laundry detergent are preferred by equal percentages of customers at a market stocking only these four brands. If this is indeed the case, the proportion preferring any given brand is .25. The goodness-of-fit test procedure presented here enables the researcher to use information on brand preference from a sample of customers in deciding whether the hypothesis of equal preference is plausible.

Let K denote the number of categories of a categorical variable, and let $\pi_1, \pi_2, \ldots, \pi_K$ represent the true proportions for categories $1, 2, \ldots, K$, respectively (so that $\pi_1 + \pi_2 + \cdots + \pi_K = 1$). The hypotheses to be tested are

$$
\begin{aligned}
H_0 : \pi_1 &= \text{hypothesized proportion for category 1} \\
\pi_2 &= \text{hypothesized proportion for category 2} \\
&\vdots \\
\pi_K &= \text{hypothesized proportion for category } K
\end{aligned}
$$

H_a : H_0 is not true—at least one of the true category proportions differs from the corresponding hypothesized value

For the example involving laundry detergents, the proportions of all shoppers preferring brands 1, 2, 3, and 4 are π_1, π_2, π_3, and π_4, respectively. The hypothesis of equal preferences is $H_0 : \pi_1 = .25,\ \pi_2 = .25,\ \pi_3 = .25$, and $\pi_4 = .25$.

The hypotheses will be tested by first selecting a random sample of size n and then classifying each sample member into one of the K possible categories. The resulting frequencies, or observed cell counts, are usually displayed in a one-way frequency table. In order to decide whether the sample data is compatible with the null hypothesis, the observed cell counts are compared to the cell counts that would have been expected when the null hypothesis is true. In general, the expected cell counts are $n\pi_1$ for category 1, $n\pi_2$ for category 2, etc. The expected cell counts when H_0 is true result from substituting the corresponding hypothesized proportion for each π.

EXAMPLE 8

The paper "Birth Order and Political Success" (*Psych. Reports* (1971):1239–42) reported that in a sample of 31 candidates for political office, 12 were firstborn children, 11 were middleborn, and 8 were lastborn. Since birth position is related to family size, all 31 candidates considered were from families with exactly 4 children. The author of this paper was interested in ascertaining whether any ordinal position was overrepresented, as this would indicate that those with certain birth orders are more likely to enter political life. In fact, the author thought that first- and lastborns would be overrepresented. Let's suppose that first-, middle-, and lastborns are equally likely to run for political office. In this case, 25% of the candidates from families with 4 children should be firstborn, 25%, lastborn, and 50%, middleborn. With π_1, π_2, and π_3 representing the true proportion of candidates for political office from families with 4 children who are first-, middle-, and lastborn, respectively, the hypotheses of interest are

$$H_0 : \pi_1 = .25,\ \pi_2 = .5,\ \pi_3 = .25$$
$$H_a : H_0 \text{ is not true}$$

Based on the hypothesized values of π_1, π_2, and π_3, the expected cell counts for cell 1 (firstborn), cell 2 (middleborn), and cell 3 (lastborn) are $31(.25) = 7.75$, $31(.5) = 15.5$, and $31(.25) = 7.75$, respectively. Observed and expected cell counts are given in the accompanying table.

Cell	Observed Cell Count	Expected Cell Count
(1) Firstborn	12	7.75
(2) Middleborn	11	15.50
(3) Lastborn	8	7.75

Since the observed cell counts are based on a *sample* of candidates from families with 4 children, even when H_0 is true, we wouldn't expect to see exactly 25% falling in cell 1, 50% in cell 2, and 25% in cell 3. If the differences between the observed and expected cell counts can reasonably be attributed to sampling variation, the data would be considered compatible with H_0. On the other hand, if the discrepancy between the observed and expected cell counts is too large to be attributed solely to chance differences from one sample to another, H_0 should be rejected in favor of H_a. Thus we need an assessment of how different the observed and expected counts are. ■

The goodness-of-fit statistic, denoted by X^2, is a quantitative measure of the extent to which the observed counts differ from those expected when H_0 is true.*

The **goodness-of-fit statistic,** X^2, results from first computing the quantity

$$\frac{(\text{observed cell count} - \text{expected cell count})^2}{\text{expected cell count}}$$

for each cell, where

$$\begin{matrix}\text{expected cell} \\ \text{count}\end{matrix} = n\begin{pmatrix}\text{hypothesized value of corresponding} \\ \text{population proportion}\end{pmatrix}$$

Then X^2 is the sum of these quantities for all K cells:

$$X^2 = \sum_{\text{all cells}} \frac{(\text{observed cell count} - \text{expected cell count})^2}{\text{expected cell count}}$$

The value of the X^2 statistic reflects the magnitude of the discrepancies between observed and expected cell counts. When the differences are sizable, the value of X^2 tends to be large. Therefore, large values of X^2 suggest rejection of H_0. A small value of X^2 (it can never be negative) occurs when the observed cell counts are quite similar to those expected when H_0 is true and so would lend support to H_0.

EXAMPLE 9

(*Example 8 continued*) For the birth-order data,

$$\begin{aligned} X^2 &= \frac{(12 - 7.75)^2}{7.75} + \frac{(11 - 15.5)^2}{15.5} + \frac{(8 - 7.75)^2}{7.75} \\ &= 2.33 + 1.31 + .01 = 3.65 \end{aligned}$$

To determine whether an X^2 value of 3.65 should lead to rejection of $H_0 : \pi_1 = .25,\ \pi_2 = .50,\ \pi_3 = .25$, we need to know whether a value this large could reasonably have resulted when H_0 is true. ■

In general H_0 will be rejected when X^2 exceeds a specified critical value (so the test is upper-tailed). As with previous test procedures, the critical value is chosen to control the probability of a type I error (rejecting H_0 when H_0 is true). This requires information concerning the sampling distribution of X^2 when H_0 is true. A key result is that when the null hypothesis is correct,

* The Greek letter χ (chi) is often used in place of X. X^2 is referred to as the chi-squared (χ^2) statistic. In using X^2 rather than χ^2, we are adhering to our convention of denoting sample quantities by Roman letters.

the behavior of X^2 (in most cases), is described approximately by what statisticians call a **chi-squared distribution**. The chi-squared curve has no area associated with negative values and is asymmetric with a longer tail on the right. There are actually many chi-squared distributions, each one identified with a different number of df. Curves corresponding to several chi-squared distributions are shown in Figure 1.

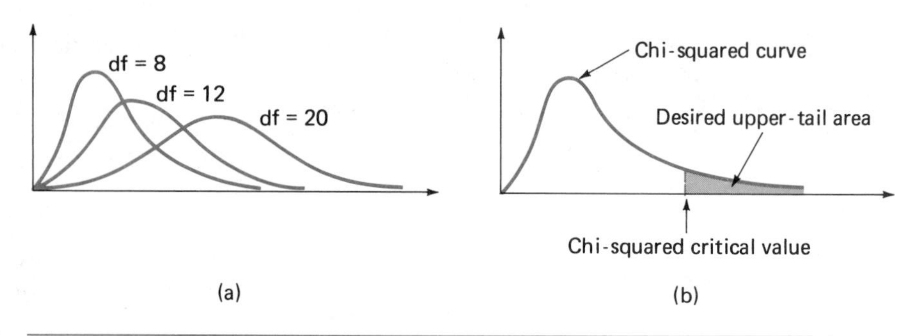

FIGURE 1 CHI-SQUARED CURVES AND AN UPPER-TAILED CRITICAL VALUE

Upper-tailed chi-squared critical values, which capture specified upper-tail areas under the corresponding chi-squared curves, are given in Table XII in the appendices. The columns of this table are headed by various areas captured in the upper tail. For example, the critical value 5.99 captures upper-tail area .05 under the chi-squared curve with 2 df.

> Summary of the Goodness-of-Fit Test Procedure
>
> $H_0 : \pi_1 =$ hypothesized proportion for cell 1, . . . , $\pi_K =$ hypothesized proportion for cell K.
>
> $H_a : H_0$ is not true.
>
> *Test statistic:* X^2.
>
> *Rejection region:* As long as none of the expected cell counts are too small, the X^2 goodness-of-fit statistic has approximately a chi-squared distribution with $K - 1$ df. It is generally agreed that the chi-squared distribution is appropriate as long as every expected cell count is at least 5. We thus reject H_0 if $X^2 >$ chi-squared critical value, where the chi-squared critical value is obtained from the $K - 1$ df row of Table XII in the column corresponding to the desired level of significance α.

If any of the expected cell frequencies are less than 5, categories may be combined in a sensible way to create acceptable expected cell counts. When this is done, however, remember to compute df based on the reduced number of categories.

EXAMPLE 10

The birth-order data of Example 8 will be used to test the researcher's hypothesis. Let's employ a .05 level of significance and use the seven-step hypothesis-testing procedure illustrated in earlier chapters.

1. Let π_1, π_2, and π_3 denote the true proportions of political candidates who are first-, middle-, and lastborn, respectively (in families with 4 children).

2. $H_0 : \pi_1 = .25, \pi_2 = .5, \pi_3 = .25$.

3. $H_a : H_0$ is not true.

4. Test statistic: X^2.

5. Rejection region: Since all expected cell counts are at least 5, the chi-squared critical value that specifies the appropriate rejection region is 5.99 (from the .05 column and the $K - 1 = 3 - 1 = 2$ df row of Table XII). H_0 will be rejected if $X^2 > 5.99$.

6. From Example 9, the computed value of X^2 is 3.65.

7. Since 3.65 is less than 5.99, H_0 is not rejected. There is not substantial evidence to suggest that any birth position is overrepresented among political candidates. The data does not support the researcher's premise that first- and lastborn children would be more likely to enter politics than those who are middleborn. ■

EXAMPLE 11

The paper "Environmentalism, Values, and Social Change" (*Brit. J. Soc.* (1981):103) investigated characteristics that distinguish environmentalists from the general public. Each member of a sample of 437 environmentalists was classified into one of nine occupational categories. The resulting data appears in the accompanying table. With the nine categories ordered as in the table, the proportions of the general public falling in the nine categories were given as .140, .116, .031, .117, .311, .088, .155, .022, and .020. If the same proportions hold for environmentalists, the corresponding expected cell counts are 437(.140) = 61.18, 437(.116) = 50.69, . . . , 437(.020) = 8.74. These expected cell counts have been entered in the accompanying table.

The X^2 goodness-of-fit test and a .01 level of significance will be used to test the null hypothesis that the true proportion of environmentalists falling into each of the nine categories is the same as that for the general population.

Cell	Occupation	Observed Cell Count	Expected Cell Count
1	Professional	67	61.18
2	Clerical	31	50.69
3	Self-employed	42	13.55
4	Service, welfare	190	51.13
5	Manual	23	135.91
6	Retired	11	38.46
7	Housewife	33	67.74
8	Unemployed	6	9.61
9	Student	34	8.74

1. Let $\pi_1, \pi_2, \ldots, \pi_9$ denote the true proportions of all environmentalists falling into the nine occupational categories.

2. $H_0 : \pi_1 = .140, \pi_2 = .116, \pi_3 = .031, \pi_4 = .117, \pi_5 = .311, \pi_6 = .088,$
$\pi_7 = .155, \pi_8 = .022, \pi_9 = .020.$

3. $H_a : H_0$ is not true.

4. Test statistic:

$$X^2 = \sum \frac{(\text{observed count} - \text{expected count})^2}{\text{expected count}}$$

5. Rejection region: All expected cell counts exceed 5, so the rejection region can be based on a chi-squared distribution with $9 - 1 = 8$ df. With a .01 significance level, the appropriate critical value is obtained from Table XII as 20.09. Therefore, H_0 will be rejected if $X^2 > 20.09$.

6. Computations:

$$X^2 = \frac{(67 - 61.18)^2}{61.18} + \frac{(31 - 50.69)^2}{50.69} + \cdots + \frac{(34 - 8.74)^2}{8.74}$$

$$= .554 + 7.648 + \cdots + 73.005 = 650.696$$

7. Since $650.696 > 20.09$, H_0 is rejected. There is strong evidence to indicate that at least one of the true cell proportions for environmentalists differs from that of the general public. ■

The goodness-of-fit test procedure developed in this section can easily be adapted for use with bivariate categorical data sets, as will be seen in the following section.

EXERCISES

14.10 A certain genetic characteristic of a particular plant can appear in one of three forms (phenotypes). A researcher has developed a theory, according to which the hypothesized proportions are $\pi_1 = .25$, $\pi_2 = .50$, and $\pi_3 = .25$. A sample of 200 plants yields $X^2 = 4.63$.

a. Carry out a test of the null hypothesis that the theory is correct using level of significance $\alpha = .05$.

b. Suppose that a sample of 300 plants had resulted in the same value of X^2. How would your analysis and conclusion differ from the analysis and conclusion in (a)?

14.11 When public opinion surveys are conducted by mail, a cover letter explaining the purpose of the survey is usually included. To determine if the wording of the cover letter influences the response rate, three different cover letters were used in a survey of students at a Midwestern university ("The Effectiveness of Cover-Letter Appeals" *J. Soc. Psych.* (1984):85–91). Suppose that each of the three cover letters accompanied questionnaires sent to an equal number of students. Returned questionnaires were then classified according to the type of cover letter (I, II, or III). Use the given data to test the hypothesis that $\pi_1 = \frac{1}{3}$, $\pi_2 = \frac{1}{3}$, and $\pi_3 = \frac{1}{3}$, where π_1, π_2, and π_3 are the true proportion of all returned questionnaires accompanied by cover letters I, II, and III, respectively. Use a .05 significance level.

	Cover-Letter Type		
	I	II	III
Frequency	48	44	39

14.12 The *Los Angeles Times* (October 17, 1984) reported that the color distribution for plain M&M's was: 40% brown, 20% yellow, 20% orange, 10% green, and 10% tan.

a. Each piece of candy in a random sample of 100 plain M&M's was classified according to color, resulting in the given data. Using a significance level of .05, test to determine if the data suggests that the published color distribution is incorrect. What can be said about the P-value?

Color	Brown	Yellow	Orange	Green	Tan
Frequency	45	13	17	7	18

b. The newspaper article also stated that peanut M&M's resulted from producing equal proportions of brown, yellow, orange, and green candies. Purchase a bag of M&M's and classify the pieces according to color. Is the resulting data compatible with the published color distribution? Test using a significance level of .01. Does this exercise leave you with a good taste for statistics?

14.13 The paper "Linkage Studies of the Tomato" (*Trans. Royal Canad. Inst.* (1931):1–19) reported the accompanying data on phenotypes resulting from a cross of tall cut-leaf tomatoes with dwarf potato-leaf tomatoes. There are four possible phenotypes: (1) tall cut-leaf, (2) tall potato-leaf, (3) dwarf cut-leaf, and (4) dwarf potato-leaf. Mendel's laws of inheritance imply that $\pi_1 = \frac{9}{16}$, $\pi_2 = \frac{3}{16}$, $\pi_3 = \frac{3}{16}$, and $\pi_4 = \frac{1}{16}$. Is the data from this experiment consistent with Mendel's laws? Use a .01 significance level.

	Phenotype			
	1	2	3	4
Frequency	926	288	293	104

14.14 A number of different terms are used today to refer to people of Spanish origin. A recent *Los Angeles Times* poll (July 25, 1983) asked each of 568 people of Spanish origin which label he or she preferred. Results are summarized in the given table.

	Label				
	Mexicano	Mexican-American	Latino	Hispanic	Chicano or Other Label
Frequency	142	131	102	80	113

Does the data suggest that the true proportions preferring each label are not all equal? Use a significance level of .01.

14.15 It is hypothesized that when homing pigeons are disoriented in a certain manner, they will exhibit no preference for any direction of flight after take-off. To test this, 120 pigeons are disoriented, let loose, and the direction of flight of each is recorded. The resulting data is given. Use the goodness-of-fit test with significance level .10 to determine if the data supports the hypothesis.

Direction	0–45°	45–90°	90–135°	135–180°
Frequency	12	16	17	15

Direction	180–225°	225–270°	270–315°	315–360°
Frequency	13	20	17	10

14.16 A college bookstore stocks three different paperpack versions of *Hamlet*. It is hypothesized that the three versions are purchased in proportions $\pi_1 = .50$, $\pi_2 = .30$, and $\pi_3 = .20$.

a. Suppose that a sample of 100 purchases yields observed counts 46, 28, and 26 for versions 1, 2, and 3, respectively. What is the value of X^2, and what do you conclude about the validity of the hypothesized proportions at level of significance .05?

b. Now suppose that a sample of 1000 purchases yields counts of 460, 280, and 260, respectively (and thus the same sample category proportions as in (a)). How do the value of X^2 and the conclusion at level .05 differ from those of (a)?

c. In general, what happens to the value of X^2 if a doubling of the sample size results in every observed cell count being doubled? Explain your reasoning (which should not depend on the number of categories under consideration).

14.3 Hypothesis Testing Using a Two-Way Frequency Table

Two-way frequency tables, first introduced in Section 14.1, arise in two different settings. One involves a comparison of two or more populations. Each individual or object in any of the populations is assumed to fall in exactly one of several possible categories. An investigator will usually wish to know whether the category proportions are the same for all populations under study. For this purpose a separate random sample is obtained from each population. The sample data can be displayed as frequencies or counts in a rectangular table in which each row refers to the sample from a different population and the columns are labeled with the possible categories (sometimes the roles of rows and columns are reversed).

Two-way frequency tables also provide a convenient way of summarizing data that results when the values of two different categorical variables are observed for each member of a single sample. The rows of the table then correspond to the possible categories of one variable and the columns to those of the second variable. In this situation, one is generally interested in deciding whether an association exists between the two variables defining the table. An X^2 statistic can be used effectively to analyze the data in a two-way frequency table arising in either setting.

Comparing Two or More Populations

The comparison of two or more populations with respect to values of a categorical variable was discussed informally in Section 14.1. Here we consider a formal procedure for testing the null hypothesis that for each category, the proportion falling in the category is the same for all populations under study. Suppose, for example, that four different candidates are running for state assembly in a district made up of three counties. There are then three different populations, each one consisting of voters from a single county. A voter belongs in one of four categories according to which candidate he or she prefers. One way for the null hypothesis to be true would be for the proportions preferring candidates 1, 2, 3, and 4 to be .40, .25, .20 and .15, respectively, in each one of the three counties. The three populations are then described as *homogeneous* with respect to the categories (candidate preferences). If, however, the proportion of voters preferring the first candidate is .45 in the first county and only .35 in the two other counties, then H_0 would be false. There are, of course, many other ways in which H_0 could be true and many in which it could be false.

Recall that when the proportion of successes in a population consisting of successes and failures is π and a random sample of size n is selected from the population, the expected, or average, number of successes is $n\pi$. More generally, if a random sample is taken from each of several populations, then

$$\begin{pmatrix}\text{expected count for any}\\ \text{particular category for}\\ \text{a specified sample}\end{pmatrix} = \begin{pmatrix}\text{sample}\\ \text{size}\end{pmatrix}\cdot\begin{pmatrix}\text{population proportion in}\\ \text{corresponding category}\end{pmatrix}$$

Even when the population proportions for any given category are assumed equal for all populations, the values of these proportions are still unknown to the investigator. However, they can be estimated from the sample data:

$$\begin{pmatrix}\text{estimate of population proportion}\\ \text{falling in a particular category}\end{pmatrix} = \frac{\text{total number of observations in all samples falling in the category}}{\text{total of all sample sizes}}$$

Thus if voters in three different counties are sampled with sample sizes of 200, 300, and 250 and the numbers of sampled voters preferring candidate 1 are 80, 112, and 93, respectively, then

$$\begin{pmatrix}\text{estimated proportion}\\ \text{preferring candidate 1}\end{pmatrix} = \frac{80 + 112 + 93}{200 + 300 + 250} = \frac{285}{750} = .380$$

Now suppose that we replace each population proportion in the above expression for expected counts by its estimate. The result is an expression for estimated expected counts (obtained by assuming the equality of category proportions across populations):

$$\begin{matrix}\text{estimated}\\ \text{expected count}\end{matrix} = \frac{(\text{sample size})(\text{total number of observations in category})}{\text{total sample size}}$$

This is exactly the formula given in Section 14.1 for expected counts. In the numerical example of the previous paragraph, the three estimated expected counts are $(200)(285)/750 = 200(.38) = 76$, $(300)(285)/750 = 114$, and $(250)(285)/750 = 95$, respectively. A test for equality of proportions across populations is based on the discrepancy between estimated expected counts and the corresponding observed counts. The procedure description given in the box assumes that the counts have been entered into a rectangular table, as described earlier.

Comparing Two or More Populations Using the X^2 Statistic

Null hypothesis: H_0 : The true category proportions are the same for all of the populations (homogeneity of populations).

Alternate hypothesis: H_a : The true category proportions are not the same for all of the populations.

Test statistic:

$$X^2 = \sum_{\substack{\text{all cells}\\ \text{in the}\\ \text{table}}} \frac{(\text{observed cell count} - \text{expected cell count})^2}{\text{expected cell count}}$$

The expected cell counts are estimated from the sample data (assuming that H_0 is true) using the formula

$$\text{expected cell count} = \frac{(\text{row marginal total})(\text{column marginal total})}{\text{grand total}}$$

Rejection region: When H_0 is true and all expected cell counts are at least 5, X^2 has approximately a chi-squared distribution with

$$\text{df} = (\text{number of rows} - 1)(\text{number of columns} - 1).$$

Therefore, H_0 should be rejected if $X^2 >$ chi-squared critical value. (If some expected cell counts are less than 5, rows or columns of the table may be combined to achieve a table with satisfactory expected counts.) The critical value comes from the column of Table XII corresponding to the desired level of significance.

EXAMPLE 12

Many surgeons and internists feel that they are spending an increasing proportion of their time treating psychiatric problems. The paper "Psychiatric Components of Medical and Surgical Practice, II: Referral and Treatment of Psychiatric Disorders" (*Amer. J. Psych.* (1983):760–63) reported the results of a survey asking both internists and surgeons how they would treat patients suffering from depression. The possible responses were : (1) treat the patients themselves, (2) request consultations, and (3) refer the patients to psychiatrists. Data is given in the accompanying two-way frequency table. The author of the paper was interested in whether the population of all internists has the same response proportions as the population of all surgeons.

The hypotheses to be tested are

H_0 : no difference between surgeons and internists with respect to the true proportion falling into each of the three response categories

H_a : surgeons and internists differ with respect to true response proportions

The estimated expected cell count for the cell in row 1 and column 1 is $(143)(63)/(185) = 48.7$ Other expected cell counts are computed in a similar fashion and appear in parentheses in the table.

	Treat	Consult	Refer	Row Marginal Total
Internists	54 (48.7)	57 (58.0)	32 (36.3)	143
Surgeons	9 (14.3)	18 (17.0)	15 (10.7)	42
Column Marginal Total	63	75	47	185

All expected cell counts exceed 5, so the X^2 test statistic is appropriate. This table has two rows and three columns, so df = $(2 - 1)(3 - 1) = 2$. We will use a .05 significance level. From Table XII, the appropriate critical value is 5.99. H_0 will be rejected if $X^2 > 5.99$. The computed value of X^2 is

$$X^2 = \frac{(54 - 48.7)^2}{48.7} + \cdots + \frac{(15 - 10.7)^2}{10.7} = 4.85$$

Since 4.85 does not exceed the critical value 5.99, H_0 is not rejected. There is not sufficient evidence to indicate that the true proportion in each response category differs for the two groups. ■

EXAMPLE 13

The results of an experiment to assess the effect of crude oil on fish parasites were described in the paper "Effects of Crude Oils on the Gastrointestinal Parasites of Two Species of Marine Fish" (*J. Wildlife Diseases* (1983): 253–58). Three treatments (corresponding to populations in the procedure described) were compared: (1) no contamination, (2) contamination by 1-year-old weathered oil, and (3) contamination by new oil. For each treatment condition, a sample of fish was taken and then each fish was classified as either parasitized or not parasitized. Data compatible with that in the paper is given; expected cell counts (computed under the hypothesis of no treatment differences) appear in parentheses. Does the data strongly indicate that the three treatments differ with respect to the true proportion of parasitized and nonparasitized fish? A significance level of .01 will be used to test the relevant hypotheses.

Treatment	Parasitized	Nonparasitized	Row Marginal Total
Control	30 (23.0)	3 (10.0)	33
Old oil	16 (16.7)	8 (7.3)	24
New oil	16 (22.3)	16 (9.7)	32
Column Marginal Total	62	27	89

H_0 : proportions of parasitized and of nonparasitized fish are the same for all three treatments

H_a: H_0 is not true

Since all expected cell counts are at least 5, the X^2 statistic can be used. The appropriate critical value, from the .01 column and $(3 - 1)(2 - 1) = 2$ df row of Table XII, is 5.99. Therefore, H_0 will be rejected if $X^2 > 5.99$. The computed value of X^2 is

$$X^2 = \frac{(30 - 23.0)^2}{23.0} + \cdots + \frac{(16 - 9.7)^2}{9.7} = 13.0$$

Since $13.0 > 5.99$, H_0 is rejected. The data strongly indicates that the proportions of parasitized and nonparasitized fish are not the same for all three treatments. ■

Testing for Independence of Two Categorical Variables

The X^2 test statistic and test procedure can also be used to investigate association between two categorical variables in a single population. Suppose that each population member has a value of a first categorical variable and also of a second such variable. As an example, television viewers in a particular city might be categorized both with respect to preferred network (ABC, CBS, NBC, or PBS) and with respect to favorite type of programming (comedy, drama, or information-news). The question of interest is often whether or not knowledge of one variable's value provides any information about the value of the other variable—that is, are the two variables independent? Continuing the example, suppose that those who favor ABC prefer the above three types of programming in proportions .4, .5, and .1 and that these proportions are also correct for individuals favoring any of the other three networks. Then learning an individual's preferred network provides no (extra) information concerning that individual's favorite type of programming. The categorical variables *preferred network* and *favorite program type* are independent.

To see how the expected counts are obtained in this situation, first recall from our probability discussion in Chapter 5 the condition for independence of two events. The events A and B were said to be independent if $P(A \text{ and } B) = P(A) \cdot P(B)$, so that the proportion of time that they occurred together in the long run is the product of the two individual long-run relative frequencies. Similarly, two categorical variables are independent in a population if for *any* particular category of the first variable and *any* particular category of the second variable,

$$\begin{pmatrix}\text{population proportion}\\ \text{of individuals falling}\\ \text{in both categories}\end{pmatrix} = \begin{pmatrix}\text{population proportion}\\ \text{in specified category}\\ \text{of first variable}\end{pmatrix} \cdot \begin{pmatrix}\text{population proportion}\\ \text{in specified category}\\ \text{of second variable}\end{pmatrix}$$

Thus if 30% of all viewers prefer ABC, the proportions of preferred programming types are as previously given, and the two variables are independent, then the proportion of individuals who both favor ABC and prefer comedy is $(.3)(.4) = .12$ (or 12%).

Multiplying the right-hand side of this expression by the sample size gives the expected number of individuals in the sample who fall in the specified categories of both variables when the variables are independent. However, these expected counts cannot be calculated because the individual population proportions are not known. The resolution of this dilemma is, of course, to estimate each population proportion by the corresponding sample proportion:

$$\begin{pmatrix}\text{estimated expected number} \\ \text{falling in specified categories} \\ \text{of the two variables}\end{pmatrix}$$

$$= \begin{pmatrix}\text{sample} \\ \text{size}\end{pmatrix} \cdot \left(\frac{\text{observed number in category of first variable}}{\text{sample size}}\right) \cdot \left(\frac{\text{observed number in category of second variable}}{\text{sample size}}\right)$$

$$= \frac{\begin{pmatrix}\text{observed number} \\ \text{in category of} \\ \text{first variable}\end{pmatrix} \cdot \begin{pmatrix}\text{observed number} \\ \text{in category of} \\ \text{second variable}\end{pmatrix}}{\text{sample size}}$$

Suppose that the observed counts are displayed in a rectangular table in which rows refer to the different categories of the first variable and columns correspond to the categories of the second variable. Then the numerator in the above expression for estimated expected counts is just the product of the row and column marginal totals. This is exactly how estimated expected counts were computed in the test for homogeneity of several populations.

Testing for Independence of Two Categorical Variables

Null hypothesis: H_0 : The two variables are independent.

Alternative hypothesis: H_a : The two variables are not independent.

Test statistic:

$$X^2 = \sum_{\substack{\text{all cells} \\ \text{in the} \\ \text{table}}} \frac{(\text{observed cell count} - \text{expected cell count})^2}{\text{expected cell count}}$$

The expected cell counts are estimated (assuming H_0 is true) by the formula

$$\begin{matrix}\text{expected} \\ \text{cell count}\end{matrix} = \frac{(\text{row marginal total})\ (\text{column marginal total})}{\text{grand total}}$$

Rejection region: When H_0 is true and all expected cell counts are at least 5, X^2 has approximately a chi-squared distribution with df = (number of rows − 1)(number of columns − 1). H_0 should be rejected if $X^2 >$ chi-squared critical value. Chi-squared critical values are given in Table XII.

EXAMPLE 14

The paper "Impulsive and Premeditated Homicide: An Analysis of Subsequent Parole Risk of the Murderer" (see Example 5) investigated the relationship between the circumstances surrounding a murder and the subsequent parole success of the murderer. A sample of 82 convicted murderers was selected, and each murderer was categorized according to type of murder (impulsive, premeditated) and parole outcome (success, failure). The resulting data is given in the 2 × 2 table below. The authors were interested in determining whether there is an association between type of murder and parole outcome. Using a .05 level of significance, we will test

H_0 : type of murder and parole outcome are independent

H_a : the two variables are not independent

Expected cell counts are as follows:

Cell		
Row	Column	Expected Cell Count
1	1	$\frac{(35)(42)}{82} = 17.93$
1	2	$\frac{(47)(42)}{82} = 24.07$
2	1	$\frac{(35)(40)}{82} = 17.07$
2	2	$\frac{(47)(40)}{82} = 22.93$

	Success	Failure	Row Marginal Total
Impulsive	13 (17.93)	29 (24.07)	42
Premeditated	22 (17.07)	18 (22.93)	40
Column Marginal Total	35	47	82

Since all expected cell counts exceed 5, the X^2 statistic can be used. With level of significance $\alpha = .05$ and df $= (2 - 1)(2 - 1) = 1$, the chi-squared critical value is 3.84. H_0 will be rejected if $X^2 > 3.84$.

The computed value of X^2 is

$$X^2 = \frac{(13 - 17.93)^2}{17.93} + \cdots + \frac{(18 - 22.93)^2}{22.93} = 4.85$$

Since $4.85 > 3.84$, H_0 is rejected. The data support the existence of an association between type of murder and parole outcome. ■

EXAMPLE 15

The accompanying two-way frequency table appeared in the paper "Marijuana Use in College" (*Youth and Society* (1979):323–34). Four hundred and forty-five college students were classified according to both frequency of marijuana use and parental use of alcohol and psychoactive drugs. Expected cell counts (computed under the assumption of no association between marijuana use and parental use) appear in parentheses in the given table.

Parental Use of Alcohol and Drugs	Student Level of Marijuana Use: Never	Occasional	Regular	Row Marginal Total
Neither	141 (119.3)	54 (57.6)	40 (58.1)	235
One	68 (82.8)	44 (39.9)	51 (40.3)	163
Both	17 (23.9)	11 (11.5)	19 (11.6)	47
Column Marginal Total	226	109	110	445

The X^2 test with a .01 significance level will be used to determine if there is an association between marijuana use and parental use of drugs and alcohol.

H_0 : student marijuana use is independent of parent's drug and alcohol use

H_a : the two variables are not independent

Since all expected cell counts are at least 5, the X^2 test can be used. With $\alpha = .01$ and df $= (3 - 1)(3 - 1) = 4$, the chi-squared critical value is 13.28 and H_0 will be rejected if $X^2 > 13.28$. The computed value of X^2 is

$$X^2 = \frac{(141 - 119.3)^2}{119.3} + \cdots + \frac{(19 - 11.6)^2}{11.6} = 22.45$$

Since $22.45 > 13.28$, H_0 is rejected. There does appear to be an association between use of marijuana and parental use of alcohol and drugs.

Most statistical computer packages will calculate both estimated expected cell counts and the value of the X^2 test statistic. MINITAB output for this data follows. (The discrepancy between $X^2 = 22.45$ and $X^2 = 22.37$ is due to rounding).

```
EXPECTED FREQUENCIES ARE PRINTED BELOW OBSERVED FREQUENCIES
        I  C1    I  C2    I  C3    ITOTALS
-------I-------I-------I-------I-------
     1  I  141   I   54   I   40   I    235
        I  119.3I   57.6I   58.1I
-------I-------I-------I-------I-------
     2  I   68   I   44   I   51   I    163
        I   82.8I   39.9I   40.3I
-------I-------I-------I-------I-------
     3  I   17   I   11   I   19   I     47
        I   23.9I   11.5I   11.6I
-------I-------I-------I-------I-------
TOTALS I   226  I   109  I   110  I    445

TOTAL CHI SQUARE =

          3.93 + 0.22 + 5.63 +
          2.64 + 0.42 + 2.85 +
          1.98 + 0.02 + 4.69 +

                = 22.37

DEGREES OF FREEDOM = ( 3 - 1) x ( 3 - 1) = 4
```

■

In some investigations, values of more than two categorical variables are recorded for each individual in a sample. For example, in addition to the variables *student marijuana use* and *parental use of drugs and alcohol,* the researchers in the study referenced in the last example might also have recorded political affiliation for each student in the sample. A number of interesting questions could then be explored: Are all three variables independent of one another? Is it possible that student use and parental use are dependent but that the relationship between them does not depend on political affiliation? For a particular political affiliation, are student use and parental use independent?

The X^2 test procedure described in this section for analysis of bivariate categorical data can be extended for use with multivariate categorical data. Appropriate hypothesis tests could then be used to provide insight into the relationships between variables. However, the computations required to calculate estimated expected cell counts and to compute the value of X^2 are quite tedious and so are seldom done without the aid of a computer. Several statistical computer packages (including BMDP and SAS) will perform this type of analysis. The chapter references can be consulted for further information.

A Note Concerning Degrees of Freedom

Let r and c denote the number of row categories and number of column categories, respectively, in the two-way frequency table under investigation. Then the X^2 test for independence and for homogeneity of populations are both based on $(r - 1)(c - 1)$ df. The number of df for each test is a special case of a more general rule for df in chi-squared tests. The rule involves both the number of freely determined cell counts in the table and the number of population characteristics estimated in order to compute expected cell counts. Consider first an investigation involving customers of three different grocery stores. Suppose that 100 customers who shop at store 1 are selected, 200 customers of store 2 are selected, and 150 customers of store 3 are chosen. Each customer is asked which of four characteristics—location, price, service, or store design—was most important in selecting a store. The data from this study can be displayed in a two-way table with three rows and four columns, as pictured.

	Location	Price	Service	Design	
Store 1					100
Store 2					200
Store 3					150

By design, the three row totals are fixed. Therefore, once any three of the four counts in the first row are known, the fourth is automatically determined. There are only three freely determined counts in the first row. Similarly, there are only three freely determined counts in each of the other two rows, so the total number of freely determined counts is $3 + 3 + 3 = 9$. Now the null hypothesis of homogeneity says that the proportions of customers falling in the four column categories are the same for each store. That is, when H_0 is true, the true proportions π_1, π_2, π_3, and π_4 are the same for each store. Expected counts are then obtained by estimating these four π's.

However, since $\pi_1 + \pi_2 + \pi_3 + \pi_4 = 1$ and the same must be true of the estimates, only three of the π's are independently estimated. For example, if π_1, π_3, and π_4 are estimated as .2, .3, and .1, respectively, then the estimate of π_2 must be $1 - (.2 + .3 + .1) = .4$.

General Rule for Degrees of Freedom in a Chi-Squared Test

$$\text{df} = \begin{pmatrix} \text{number of} \\ \text{freely determined} \\ \text{cell counts} \end{pmatrix} - \begin{pmatrix} \text{number of independently} \\ \text{estimated population} \\ \text{characteristics} \end{pmatrix}$$

For the given example, this rule would give df = 9 − 3 = 6, which is exactly $(r - 1)(c - 1) = (2)(3)$. More generally, if a test of homogeneity is based on an $r \times c$ frequency table (the same c categories for each of r populations), then there are $c - 1$ freely determined counts in each row and, therefore, a total of $r(c - 1)$ freely determined counts. According to H_0, the population proportions $\pi_1, \pi_2, \ldots, \pi_c$ are the same for all populations, but since these have a sum of 1, only $c - 1$ are independently estimated. Thus

$$\text{df} = r(c - 1) - (c - 1) = rc - r - c + 1 = (r - 1)(c - 1)$$

The specialization of the general rule to the test for independence involves slightly different reasoning. In this situation only the total sample size is fixed (total of the counts for all rc cells), so there are $rc - 1$ freely determined cell counts. To obtain expected counts, the r population row proportions and the c population column proportions must be estimated. But since each set of proportions has a sum of 1, only $(r - 1) + (c - 1)$ are independently estimated. Thus $\text{df} = rc - 1 - [(r - 1) + (c - 1)] = rc - 1 - r + 1 - c + 1 = rc - r - c + 1 = (r - 1)(c - 1)$.

The general rule can be used to justify the number of df for the goodness-of-fit tests discussed in the next section. It is also used when relationships between more than two variables or factors are studied (e.g., political preference, religious preference, and social class using a multidimensional frequency table analysis).

EXERCISES

14.17 A particular state university system has six campuses. On each campus a random sample of students will be selected and each student will be categorized with respect to political philosophy as liberal, moderate, or conservative. The null hypothesis of interest is that the proportion of students falling in these three categories is the same at all six campuses.

a. On how many degrees of freedom will the resulting X^2 test be based, and what is the rejection region for significance level .01?

b. How do your answers in (a) change if there are seven campuses rather than six?

c. How do your answers in (a) change if there are four rather than three categories for political philosophy?

14.18 The following data on salary satisfaction for pharmaceutical salespeople appeared in the article "Some Job-Different Views: Women and Men in Industrial

Sales" (*J. Marketing* (1978):92–98). Estimated expected counts under the hypothesis that the population proportions in the three satisfaction categories are the same for each sex are also given. Carry out the X^2 test using significance level .10.

	Observed Satisfaction Level			Expected Satisfaction Level		
	Low	Medium	High	Low	Medium	High
Male	46	61	53	45.71	59.26	55.03
Female	8	9	12	8.29	10.74	9.97

14.19 Suppose that a random sample of college graduates is selected and each graduate is categorized according to both decade of graduation (50s, 60s, 70s, or 80s) and present level of drug use (none, infrequently, frequently). Test for independence using significance level .01 if the resulting value of X^2 is 21.35.

14.20 An increasing number of people are spending their working hours in front of a video display terminal (VDT). The paper "VDT Workstation Design: Preferred Settings and Their Effects" (*Human Factors* (1983):161–75) summarized a study of adjustable VDT screens. Sixty-five workers using nonadjustable screens and 66 workers using adjustable screens were asked if they experienced annoying reflections from the screens. The resulting data is given in the accompanying table.

	Annoying Reflection	
Screen Type	No	Yes
Nonadjustable	15	50
Adjustable	28	38

a. The investigators were interested in whether the proportion experiencing annoying reflections was the same for both types of VDT screens. Does this problem situation involve comparing two populations or testing for independence? Explain.

b. Use a .05 significance level and the X^2 statistic to test the appropriate hypotheses.

c. Can you think of another test statistic that could be used to answer the researchers' question? (*Hint:* See Chapter 9.) If the researchers were interested in determining whether the proportion experiencing annoying reflection was smaller for the adjustable VDT screens, which test statistic would you recommend? Explain. (*Hint:* Look at the alternative hypothesis for the X^2 test.)

14.21 The paper "Consumer Reaction to Variable Rate Mortgages" (*J. Consumer Affairs* (1979):262–81) investigated the relationship between type of mortgage (variable rate (VR) or fixed rate (FR)) and the number of financial institutions contacted prior to obtaining a home loan. A sample of 292 new home buyers was selected, resulting in the accompanying 6×2 frequency table.

Number of Institutions	Type of Mortgage VR	FR
1	4	2
2	24	28
3	46	55
4	21	25
5	14	20
6 or more	29	24

a. If you were interested in determining whether a relationship exists between the type of mortgage and the number of financial institutions contacted, what null and alternative hypotheses would you test?

b. Compute the expected count for each cell in the table.

c. Since the expected counts for row 1 of the table aren't at least 5, rows 1 and 2 can be combined to form a 5×2 table prior to conducting the hypothesis test. Use the resulting table to test the hypotheses in (a).

14.22 Elbow dislocation is a common injury. The given frequency table appeared in the paper "Elbow Dislocations" (*J. Amer. Med. Assoc.* (1965):113). Samples of 61 men and 44 women suffering elbow dislocations were classified into 6 categories according to age. Does the data suggest that the true age distribution (as given by the proportions falling into each of the six age categories) differs for men and women who dislocate an elbow? Use a significance level of .01.

				Age		
Sex	1–10	11–20	21–30	31–40	41–50	Over 50
Male	7	21	9	13	6	5
Female	7	6	2	3	10	16

14.23 Are the educational aspirations of students related to family income? This question was investigated in the article "Aspirations and Expectations of High School Youth" (*Int. J. Comparative Soc.* (1975):25). The given 4×3 table resulted from classifying 273 high school students according to level of education expected and family income. Does the data indicate that educational aspiration and family income are not independent? Use a .10 level of significance.

		Income	
Aspired Level	Low	Middle	High
Some high school	9	11	9
High school graduate	44	52	41
Some college	13	23	12
College graduate	10	22	27

14.24 The *Los Angeles Times* (July 29, 1983) conducted a survey to find out why some Californians don't register to vote. Random samples of 100 Latinos, 100 Anglos, and 100 Blacks who were not registered to vote were selected. The resulting data is summarized in the accompanying table. Does the data suggest that the true proportion falling into each response category is not the same for Latinos, Anglos, and Blacks? Use a .05 significance level.

Reason for Not Registering	Latino	Anglo	Black
Not a citizen	45	8	0
Not interested	19	33	19
Can't meet residency requirements	9	35	23
Distrust of politics	5	10	8
Too difficult to register	10	10	27
Other reason	12	4	23

14.25 The accompanying frequency table appeared in the paper "Commitment to Work in Immigrants: Its Functions and Peculiarities" (*J. Vocational Behavior* (1984):329–39). The data resulted from classifying 175 workers according to two variables: *job type* (with 3 categories: immigrant (I), white-collar nonimmigrant (W), and executive nonimmigrant (E)), and *attitude toward authority* (with two categories: positive (P) and negative (N)). Does the data support the theory that there is an association between attitude toward authority and job type? Use a significance level of .01.

	Attitude	
Job Type	P	N
I	51	23
W	34	32
E	25	10

14.26 The effect of copper on earthworms was investigated in the paper "Sublethal Toxic Effects of Copper on Growth, Reproduction and Litter Breakdown Activity in the Earthworm *Lumbricus rubellus*, with Observations on the Influence of Temperature and Soil pH" (*Environ. Pollution* (1984):207–19). Each of four concentrations of copper (14, 54, 131, and 372 mg/kg soil) was applied to a sandy soil containing a known number of worms. After 6 weeks, the number of surviving worms was recorded. Data compatible with that given in the paper appears in the accompanying table. Does this data strongly suggest that the mortality rate (proportion not surviving) differs for the four concentrations? Use a .01 significance level.

	Concentration Level			
	I	II	III	IV
Survived	80	74	78	66
Died	0	6	2	14

14.27 In a study of 2989 cancer deaths, the location of death (home, acute-care hospital, or chronic-care facility) and age at death were recorded, resulting in the given two-way frequency table. ("Where Cancer Patients Die" *Public Health Reports* (1983):173). Using a .01 significance level, test the null hypothesis that age at death and location of death are independent.

	Location		
Age	Home	Acute Care	Chronic Care
15–54	94	418	23
55–64	116	524	34
65–74	156	581	109
Over 74	138	558	238

14.28 The relative importance attached to work and home life by high school students was examined in "Work Role Salience as a Determinant of Career Maturity in High School Students" (*J. Vocational Behavior* (1984):30–44). Does the data summarized in the accompanying two-way frequency table suggest that sex and relative importance assigned to work and home are not independent? Test using a .05 level of significance.

	Relative Importance		
Sex	Work > Home	Work = Home	Work < Home
Female	68	26	94
Male	75	19	57

14.4 Goodness-of-Fit Tests for a Population Model

Many of the small-sample inferential procedures introduced in earlier chapters required the assumption of a normal population distribution. Normal probability plots provided one way of assessing the reasonableness of this assumption. The goodness-of-fit procedure can also be used to decide whether it is plausible that the population distribution is normal. The X^2 test statistic again compares observed and expected cell counts for a specified set of categories. Since a variable with a normal distribution is numerical, we begin by constructing a frequency distribution whose class intervals and corresponding frequencies serve as categories or cells and observed cell counts, respectively. The expected counts are particularly easy to compute when H_0 specifies the values of μ and σ for the normal population, so we consider this case first.

> X^2 Goodness-of-Fit Test for a Particular Normal Distribution (Based on a Frequency Distribution with K Classes)
>
> *Null hypothesis:* H_0 : The population distribution is normal with specified values of μ and σ.
>
> *Alternative hypothesis:* H_a : The population distribution is not the specified normal distribution.
>
> *Test statistic:* X^2 goodness-of-fit statistic.
>
> *Rejection region:* As long as all expected counts are at least 5, reject H_0 if $X^2 >$ chi-squared critical value. If some expected counts are smaller than 5, adjacent class intervals can be combined. The chi-squared critical value is obtained from the appropriate column of Table XII and the row corresponding to df = (number of class intervals − 1).

EXAMPLE 16

The article "A Probabilistic Analysis of Dissolved Oxygen—Biochemical Oxygen Demand Relationship in Streams" (*J. Water Resources Control Fed.* (1969):73–90) reported data on the rate of oxygenation in streams at 20°C in a certain region. Based on the accompanying frequency distribution, can we conclude that the true distribution of oxygenation rates is normal with $\mu = .2$ and $\sigma = .05$? A .01 significance level has been selected.

Rate (Per Day)	Frequency
Below .100	12
.100–.150	20
.150–.200	23
.200–.250	15
Above .250	13
	83

The appropriate hypotheses are:

H_0 : the oxygenation rate distribution is normal with $\mu = .2$ and $\sigma = .05$

H_a : the oxygenation rate distribution is not normal with $\mu = .2$ and $\sigma = .05$ (H_0 is not true)

If H_0 is rejected, we will have ruled out only this one particular normal distribution as a population model. It is possible that a different normal distribution (different μ and/or σ) might be appropriate.

The next step in the analysis is to compute the expected cell counts for the five categories. When H_0 is true, the population distribution is normal with $\mu = .2$ and $\sigma = .05$. The proportion of oxygenation rates below .100 is the area under the corresponding normal curve to the left of .100, as pictured. To find this area, we compute $z = (.1 - .2)/.05 = -2$ and refer to Table I. This gives .0228 as the proportion of oxygenation rates below .100 when H_0 is true. Since the sample size was 83, the expected cell count for the first cell is $83(.0228) = 1.89$.

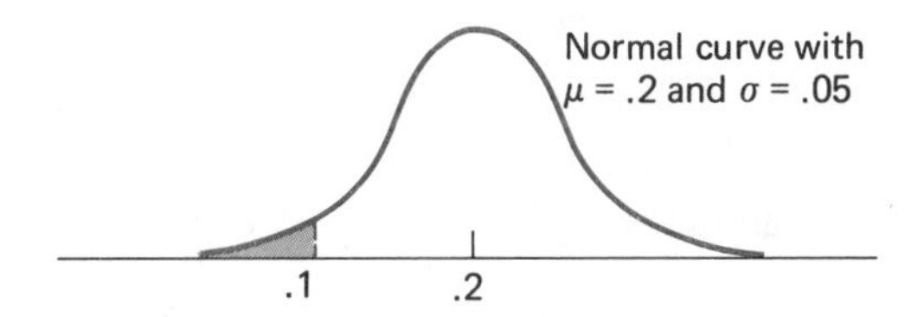

The proportion of oxygenation rates for cell 2 (.100 − .150) is the shaded area in the accompanying picture. This is equal to the area above the interval from $(.1 - .2)/.05 = -2$ to $(.15 - .2)/.05 = -1$ under the standard normal curve. From Table I, this area is $.1587 - .0228 = .1359$, and so the corresponding expected cell count is $83(.1359) = 11.28$. Expected counts for the remaining three cells are computed in a similar manner.

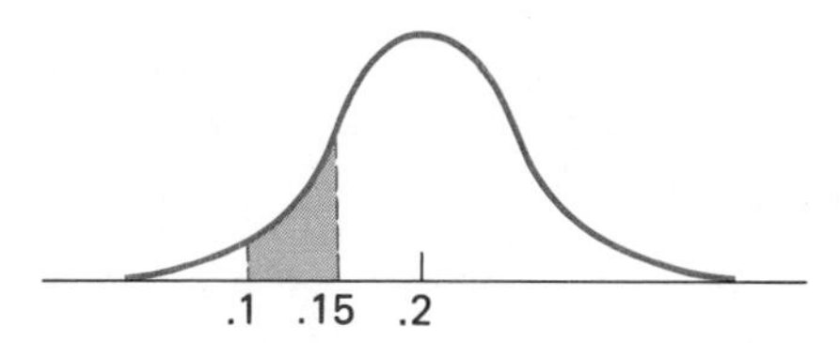

Cell	Interval	Observed Cell Count	Expected Cell Count
1	Below .100	12 } 32	1.89 } 13.17
2	.100–.150	20	11.28
3	.150–.200	23	28.33
4	.200–.250	15	28.33
5	Above .250	13	13.17

Since the expected count for the first cell is less than 5, let's combine the first two cells. The subsequent analysis is based on 4 cells and, therefore, $K - 1 = 3$ df.

The X^2 goodness-of-fit statistic will be used. With a significance level of .01, the chi-squared critical value in the 3 df row of Table XII is 11.34. H_0 will be rejected if $X^2 > 11.34$. Otherwise, H_0 will not be rejected.

The computed value of X^2 is

$$X^2 = \frac{(32 - 13.17)^2}{13.17} + \cdots + \frac{(13 - 13.17)^2}{13.17} = 34.20$$

Since $34.20 > 11.34$, H_0 is rejected. The data strongly indicates that the true distribution of oxygenation rate is not normal with $\mu = .2$ and $\sigma = .05$. ■

In Example 16, the null hypothesis specified a particular normal distribution ($\mu = .2$ and $\sigma = .05$). Much more frequently, an investigator is interested in whether the population distribution is normal in shape but doesn't have a specific normal distribution in mind. In this case, μ and σ are estimated from the data by $\overline{x}$ and s. The estimation of the population characteristics μ and σ results in a loss of 2 df, so the test is now based on $K - 3$ df (recall the general rule for df from the previous section).

X^2 Goodness-of-Fit Test for Normality (μ and σ not specified) Based on a Frequency Distribution with K Classes

When the X^2 goodness-of-fit test is used to test

H_0 : the population distribution is normal

versus

H_a : the population distribution is not normal

the sample mean, $\overline{x}$, and sample standard deviation, s, are used in place of μ and σ to compute (estimated) expected cell counts. The chi-squared critical value for the rejection region is based on $K - 3$ df.

EXAMPLE 17

The paper discussed in Example 16 reported the values $\overline{x} = .173$ and $s = .066$. Is it reasonable to think that the true distribution of oxygenation rate is normal? A .05 significance level will be used.

H_0 : the oxygenation rate distribution is normal

H_a : the oxygenation rate distribution is not normal

Expected cell counts are computed in a manner analogous to that of Example 16. The expected proportion of oxygenation rates less than .100 is the area to the left of .1 under the normal curve with mean .173 and standard deviation .066. The corresponding z value is $z = (.1 - .173)/.066 = -1.1$. From Table I, the desired area is .1357. The expected count for the first cell is then $83(.1357) = 11.26$. The remaining expected counts have been computed in a similar fashion and appear in the accompanying table.

Cell	Interval	Observed Cell Count	Expected Cell Count
1	Below .100	12	11.26
2	.100–.150	20	20.45
3	.150–.200	23	22.68
4	.200–.250	15	19.05
5	Above .250	13	9.55

All expected cell counts are greater than 5, so there is no need to combine cells. With $5 - 3 = 2$ df and a .05 significance level, the chi-squared critical value (from Table XII) is 5.99. Therefore, H_0 will be rejected if $X^2 > 5.99$.

The computed value of X^2 is

$$X^2 = \frac{(12 - 11.26)^2}{11.26} + \cdots + \frac{(13 - 9.55)^2}{9.55} = 2.17$$

Since $2.17 < 5.99$, H_0 is not rejected. The data supports the hypothesis of normality. ■

The goodness-of-fit test procedure can also be used to compare observed cell counts to those expected under hypothesized population distributions other than the normal. Several examples are included in the exercises.

EXERCISES

14.29 The accompanying frequency distribution of protein level in alfalfa clones appeared in the paper "Variability of Fraction 1 Protein and Total Phenolic Constituents in Alfalfa" (*Agronomy J.* (1974):384–86). Use the goodness-of-fit test with a .05 significance level to test the null hypothesis that the true distribution of protein level in alfalfa clones is normal with mean 4.3 and standard deviation .8.

Protein Level (% of Oven-Dry Weight)	Frequency
Less than 2.4	8
2.4–<3.0	65
3.0–<3.6	180
3.6–<4.2	328
4.2–<4.8	408
4.8–<5.4	284
5.4–<6.0	83
6.0–<6.6	13
≥ 6.6	3

14.30 The Institute of Central America and Panama has carried out extensive dietary studies and research projects in Central America. In one study, reported in "The Blood Viscosity of Various Socioeconomic Groups in Guatemala" (*Amer. J. of Clin. Nutr.* (November 1964)), serum total cholesterol measurements for a sample of 49 rural low-income Indians were reported as follows (in mg/L):

204	108	140	152	158	129	175	146	157	174	192	194	144
152	135	223	145	231	115	131	129	142	114	173	226	155
166	220	180	172	143	148	171	143	124	158	144	108	189
136	136	197	131	95	139	181	165	142	162			

a. Use the cholesterol data to construct a frequency distribution. Note: The sample mean and sample standard deviation are 157.02 and 31.75, respectively.

b. Using the frequency distribution in (a), conduct a goodness-of-fit test to determine if it is plausible that the population distribution of total serum cholesterol is normal. Use a .05 significance level.

14.31 In a study similar to that of Exercise 14.30, the blood cholesterol measurements of 500 American males were recorded and summarized in the accompanying frequency distribution (*Primer for the Biomedical Sciences* by O. J. Dunn (New York: John Wiley, 1977)). The corresponding sample mean and standard deviation were calculated to be 275 and 47, respectively.

Cholesterol Level	Frequency
Under 189.5	17
189.5–<219.5	41
219.5–<249.5	80
249.5–<279.5	134
279.5–<309.5	118
309.5–<339.5	71
339.5–<369.5	24
369.5–<399.5	13
≥ 399.5	2

a. Use the X^2 test to determine whether a normal distribution provides an adequate description of the population. Use a .05 level of significance.

b. Do you think the same normal distribution could be used to describe the population of cholesterol measurements of Guatemalan Indians (see Exercise 14.30) and American males? Explain.

14.32 In a Swedish study, the number of males among the first seven children was recorded for each of 1334 families with seven or more children. The accompanying observed counts appeared in the article "Distribution and Sequences of Sexes in a Selected Sample of Swedish Families" (*Ann. Human Genetics* (1960):245–52). The expected counts have been computed under the assumption that the variable x = *number of males in first seven children* has a binomial distribution with $n = 7$ and $\pi = .5$ (for a discussion of the binomial distribution, see Section 5.6). Use the goodness-of-fit test and a .01 level of significance to test the hypotheses

H_0 : x has a binomial distribution with $n = 7$ and $\pi = .5$

H_a : H_0 is not true

	Number of Males							
	0	1	2	3	4	5	6	7
Observed count	6	57	206	362	365	256	69	13
Expected count	10.4	73.0	218.8	364.8	364.8	218.8	73.0	10.4

14.33 The paper "Some Studies on Tuft Weight Distribution in the Opening Room" (*Textile Res. J.* (1976):567–73) reported the accompanying observed counts for output tuft weight (mg) of cotton fibers. The authors hypothesized that the true distribution of tuft weights was described by a distribution called a *truncated exponential distribution* and computed the given expected counts in order to test

H_0 : tuft weight has a particular truncated exponential distribution

H_a : H_0 is not true

Use the goodness-of-fit test with a .01 significance level to test these hypotheses.

Tuft weight	0–<8	8–<16	16–<24	≥24
Observed count	20	8	7	5
Expected count	18.0	9.9	5.5	6.6

14.34 The authors of the paper "Some Sampling Characteristics of Plants and Arthropods of the Arizona Desert" (*Ecology* (1962):567–71) hypothesized that an important type of probability distribution called a *Poisson distribution* could be used to describe the distribution of a number of *Larrea divaricata* plants found in sampling regions of fixed size. The Poisson distribution has a single characteristic, whose value was estimated from the sample data, resulting in the loss of 1 df. Use a .05 significance level and the observed and expected counts given to test

H_0 : number of *Larrea divaricata* plants follows a Poisson distribution

H_a : H_0 is not true

Number of plants	0	1	2	3	4 or More
Observed count	9	9	10	14	6
Expected count	5.9	12.3	13.0	9.0	7.8

14.35 A certain vision test involves marking a measurement scale on a horizontal axis, placing a point at some distance above the axis, and asking the subject to locate the point on the marked scale. Suppose it is hypothesized that errors in location have a normal distribution with mean 0 and standard deviation 1 (a standard normal distribution). A goodness-of-fit test is to be based on the six class intervals $(-\infty, -b)$, $(-b, -a)$, $(-a, 0)$, $(0, a)$, (a, b), and (b, ∞). What values of a and b result in these class intervals having equal probability under the null hypothesis?

KEY CONCEPTS

One-way frequency table A compact way of summarizing data on a categorical variable. It gives the number of times (frequency) that each of the possible categories occurred in the data set. (p. 621)

Two-way frequency table A rectangular table used to summarize a bivariate categorical data set. Two-way tables are used to compare several groups on the basis of a categorical variable or to identify whether an association exists between two categorical variables. (p. 621)

Row proportions, column proportions, and joint relative frequencies They provide insight into the structure of a two-way frequency table. Row and column proportions, obtained by dividing each cell count by the corresponding row or column marginal total, are used to examine the way in which groups might differ with respect to a categorical variable. Joint relative frequencies are obtained from the cell counts through division by the total of all counts in the table. (p. 624–25)

Index of association i A measure of the extent of association in a 2×2 table. It ranges between -1 and $+1$, with a value of zero corresponding to independence. (p. 632)

X^2 statistic A comparison of observed and expected cell counts, it is given by

$$X^2 = \sum \frac{\left(\begin{matrix}\text{observed} \\ \text{cell count}\end{matrix} - \begin{matrix}\text{expected} \\ \text{cell count}\end{matrix}\right)^2}{\text{expected cell count}}$$

When all expected cell counts are at least five, X^2 has (approximately) a chi-squared distribution. The null hypothesis being tested is then rejected if $X^2 >$ chi-squared critical value, where the chi-squared critical value is obtained using the appropriate df. (p. 639)

X^2 test in a one-way frequency table The X^2 statistic is used to test a null hypothesis that specifies the population proportion corresponding to each cell of a one-way table. (p. 640)

X^2 test in a two-way frequency table The X^2 statistic is used to test for homogeneity of several populations based on a sample from each one or to test for the independence of two categorical variables in a single population based on a sample from that population. (p. 646, 649)

X^2 goodness-of-fit test for a population model The X^2 statistic can be used to decide whether it is plausible that a sample was selected from a particular population distribution. It compares the observed counts in a frequency distribution to those expected when the hypothesized population distribution is correct. (p. 657)

SUPPLEMENTARY EXERCISES

14.36 Each driver in a sample of size 1024 was classified according to both seat-belt usage and sex to obtain the accompanying 2 × 2 frequency table ("What Kinds of People Do Not Use Seat Belts" *Amer J. Public Health* (1977):1043–49). Does the data strongly suggest an association between sex and seat belt usage? Use a .05 significance level.

	Seat-Belt Usage	
Sex	Don't Use	Use
Male	192	272
Female	284	276

14.37 One important factor that affects the quality of sorghum, a major world cereal crop, is presence of pigmentation. The paper "A Genetic and Biochemical Study on Pericarp Pigments in a Cross Between Two Cultivars of Grain Sorghum, Sorghum Bicolor" (*Heredity* (1976):413–16) reported on a genetic experiment in which three different pigmentations—red, yellow, or white—were possible. A particular genetic model predicted that these colors would appear in the ratios 9:3:4 (i.e., proportions $\frac{9}{16}$, $\frac{3}{16}$, and $\frac{4}{16}$). The experiment yielded 195 seeds with red pigmentation, 73 with yellow, and 100 with white. Does this data cast doubt on the appropriateness of this genetic theory? Carry out a goodness-of-fit test at significance level .10.

14.38 The Japanese farming community of Achihara was the focus of a study described in the article "Part-Time Farming: A Japanese Example" (*J. Anthro. Res.* (1984):293–305). A random sample of farms for which the head of household was between 45 and 72 years old was selected, and each member of the sample was classified according to size of farm and residence of the farmer's oldest child. Data compatible with that given in the paper is summarized in the two-way table below. Using a .05 significance level, test the null hypothesis of no association between farm size and child's residence.

	Oldest Child's Residence	
Size (ha)	With Parents	Separate Residence
0–<.3	28	8
.3–<1.0	10	8
1.0–<9.0	18	20

14.39 The accompanying 2 × 2 frequency table is the result of classifying random samples of 112 librarians and 108 faculty members of the California State University System with respect to sex ("Job Satisfaction Among Faculty and Librarians: A Study of Gender, Autonomy, and Decision Making Opportunities" *J. Library Admin.* (1984):43–56). Does the data strongly suggest that librarians and faculty members differ with respect to the proportion of males and females? Test using a .05 level of significance.

	Faculty	Librarians
Male	56	59
Female	52	53

14.40 Is there any relationship between age of an investor and the rate of return that the investor expects from an investment? A sample of 972 common stock investors was selected and each was placed in one of four age categories and in one of four *rate believed attainable* categories ("Patterns of Investment Strategy and Behavior Among Individual Investors" *J. Business* (1977):296–333). The resulting data is given. Does there appear to be an association between age and rate believed attainable? Test the appropriate hypotheses using a .01 significance level.

	Rate Believed Attainable			
Investor Age	0–5%	6–10%	11–15%	Over 15%
Under 45	15	51	51	29
45–54	31	133	70	48
55–64	59	139	35	20
65 or over	84	157	32	18

14.41 The given 2 × 5 table is the result of classifying each seed in samples of five different types of fir seed according to whether or not the seed germinated within 5 weeks of planting. The data appeared in the paper "Nondestructive Optical Methods of Food Quality Evaluation" (*Food Sci. and Nutr.* (1984):232–79). Do the five seed types differ with respect to the true proportion that germinate? Test using a .01 significance level. What can be said about the P-value?

	Type of Seed				
	A	B	C	D	E
Germinated	31	57	87	52	10
Failed to germinate	7	33	60	44	19

14.42 The paper "Participation of Senior Citizens in the Swine Flu Inoculation Program" (*J. Gerentology* (1979):201–08) described a study of the factors thought to influence a person's decision to obtain a flu vaccination. Each member of a sample of 122 senior citizens was classified according to belief about the likelihood of getting the flu and vaccine status to obtain the given two-way frequency table. Using a .05 significance level, test to determine if there is an association between belief and vaccine status.

Belief	Vaccine Status: Received Vaccine	Didn't Receive Vaccine
Very unlikely	25	24
Unlikely	30	11
Likely	6	8
Don't know	5	13

14.43 The paper "An Instant Shot of 'Ah': Cocaine Use Among Methadone Clients" (*J. Psychoactive Drugs* (1984):217–27) reported the accompanying data on frequency of cocaine use for individuals in three different treatment groups. Does the data suggest that the true proportion of individuals in each of the different cocaine-use categories differs for the three treatments? Carry out an appropriate test at level .05.

	Treatment		
Cocaine Use	A	B	C
None	149	75	8
1–2 times	26	27	15
3–6 times	6	20	11
At least 7 times	4	10	10

14.44 The paper "Identification of Cola Beverages" (*J. Appl. Psych.* (1962): 356–60) reported on an experiment in which each of 79 subjects was presented with glasses of cola in pairs and asked to identify which glass contained a specific brand of cola. The accompanying data appeared in the paper. Does this data suggest that individuals' abilities to make correct identifications differ for the different brands of cola?

	Number of Correct Identifications			
Cola	0	1	2	3 or 4
Coca·Cola	13	23	24	19
Pepsi Cola	12	20	26	21
Royal Crown	18	28	19	14

14.45 A variety of different probability distributions have been fit to data sets consisting of scores of games in various sports (this is what statisticians do for recreation). The paper "Collegiate Football Scores and the Negative Binomial Distribution" (*J. Amer. Stat. Assoc.* (1973):351–52) proposed a probability distribution called the *negative binomial distribution* for describing the points scored per game by an individual team. Observed and expected counts taken from the paper are given. The expected counts were computed by first estimating two population characteristics (analogous to estimation of μ and σ for the normal distribution); this reduces the number of df for the goodness-of-fit test by 2. Does this distribution provide a good fit to the data?

Points	0–5	6–11	12–17	18–24	25–31	32–38
Observed	272	485	537	407	258	157
Expected	278.7	490.2	509.1	406.6	275.9	167.3

Points	39–45	46–52	53–59	60–66	Over 66
Observed	101	57	23	8	11
Expected	93.5	49.0	24.4	11.7	9.7

14.46 Many shoppers have expressed unhappiness over plans by grocery stores to stop putting prices on individual grocery items. The paper "The Impact of Item Price Removal on Grocery Shopping Behavior" (*J. Marketing* (1980):73–93) reported on a study in which each shopper in a sample was classified by age and by whether or not he or she felt the need for item pricing. Based on the accompanying data, does the need for item pricing appear to be independent of age? (*Hint:* Construct the appropriate two-way frequency table.)

	Age				
	<30	30–39	40–49	50–59	≥60
Number in sample	150	141	82	63	49
Number who want item pricing	127	118	77	61	41

14.47 The paper "Social Class and Corporal Punishment in Childrearing: A Reassessment" (*American Soc. Review* (1974):68–85) reported on a study in which a random sample of 851 adults was obtained and each was classified with respect to three categorical variables: (1) frequency of spanking as a child (infrequently or frequently), (2) age group (18–50 or 51+), and (3) social class (middle class or working class). The data is summarized in the accompanying table.

	Middle Class		Working Class	
	18–50	50+	18–50	50+
Infrequently	145	67	234	140
Frequently	39	28	105	93

Consider the hypothesis that the three variables are independent. Estimated expected cell counts in this case are computed as

$$\frac{\begin{pmatrix}\text{appropriate marginal}\\ \text{total for first}\\ \text{variable}\end{pmatrix}\begin{pmatrix}\text{appropriate marginal}\\ \text{total for second}\\ \text{variable}\end{pmatrix}\begin{pmatrix}\text{appropriate marginal}\\ \text{total for third}\\ \text{variable}\end{pmatrix}}{(\text{sample size})^2}$$

For example, the number infrequently spanked is 586, the number whose age is 18–50 is 523, and the number of middle class individuals is 279, so the estimated expected cell count for the infrequently/18–50/middle class cell is $(586)(523)(279)/(851)^2 = 118.07$, whereas the observed count is 145.

a. Use the above expression to compute the remaining seven estimated expected cell counts for the independence hypothesis.

b. The X^2 statistic is Σ(observed − expected)^2/expected, where the sum is over all cells. In general, if there are c_1, c_2, and c_3 categories for the three variables, the number of df for the chi-squared test is $c_1c_2c_3 - c_1 - c_2 - c_3 + 2$. Carry out the test for the above data using significance level .05.

14.48 The May 13, 1985, issue of the (world-famous) Cal Poly *Mustang Daily* reported that among 293 students asked to taste a sample of vanilla ice cream from each of three unmarked tubs and state a preference, 40% preferred Carnation, 26% preferred Knudsen, and 34% preferred Burnardo'z (a brand favored by local Yuppies). Does this data suggest that students are not equally divided in their preferences for these three brands? State and test the relevant hypotheses using $\alpha = .05$ (and if you've come this far in our book, treat yourself to an ice cream cone!).

REFERENCES

Agresti, Alan, and Barbara Agresti. *Statistical Methods for the Social Sciences.* New York: Dellen 1979. (This book includes a good discussion of measures of association for two-way frequency tables.)

Everitt, B. S. *The Analysis of Contingency Tables.* New York: Halstead Press, 1977. (A compact but informative survey of methods for analyzing categorical data.)

Mosteller, Frederick, and Robert Rourke. *Sturdy Statistics.* Reading, MA.: Addison-Wesley, 1973. (Contains several very readable chapters on the varied uses of the chi-squared statistic.)

APPENDIX STATISTICAL TABLES

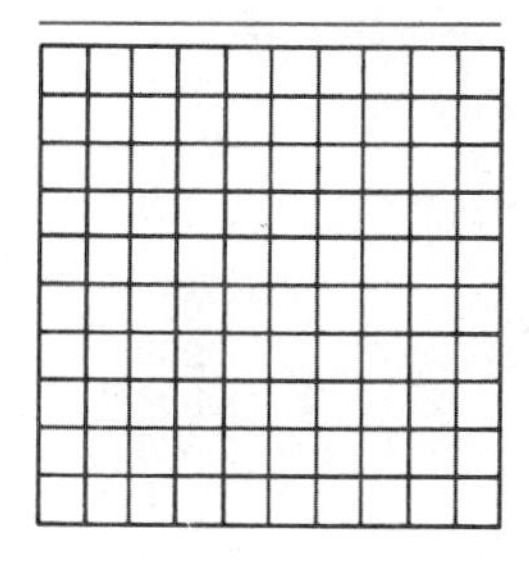

TABLE I STANDARD NORMAL (z) CURVE AREAS

z	Area to the Left of z	Area to the Right of z	z	Area to the Left of z	Area to the Right of z
−5.0	*	**	0	.5000	.5000
−4.5	.000004	.999996	.1	.5398	.4062
−4.0	.000033	.999967	.2	.5793	.4207
−3.5	.0002	.9998	.3	.6179	.3821
−3.4	.0003	.9997	.4	.6554	.3446
−3.3	.0005	.9995	.5	.6915	.3085
−3.2	.0007	.9993	.6	.7257	.2743
−3.1	.0010	.9990	.7	.7580	.2420
−3.0	.0013	.9987	.8	.7881	.2119
−2.9	.0019	.9981	.9	.8159	.1841
−2.8	.0026	.9974	1.0	.8413	.1587
−2.7	.0035	.9965	1.1	.8643	.1357
−2.6	.0047	.9953	1.2	.8849	.1151
−2.58	.0050	.9950	1.28	.9000	.1000
−2.5	.0062	.9938	1.3	.9032	.0968
−2.4	.0082	.9918	1.4	.9192	.0808
−2.33	.0100	.9900	1.5	.9332	.0668
−2.3	.0107	.9893	1.6	.9452	.0548
−2.2	.0139	.9861	1.645	.9500	.0500
−2.1	.0179	.9821	1.7	.9554	.0446
−2.0	.0228	.9772	1.8	.9641	.0359
−1.96	.0250	.9750	1.9	.9713	.0287
−1.9	.0287	.9713	1.96	.9750	.0250
−1.8	.0359	.9641	2.0	.9772	.0228
−1.7	.0446	.9554	2.1	.9821	.0179
−1.645	.0500	.9500	2.2	.9861	.0139
−1.6	.0548	.9452	2.3	.9893	.0107
−1.5	.0668	.9332	2.33	.9900	.0100
−1.4	.0808	.9192	2.4	.9918	.0082
−1.3	.0968	.9032	2.5	.9938	.0062
−1.28	.1000	.9000	2.58	.9950	.0050
−1.2	.1151	.8849	2.6	.9953	.0047
−1.1	.1357	.8643	2.7	.9965	.0035
−1.0	.1587	.8413	2.8	.9974	.0026
−.9	.1841	.8159	2.9	.9981	.0019
−.8	.2119	.7881	3.0	.9987	.0013
−.7	.2420	.7580	3.1	.9990	.0010
−.6	.2743	.7257	3.2	.9993	.0007
−.5	.3085	.6915	3.3	.9995	.0005
−.4	.3446	.6554	3.4	.9997	.0003
−.3	.3821	.6179	3.5	.9998	.0002
−.2	.4207	.5793	4.0	.999967	.000033
−.1	.4602	.5398	4.5	.999996	.000004
0	.5000	.5000	5.0	**	*

* = .00000029, or zero to six decimal places

** = .99999971, or one to six decimal places

TABLE II

EXPECTED NORMAL SCORES ($\mu = 0, \sigma = 1$)

	n			
Ordered Position	10	20	25	30
1	−1.539	−1.867	−1.965	−2.043
2	−1.001	−1.408	−1.524	−1.616
3	−.656	−1.131	−1.263	−1.365
4	−.376	−.921	−1.067	−1.179
5	−.123	−.745	−.905	−1.026
6	.123	−.590	−.764	−.894
7	.376	−.448	−.637	−.777
8	.656	−.315	−.519	−.669
9	1.001	−.187	−.409	−.568
10	1.539	−.062	−.303	−.473
11		.062	−.200	−.382
12		.187	−.100	−.294
13		.315	.000	−.209
14		.448	.100	−.125
15		.590	.200	−.041
16		.745	.303	.041
17		.921	.409	.125
18		1.131	.519	.209
19		1.408	.637	.294
20		1.867	.764	.382
21			.905	.473
22			1.067	.568
23			1.263	.669
24			1.524	.777
25			1.965	.894
26				1.026
27				1.179
28				1.365
29				1.616
30				2.043

TABLE III BINOMIAL PROBABILITIES

$n = 10$

x \ π	0.05	0.1	0.2	0.25	0.3	0.4	0.5	0.6	0.7	0.75	0.8	0.9	0.95
0	.599	.349	.107	.056	.028	.006	.001	.000	.000	.000	.000	.000	.000
1	.315	.387	.268	.188	.121	.040	.010	.002	.000	.000	.000	.000	.000
2	.075	.194	.302	.282	.233	.121	.044	.011	.001	.000	.000	.000	.000
3	.010	.057	.201	.250	.267	.215	.117	.042	.009	.003	.001	.000	.000
4	.001	.011	.088	.146	.200	.251	.205	.111	.037	.016	.006	.000	.000
5	.000	.001	.026	.058	.103	.201	.246	.201	.103	.058	.026	.001	.000
6	.000	.000	.006	.016	.037	.111	.205	.251	.200	.146	.088	.011	.001
7	.000	.000	.001	.003	.009	.042	.117	.215	.267	.250	.201	.057	.010
8	.000	.000	.000	.000	.001	.011	.044	.121	.233	.282	.302	.194	.075
9	.000	.000	.000	.000	.000	.002	.010	.040	.121	.188	.268	.387	.315
10	.000	.000	.000	.000	.000	.000	.001	.006	.028	.056	.107	.349	.599

$n = 20$

x \ π	0.05	0.1	0.2	0.25	0.3	0.4	0.5	0.6	0.7	0.75	0.8	0.9	0.95
0	.358	.122	.012	.003	.001	.000	.000	.000	.000	.000	.000	.000	.000
1	.377	.270	.058	.021	.007	.000	.000	.000	.000	.000	.000	.000	.000
2	.189	.285	.137	.067	.028	.003	.000	.000	.000	.000	.000	.000	.000
3	.060	.190	.205	.134	.072	.012	.001	.000	.000	.000	.000	.000	.000
4	.013	.090	.218	.190	.130	.035	.005	.000	.000	.000	.000	.000	.000
5	.002	.032	.175	.202	.179	.075	.015	.001	.000	.000	.000	.000	.000
6	.000	.009	.109	.169	.192	.124	.037	.005	.000	.000	.000	.000	.000
7	.000	.002	.055	.112	.164	.166	.074	.015	.001	.000	.000	.000	.000
8	.000	.000	.022	.061	.114	.180	.120	.035	.004	.001	.000	.000	.000
9	.000	.000	.007	.027	.065	.160	.160	.071	.012	.003	.000	.000	.000
10	.000	.000	.002	.010	.031	.117	.176	.117	.031	.010	.002	.000	.000
11	.000	.000	.000	.003	.012	.071	.160	.160	.065	.027	.007	.000	.000
12	.000	.000	.000	.001	.004	.035	.120	.180	.114	.061	.022	.000	.000
13	.000	.000	.000	.000	.001	.015	.074	.166	.164	.112	.055	.002	.000
14	.000	.000	.000	.000	.000	.005	.037	.124	.192	.169	.109	.009	.000
15	.000	.000	.000	.000	.000	.001	.015	.075	.179	.202	.175	.032	.002
16	.000	.000	.000	.000	.000	.000	.005	.035	.130	.190	.218	.090	.013
17	.000	.000	.000	.000	.000	.000	.001	.012	.072	.134	.205	.190	.060
18	.000	.000	.000	.000	.000	.000	.000	.003	.028	.067	.137	.285	.189
19	.000	.000	.000	.000	.000	.000	.000	.000	.007	.021	.058	.270	.377
20	.000	.000	.000	.000	.000	.000	.000	.000	.001	.003	.012	.122	.358

TABLE IV *t* CRITICAL VALUES

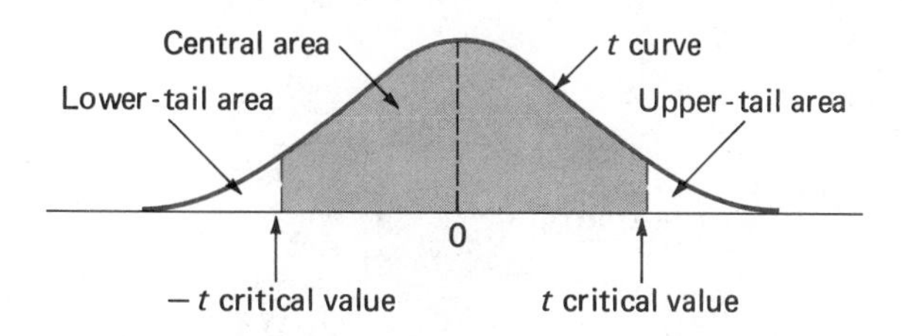

Central Area Captured Confidence Level	.80 80%	.90 90%	.95 95%	.98 98%	.99 99%	.998 99.8%	.999 99.9%
Degrees of Freedom 1	3.08	6.31	12.71	31.82	63.66	318.31	636.62
2	1.89	2.92	4.30	6.97	9.93	23.33	31.60
3	1.64	2.35	3.18	4.54	5.84	10.21	12.92
4	1.53	2.13	2.78	3.75	4.60	7.17	8.61
5	1.48	2.02	2.57	3.37	4.03	5.89	6.86
6	1.44	1.94	2.45	3.14	3.71	5.21	5.96
7	1.42	1.90	2.37	3.00	3.50	4.79	5.41
8	1.40	1.86	2.31	2.90	3.36	4.50	5.04
9	1.38	1.83	2.26	2.82	3.25	4.30	4.78
10	1.37	1.81	2.23	2.76	3.17	4.14	4.59
11	1.36	1.80	2.20	2.72	3.11	4.03	4.44
12	1.36	1.78	2.18	2.68	3.06	3.93	4.32
13	1.35	1.77	2.16	2.65	3.01	3.85	4.22
14	1.35	1.76	2.15	2.62	2.98	3.79	4.14
15	1.34	1.75	2.13	2.60	2.95	3.73	4.07
16	1.34	1.75	2.12	2.58	2.92	3.69	4.02
17	1.33	1.74	2.11	2.57	2.90	3.65	3.97
18	1.33	1.73	2.10	2.55	2.88	3.61	3.92
19	1.33	1.73	2.09	2.54	2.86	3.58	3.88
20	1.33	1.73	2.09	2.53	2.85	3.55	3.85
21	1.32	1.72	2.08	2.52	2.83	3.53	3.82
22	1.32	1.72	2.07	2.51	2.82	3.51	3.79
23	1.32	1.71	2.07	2.50	2.81	3.49	3.77
24	1.32	1.71	2.06	2.49	2.80	3.47	3.75
25	1.32	1.71	2.06	2.49	2.79	3.45	3.73
26	1.32	1.71	2.06	2.48	2.78	3.44	3.71
27	1.31	1.70	2.05	2.47	2.77	3.42	3.69
28	1.31	1.70	2.05	2.47	2.76	3.41	3.67
29	1.31	1.70	2.05	2.46	2.76	3.40	3.66
30	1.31	1.70	2.04	2.46	2.75	3.39	3.65
40	1.30	1.68	2.02	2.42	2.70	3.31	3.55
60	1.30	1.67	2.00	2.39	2.66	3.23	3.46
120	1.29	1.66	1.98	2.36	2.62	3.16	3.37
z critical values	1.28	1.645	1.96	2.33	2.58	3.09	3.29
Level of significance for a *two*-tailed test	.20	.10	.05	.02	.01	.002	.001
Level of significance for a *one*-tailed test	.10	.05	.025	.01	.005	.001	.0005

TABLE V CURVES OF $\beta = P$(TYPE II ERROR) FOR t TESTS

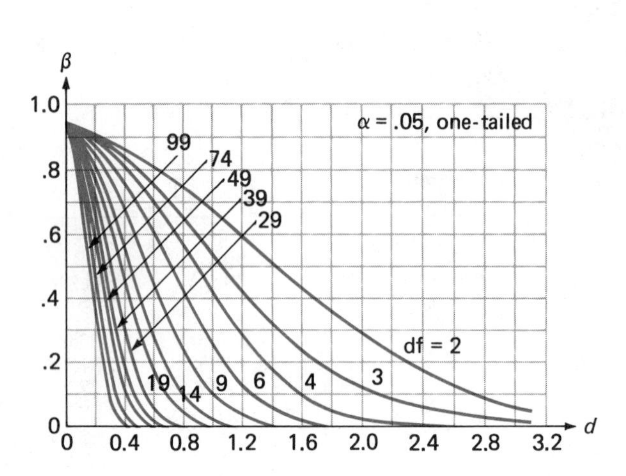

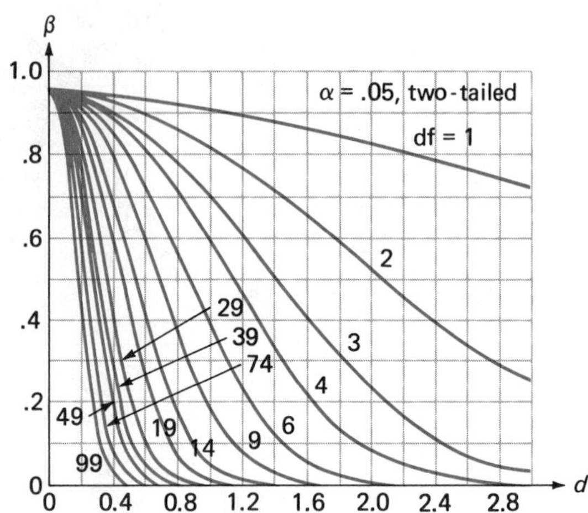

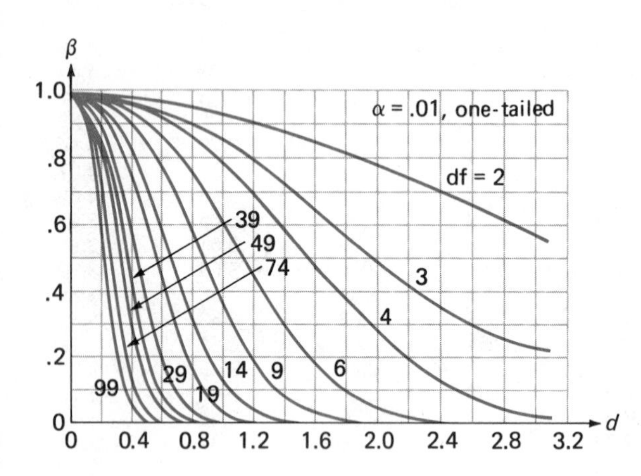

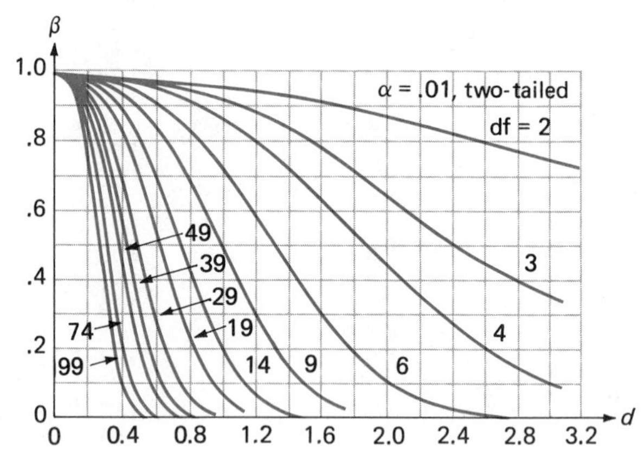

TABLE VI

CRITICAL VALUES FOR THE RANK SUM TEST*

		Upper-Tailed Test		Lower-Tailed Test		Two-Tailed Test	
n_1	n_2	$\alpha = .05$	$\alpha = .01$	$\alpha = .05$	$\alpha = .01$	$\alpha = .05$	$\alpha = .01$
3	3	15	—	6	—	—	—
3	4	17	—	7	—	18,6	—
3	5	20	21	7	6	21,6	—
3	6	22	24	8	6	23,7	—
3	7	24	27	9	6	26,7	27,6
3	8	27	29	9	7	28,8	30,6
4	3	21	—	11	—	—	—
4	4	24	26	12	10	25,11	—
4	5	27	30	13	10	29,11	30,10
4	6	30	33	14	11	32,12	34,10
4	7	33	36	15	12	35,13	37,11
4	8	36	40	16	12	38,14	41,11
5	3	29	30	16	15	30,15	—
5	4	32	35	18	15	34,16	35,15
5	5	36	39	19	16	37,18	39,16
5	6	40	43	20	17	41,19	44,16
5	7	43	47	22	18	45,20	48,17
5	8	47	51	23	19	49,21	52,18
6	3	37	39	23	21	38,22	—
6	4	41	44	25	22	43,23	45,21
6	5	46	49	26	23	47,25	50,22
6	6	50	54	28	24	52,26	55,23
6	7	54	58	30	26	56,28	60,24
6	8	58	63	32	27	61,29	65,25
7	3	46	49	31	28	48,29	49,28
7	4	51	54	33	30	53,31	55,29
7	5	56	60	35	31	58,33	61,30
7	6	61	65	37	33	63,35	67,31
7	7	66	71	39	34	68,37	72,33
7	8	71	76	41	36	73,39	78,34
8	3	57	59	39	37	58,38	60,36
8	4	62	66	42	38	64,40	67,37
8	5	68	72	44	40	70,42	73,39
8	6	73	78	47	42	76,44	80,40
8	7	79	84	49	44	81,47	86,42
8	8	84	90	52	46	87,49	92,44

*As explained in Chapter 9, the significance levels .05 and .01 cannot be achieved exactly for most sample sizes. For example, in the case $n_1 = 3$, $n_2 = 6$, the actual significance level for the test that rejects H_0 if *rank sum* ≥ 24 is .012, as close to .01 as can be achieved. Whenever a critical value is missing (e.g., $n_1 = 3$, $n_2 = 4$, $\alpha = .01$), it is not possible to get close to the desired α.

TABLE VII VALUES OF d FOR THE RANK SUM CONFIDENCE INTERVAL (dth SMALLEST DIFFERENCE, dth LARGEST DIFFERENCE)

Sample Sizes*	d for 90% Confidence	d for 95% Confidence	d for 99% Confidence
5, 5	5	4	1
5, 6	6	5	2
5, 7	8	6	3
5, 8	9	7	4
5, 9	11	8	5
5, 10	12	10	5
5, 11	13	11	6
5, 12	15	12	7
6, 6	8	6	3
6, 7	10	8	4
6, 8	12	9	5
6, 9	13	11	6
6, 10	15	13	8
6, 11	17	14	9
6, 12	19	16	10
7, 7	12	10	6
7, 8	14	12	7
7, 9	16	14	8
7, 10	19	16	10
7, 11	21	17	12
7, 12	23	19	13
8, 8	17	14	9
8, 9	19	16	11
8, 10	22	19	12
8, 11	24	21	14
8, 12	27	23	16
9, 9	22	19	13
9, 10	25	22	15
9, 11	28	24	17
9, 12	31	27	19
10, 10	29	25	17
10, 11	32	28	20
10, 12	35	31	22
11, 11	36	31	23
11, 12	40	35	25
12, 12	44	39	29

*When $n_1 \neq n_2$, the smaller sample size is listed first.

TABLE VIII

CRITICAL VALUES FOR THE SIGNED-RANK TEST

n	Significance Level for One-Tailed Test	Significance Level for Two-Tailed Test	Critical Value
5	.031	.062	15
	.062	.124	13
	.094	.188	11
6	.016	.032	21
	.031	.062	19
	.047	.094	17
	.109	.218	13
7	.008	.016	28
	.023	.046	24
	.055	.110	20
	.109	.218	16
8	.012	.024	32
	.027	.054	28
	.055	.110	24
	.098	.196	20
9	.010	.020	39
	.027	.054	33
	.049	.098	29
	.102	.204	23
10	.010	.020	45
	.024	.048	39
	.053	.106	33
	.097	.194	27
11	.009	.018	52
	.027	.054	44
	.051	.102	38
	.103	.206	30
12	.010	.020	58
	.026	.052	50
	.046	.092	44
	.102	.204	34
13	.011	.022	65
	.024	.048	57
	.047	.094	49
	.095	.190	39
14	.010	.020	73
	.025	.050	63
	.052	.104	53
	.097	.194	43
15	.011	.022	80
	.024	.048	70
	.047	.094	60
	.104	.208	46
16	.011	.022	88
	.025	.054	76
	.052	.104	64
	.096	.192	52

continued

TABLE VIII CRITICAL VALUES FOR THE SIGNED-RANK TEST (continued)

n	Significance Level for One-Tailed Test	Significance Level for Two-Tailed Test	Critical Value
17	.010	.020	97
	.025	.050	83
	.049	.098	71
	.103	.206	55
18	.010	.020	105
	.025	.048	91
	.049	.098	77
	.098	.196	61
19	.010	.020	114
	.025	.050	98
	.052	.104	82
	.098	.196	66
20	.010	.020	124
	.024	.048	106
	.049	.098	90
	.101	.202	70

TABLE IX VALUES OF d FOR THE SIGNED-RANK CONFIDENCE INTERVAL

n	Confidence Level	d	n	Confidence Level	d
5	93.8	1	14	99.1	13
	87.5	2		95.1	22
				89.6	27
6	96.9	1			
	90.6	3	15	99.0	17
				95.2	26
7	98.4	1		90.5	31
	96.9	2			
	89.1	5	16	99.1	20
				94.9	31
8	99.2	1		89.5	37
	94.5	5			
	89.1	7	17	99.1	25
				94.9	36
9	99.2	2		90.2	42
	94.5	7			
	90.2	9	18	99.0	29
				95.2	41
10	99.0	4		90.1	48
	95.1	9			
	89.5	12	19	99.1	33
				95.1	47
11	99.0	6		90.4	54
	94.6	12			
	89.8	15	20	99.1	38
				95.2	53
12	99.1	8		90.3	61
	94.8	15			
	90.8	18			
13	99.0	11			
	95.2	18			
	90.6	22			

TABLE X BONFERRONI 95% t CRITICAL VALUES

Number of df	Number of Intervals						
	2	3	4	5	6	10	15
2	6.21	7.65	8.86	9.92	10.89	14.09	17.28
3	4.18	4.86	5.39	5.84	6.23	7.45	8.58
4	3.50	3.96	4.31	4.60	4.85	5.60	6.25
5	3.16	3.53	3.81	4.03	4.22	4.77	5.25
6	2.97	3.29	3.52	3.71	3.86	4.32	4.70
7	2.84	3.13	3.34	3.50	3.64	4.03	4.36
8	2.75	3.02	3.21	3.36	3.48	3.83	4.12
9	2.69	2.93	3.11	3.25	3.36	3.69	3.95
10	2.63	2.87	3.04	3.17	3.28	3.58	3.83
11	2.59	2.82	2.98	3.11	3.21	3.50	3.73
12	2.56	2.78	2.93	3.05	3.15	3.43	3.65
13	2.53	2.75	2.90	3.01	3.11	3.37	3.58
14	2.51	2.72	2.86	2.98	3.07	3.33	3.53
15	2.49	2.69	2.84	2.95	3.04	3.29	3.48
16	2.47	2.67	2.81	2.92	3.01	3.25	3.44
17	2.46	2.66	2.79	2.90	2.98	3.22	3.41
18	2.45	2.64	2.77	2.88	2.96	3.20	3.38
19	2.43	2.63	2.76	2.86	2.94	3.17	3.35
20	2.42	2.61	2.74	2.85	2.93	3.15	3.33
21	2.41	2.60	2.73	2.83	2.91	3.14	3.31
22	2.41	2.59	2.72	2.82	2.90	3.12	3.29
23	2.40	2.58	2.71	2.81	2.89	3.10	3.27
24	2.39	2.57	2.70	2.80	2.88	3.09	3.26
25	2.38	2.57	2.69	2.79	2.86	3.08	3.24
26	2.38	2.56	2.68	2.78	2.86	3.07	3.23
27	2.37	2.55	2.68	2.77	2.85	3.06	3.22
28	2.37	2.55	2.67	2.76	2.84	3.05	3.21
29	2.36	2.54	2.66	2.76	2.83	3.04	3.20
30	2.36	2.54	2.66	2.75	2.82	3.03	3.19
40	2.33	2.50	2.62	2.70	2.78	2.97	3.12
60	2.30	2.46	2.58	2.66	2.73	2.91	3.06
120	2.27	2.43	2.54	2.62	2.68	2.86	3.00
∞	2.24	2.39	2.50	2.58	2.64	2.81	2.94

TABLE XI(a) *F* DISTRIBUTION CRITICAL VALUES FOR TESTS WITH SIGNIFICANCE LEVEL .05

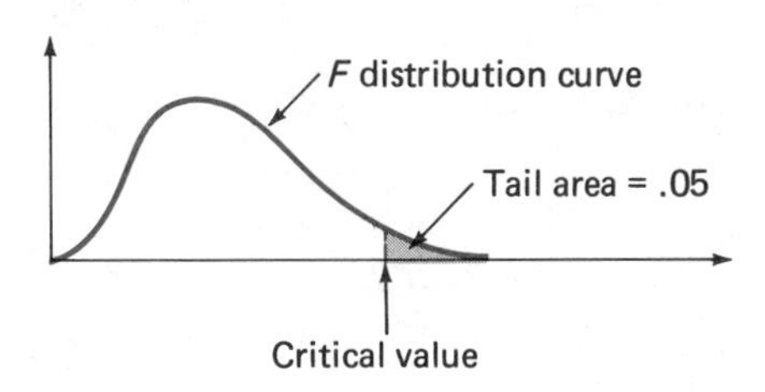

Denominator Degrees of Freedom	Numerator Degrees of Freedom 1	2	3	4	5	6	7	8	9	10
1	161.4	199.5	215.7	224.6	230.2	234.0	236.8	238.9	240.5	241.9
2	18.51	19.00	19.16	19.25	19.30	19.33	19.35	19.37	19.38	19.40
3	10.13	9.55	9.28	9.12	9.01	8.94	8.89	8.85	8.81	8.79
4	7.71	6.94	6.59	6.39	6.26	6.16	6.09	6.04	6.00	5.96
5	6.61	5.79	5.41	5.19	5.05	4.95	4.88	4.82	4.77	4.74
6	5.99	5.14	4.76	4.53	4.39	4.28	4.21	4.15	4.10	4.06
7	5.59	4.74	4.35	4.12	3.97	3.87	3.79	3.73	3.68	3.64
8	5.32	4.46	4.07	3.84	3.69	3.58	3.50	3.44	3.39	3.35
9	5.12	4.26	3.86	3.63	3.48	3.37	3.29	3.23	3.18	3.14
10	4.96	4.10	3.71	3.48	3.33	3.22	3.14	3.07	3.02	2.98
11	4.84	3.98	3.59	3.36	3.20	3.09	3.01	2.95	2.90	2.85
12	4.75	3.89	3.49	3.26	3.11	3.00	2.91	2.85	2.80	2.75
13	4.67	3.81	3.41	3.18	3.03	2.92	2.83	2.77	2.71	2.67
14	4.60	3.74	3.34	3.11	2.96	2.85	2.76	2.70	2.65	2.60
15	4.54	3.68	3.29	3.06	2.90	2.79	2.71	2.64	2.59	2.54
16	4.49	3.63	3.24	3.01	2.85	2.74	2.66	2.59	2.54	2.49
17	4.45	3.59	3.20	2.96	2.81	2.70	2.61	2.55	2.49	2.45
18	4.41	3.55	3.16	2.93	2.77	2.66	2.58	2.51	2.46	2.41
19	4.38	3.52	3.13	2.90	2.74	2.63	2.54	2.48	2.42	2.38
20	4.35	3.49	3.10	2.87	2.71	2.60	2.51	2.45	2.39	2.35
21	4.32	3.47	3.07	2.84	2.68	2.57	2.49	2.42	2.37	2.32
22	4.30	3.44	3.05	2.82	2.66	2.55	2.46	2.40	2.34	2.30
23	4.28	3.42	3.03	2.80	2.64	2.53	2.44	2.37	2.32	2.27
24	4.26	3.40	3.01	2.78	2.62	2.51	2.42	2.36	2.30	2.25
25	4.24	3.39	2.99	2.76	2.60	2.49	2.40	2.34	2.28	2.24
26	4.23	3.37	2.98	2.74	2.59	2.47	2.39	2.32	2.27	2.22
27	4.21	3.35	2.96	2.73	2.57	2.46	2.37	2.31	2.25	2.20
28	4.20	3.34	2.95	2.71	2.56	2.45	2.36	2.29	2.24	2.19
29	4.18	3.33	2.93	2.70	2.55	2.43	2.35	2.28	2.22	2.18
30	4.17	3.32	2.92	2.69	2.53	2.42	2.33	2.27	2.21	2.16
40	4.08	3.23	2.84	2.61	2.45	2.34	2.25	2.18	2.12	2.08
60	4.00	3.15	2.76	2.53	2.37	2.25	2.17	2.10	2.04	1.99
120	3.92	3.07	2.68	2.45	2.29	2.17	2.09	2.02	1.96	1.91
∞	3.84	3.00	2.60	2.37	2.21	2.10	2.01	1.94	1.88	1.83

TABLE XI(b): *F* DISTRIBUTION CRITICAL VALUES FOR TESTS WITH SIGNIFICANCE LEVEL .01

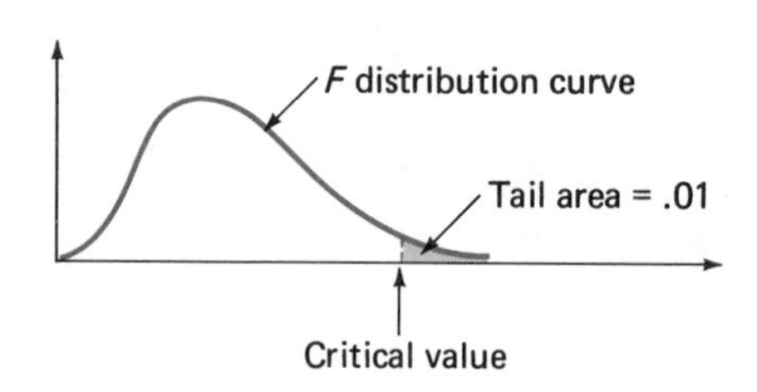

Denominator Degrees of Freedom	Numerator Degrees of Freedom 1	2	3	4	5	6	7	8	9	10
1	4,052	4,999.5	5,403	5,625	5,764	5,859	5,928	5,982	6,022	6,056
2	98.50	99.00	99.17	99.25	99.30	99.33	99.36	99.37	99.39	99.40
3	34.12	30.82	29.46	28.71	28.24	27.91	27.67	27.49	27.35	27.23
4	21.20	18.00	16.69	15.98	15.52	15.21	14.98	14.80	14.66	14.55
5	16.26	13.27	12.06	11.39	10.97	10.67	10.46	10.29	10.16	10.05
6	13.75	10.92	9.78	9.15	8.75	8.47	8.26	8.10	7.98	7.87
7	12.25	9.55	8.45	7.85	7.46	7.19	6.99	6.84	6.72	6.62
8	11.26	8.65	7.59	7.01	6.63	6.37	6.18	6.03	5.91	5.81
9	10.56	8.02	6.99	6.42	6.06	5.80	5.61	5.47	5.35	5.26
10	10.04	7.56	6.55	5.99	5.64	5.39	5.20	5.06	4.94	4.85
11	9.65	7.21	6.22	5.67	5.32	5.07	4.89	4.74	4.63	4.54
12	9.33	6.93	5.95	5.41	5.06	4.82	4.64	4.50	4.39	4.30
13	9.07	6.70	5.74	5.21	4.86	4.62	4.44	4.30	4.19	4.10
14	8.86	6.51	5.56	5.04	4.69	4.46	4.28	4.14	4.03	3.94
15	8.68	6.36	5.42	4.89	4.56	4.32	4.14	4.00	3.89	3.80
16	8.53	6.23	5.29	4.77	4.44	4.20	4.03	3.89	3.78	3.69
17	8.40	6.11	5.18	4.67	4.34	4.10	3.93	3.79	3.68	3.59
18	8.29	6.01	5.09	4.58	4.25	4.01	3.84	3.71	3.60	3.51
19	8.18	5.93	5.01	4.50	4.17	3.94	3.77	3.63	3.52	3.43
20	8.10	5.85	4.94	4.43	4.10	3.87	3.70	3.56	3.46	3.37
21	8.02	5.78	4.87	4.37	4.04	3.81	3.64	3.51	3.40	3.31
22	7.95	5.72	4.82	4.31	3.99	3.76	3.59	3.45	3.35	3.26
23	7.88	5.66	4.76	4.26	3.94	3.71	3.54	3.41	3.30	3.21
24	7.82	5.61	4.72	4.22	3.90	3.67	3.50	3.36	3.26	3.17
25	7.77	5.57	4.68	4.18	3.85	3.63	3.46	3.32	3.22	3.13
26	7.72	5.53	4.64	4.14	3.82	3.59	3.42	3.29	3.18	3.09
27	7.68	5.49	4.60	4.11	3.78	3.56	3.39	3.26	3.15	3.06
28	7.64	5.45	4.57	4.07	3.75	3.53	3.36	3.23	3.12	3.03
29	7.60	5.42	4.54	4.04	3.73	3.50	3.33	3.20	3.09	3.00
30	7.56	5.39	4.51	4.02	3.70	3.47	3.30	3.17	3.07	2.98
40	7.31	5.18	4.31	3.83	3.51	3.29	3.12	2.99	2.89	2.80
60	7.08	4.98	4.13	3.65	3.34	3.12	2.95	2.82	2.72	2.63
120	6.85	4.79	3.95	3.48	3.17	2.96	2.79	2.66	2.56	2.47
∞	6.63	4.61	3.78	3.32	3.02	2.80	2.64	2.51	2.41	2.32

TABLE XII CHI-SQUARED DISTRIBUTION CRITICAL VALUES

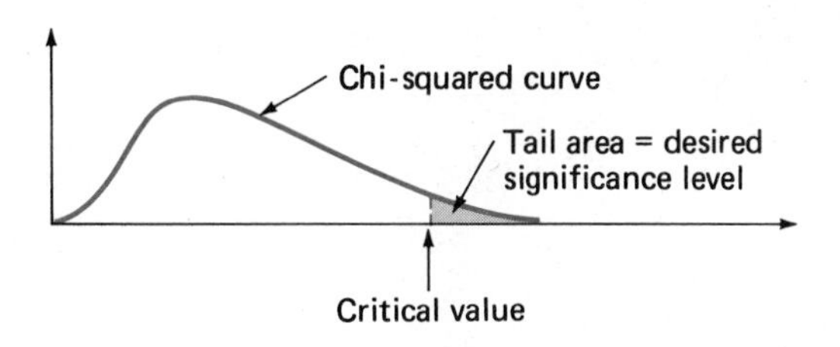

		Significance Level			
		.10	.05	.01	.001
Degrees of Freedom	1	2.71	3.84	6.64	10.83
	2	4.61	5.99	9.21	13.82
	3	6.25	7.82	11.34	16.27
	4	7.78	9.49	13.28	18.47
	5	9.24	11.07	15.09	20.52
	6	10.64	12.59	16.81	22.46
	7	12.02	14.07	18.48	24.32
	8	13.36	15.51	20.09	26.12
	9	14.68	16.92	21.67	27.88
	10	15.99	18.31	23.21	29.59
	11	17.28	19.68	24.72	31.26
	12	18.55	21.03	26.22	32.91
	13	19.81	22.36	27.69	34.53
	14	21.06	23.68	29.14	36.12
	15	22.31	25.00	30.58	37.70
	16	23.54	26.30	32.00	39.25
	17	24.77	27.59	33.41	40.79
	18	25.99	28.87	34.81	42.31
	19	27.20	30.14	36.19	43.82
	20	28.41	31.41	37.57	45.31

ANSWERS TO SELECTED EXERCISES

CHAPTER TWO

2.1 **a.** categorical **b.** categorical **c.** numerical (discrete) **d.** numerical (continuous) **e.** categorical **f.** numerical (continuous)
2.7 no
2.11 **b.** .227 **c.** .272, .136
2.15 **b.** .32
2.17 **d.** .8651 **e.** .5506 **f.** 20.46 months **g.** 29.21 months
2.23 **b.** .0908 **c.** .2382
2.31 (i) symmetric (ii) positively skewed (iii) bimodal, approximately symmetric (iv) bimodal (v) negatively skewed
2.33 **c.** both are positively skewed
2.37 **a.** positively skewed
2.39 **a.** .3167
2.47 **b.** positively skewed **d.** (i) .2979 (ii) .4043 (iii) .2978

CHAPTER THREE

3.1 **a.** 54.59 **b.** 55 **c.** 91.18
3.3 .7
3.5 **a.** $\bar{x} = 36.33$, median = 35 **b.** 35.42
3.9 $1738.10
3.13 **a.** 12.93 **c.** $s^2 = .16204$, $s = .4025$
3.15 **a.** 48.364 **b.** $s^2 = 327.05$, $s = 18.08$
3.17 **a.** $s^2 = 9.061$, $s = 3.01$ **b.** 13.05, 7.95
3.19 $\bar{x} = 10.02$, $s^2 = .044$, $s = .2098$
3.21 $\bar{x} = 483.63636$, $s^2 = 32705.4545$, $s = 180.846$
3.23 **a.** sample 1: $\bar{x} = 7.81$, $s = .39847$ sample 2: $\bar{x} = 49.68$, $s = 1.73897$ **b.** 5.10, 3.50
3.25 .16
3.27 **a.** 11, 18.5, 7.5 **b.** no
3.31 **a.** no **b.** at most .28
3.35 at least .75
3.37 **a.** −1.86 **b.** 0.36 **c.** −0.43 **d.** 2.29
3.39 0.086
3.41 $\bar{x} = 1.66808$, median = 1.285, $s = 1.0925$
3.43 **a.** $\bar{x} = 103.83$, median = 82.5, $s^2 = 3497.8091$, $s = 59.14$ **b.** 58, 137.5, 79.5, no
3.45 **a.** $\bar{x} = 8.006$, $s^2 = .0824$. $s = .287$ **b.** 10% trimmed mean = 7.976, median = 7.91 **c.** 8.01, 7.82, 0.19
3.47 **a.** $\bar{x} = 22.15$, $s = 11.366$ **b.** 19.4375 **c.** 20.5, 18, 2.5 **d.** 25 and 28 are mild outliers, 69 is an extreme outlier

CHAPTER FOUR

4.5 **a.** linear with a positive slope **b.** no
4.9 .9963
4.11 .75, moderate positive
4.13 .682, moderate
4.15 .355
4.17 **b.** .981 **c.** 1
4.19 **a.** (i) .822 (ii) .637 (iii) .593 **b.** students and nonstudents
4.23 **a.** .95
4.27 **a.** $1.415 - .0074x$ **b.** $-.0074$, $-.074$
4.29 **b.** positive **c.** $-1.61 + 13.983x$ **d.** 5.52
4.31 **a.** $22.13 + .1999x$ **b.** .1999
4.33 **a.** $4027.083 - 577.895x$ **b.** negative
4.37 **a.** yes **b.** $112.58x$, 28.145
4.41 **b.** 630.02 **c.** larger
4.43 **a.** $11.849 - 1.5122x$ **b.** SSResid = 29.80933
4.45 $94.33 - 15.389x$, SSResid = 285.66
4.47 .538
4.49 $r^2 = .693$, $s_e = 7.085$
4.59 **d.** $\log y$ and $\log x$
4.67 **a.** $291.3 - 3.3755x$ **b.** $r^2 = .793$
4.69 **a.** .688 **b.** .5714
4.71 Model A: SSResid = 6.67, Model B: SSResid = 4.11
4.73 **b.** .0686 **d.** $-.246$
4.75 .805
4.77 .132, no
4.79 **a.** 0
4.81 **a.** $-.981$, yes

CHAPTER FIVE

5.3 **a.** .0119 **b.** .00000238 **c.** .012
5.5 **a.** .91 **b.** at least 4
5.7 **b.** (1, 2, 4, 3), (1, 4, 3, 2), (1, 3, 2, 4), (2, 1, 3, 4), (3, 2, 1, 4) and (4, 2, 3, 1); 6/24 = .25 **c.** .33 **d.** 0 **e.** .2917
5.9 **a.** .000495, .00198 **b.** .02432 **c.** .02531 **d.** .14592
5.11 **a.** .12, .88 **b.** .18, .38
5.13 **a.** .82 **b.** .18 **c.** .65, .27

5.15 **b.**

x	0	1	2
$p(x)$	.3	.6	.1

5.19 **a.**

x	0	1	2	3	4
$p(x)$	.1296	.3456	.3456	.1536	.0256

b. 1 and 2

5.21 **b.** .5, .125, .125 **c.** .75
5.23 **a.** 46.5 **b.** 890
5.27 3.114, .6364
5.29 **a.** 1.8 **b.** 1.56, 1.249
5.33 **a.** 1.3 **b.** -1.3 **c.** .7 **d.** $-.7$ **e.** 1.645
5.35 31.5, 17.0
5.37 **a.** .6826 **b.** .5000 **c.** .0013 **d.** 5.28 **e.** .6826

5.39 yes
5.41 660 hours
5.43 second machine
5.45 yes
5.55 **a.** .735 **b.** .392 **c.** .07
5.57 200, 13.42
5.59 **a.** .166 **b.** .633 **c.** .952
5.61 **a.** .25 **b.** .29
5.63 20/36
5.65 **a.** .9000 **b.** .3118

5.67 **a.**

x	1	2	3	4
$p(x)$	.4	.3	.2	.1

b. 2, 1, 1

5.69 **a.**

y	3	4	5
$p(y)$	.2800	.3744	.3456

b. 4.0656

5.71 **a.** .250 **b.** .003 **c.** .992
5.75 **a.** .05 **b.** .10 **c.** .10 **d.** .40 **e.** .4, .1

f.

x	0	1	2	3
$p(x)$	.4	.3	.2	.1

CHAPTER SIX

6.3 The sample means for the 10 samples are: 29.83, 29.9, 30.1, 29.93, 30.13, 30.2, 30.03, 30.23, 30.3, and 30.33. Each of these values has probability .1.
6.9 **b.** 3.5, 1.7078
6.13 **a.** approximately normal with mean 50 and standard deviation 1.875 **b.** .4452 **c.** .007
6.15 **a.** .8185, .0013 **b.** .977167, 0
6.17 **a.** .9544 **b.** 2, 3
6.19 **a.** .5, .072 **b.** .5, .041
6.25 **a.** when $\pi = .5$, $\mu_p = .5$ and $\sigma_p = .0333$ when $\pi = .6$, $\mu_p = .6$ and $\sigma_p = .0327$ **b.** .0013, .5
6.27 **a.** $s = 2.04$ **b.** sample median = 11.35 **c.** 10% trimmed mean = 11.83 **d.** 14.70
6.29 111.5

6.31 **c.**

$\bar{x}$	12	13.5	15	16	17.5	20
$p(\bar{x})$	.25	.20	.04	.30	.12	.09

6.33 **a.** .4 **b.** .16
6.35 **a.** minimum of the observed values **b.** 152
6.37 **a.** approximately normal with $\mu = 50$ and $\sigma = .1$ **b.** .9876 **c.** .5
6.39 .1587, .0013
6.41 73.2

CHAPTER SEVEN

7.1 **a.** 1.96 **b.** 1.645 **c.** 2.58 **d.** 1.28 **e.** approximately 1.45
7.3 (85.08, 92.92)
7.5 (349.42, 354.58), yes
7.7 246
7.9 **a.** for boys: (99.486, 103.914) for girls: (99.423, 104.177) **b.** (83.05, 90.35)

7.11 **b.** (9.039, 9.361)
7.13 (35.90, 43.70), no
7.17 **a.** 2.12 **b.** 1.80 **c.** 2.81 **d.** 1.71 **e.** 1.78 **f.** 2.26
7.19 (84.7, 113.3)
7.21 (5.77, 28.23)
7.23 **a.** (99.04, 100.16)
7.25 (73.16, 86.84), yes
7.27 (128.98, 199.02)
7.31 **a.** yes **b.** no **c.** yes **d.** no **e.** yes **f.** yes **g.** yes **h.** yes
7.33 (.1777, .3157)
7.35 (.1612, .1836)
7.37 **a.** (.356, .684) **b.** (.5313, .7845) **c.** (.0606, .2552)
7.39 (.822, .998)
7.43 **a.** (.7200, .7652)
7.45 385
7.47 385
7.49 (.6218, .6782)
7.51 (.4702, .5382)
7.53 (.455, .517)
7.55 **a.** (6.88, 9.38) **b.** no
7.57 **a.** (.96, 1.02) **b.** (1.28, 1.66) **c.** (.90, 1.08)
7.61 (4.20, 7.65)

CHAPTER EIGHT

8.5 $\mu = 12$ vs. $\mu < 12$
8.7 $\mu = 40$ vs. $\mu \neq 40$
8.9 **a.** $\mu = 1$ vs. $\mu \neq 1$ **b.** $\sigma = .15$ vs. $\sigma > .15$
8.11 **a.** $z < -1.645$ **b.** $z = -3.4$, reject
8.13 **a.** $\pi = .5$ vs. $\pi \neq .5$ **b.** .5, .05 **c.** .8
8.17 **c.** .0359 **e.** .3821 **f.** 0 **g.** do not reject, type II
8.19 **a.** .0287 **b.** smaller **c.** greater **d.** larger for $\mu = 2.5$
8.21 **a.** $z > 2.33$ **b.** $z = 1.645$ **c.** $z > 1.28$ **d.** $z > 1.1$ (approximately)
8.23 **a.** $z < -2.33$ **b.** $z < -1.645$ **c.** $z < -1.28$
8.25 $z = 11.82$, reject
8.27 $z = -4.18$, reject
8.29 **a.** $z = -1.80$, do not reject
8.31 $z = -5$, reject
8.33 reject only in (d)
8.35 **a.** .0358 **b.** .0892 **c.** .6170 **d.** .1616 **e.** 0
8.37 **a.** $.05 > P\text{-value} > .02$ **b.** $.10 > P\text{-value} > .05$ **c.** $P\text{-value} > .20$ **d.** $.20 > P\text{-value} > .10$ **e.** $.001 > P\text{-value}$
8.39 $z = -5.84$, $P\text{-value} < .001$, reject
8.41 $z = 15$, $P\text{-value} < .0005$, reject
8.43 **a.** reject if $t > 1.83$ **b.** reject if $t > 2.57$ **c.** reject if $t > 3.47$ **d.** reject if $t > 1.30$
8.45 no, yes, yes
8.47 **a.** $.10 > P\text{-value} > .05$ **b.** $.01 > P\text{-value} > .002$ **c.** $.001 > P\text{-value}$ **d.** $P\text{-value} > .20$
8.49 $t = -0.21$, do not reject
8.51 $t = 1.94$, do not reject
8.53 **a.** $t = 15.55$, reject **b.** $P\text{-value} < .0005$
8.55 **a.** $t = -33.54$, do not reject (upper tail test) **b.** $P\text{-value} \approx 1$
8.57 $z = -0.92$, do not reject
8.59 $z = 3.55$, reject

8.61 $z = -3.27$, P-value $= .0005$, reject
8.63 $z = -2.41$, reject
8.65 $z = -6.02$, reject
8.67 $z = 4.13$, reject
8.69 **a.** $t = 0.47$, do not reject **b.** $\beta \approx .75$
8.71 **a.** $z = 1.70$, reject **b.** .6554
8.73 **a.** .8643 **b.** .5398, .0287
8.75 $z = .198$, do not reject
8.77 **a.** $z = 4.19$, reject **b.** P-value $< .000033$
8.79 $z = -6.62$, reject
8.81 $t = 0.16$, do not reject
8.83 $x = -21.6$, reject
8.85 **a.** $z = 6.43$, reject **b.** P-value $< .000004$, reject
8.87 $z = -3.99$, reject

CHAPTER NINE

9.1 50, 22.36
9.3 **a.** 3200 **b.** 2830.19 **c.** 0, 2121.32
9.5 **a.** $\mu_x = 3.2$, $\mu_y = 2.5$, $\sigma_x = 1.4$, $\sigma_y = 1.118$ **b.** .7, 1.792
9.7 approximately normal with mean 5 and standard deviation .529
9.9 **a.** .25 **b.** .0726, no, four times as large **c.** $< .0003$
9.11 $z = 6.41$, reject
9.13 $(-2.176, -1.84)$, no (because of large sample sizes)
9.15 **a.** $z = -1.44$, do not reject **b.** $z = -2.85$, reject **c.** $z = 0.145$, do not reject **d.** $z = -1.72$, P-value $= .0892$, do not reject
9.17 $z = -1.31$, do not reject
9.19 $z = 2.55$, do not reject
9.23 $t = -1.20$, do not reject
9.25 $t = 3.48$, reject
9.27 $t = -23.51$, reject
9.29 $t = 2.31$, reject
9.31 **b.** $t = 2.39$, reject **c.** $.025 > P$-value $> .01$, do not reject at level .01
9.33 $t = 2.629$, reject
9.35 $(-.033, -.021)$
9.37 $t = 3.04$, reject
9.39 $z = -4.18$, reject
9.41 $z = 4.17$, reject
9.43 **a.** $z = 1.53$, do not reject **b.** .0668
9.45 **a.** $z = 2.36$, reject **b.** $.0164 < P$-value $< .02$, larger than reported value
9.47 **a.** $z = 1.66$, do not reject **b.** $.0892 < P$-value $< .10$, same conclusion
9.49 $(-.374, -.226)$
9.51 $z = -.403$, do not reject
9.53 **a.** $z = 2.41$, $.0124 < P$-value $< .0164$, reject for any $\alpha > .0164$
9.55 rank sum $= 65$, do not reject
9.57 **b.** rank sum $= 151$, $z = 3.17$, reject
9.59 rank sum $= 39.5$, do not reject
9.61 rank sum $= 346.5$, $z = 2.49$, reject
9.63 (16, 96)
9.65 **a.** $t = -5.03$, reject

9.67 $z = 1.79$, do not reject
9.69 $(-42.18, 69.98)$
9.71 **a.** $z = -10.35$, reject **b.** P-value $< .00000058$, no
9.73 $t = 4.92$, reject
9.75 **a.** $t = -1.94$, do not reject **b.** $(-5.3, .5)$
9.77 **a.** $t = -1.44$, do not reject
9.79 **a.** $z = -6.43$, reject **b.** $t = 1.12$, do not reject
9.81 $t = -1.21$, do not reject

CHAPTER TEN

10.9 $t = .86$, do not reject
10.11 **a.** no, because the number of differences is large **b.** $t = 3.95$, reject **c.** $t = -1.17$, do not reject
10.13 **a.** $(-.1861, .1111)$
10.15 $z = 8.07$, reject
10.17 $(59.25, 275.17)$
10.21 signed-rank sum $= -15$, reject
10.23 signed-rank sum $= 105$, reject
10.25 signed-rank sum $= 12$, do not reject
10.27 $(3.65, 5.55)$
10.29 **a.** $t = 2.17$, reject **b.** $t = 9.23$, reject **c.** $(25.15, 41.80)$
10.31 $t = 0.51$, do not reject
10.33 **a.** $t = 2.63$, reject

CHAPTER ELEVEN

11.1 **a.** 3.6, 4.3 **b.** .1151 **c.** .5, .1587
11.5 **a.** $2.7998 + .0265x$ **b.** 3.065 **c.** 3.065 **d.** no
11.7 **a.** $30.816 - .9105x$ **b.** 8.0535 **c.** 4.4115
11.9 **b.** 1400.9, 51.89 **c.** .342
11.13 **a.** $(.4, 1.2)$ **b.** yes
11.15 $(-.07, -.042)$
11.17 **a.** $a = 11.85$, $b = -1.51$ **b.** $t = -3.58$, reject, yes **c.** $.01 > P\text{-value} > .002$
11.19 **a.** $a = 100.79$, $b = -.778$ **b.** $t = -2.80$, reject, yes **c.** 77.45 **d.** $(-1.306, -.250)$
11.21 $t = 2.64$, reject
11.23 **a.** $t = 3.10$, reject, yes **b.** $.02 > P\text{-value} > .01$ **c.** $(.006, .0074)$
11.25 **a.** .253 **b.** no, it is .178 **c.** 4, .1265
11.27 yes, since s_b would be smaller
11.29 **a.** $(12.82, 13.54)$ **b.** $(13.63, 17.33)$ **c.** narrower
11.31 **a.** $(17.669, 26.465)$ **b.** narrower
11.33 interval with $x^* = 17$ would be wider
11.35 **a.** $-8.779 + 18.874x$ **b.** $t = 21.22$, reject, yes **c.** $(318.75, 343.17)$
11.37 $t = .26$, do not reject
11.41 $t = 2.47$, do not reject
11.43 $t = 2.44$, do not reject
11.45 **b.** $r = .913$, $t = 6.33$, reject

11.47 **a.** .8777 **b.** $t = 6.07$, reject **c.** P-value $< .0005$
11.49 $t = 2.2$, reject
11.59 **b.** $2.2255 + .1521x$ **c.** $t = 8.17$, reject **d.** .881, .3445 **e.** (4.903, 5.632)
11.61 $t = .73$, do not reject
11.63 **a.** $r = -.0164$, $t = -.06$, do not reject. **b.** $r = .747$, $t = 4.35$, reject **c.** $a = 3.078$, $b = .0116$ **d.** (2.9776, 4.3396) **e.** (3.619, 4.162)
11.67 **a.** $57.964 + .0357x$ **b.** $t = .023$, do not reject

CHAPTER TWELVE

12.3 **b.** .237 **c.** 1.213
12.5 **a.** 9.364 **b.** increase
12.7 **b.** 67948.5564 **c.** smaller for May 22
12.13 three dummy variables would be needed
12.15 **b.** increase
12.17 $F = .638$, do not reject
12.19 **a.** $86.85 - .12297x_1 + 5.090x_2 - .07092x_3 + .001538x_4$ **b.** $F = 64.38$, reject
12.21 **a.** $F = 6$, reject **b.** .4898, .894
12.23 $F = 96.64$, reject
12.27 $F = 18.95$, reject
12.29 **a.** (−18.438, −.318) **b.** (−4.049, −.309)
12.31 **a.** $F = 106.12$, reject **b.** (.015, .067) **c.** $t = 5.86$, reject **d.** 99.514
12.33 **a.** $F = 5.47$, reject **b.** $t = .04$, fail to reject
12.35 **a.** $F = 339.34$, reject **b.** (i) $t = 6.95$, reject (ii) $t = 18.51$, reject **d.** (40.21, 41.52) **e.** (54.553, 56.34) **f.** (55.37, 57.19)
12.37 **a.** $F = 30.81$, reject **b.** for linear term, $t = 6.57$, reject; for quadratic term, $t = -7.69$, reject **c.** (43.0, 47.92) **d.** (44.0, 47.92)
12.41 temperature (x_1) can be eliminated
12.45 Adjusted R^2 will be negative when $R^2 < .5$
12.49 **b.** $F = 21.25$, reject **c.** $t = 6.52$, reject, cannot be eliminated
12.51 **d.** (12.684, 18.734)
12.53 **a.** $F = 8.64$, reject **b.** x_3 **c.** sex (x_1) can be eliminated
12.55 **a.** $F = 2.4$, do not reject **c.** $R^2 > .956$

CHAPTER THIRTEEN

13.3 **a.** $F = 4.12$, reject
13.5 $F = 1.85$, do not reject
13.7 $F = 4.01$, do not reject
13.9 $F = 2.46$, do not reject
13.11 $F = 4.1$, do not reject
13.13 $F = .83$, do not reject

13.15

Source	df	SS	MS	F
Treatments	3	75081.72	25027.24	1.70
Error	16	235419.04	14713.69	
Total	19	310500.76		

Since 1.70 < 3.24, do not reject

13.17

Source	df	SS	MS	F
Treatments	3	136.14	45.38	5.12
Error	60	532.26	8.871	
Total	63	668.40		

Since 5.12 > 4.13, reject

13.19

Source	df	SS	MS	F
Treatments	4	.92928	.23232	2.17
Error	25	2.68072	.10722	
Total	29	3.61000		

Since 2.17 < 2.76, do not reject

13.21

Source	df	SS	MS	F
Treatments	2	1.6297	.8148	4.589
Error	21	3.7289	.1776	
Total	23	5.3586		

Since 4.589 > 3.47, reject

13.25

Group	Simultaneous	Sequential	Control
Mean	$\bar{x}_3$	$\bar{x}_2$	$\bar{x}_1$

13.27 **a.** Since 5.56 > 2.81, reject **b.** (−5.61, 1.43), (−9.30, −2.26), (−5.47, 1.57), (−7.21, −0.17), (−3.38, 3.66), (0.31, 7.35)

c.

Group	Apathetic	Conservative	Erratic	Strategic
Mean	2.96	4.91	5.05	8.74

13.29 **a.**

Source	df	SS	MS	F
Treatments	5	108.185	21.637	79.264
Error	20	5.460	.273	
Total	25	113.645		

Since 79.264 > 2.71, reject

c.

Brand	2	4	3	1	5	6
Mean	12.8	13.1	13.825	14.1	17.14	18.1

13.33 **a.**

Source	df	SS	MS	F
Treatments	2	1835.2	917.60	35.62
Blocks	4	93.1	23.28	
Error	8	206.1	25.76	
Total	14	2134.4		

b. Since 35.62 > 4.46, reject

13.35 **a.**

Source	df	SS	MS	F
Treatments	2	28.78	14.39	1.04
Blocks	17	2977.67	175.16	
Error	34	469.55	13.81	
Total	53	3476.00		

Since 1.04 < 3.23, do not reject

13.37 **a.**

Source	df	SS	MS	F
Treatments	1	289	289.00	1.09
Blocks	7	623	89.00	
Error	7	1858	265.43	
Total	15	2770		

Since 1.09 < 5.59, do not reject

b. $t = 1.04$, do not reject

13.39

Source	df	SS	MS	F
Treatments	3	3141153.5	1040751.17	2.276
Blocks	3	19470550.0	6490183.33	
Error	9	4141165.5	460129.50	
Total	15	26752869.0		

Since 2.276 < 3.86, do not reject

13.41 **a.**

Source	df	SS	MS	F
Treatments	4	4195.603	1048.901	10.932
Blocks	3	336.354	112.118	
Error	12	1151.389	95.949	
Total	19	5683.348		

Since 10.932 > 3.26, reject

b.

Source	5	2	1	4	3
Mean	51.975	53.525	67.575	77.35	90.25

13.45 **a.**

Source	df	SS	MS	F
A main effects	7	102526	14646.57	384.05
B main effects	4	95	23.75	.62
AB interaction	28	315	11.25	.29
Error	80	3051	38.138	
Total	119	105987		

b. Since .29 < 1.62, do not reject
c. For factor A, since 384.05 > 2.14, reject
For factor B, since .62 < 2.51, do not reject

13.47 For AB interaction, since $F_{AB} < 1 < 3.07$, do not reject
For factor A, since 4.99 > 3.07, reject
For factor B, since 4.81 > 3.92, reject

13.49

Source	df	SS	MS	F
Diet	2	18138	9069.0	28.92
Temperature	2	5182	2591.0	8.26
Interaction	4	1737	434.25	1.38
Error	36	11291	313.64	
Total	44	36348		

For interaction, since 1.38 < 2.63, do not reject
For diet main effect, since 28.92 > 3.26, reject
For temperature main effect, since 8.26 > 3.26, reject

13.51 KW = .17, do not reject

13.53 KW = 17.86, reject

13.55 F_r = 28.92, reject

13.57 F_r = 15, reject

13.59

Source	df	SS	MS	F
Treatments	4	14.962	3.741	36.7
Blocks	8	9.696	1.212	
Error	32	3.262	0.102	
Total	44	27.920		

Since 36.7 > 2.69, reject

13.61

Source	df	SS	MS	F
Treatments	4	0.26	.065	2.24
Error	20	0.58	.029	
Total	24	0.84		

Since 2.24 < 2.87, do not reject (based on α = .05)

13.63

Source	df	SS	MS	F
Treatments	4	290.96	72.74	8.07
Error	30	270.58	9.02	
Total	34	561.54		

Since 8.07 > 2.69, reject

13.65 KW = 10.38, reject

13.69 F_r = 6.5, reject

13.71

Source	df	SS	MS	F
A main effects	1	322.667	322.667	980.38
B main effects	3	35.623	11.874	36.08
Interaction	3	8.557	2.852	8.67
Error	16	5.266	0.329	
Total	23	372.113		

Since 8.67 > 3.24, there is a significant interaction.

CHAPTER FOURTEEN

14.1

1970	.44	.24	.32
1980	.42	.28	.30

14.5

38 (23.5)	9 (23.5)
26 (22.5)	19 (22.5)
24 (19.5)	15 (19.5)
12 (13.5)	15 (13.5)
12 (16.5)	21 (16.5)
10 (26.5)	43 (26.5)

14.7

19 (15.5)	24 (27.0)	57 (57.5)
12 (15.5)	30 (27.0)	58 (57.5)

14.9 $i = .4375$
14.11 $X^2 = .932$, do not reject
14.13 $X^2 = 1.47$, do not reject
14.15 $X^2 = 4.8$, do not reject
14.17 **a.** df = 10, reject if $X^2 > 23.21$ **b.** df = 12, reject if $X^2 > 26.22$ **c.** df = 15, reject if $X^2 > 30.58$
14.19 Since $21.35 > 16.81$, reject

14.21 **c.**

28 (27.41)	30 (30.59)
46 (47.73)	55 (53.27)
21 (21.74)	25 (24.26)
14 (16.07)	20 (17.93)
29 (25.05)	24 (27.95)

$X^2 = 1.88$, do not reject
14.23 $X^2 = 8.87$, do not reject
14.25 $X^2 = 5.90$, do not reject
14.27 $X^2 = 197.62$, reject
14.29 The nine expected counts are 11.25, 63.94, 177.40, 378.81, 364.27, 265.48, 86.30, 21.95, 2.61. The last two are combined since $2.61 < 5$. $X^2 = 17.5$, reject
14.31 **a.** The nine expected counts are 17.95, 39.60, 96.70, 115.65, 109.10, 80.60, 29.00, 9.05, 2.35. The last two are combined since $2.35 < 5$. $X^2 = 9.76$, do not reject **b.** no, $z = -46.86$
14.33 $X^2 = 1.38$, do not reject
14.35 $b = 1, a = .4$
14.37 $X^2 = 1.63$, do not reject
14.39 $X^2 = .02$, do not reject
14.41 $X^2 = 16.86$, reject
14.45 $X^2 = 6.87$, do not reject
14.47 $X^2 = 21.6$, reject

INDEX